CHEMISTRY FOR THE PROTECTION OF THE ENVIRONMENT

ENVIRONMENTAL SCIENCE RESEARCH

Recent Volumes in this Series

Volume 34 – ARCTIC AND ALPINE MYCOLOGY II
Edited by Gary A. Laursen, Joseph R. Ammirati, and Scott A. Redhead

Volume 35 – ENVIRONMENTAL RADON
Edited by C. Richard Cothern and James E. Smith, Jr.

Volume 36 – SHORT-TERM BIOASSAYS IN THE ANALYSIS OF COMPLEX ENVIRONMENTAL MIXTURES V
Edited by Shahbeg S. Sandhu, David M. DeMarini, Marc J. Mass, Martha M. Moore, and Judy L. Mumford

Volume 37 – HAZARDS, DECONTAMINATION, AND REPLACEMENT OF PCB: A Comprehensive Guide
Edited by Jean-Pierre Crine

Volume 38 – *IN SITU* EVALUATION OF BIOLOGICAL HAZARDS OF ENVIRONMENTAL POLLUTANTS
Edited by Shahbeg S. Sandhu, William R. Lower, Frederick J. de Serres, William A. Suk, and Raymond R. Tice

Volume 39 – GENETIC TOXICOLOGY OF COMPLEX MIXTURES
Edited by Michael D. Waters, F. Bernard Daniel, Joellen Lewtas, Martha M. Moore, and Stephen Nesnow

Volume 40 – NITROARENES: Occurrence, Metabolism, and Biological Impact
Edited by Paul C. Howard, Stephen S. Hecht, and Frederick A. Beland

Volume 41 – ENVIRONMENTAL BIOTECHNOLOGY FOR WASTE TREATMENT
Edited by Gary S. Sayler, Robert Fox, and James W. Blackburn

Volume 42 – CHEMISTRY FOR THE PROTECTION OF THE ENVIRONMENT
Edited by L. Pawlowski, W. J. Lacy, and J. J. Dlugosz

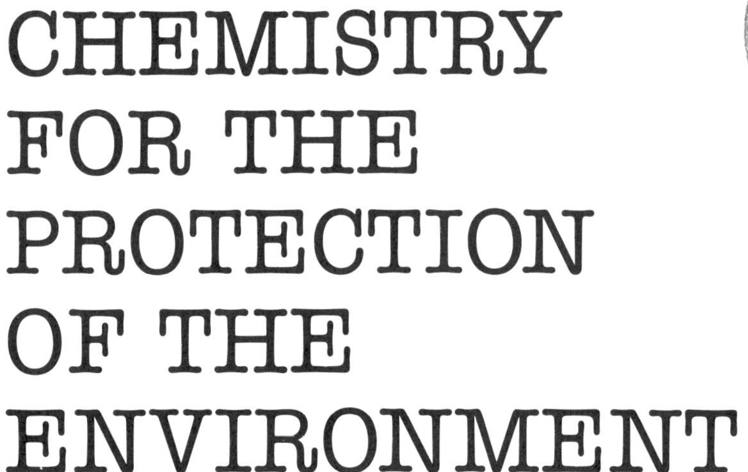

CHEMISTRY FOR THE PROTECTION OF THE ENVIRONMENT

Edited by

L. Pawlowski

Technical University of Lublin
Lublin, Poland

W. J. Lacy

President, Lacy and Associates
Alexandria, Virginia

and

J. J. Dlugosz

U.S. Environmental Protection Agency
Las Vegas, Nevada

PLENUM PRESS • NEW YORK AND LONDON

Library of Congress Cataloging-in-Publication Data

International Conference on Chemistry for the Protection of the Environment (7th : 1989 : Lublin, Poland)
Chemistry for the protection of the environment / edited by L. Pawlowski, W.J. Lacy, and J.J. Dlugosz.
p. cm. -- (Environmental science research ; v. 42)
"Proceedings of the Seventh International Conference on Chemistry for the Protection of the Environment, held September 4-7, 1989, in Lublin, Poland"--T.p. verso.
Includes bibliographical references and index.
ISBN 0-306-43904-2
1. Environmental chemistry--Congresses. 2. Sewage--Purification--Congresses. 3. Water--Pollution--Congresses. I. Pawlowski, Lucjan. II. Lacy, W. J. (William J.) III. Dlugosz, J. J.
IV. Title. V. Series.
TD193.I58 1989
628--dc20 91-20150
CIP

Proceedings of the Seventh International Conference on Chemistry for the Protection of the Environment, held September 4-7, 1989, in Lublin, Poland

ISBN 0-306-43904-2

Printed in the United States of America

ANDRE VAN HAUTE
(1930-1989)

Andre Van Haute, a member of International Committee on "Chemistry for Protection of the Environment," professor at the Faculty of Engineering of the Catholic University of Leuven (Belgium), died on October 22, 1989, after a hard and brave fight against a fatal disease.

A. Van Haute was born in Hamme on April 28, 1930. He obtained a degree in chemical engineering from the University of Leuven in 1953 and a master's degree in chemical engineering from California Institute of Technology in Pasadena in 1954. Back in Leuven, he began teaching at the Institute of Industrial Chemistry and Chemical Engineering. Soon he replaced Prof. Eugene Mertens de Wilmars in teaching industrial chemistry, both in Flemish and in French.

He also taught industrial chemistry (inorganic chemical processes, polymers, materials, and glass), industrial electrochemistry, and corrosion. In the mid-sixties, he turned his attention to water treatment. The subject of his first research was water desalination by membranes and hydration processes. He started a new specialization in environmental management and was a leading figure in this field for about twenty years. Many government projects, the books Dr. Van Haute wrote about environmental problems, and many doctorates related to a wide range of aspects of wastewater purification resulted from this work. In the late 1970s and early 1980s, Prof. Van Haute founded a program for Sanitary Engineering at the Institute of Technology in Surabaya.

He was a member of the Royal Academy of Overseas Sciences, member of the Flemish Council for the Environment, administrator of the Environment Foundation, president of the Belgian National Committee of the International Association of Water Pollution Research and Control, and Belgian representative in the working party of the European Federation of Chemical Engineering.

Prof. P. Van Rompay

PREFACE

Over the last decade and a half, an environmental conference series has emerged to become one of the major international forums on the chemical aspects of environmental protection. The forum is called Chemistry for the Protection of the Environment (CPE). The sponsors of this CPE series have included the Chemical Societies of Poland, France, Belgium, Italy, and the U.S.A., the European Federation of Chemical Societies, the American Institute of Chemical Engineers, the American Society of Testing and Materials, the International Ozone Association, the United Nations Industrial Development Organization, the Ministries of the Environment of Poland, France, Belgium, and Italy, US Environmental Protection Agency, more than twenty universities and institutes of higher learning, and five academies of sciences.

The first meeting in this series was organized in 1976 at the Marie Curie-Sklodowska University in Lublin, Poland. The conference dealt with various physicochemical methodologies for water and wastewater treatment research projects that were jointly sponsored by US EPA and Poland.

The great interest expressed by the participants led the organizers to expand the scope of the second conference, which was also held in Poland in 1979. The third and enlarged symposium was again successfully held in 1981 in Lublin, Poland. At that time the participating scientists and engineers expressed their desire to broaden the coverage as well as the title of the conference series. The International Committee, the governing body of the symposium, approved the title "Chemistry for the Protection of the Environment" and designated the date of the fourth conference, CPE IV, which was convened in September 1983 at the Paul Sabatier University in Toulouse, France, and was hosted and arranged by Prof. A. Verdier. The international scientific community expressed the need to have an independent, nonpolitical forum for chemists, chemical engineers, biologists, environmental scientists, and other professionals involved in environmental protection. Furthermore, this conference series includes participants from various government agencies, academia, and the private sector, representing industrialized countries as well as emerging nations, both East and West.

The central goals of CPE are to improve technology transfer and scientific dialogue, thereby leading to a better comprehension of and solution to a broad spectrum of environmentally related problems. The fifth conference was held in September 1985 at the Catholic University in Leuven, Belgium. It was hosted by Profs. A. Van Haute and G. Alaerts. CPE V covered topics dealing with treatment technologies and phenomena related to hazardous waste and the utilization of fossil fuels. It provided an opportunity for interdisciplinary discussions and encouraged the exchange of ideas among international specialists from diverse fields and backgrounds. Under the leadership of Profs. E. Mentasti and C. Sarzanini and with the able assistance of Dr. M. Gennero, CPE VI, which was held in September 1987 at the University of Turin in Italy, was a success. Over 150 selected scientific papers and posters were presented to an audience of specialists from 32 nations. This assemblage comprised in equal measure scientists from Europe, the New World, and developing nations.

CPE VII, of which these are the proceedings, was convened at the Catholic University in Lublin, Poland. The exchange of information by approximately 200 scientists and engineers made this a most memorable scientific occurrence. The scientific committee selected presentors of high intellectual and technical merit. The distinguished participants of CPE VII included Poland's Minister and Deputy Minister for Environmental Protection, U.S. Scientific Council, Israel's Deputy Minister of the Environment, presidents and vice presidents of five universities, representatives of the academies of sciences for Chechoslovakia, France, Italy, Poland, and the U.S.S.R., as well as many department heads and acclaimed scientists.

162 interesting, original, and informative papers and posters were presented on the following topics: adsorption, analytical methods, chemical/biological/physical/treatment, groundwater studies, ion exchange, modeling, ozonation, photodegradation, risk assessment, sludge treatment, waste minimization, and innovative technology.

Some of the major benefits for the participants were technology transfer and exchange of innovative and alternative treatment methods and information about activities in other countries related to environmental problems. The proceedings of all these CPE conferences have been published in either hard-bound books or in selected peer review journals.

The venue for the next conference, CPE VIII, will be selected during the spring 1990 meeting of the International Committee. In his closing remarks at CPE VII, Dr. W.J. Lacy stated that all scientists and engineers need to be aware of what is being done in other countries, since even scientists from the smallest developing nation could have the solution to the most vexing environmental problems. Prof. L. Pawlowski agreed that as long as scientists from around the globe wish to meet in this type of open, technical forum and exchange ideas and information on environmental protection, the Chemistry for Protection of the Environment series will continue.

L. Pawlowski, W.J. Lacy, and J. Dlugosz

CONTENTS

GENERAL PROBLEMS

MONITORING METHODS FOR SURFACE AND GROUND WATER AND ANALYSIS OF POLLUTANTS

PHYSICOCHEMICAL TREATMENT: ION EXCHANGE

PHYSICOCHEMICAL TREATMENT: COAGULATION, FLOCCULATION AND SORPTION

PHYSICOCHEMICAL TREATMENT: OXIDATION-REDUCTION PROCESSES

PHYSICOCHEMICAL TREATMENT: MEMBRANE PROCESSES

MISCELLANEOUS METHODS FOR REMOVAL OF POLLUTANTS

GENERAL PROBLEMS

GROUND WATER POLLUTION

PLENARY LECTURE

A.L. KOWAL
Institute of Environmental Protection Engineering
Technical University of Wroclaw, Poland

Water supply in Poland is based (in equal proportion) on surface and ground water.

Central water supply exploits usually deep ground water or shallow infiltration water.

Private wells used for individual houses or farms take in shallow quaternary water, generally not covered by an impermeable layer and prone to pollution.

Ground water may be polluted from various pollution sources. Industrial and agricultural activities as wel as municipalities, suburbs and villages may add to ground water pollution.

Industries can affect the ground water by industrial sewerage systems, washout from industrial areas, storage of raw materials, sewage lagoons and damping sites.

In many industrial plants and their vicinity there were observed substantial changes in ground water pH, varying from pH=3 up to over 10, which could cause damage even to structures and installations.

Ground water in urban areas may be polluted by leakage from sewerage systems and by salts used during the winter season.

Suburban areas, without sewerage systems, using septic tanks for domestic sewage disposal, reveal increased pollution of ground water, which overtakes large areas. Pollution by domestic sewage is easily manifested by the presence of detergents.

Chemistry for the Protection of the Environment
Edited by L. Pawlowski *et al.*, Plenum Press, New York, 1991

Ground water in village areas is affected by domestic sewage as well as by inadequately stored manure in piles. The analysis of private well water shows that over 50% of wells supply water of improper quality polluted chemically and bacteriologically. After construction of central water supply systems for villages, the public usually neglects the ground water pollution control, and sometimes the dug wells are illegally converted to septic tanks. In small farms the manure pile is usually located in the farm yard. The seepage from manure affects ground and surface water. Manure from small farms affects only the nearest wells, whereas wells located at a distance of hundred meters remain unaffected.

A real hazard is created by uncontrolled storage of fertilizers which due to weather condition may be dissolved and in the form of strong solution pass to the ground water.

Well water analyses in a village in the vicinity of Wroclaw have shown concentrations of nitrates up to 100 mg NO /l and potassium up to 300 mg K/l.

Local pollution sources and pollution caused by overfertilization of the land contibuted to ground water pollution.

The investigation of the surface water pollution due to land fertilization in the region of Legnica (Lower Silesia) shows overfertilization of the land, which was not in accordance with plants demand and crops.

The farmers treat overfertilization as substitute for improper land cultivation and weather conditions.

In many areas of the country the concentration of nitrates in ground water exceeds the permissible concentration of nitrates in drinking water.

There is also substantial increasing in total dissolved solids concentration (TDS) in ground water. The concentration of chlorides, sulphates, sodium, potassium, calcium and magnesium is evidently higher than in ground water of adjacent forested areas.

The soluble parts of fertilizers are intruding to ground water, since the excessive doses of readily soluble fertilizer are not used by the root system of plants or adsorbed in the soil. The fertilization affects mainly quaternary ground water, whereas deeper tertiary water remains usually unpolluted. So it is in Poland, but in some other countries like Hungary, France, FRG, the ground water up to 80 m deep is polluted by nitrates.

The pesticides applied in agriculture are more often present in minute concentration in surface water than in ground water. 20 to 70% of pesticides applied remain persistently bound in the soil colloids. Release of bound residue proceeds by slow microbiological degradation. It is very hard to characterize and monitor the pesticides residue in the soil.

Pollution of water by biodegradable organic substances creates smaller hazard than the pollution by refractory substances, which are incidentally spilled or discharged to the ground. Very hazardous to ground water are leaks from gazoline or diesel oil tanks. These kinds of pollution maymake it necessary to abandon the water intake.

Manure applied in agriculture can be totally absorbed in the soil if the doses are relatively low. It is commonly recognized that 30 to 50 tons of manure can be absorbed on 1 hectar of land without causing damage to ground water.

Nitrogen compounds in manure undergo complicated biochemical decomposition.

Urea is biodegraded in the presence of urease to ammonia and carbon dioxide

$$N_2H - CO - NH_2 + H_2O \xrightarrow{\text{urease}} 2NH_3 + CO_2$$

Ammonia can be also released from organic compounds by hydrolythic ammonification

$$R - NH_2 + H_2O \longrightarrow R - OH + NH_3$$

or oxidative ammonification

$$R - CHNH_2 - COOH + H_2O \longrightarrow R - CO - COOH + 2[H^+] + NH_3$$

and also reduced ammonification

$$R - CHNH_2 - COOH + 2[H^+] \longrightarrow R - CH_2 - COOH + NH_3$$

Nitrogen is released from organic compounds in the form of ammonia, urea and amino acids. The latter are converted to ammonia directly or by transammonification with transitional formation of glutamin acid. If the reaction does not proceed to a final state the transitional products remain in the water.

Ammonia has a good affinity to soil colloids and is well adsorbed. Excess of ammonia can be washed down to lower parts of the ground, where it can be nitrified, and denitrified after depletion of oxygen.

By manuring of land a vast number of microorganisms are introduced into the soil and subsoil, and the formation of a nitrification zone in the ground is enhanced.

With only mineral fertilization of land the zones may not be formed, and if they are, they will be overloaded and nitrates will pass through and reach the ground water table. Thus the manuring--fertilization of land--has to be considered as a very important process in ground water pollution control, which has special importance for light sandy soils predominating in Poland.

In order to decrease ground water pollution by nitrates the quality of fertilizers and methods of their application should be changed. It can be done by production of low soluble fertilizers and by division of the fertilizers dose in two parts, according to plants' demand for nitrogen. The last procedure is advised in France. These will diminish the fertilizer losses and decrease the ground water pollution and surface water eutrophication.

The economical aspects of these procedures should be solved. The ground water pollution by nitrates can be also limited by supplementing application of manure on farmland, which supports the formation of denitrification zones in the ground.

The decrease of existing nitrates concentration in quaternary water will take a long time. In the meantime the shallow water sources will have to be used for drinking water supply, especially for private houses and farms.

The treatment processes for nitrates removal in private water supply may rely on ion exchange methods.

For economic reasons only the water for consumption should be treated by the exchange methods.

Fortunately there is no public water supply system in Poland which intakes the ground water polluted by nitrates. Thus nitrates removal has to be solved only for private water supply.

In public water supply systems the biological denitrification with or without a ground passage of the water will be more economical. The application of surface water infiltration for ground water restoration and recharge is also possible.

HYDROLOGIC AND GEOCHEMICAL IMPLICATIONS OF WASTE DISPOSAL SCENARIOS

M. S. BEDINGER

Environmental Research Center
University of Nevada-Las Vegas
Las Vegas, NV 89154[1]

ABSTRACT

Hazardous wastes are commonly disposed of by land burial with the intention of isolating the materials from man's environment. The waste disposal site may or may not provide an environment in which the wastes can degrade. Burial sites vary greatly in their chemical and hydrologic environment and in their capacity to retain waste in or near the burial area. They provide an environment in which the waste may degrade to innocuous levels or become immobilized. Failure of waste disposal repositories may arise from inadequate design and construction because of insufficient knowledge of the required disposal environment for specific wastes. Repository design and siting requirements are specific to the chemical and physical properties of waste to be disposed. Consider, for example, four broad types of waste:

1) high-level radioactive waste

2) low-level radioactive waste

3) hazardous organic wastes

4) mixed municipal wastes

[1]Work done in cooperation with U. S. Environmental Protection Agency, Environmental Monitoring Systems Laboratory, Las Vegas, NV.

Chemistry for the Protection of the Environment
Edited by L. Pawlowski *et al.*, Plenum Press, New York, 1991

The half-lives of radionuclides range from a fraction of a second to millions of years. Because of the extreme range in the duration of toxicity of radionuclides, there are different requirements for siting and design of repositories for long- and short-lived radioactive wastes. Radionuclides decay at rates independent of their chemical environment. However, sorption of radionuclides by material within a repository or by natural material in the ground-water flow systems may effectively isolate radionuclides until they decay to innocuous levels. Sorption of radionuclides is a function of the geochemistry of the environment and the nature of the porous media in which the radionuclides are enclosed. The degradation rate of organic compounds is related to the biological, chemical, and physical characteristics of their environment. The waste of municipal landfills is composed largely of non-hazardous organic and miscellaneous materials. The chemical changes in municipal landfills are dominated by the degradation of the organic materials and the generation of methane and carbon dioxide.

INTRODUCTION

The primary goal of land disposal of waste is to provide an environment that will isolate the waste from the biosphere. If is desirable for the repository to provide an environment that will degrade the hazardous waste to relatively innocuous products. Each type of hazardous waste presents requirements for disposal to provide adequate protection of the environment. Requirements for satisfactory disposal of different waste materials may include factors such as depth of burial, physical and chemical nature of the backfill and host media, moisture content of the repository, saturated or unsaturated conditions, infiltration into the repository, and length of ground-water flow path to the accessible environment.

RADIOACTIVE WASTE REPOSITORIES

The rationale for disposal of radioactive waste is to isolate it from the biosphere until radioactive decay renders the radionuclides no longer hazardous to life. The isolation time required for such a waste is a function of the half-lives of the radionuclides and their daughter products in the waste. A rule of thumb commonly used in the nuclear industry [1] is that radioactive waste should be confined to at least 10 times the half-life of the longest dominant isotopes. The rate of radioactive decay is independent of the geochemical environment of the repository. However, the repository environment should not degrade the waste form and mobilize the radionuclides or transport them beyond the repository. In the United States, high-level radioactive waste will be disposed in deep mined repositories and may need to be isolated for tens or hundreds of thousands of years. Low-level radioactive waste may need to be isolated for 300 to 500 years.

High-level radioactive waste repositories

High-level radioactive wastes are composed of nuclides of long half-lives, low- to highheat output, and intermediate to high radiotoxicity. High-level radioactive waste includes spent fuel from commercial reactors and waste generated during the separation of uranium and plutonium from the fission products and transuranic elements in the spent fuel. High-level radioactive waste generates much heat and requires heavy shielding to control penetrating radiation.

Transuranic waste comes from the reprocessing of spent fuel and from the use of plutonium in the fabrication of nuclear weapons. It is defined as waste contaminated with alpha-emitting radionuclides of atomic number greater than 92 (that is, uranium, hence the term transuranic). Because of the long half-lives of some transuranic elements, transuranic waste requires the same long-term isolation as that required for high-level radioactive waste.

High-level radioactive waste will be disposed in the United States and in many other countries in deep mined repositories. In the United States, the characteristics sought in a high-level radioactive waste repository include:

(1) disposal in a deep mined repository;

(2) capability to retrieve the waste for as much as 50 years after closure of the repository;

(3) confidence of isolation of the waste from the accessible environment.

Isolation of the waste will need to be assured using geologic, geochemical, and hydrologic conditions that:

(1) minimize risk of inadvertent future intrusion by man;

(2) minimize the possibility of disturbance by processes that would expose the waste or increase its mobility; and

(3) provide a system of natural barriers to the migration of waste by ground water.

The criteria used to identify suitable repository sites include the following categories: (1) potential host media, (2) ground water conditions, (3) tectonic conditions, and (4) occurrence of natural resources. The host medium constitutes the first natural barrier to migration of radionuclides.

The host medium ideally should be a rock type that prevents or retards the dissolution and transport of radionuclides. Rocks in both the unsaturated and saturated zones may have desirable characteristics for host media. Rocks–other than the host–in the ground-water flow path from the repository ideally should be major barriers to radionuclide migration. Confining beds of low permeability might be present to retard the rate of flow between more permeable beds. Additionally, sorption of radionuclides by materials such as clays and zeolites in the flow path can further retard the flow of radionuclides by several orders of magnitude. Tectonic conditions in an area should not present a probable cause for exhumation or increased mobility of radioactive waste. Natural resources are a factor for consideration because of the problem of future human intrusion and exposure to radioactivity in the quest for minerals, oil, gas, water, and geothermal resources.

Chemical reactions between a nuclide and the solid phase tend to decrease the concentration of radionuclide in solution. The important reactions are dissolution, precipitation, adsorption, and ion exchange. Such nuclide-solid reactions are considered by most workers to be more important in nuclide retardation than nuclide-solute interactions.

The guidelines discussed in the preceding paragraph may be illustrated by a hypothetical setting for a high-level radioactive waste repository shown in Fig. 1. The repository is in a mesa overlying a folded and faulted sequence of carbonates, shales, and sandstone of Paleozoic age. Depth to water is about 700 meters below the surface. A repository in the unsaturated zone could be in a fractured or jointed basalt or densely welded tuff or in a well sorted sand that would provide drainage from the repository if recharge from the surface reached the repository.

In an arid environment–one in which precipitation is greatly exceeded by potential evapotranspiration–recharge rates can be expected to be extremely low or nil. Leaching of the waste material, if it occurs, would presumably be at a significantly slower rate than in the saturated zone. However, assuming leaching of the waste occurs, a regional flow system in the Paleozoic rocks, discharging 75 kilometers away from the mesa under an average gradient of 6×10^{-4} would provide a conservatively estimated travel time in the saturated zone of about 20,000 years from the mesa to discharge area. Transport time of sorbed radionuclides will be one to three orders of magnitude greater because of sorption by zeolitized tuff, clayey siltstone, and a lacustrine sequence overlying the Paleozoic rocks. Dilution of the radionuclide concentration would be affected by the greater flow volume and dispersion in Paleozoic rocks. The risk of release of radionuclides by tectonic hazards would need to be considered in this and other waste disposal scenarios.

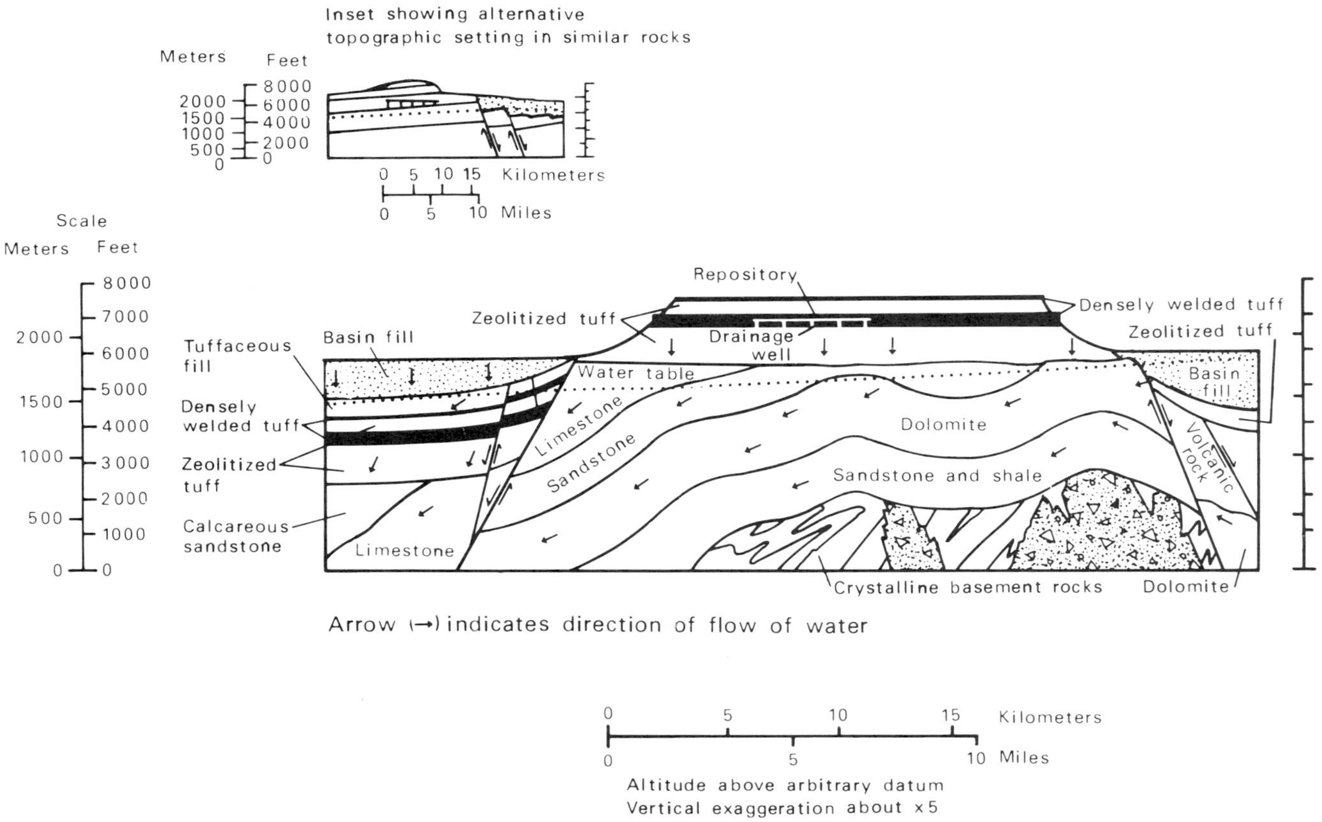

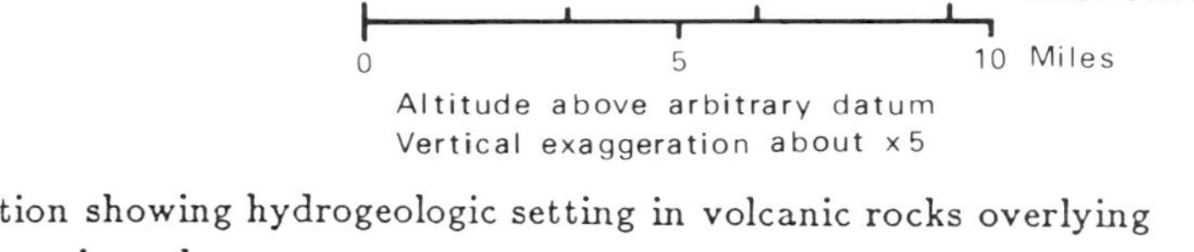

Fig. 1. Section showing hydrogeologic setting in volcanic rocks overlying Paleozoic rocks.

Low-level radioactive waste

Low-level radioactive wastes are composed primarily of nuclides of relatively short half-lives. Low-level radioactive wastes are produced in the nuclear-fuel production cycle, as well as from the use of radionuclides in research, medicine, and industry. The half-lives of principal components of low-level radioactive waste range from 3 to 30 years.

Land disposal methods and engineered repositories have been proposed or used in efforts to isolate low-level radioactive waste from the biosphere during its hazardous life. Trench disposal facilities for low-level waste are near-surface, within 50 meters of the land surface, and may include engineered barriers. Summaries of shallow-land trench burial sites in the United States are given in Fischer and Robertson [2].

Many first-generation low-level radioactive waste disposal sites in the United States were excavated in deposits of low permeability. Such siting resulted in accumulation of water in the trenches, saturation and accelerated leaching of the waste. This phenomenon, commonly referred to as the"bathtub effect", has occurred at low-level radioactive disposal sites at West Valley, New York, Maxey Flats, Kentucky, and Oak Ridge National Laboratory, Tennessee.

Another problem that has occurred at low-level radioactive waste sites was collapse of trench caps caused principally by compaction of the waste packages and backfill material. Trench cap stability has been improved by compacting the clay capping materials with heavy machinery and applying and overburden of earth material for a limited period of time. At Barnwell, South Carolina, the waste packages are stacked systematically to allow backfill to fill the voids. Furthermore, sand has been used as the backfill material which minimizes compaction.

Accompanying the problems of trench instability and leaching of waste was migration of radionuclides from many of the burial sites. It is recognized that the waste cannot be retained perpetually within the repository, but efforts are made to minimize the transport of radionuclides from the repository by employing hydraulic barriers, both natural and engineered, in design of trench repositories to minimize water contact with the waste. In the unsaturated zone, flow may be retarded by capillary barriers in which a fine-grained layer overlies a coarse-grained layer. The barrier is created by the lower permeability of the coarse-grained layer at high moisture tension underlying a more permeable fine-grained bed.

Low permeability materials are severely limited as suitable host media for low-level radioactive waste repositories in the unsaturated zone. Capping materials are not impermeable, and experience has shown that they will become more permeable with time due to weathering, dessication, and biologic

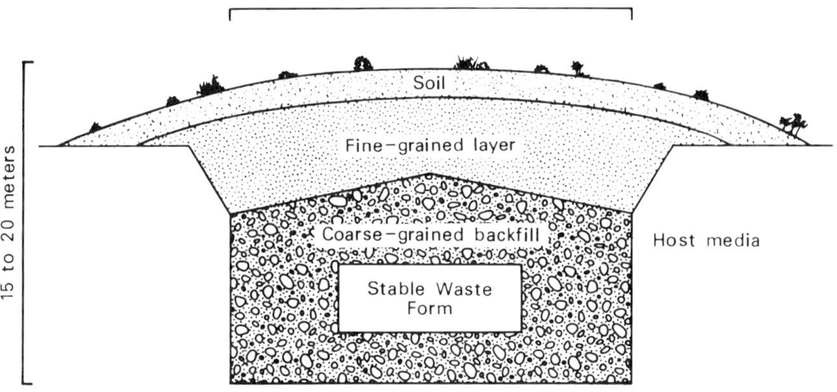

Fig. 2. Diagrammatic cross-section of a trench repository incorporating a capillary barrier.

activity. The resulting phenomenon is increased infiltration through the trench cap. A repository in low permeability host media will not permit the repository to drain as rapidly as water infiltrates the repository. Though the severe restrictions of low permeability material as a suitable host media have been amply demonstrated by the first generation of trench repositories, recent screening studies for low-level radioactive waste sites have been conducted that have targeted low permeability material as the host medium. Meyer [3] questioned if a more permeable host medium would not avoid the problems of the bathtub effect. Fischer and Robertson [2] recommended that the host medium be permeable.

Barriers to radionuclide migration should also be present in the ground-water flow path in the region between the repository and the natural discharge area. Factors which present major barriers to radionuclide transport include: (1) long flow paths, low hydraulic gradients and large effective porosity, and (2) those which decrease concentration of radionuclides in solution, such as sorption and low solubility of waste. Dispersion, diffusion, and dilution are processes which also reduce point concentrations of contaminants but do not reduce the amount of contaminant in solution. A diagrammatic cross section through a trench repository incorporating a capillary barrier is shown in Fig. 2.

Low-level radioactive waste may contain non-radioactive hazardous constituents (General Research Corp., [4]). Such constituents expected to be found in low-level radioactive wastes would include liquid organic compounds, lead, and chromium (Bowerman and others, [5]). Goode [6] sampled

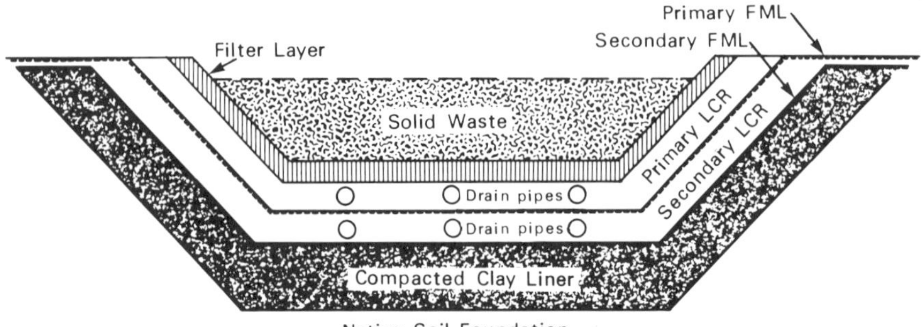

Fig. 3. Diagrammatic section through a hazardous waste landfill incorporating two flexible membrane liners (FML) and two leachate collection and removal systems (LCR).

ground water at several low-level radioactive waste sites in the United States and summarized the results of sampling done by other investigators. At the waste burial site at Sheffield [1], Goode found that organic-constituent concentrations correlated with the occurrence of tritium, supporting the hypothesis that the organic chemicals are associated with the tritium source.

LAND DISPOSAL OF SOLID WASTES

Landfills are often used to dispose of various types of solid hazardous and non-hazardous wastes. The regulations for land disposal of hazardous and non-hazardous wastes are contained in the Resource Conservation and Recovery Act (RCRA), U.S. Code of Federal Regulations, 40, Parts 240-280. The general provisions for siting and operating landfills apply to both non-hazardous and hazardous waste. Specific provisions for hazardous waste disposal are more stringent than for non-hazardous waste, such as requiring impermeable liners and leachate collection systems in hazardous waste landfills.

Hazardous organic waste

Guidance developed to implement the Resource Conservation and Recovery Act requires that hazardous wastes be disposed in landfills constructed with a flexible membrane liner and a leachate collection system. A diagrammatic section through a hazardous waste landfill incorporating two flexible membrane liners (FML) and two leachate collection and removal systems (LCR) is shown in Fig. 3. Ground-water monitoring to detect contaminants that may be migrating from the landfill is required.

Organic compounds comprise a large and complex group of chemicals having widely varying characteristics, reactions, and persistence to chemical and biological transportation. The major processes that are of significance in the degradation of organic compounds are sorption and degradation by oxidation, reduction, hydrolysis, and biological transformation. Although not well known, transformation of organic compounds by reduction is a potentially significant process under low-Eh conditions.

Many of the chemical reactions involving organic compounds are influenced by microorganisms. The cultivation of microorganisms for the transformation of specific organic compounds and groups of compounds is an area of intense research. However, the use of microorganisms in the control of organic compounds is still undergoing research and is not fully operational as a control method. It is expected that efforts to control the degradation of particular contaminants by the introduction of specific microorganisms cultures will continue. Preliminary data in Table 1 from Wilson and McNabb [12] indicate that biotransformation of organic compounds may be a function of aerobic and anaerobic conditions and the concentration of the individual organic compound. RCRA regulations require that leachate be withdrawn to maintain the leachate depth to less than a meter. In effect, the landfill is thus unsaturated and aerobic conditions probably will be prevalent. Table 1 shows that organic compounds are possibly transformed by microorganisms under either aerobic or anaerobic conditions. Therefore, it would appear that some organic compounds, such as some halogenated aliphatic hydrocarbons, may not be degraded in landfill repositories operated in accordance with RCRA regulations.

Transformation of organic compounds by hydrolysis refers to the reaction of the compound with water in which bonds are broken and new bonds with HO- and H- are formed. Hydrolysis is pH and metal-ion-concentration dependent and may vary with the degree of saturation. A sampling of half-lives of 54 organic chemicals based on degradation by hydrolysis (Ellington and others [7]) ranged from 1.7 hours to >197,000 years.

Treatment of organic chemicals by incineration of those amenable to the process would seem to be an efficient manner of treatment for many organic wastes intended for landfill disposal. Some organic compounds are resistant to known biological transformation and are not broken down to nonhazardous components by incineration. Organic compound which cannot be degraded by known means may have to be disposed by deep burial, as is proposed for high-level radioactive waste.

Table 1. Prospect of Biotransformation of Selected Organic Pollutants in Water-Table Aquifers[a]

Class of Compounds	Aerobic Water, Concentration of Pollutant ($\mu g/l$) 100	10	Anaerobic Water
Halogenated aliphatic hydrocarbons			
Trichloroethylene	None	None	Possible[b]
Tetrachloroethylene	None	None	Possible[b]
1,1,1-Trichloroethane	None	None	Possible[b]
Carbon tetrachloride	None	None	Possible[b]
Chloroform	None	None	Possible[b]
Methylene chloride	Possible	Improbable	Possible
1,2-Dichloroethane	Possible	Improbable	Possible
Brominated methanes	Improbable	Improbable	Probable
Chlorobenzenes			
Chlorobenzene	Probable	Possible	None
1,2-Dichlorobenzene	Probable	Possible	None
1,4-Dichlorobenzene	Probable	Possible	None
1,3-Dichlorobenzene	Improbable	Improbable	None
Alkylbenzenes			
Benzene	Probable	Possible	None
Toluene	Probable	Possible	None
Dimethylbenzenes	Probable	Possible	None
Styrene	Probable	Possible	None
Phenol and alkyl phenols	Probable	Probable	Probable[c]
Chlorophenols	Probable	Possible	Possible
Aliphatic hydrocarbons	Probable	Possible	None
Polinuclear aromatic hydrocarbons			
Two and three rings	Possible	Possible	None
Four or more rings	Improbable	Improbable	None

[a]From Wilson and McNabb [12]

[b]Possible but probably incomplete

[c]Probable but at high concentration

Municipal waste

The landfill should be sited and designed to minimize its impact on the ground-water resource. The typical municipal landfill receiving non-hazardous solid waste is operated by spreading waste in thin layers, then compacting and covering the waste with earth materials. When successive layers are added, the total height may be as great as 15 to 30 meters. The completed landfill is covered with compacted earth material and contoured to promote drainage and minimize infiltration. Vegetative growth at the surface is promoted to minimize infiltration. Vegetative growth at the surface

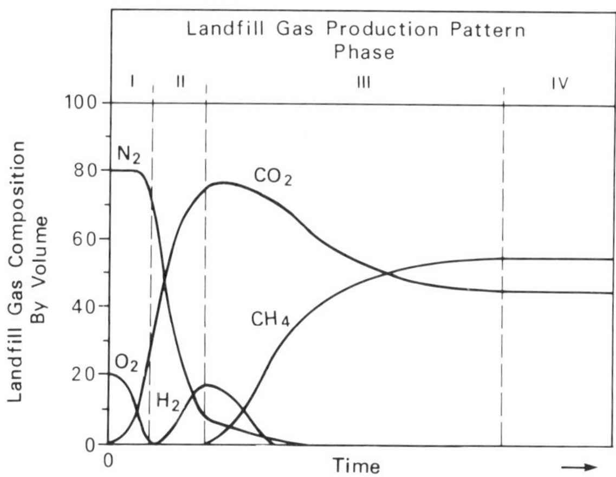

Fig. 4. Sanitary landfill gas production pattern (from Farquhar and Rovers [8]).

is promoted to minimize erosion and improve appearance. Sanitary landfills accept waste from local residential and commercial establishments. Impermeable liners, leachate collection systems, and monitoring systems are not required for sanitary landfills at the present time (1989), but some landfills are constructed with such features.

Farquhar and Rovers [8] have proposed four phases of landfill evolution based on the generation of gas in the landfill (Fig. 4).

Kjeldsen [9] discusses the processes involved in the degradation of the organic waste during the four stages. Phase 1 is an aerobic phase characterized by the rapid oxidation of readily degradable compounds, the buildup of heat and generation of carbon dioxide and nitrogen. Phase 2 is an anaerobic phase characterized by a decline in nitrogen production, an increase in carbon dioxide production, formation of ammonia and fatty acids, and lowering of the pH. Phase 3 is an anaerobic phase characterized by the steady production of methane and continued production of carbon dioxide. During this phase fatty acids are readily degraded to methane and carbon dioxide, resulting in low levels of fatty acids in the leachate. Phase 1 is a short period of a few weeks duration. Phase 2 may last from a few to several years. The duration of phases 3 and 4 is the longest and may last for decades.

Characteristics of municipal landfill leachate and waste water are shown in Table 2. As indicated by the analyses, municipal wastes contain materials other than organic wastes. Constituents such as metals in the landfill may be mobilized at certain phases of the landfill evolution.

The rates of waste degradation and gas generation are controlled by the composition of the waste, moisture, temperature, alkalinity, and pH. The optimum pH range for methanogenic bacterial activity is 6 to 8.5. A pH below 6 will lead to a much reduced methanogenic production. Sulfate and methanogenic bacteria compete for the same substances; therefore, methane production will be decreased in the presence of high sulfate concentrations. Nutrients, commonly present in the waste, are needed to sustain the activity of bacteria in the landfill. Anaerobic bacterial activity increases with temperature, with an optimal temperature for methanogenic bacterial activity at about 40 degrees Celsius. Methane production in the unsaturated zone increases with increased water content; however, saturation of the waste with water will cause methanogenesis to cease.

The primary objective of the landfill operation is to isolate the waste from human contact. It is also desirable to provide an environment where the waste will degrade to innocuous levels. The most likely means by which the waste products can be released from the landfill is by migration of leachate from the landfill. Degradation rates within the landfill are dependent upon adequate moisture to promote the degradation processes. If infiltration and leachate production is minimal, the processes of degradation will be decreased and the duration of potential ground-water contamination from the landfill be extended.

Kjeldsen [9] suggested the following landfill operational factors to aid in the degradation of the waste: 1) constructing a tall landfill with a low volume-to-surface area ratio decreases the heat loss from the landfill and decreases the infiltration required to provide adequate moisture to the landfill; 2) collection of leachate from drains at the base of the landfill and recirculation of the leachate through the landfill; 3) use of permeable materials for layering the waste during the build-up of the waste pile providing for rapid leaching of the landfill; and 4) mixing of the waste to form a homogeneous matrix and digested sewage sludge to provide nutrients to enhance the methanogenic processes.

CONCLUDING REMARKS

Trench disposal of low-level radioactive waste is similar to landfill disposal of municipal and hazardous solid wastes. U.S. Nuclear Regulatory Commission standards do not require liners in low-level radioactive waste repositories (U.S. Nuclear Regulatory Agency, 1982). The U.S. Environmental Protection Agency requires liners and leachate collection systems in hazardous waste landfill repositories but not in sanitary landfill (U.S. Environmental Protection Agency, 1986). Such regulatory differences impose

Table 2. Characteristics of Leachate from Municipal Solid Waste

Constituent	Range[a] (mg/L)	Range[b] (mg/L)
Chloride (Cl)	34–2,800	600–800
Iron (Fe)	0.2–5,500	210–325
Manganese (Mn)	0.06–1,400	75–125
Zinc (Zn)	0–1,000	10–30
Magnesium (Mg)	16.5–15,600	160–250
Calcium (Ca)	5–4,080	900–1,700
Potassium (K)	2.8–3,770	295–310
Sodium (Na)	0–7,700	450–500
Phosphate (P)	0–154	–
Copper (Cu)	0–9.9	0.5
Lead (Pb)	0–5.0	1.6
Cadmium (Cd)	–	0.4
Sulfate (SO_4)	1–1,826	400–650
Total N	0–1,416	–
Conductivity (μmhos)	–	6,000–9,000
Total dissolved solids	0–42,276	10,000–14,000
Total suspended solids	0–2,685	100–700
pH	0.7–8.5	5.2–6.4
Alkalinity as $CaCO_3$	0–20,850	800–4,000
Hardness total	0–22,800	3,500–5,000
Biological oxygen demand	9–54,610	7,500–10,000
Chemical oxygen demand	0–89,520	16,000–22,000

[a]U.S. Environmental Protection Agency [10]
[b]U.S. Environmental Protection Agency [11]

differences in the design, siting, and operational procedures for the various categories of waste and may indirectly result in differences in the hydrologic and geochemical environment of the repositories. An unlined radioactive waste repository would be designed to minimize infiltration by the use of capillary barriers and low permeability layers in the trench cap and would be located where long travel time of ground water would isolate the waste from the surface environment. A hazardous waste landfill would also be designed to minimize infiltration and would require maintenance to withdraw and treat the leachate from the collection sump. Both repositories would require monitoring to detect movement of hazardous constituents from the repository.

Most wastes are "mixed" waste; that is, they consist of variety of chemical products with different characteristics. The nature of leachates from low-level radioactive waste disposal facilities clearly indicates the "mixed" waste nature of some low-level wastes and the mobility of non-radioactive components of waste in the low-level radioactive waste. The inclusion of a mixed waste in a repository designed for a waste of a particular physical or chemical character may compromise the efficiency and integrity of a waste repository.

Burial sites vary greatly in their chemical and hydrologic environment and in their capacity to retain waste in or near the burial area and to provide an environment in which the waste will degrade. Most repositories are designed to provide a combination of engineered and natural barriers to retard release of waste from the repository and to delay the transport of waste from the repository to the biosphere. However, it is unrealistic to expect that hazardous waste or its products will be permanently contained in the waste disposal area.

Monitoring of the ground-water flow system at hazardous landfill sites is required by the Resource Conservation and Recovery Act. Based on the composition of observed leachate from sanitary landfills, it also would appear desirable to provide impermeable membrane liners, leachate collection systems, and ground-water monitoring networks at municipal landfill sites.

NOTICE

The information in this document has been funded in part by the United States Environmental Protection Agency under Cooperative Agreement #CR814701-01 to the University of Nevada's Environmental Research Center. It has been subjected to agency review and approved for publication.

REFERENCES

[1] Sun, R.J., 1982, Selection and investigation of sites for the disposal of radioactive wastes in hydraulically induced subsurface fractures: U.S. Geological Survey Professional Paper 1215, 87 p.

[2] Fischer, J.N., and Robertson, J.B., 1984, Geohydrologic problems at low-level radioactive waste sites in the United States of America, in Radioactive waste management, an International Conference, Seattle, Wash., 1983, Proceedings: Vienna, International Atomic Energy Agency, v. 3, p. 537-547.

[3] Meyer, G.L., 1979, Problems and issues in the ground disposal of low-level radioactive wastes, 1977, in Carter, M.W., Moghissi, A.A., and Kahn, Bernd, eds., Management of low-level radioactive waste: New York, Pergamon Press, v. 2, p. 637-664.

[4] General Research Corp., 1980, Study of chemical toxicity of low-level wastes: U.S. Regulatory Commission Report NUREG/CR-1973, v. 1, p. 253.

[5] Bowerman, B.S., Kempf, C.R., MacKenzie, D.R., Siskind, B., and Piciulo, P.L., 1985, An analysis of low-level wastes - Review of hazardous waste regulations and identification of radioactive mixed wastes: U.S. Nuclear Regulatory Commission Report NUREG/CR-4406, p. 159.

[6] Goode, D.J., 1989, Mixed waste leachates in ground-water at low-level radioactive waste repository sites: in Bedinger, M.S., and Stevens, P.R., eds., 1989, Proceeding

of the U.S. Geological Survey Workshop on Low-Level Radioactive-Waste Disposal, Big Bear Lake, California, July 11-16, 1987, p. 146-166.

[7] Elligton, J.J., Stancil, F.E., Jr., Payne, W.D., and Trusty, C., 1987, Measurements of hydrolysis rate constants for evaluation of hazardous waste land disposal: Volume 2. Data on 54 Chemicals: U.S. Environmental Protection Agency, Project Summary EPA/600/S3-87/019, p. 6.

[8] Farquhar, G.J., and Rovers, F.A., 1975, Leachate attenuation in undisturbed and remoulded soils, Proc. Res. Symp. Gas and Leachate from Environmental Protection Agency, Cincinnati, Ohio.

[9] Kjeldsen, Peter, 1988, Degradation within waste disposal sites: Proceedings UNESCO Workshop on Impact of Waste Disposal on Groundwater and Surface Water, August 15-19, 1988, Copenhagen, Denmark, p. 12.

[10] U.S. Environmental Protection Agency, 1973, An environmental assessment of potential gas and leachate problems at land disposal sites: U.S. Environmental Protection Agency, Cincinnati, Ohio, Open-File Report SW-110, p. 33.

[11] U.S. Environmental Protection Agency, 1975, Gas and leachate from land disposal of municipal solid waste: U.S. Environmental Protection Agency, Cincinnati, Ohio.

[12] Wilson, J.T., and McNabb, J.F., 1983, Biological transformation of organic pollutants in groundwater: EOS, Trans. Am. Geophys. Union, vol. 64, p. 505.

CONSIDERATION OF HYDROGEOLOGIC FACTORS IN DESIGNING WELLHEAD PROTECTION AREAS

M. S. BEDINGER

Environmental Research Center,
University of Nevada–Las Vegas, Las Vegas, NV 89154

S. P. GARDNER

U.S. Environmental Protection Agency Environmental
Monitoring Systems Laboratory, Las Vegas, NV 89119

ABSTRACT

Legislation was enacted in 1986 to establish the Wellhead Protection Program in the United States. This is a Federal program directing the U.S. Environmental Protection Agency to assist the states in designating and protecting areas surrounding public water supply wells from contaminants that may adversely affect human health. Under this program, the states will delineate the Wellhead Protection Area (WHPA) for each public well or wellfield. The WHPA is defined as the surface and subsurface area surrounding a well or wellfield through which contaminants can reasonably be expected to move toward and reach the well. The monitoring of WHPAs can be an integral component of the wellhead protection program. The components of monitoring network design in wellhead protection areas include geohydrologic and geochemical factors of the aquifer flow system; the hydraulics

Chemistry for the Protection of the Environment
Edited by L. Pawlowski *et al.*, Plenum Press, New York, 1991

of the wellfield; external hydraulic stresses on the flow system, that is, stresses other than those of the wellfield; anthropogenic and natural sources of potential contaminants; and the chemical and organic nature of potential contaminants. The monitoring of wellhead protection areas should be designed to determine sources of potential contaminants in the WHPA and to detect migration of contaminants toward the wellfields in time to provide remedial or mitigating action to protect the water supply, or in time to provide alternative public water supplies before the source is contaminated. Another role of monitoring is to refine the delineation of the WHPA. The contributing area to the wellfield is not static, but subject to change as it is influenced by increase or decrease in withdrawal from the public supply and as the flow system is influenced by other stresses, including climatic changes that affect recharge from infiltration of precipitation and streamflow, and withdrawal for agricultural or industrial supplies. These factors and others may influence the size and location of the zone of contribution to the well or wellfield. It is therefore important to initially delineate a WHPA that is large enough to adequately protect the water supply under different discharge-recharge conditions.

INTRODUCTION

Nearly half the population of the United States obtains drinking water from wells and springs [2]. Frequently, these water supplies are contaminated as the result of improper management of the wastes produced by human activity. The U.S. Environmental Protection Agency's Wellhead Protection Program, a provision of the 1986 Amendments to the Safe Drinking Water Act (SDWA), is one of several methods intended to solve such drinking-water contamination problems. The Wellhead Protection Program is designed to assist each state in delineating and monitoring WHPAs. These areas will be managed by state and local authorities in order to prevent contamination to the public water supplies. By providing a strategy for delineating and monitoring WHPAs, the Wellhead Protection Program offers the States a mechanism to ensure early warning of contamination and to provide time to take preventative or defensive action against contamination.

Designing an effective Wellhead Protection Program for a particular water supply involves several major steps. First, the boundaries of the WHPA must be defined. The WHPA is a zone that is to be protected against the introduction and migration of contaminants that may adversely affect human health (SDWA, Section 1428(a)). Therefore, the WHPA should surround the well or wellfield and should encompass all the areas through which contaminants can reasonably be expected to move toward and reach the well. Second, potential sources of contamination within the WHPA must be identified. These sources can include agricultural pesticide application and runoff, leaking underground storage tanks, accidental releases of hazardous chemicals, road-salting operations, illegal dumping, septic systems, and leaking sewers. Third, a WHPA management plan must be formulated. A common element of this plan is implementation of land-use controls on industries that have the potential to cause pollution. Another common element is the implementation of a ground–water monitoring program. Fourth, a contingency plan must be formulated to protect against the effects of an accident such as an overturned tanker truck on a highway or a fire at a chemical factory. Contingency planning can be critical: In one case in Ohio, a warehouse containing paint and chemicals was allowed to burn. If, instead, water had been sprayed on the fire, contaminants would have been washed into the aquifer. Fifth, the need for new water supply wells must be addressed, and appropriate locations should be identified. The installation, testing and implementation of a ground-water monitoring network can provide hydrogeologic information that is useful in refining the delineation of WHPAs and in identifying suitable locations for new wells or wellfields. In addition, the monitoring network can assist in detecting contamination and in identifying unknown sources of contamination. The factors involved in delineating WHPAs, the hydraulic principles used to determine the ground-water zone of contribution to the wellhead, and the factors to be considered in developing strategies for monitoring WHPAs are the subjects of this paper.

DELINEATING WELLHEAD PROTECTION AREAS

Amendments to the Safe Drinking Water Act mandated that the Administrator of the Environmental Protection Agency issue technical guidance that the States may use to determine the extent of each WHPA (SDWA, Section 1428(e)). Criteria used for delineating WHPAs are distance from the well, drawdown of the aquifer, time of travel of water to the well, hydrogeologic boundaries of the aquifer, and the capacity of the aquifer to assimilate contaminants. Methods proposed by the U.S. Environmental Protection Agency [3] to implement these criteria for the delineation of WHPAs include (1) defining an arbitrary radius around the well to be protected, (2) defining a fixed radius based on a volume of water withdrawn by the well, (3) defining a variably shaped zone, that is a zone that is not circular, based on hydraulic properties of the aquifer, the flow regime of the aquifer, and the time of travel to the well, (4) using analytical methods to define the area, based on the hydrogeologic properties of the aquifer and on the slope of the water table, (5) conducting hydrogeologic mapping to define the radius from the well, based on the flow boundaries of the aquifer, (6) developing deterministic flow models of the aquifer system, and (7) developing fate and transport models of contaminant movement in the subsurface. These methods represent a general progression of sophistication from practically no knowledge of the flow system for the first method to a detailed knowledge for the sixth method. Delineation of the WHPA should be based primarily on hydrogeologic characterization of the site. Analytical and modeling methods may be best used as tools to complement a site characterization study. The adequacy of a network for monitoring WHPAs is as good as the basis for delineating the WHPA, and the quality of both elements is a function of the quality of site characterization.

In the more sophisticated approaches to delineation, the WHPA is all or part of the zone of contribution to the well. In the most sophisticated approach, using the assimilative capacity criterion, a fate and transport model predicting contaminant movement within the zone of contribution would be implemented to study the predicted concentration of contaminants at the wellhead. The delineation of the WHPA will depend upon the time of travel within the zone of contribution to the well or wellfield and the nature and persistence of the potential contaminants within the aquifer. Figure 1A shows the water table, the zone of contribution to a pumping well, and the contours showing time of travel to the pumping well. The time of travel represents the time it would take a drop of water or a conservative (nonreactive) chemical constituent to reach the pumping well.

ZONE OF CONTRIBUTION TO WELLS

The hydraulic response of an aquifer to withdrawal of water from wells was first comprehensively expressed by Theis [1]. He recognized that when a well is pumped or otherwise discharged, water is initially withdrawn from storage in the aquifer near the well. With continued pumping, the water discharged is eventually balanced by an increase in natural recharge to the aquifer or a decrease in natural discharge, or both. If withdrawal from a well or wellfield continues for a long period of time at a steady rate, the flow system will reach a state of dynamic equilibrium. At equilibrium, the zone of contribution to the well will be stabilized. Wellhead protection is concerned with the source of the water contributed to the well or wellfield. These sources may include infiltration of recharge; induced infiltration from a surface water body, such as a lake or stream; or induced leakage from or through a confining bed.

The extent of the zone of contribution is a function of (1) the rate and distribution of withdrawal and (2) hydrogeologic site factors. The site factors include the physical and hydraulic characteristics of the aquifer flow system and the nature, geometry, and flow conditions of the boundaries of the aquifer. The zone of contribution may be approximately determined by superposing the cone of depression on the natural head in the flow system. A cross section and plan view of an unconfined aquifer which illustrate the cone of depression about a pumping well and the zone of contribution to the well, are shown in Figure 1. In small aquifers, the total zone of contribution may become the WHPA. In larger aquifers, some time-of-travel criterion, such as a five- to ten-year travel time, may be applied within the zone of contribution to define the WHPA.

MONITORING WELLHEAD PROTECTION AREAS

The ultimate purpose of monitoring WHPAs is to detect the presence of contaminants that threaten to contaminate the water supply. Elements that should be considered in devising a wellhead monitoring network include (1) identification of potential sources of contamination before contaminants are introduced to the aquifer, (2) identification of potential contaminants in the subsurface before they reach the wellhead, (3) chemical, biological, and physical nature of the potential contaminants, and (4) the hydrogeologic nature of the ground-water flow system. Some monitoring methods may be entirely external to the aquifer. For example, in an aquifer subject to contamination by infiltration from the surface, the monitoring methods may include development of a history of land use, such as agricultural, industrial, and commercial activities in the area, beginning before the development of a wellfield. Such developmental histories and research on the potential for contamination of the aquifer may be aided greatly by use of historic and specially acquired aerial photographs and their interpretation. The documentation of changes in land use in the WHPA should be continued as a part of the wellhead monitoring program. The chemical, biological, and physical nature of the potential wastes may determine to great extent the method of data collection, the frequency of data collection, method of sample collection and field treatment, and the analytical treatment and procedures to be used on the samples. The hydrogeologic nature of the ground-water flow system is used in combination with the preceding factors in determining the areal and vertical placement of sampling points and the frequency of sample collection. The hydrogeologic factors determine the nature of the vulnerability of the aquifer to contamination. The hydrogeology also influences the rate and direction of movement of contaminants and the fate of contaminants in the aquifer.

In addition to the primary objective of wellhead monitoring, the wellhead monitoring network has other useful purposes. Wellhead monitoring

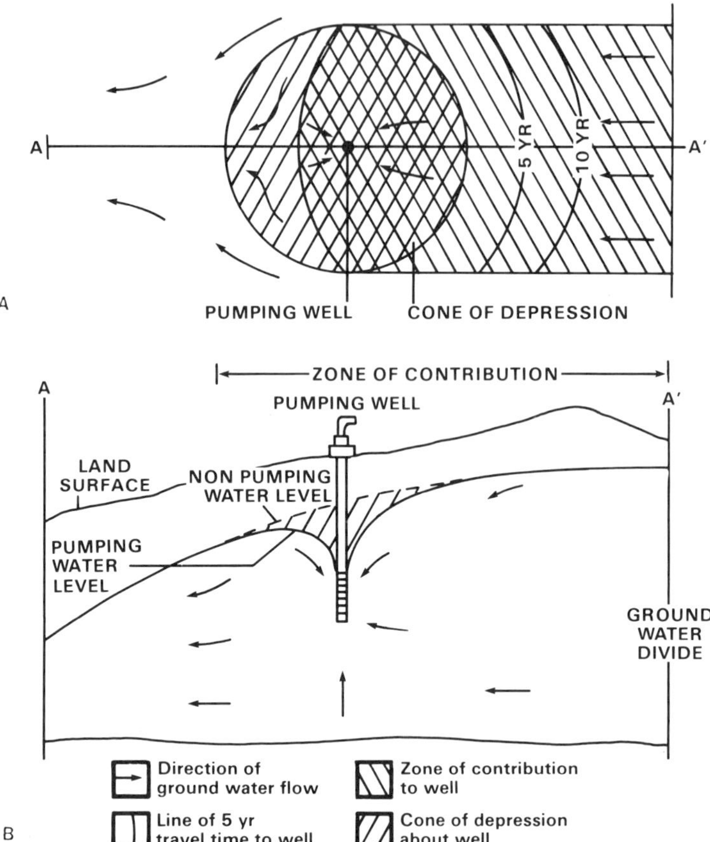

Fig. 1. Plan view and cross section of a pumping well in an unconfined aquifer. A. Plan view of cone of depression and resulting zone of contribution to a pumping well. B. Cross section showing cone of depression about a pumping well in an unconfined aquifer.

is useful in obtaining hydrologic data to further define the zone of contribution. Data obtained can also assist in refining the extent of changes in the zone of contribution in response to changes in withdrawal from the well or wellfield and in response to changes in discharge or natural stresses acting on the aquifer. WHPA monitoring is a useful aid in managing the wellfield efficiently and economically. The information gained may also be an aid in expanding the wellfield.

The Environmental Monitoring Systems Laboratory, Las Vegas, Nevada (EMSL–LV) of the U.S. Environmental Protection Agency is conducting research to develop innovative and cost–effective monitoring strategies in different hydrologic and contaminant source settings. EMSL-LV is funding research in cooperation with local governments at five sites in different hydrogeologic settings to develop case studies of design, implementation, and evaluation of WHPA monitoring programs. The five hydrogeologic settings include (1) an unconfined aquifer, (2) an aquifer in connection with a perennial stream, (3) a confined aquifer, (4) a karst aquifer, and (4) a fractured bedrock aquifer.

UNCONFINED AQUIFER

Consider an unconfined, porous media, areally extensive aquifer, as shown in Figure 1. Recharge to the aquifer is by infiltration of precipitation; discharge is by evapotranspiration (in areas where water levels are shallow) or as liquid discharge to springs, seeps, or streams. We assume that the aquifer is areally extensive, such that the effects of ground–water discharge from the well do not induce flow directly from a stream or surface water body to the well. The zone of contribution to the well is within the recharge area of the aquifer. The zone of contribution extends to the ground-water divide; the natural discharge of the aquifer is reduced by the amount withdrawn from the well. Within the zone of contribution there is potential for contaminants that enter the aquifer to reach the wellhead.

Monitoring for wellhead protection is concerned with infiltration of contaminants from the surface and near surface. Unconfined aquifers are commonly at shallow depths and receive recharge by infiltration of precipitation. Thus, unconfined aquifers are subject to contamination by infiltration of a wide variety of soluble and liquid contaminants resulting from numerous human activities. Among the most prominent sources of contamination are municipal and commercial waste (sanitary) landfills, hazardous waste landfills, radioactive waste repositories, accidental spills, unregulated dumping of wastes, underground storage tanks, surface and subsurface mining operations, septic tanks and cesspools, waste water and waste liquid storage lagoons, beef cattle feed lots, municipal sanitary sewer systems, urban runoff, land application of wastes, and agricultural application of herbicides and pesticides. Unconfined aquifers may also be contaminated by subsurface injection of wastes, either by direct injection or by leakage from poorly constructed or deteriorated wells.

Because contaminants can enter an unconfined aquifer from any point along its upper surface, the cost of drilling monitoring wells and performing routine water analyses may become a large factor in the monitoring network design. Although timely detection of contamination is essential, it may be too expensive to drill monitoring wells on a closely spaced grid. Geostatistical approaches such as kriging, and the use of ground-water models may aid in determining optimum well locations in such situations. Monitoring wells should also be placed next to known potential sources of contamination. In the city of Littleton, Massachusetts, the water department is requiring the industries that could potentially pollute the aquifer to provide their own monitoring networks around their facilities. Routine water analysis is paid for by the industries themselves.

AQUIFER IN CONNECTION WITH A STREAM

Consider an unconfined aquifer in which discharge from the well causes a reversal of the natural hydraulic gradient, that is, causes the gradient to slope from the stream to the well (Fig. 2). The zone of contribution to the well includes the stream, which contributes directly to the well, and recharge by infiltration of precipitation.

The monitoring network, as in the previous case, must account for all surface activities which may release liquid or soluble contaminants. These contaminants may infiltrate to the water table and waste injection wells. In addition, the stream, the contaminant sources feeding into the stream, and the ground-water flow paths from the stream to the well are of primary importance to monitoring.

In some cases, the aquifer system may be able to assimilate some of the contaminants induced into the aquifer from the stream. Hydrocarbon compounds of fuels that are soluble in water do not persist long in the presence of oxygen due to aerobic bacteria. Chlorinated hydrocarbons used as solvents do not readily oxidize and can persist for long periods of time in the subsurface. Fate and transport computer models may have to be developed for the contaminants of concern in the surface water body to determine if induced recharge will affect water quality at the wellhead.

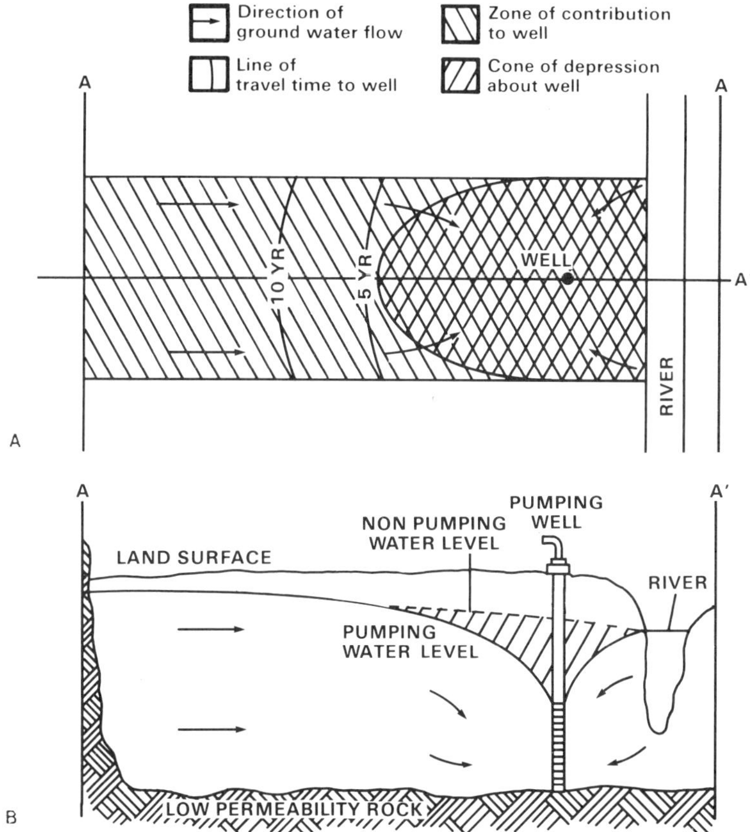

Fig. 2. Plan view and cross section of a pumping well inducing recharge from a stream. A. Plan view of cone of depression and resulting zone of contribution to a pumping well including recharge from a stream. B. Cross section showing cone of depression about a pumping well including recharge from a stream.

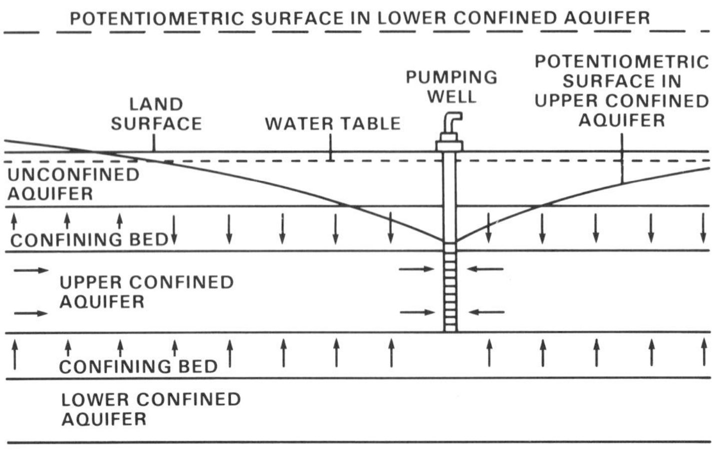

Fig. 3. Confined aquifer flow system.

CONFINED AQUIFER

Consider the confined hydrogeologic setting shown in Figure 3. The upper surface of the aquifer is formed by a bed of low permeability; the water level in a well tapping the aquifer will rise above the aquifer. The water discharged by a well tapping an artesian aquifer is derived initially by expansion of the water and from storage in the vicinity of the well. As pumping continues, water may be drawn from greater distances within the aquifer and from overlying and underlying confining beds. Commonly, this distance is so great that the outcrop area is well beyond the WHPA. Because the permeability of the overlying confining bed is relatively low, there is less opportunity for infiltration of recharge and contaminants from the surface in the immediate vicinity of the wellfield. However, if the head in the confined aquifer is lowered beneath the head in the overlying or underlying beds, flow will be induced from the adjacent beds to the confined aquifer being pumped.

The zone of contribution for a deep artesian aquifer may be a volume totally beneath the land surface. Deep artesian aquifers may be many tens or hundreds of miles from the outcrop of the formation comprising the aquifer. On the other hand, the artesian aquifer may be at relatively shallow depth, and the beds may be of moderate hydraulic conductivity. Thus, the sequence of beds above and below the aquifer, their hydraulic conductivity and thickness, and the head in overlying and underlying beds is significant in delineating the WHPA and in designing the wellhead monitoring network. The potential for contamination from upward leakage in deep artesian aquifers may be a greater threat to the wellhead than to shallow water-table aquifers. Deep artesian aquifers may be near underlying beds that contain saline water or near beds into which wastes have been injected. Abandoned and improperly constructed supply and waste injection wells provide potential pathways for contamination from adjacent beds.

The monitoring requirements for each artesian aquifer must be assessed individually on the basis of the hydrogeologic setting and the properties of the rocks. In some cases, the need to monitor the water quality in a confined zone must be weighed against the possibility of introducing contamination by drilling through a polluted surface aquifer. Natural and man-made breaks in confining layers represent one of the largest threats for introducing contaminants in confined aquifers.

AQUIFER IN KARST TERRANE

Karst terrane is typically formed in areas underlain by limestone, dolomite, or gypsum. Such rocks are commonly aquifers by virtue of solutional enlargement of joints, fractures, and bedding-plane openings. Karst regions typically contain sinkholes, caverns, and springs. Consider the karst terrane aquifer shown in Figure 4. The karst aquifer is subject to contamination from surface sources, as is the unconfined aquifer. The spring serves as the source of ground water for a public water supply. The exact nature and extent of the flow system contributing to the spring is uncertain. The flow of water from a sinkhole to the spring is relatively rapid; the flow paths from the sinkholes to the spring may be difficult or impossible to locate and tap by monitoring wells. Because water in a karst setting travels rapidly, WHPAs should probably be delineated on the basis of the hydrologic flow boundaries. Dye–tracing may offer a method of determining flow boundaries in karst terrane. The monitoring of the WHPA should be keyed primarily to the monitoring of potential surface sources of contaminants that may enter the flow system by way of the sinkholes or by infiltration at other points. Another key point of monitoring in a spring system is the spring itself. Continuous monitoring devices may be used at points along the flow system, if it can be traced. Because of the rapid flow rates in karst, options after detecting contamination are limited. Shutting off a city's water intake until the pulse of contamination has passed is one option, though not an attractive one if the spring is the only source of supply.

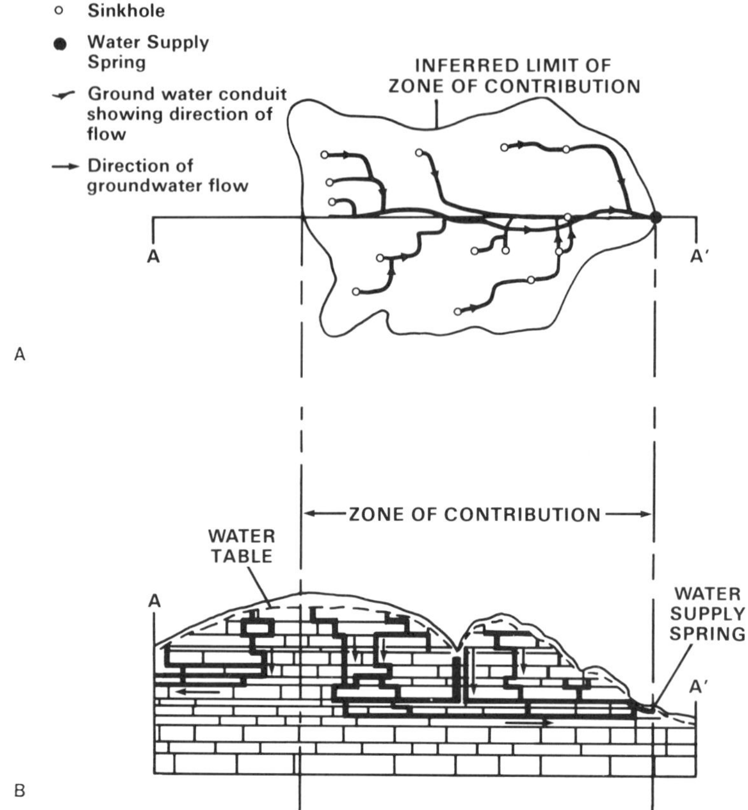

Fig. 4. Diagrammatic plan and cross sectional views of a flow system in a karst aquifer. [3] A. Plan view showing location of sinkholes, springs and inferred ground water conduits. B. Cross section illustrating principal flow paths in solution channels.

FRACTURED BEDROCK AQUIFER

Many dense rocks such as those of igneous and metamorphic origin may have negligible interstitial porosity and permeability. Such rocks might be potential aquifers solely by virtue of fractures and joints (Fig. 5). Depending on the degree of heterogeneity of the aquifer, the zone of contribution to the wellhead may be very difficult to define. Field observations and geophysical methods can sometimes be used to determine fracture locations and orientations. Dye–tracing techniques used to identify flow patterns between wells may aid in defining the flow system and rates of ground-water movement.

Fractured bedrock aquifers may be confined or unconfined. If uncon-

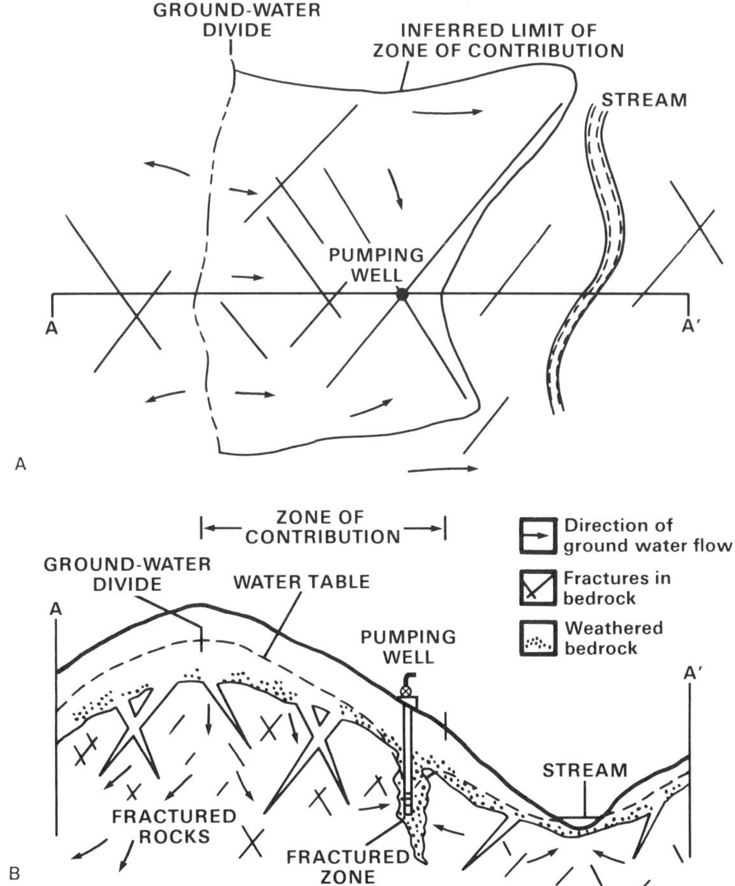

Fig. 5. Diagrammatic plan view and cross section of a pumping well in a fractured rock aquifer. [3] A. Plan view showing pumping well and inferred zone of contribution. B. Cross section showing pumping well.

fined, the aquifer is subject to contamination from surface sources. In many fractured aquifer systems, movement from the source of contamination to the wellhead will be rapid. The keys to monitoring in many fractured-rock aquifers will be to monitor potential surface sources of contamination and to monitor the discharge from the well. Monitoring wells may be situated at points along the fractures that feed the wells to determine water quality and to provide information on identifying new wells.

CONCLUDING REMARKS

The zone of contribution to a well or wellfield may change with time thereby affecting the potential zone of contamination. The delineated WHPA should be large enough initially to include these changes in the zone of contribution. Otherwise, activities that are conducted legally outside the WHPA but have high potential for contributing to contamination may be brought inside the management zone over time, causing expensive relocation of facilities. Data collected from a monitoring network should include observations of changes in active well locations and discharge in addition to routine and periodic synoptic measurements of water level in the aquifer. These data will assist in monitoring changes in the zone of contribution and redefinition of time-of-travel zones relative to the wellfield.

The delineation of the WHPA and the design of the monitoring network are dependent upon the characterization of the site. Lack of site characterization data and arbitrary delineation of the zone of contribution to the wellfield will result in a correspondingly weak, possibly erroneous and self-defeating basis from which to design an effective monitoring network.

Target zones for wellhead monitoring include (1) potential sources of contamination, (2) spatially distributed points within the WHPA, and (3) the discharge from the wellfield or spring. The relative emphasis and importance of each target zone for a given wellhead monitoring network is a function of the nature of the flow system and the quality of the site characterization.

NOTICE

The information in this document has been funded by the United States Environmental Protection Agency under Cooperative Agreement #CR814701-01 to the University of Nevada's Environmental Research Center. It has been subjected to Agency review and approved for publication.

REFERENCES

[1] Theis, C.V., 1940, The source of water derived from wells: Civil Engineering, vol. 10, no. 5, pp. 277-280.

[2] U.S. Geological Survey, 1984, National Water Summary: U.S. Geological Survey Water Supply Paper 2275.

[3] U.S. Environmental Protection Agency, 1987, Guidelines for Delineation of Well-head Protection Areas: U.S. Environmental Protection Agency, EPA 440/6-87-010, 131 p.

AN EXPERT SYSTEM APPROACH FOR SELECTION OF SAMPLING METHODS FOR GROUND-WATER CONTAMINATION AT HAZARDOUS SITES

R. E. CAMERON, R. A. OLIVERO, K. J. CABBLE, C. CARLSEN, M. D. TEUBNER, D. W. BOTTRELL[1] and M. T. HOMSHER[2]

Lockheed Engineering & Sciences Company
Environmental Programs Office 050 East Flamingo Road
Las Vegas, NV 89119, USA
[1] U.S. Environmental Protection Agency
Environmental Monitoring Systems Laboratory-Las Vegas
P.O. Box 93478, Las Vegas, NV 89193-3478, USA
[2] National Sanitation Foundation
P.O. Box 1468 Ann Arbor, MI 48106, USA

ABSTRACT

The development of an expert system to assist in sampling plan preparation for ground water contamination assessment projects is described. The modules in the current prototype provide information to guide the choice of methods for sampling, sample handling, quality assurance/quality control, safety measures, and documentation. Operation of the environmental sampling expert system is presented, including an example of the sample handling module for the pesticide Aldicarb. Some advantages and limitations of the expert system approach are discussed.

Chemistry for the Protection of the Environment
Edited by L. Pawlowski *et al.*, Plenum Press, New York, 1991

INTRODUCTION

A worldwide growing awareness and concern about the environmental impact of human activity has created a need for the development of better and, at the same time, cost-effective ways to assess the condition and quality of the air, surface water, ground water, sediment, soil, and biota that constitute most of our natural environment. The protection of vital water resources is of particular interest. For example, in the United States the first two major pieces of environmental legislation were aimed at the preservation and improvement of the aquatic environment (the Clean Water Act of 1972 and the Safe Drinking Water Act of 1974) [1]. One of the areas receiving increased attention is the monitoring of subsurface reservoirs, on which a large segment of the population depends for potable and agricultural water use.

Ground-water protection is a primary environmental concern in the United States as well as an international issue. Common objectives in ground-water monitoring are: to determine the water quality and the chemistry of the ground-water environment in a particular region; to determine the water quality and the chemistry of a specific water supply, well, or well field for either drinking or agricultural use; to determine the extent of ground-water contamination from a known source; or to monitor a potential source of contamination to protect the aquatic environment [2]. Ground water may become contaminated in several ways. Leachates from industrial areas, mining, and drill sites; urban runoff; landfills; chemical deposits and spills; and applications of agrochemicals can percolate from the land surface into the ground water and affect the water quality [3]. Industrial areas, landfills, and chemical spills can introduce volatile organics, semi-volatile organics, and metals into the ground water [4]. Agricultural operations and applications can introduce pesticides, hydrocarbon carriers, nutrients from fertilizers (particularly nitrates), other salts, and microorganisms to the ground water [5].

Chemical analyses and the associated sampling play a critical role in determining our ability to monitor and protect the environment. Obtaining reliable data in a cost-effective manner requires careful planning and the use of sound methodology. In particular, sampling and monitoring of ground water are especially difficult tasks that demand great expertise due to the slow rate of movement of ground water and the various routes the contaminated water can take. Monitoring wells are installed to determine ground-water quality of the aquifer at a particular location [6]. The main items to consider when sampling ground water are the sampling device that is to be used, purge volumes, field determinations, sample handling, sampling frequency, and quality assurance and quality control (QA/QC) that will meet the data quality objectives of the project [7]. The use of inappropriate techniques and equipment may result in the generation of unreliable data, a situation that has been found to be rather common. The problem in many occasions is compounded by lack of expertise on the part of managers, as well as field personnel, which results in the use of unknown-quality data for decision making.

Ground-water sampling expertise can be made more widely available for planning, consulting, evaluation, reviewing, and training purposes through the implementation of computerized advisory systems called "expert systems." This work describes an expert system prototype for sampling of ground-water contaminants being designed and developed as an extension to a previous Environmental Sampling Expert System prototype which provides advice for certain aspects of soil sampling [8].

ENVIRONMENTAL SAMPLING EXPERT SYSTEM

Expert systems (or knowledge-based systems) are computer programs that provide users with advice in specific areas of expertise by following a process similar to the one employed by human experts [9]. In most expert systems the process employed is a chain of logical decisions that are made by using a set of rules with antecedents and conclusions (Fig. 1). The knowledge about the problem domain (ground-water sampling in this case) is stored in the rules and in a collection of facts about the domain. When the user's information is input into the program, the result of the inference process is a set of advice statements containing solutions relevant to the specific problem described by the user. An important feature of expert systems is that they can only be applied to specific, and many times narrow, problem domains because it is not possible at present to provide a computer with an "artificial common sense" that would allow it to reason generally about broad subjects.

Two expert system prototypes currently under development at the U.S. Environmental Protection Agency's Environmental Monitoring Systems Laboratory-Las Vegas (USEPA EMSL-LV), are intended to provide advice in the preparation or review of sampling plans and to serve as training aids in this area. The Environmental Sampling Expert System-Soil Metals (ESES-SM) program [10] is a predecessor to the Ground Water module (ESES-GW). ESES-SM assists in sampling for the determination of the extent of metal pollution in soil. It queries the user for information on the project data quality objectives and site and soil characteristics. It then provides advice on appropriate statistical designs, quality assurance and quality control procedures, sampling techniques, sample handling, budget requirements, personnel safety measures, and documentation. ESES-GW, the ground-water

```
IF analyte-name INCLUDES aldicarb
THEN analyte-class INCLUDES pesticide

IF analyte-class INCLUDES pesticide
AND pesticide-carrier IS hydrocarbon
THEN analyte-class INCLUDES volatile_organics

IF analyte-class INCLUDES volatile_organics
AND bladder_pump-viability IS NOT negative
THEN pump-type IS bladder

IF analyte-class INCLUDES volatile_organics
AND bladder_pump-viability IS negative
AND bailer_pump-viability IS NOT negative
THEN pump-type IS bailer

IF analyte-class INCLUDES volatile_organics
AND bladder_pump-viability IS negative
AND bailer_pump-viability IS negative
AND submersible_pump-viability IS NOT negative
THEN pump-type IS submersible

IF pump-type IS bladder
AND ambient_air-quality IS good
THEN purge_gas-type IS ambient_air

IF pump-type IS bladder
AND ambient_air-quality IS poor
THEN purge_gas-type INCLUDES compressed_air
AND purge_gas-type INCLUDES nitrogen

IF well_casing-diameter IS LESS THAN 2 [inches]
OR sampling-depth IS GREATER THAN 500 [feet]
THEN bladder_pump-viability IS negative

IF water-turbidity IS high
THEN bladder_pump-viability IS negative

IF well_casing-diameter IS EQUAL OR GREATER THAN 2 [inches]
OR sampling-depth IS GREATER THAN 100 [feet]
THEN bailer_pump-viability IS negative

IF well_casing-diameter IS LESS THAN 3 [inches]
THEN submersible_pump-viability IS negative

IF analyte_class INCLUDES volatile_organics
THEN peristaltic_pump-viability IS negative
AND air_lift_pump-viability IS negative
```

Figure 1. Example of inference rules. Pump selection for ground-water pesticide sampling.

counterpart, has an extended analyte coverage that includes organics in addition to metals.

GROUND-WATER SAMPLING EXPERT SYSTEM DEVELOPMENT

A ground-water sampling project has characteristic design features that depend on the overall objectives of the study and the conditions in the field. In general, each project requires that decisions be made about the number of wells, the location of each well, well design and construction, well-drilling method, well development, sampling, field determinations, and sample handling. Related aspects are QA/QC, personnel safety, budgeting, documentation, and certain regulatory aspects. Figure 2 is a schematic representation of the ground-water monitoring process.

The goal of the first phase for the ESES-GW prototype is to provide sampling planners and reviewers with a computerized tool for the sampling phase. The system can help decide what types of ground-water sampling devices are appropriate to use under given site conditions. Advice is also given on proper sample handling, field determinations, QA/QC procedures, personnel safety, and documentation. The aspects covered in the current ESES-GW prototype are denoted by double-frame boxes in Figure 2. The current prototype does not cover well placement and well installation. The system makes the following assumptions: the initial site characterization has been performed for the hydrogeological formation of interest and the wells to be sampled are already in place. Information regarding project objectives and specific site conditions is entered into the system in response to questions asked from the user in an interactive session. The system then leads the user through various aspects of a sampling program and selects the most appropriate options based on the expert knowledge coded into it.

Project objectives and goals for monitoring or remediation efforts are supported by data quality objectives. Data quality objectives (DQOs) are qualitative and quantitative statements that specify the quality of data required to support an environmental decision [11]. For the sampling plans, a primary DQO concern is the representativeness of a ground-water sample. For the selection of analysis options, precision and accuracy are primary DQO concerns. Completeness, comparability, and detectability are other major DQO issues. The development of QA/QC in the expert system focuses on the DQOs for the specific project. DQO information is requested from the user to identify the level of uncertainty that a project manager is willing to accept in the results obtained from the data collected. Without DQOs, it is not possible to assure that the quality of the data will be high enough to satisfy project objectives, although the quality of data can be documented.

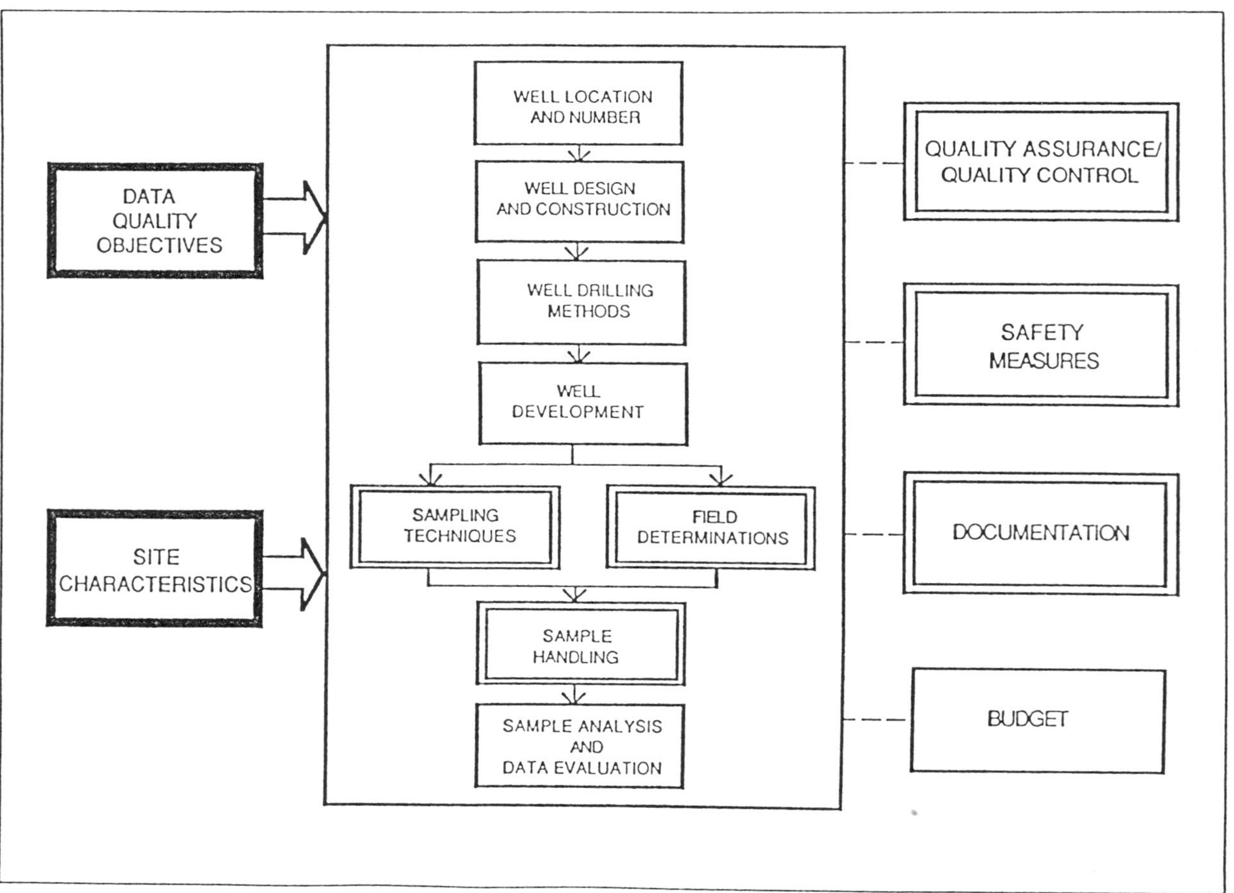

Figure 2. Elements of ground-water sampling.

Information on site and well characteristics is requested from the user on specific aspects, including: the chemical constituents of interest (e.g., volatile organics, pesticides, and metals), the hydraulic conductivity of the aquifer from which the samples will be collected, the diameter of the borehole to be sampled, the location of the top of the screened interval, ground-water turbidity, and the ambient air quality [12]. Information about each of these site characteristics, in addition to a number of others, is needed by the expert system to help identify the most appropriate sampling device.

The current modules, the information that is needed from the user in each module, and the type of information provided back to the user are explained below.

(i) Sampling Techniques

The characteristics of various sampling devices are defined within the system, and the conditions for which each device is appropriate are stored in the system knowledge base. Once the user has identified the study objectives and the specific site conditions, the system proceeds to identify the most appropriate methods for well purging, field determinations, sample collection, and if needed, well redevelopment.

The sampling devices currently included in the expert system prototype are air-lift pumps, bailers, bladder pumps, suction-lift (peristaltic) pumps, and submersible pumps [13]. In air-lift pumps, air is injected into a well and water is lifted to the top of the casing. Bailers are cylindrical devices that have a check valve on the bottom for discharging the water sample. The bailer is lowered into the monitoring well by a cable or rope. Bladder pumps are positive-displacement devices that use a pulse of gas or air to push the sample to the surface. Suction-lift pumps draw the ground water to the surface by creating a negative pressure at the land surface. In submersible pumps, ground water is drawn through the intake of the pump, moved through the various stages (impellers, bowls, or helical rotors driven by an electric motor), and discharged at the ground surface [2, 6].

(ii) Sample Handling

The Sample Handling module provides advice on the correct procedures to be followed after the sample is collected. This advice covers the container type to be used, preservation sample, filtration, maximum holding time, and shipping procedures. The information provided by the system is based upon the type of contaminants being sampled and the analytical techniques to be used in the analysis of the samples.

(iii) Quality Assurance and Quality Control

The Quality Assurance and Quality Control module is designed to help the user measure and obtain the required precision, accuracy, comparability, representativeness, and completeness of the data. The minimum required levels for these parameters are established in the data quality objectives of the project. The system recommends the proper type and number of blanks, duplicates, standards, and other QA/QC samples to be collected.

(iv) Safety Measures

The proper levels of personnel protection, as well as safety and screening instruments recommended by the USEPA Office of Emergency and Remedial Response [14] and other agencies is contained in the health and safety module. The user is asked by the system to supply levels and probabilities of expected or potential hazards at the site. Based on this information the appropriate protective gear and monitoring instruments are recommended for sampling personnel.

(v) Documentation

The documentation module provides documentation and chain-of-custody procedures to insure the integrity of the sample and accuracy of reporting and support the generation of legally admissible data. The specific format followed is the one recommended for standard USEPA work. The system provides illustrations of information required for logbooks, sample labels, well logs, chain-of-custody forms, calibration forms, and other documentation.

ESES-GW OPERATION

The ESES software runs on microcomputers operating under MS-DOS (or PC-DOS) and is written in the KnowledgePro [15] expert system development shell. Some of the statistics, graphics, and string manipulation routines are written in Pascal language [16]. The delivered program is a stand-alone software package and runs on computers with 640 kilobytes of memory, with color or monochrome monitors. A hard-disk drive and at least an 8-MHz processor clock speed are recommended for efficient operation. A mouse pointing device is also supported, if available.

During operation the system asks the user a series of questions in an interactive consultation and uses its stored knowledge and rules to produce appropriate recommendations on the various aspects covered. Specific technical terms appear highlighted on the computer screen and the user can choose to view a definition, explanation, or reference to those terms and concepts. This feature is implemented by using the hypertext technique, which is a method to present multilayered information on a computer screen. The amount and depth of the information received is controlled by the user, which makes the program suitable for efficient use by people with various degrees of training, ranging from the novice to the moderately knowledgeable person, in ground-water sampling.

Figure 3 is an example of how the Sample Handling module operates. In this example, the user has selected Aldicarb from a list provided by the system as a pesticide of interest. ESES-GW then shows the Figure 3 "window" illustrated with some of the sample-handling recommendations [17]. The rest of the information for samples to be analyzed for Aldicarb can be obtained by pressing a designated key on the computer keyboard. The program includes options to go back to previous modules and to check related information and previous recommendations, print the currently shown information, reference data that may have been entered as answers to previous questions, and receive an explanation of how the program arrived at the recommendations provided.

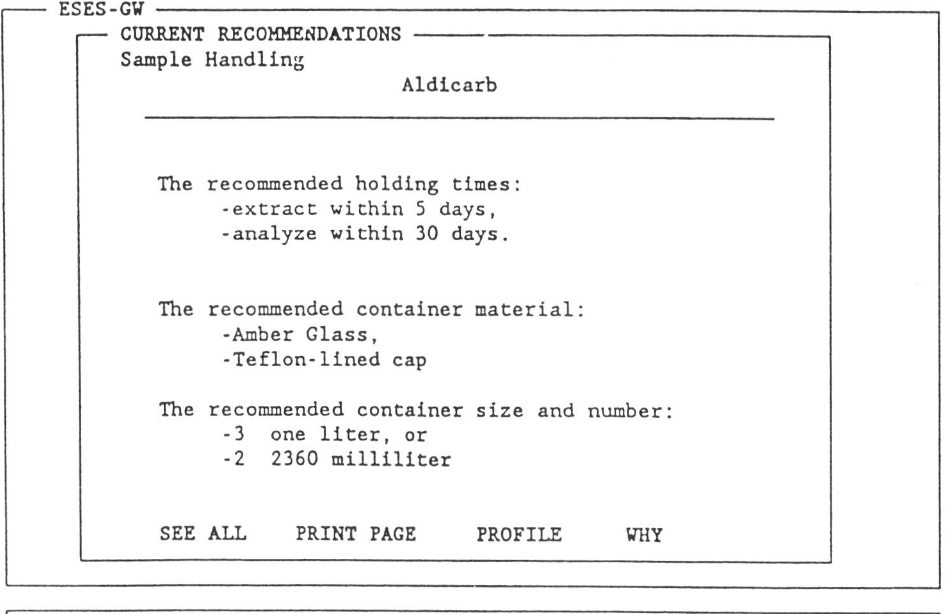

Figure 3. Example of recommendation screen. Sample handling for groundwater pesticide sampling.

User input can be varied to explore the impact of different circumstances on the recommendations. This feature provides an opportunity to optimize the sampling plan without committing resources.

By 6.

CONCLUSION

It should be kept in mind that automated systems, like the one described, have many limitations and must be used properly. There are many factors to consider when collecting samples and the contribution of human experts, such as hydrogeologists and environmental scientists, is always vital. The use of an expert system helps anticipate or prevent many of the problems that may be encountered before actual field sampling is undertaken. Also, the system can complement expert advice by identifying items that might otherwise be overlooked. Identified requirements can then be included in the sampling plan for the project. Expert systems at this stage of development cannot substitute for human experts, but they do have a place in the planning process in the role of automated trainers and consultants.

The current prototype for ground-water sampling which is under development is being readied for demonstration and the first round of external review. Subsequently, a full prototype will be developed as a predecessor to the finished production system. One important advantage of the availability of the system described in this work is the accessibility to a single, relatively easy to use, compiled source of information on sampling methodology. Also, the information provided is useable by people with a wide range of expertise. This is particularly important in the environmental field, since the present knowledge in this area is distributed among different organizations and across different countries. The next module of ESES to be developed is a sampling advisor for surface water. The experience gained from the development of the two current prototype systems, and contributions from other environmental sampling experts, will be applied during the next phase of the overall development of the Environmental Sampling Expert System.

ACKNOWLEDGMENTS

The authors want to acknowledge Kenneth W. Brown (EMSL-LV, Exposure Assessment Research Division) for his support of the initial expert system prototypes; Joseph J. Dlugosz (EMSL-LV, Aquatic and Subsurface Monitoring Branch) for his encouragement of this work; and Kelly R. York (Lockheed Engineering & Sciences Company) for the prototype programming.

NOTICE

Although the research described in this article has been funded wholly or in part by the United States Environmental Protection Agency through contract number 68-03-3249 to Lockheed Engineering & Sciences Company, it has not been subjected to Agency review and therefore does not necessarily reflect the views of the Agency and no official endorsement should be inferred. Mention of trade names or commercial products does not constitute endorsement or recommendation for use.

REFERENCES

[1] Fisk J. F., The Driving Force: Environmental Legislation, Environmental Lab, 1(1), pp. 30-35, March 1989.

[2] Fetter C. W., Applied Hydrogeology, 2nd edition, Merrill Publishing Company, Columbus, Ohio, 1988.

[3] Collins A. G. and Johnson A. I., Eds., Groundwater Contamination Field Methods, ASTM Special Technical Publication 963, Philadelphia, Pennsylvania, 1988.

[4] Sources of Contamination, Chapter 4, in Groundwater Protection, The Conservation Foundation, Washington, D.C., pp. 105-169, 1987.

[5] Agriculture and Groundwater Quality, Council for Agricultural Science and Technology, Report No. 103, Ames, Iowa, 1985.

[6] Driscoll F. G., Groundwater and Wells, 2nd edition, Johnson Division, St. Paul, Minnesota, 1986.

[7] U.S.Environmental Protection Agency, RCRA Ground-Water Monitoring Technical Enforcement Guidance Document, Office of Waste Programs Enforcement and Office of Solid Waste and Emergency Response, September 1986.

[8] Olivero R. A., York K. R., Homsher M. T. and Cabble K. J., A Hypertext-Based System for Planning of Environmental Sampling Projects, in Proceedings of the Fourth Annual Lockheed Artificial Intelligence Symposium, Calabasas, California, pp. 4-81 - 4-91, 1989.

[9] Waterman D. A., A Guide to Expert Systems, Addison-Wesley Publishing Company, Reading, Massachusetts, 1986.

[10] Olivero R. A., Cameron R. E., Cabble K. J., Homsher M. T., Stapanian M. A. and Brown K. W., Environmental Field Sampling Expert System - Development of a Soil Sampling Advisor, in Proceedings of the First International Symposium on Field Screening Methods for Hazardous Waste Site Investigations, Las Vegas, Nevada, pp. 325-339, 1988.

[11] U.S. Environmental Protection Agency, Data Quality Objectives for Remedial Response Activities-Development Process, EPA/6-87/003, Office of Solid Waste and Emergency Response, Washington, D.C., 1987.

[12] U.S. Environmental Protection Agency, Handbook; Ground Water, EPA/625/6-87/016, Robert S. Kerr Environmental Research Laboratory, Ada, Oklahoma, 1987.

[13] Barcelona M. J., Gibb J. P., Helfrich J. A. and Garski E. E., Practical Guide for Ground-Water Sampling, EPA 600/S2-85/104, Robert S. Kerr Environmental Research Laboratory, Ada, Oklahoma, 1985.

[14] U.S. Environmental Protection Agency, Standard Operating Safety Guides, Office of Emergency and Remedial Response, 1988.

[15] Thompson B. and Thompson B., KnowledgePro User Manual, Version 1.3, Knowledge Garden, Nassau, New York, 1988.

[16] Turbo Pascal Version 3.0 Reference Manual, Borland International, Inc., Scotts Valley, California, 1985.

[17] U.S. Department of Energy, The Environmental Survey Manual. Appendix I, DOE/EH-0053, Office of the Assistant Secretary-Environmental, Safety, and Health and Office of Environmental Audit, Washington, D.C., 1987.

CLASSIFICATION AND CHARACTERISTICS OF AIR POLLUTION MODELS

K. JUDA-REZLER

Institute of Environmental Engineering
Technical University of Warsaw
Nowowiejska 20, 00-635 Warsaw, Poland

ABSTRACT

The classification of air pollution models as well as a comparison between different modelling approaches is given. Each class of models is discussed briefly and major model's attributes are described.

INTRODUCTION

The fundamental problem of air pollution modelling is to calculate air concentration of one or more species in space and time as related to the independent variables such as emissions into the atmosphere (E), the meteorological variables (M), and parameters which describe removal and transformation processes (P). Given E, M and a set of additional parameters P (see Fig. 1) the air quality (AQ) modelling problem can be represented in general way by the mapping [1]:

$$F(P, M) \; : \; E \longrightarrow AQ \tag{1}$$

where the function F, or air quality model, denotes the means of relating

Chemistry for the Protection of the Environment
Edited by L. Pawlowski *et al.*, Plenum Press, New York, 1991

Table 1. Classification of Air Pollution Models by their Applications

Category of Use	Purpose
Scientific Research	To acquire knowledge about basic atmospheric processes To prove new theories To educate
Air Management and Decision Making	To incorporate air quality constrains into the planning system of land-use and transportation
Air Pollution Control	To develop control strategies To forecast and warn To extend the information system
Environmental Impact	To provide environmental modelling with air pollution sub-model To study the impact of air pollution from new sources
Air Pollution Episodes	To create an alert system for air pollution episodes (in Japan an alert system is in operation in several areas) To specify endangered areas in case of air pollution incidents (e.g., Chernobyl accident)

changes in contaminant emissions to the resulting air quality (AQ).

MODEL PARAMETERS
(P)

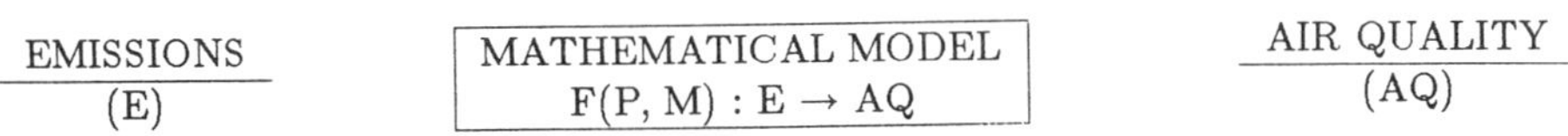

METEOROLOGY
(M)

Figure 1. Schematic structure of an Air Quality Model [1].

Nowadays, when increasing environmental pollution is becoming a world-wide problem, models provide an important tool for making decisions concerning environmental protection. A variety of different air pollution models have been developed in the last twenty years, ranging from those based upon simple to the most sophisticated approaches. They address various atmospheric problems thus in choosing an appropriate model the following question should first of all be posed: What will the model be used for?

Classification of air pollution models by their purpose is given in Table 1. Having determined the goal of a model, the best modelling approach for a given problem may be decided upon. An "ideal" model should be:

(a) Physically realistic and accurate;

(b) Universal, suitable for various:

- temporal and spatial scales
- meteorological and topographical conditions
- emission sources (point, area, line)
- air pollution species;

(c) Supported by readily available input data;

(d) Easy to understand and use;

(e) Easy to adopt for running on typical computers;

(f) Computationally fast;

(g) Interactive with user;

(h) Well documented;

(i) Well evaluated against observations;

(j) Fully validated on real data (for different cases).

Unfortunately, such an "ideal" model does not exist, because some of its features are mutually exclusive (e.g., a universal model will probably not be easy to use and efficient). That is why each model is a compromise between simplicity and sophistication, accuracy and practicality. It is very important to remember that accuracy of the model output is dependent on its input data, thus output resolution can never be better than input resolution.

Therefore, when the purpose of modelling is known, the next important question is: What input data are available?

Answering this should be the first step in choosing a proper model from the class of models appropriate for a given purpose. A sophisticated approach will be absolutely useless if we cannot collect all necessary input data for it.

APPROACHES TO MODEL CLASSIFICATION

A large number of air pollution models have been developed, and still research projects in this area are undertaken. Therefore, there is a need for presenting a clear classification of the existing models, as well as a comprehensive discussion on specific model characteristics. This is not an easy task, because models differ from one another in all possible aspects, and different model attributes can be used for their classification. Among them the most important are:

(1) Basic model structure

- deterministic or non-deterministic

(2) Time resolution

- steady-state or time-dependent

(3) Frame or reference

- Eulerian
- Lagrangian

(4) Dimensionality of computational domain (1D, 2D, 3D, multilevel)

(5) Methods of model equations resolution

- analytical
- numerical (various methods).

Equally important are the characteristics of the system being studied:

(1) Size (local, regional, national, global)

(2) Time horizon (hour, day, month, year, decade)

(3) Pollutant of concern (sulphur dioxide, nitrogen oxides, hydrocarbons, particulate matter, photochemical oxidants, secondary pollutants).

The most significant factor, however, which divides air pollution models into two independent groups is the *basic model structure*. This will be used here for first classification into deterministic and nondeterministic approach.

Deterministic models calculate the concentrations from an emission inventory and other independent, mostly meteorological, variables according to the solution of various equations which represent the relevant physical processes. In most cases, they use solutions of the turbulent diffusion equation derived in several ways and under different assumptions.

Nondeterministic models can be further divided into two groups:

(1) Statistical models

(2) Physical (fluid) models.

The *statistical* model calculates concentration by statistical methods from meteorological and other parameters after the statistical relationship has been obtained empirically from measured concentrations. The *physical* model is one in which nature is simulated on a smaller scale in the laboratory, e.g., in a wind-tunnel.

The deterministic models are most suitable for long-term planning decisions. The statistical are very useful for short-term forecast of concentrations, and the physical models are of use if specific processes are being considered, e.g., influences of topography on the mean airflow.

DETERMINISTIC MODELLING

Major Classes of Models

Deterministic models form the basis for solving the majority of problems within an air pollution research field. They can be categorized into several groups based on the specific model's attributes. The classification proposed here is based on a fundamental approach used for the resolution of the basic diffusion equation. According to that approach, deterministic models are divided into two generic classes:

(1) Closed-form analytical (Gaussian plume/puff)

(2) Numerical.

The basic turbulent diffusion equation can be derived from the mass conservation principle and has the following form:

$$\frac{\partial C}{\partial t} = -\vec{U}\nabla C - \nabla \vec{F_t} + Q + R \tag{2}$$

where: C = pollutant concentration
t = time
$\vec{U}$ = wind vector, $U[u, v, w]$
$\vec{F_t}$ = turbulent flux of pollutants,

$$\vec{F_t}[\overline{u'C'}, \overline{v'C'}, \overline{w'C'}]$$

Q = source term
R = removal term

$$\nabla = \vec{i}\frac{\partial}{\partial x} + \vec{j}\frac{\partial}{\partial y} + \vec{k}\frac{\partial}{\partial z}$$

Note, that C and $\vec{U}$ in Equation (2) are averages over a time interval which is large in comparison with the dominant time scale of turbulent fluctuations, but small in comparison with the time scale of variations of mean concentration and wind speed.

The diffusion equation (Eq. (2)) may be solved only by means of numerical methods, as it is done in numerical models. However, under a set of simplifying assumptions, the analytical solutions of this equation can be obtained and such a solution is used in Gaussian plume/puff models.

The main problem of deterministic modelling is that the $\vec{F}_t$ term in Equation (2) is unknown, so the diffusion equation is not closed. The *numerical* models can be further classified according to the method used for the closure of basic diffusion equation as follows:

(a) First-order-closure (K-theory) models

- Eulerian grid
- Lagrangian trajectory
- Hybrid Lagrangian-Eulerian (Particle-in-Cell, PIC)
- Random-Walk (Monte-Carlo) trajectory particle

(b) Second-order-closure models.

If the short and comprehensive comparison between analytical and numerical models is needed, it can be presented as follows.

Gaussian plume model formula was obtained under a set of simplifying assumptions, from which the most important are the steady-state and homogeneous flow assumptions. That implies that the Gaussian plume model can be applied only for shorter distances (order of 10 km) and of shorter travel times (order of 2 hours).

Numerical models are time-dependent; their structure allows space and time variations in the field of meteorological parameters, as well as the concentration field. Therefore, numerical models are capable of handling a much wider variety of air pollution problems than the analytical ones. On the other hand, the Gaussian models enjoy a major advantage in having a simple structure and low computational costs.

Main characteristics of all mentioned types of models are as follows:

Gaussian Plume/Puff Models

The first analytical closed-form solution of the basic diffusion equation (Eq. (2)) for the steady-state concentration downwind from a *continuous point source* was presented in [2], and further developed in [3] and [4]. This solution is commonly known as Gaussian plume model. The concentration distribution perpendicular to the plume axis is assumed to be Gaussian. The plume travels with a uniform wind velocity ($\bar{u}$) down-wind from the source. Its dimensions perpendicular to the wind direction are described by dispersion coefficients as a function of distance or travel time from the source. The standard *Gaussian plume model equation* for the continuous point source is given by:

$$C(x, y, z; H) = \frac{Q}{2\pi\sigma_y\sigma_z\bar{u}} \cdot \exp\left[-\frac{1}{2}\left(\frac{y}{\sigma_y}\right)^2\right] \cdot$$

$$\cdot \left\{\exp\left[-\frac{1}{2}\left(\frac{z-H}{\sigma_z}\right)^2\right] + \exp\left[-\frac{1}{2}\left(\frac{z+H}{\sigma_z}\right)^2\right]\right\} \qquad (3)$$

where C denotes the pollutant concentration (units/m^3); Q is the emission rate from the point source (units/sec); and x and y denote the down-wind and lateral distances from the pollutant source to the receptor point (m); $\bar{u}$ is the mean horizontal wind speed (m/sec); H is the effective stack height (m) – the sum of the physical stack height (h) and the plume rise (Δh); and σ_y, σ_z are horizontal and vertical dispersion coefficients, respectively (m).

The Gaussian plume model is based on the following assumptions:

(a) Steady-state conditions, which imply that all variables and parameters are constant in time;

(b) Homogeneous flow, which implies no spatial variability in the meteorological parameters (in particular wind-direction shear is not considered);

(c) An inert passive pollutant, i.e., no atmospheric chemical reactions and no gravity fall-out;

(d) Perfect reflection of the plume at the underlying surface, i.e., no ground absorption;

(e) The turbulent diffusion in the x-direction is neglected relative to advection in the transport (x) direction, which implies that model should be applied for average wind speeds of more than 1 m/s ($\bar{u} > 1$ m/s);

(f) The coordinate system is directed with its x-axis into the direction of the flow, the v (lateral) and w (vertical) components of the time averaged wind vector are set to zero;

(g) The terrain underlying the plume is flat;

(h) All variables are averaged over a period of about 10 min, which implies that for different averaging times, corrections to Equation (3) have to be made.

The steady-state assumption (a) implies that the Gaussian plume equation can be applied only for shorter travel times (order of 2 hours). Despite their disadvantages arising from accepted assumptions, the Gaussian plume models are still the most commonly applied method for calculating dispersion of pollution from the point sources. The reasons for that are the following:

(a) Much experience has been gained since first model formulation (in particular in the field of dispersion coefficients estimating);

(b) Model is easy to understand and use, is efficient in computer running time;

(c) Results agree with experimental data quite well.

A collection of information about Gaussian plume models is given in [5] and [6]. A number of applied Gaussian models is fully presented in [7] and [8]. Gaussian plume models are based, in general, on Equation (3) or its variations, most of which arise from different prescriptions of the four physical parameters of model – $\bar{u}$, H, σ_y, σ_z. Also, Gaussian plume model equations differ according to specific model applications (short-term or long-term model, point, line or area sources, multiple-source model, inversion situations, maximum concentration calculations).

In the Gaussian plume model, it is assumed that the mass of air pollutants is conserved throughout the transport process. The estimates of physical or chemical transformations cannot be taken into account without violation of the Gaussian hypothesis. However, the exponential correction factor for some of the removal processes can be introduced to the basic plume equation (Eq. (3)). This includes dry and wet deposition of pollutants (e.g., see [9]). Chemical transformation of primary air pollutants (as SO_2, CO and NO_2) is a complex process, and thus, difficult to incorporate in Gaussian plume models. Secondary pollutants (which are created through chemical reactions involving the primary pollutants) cannot be modelled by simple plume models. The most important transformation process in the atmosphere, which can be included in the Gaussian plume model, is the oxidation of sulphur dioxide to sulphates $SO_2 \rightarrow SO_4^{2-}$ (e.g., see [10]). This process may be, as in case of deposition, estimated by a first-order decay function.

The Gaussian plume model is applicable only for sources that are emitting continuously or for time periods equal to or greater than the travel times from the source to the receptor. However, the solution also is needed for the cases of instantaneous release of pollutants. Such an alternative approach has been developed and is known as a "puff" model (for the theory presentation see [6]).

The puff model applies to *instantaneous sources*, where the release time is short compared with the travel time. The centre of the puff is moving along wind trajectory. The concentration distribution inside the puff is assumed to be Gaussian. The dimensions of the puff are described by dispersion coefficients as a function of travel time of the puff. The *Gaussian puff* formula for the ground-level concentration down-wind from a release from height H reads:

$$C(x, y, 0, t; H) = \frac{2M}{(2\pi)^{3/2}\sigma_z\sigma_h^2} \cdot \exp\left\{-\frac{1}{2}\left[\frac{(x-\bar{u}t)^2 + y^2}{\sigma_y^2}\right]\right\} \cdot \exp\left[-\frac{1}{2}\left(\frac{H}{\sigma_z}\right)^2\right] \quad (4)$$

where the symbols have the same meanings as in Equation (3), with two important exceptions: M ($M = Q \cdot t$) denotes the total mass of the pollutant release; and σ_h, σ_z are standard deviations of the puff material distribution in horizontal and vertical directions, respectively, but they are not those σ's evaluated with respect to the dispersion of a continuous source; t is the transport time from the source to the receptor point. In Equation (4) horizontal dispersion of the puff is assumed to be isotropic ($\sigma_y = \sigma_x = \sigma_h$).

The puff formulation overcomes some problems connecting with the plume models. First, the puff model enables calculations of pollutant concentration in near-calm conditions. Second, both small-scale and large-scale diffusion can be described. However, this method has large computational requirements (tracking the puff in space involves a laborious procedure), which limit its applications.

First-Order-Closure (K-Theory) Models

First-order-closure models have their common roots in the K-theory diffusion equation derived by using a K-theory approximation for the closure of the basic diffusion equation (Eq. (2)). In such models unknown turbulent flux of pollutant $\vec{F}_t$ is parametrized by the product of an eddy diffusivity and the local spatial gradient of the quantity being transported. For the pollution concentration, this approximation reads:

$$\vec{F}_t = \overline{\vec{U}'C'} = -\hat{K} \cdot \nabla C \quad (5)$$

where $\hat{K}$ denotes a diffusivity tensor. In most first-order-closure models used in planetary boundary layer applications, the diffusivity tensor $\hat{K}$, can generally be simplified by employing the isotropic argument. In this case, the off-diagonal components can be represented by a horizontal term K_H ($K_{xx} = K_{yy} = K_H$) and a vertical term K_z.

Finally, the K-theory diffusion equation, called also advection-diffusion equation, has the following form:

$$\frac{\partial C}{\partial t} = -\left\{u\frac{\partial C}{\partial x} + v\frac{\partial C}{\partial y} + w\frac{\partial C}{\partial z}\right\} + \frac{\partial}{\partial x}K_H\frac{\partial C}{\partial x} +$$

$$+\frac{\partial}{\partial y}K_H\frac{\partial C}{\partial y} + \frac{\partial}{\partial z}K_z\frac{\partial C}{\partial z} + Q + R \quad (6)$$

where C is the pollutant concentration (mass/volume); $\vec{U}(u, v, w)$ is the wind vector (m/sec); K_H and K_z are horizontal and vertical turbulent exchange coefficients (called also eddy diffusivities), (m^2/sec); Q and R are the emission and the removal terms (mass/volume · time).

The first-order-closure models have common limitations, arising from employing the K-theory for the closure of basic diffusion equation (see [11]). The most important limitations are:

(1) The K-theory diffusion equation is valid only if the size of the plume of puff of pollutants is greater than the size of the dominant turbulent eddies;

(2) The K-model assumption (Equation (5)) is not valid for the convective boundary layer under strong instability.

The first constraint implies that for the greater point release heights (tall stacks) the advection-diffusion equation should not be used until the pollutant of interest in spread out over several hundred meters.

Advantages of this group of models are the following:

(1) K-theory models are time dependent, so all variables can be functions of time;

(2) Equation (6) allows space variations in the fields of meteorological parameters, as well as the concentration field;

(3) The nonlinear chemistry (e.g., formation of photochemical smog) can usually be included;

(4) A widevariety of problems, ranging from microscale up to the long range transport (LRT) of air pollutants, can be modelled.

At a greater level of detail, first-order-closure models can be further categorized as follows:

1) Eulerian grid model

2) Lagrangian trajectory model

3) Hybrid Lagrangian-Eulerian model (Particle-in-Cell)

4) Random-Walk (Monte-Carlo) trajectory particle model.

In the following a short description of each class of K-models will be given.

The Eulerian Grid Model employs a coordinate system which is fixed with respect to the ground. The region of interest is subdivided into a two- or three-dimensional array of grid cells, and the polluted air is simulated as it passes from cell to cell.

The governing advection-diffusion equation (Eq. (6)) is solved numerically to yield the desired time-dependent concentration distribution.

In principle, the Eulerian grid model is capable of incorporating more physical realism than Gaussian and Lagrangian trajectory models. The major advantages of this approach are:

(1) All the various terms in the advection-diffusion equation (Eq. (6)) can be accommodated;

(2) Sophisticated three-dimensional treatment of the air pollution problem is possible;

(3) Any scale of dispersion may be modeled;

(4) The nonlinear chemistry (e.g., a formation of photochemical smog) can be included.

However, these attributes are offset in some applications by the model disadvantages, which are the following:

(1) Large computational costs (time and storage);

(2) Requirement of large amounts of input data;

(3) The problem of artificial diffusion.

The artificial diffusion is generated by finite-difference representation of the advection-diffusion equation. Various numerical techniques for solving the diffusion equation have been developed to minimize this effect, such as the higher-order difference schemes, the moment method, the pseudospectral method, the cubic spline and chapeau-function (Galerkin) method.

One of the biggest problems in Eulerian grid models concerns supplying necessary input data. Since the output resolution can never be better than the input resolution, the question of sophistication versus practicality is specially important here. The best way of providing input data is by coupling the numerical solution of the 3D diffusion equation with the predicated characteristics of the planetary boundary layer (3D PBL model). This approach has been adopted in [12].

The problem of treating the subgrid scale dispersion from high point sources can be solved by applying the Lagrangian plume trajectory submodel near such sources (e.g., see [13]).

There is a relatively small number of implementations of three-dimensional (3D) air pollution models in real cases, in comparison with a conspicuous number of theoretical contributions. However, the most often applied 3D or multilevel models, are the Eulerian grid ones [12], [13], [14], [15], [16], [17], [18], [19], [20] and [21].

In contrast to the 3D approaches, the 2D Eulerian grid models are widely implemented. In such models, the model parameters preserve their spatial variability in the horizontal plane (x, y), but their values are averaged in the vertical plane. Thus, some important features such as wind-direction shear with height cannot be incorporated.

A simplified version of the two-dimensional Eulerian grid model is the so-called *Multi-Box Model*. The region of interest is divided into a grid of boxes. The basic advection-diffusion equation (Eq. (6)) is simplified by the following assumptions:

(1) Complete vertical mixing within the box (the vertical diffusion term is dropped);

(2) No diffusion between the boxes (the horizontal diffusion terms are neglected);

(3) First-order, explicit, finite-difference solution to the diffusion equation.

This approach is specially useful for urban diffusion modelling. The box model can also be applied as a street-level submodel. A large number of box and multi-box models for various applications and of different levels of complexity are presented in [22].

The Lagrangian Trajectory Model "attaches" its coordinate system to a fictitious vertical air column which moves horizontally with the advective wind. The concentration distribution within the air column is obtained by solving the advection-diffusion equation (Eq. (6)) without advection term:

$$\frac{\partial C}{\partial \tau} = -w\frac{\partial C}{\partial z} + \frac{\partial}{\partial x}K_H\frac{\partial C}{\partial x} + \frac{\partial}{\partial y}K_H\frac{\partial C}{\partial y} + \frac{\partial}{\partial z}K_z\frac{\partial C}{\partial z} + Q + R \qquad (7)$$

where C is the concentration within the air column and τ is the transport time.

The following simplifying assumptions are usually employed with Equation (7):

(1) Vertical advection is neglected $(-w \cdot \frac{\partial C}{\partial z} = 0)$;

(2) Vertical variability of the horizontal wind is suppressed (no wind shear);

(3) The horizontal diffusion term is either highly parametrized or neglected.

The approximations necessary to solve the Lagrangian formulation limit its applicability and the accuracy of the solution. Although some trajectory models for the urban scale have been developed (e.g., see [23]), this type of model has found its main application in long-range transport (LRT) of air pollutants.

In the Lagrangian LRT modelling further simplification to Equation (7) is introduced:

(1) Complete mixing within each column up to some specified height (H) is assumed.

Thus the simplest form of the basic equation for the Lagrangian LRT models is:

$$\frac{dC}{dt} = Q + R(C) \tag{8}$$

where $\frac{d}{dt}$ is the total (Lagrangian) derivative; Q is the emission term and $R(C)$ is the removal term, including dry and wet deposition and first-order chemical decay.

A comparison of various LRT models is presented in [24]. Further information concerning LRT modelling can also be found in [25] and [26]. In general, the Lagrangian trajectory model has the following advantages [24]:

(1) Relatively inexpensive to run on a computer;

(2) Easy to keep track of pollutant mass balances;

(3) No artificial diffusion;

(4) Individual sources or receptors can be run separately.

Disadvantages of these models are [24]:

(1) Extension to 3D not straightforward;

(2) Nonlinear chemistry may be difficult to incorporate;

(3) Horizontal and vertical diffusion neglected or highly parametrized;

(4) Errors can be introduced in interpolating results into Eulerian grid.

The Particle-in-Cell (PIC) is the *Hybrid Eulerian-Lagrangian Model* using the most desirable features of both fixed- and moving-coordinate approaches. In this method, the mass of pollutant is separated into discrete masses (particles). Particles follow a Lagrangian trajectory under the influence of the prevailing wind and a velocity representing turbulent diffusion. Each particle is tracked on a Eulerian grid, and the concentrations are computed by counting the total number of particles in a given cell.

A three-dimensional numerical model of this type has been developed in [27]. This model, called ADPIC, solves the advection-diffusion equation (Eq. (6)) in its flux conservative form (pseudovelocity technique) for a given mass-consistent advective wind field by a finite-difference approximation in an Eulerian grid. The method is based on the PIC technique [28].

The PIC model has all the advantages of the Eulerian grid approach, but it eliminates the artificial diffusion (because advection takes place during the Lagrangian step of the computational cycle). This determines the great importance of the PIC model.

However, a large number of particles are required to yield accurate averaging statistics for estimating the concentrations. Thus, the PIC method has a long execution time, which limits its practical applications.

In the *Random-Walk (Monte-Carlo) Particle Trajectory Model* the dispersion is described in terms of particles, each representing a specified mass of pollutant. Particles are released in numbers proportional to the strength of a given source, and they move under the influence of the mean wind and the atmospheric turbulence. The effect of turbulence is simulated by random movements of the particles. The concentration at any given point is provided by the local density of particles.

A large number of particles are released at some point for simulating the process described by Equation (6). Their trajectories are calculated independently, according to the meteorological conditions found at the position of each particle. The trajectory equations read:

$$\begin{aligned} X(t+\Delta t) &= X(t) + (u+u')\cdot \Delta t \\ Y(t+\Delta t) &= Y(t) + (v+v')\cdot \Delta t \\ Z(t+\Delta t) &= Z(t) + (w+w')\cdot \Delta t \end{aligned} \tag{9}$$

where (u, v, w) are components of the mean wind vector and (u', v', w') are turbulent components.

In some random-walk models (see [29] and [30]), a set of turbulent particle velocities (u', v', w') is chosen randomly from a Gaussian distribution with a zero mean. More complicated models are based on a Markov chain principle of correlations between component particle velocities at successive time steps (see [9] and [31]). The calculations of (u', v', w') require an estimate of the Lagrangian integral time scale of turbulence T_L (see [31] for a comparison of various Markovian random-walk models).

The random-walk particle trajectory model has the following advantages:

(1) The diffusion calculations are related directly to basic turbulence characteristics;

(2) Great generality and flexibility for handling complex and time-varying emissions and atmospheric conditions are obtainable;

(3) Any scale of dispersion may be simulated.

However, practical applications of such models are limited by:

(1) Fairly high computational cost;

(2) Computer programing complexity;

(3) Required input data, which can be difficult to obtain.

In general, the random-walk models will be best applicable to difficult situations such as sea breeze or complex terrain, in which other models do not give satisfactory results.

Second-Order-Closure Models

The first-order-closure assumption introduces some limitations, which can be eliminated by invoking a higher-order closure to the basic diffusion equation (Eq.(2)).

The second-order-closure models are based on deriving prognostic equations for each of the unknown correlations $(\overline{u'C'}, \overline{v'C'}, \overline{w'C'})$. That can be done formally by multiplying the basic diffusion equation (Eq. (2)) by u', v', w' successively, and then averaging. The result is a set of equations, as e.g.:

$$\frac{\partial(\overline{w'C'})}{\partial t} = F \tag{10}$$

where F is a function of mean quantities, second-order correlations and triple correlations. The triple correlations like $\overline{w'w'C'}$ are now terms that have to be specified. By expressing these in some way in terms of the mean quantities and second-order correlations a "second-order" closure is achieved; as e.g.:

$$\overline{w'w'C'} = \sigma_w \Lambda \frac{\partial(\overline{w'C'})}{\partial z} \tag{11}$$

where Λ is an unknown length scale.

The derivation of a set of second-order-closure model equation is presented in [32]. Here are given the methods of estimating Λ which turns out to be less variable than $\hat{K}$ in the first-order-closure models.

Currently, the number of second-order-closure models specifically applicable to atmospheric problem is quite limited. A three-dimensional second-order-closure model has been developed at Argonne National Laboratory [33]. The model has been used to simulate the behavior of buoyant plumes from a large cooling pond.

The second-order-closure models are the most sophisticated air pollution models discussed in this work, however, model attributes as:

(1) Large computer time and core requirements;

(2) Computer programing complexity;

(3) Input data are not easily available.

are prohibitive for using them in operational applications.

On the other hand, second-order-closure models generate a multitude of data that can be used for designing more efficient parametrization schemes, for example, K-formulation. Therefore, at the present time, the second-order-closure models are used, above all, as a research tool.

NONDETERMINISTIC MODELLING

Statistical Models

Most statistical models proposed in air pollution are applications of well-known statistical methods used in meteorology. They vary from simple contingency tables through regression and multiple regression models to the time-series techniques (see [27] for information about all these methods).

The statistical models are essentially *empirical*, because even the most complex ones are based on a group of observations. The basic simple statement encompasses all the limitations of this type of model [22].

In the last decade, the time series analysis techniques (see [34]) have been widely used to describe the dispersion of air pollutants on a local scale. Especially the ARIMA and ARIMAX stochastic models have been adopted in many applications.

In an ARIMA (Auto Regressive Integrated Moving Average) model, the concentrations at a certain instant are expressed as linear combinations of previous concentration values and random terms (noise), which are specified in a statistical sense (i.e., are properly described in terms of a random process). Thus, in ARIMA models the physical causes of phenomena (meteorological variables and emission rates of the sources) are not distinguished in the input. Such models represent a "black-box" approach. All possible uncertainties of the model are taken into account by a "noise" variable with assigned statistical properties. The three-variant ARIMA model can be described as follows [35]:

$$C(k+1) = \sum_{j=1}^{p} \phi_j C(k-j+1) + \varepsilon(k+1) + \sum_{m-1}^{q} \psi_m \cdot \varepsilon(k-m+1) \quad (12)$$

where ψ_j, ϕ_m are 3×3 matrices of model parameters; p, q are the model orders; and $\{\varepsilon(k)\}_k$ is a three-variate, purely random, zero mean process (white noise), namely a Gaussian random process.

The most suitable in the ARIMA class of models has turned out to be the AR(1) (Auto Regressive of order 1), which can be presented as [35]:

$$C(k+1) = \phi C(k) + \varepsilon(k+1) \quad (13)$$

The ARIMAX (ARIMA with Exogeneous input) model represents a "grey-box" approach. In the ARIMAX model, the pollutant concentrations at a certain instant are expressed as a linear combination of previous concentrations and a linear combination of present and previous physical inputs plus the noise term.

In accordance with the techniques described in [34], ARIMA and ARIMAX models can be employed for supplying real-time forecast of pollutant concentrations, predicting, at intervals, future concentration levels on the basis of data recorded in previous periods. A better real-time episode predictor can be derived from ARIMAX stochastic mathematical representation, to which physical inputs are introduced. Such a technique has already been used for air pollution forecasting (SO_2 concentration) in many urban areas, e.g.: the Venetian Lagoon Area, Italy [36], Vienna, Austria [37], Milan, Italy [35], [38], and Madrid, Spain [39].

A different and more complex type of real-time predictor is *the Kalman predictor* (e.g., see [40]). In this method, a discretized advection-diffusion equation (see Eq. (6)) is subsequently turned into a stochastic system by introducing properly defined noise terms. Then, an adaptive Kalman predictor is derived from stochastic system. Such a technique has been applied to the real-time forecasting of air pollution episodes in the Venetian Region [41]. The Kalman predictor exhibits a higher degree of reliability than ARIMA and ARIMAX predictors; however, the computational effort increases considerably.

In general, the statistical models are best applicable for short-term forecasts of concentrations on a local scale. Their simplicity forms an alternative to the deterministic models. However, statistical models have a number of disadvantages as:

(1) Requirement of long historical data sets;

(2) Lack of physical interpretations.

Moreover, it should be remembered that such models have to be developed for each region individually, since their empirical nature does not allow a universal model structure.

Physical (Fluid) Models

In a physical model a real process is simulated on a smaller scale in the laboratory, by a physical experiment, which models the important features of the original process being studied.

The pollutant concentrations measured within the physical model can be converted to equivalent atmospheric concentrations through the use of appropriate scaling relationships. The most significant problem in fluid modelling is that of ensuring that the scaling (or similarity) criteria used are suitable to permit a realistic simulation of the atmosphere.

The application of wind tunnel modelling to air flow and diffusion has been discussed in [42] (and the 39 references therein) and [43] (and the 229 references therein).

Recently (see [44]), the discussion of simulation techniques showed that in many areas, there is no consensus view of the appropriate methods to be used. This is particularly true for the modelling of emission buoyancy and momentum effects, where the need for sensitivity studies and the benefits of large facilities were stressed. However, in other areas, notably flow field modelling, progress was noted.

In general, the physical model should be used as a research tool to study specific atmospheric processes. Also, in case of complex air pollution situations, when detailed deterministic models and/or experimental field measurements become very costly, laboratory simulation using scaled-down models in wind tunnels or water channels is often the best approach. As the scale of the prototype atmospheric situation decreases, the accuracy of fluid model simulation increases.

The most important advantage of fluid models is that the scale-model geometry as well as flow speeds and other essential variables can be easily changed and controlled. Their disadvantages are connected with the costs of building new facilities.

CONCLUDING REMARKS

Major types of air pollution models have been classified and presented together with their applications, advantages and disadvantages. The above presentation shows that a universal model does not exist. The value of each model type may only be evaluated in accordance with the given situation. Physical models form a separate group to be used as a research tool to study special atmospheric processes. Statistical models are of use if short-term forecast of concentrations is needed. Second-order- (or higher-order-)

Table 2. Deterministic air pollution models – their scales of applications, capabilities, required input data, and costs

Model	Scale of Application	Capability of Modelling Non-stationary and Nonhomogeneous Flow	Required Input Data	Computational Cost
Gaussian Plume	Local (point sources)	Limited	Easily available	Low
Box	Urban	Very limited	Easily available	Low
Eulerian Grid	All scales	Complete	May be difficult to obtain (depending on chosen parametrization schemes)	May be considerable (depending on applied numerical solution technique)
Lagrangian Trajectory	Urban, Regional, Long Distance	Limited	Available	Relatively low (depending on the number of calculated trajectories)
Particle-in-Cell	All scales	Complete	May be difficult to obtain (depending on chosen parametrization schemes)	High
Random-Walk Particle Trajectory	All scales	Complete	Difficult to obtain	High

From *Encyclopedia of Environmental Control Technology*, Volume 2, Air Pollution Control, edited by Paul N. Cheremisinoff.

closure models are sophisticated solutions which are seldom applied to real air pollution problems.

The comparison of main attributes, as scale of application, capabilities, required input data and computational costs of analytical and first-order-closure models is given in Table 2. As indicated there, the advantages of a given method always cost some disadvantages. Therefore the superiority of one modelling technique over another can only be decided by the user after answering the following questions:

(1) What will the model be used for?

(2) What input data are available?

(3) What meteorological and topographical situation is to be modelled?

(4) What computer equipment is available?

(5) What funds can be assigned to the modelling project?

Answering these questions and being acquainted with individual model's attributes can in consequence, give, an appropriate decision concerning the choice of the best model type for the intended application.

REFERENCES

[1] G.J. McRae, W.R. Goodin and J.H. Seinfeld, "Mathematical Modelling of Photochemical Air Pollution", *Environmental Quality Laboratory, Report* No. 18, 1982.

[2] O.G. Sutton, "The Problem of Diffusion in the Lower Atmosphere", *Quart. J. Roy. Met. Soc.*, **73**, 257, 1947.

[3] F. Pasquill, "The Estimation of the Dispersion of Windborne Material", *Meteorol. Mag.*, **90**, 33, 1961.

[4] F.A. Gifford, "Uses of Routine Meteorological Observations for Estimating Atmospheric Dispersion", *Nuclear Safety*, **2**, 47, 1961.

[5] D.B. Turner, Workbook of Atmospheric Dispersion Estimates. AP-26, Office of Air Programs, U.S. Environmental Protection Agency, Research Triangle Park, North Carolina, 1970, p. 84.

[6] F. Pasquill, Atmospheric Diffusion, Wiley, New York, 1974, p. 429.

[7] EPA, Guideline on Air Quality Models, EPA-450/2-78-027, U.S. Environmental Protection Agency, Research Triangle Park, North Carolina, 1978.

[8] EPRI, Survey of Plume Models for Atmospheric Application, EPRI EA-2243, Interim Report, Electric Power Research Institute, California, 1982.

[9] S.R. Hanna, G.A. Briggs and R.P. Hosker,Jr., Handbook on Atmospheric Diffusion, DOE/TIC-11223,*Tech. Inf. Center, U.S. Department of Energy*, 1982, p. 102.

[10] W.E. Wilson, "Sulphates in the Atmosphere: A Progress Report on Project MISTT", *Atmos. Environ.*, **12**, 537, 1978.

[11] S. Corrsin, "Limitations of Gradient Transport Models in Random Walk and in Turbulence", *Adv. Geophys.*, **18A**, 25, 1974.

[12] R. Bornstein et al., "Application of Linked Three-Dimensional PBL and Dispersion

Models to New York City", *Proceedings of 15th NATO/CCMS ITM on Air Pollution Modelling, St. Louis, Missouri, USA*, 1985.

[13] K. Juda, "Modelling of the Air Pollution in the Cracow Area", *Atmos. Environ.*, **20**, 2449, 1986.

[14] D. Randerson, "A Numerical Experiment in Simulating the Transport of SO_2 through the Atmosphere", *Atmos. Environ.*, **4**, 615, 1970.

[15] S.D. Reynolds, P.M. Roth and J.H. Seinfeld, "Mathematical Modelling of Photochemical Air Pollution - I Formulation of the Model", *Atmos. Environ.*, **7**, 1033, 1973.

[16] C.C. Shir and L.J. Shieah, "A Generalized Urban Air Pollution Model and Its Applications to the Study of SO_2 Distributions in the St. Louis Metropolitan Area", *J. Appl. Meteor.*, **13**, 185, 1974.

[17] G.J. McRae and J.H. Seinfeld, "Development of a Second-Generation Mathematical Performance", *Atmos. Environ.*, **17**, 501, 1983.

[18] Z. Zlatew, R. Berkowicz and L.P. Prahm, "Three-Dimensional Advection - Diffusion Modelling for Regional Scale", *Atmos. Environ.*, **17**, 491, 1983.

[19] N.D. van Egmond and H. Kesseboom, "Mesoscale Air Pollution Dispersion Models - I. Eulerian Grid Model", *Atmos. Environ.*, **17**, 257, 1983.

[20] G.R. Carmichael and L.K. Peters, "An Eulerian Transport (Transformation) Removal Model for SO_2 and Sulphate - I Model Development; - II Model Calculations of SO_x Transport in the Eastern United States", *Atmos. Environ.*, **18**, 937 (Part I), 953 (Part II), 1984.

[21] K.E. Grønskei, Oxident Formation Described in a Three-Level Grid, Air Pollution Modelling and Its Applications III (edited by C. De Wispelaere), **5**, pp. 583 - 602. Plenum Press, New York, 1984.

[22] M.M. Benarie, Urban Air Pollution Modelling, McMillan Press, London and Basingstoke, p. 405, 1980.

[23] A.Q. Eschenroeder and J.R. Martinez, "Concepts and Applications of Photochemical Smog Models", *Adv. Chem.*, **113**, 101, 1972.

[24] W.B. Johnson, Interregional Exchanges of Air Pollution: Model Types and Applications, Air Pollution Modelling and Its Application I (edited by C. de Wispelaere), **1**, pp. 3 - 41, Plenum Press, New York, 1981.

[25] A. Eliassen, "Aspects of Lagrangian Air Pollution Modelling", Proceedings of 13th NATO/CCMS ITM on Air Pollution Modelling, Ile des Embiez, France, 1982.

[26] B. Ottar, The Long Range Transport of Air Pollutants, Norwegian Institute for Air Research, NILU, TN 4/83, Lillestrøm, p. 53, 1983.

[27] R. Lange, "ADPIC - A Three-Dimensional Transport-Diffusion Model for the Dispersal of Atmospheric Pollutants and its Validation Against Regional Tracer Studies", *J. Appl. Meteor.*, **17**, 320, 1978.

[28] R.C. Sklarew, A.J. Fabrick and J.E. Prager, "Mathematical Modelling of Photochemical Smog Using the PIC Method", J. Air Pollut. Control Assoc., 22: 865 (1972).

[29] R.C. Joynt and R. Blackman, "A Numerical Model of Pollutant Transport". Atoms. Enviro., 10: 433 (1976).

[30] J.E. Lehmhaus, E. Roeckner, I. Bernhardt and J. Pankrath, "A Monte-Carlo Model for the Simulation of Long-Range Transport of Air Pollutants", Proceedings of 13th NATO/CCMS ITM on Air Pollution Modelling, Ile des Embiez, France 1982).

[31] A.J. Ley, "A Random-Walk Simulation of Two-Dimensional-Turbulent Diffusion in the neutral Surface Layer", Atmos. Environ., 16: 2799 (1982).

[32] C. du P. Donaldson, Construction of a Dynamic Model of the Production of Atmospheric Turbulence and the Dispersal of Atmospheric Pollutants, Workshop on Micrometeorology (edited by D.A. Haugen), pp. 313-319, Am. Meteor. Soc., Boston (1973).

[33] T. Yamada, "An Application of a Three-Dimensional Simplified Second-Moment Closure Numerical Model to Study Atmospheric Effects of a Large Cooling-Pond", Atmos. Environ., 13: 693 (1979).

[34] G.E.P. Box and G.M. Jenkins, Time Series Analysis, Forecasting and Control, Holden-Day, San Francisco, California, 1970, p. 544.

[35] G. Finzi, G. Fronza and A. Spirito, "Multivariate Stochastic Models of SO_2 Pollution in an Urban Area", APCA, 30: 1212 (1980).

[36] G. Finzi, P. Zannetti, G. Fronza and S. Rindaldi, "Real-Time Prediction of SO_2 Concentration in the Venetian Lagoon Area", Atmos. Environ., 13: 1249 (1979).

[37] P.G. Bolzern, G. Fronza, E. Runca and C. Überhuber, Statistical Analysis of Winter SO_2 Concentration Data in Vienna, International Inst. for Applied Systems Analysis, IIASA Collaborative Paper CP-81-25, Laxenburg, Austria, 1981.

[38] G. Finzi and G. Tebaldi, "A Mathematical Model for Air Pollution Forecast and Alarm in an Urban Area", *Atmos. Environ.,* **16**, 2055, 1982.

[39] G. Finzi, E. Hernandez and R. Garcia, "The SO_2 Pollution in Madrid. II - A Comparison Between Two Stochastic Models For Real-Time Forecast Purposes", *Il Nuovo Climento,* **6**, 605, 1983.

[40] A.H. Jazwinski, Stochastic Processes and Filtering Theory, Academic Press, New York, 1970.

[41] G. Fronza, A. Spirito and A. Tonielli, Real-Time Forecasting of Air Pollution Episodes in the Venetian Region. Part II: The Kalman Predictor, International Inst. for Applied Systems Analysis, RR-79-11, Laxenburg, Austria, 1979.

[42] J.E. Cermak, "Laboratory Simulation of Atmospheric Boundary Layer", *Aircraft Ind, Assoc. Am. J.*, **9**, 1746, 1971.

[43] W.H. Snyder, Fluid Models for the Study of Air Pollution Meteorology: Similarity, Facilities, Review of Literature and Recommendations, *Div. Meteorol., U.S. Environ. Prot. Agency, Publ.*, p. 106, 1972.

[44] EURASAP, Proceedings of 3rd EURASAP Meeting: "Third International Workshop on Wind and Water Tunnel Modelling of Atmospheric Flow and Dispersion", Lausanne, Switzerland, September 1986 (in press, contact Dr. J.A. Hertig EPEL, Department de Genie Civil IENER, CH-1015 Lausanne, Switzerland).

MONITORING METHODS FOR SURFACE AND GROUND WATER AND ANALYSIS OF POLLUTANTS

REMOTE WATER QUALITY MONITORING WITH AN AIRBORNE LASER FLUOROSENSOR

M. BRISTOW

U.S. Environmental Protection Agency
P.O. Box 93478
Las Vegas, NV 89193, USA

R. ZIMMERMANN

Environmental Research Center
University of Nevada - Las Vegas
Las Vegas, NV 89154, USA

ABSTRACT

Remotely measured laser-induced fluorescence spectra of surface waters, whether fresh, estuarine or marine, contain information on a multiplicity of water quality parameters and related phenomena. These include chlorophyll a, dissolved organic carbon (DOC), the optical attenuation coefficient and the density of blue-green algal blooms. In addition, both DOC and the optical attenuation coefficient can be used to map the degree and extent of water body mixing and estuarine salinity. Airborne laser fluorosensors have also been used to detect and map fluorescent tracing dyes, petroleum oil spills, pulp mill effluents and certain phycoerythrin-bearing marine phytoplankton.

The relationship between these water quality parameters and their fluores-

Chemistry for the Protection of the Environment
Edited by L. Pawlowski *et al.*, Plenum Press, New York, 1991

cence analogues is one of simple proportionality. By performing linear regression between "ground–truth" measurements for these parameters obtained at a small number of selected sites under the sensor flight path and the concurrent fluorescence measurements, it is possible to convert the airborne data into extensive regional-scale profiles for these parameters. Moreover, as these parameters are all measured on the same laser-irradiated surface water sample volume, it is possible to perform cross correlation, visual or otherwise, between the various profiles as an aid to revealing anomalies or eliminating ambiguities encountered in the interpretation of any one profile.

INTRODUCTION

Airborne laser fluorosensors have been successfully used to measure the concentration of a number of fluorescent substances and related optical properties in the near-surface water column of fresh, estuarine, and marine waters. A broad range of application areas can now be addressed: phytoplankton biomass from chlorophyll a fluorescence [35,12,3,4]; algal diversity using multiple laser excitation of chlorophyll a fluorescence [5]; the distribution of phycoerythrin-bearing algae specific to certain cyanophytes, cryptophytes, and rhodophytes in marine and estuarine environments using phycoerythrin fluorescence [6,7,8]; water movement studies using fluorescent tracing dyes [9,10]; water body mixing, including a potential for estuarine salinity measurements, using as indicators either dissolved organic matter (DOM) fluorescence [11,12,4] or water Raman emission [11,12]; optical transmission as represented by the beam attenuation coefficient using water Raman emission [35,12,4]; depth-resolved optical attenuation using water Raman emission [37,13]; dissolved organic carbon (DOC) using DOM fluorescence [12], and pulp and paper mill effluents using lignin sulphonate fluorescence [14]. In addition, these airborne systems have been used to determine the type, extent and thickness of surface-distributed petroleum oil spills (see e.g., Measures [36] and references cited therein).

A significant advance in the laser fluorosensor method occurred when it was realized that changes in fluorescence intensity caused not by changes in fluorescent-substance concentration but by point-to-point changes in optical attenuation in the water column could be eliminated by normalizing the fluorescence signal with the concurrent laser-induced water Raman signal [35]. These changes in optical attenuation are produced by changes in concentration of either particulate or dissolved substances regardless of their fluorescence properties. It should be pointed out that an absolute calibration of the laser fluorosensor such that the airborne data can be read out in real time in units of substance concentration is generally not possible.

This situation exists because the quantum yield of the waterborne fluorescent substances are, with the possible exception of fluorescent tracing dyes, not known and, in any case, can vary on a seasonal and regional basis. Consequently, system calibration is usually achieved by performing a linear regression between the airborne data and concurrent ground-truth data obtained at a few key sites under the sensor flight path during a particular survey.

The performance and capabilities of the airborne laser fluorosensor for monitoring substances and properties in the near-surface water column are exemplified here by describing the results of two separate projects conducted by the U.S. Environmental Protection Agency (EPA). The first involves a system demonstration project performed on a small eutrophic bay in Lake Mead, Nevada [35], and the second describes a survey conducted over a 734-km length of the Snake and Columbia Rivers [12]. The airborne data consists of essentially continuous profiles of chlorophyll a fluorescence representing chlorophyll a concentration (μg/L), water Raman emission representing the reciprocal of the beam attenuation coefficient (m^{-1}), and, in the case of the latter survey, DOM fluorescence as a measure of DOC (mg/L). In addition, developments aimed at using a modified version of this system to monitor the distribution and concentration of blue-green algae are briefly discussed.

MODE OF OPERATION

The theoretical basis of the airborne laser fluorosensor as applied to nondepth resolved, near-surface water column monitoring has been discussed in detail by Browell [15] and Bristow et al. [35,12], so that only a brief discussion is presented here.

An idealized fluorescence emission spectrum excited at 487 nm and typical of many fresh surface waters is composed of several components (Figure 1). Excitation in the blue-green spectral region is a compromise based on the consideration of numerous conflicting factors and has been discussed elsewhere [35,12]. The broad fluorescence band peaked in the region of 540 nm is due to natural dissolved organic substances known collectively as dissolved organic matter (DOM). Numerous studies have indicated that the intensity of this DOM fluorescence is related to sample DOC concentration, but with the regression coefficients and the degree of correlation varying between studies [1]. It is generally accepted that, if a high degree of linear correlation between DOM fluorescence and measured DOC concentrations can be established for a given water body, lake system or watershed at a given time, DOM fluorescence can be used as a DOC indicator. In addition, it is suggested that the detection of DOM fluorescence anomalies can be

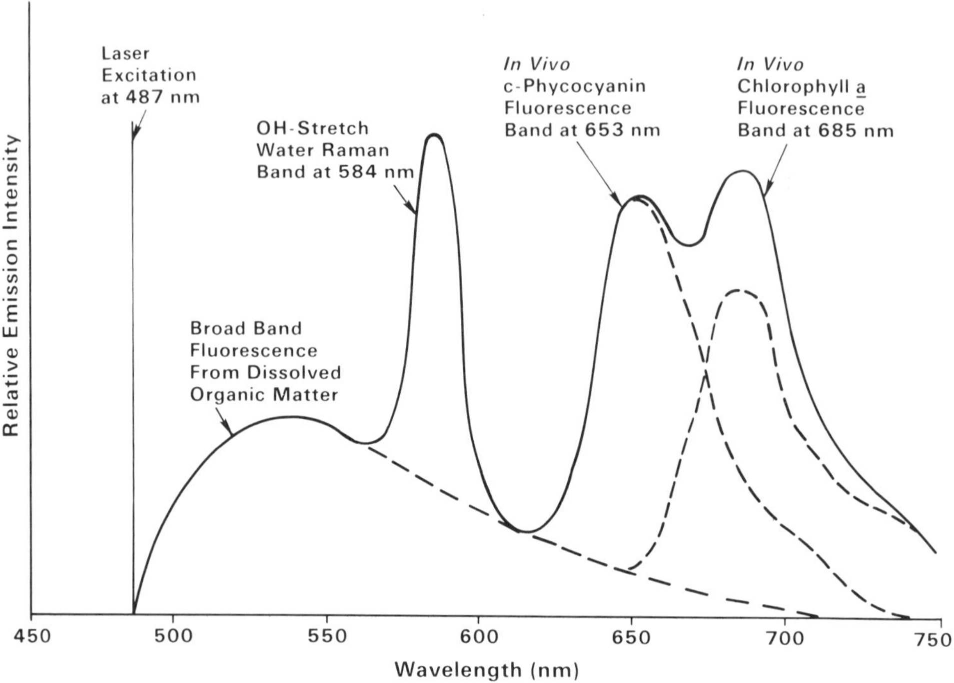

Fig. 1. Schematic showing spectral emission features typical of many fresh waters produced by excitation at 487 nm.

used as a general screening tool for designating sites and regions for more intensive investigation by established in situ sampling and analytical methods. A potential application for this measurement is the detection of either point or nonpoint sources of organic pollutants that can be expected to exhibit anomalous fluorescence properties. Further, as DOM fluorescence is a conservative property as indicated by its inverse linear relationship with salinity in estuarine environments, it can be used as a tracer to monitor the movement and mixing of different water bodies.

The Raman emission band (Figure 1) for the O-H vibrational stretching mode of liquid water is located at 584 nm when excited at 487 nm. As changes in intensity in the remotely measured water Raman signal are due solely to changes in optical attenuation in the water column, they can be used to provide a relative measure of the depth of penetration of the laser beam and, hence, of the size of the water sample being interrogated by the airborne sensor. The water Raman signal can therefore be used as a built-in reference standard with which to correct the intensity of the various fluorescence features for changes produced by variations in optical attenuation [35]. In addition, remote measurements of the reciprocal of the water Raman signal were shown to be a sensitive indicator of the beam attenuation coefficient [35]. In the measurements described here, DOM fluorescence was measured, with little loss in sensitivity, at the water Raman wavelength (584 nm) rather than at the wavelength of maximum DOM emission. In practice, emission measurements were made at the water Raman wavelength and at wavelengths about 40 nm above and below this wavelength. The DOM fluorescence and Raman components at the water Raman wavelength were then separated by using the DOM fluorescence data at the other two wavelengths to determine the DOM fluorescence value at the water Raman wavelength by linear interpolation.

The intensity of the fluorescence band at 685 nm (Figure 1) varies as the concentration of algal chlorophyll a concentration and is widely accepted as a measure of phytoplankton biomass on the assumption that the fluorescence to extracted chlorophyll a ratio remains constant within a given water system and time frame. Many factors can influence this ratio and include water temperature, nutrient availability, excitation wavelength, relative concentration of phytoplankton from different algal color groups, solar history and present solar intensity at the water surface [35] and references cited therein).

The band at 653 nm, though not always present, is due to fluorescence emission from c-phycocyanin, an accessory phytopigment that can be used to indicate the presence, extent and, possibly, concentration of certain bloom-forming phytoplankton. This pigment together with c-phycoerythrin form a group of autofluorescent substances known as phycobilins. These substances are the primary light–harvesting pigments responsible for coupling energy into the chlorophyll a photosynthetic system of blue-green algae (cyano-bacteria), red algae (rhodophytes), and cryptophytes. Direct absorption by the chlorophyll a in these phytoplankton, which has a maximum absorption in the region of 440 nm, is relatively weak. c-Phycocyanin and c-phycoerythrin have excitation (absorption) peaks at about 620 nm and 550 nm, respectively, and emission peaks at about 650 nm and 570 nm, respectively. As such, the fluorescence emission from the in vivo excitation of these phytopigments is a manifestation of incomplete coupling to chlorophyll a.

Although the fluorescence excitation cross section at 487 nm for c-phycocyanin emission at 653 nm in *Microcystis aeruginosa* is generally about 1% of the peak value at 620 nm [16], excitation at 487 nm results in c-phycocyanin and chlorophyll a emission bands of about equal amplitude. The relative magnitude of these two bands is dependent on many factors including genus type, nutrient status, light history and the presence of free c-phycocyanin that is released from dying cells during the late stages of a bloom. In fresh water environments, blue-green algae are the only phycobilin-bearing phytoplankton present in significant concentrations, and many of the ubiquitous surface bloom-forming species such as *Microcystis aeruginosa, Anabaena* flos-aquae and *Aphanizomenon* flos-aquae contain only c-phycocyanin. However, species in other blue-green algal families such as Oscillatoria may contain varying amounts of both pigments or only phycoerythrin [17]. The remote detection (in a freshwater environment) of phycoerythrin emission, exclusively or in addition to that of phycocyanin, would therefore constitute a positive indication of the presence of an Oscillatoria bloom.

INSTRUMENTATION

A full description of the development of the EPA laser fluorosensor has been given by Bristow et al. [35,12]. The sensor consisted of a coaxial flashlamp-pumped dye laser operating at 475 nm attached, in a noncoaxial configuration, to a 30-cm diameter Fresnel-lens telescope receiver. The return signal was split into four components via a series of beam splitters, which were then directed through interference filters (531 nm, 567 nm, 603 nm, and 685 nm) onto four gated photomultipliers (567 nm corresponds to the water Raman band wavelength when excited at 475 nm). The detector signals were then digitized (8 bits at 30 MS/s), displayed and recorded under the control of a microcomputer (DEC LSI 11-23). An important adjunct to the sensor system was a down-looking color video camera boresighted with the laser beam. The realtime video image provided a valuable in-flight navigation and target evaluation aid for both the system operator and pilot, whereas the video record of the sensor flight path, imprinted with alphanumeric time data, was an indispensable aid to interpreting the laser fluorosensor data. In fact, many anomalies in the fluorescence data could only be resolved by careful examination of the concurrent video record. For the measurements obtained from Lake Mead [35], the system was flown on a Bell UH 1D/H (Huey) helicopter, and for those obtained from the Columbia and Snake Rivers [12] on an Aero Commander 680-V twin-engined turboprop fixed-wing aircraft in a nonpressurized condition.

DISCUSSION OF AIRBORNE MEASUREMENTS

(a) Lake Mead, Nevada

The site of these measurements was a 10-km long inlet known as Las Vegas Bay, which is fed by a stream that is partly surface runoff and partly

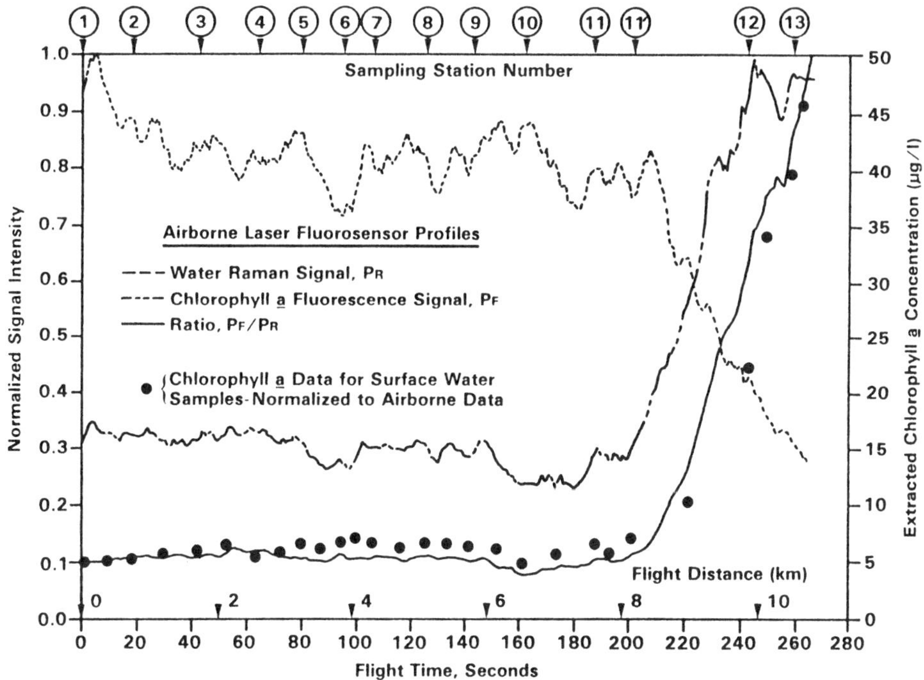

Fig. 2. Airborne laser fluorosensor profiles of chlorophyll a fluorescence $P_F(Ca)$, water Raman emission P_R and $P_F(Ca)/P_R$ for the Las Vegas Bay region of Lake Mead, Nevada, on June 7, 1979. Also shown are the chlorophyll a values for 28 ground-truth samples, which have been normalized to the profile for P_F $(Ca)/P_R$.

a combination of treated and partially treated waste water from the city of Las Vegas. This nutrient-rich stream is known to create a steep chlorophyll a gradient between its entry point into Las Vegas Bay and the central region of Lake Mead, which is fed principally by the relatively sterile waters of the Colorado River. Concurrent ground-truth coverage in the form of grab sampling for chlorophyll a analyses and in situ beam transmissometry was performed at and approximately halfway between thirteen navigation marker buoys that define the inlet center line.

The quasi-continuous airborne data in the form of the chlorophyll a fluorescence profile $P_F(Ca)$, the water Raman profile P_R and the Raman normalized chlorophyll a fluorescence profile $P_F(Ca)/P_R$ are shown in Figure 2, together with the discrete chlorophyll a ground-truth data. The rapid rise in $P_F(Ca)$ as the profile approaches the stream entry point is clearly visible and is matched by a concurrent fall in P_R, which is a measure of the

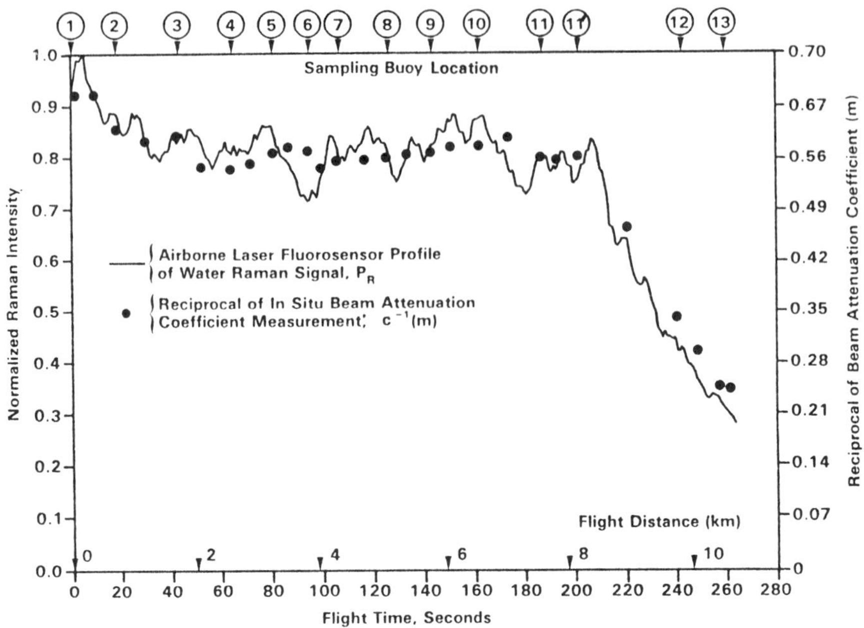

Fig. 3. Airborne laser fluorosensor profile of water Raman emission signal P_R for flight of June 7, 1979 overlayed with normalized values for the reciprocal of the beam attenuation coefficient obtained from in situ measurements at the 28 sampling sites in Las Vegas Bay at the time of the overflight.

increased optical attenuation in the water column, due, in this case, to the increased phytoplankton concentration. The observed variation in chlorophyll a concentration covers a 10:1 range, in agreement with the range of variation in $P_F(Ca)/P_R$, whereas the uncorrected signal $P_F(Ca)$ varies over only a 3:1 range. Figure 3 shows the same profile of P_R overlayed with the reciprocal of the discrete beam attenuation coefficient data obtained from the in situ beam transmissometer measurements. The agreement between the airborne and ground–truth data is consistent with a simple theoretical analysis [35].

(b) Snake and Columbia Rivers

These measurements were obtained from an extensive riverine-estuarine continuum consisting of a 734-km segment of the lower Snake and

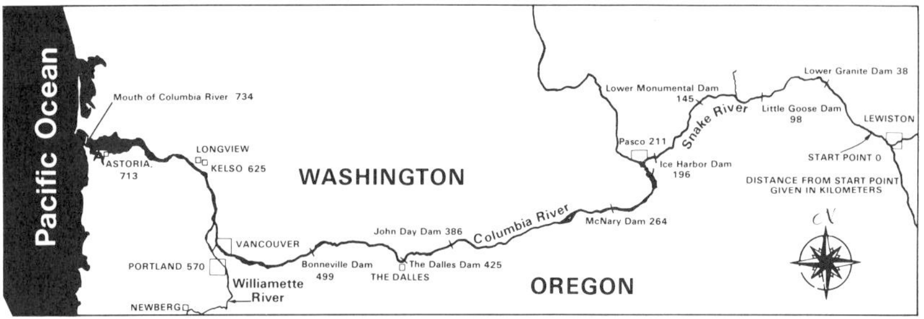

Fig. 4. Map showing region of Snake and Columbia Rivers surveyed by airborne laser fluorosensor. Key features, including the eight hydroelectric facilities of the Columbia-Snake Inland Waterway, are indicated together with their distance from the start point in kilometers.

Columbia Rivers during the period of peak snow-melt runoff. This part of the river system is more commonly known as the Columbia-Snake Inland Waterway, and includes eight impoundments formed by hydroelectric dam/lock facilities (see Figure 4).

The airborne data in the form of the Raman-normalized chlorophyll a fluorescence signal, $P_F(Ca)/P_R$, the reciprocal of the water Raman signal, P_R^{-1}, and the Raman-normalized DOM fluorescence signal, $P_F(DOM)/P_R$, were calibrated in terms of, respectively, extracted chlorophyll a, beam attenuation coefficient and DOC. These calibrations were obtained from linear regression plots produced by comparing airborne and concurrent ground-truth data collected from 12 sites in the Columbia River estuary. As such, the validity of these calibrations at upstream sites remote from the estuary may be questionable because of the presence of a different type of DOM or algal community, but, nevertheless, they do provide a general order-of-magnitude indication of the values for these parameters.

The profiles of the airborne data for chlorophyll a, beam attenuation coefficient and DOC are shown in Figures 5a, b, and c, respectively. A number of significant features are apparent in each of these profiles.

A marked feature of the chlorophyll a profile (Figure 5a) is the series of sharp steps occurring at each dam location. These steps are interpreted as being due to a high subsurface concentration of nonmotile phytoplankton on the immediate upstream side of each dam. Due to an absence of turbulent mixing as the river water progresses across each impoundment, the phytoplankton sink below the surface layer that is visible to the laser

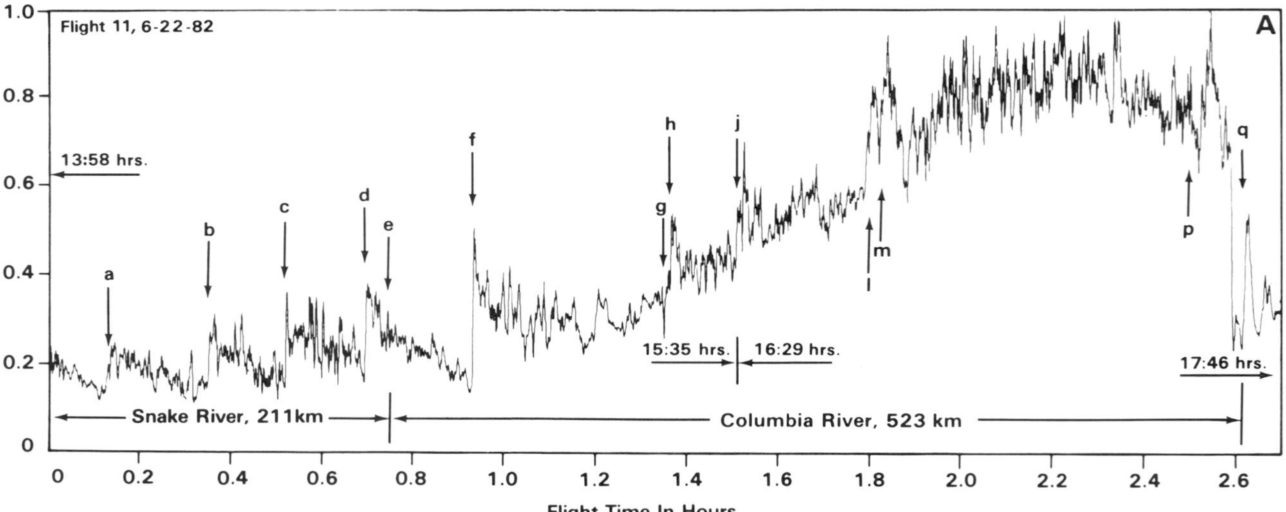

Fig. 5a. Airborne laser fluorosensor profiles of (a) chlorophyll a fluorescence ($P_F(Ca)/P_R$), (b) reciprocal of water Raman emission (P_R^{-1}), and (c) DOM fluorescence ($P_F(DOM)/P_R$ as a function of aircraft flight time for flight of June 22, 1982. Equivalent full-scale ground-truth values are, respectively, n(Ca) = 27.7 μg/L, c = 2.50 m^{-1}, and n(DOC) = 3.24 mgC/L. Symbols to key locations: a, Lower Granite Dam; b, Little Goose Dam; c, Lower Monumental Dam; d, Ice Harbor Dam; e, confluence of Snake and Columbia rivers; f, McNary Dam; h, John Day Dam; j, The Dalles Dam; m, Bonneville Dam, p, ground-truth sampling site; q, mouth of Columbia River.

B
Flight 11, 6-22-82
Normalized Value
1.0
0.8
0.6
0.4
0.2
0
0
0.2
0.4
0.6
0.8
1.0
1.2
1.4
1.6
1.8
2.0
2.2
2.4
2.6
Flight Time In Hours
a
b
c
d
e
f
h
j
m
p
q
Snake River
Columbia River

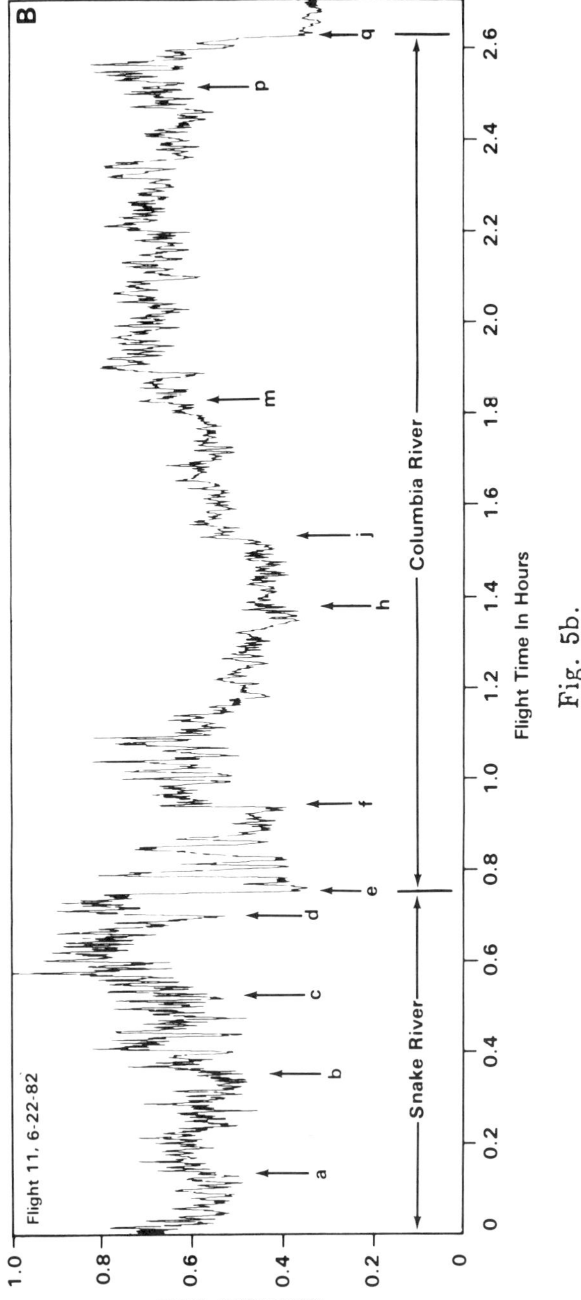

Fig. 5b.

C
Flight 11, 6-22-82
a
b
c
d
e
f
g h
j
m
p
q
Snake River
Columbia River
Normalized Value
1.0
0.8
0.6
0.4
0.2
0
0
0.2
0.4
0.6
0.8
1.0
1.2
1.4
1.6
1.8
2.0
2.2
2.4
2.6
Flight Time In Hours

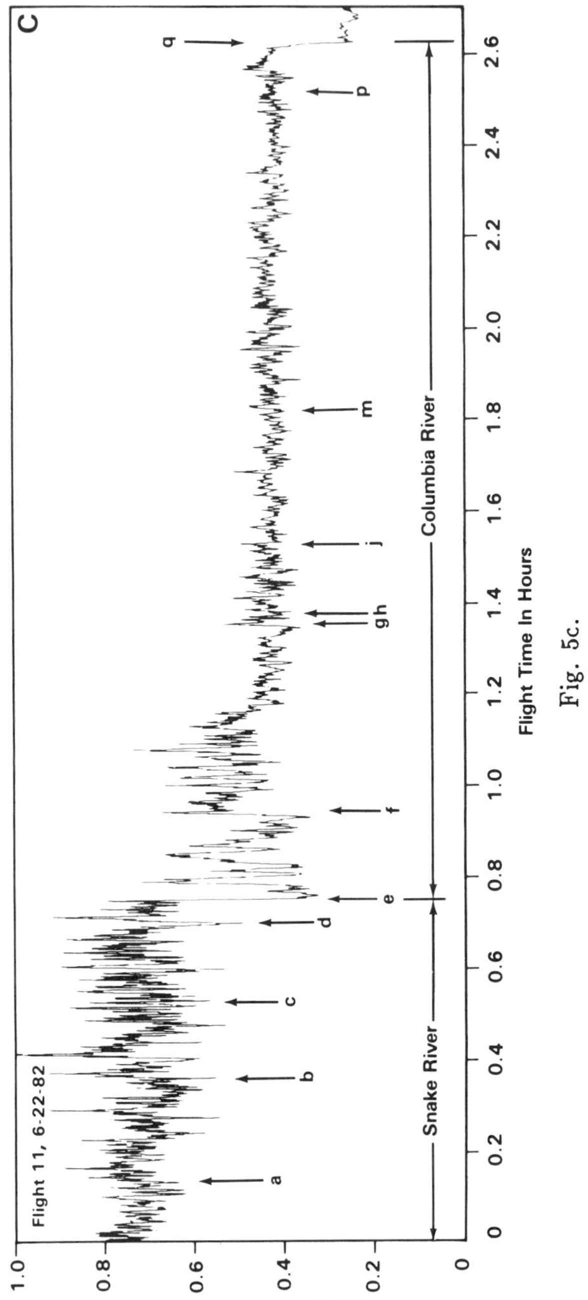

Fig. 5c.

fluorosensor. As the water passes through the turbines or over the spillways, the phytoplankton become redistributed throughout the water column and hence visible to the laser fluorosensor. This behavior is then repeated across each subsequent impoundment. Surprisingly, the chlorophyll a level did not change at the confluence, implying that, at this time, both rivers had similar chlorophyll a concentrations upstream of the confluence. The slow increase in chlorophyll a concentration to a maximum in the lower region of the Columbia River is believed to be due to a combination of factors that include high levels of nutrients, a warming trend introduced by the shallower, slower moving water that results from a broadening of the river, and increased light penetration due to some settling-out of suspended sediments.

The beam attenuation profile (Figure 5b) shows a steady rise as the Snake River approaches the confluence. This is considered to be due to the scouring action of the peak runoff flow in resuspending sediments into the water column. Whetten et al. [18] have shown that in this part of the river system sediments deposited throughout the year are often removed to the point that bedrock is exposed. The behavior at and below the confluence implies a significantly higher suspended sediment levels for the Snake River in contrast to the upstream Columbia River, and indicates a mixing region that extends for about 100 km downstream of the confluence. After a short region in which the beam attenuation coefficient levels out, it begins to rise in the lower Columbia River due, presumably, to the increased phytoplankton population indicated in Figure 5a.

The DOC profile is shown in Figure 5c and suggests that the levels in the Snake River are approximately twice those of the Columbia River upstream of the confluence. Again, the DOC profile, like that for the beam attenuation coefficient, provides a clear indication of the mixing region downstream of the confluence. The existence of a steady DOC level below this region and extending out to the river mouth was a little surprising in view of the extensive logging, timber, and industrial activity in this region of the river, which includes an extensive urban region centered on the city of Portland. However, an independent study by Dahm et al. [19] has confirmed this behavior.

FEASIBILITY OF REMOTE SENSING FRESHWATER BLUE-GREEN ALGAE

A need exists for a remote sensing technique that can measure the distribution and density of blue-green algae in lakes and impoundments. From such synoptic measurements in conjunction with other, readily available independent data such as water temperature, nutrient levels, wind velocity, etc., it should be possible to predict whether conditions exist for widespread bloom formation. Blue-green algae are notorious for forming extensive surface blooms that can create toxic and malodorous water conditions leading, in extreme situations, to severe oxygen depletion. This is a problem that occurs at all latitudes, most frequently in late summer. In fresh waters, toxic conditions are produced by blue-green algae, whereas in estuarine and marine waters dinoflagellates are the offending algae type, being responsible for the so-called red tides. The nature of nuisance blooms and the conditions leading to their formation has recently been discussed by Paerl [20].

Near-IR photography [21] and near-IR satellite multispectral imagery [22] have proved successful in mapping the distribution of different blooms of *Aphanizomenon* flos-aquae. However, in these situations the dramatic imagery was, for the most part, a record of the large surface accumulations of dead cell material of decaying blooms. With few exceptions, passive airborne or satellite remote–sensing techniques have not been successful in measuring nonbloom concentrations of chlorophyll a or blue-green algae in fresh–waters because of spectral interference from suspended sediments and dissolved organics. Airborne laser fluorosensors, on the other hand, have demonstrated high sensitivity for measuring chlorophyll a under these (freshwater) conditions [12]. In addition, they have also been used to measure the concentration of dissolved organics (as DOC) and the optical attenuation coefficient, which is often a good indicator of suspended sediment levels [12]. It is therefore proposed that, as the in vivo phycocyanin fluorescence emission excited at 487 nm is at least as intense as that for chlorophyll a, the airborne laser fluorosensor be evaluated as a device for monitoring blue-green algae at both preemergent and bloom concentrations in addition to the parameters discussed earlier. To meet this objective, a laboratory feasibility study is being conducted and the existing laser fluorosensor modified so that the optical receiver can measure fluorescence emission spectra on a quasi-continuous spectral basis. Preliminary results of these preparations are described.

The laboratory study consisted of a laser fluorosensor simulation experiment to measure the laser-induced fluorescence emission spectra of a culture of a common bloom-forming species using the same multichannel spectrograph as will be incorporated into the modified airborne system. Samples of *Microcystis aeruginosa* Kütz. obtained from the University of Texas Collection were cultured under white fluorescent lighting in a BGJ/84 medium and harvested at near-bloom concentrations. Fluorescence excitation was achieved with a N_2 laser-pumped tunable dye laser (Laser Science, Inc.) operating in the Littrow configuration. Laser beam and pulse characteristics were as follows: wavelength, 487 nm; bandwidth, 0.3 nm; pulsewidth, 3 nsec; pulse energy, 20 μ J; polarization, linear; beam dimensions, 1 x 3 mm; repetition rate, 12 Hz. Samples were diluted (3:1 to 10:1) with deionized water so as to make the fluorescence band intensities approximately equal to that of the water Raman band. In regard to these dilution ratios, it should be noted that the fluorescence excitation crosssection at 487 nm for chlorophyll a emission at 685 nm for marine blue-green algae is typically four- to five-fold smaller than that of marine green algae [15]. Consequently, the density of blue-green algae (as characterized by chlorophyll a concentration data) needed to achieve the same emission intensity at 685 nm must be four to five times larger than that for green algae. A possible penalty for using 487 nm as the excitation wavelength might then be that the fluorescence emission for the blue-green algae is dominated by that from all other algal color groups except under bloom conditions. This situation would then necessitate the use of a longer excitation wavelength closer to the peak excitation wavelength for c-phycocyanin (620nm).

The phytoplankton samples, held in an open-top 1-cm pathlength fused silica cuvette, were illuminated from above by the unfocused laser beam so that the irradiated cylindrically-shaped sample volume is parallel to the spectrograph entrance slit and to the individual 2.5-mm high diodes of the array detector. A lens with angular aperture matched to that of the f/2 spectrograph collected the fluorescence emitted at 90° to the laser beam and focused a slit-sized image of the excited sample volume onto the entrance slit. Because of the linearly polarized state of the laser beam, the 90°-scattering geometry, and the polarization sensitivity of the spectrograph, it is possible to encounter polarization artifacts that result in emission spectra different from those observed in a remote sensing configuration. This occurs because the polarization state of the Raman emission is highly dependent both on that of the excitation source and on the scattering geometry [23], whereas the fluorescence emission from algae is essentially depolarized regardless of scattering geometry. This phenomenon, which influences the 90°-scattering geometry of the laboratory measurements differently than the 180°-scattering geometry of the airborne sensor, can be avoided by using a depolarizer [1]. For the laboratory study, a large aperture quartz-wedge type depolarizer was inserted into the emission beam just after the collection lens. The entry of scattered laser light into the spectrograph was blocked by placing a long-wave-pass absorption filter (Corning 3-69) having low inherent fluorescence [24] into the emission beam just prior to the entrance slit.

A description of the evaluation and calibration of this spectrograph-detector subsystem has been given elsewhere [25], so only a brief description is given here. The f/2, 200-mm focal length flat-field spectrograph uses a custom, concave holographic grating to disperse and focus the fluorescence spectra onto a 25-mm wide image intensifier that is phosphor coupled to a 1024-element linear CCD array detector (Princeton Applied Research Model 1421 R). The image intensifier was gated on for about 100 ns in order to

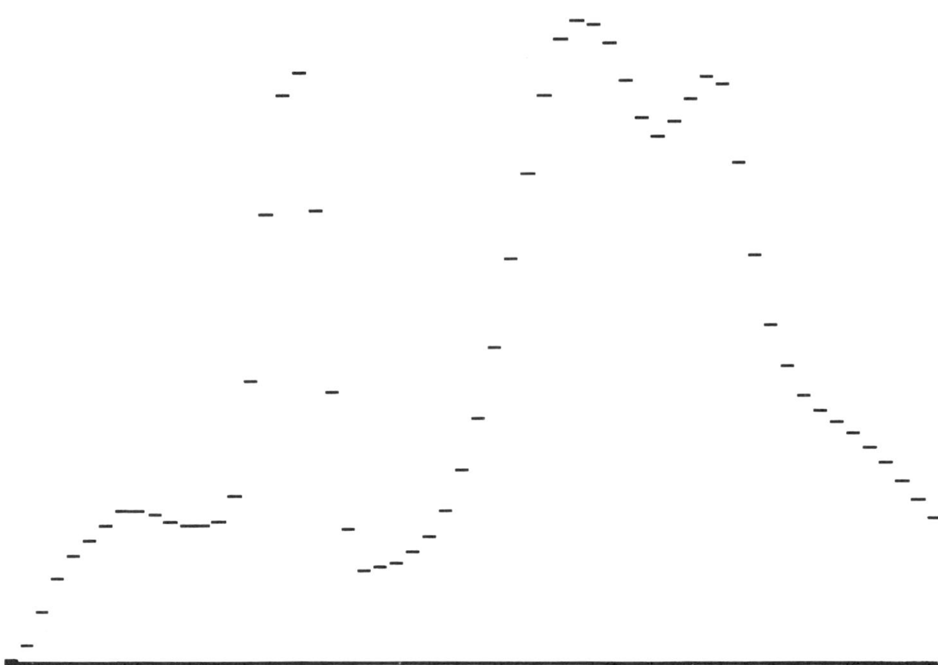

Fig. 6. Uncorrected fluorescence emission spectrum of a dark-adapted sample of *Microcystis aeruginosa* excited at 487 nm. Spectral coverage is from 516 nm to 732 nm with shorter wavelengths at the left.

prevent integration by the device of extraneous background light during a series of measurements. During both laboratory and airborne operation of this detector, the signals from the individual diodes of the CCD array are ganged together into 64 groups of 16 consecutive diodes per group for reasons related to the airborne application of the device [25]. Under these circumstances, the overall system, which has an effective spectral resolution of 7.5 nm, reproduces the water Raman band without loss in amplitude. Good spectral signal-to-noise ratios were achieved by (digitally) summing a series of spectra obtained at a laser pulse rate of 12 Hz for which each spectrum of the series has had its (internal) independently measured background signal removed by simple subtraction. The alternate approach of integrating the series of spectra directly on the CCD array results in unacceptably large internal dark signal and noise components because of the relatively large time period between each laser stimulated event.

A typical fluorescence emission spectrum of a 5:1 dilution of the *Microcystis aeruginosa* culture excited at 487 nm under dark-adapted conditions is shown in Figure 6. This spectrum, uncorrected for system spectral artifacts and composed of the sum of 450 individual spectra, covers the spectral region from 516 nm to 723 nm. Both the first and last three elements of the spectrum have been omitted because of intensifier edge effects. The slow rise in the dissolved organics part of the spectrum at shorter wavelengths is due to the influence of the long-wave-pass filter (Corning 3-69) and can be eliminated during spectral correction of the data. With excitation at 487 nm, the water Raman band centered at 584 nm just avoids overlapping the c-phycocyanin emission band centered at 653 nm. By this arrangement, it is possible to separate (by interpolation) the Raman band from the underlying DOM fluorescence component. Separation of the overlapped c-phycocyanin and chlorophyll a bands centered, respectively, at 653 nm and 685 nm is more complicated. It is proposed that this separation be achieved by using the procedure of curve fitting [26,27,28]. In this approach the functional form of the chlorophyll a and c-phycocyanin emission bands are predetermined by performing in vivo fluorescence emission spectroscopy on a culture of green algae such as that of *Chlorella* and on a pure c-phycocyanin extract, respectively. The form of the underlying DOM fluorescence background signal, which does not change rapidly with wavelength and which is usually stable within a given watershed, can be obtained by performing similar spectroscopy on a small number of representative field samples that have been filtered (e.g., 0.2-μm pore size) to remove the phytoplankton fluorescence contribution.

Having separated the phycocyanin emission band from the other spectral components, it still remains to interpret this fluorescence signal in terms of its equivalent (extracted) c-phycocyanin pigment concentration. In a remote sensing scenario, this would be performed in an analogous manner to that described earlier for chlorophyll a, DOC and the beam attenuation coefficient, where a small number of representative "ground truth" samples are collected under the sensor flight path at the time of the overpass. These samples are then analyzed and, through simple linear regression, used to convert the airborne data to equivalent water quality data. This procedure would then be extended to include analyses for c-phycocyanin content by performing absorption spectroscopy on the c-phycocyanin pigment extracted from a specific sample volume [29,30,2].

For remotely sensed fluorescence emission spectra containing a c-phycocyanin emission band, interpretation of the intensities of this and the chlorophyll a band in terms of pigment concentrations may be complicated by a number of factors. First, because the chlorophyll a fluorescence cross section for blue-green algae excited at 487 nm is, typically, lower by a factor of 5:1 then that for other algal color groups, this signal will tend to underestimate the chlorophyll a content of the water column if a significant concentration of non blue-green algae are present. This situation can be further exacerbated by water surface solar irradiation levels that change significantly over short times or distances due to changes in cloud cover. Most phytoplankton exhibit photoinhibition of fluorescence emission when solar irradiation levels exceed some algal-class dependent value [31]. This effect has already been observed in airborne laser fluorosensor measurements [12]. Blue-green algae, however, often exhibit the opposite effect whereby an increase in solar radiation induces an increase in the independently-stimulated chlorophyll a fluorescence emission [32]. In the present study, this effect was simulated by performing laser-excited fluorescence measurements on both dark-adapted samples (as illustrated by the spectrum in Figure 6) and on samples exposed to a bright steady white light source for several minutes (not shown). The effect of this background light was to increase the intensity of the laser-excited in vivo chlorophyll a emission by about 40% without changing the intensity of the c-phycocyanin band. Similar percentage increases have been observed in Kautsky-type studies (where the background and excitation sources are one and the same) performed in vivo on the marine blue-green alga, *Anacystis nidulans* [33,34]. Finally, remotely measured c-phycocyanin fluorescence may originate, either partly or wholly, from free rather than cell-bound pigment, a situation that will occur at the mature stage of a bloom when cells are dying and rupturing in large numbers. Under these circumstances the water surface can take on a markedly blue appearance. It should be noted that the sample used to produce the spectrum in Figure 6 contained no free c-phycocyanin as confirmed by running a similar spectrum on a sample that had been passed through an 0.2-μm membrane filter. However, situations in which high concentrations of free c-phycocyanin are present are not likely to produce ambiguous data, particularly as the c-phycocyanin signal will be extremely large. It is planned to continue this laboratory feasibility study, but it is likely that these questions and anomalies may only be resolved by conducting a series of remote sensing measurements over a bloom-impacted lake or impoundment.

Modifications to the airborne laser fluorosensor used in the earlier field studies and described by Bristow et al. [35,12] primarily involve construction of a new receiver system capable of producing essentially continuous spectra. This system is composed of an f/2 parabolic Newtonian telescope coupled to a custom f/2 flat-field, concave-holographic-grating spectrograph, which disperses the spectrum across the face of a gated, intensified linear CCD array detector. This is the same spectrograph-detector combination that was used to produce the spectrum in Figure 6. The evaluation, calibration and, in particular, the nontrivial procedure needed to remove the solar background signal from the laser-induced fluorescence signal are described by Bristow et al. [25]. The relationship of this receiver subsystem to the overall laser fluorosensor is shown schematically in Figure 7. Optimum signal-to-noise

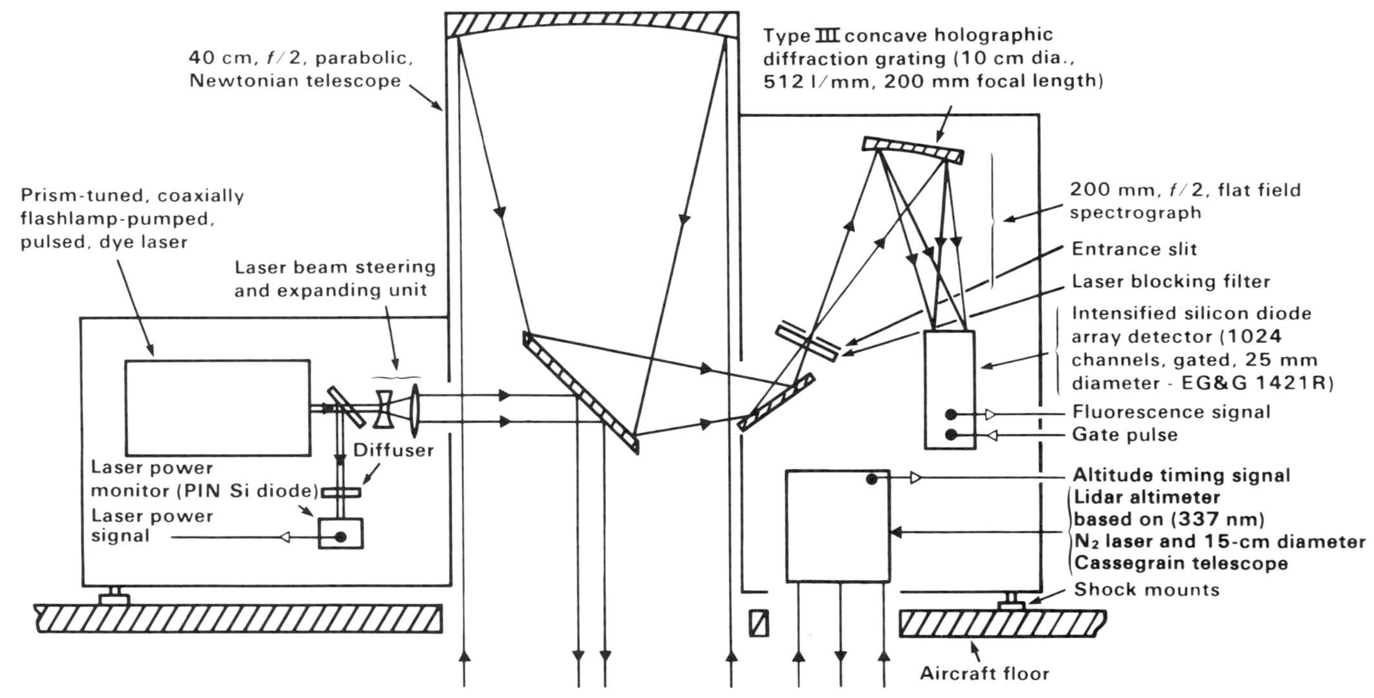

Fig. 7. Optical schematic of modified airborne laser fluorosensor with Newtonian telescope, spectrograph and multichannel detection system.

ratios are achieved by (temporally) shaping the signal used to gate the image intensifier of the multichannel detector so that it exactly matches that of the returned fluorescence signal. The precise gate timing needed for this purpose is achieved by using an independent laser altimeter based on a N_2 laser operating at 337 nm.

CONCLUSIONS

The airborne laser fluorosensor is shown to be capable of measuring several parameters in the water column at the same time. Moreover, in situations where interrelationships between these parameters are known to exist such as with the water quality measurements discussed here, it becomes possible to cross-correlate between concurrent profiles to reveal anomalies, eliminate ambiguities and enhance the interpretation of any one series.

In addition, a feasibility study has demonstrated that a potential exists for using the airborne laser fluorosensor to monitor the distribution and concentration of blue-green algae in fresh surface waters.

NOTICE

Although the information in this document has been funded wholly or in part by the United States Environmental Protection Agency under Cooperative Agreement #CR814002-01 to the University of Nevada's Environmental Research Center, it does not necessarily reflect the views of the Agency and no official endorsement should be inferred.

REFERENCES

[1] Bristow, M. P. F. and D. Nielsen, 1981. Remote Monitoring of Organic Carbon in Surface Waters. Project Report EPA-600/4-81-001, U.S. Environmental Protection Agency, Las Vegas, Nevada. Available from NTIS as PB81-168965, Springfield, Virginia, 1981.

[2] Beer, S. and A. Eshel, 1985. Determining Phycoerythrin and Phycocyanin Concentrations in Aqueous Crude Extracts of Red Algae. *Aust. J. Mar. Freshw. Res.* **36**, 785-792, 1985.

[3] Hoge, F. E. and R. N. Swift, Airborne Simultaneous Spectroscopic Detection of Laser-Induced Water Raman Backscatter and Fluorescence from Chlorophyll a and Other Naturally Occurring Pigments. *Appl. Opt.* **20**, 3197-3205, 1981.

[4] Diebel-Langohr, D., T. Hengstermann, and R. Reuter, Identification of Hydrographic Fronts by Airborne Lidar Measurements of Gelbstoff Distributions. In: Dynamic Biological Processes of Marine Physical Interfaces. J. C. J. Nihoul (ed.), Proc. 17. Int. Liege Coll. on Ocean Hydrodynamics. May 13-17, 1985. Elsevier, Amsterdam, Holland, 1986.

[5] Farmer, F. H., G. A. Vargo, C. A. Brown, and O. Jarrett, Spatial Distributions of Major Phytoplankton Community Components in the Narragansett Bay at the Peak of the Winter-Spring bloom. *J. Mar. Res.* **40**, 593-614, 1982.

[6] Hoge, F. E. and R. N. Swift, Airborne Dual Laser Excitation and Mapping of Phytoplankton Photopigments in a Gulf Stream Warm Core Ring. *Appl. Opt.* **22**, 2272-2281, 1983.

[7] Hoge, F. E. and R. N. Swift, Airborne Mapping of Laser-Induced Fluorescence of Chlorophyll a and Phycoerythrin in a Gulf Stream Warm Core Ring. In: Mapping Strategies in Chemical Oceanography. Alberto Zirino (ed.). Advances in Chemistry Series. No. 209. American Chemical Society. Washington D.C., 1985.

[8] Houghton, W. M., R. J. Exton, and R. W. Gregory, 1983. Field Investigation of Techniques for Remote Laser Sensing of Oceanographic Parameters. *Remote Sensing Environ.* **13**, 17-32, 1983.

[9] Hoge, F. E. and R. N. Swift, Absolute Tracer Dye Concentration Using Airborne Laser-Induced Water Raman Backscatter. *Appl. Opt.* **20**, 1191-1202, 1981.

[10] Franz, H., V. Gehlhaar, K. P. Gunther, A. Klein, J. Luther, R. Reuter, and H. Weidmann, Airborne Fluorescence LIDAR Monitoring of Tracer Dye Patches - A Comparison with Shipboard Measurements. *Deep Sea Res.* **29**, 893-901, 1982.

[11] Hoge, F. E. and R. N. Swift, Delineation of Estuarine Fronts in the German Bight Using Laser-Induced Water Raman Backscatter and Fluorescence of Water Column Constituents. *Int. J. Remote Sensing.* bf 3, 475-495, 1982.

[12] Bristow, M. P. F., D. H. Bundy, C. M. Edmonds, P. E. Ponto, B. E. Frey, and L. F. Small, Airborne Laser Fluorosensor Survey of the Columbia and Snake Rivers: Simultaneous Measurement of Chlorophyll, Dissolved Organics and Optical Attenuation. *Int. J. Remote Sensing.* **6**, 1707-1734, 1985.

[13] Diebel-Langohr, D., T. Hengstermann, and R. Reuter, Water Depth Resolved Determination of Hydrographic Parameters from Airborne Lidar Measurements. In: Dynamic Biological Processes at Marine Physical Interfaces. J. C. J. Nihoul (ed.), Proc. 17. Int. Liege Call. on Ocean Hydrodynamics. May 13-17, 1985. Elsevier, Amsterdam, Holland, 1986.

[14] Bristow, M. P. F., 1978. Airborne Monitoring of Surface Water Pollutants by Fluorescence Spectroscopy. *Remote Sensing Environ.* **7**, 105-127, 1978.

[15] Browell, E. V., Analysis of Laser Fluorosensor Systems for Remote Algae Detection and Quantification. NASA Technical Note TND-8447. Washington, D.C., 1977

[16] Watras, C. J. and A. L. Baker, Detection of Planktonic Cyanobacteria by Tandem In Vivo Fluorometry. *Hydrobiol.* **169**, 77-84, 1988.

[17] Chapman, D. J., Biliproteins and Bile Pigments. In: The Biology of Blue-Green Algae. N. G. Carr and B. A. Whitton (ed). University of California Press, Berkeley, 1973.

[18] Whetten, J. T., J. C. Kelley, and L. G. Hanson, Characteristics of Columbia River Sediment and Sediment Transport. *J. Sediment. Petrol.* **39**, 1149-1166, 1969.

[19] Dahm, C. N., S. V. Gregory, and P. K. Park, 1981. Organic Carbon Transport in the Columbia River. *Estuar. Coastal Shelf Sci.* **13**, 645-658, 1981.

[20] Paerl, H. W., Nuisance Phytoplankton Blooms in Coastal, Estuarine and Inland Waters. *Limnol. Oceanogr.* **33**, 823-847, 1988.

[21] Wrigley, R. C. and A. J. Horne, Remote Sensing and Lake Eutrophication. *Nature* **25**, 213-214, 1974.

[22] Strong, A. E., 1974. Remote Sensing of Algae Blooms by Aircraft and Satellite in Lake Erie and Utah Lake. *Rem. Sens. Environ.* **3**, 99-107, 1974.

[23] Gilson, T. R. and P. J. Hendra, Laser Raman Spectroscopy. Wiley - Interscience, London. 1970.

[24] Bristow, M. P. F., 1979. Fluorescence of Short Wavelength Cutoff Filters. *Appl. Opt.* **18**, 952-955, 1979.

[25] Bristow, M. P. F., R. M. Turner, C. M. Edmonds, and D. H. Bundy, Short- and Long-Term Memory Effects in Intensified Array Detectors: Influence on Airborne Laser Fluorosensor Measurements. *Appl. Opt.* **28**, 472-480, 1988.

[26] Gans, P. and J. B. Gill, On the Analysis of Raman Spectra by Curve Resolution. *Appl. Spectrosc.* bf 31, 451-455, 1977.

[27] Maddams, W. F., The Scope and Limitations of Curve Fitting. *Appl. Spectrosc.* **34**, 245-267, 1980.

[28] Barker, B. E. and M. F. Fox, 1980. Computer Resolution of Overlapping Electronic Absorption Bands. *Chem. Soc. Reviews.* **9**, 143-184, 1980.

[29] Siegelman, H. W. and J. H. Kycia, Algae Bile Proteins. In: Handbook of Phycological Methods. J. A. Helleburst and J. S. Craigie (ed.). Cambridge University Press, Cambridge, 1978.

[30] Stewart, D. E. and F. H. Farmer, Extraction, Identification, and Quantitation of Phycobiliprotein Pigments from Phototropic Plankton. *Limnol. Oceanogr.* **29**, 392-397 1984.

[31] Harris, G. P., Photosynthesis, Productivity and Growth: The Physiological Ecology of Phytoplankton. *Arch. Hydrobiol. Beih.* **10**, 1-171, 1978.

[32] Heaney, S. I., Some Observations on the Use of the In Vivo Fluorescence Technique to Determine Chlorophyll a in Natural Populations and Cultures of Freshwater Phytoplankton. *Freshwater Biol.* **2**, 115-126, 1978.

[33] Papageorgiou, G. and Govindjee, Light Induced Changes in the Fluorescence Yield of Chlorophyll a In Vivo. I. Anacystis Nidulans. *Biophys. J.* **8**, 1299-1315, 1968.

[34] Goedheer, J. C., Spectral Properties of the Blue-Green Alga Anacystis Nidulans Grown Under Different Environmental Conditions. *Photosynthetica.* **10**, 411-422, 1976.

[35] Bristow, M. P. F., D. Nielsen, D. Bundy, and R. Furtek, 1981. Use of Water Raman Emission to Correct Airborne Laser Fluorosensor Data for Effects of Water Optical Attenuation. *Appl. Opt.* **20**, 2889-2906, 1981.

[36] Raymond M., Laser Remote Sensing Measures. Wiley, New York., 1984.

[37] Hoge, F. E. and R. N. Swift, Airborne Detection of Oceanic Turbidity Cell Structure Using Depth-Resolved Laser-Induced Water Raman Backscatter. *Appl. Opt.* **22**, 3778-3786, 1983.

AN OVERVIEW OF ADVANCED SPECTROSCOPIC FIELD SCREENING AND IN-SITU MONITORING INSTRUMENTATION AND METHODS

D. EASTWOOD, R.L. LIDBERG, and S. J.SIMON

Lockheed Engineering and Sciences Co.
Advanced Technology Project Office
Las Vegas, Nevada 89119, USA

T. VO-DINH

Oak Ridge National Laboratory
Advanced Monitoring Development Group
Health and Safety Division
Oak Ridge, Tennesseev 37831 USA

ABSTRACT

The United States Environmental Protection Agency (U.S. EPA) is interested in screening suspected hazardous waste sites for a variety of environmental pollutants in surface water, ground water and soil. To obtain rapid response and to reduce the total cost of environmental assessment, it is desirable to use portable or field deployable spectroscopic methods and instruments to rapidly screen samples in situ or with little or no sample preparation. A more detailed and expensive chemical characterization is then required only on a select subset of the samples.

Lockheed Engineering & Sciences Company, as the prime in-house EPA

Chemistry for the Protection of the Environment
Edited by L. Pawlowski *et al.*, Plenum Press, New York, 1991

contractor in Las Vegas, and Oak Ridge National Laboratory (ORNL), through an inter-agency agreement with the EPA, are providing research and development support for the EPA in a variety of areas including state-of-the-art portable fluorescence analyzers, room temperature phosphorescence for polychlorinated and polybrominated biphenyls and other polycyclic compounds, fiber optic chemical sensors for deep well groundwater monitoring, a fiber optic luminoscope for surface water, surface-enhanced Raman spectroscopy, and fluoroimmunoassay. An overview of several specific examples of on-going spectroscopic instruments and methods development is presented to illustrate the usefulness of these techniques.

INTRODUCTION

For many environmental monitoring and assessment applications, there is interest by the U.S. Environmental Protection Agency (EPA) in field screening hazardous waste sites for pollutants in surface and ground water as well as in soil.

Spectroscopic methods and instruments that are field deployable or portable permit samples to be screened and prioritized in the field with little or no sample preparation. These screening techniques permit rapid response and considerable cost savings since more detailed analyses are required only on a select subset of samples.

This presentation provides an overview of the advanced spectroscopic techniques and state-of-the-art instrumentation that Lockheed as the prime EPA contractor in Las Vegas and Oak Ridge National Laboratory (ORNL) through an interagency agreement with the EPA are investigating and evaluating for field screening purposes. The detection, identification and quantitation are important, but for field screening purposes for complex mixtures, characterization including classification, and semiquantitation may also be of interest.

One of the most mature, although often neglected, spectroscopic techniques is luminescence because of its potentially high sensitivity and selectivity especially with new laser sources and detectors recently developed. A number of luminescence techniques [emission, synchronous, room temperature phosphorescence (RTP) and contour] are discussed along with currently available instrumentation. The power of fluorescence techniques can also be extended by chemical derivatization, by use with liquid chromatography, and by fluoroimmunoassay.

Ultraviolet-visible (UV-vis) absorption spectroscopy and colorimetric techniques are, of course, complementary techniques to UV-vis luminescence and have also been used for field testing. These methods have disadvantages, however, relative to fluorescence in sensitivity and selectivity.

Although infrared spectroscopy (IR) is frequently used in field laboratories because of its specificity and capability to classify unknown chemicals using group frequencies, disadvantages to its use have included its relatively low sensitivity compared to fluorescence and interferences due to the presence of water that complicate sample preparation.

Laser Raman spectroscopy, complementary to infrared in the vibrational information obtained, has been neglected for field applications because of relative lack of sensitivity, interferences and complicated instrumentation. Advances in Raman instrumentation and techniques now make this approach more promising. Surface-enhanced Raman spectroscopy (SERS) is of special interest because, in some cases, it may have sensitivities rivaling luminescence, and it is potentially applicable to a larger group of pollutants.

Fiber optic chemical sensors of several types (colorimetric, fluorometric, refractive index, fluoroimmunoassay, etc.) are gaining importance for in situ monitoring of trace pollutants in groundwater wells.

For all of these spectroscopic approaches, trade-offs must be considered among sensitivity, selectivity, cost-effectiveness, simplicity and ruggedness under field conditions.

METHODS AND INSTRUMENTATION

Luminescence screening techniques

Luminescence spectroscopy, usually as room temperature fluorescence emission, has long been used for field and laboratory screening applications because of its sensitivity, selectivity, freedom from interferences by water and many nonfluorescing chemicals, ease of sample preparation, and relative availability of fielddeployable instrumentation [1-9]. Luminescence excitation spectra have also been used to permit comparison with absorption spectra when reference fluorescence spectra were unavailable. Luminescence has proven especially useful for highly fluorescent species such as aromatics and polynuclear aromatic hydrocarbons (PNAs) and their derivatives and for petroleum oils, although it is also useful for pollutants such as phenols, (see Figure 1), polychlorinated biphenyls (PCBs) and other halogenated polyaromatics, some pesticides and heterocyclic compounds as well as some inorganics, such as uranium [10] and metals as organometallics.

The utility of luminescence for analyzing complex mixtures has been improved by the use of synchronous spectroscopy [11-13], where the luminescence signal is recorded while the excitation and emission wavelengths are scanned simultaneously, usually with a constant wavelength interval ( ) between them. This procedure improves selectivity by narrowing spectral

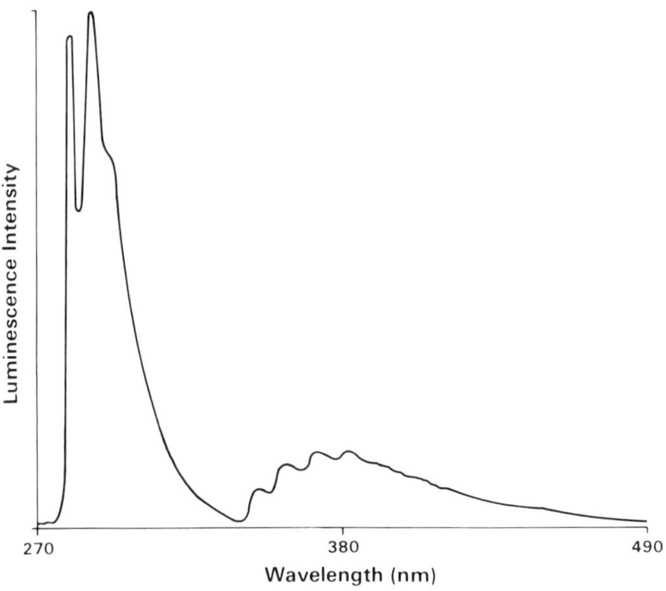

Fig. 1. Low temperature luminescence (77 K) of phenol in 1:1 ethanol:methanol. Excitation wavelength = 265 nm. Excitation bandpass = 4.5 nm. Emission bandpass = 0.9 nm. Phenol 11μg/mL. 1:1 EtOH/MeOH. Ex = 265 nm. 77 K.

bands to reduce emission from interfering compounds. For PNAs, it has the further advantage of roughly separating them on the basis of the number of aromatic rings (in a homologous series). Figure 2 gives an example of the use of synchronous fluorescence for components in a coal liquid from a synfuel plant. Another example for a heavy crude petroleum oil is given in Figure 3. For synchronous fluorescence, the optimum wavelength interval depends on the Stokes shift between the absorption and fluorescence bands and is typically 3-5 nm. For synchronous phosphorescence the optimum interval is determined by the singlet-triplet energy difference and might range from 100 to 300 nm.

Emission, excitation and synchronous spectra can all be considered crosssections of the total or contour luminescence (excitation-emission matrix) spectrum that can be computer-generated if further information is needed for an unknown mixture. Fluorescence lifetimes can also be used to increase spectral sensitivity, but so far this approach has seldom been used for routine field measurements.

Low-temperature phosphorescence, utilizing low-temperature rigid

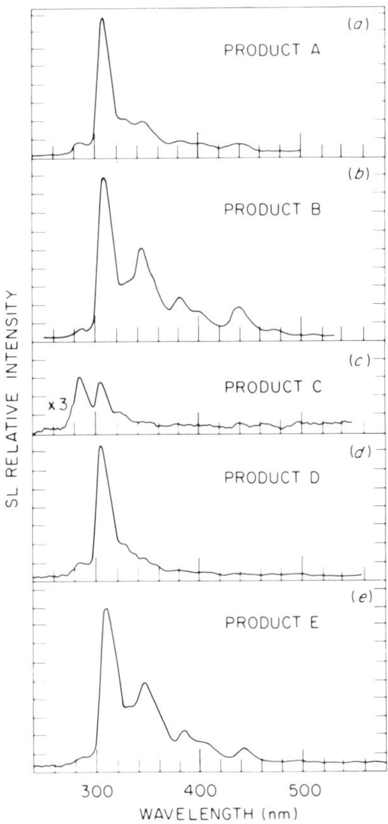

Fig. 2. Synchronous fluorescence of different coal liquid products from a synfuel plant. Wavelength interval between excitation and emission monochromators = 5 nm. Coal liquid products synchronous emission $\Delta\lambda = 5$ nm.

matrices to reduce collisional quenching and other deactivation processes, has frequently been used in laboratory analysis of chemicals with relatively high phosphorescence quantum yields. Phosphorescence has an advantage in that many interferences from fluorescent chemicals can be eliminated either by wavelength or lifetime selectivity. The additional complexity of required cryogenic equipment and refrigerant has limited the usefulness of this approach for field screening. The use of organized media such as solid substrates, micelle or cyclodextrin solutions permits room–temperature phosphorescence (RTP) to be observed with simpler apparatus [15-21]. Figure 4 shows the RTP of a PNA mixture spotted onto filter paper by a simple procedure. RTP sensitivity can often be enhanced by mixing the sample or

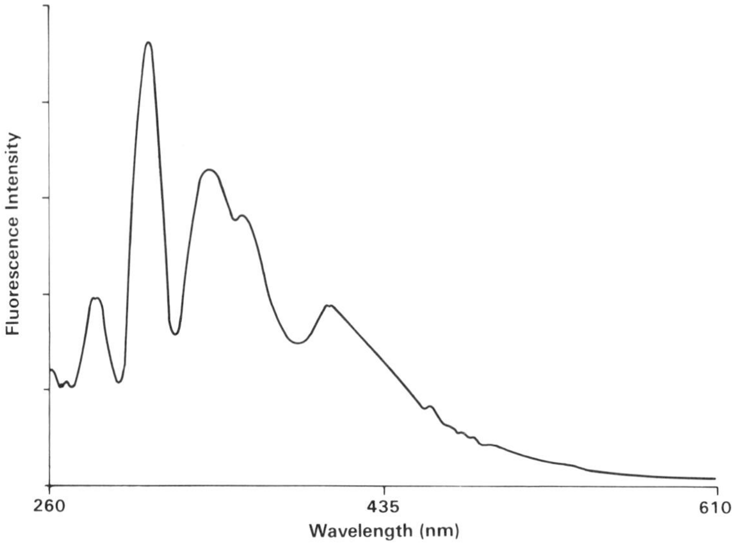

Fig. 3. Synchronous emission of Prudhoe Bay crude oil. 10 ug/mL in cyclohexane. $\Delta\lambda$=10 nm. Bandpasses = 2.5 nm. Synchronous emission Prudhoe Bay crude oil 10 μg/mL. $\Delta\lambda$ = 10 nm.

pretreating the matrix with a heavy-atom salt solution (such as thallium acetate or cesium iodide) which increases intersystem crossing between singlet and triplet states and thus may raise phosphorescence quantum yields.

Other luminescence applications

Chemical derivatization has long been proven to increase the number of pollutants which can be studied by luminescence. In some cases spectra of different derivatives of the same reagent may be similar, limiting specificity. Examples which have recently been found useful include cyanide [22-27] and aluminum [28,29].

Liquid chromatography (HPLC) [30,31] frequently employs fluorescence detection, and photodiode arrays to record the emission spectrum (or even the contour spectrum) of each chromatographic peak are now available. As liquid chromatographic techniques become more accepted for environmental analysis, their use in mobile field laboratories will be increasingly feasible, but the relative complexities of sample preparation, instrumentation and spectral data analysis so far have limited their applicability for field analysis.

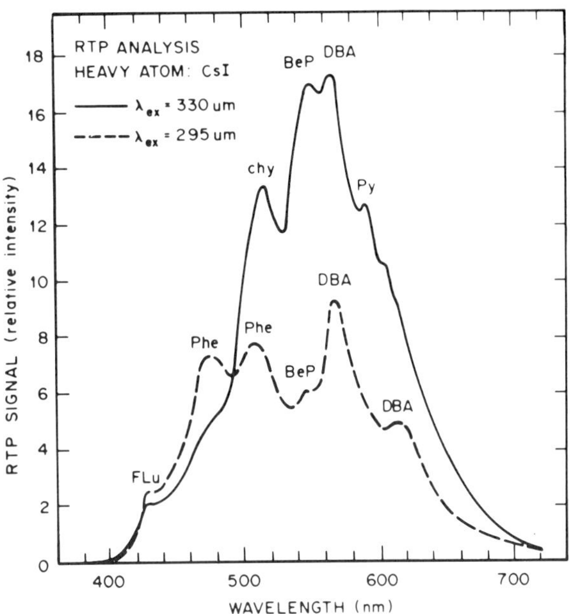

Fig. 4. Room temperature phosphorescence of PNA mixture spotted on filter paper treated with a heavy atom salt solution of cesium iodide. Spectra excited at 295 (dashed line) and 330 nm (solid line). Flu = Fluorene, Phe = Phenanthrene, Chy = Chrysene, BeP = Benzo(e)pyrene, DBA = 1,2,5,6-Dibenzanthracene.

Immunological methods offer excellent selectivity through the process of antibody-antigen reactivity, which also permits trace level sensitivity. Especially for field applications, fluoroimmunoassay would be more advantageous than radioimmunoassay in terms of cost of instrumentation, avoidance of radioisotopes, etc. These techniques are now being developed for pollutants such as benzo(a)pyrene and pesticides and offer many advantages for field screening [32-37]. One possible disadvantage is that immunoassay techniques may be too specific for field screening applications where the pollutants are unknown.

Fluorescence instrumentation

For effective field screening applications where access is limited, fluorescence instrumentation should be rugged and portable. This portability entails light-weight and battery-powered instrumentation that maintains the sensitivity and resolution needed to perform the analysis. To date only one truly portable fluorometer exists on the market. It is the Baird Field Identification Luminescence Monitor (FILM), manufactured for the US Coast Guard for oil and hazardous chemical identification. It operates off a 12 V dc battery supply and uses a mercury lamp with the 254 nm line isolated for excitation. The emission spectrum (250–600 nm) is then dispersed by a flat-field grating and recorded on Polaroid film. Another small instrument that is not fully portable because it operates off 120 V ac is the L-101A Fiberoptic Luminoscope from Environmental Systems Corporation. This instrument also uses a mercury lamp for excitation but isolates the 365 nm or other Hg line. The excitation light is focused into a bifurcated fiber-optic lightpipe and is transmitted to a probe which can be placed against soil surfaces or into water. The returning emission is then directed into a manually scanned monochromator with photomultiplier detection.

The availability of a mobile laboratory alleviates the need for portable instruments. Mobile laboratories equipped with ac power supplies can support most commercially available spectrofluorometers. The use of these instruments allows for greater selectivity and sensitivity. Instruments that have continuous sources and scanning excitation and emission monochromators allow for selective excitation, collection of total emission spectra and the use of synchronous techniques. The use of diode-arrays or charge-coupled devices (CCDs) as detectors allows for faster data acquisition and for rapid collection of contour spectra. These instruments, coupled with computer control, provide data processing and allow signal enhancement techniques to be implemented, which increases the amount of information that can be extracted from the spectral data. Computer systems also provide the opportunity to utilize pattern recognition techniques and expert systems in the analysis of data.

Dedicated fluorosensors are also of interest where the analysis is concerned with only one analyte or one class of analytes. An example of a dedicated laser fluorosensor is instrumentation developed to analyze the phosphorescence lifetimes of uranium complexes. This system uses a single excitation wavelength from a nitrogen dye laser and collects the phosphorescence signal at a predetermined emission wavelength and lifetime [38]. This design allows for the selective analysis of compounds and for the ability to correct for interferences, such as organics with short lifetimes.

Existing luminescence instrumentation is designed mainly for laboratory settings, with the exception of the two systems mentioned above. The recent advances in spectroscopic hardware offer opportunities to develop portable instrumentation that will match the performance standards of the best laboratory equipment. The use of new solid state lasers and detectors,

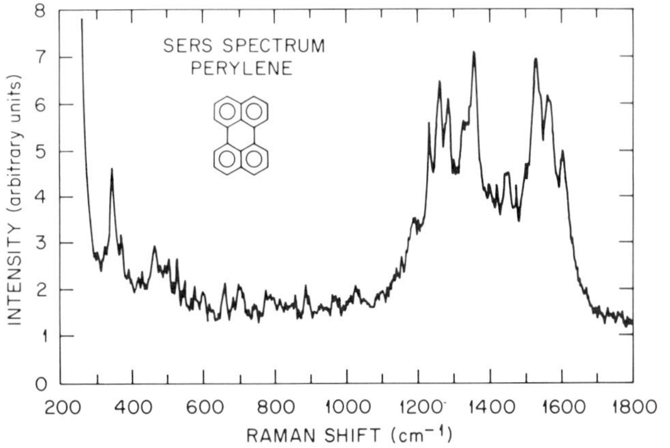

Fig. 5. Surface-enhanced Raman spectrum of perylene adsorbed on silver-coated submicron quartz microspheres as substrate.

along with increased signal processing capabilities, should allow the development of a more compact and rugged instrument with the sensitivity and selectivity needed for field screening.

Vibrational spectroscopy

Infrared spectroscopy (dispersive and Fourier transform) [39-47] has been used in field applications, especially for monitoring air pollutants with a gas cell or for characterizing oil or hazardous chemicals where structural information obtainable from group frequencies is useful and where sensitivity is not important. Some disadvantages have been the need to prepare samples and to eliminate water as an interferent, the sensitivity of some infrared cell windows and optics to water, the toxicity of some solvents used for infrared analysis, and the relative insensitivity of the techniques.

Raman spectroscopy [48-50], which also provides information about group frequencies, is not sensitive to water and can be used in the visible or near infrared region. Until recently, Raman techniques were considered to have several disadvantages for field use, such as complex instrumentation, fluorescence interferences, and relatively low sensitivity requiring the use of bulky and costly laser sources. Some of these disadvantages have been eliminated by the advent of more compact Raman spectrometers, smaller and cheaper lasers, near-infrared lasers that remove most fluorescence background signals, and special, more sensitive Raman techniques. The most promising Raman technique for field use is surface-enhanced Raman spectroscopy (SERS) in which Raman scattering efficiency can be enhanced by factors up to 10^6 -D when a chemical is absorbed on special metal (Cu, Ag, Au) surfaces [51-55]. This enhancement is believed to result from a combination of several electromagnetic and chemical interactions between the molecule and the roughened metal surface. Although this technique may be promising for future field applications, it is not yet fully understood or developed and may not apply to all chemicals. Figure 5 shows the SERS spectrum of perylene on a silver coated quartz substrate with silver-coated submicron spheres.

Fiber optic chemical sensors (FOCS)

Fiber optic chemical sensors [56-59] are gaining prominence especially for monitoring trace levels of pollutants in deep ground water wells without costs or errors introduced by sampling. The FOCS system consists of the following elements: a source which may be a lamp or laser; the sensor; a spectral dispersive element, which may be a filter or a monochromator; a detector; and a data processor. The sensor itself is usually an optical fiber which serves as a light pipe transmitting in the visible and near-infrared range, and a cell or coating containing a chemically sensitive reagent at the distal end of the fiber. Special optical materials, which tend to be expensive, are now extending the useful spectral range of these fibers into the ultraviolet and mid-infrared. The more common FOCS operate fluorometrically or colorimetrically using an immobilized reagent, but other types might include sensors that monitor fluorescence quenching, changes in the refractive index or some other optical property of the cladding, or fluoroimmunoassay. Immobilization techniques include imbedding the reagent in a polymer, using a reservoir cell with a semipermeable membrane, using porous glass or pumping fresh reagent past the tip of the fiber (rather than immobilizing it). Some FOCS, such as pH, temperature, CO_2, and alkali metals, have already proven to be rugged and usable under field conditions while others are still being developed and tested for environmental pollutants such as gasoline (see Figure 6) and chloroform. Some of these sensors appear more promising as gas phase sensors than as aqueous phase sensors because of the problems associated with aqueous-phase diffusion and leaching of the reagent. Another possible disadvantage is that many of these sensors are extremely specific, assuming that the likely pollutant is known. In any case, FOCS are being developed and improved, and they show great promise as the most cost-effective solution to many environmental monitoring problems.

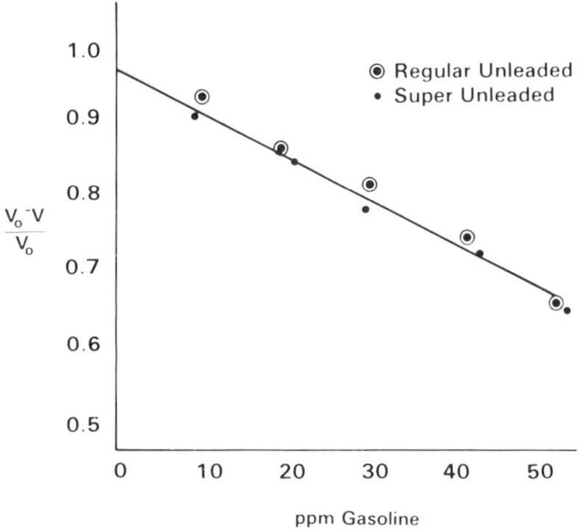

Fig. 6. Response of a refractive index fiber optic chemical sensor (FOCS) to gasoline vapors (3-hour equilibrium).

SUMMARY AND CONCLUSIONS

A variety of spectroscopic techniques are being used with increasing frequency for field screening. These techniques permit rapid response and reduce costs for environmental monitoring programs. Spectroscopic field screening also allows sampling efforts to be optimized and permits samples to be prioritized for more detailed analysis. Some of these spectroscopic methods are used in place without sampling, e.g., FOCS; others are used with portable instrumentation or, more commonly, are field deployable in a mobile laboratory. Recently, instrument developments, such as more compact and rugged spectrofluorometers, miniaturized optical hardware, more compact or solid-state lasers, new types of detectors such as charge-coupled devices, increased use of fiber optics, and better computer software for spectral data processing, have improved the utility of field spectroscopic techniques. Further improvements are needed, especially for portable instruments. Even techniques with field-ready instruments have often been tested on only a few of the applicable pollutants. For most of these techniques, detailed analytical protocols with appropriate standards, calibration criteria, and appropriate quality assurance for intended data use remain to be developed. Field spectroscopic instrumentation and methods represent a dynamic and rapidly changing approach to field screening that can greatly reduce monitoring expenditures and improve environmental analytical technology.

ACKNOWLEDGEMENTS

This work is jointly sponsored by the United States Environmental Protection Agency, under contract number 68-03-3249 to Lockheed-ESC, and under an interagency agreement, EPA No. DW89933900-0 (DOE No. 1824-B124-A1) to Oak Ridge National Laboratory, and the Office of Health and Environmental Research, U.S. Department of Energy, under contract DE-AC05-840R21400 with Martin Marietta Energy Systems, Inc.

The authors greatly appreciate the helpful suggestions and support of Larry Eccles, EPA Project Manager.

DISCLAIMER

Although the research in this paper was funded in part by the United States Environmental Protection Agency under contract number 68-03-3249 to Lockheed Engineering and Sciences Company, it has not been subjected to Agency review and therefore does not necessarily reflect the views of the Agency and no official endorsement should be inferred.

REFERENCES

[1] Parker, C.A., *Photoluminescence of Solutions*, Elsevier, New York, N.Y., 1968.

[2] Lloyd, J.B.F., *Analyst*, 99, 729, 1974.

[3] Brownrigg, T.J., Busch, D.A. and Giering, L.P., A Luminescence Survey of Hazardous Materials, Report No. DOT-CG-D-53-79, 1979.

[4] Eastwood, D., "Use of Luminescence Spectroscopy in Oil Identification", *Modern Fluorescence Spectroscopy*, Wehry E., ed., 4, 251, Plenum, New York, NY, 1981.

[5] Sogliero, G., Eastwood, D., and Ehmer, R., "Some Pattern Recognition Considerations for Low-Temperature Luminescence and Room Temperature Fluorescence Spectra", *Appl.* Spectr.DΦD, 36, 110, 1981.

[6] Eastwood, D., ed., *New Directions in Molecular Luminescence*, ASTM STP 822, 95, (and references therein) ASTM Philadelphia, PA, 1983.

[7] Sogliero, G., Eastwood, D. and Gilbert, J., "A Concise Feature Set for the Pattern Recognition of Low Temperature Luminescence Spectra of Hazardous Chemicals", in *Adv. in Luminescence Spectroscopy*, Cline-Love, L. J. and Eastwood, D., eds.,ASTM STP 863, 95, ASTM, Philadelphia, PA, 1985.

[8] Eastwood, D., Lidberg, R.L., "Application of Fluorescence and FTIR Techniques to Screening and Classifying Hazardous Waste Samples", *Proceedings of the 7th Natl. Conf. on Management of Uncontrolled Hazardous Waste Sites*, Washington D.C., 370, 1986.

[9] Schulman, S.G., *Molecular Luminescence Spectroscopy: Methods and Applications, Parts 1 and 2*, Wiley-Interscience, New York, NY, 1985, 1988.

[10] Kaminski, R., Purcell, F.J. and Russavage, E., "Uranyl Phosphorescence at the Parts-per-Trillion Level", *Anal. Chem.*, 53, 1093, 1981.

[11] Lloyd, J.B.F., *J. Forens. Sci. Soc.*, 11, 83, 135, 153, 1971.

[12] Vo-Dinh, T., "Molecular Analysis by Synchronous Luminescence Spectroscopy", *Anal. Chem.*, 50, 396-401, 1978.

[13] Vo-Dinh, T., "Synchronous Excitation Spectroscopy", Wehry, E. L., ed., *Modern Fluorescence Spectroscopy*, vol. 4, 167-192, Plenum Press, New York, NY, 1981.

[14] Vo-Dinh, T., Hooyman, R., "Selective Heavy-Atom Perturbation for Analysis of Complex Mixtures by Room Temperature Phosphorimetry", *Anal. Chem.*, 51, 1915-1921, 1979.

[15] Vo-Dinh, T., *Room Temperature Phosphorimetry for Chemical Analysis*, Wiley, New York, NY, 1984.

[16] Cline-Love, L.J., Skrilec, M. and Habarta, J.G., *Anal. Chem.*, 53, 754-759, 1980.

[17] Cline-Love, L.J., Habarta, J.G. and Skrilec, M., *Anal. Chem.*, 53, 437-444, 1981.

[18] Femia, R.A. and Cline-Love, L.J., *Anal. Chem.*, 56, 327-33, 1984.

[19] Weinberger, R., Rembish, K. and Cline-Love, L.J., "Comparison of Techniques for Generating Room Temperature Phosphorescence in Fluid Solution", in *Advances in Luminescence Spectroscopy*, Cline-Love, L.J. and Eastwood, D., eds., ASTM STP 863, 40-51, ASTM, Philadelphia, PA, 1985.

[20] Warner, I.M., Patonay, G., Rollie, M.E., Thomas, M., and Nelson, G., "Optimization of Fluorescence Measurements", in *Progress in Analytical Luminescence*, Eastwood, D. and Cline-Love, L.J., eds., ASTM STP 1009, 1988.

[21] Khasawneh, I.M. and Winefordner, J.D., *Talanta*, 35, 267-270, 1988.

[22] McKinney, G.L., Lau, H.K.Y. and Lott, P.F., "A Rapid Fluorometric Determination of Cyanide", *Microchem. J.*, 17, 375-379, 1972.

[23] White, C.E. and Argauer, R.J., *Fluorescence Analysis: A Practical Approach*, Marcel Dekker, Inc., New York, NY, 1970.

[24] Guilbault, G.G. and Kramer, D.N., "A Specific Fluorometric Method for the Detection of Cyanide", *Anal. Chem.*, 37, 918, 1965.

[25] Guilbault, G.G. and Kramer, D.N., "Specific Detection and Determination of Cyanide Using Various Quinone Derivatives", *Anal. Chem.*, 37, 1395-1399, 1965.

[26] Jeffrey, J.G. and Wiebe, L.I., "Cyanide Determination on the Fruit of Ornamental Plants Grown in Saskatchewan", *Can. J. Pharm. Sci.*, 6, 53-54, 1971.

[27] Hanker, J.S., Gelberg, A. and Witten, B., "Fluorometric and Colorimetric Estimation of Cyanide and Sulfide by Demasking Reactions of Palladium Chelates", *Anal. Chem.*, 30, 93-95, 1958.

[28] Siegel, N. and Haug, A., "Aluminum Interaction with Calmodulin - Evidence for Altered Structure and Function from Optical and Enzymatic Studies", *Biochimica et Biophysica Acta*, 744, 36-45, 1983.

[29] Weis, C. and Haug, A., "Aluminum Conformational Changes in Calmodulin Alter the Dynamics of Interaction with Melittin", *Archives of Biochemistry and Biophysics*, 254, 304-312, 1987.

[30] Cobb, W.T., Nithipatikom, K. and McGown, L.B., "Multicomponent Detection and Determination of Polycyclic Aromatic Hydrocarbons Using HPLC and a Phase-Modulation Spectrofluorometer", *Progress in Analytical Luminescence*, ASTM STP 1009, 12-25, Eastwood, D. and Cline-Love, L.J., eds., ASTM, Philadelphia, PA, 1988.

[31] Ogan, K., Katz, E. and Slavin, W., *Anal. Chem.*, 51, 1315, 1979.

[32] Smith, D.S., Hassan, M. and Nargessi, "Principles and Practice of Fluoroimmunoassay Procedures", in *Modern Fluorescence Spectroscopy, 3*, Wehry, E.L., ed., Plenum, NY, 143-191, 1981.

[33] Peterson, J.I. and Vurek, G.G., "Fiber Optic Sensors for Biomedical Applications", *Science*, 224, 123-127, 1984.

[34] Tromberg, B.J., Sepaniak, M.J., Vo-Dinh, T. and Griffin, G.D., "Fiber-Optic Chemical Sensors for Competitive Binding Fluoroimmunoassay", *Anal. Chem.*, 59, 1226-1232, 1987.

[35] Vo-Dinh, T., Tromberg, B.J., Griffin, G.D., Ambrose, K.R., Sepaniak, M.J. and Gardenhire, E.M., "Antibody-based Fiberoptic Biosensor for the Carcinogen Benzo(a)pyrene", *Appl. Spectr.*, 41, 735-738, 1987.

[36] Vo-Dinh, T., Griffin, G.D. and Ambrose, K.R., "A Portable Fiberoptic Monitor for Fluorimetric Bioassays", *Appl. Spectr.*, 40, 696-700, 1986.

[37] Tromberg, B.J., Sepaniak, M.J., Alarie, J.P., Vo-Dinh, T. and Santella, R.M., "Development of Antibody-Based Fiberoptic Sensors for Detection of a Benzo(a)pyrene Metabolite", *Anal. Chem.*, 60, 1901-1908, 1988.

[38] Bushaw, B.A., "Kinetic Analysis of Laser-Induced Phosphorescence in Uranyl Phosphate for Improved Analytical Measurements", (presented at 26th Oak Ridge Conference on Analytical Chemistry in Energy Technology, Knoxville, TN), October 1983.

[39] Colthup, N.B., Daly, L.H. and Wiberly, S.E., *Introduction to Infrared and Raman Spectroscopy*, Academic Press, New York, NY, 1964.

[40] Griffiths, P.R. and DeHaseth, J.A., *Fourier Transform Infrared Spectroscopy*, Wiley-Interscience, New York, NY, 1986.

[41] Eastwood, D. and Grant, D.F., "Field Infrared Method to Discriminate Natural Seeps from Non-Seeps", (Santa Barbara, California Area) USDOT Report No. CG-D32-77, ADA042861, 1976.

[42] Grant, D.F. and Eastwood, D., "A Field Method for Natural Seep Oil Discrimination by Infrared Spectroscopy", *Talanta*, 1983.

[43] Ahmadjian, M., Baer, C.D., Lynch, P.F. and Brown, C.W., "Infrared Spectra of Petroleum Weathered Naturally and Under Simulated Conditions", *Environ. Sci. Tech.*, 10, 777, 1976.

[44] Anderson, C.P., Killeen, T.J. Taft, J.B. and Bentz, A.P., "Improved Identification of Spilled Oils by Infrared Spectroscopy", *Environ. Sci. Tech.*, 14, 1230, 1980.

[45] Brown, C.W., "Multicomponent Analysis Using P-Matix Methods", ASTM 12, 86, 1984.

[46] Puskar, M.A., Levine, S.P. and Lowry, S.R., "Computerized Infrared Spectral Identification of Compounds Frequently Found at Hazardous Waste Sites", *Anal. Chem.*, 58, 1156, 1986.

[47] Gurka, D.F., Project Summary, "Interim Protocol for Automated Analysis of Semivolatile Organic Compounds by Gas Chromatography/Fourier Transform Infrared (GC/FTIR) Spectrometry", EPA-600/S4-84-081, 1984.

[48] Long, D.A., *Raman Spectroscopy*, McGraw Hill, London, 1977.

[49] Grasselli, J.G., Snavely, M.K. and Bulkin, B.J., *Chemical Applications of Raman Spectroscopy*, Wiley, New York, NY, 1981.

[50] Baranska, H., Labudzinska, A. and Terpinski, J., *Laser Raman Spectrometry, Analytical Applications*, Halsted, Wiley, New York, NY, 1987.

[51] Chang, R.K. and Furtak, T.E., eds., *Surface-Enhanced Raman Scattering*, Plenum, New York, NY, 1982.

[52] Vo-Dinh, T., Hiromoto, M.Y.K., Begun, G.M., and Moody, R.L., "Surface-Enhanced Raman Spectroscopy for Trace Organic Analysis", *Anal. Chem.*, 56, 1667-1672, 1984.

[53] Moody, R.L., Vo-Dinh, T. and Fletcher, W.R., "Investigations of Experimental Parameters for Surface-Enhanced Raman Spectroscopy", *Appl. Spectr.*, 41, 966-971, 1987.

[54] Alak, A. and Vo-Dinh, T., "Surface-Enhanced Raman Spectroscopy of Chlorinated Pesticides", *Anal. Chim. Acta.*, 206, 333-337, 1988.

[55] Vo-Dinh, T., Alak, A. and Moody, R.L., "Recent Advances in Surface-Enhanced Raman Spectroscopy for Chemical Analysis", *Spectrochim. Acta,* 43B, 605*615, 1988.

[56] Hirschfeld, T., Deaton, T., Milanovich, F. and Klainer, S., "The Feasibility of Using Fiber Optics for Monitoring Ground Water Contaminants", *Opt. Eng.*, 22, 527, 1983.

[57] Seitz, W.R., "Chemical Sensors Based on Fiber Optics", *Anal. Chem.*, 56, 17A*34A, 1984.

[58] Wolfbeis, O., "Analytical Chemistry with Optical Sensors", *Fresenius'Z. Anal. Chem.*, 325, 387, 1986.

[59] Wolfbeis, O., "Fiber-Optic Probes for Chemical Sensing", *Anal. Proceed.*, 24, 74, 1987.

of packed columns for instrumental analysis, and the specified GC column temperatures and other GC parameters vary from column to column.

Method improvement and standardization studies are being carried out by our laboratory with the overall goal to simplify specific GC methods, to make them more rugged and to shorten analysis time, thus reducing the cost per analysis. One part of our present studies is directed towards replacing the packed GC columns specified in SW-846 GC methods with fused-silica capillary/megabore columns and, whenever feasible, establishing identical GC conditions for both the primary and the confirmatory GC column. The use of capillary or megabore GC columns in place of packed columns results in improved resolution, and, when a dual-column, dual-detector set-up is used with identical GC conditions for both columns, the instrumental procedure is simplified, operator error is reduced, and analysis time is shortened.

In another part of our studies we have been evaluating the efficiency of conventional sample extract cleanup on glass columns packed with adsorbents (alumina, Florisil, silica gel), as specified in SW-846 Methods 3610, 3620, and 3630. These procedures specify amounts of adsorbents in excess of 10 g and large volumes of eluting solvents (e.g., a 10-g Florisil column and 100 mL of 20-percent diethyl ether in hexane are recommended for cleanup of sample extracts containing phthalate esters). Such large volumes of solvents increase the likelihood of sample contamination by impurities in the solvents. Furthermore, the adsorbent materials and the solvents are not recycled, and although such materials are not overly expensive, the time required for the preparation of the adsorbent, for the packing of the chromatographic columns, for the elution of the target analytes, and for the evaporation of solvents contributes to the overall cost of analysis. The solution to all these problems would be to use disposable cartridges known as solid-phase extraction (SPE) cartridges. Solid-phase extraction is one of the fastest-growing sample preparation methods [2], and a variety of cartridges with different adsorbents is available from several suppliers. Our efforts focused on the use of prepackaged disposable SPE cartridges with alumina, Florisil and silica gel as adsorbents. The groups of pollutants we considered were phthalate esters (Method 8060), organochlorine pesticides (Method 8080/8081[1]) and chlorinated hydrocarbons (Method 8120).

[1]Method 8081 is an as yet unpublished improved version of Method 8080.

EXPERIMENTAL

Apparatus

Gas Chromatographs – Varian 6000 with constant current/pulsed frequency ECD, interfaced with a Varian Vista 402 data system; Varian 6500 with FID, interfaced with either a Spectra Physics 4290 integrator or a Varian Vista 402 data system. For simultaneous analysis, the megabore open tubular columns were connected to either a 6-in or 8-in injection tee and to identical ECDs.

Autosampler – Varian Model 8000

GC Columns – (1) DB-5, (2) DB-608, (3) DB-1701, (4) Supelcowax-10, (5) SPB-608, (6) DB-210, (7) DB-WAX

Vacuum Manifold – VacElute Manifold SPS-24 (Analytichem International) or Visiprep (Supelco, Inc.) or equivalent, consisting of glass vacuum basin, collection rack and funnel, collection vials, replaceable stainless steel sample delivery tips, built-in vacuum bleed valve and gauge; the system is connected to a vacuum pump or water aspirator through a vacuum trap made from a 500-mL sidearm flask fitted with a one-hole stopper and tubing.

Materials

Cartridges – Florisil, alumina, and silica gel, of different sizes (0.5 g, 1.0 g, 2.0 g). The cartridges consist of serological-grade polypropylene tubes, 3 mL or 6 mL in volume; the adsorbent material is held between two polyethylene frits (20 μ pores).

Standards – Analytical reference standards of the organochlorine pesticides, the phthalate esters and the chlorinated hydrocarbons were obtained from the U.S. Environmental Protection Agency–Pesticides and Industrial Chemicals Repository, Aldrich Chemical, Ultra Scientific Inc., Chem Service, and Scientific Polymer Products. Purities were stated to be greater than 98 percent. Stock solutions of each test compound were prepared in pesticide-grade hexane at 1 mg/mL (isooctane for the chlorinated hydrocarbons). Working calibration standards were prepared by serial dilutions of a composite stock solution prepared from the individual stock solutions.

Corn Oil – Stock solution was prepared in hexane at 1.1 mg/mL.

The following materials were used as matrices:

* Sandy loam soil obtained from Soils Incorporated, Puyallup, Washington, with the following characteristics: pH 5.9 to 6.0; 89% sand, 7% silt, 4% clay; cation exchange capacity 7 meq/100 g; total organic carbon content 1,290 $\pm$ 185 mg/kg.

* A sediment sample of unknown origin. Analysis of the extract by

GC/MS indicated the presence of petroleum hydrocarbons.

* NBS SRM-1572, citrus leaves.
* NBS SRM-1632a, coal.
* NBS SRM-1633a, coal flyash.

GC Conditions

The GC operating conditions are specified in the respective tables.

Cartridge Cleanup Procedure

Each Florisil, alumina and silica gel cartridge was conditioned prior to use by washing with 4 mL hexane. Two-mL aliquots of hexane solutions containing the test compounds and the interferents were loaded onto cartridges using a micropipette. To ensure that the packing did not get dry in between cartridge conditioning and sample addition or in between collection of fractions, we always let 1 mm of the last solvent remain on top of the frit. A Supelclean vacuum manifold (Supelco, Inc.) was used to simultaneously prepare as many as 12 samples, an Analytichem International vacuum manifold (SPS-24) was used to prepare 24 samples simultaneously. When using the Visiprep Supelco vacuum manifold, the vacuum for each cartridge was adjusted manually using chemically inert screw-type valves.

Compounds were eluted with the various solvents identified in the tables that summarize the results. The volume of each fraction was adjusted to 5 mL prior to gas chromatographic analysis.

RESULTS AND DISCUSSION

Gas Chromatography

Packed gas chromatographic columns are specified for most of the SW-846 GC methods. The results of our evaluation of Method 8060 (phthalate esters), Method 8080/8081 (organochlorine pesticides) and Method 8120 (chlorinated hydrocarbons) with the specified packed columns and with fused-silica capillary and open tubular ("megabore") columns can be summarized as follows.

The analysis using the packed columns for the phthalate esters in the 16-compound standards mixture (Table 1) was not practical for quantification for several of these compounds [3]. The use of fused-silica capillary columns (DB-5 and Supelcowax-10) gave much improved resolution for all target compounds (Table 1). Simultaneous analysis of the standards mixture under identical GC conditions ("dual-column analysis") on open tubular fused-silica columns (0.53-mm ID) further improved resolution (Table 1).

The packed-column chromatographic determination of the organochlorine pesticides does not allow complete separation of several pairs of

Table 1. GC Retention Times for the Phthalates[a]

Phthalate	Retention time (min)			
	DB-5	Supelco-wax-10	DB-608	DB-1701
Dimethyl	3.42	5.62	6.72	6.73
Diethyl	3.45	6.11	8.69	8.85
Diisobutyl	6.48	7.26	12.74	13.36
Di-n-butyl	7.14	8.43	14.68	15.13
Bis(2-methoxyethyl)	7.40	12.05	17.24	16.96
Bis(4-methyl-2-pentyl)	7.96	8.14	15.76	16.73
Bis(2-ethoxyethyl)	8.17	12.41	18.93	18.80
Diamyl	8.41	10.15	17.94	18.64
Hexyl 2-ethylhexyl	8.63	11.13	19.70	19.56
Dihexyl	9.62	12.21	21.50	22.48
Benzyl butyl	9.69	16.36	24.64	23.76
Bis(2-n-butoxyethyl)	10.53	16.94	25.71	25.96
Dicyclohexyl	10.98	16.66	28.33	27.06
Bis(2-ethylhexyl)	11.13	13.31	24.94	26.35
Di-n-octyl	13.03	17.25	29.14	30.57
Dinonyl	16.00	20.73	32.97	34.71

[a] Operating conditions: DB-5 – 120°C to 260°C (hold 16 min) at 15°C/min, injector temp. 275°C, detector temp. 320°C; Supelcowax-10 – 150°C (hold 2 min) to 220°C at 15°C/min, then to 260°C (hold 16 min) at 4°C/min, injector temp. 270°C, detector temp. 270°C; DB-608 and DB-1701 – 150°C (hold 0.5 min) to 220°C at 3°C/min, then to 275°C (hold 15 min) at 5°C/min, injector temp. 250°C, detector temp. 320°C.

pesticides. The use of either a pair of fused-silica capillary columns (e.g., 30-m x 0.25-mm ID DB-5 and SPB-608) or a pair of open tubular fused-silica columns (e.g., 30-m x 0.53-mm ID DB-5, SPB-5, or RTx-35) resulted in complete separation of the 18 organochlorine pesticides except for two compound pairs on the 0.53-mm ID DB-5 [4]. The retention times for the DB-5/SPB-608 (0.25 mm ID) and the DB-5/ DB-608 (0.53-mm ID) are presented in Table 2. Again, the values for the latter column pair were generated by the dual-column GC method (simultaneous analysis under identical GC conditions).

The determination of the 22 chlorinated hydrocarbons (Table 3) proved to be especially troublesome. At 65°C (one of the temperatures specified for the packed columns), only 11 compounds eluted from the specified column within 30 min, and even at 150°C, the second specified temperature, several compounds still did not elute. Furthermore, the resolution was so

Table 2. GC Retention Times for the Organochlorine Pesticides[a]

	Retention Time (min)			
Compound	DB-5 (.25-mm ID)	SPB-608 (.25-mm ID)	DB-5 (.53-mm ID)	DB-608 (.53-mm ID)
alpha-BHC	12.29	9.46	10.94	11.43
beta-BHC	13.13	11.33	11.51	12.59
gamma-BHC (Lindane)	13.37	10.97	11.71	12.46
delta-BHC	14.14	12.73	12.20	13.69
Heptachlor	15.91	12.46	13.59	13.41
Aldrin	17.16	13.76	14.70	14.51
Heptachlor epoxide	18.60	15.98	16.05	16.62
gamma-Chlordane	19.48	16.70	17.02	17.34
Endosulfan I	19.94	17.40	17.62	18.27
4,4'-DDE	20.83	18.36	18.30	19.09
Dieldrin	20.91	18.60	18.74	19.67
Endrin	21.71	19.96	19.73	21.37
Endosulfan II	22.05	20.69	20.11[b]	22.17
4,4'-DDD	22.38	20.53	20.11[b]	21.67
Endrin aldehyde	22.75	21.90	20.85	23.78
Endosulfan sulfate	23.64	22.54	21.84[c]	24.45
4,4'-DDT	23.79	21.72	21.84[c]	23.13
4,4'-Methoxychlor	25.94	24.90	24.43	28.65

[a] DB-5 (0.25-mm ID) - 100°C (hold 2 min) to 160°C at 15°C/min, then to 270°C at 5°C/min; SPB-608 (0.25-mm ID) - 160°C (hold 2 min) to 290°C (hold 1 min) at 5°C/min; DB-5 (0.53-mm ID) and DB-608 (0.53-mm ID) - 140°C (hold 2 min) to 240°C (hold 5 min) at 10°C/min, then to 265°C (hold 18 min) at 5°C/min. [b,c] Compound pairs that are not separated under these conditions.

poor that identification was not possible. Temperature-programming did not alleviate the problem. The evaluation of a series of open tubular and capillary fused-silica columns [5] led to the selection of the pair DB-210 (30 m x 0.53-mm ID)/DB-WAX (30-mm x 0.53-mm) although on each column two compound pairs are not resolved. The retention times are listed in Table 3.

SAMPLE EXTRACT CLEANUP

Phthalate Esters

Alumina and Florisil chromatography was performed with phthalate ester standards in hexane according to Methods 3610 and 3620, respectively (Table 4). For the Florisil cartridge cleanup, various solvents and solvent combinations were tried on standards in hexane and on standards in the pres-

Table 3. Hydrocarbons[a]

Compound	Retention time (min)	
	DB-210	DB-WAX
Benzal chloride	6.86[b]	15.59
Benzotrichloride	7.85	15.10
Benzyl chloride	4.59	10.42[d]
2-Chloronaphthalene	13.45	22.70[e]
1,2-Dichlorobenzene	4.44	9.72
1,3-Dichlorobenzene	3.66	7.95
1,4-Dichlorobenzene	3.80	8.68
Hexachlorobenzene	19.23	27.44
Hexachlorobutadiene	5.77	10.08
alpha-BHC	22.21	36.86
beta-BHC	25.54	47.46
gamma-BHC	24.07	46.24
delta-BHC	26.16	31.33
Hexachlorocyclopentadiene	8.86	[f]
Hexachloroethane	3.35	8.33
Pentachlorobenzene	14.86	22.70[e]
1,2,3,4-Tetrachlorobenzene	11.90	20.33
1,2,4,5-Tetrachlorobenzene	10.18[c]	17.28
1,2,3,5-Tetrachlorobenzene	10.18[c]	17.00
1,2,4-Trichlorobenzene	6.86[b]	13.57
1,2,3-Trichlorobenzene	8.14	15.62
1,3,5-Trichlorobenzene	5.45	10.42[d]

[a]DB-210 – 65°C to 175°C (hold 20 min) at 40°C/min; DB-WAX – 60°C to 170°C (hold 30 min) at 4°C/min.

[b–e]Pairs not resolved on respective column.

[f]Compound decomposes on column.

ence of organochlorine pesticides. It was found that the organochlorine pesticides can be removed efficiently from the cartridges with hexane/methylene chloride (4:1); under these conditions, the phthalate esters are retained on the Florisil cartridge and can be recovered with hexane/acetone (9:1). The recoveries are presented in Table 4. Additional details on the Florisil cartridge cleanup method can be found in Reference 6. The Florisil cartridge method was tested for the 16 phthalate esters with five environmental materials. The recoveries (Table 4) were > 74 percent for all phthalate esters but bis(2-ethoxyethyl) phthalate. Similar results (not listed) were obtained with alumina cartridges.

Table 4. Percent Recoveries of Phthalate Esters from Extract Cleanup

	Standards in hexane				Environmental Materials (SPE Cleanup)[c]				
			Florisil						
	Alumina		Cartr.[b]		Sandy Loam	Sedi-		NBS SRM	
Phthalate	Column[a]		Fr. 1	Fr. 2	Soil	ment	1572	1632[a]	1633[a]
Dimethyl	40	65	0	130	78	75	80	76	82
Diethyl	57	62	0	88	79	79	89	79	84
Diisobutyl	80	77	0	118	79	82	90	108	86
Di-n-butyl	85	77	12	121	74	78	84	83	83
Bis(2-methoxyethyl)	0	70	0	32	d	d	d	d	d
Bis(4-methyl-2-pentyl)	85	89	0	123	77	84	102	91	86
Bis(2-ethoxyethyl)	0	67	0	82	37	24	62	32	33
Diamyl	82	75	3.3	94	82	86	100	76	89
Hexyl 2-ethylhexyl	105	91	0	126	80	88	95	93	81
Dihexyl	74	73	0	62	78	88	86	92	80
Benzyl butyl	90	87	0	98	82	99	114	102	98
Bis(2-n-butoxyethyl)	0	63	0	135	86	94	98	106	98
Dicyclohexyl	84	84	0	106	91	96	106	98	95
Bis(2-ethylhexyl)	82	91	0	110	74	85	108	88	112
Di-n-octyl	115	108	0	123	80	92	104	95	88
Dinonyl	73	71	0	102	84	96	106	111	92

[a]Average of two determinations.

[b]Average of three determinations. Fraction 1 was eluted with 5 mL hexane/methylene chloride (4:1), Fraction 2 with 5 mL hexane/acetone (9:1).

[c]Spiking level is 50 ng/mL for each compound. The data shown are for Fraction 2 (hexane/acetone 9:1); Fraction 1 (5 mL hexane/methylene chloride 4:1) was discarded.

[d]Values not available.

ORGANOCHLORINE PESTICIDES

EPA Method 8080/8081 recommends Florisil cleanup (Method 3620) for extracts that are analyzed for organochlorine pesticides and PCBs. The target compounds are eluted from the Florisil column with 6, 15, and 50 percent diethyl ether in hexane. The PCBs are not separated from the organochlorine pesticides under these conditions. Nonetheless, we found that the Florisil procedure gave quantitative recoveries (<75 percent) for 17 organochlorine pesticides, toxaphene, technical chlordane, and 7 Aroclors [4]. The use of a 3-g silica gel column (the silica gel was deactivated with water at 3.3 percent by weight) allowed the separation of PCBs from most of the organochlorine pesticides. However, the column had to be eluted with large volumes of hexane (130 mL) and methylene chloride (15 mL) in order to allow quantitative recovery of the PCBs and organochlorine pesticides.

Table 5a. Elution Patterns and Average Percent Recoveries of the Organochlorine Pesticides Using Florisil and Silica Gel Cartridges[a,b]

	0.5 μg/compound					
	Fract. 1		Fract. 2		Fract. 3	
Compound	F[c]	SG[c]	F	SG	F	SG
alpha-BHC	82.6		26.3	111.0		
beta-BHC			102.0	109.0		
gamma-BHC (Lindane)	35.8		77.5	110.0		
delta-BHC			99.6	106.0		
Heptachlor	94.4	98.4				
Aldrin	93.1	96.6				
Heptachlor epoxide			102.0	109.0		
gamma-Chlordane	47.6		65.6	105.0		
Endosulfan I			101.0	111.0		
4,4'-DDE	94.5	104.0				
Dieldrin			101.0	110.0		
Endrin			57.2	d		
Endosulfan II			58.2		61.1	111.0
4,4'-DDD	38.5		68.2	111.0		
Endrin aldehyde			36.3	48.9	78.7	47.7
Endosulfan sulfate[e]						
4,4'-DDT[e]	49.6	40.1	11.2	16.7	59.4	63.4
4,4'-Methoxychlor			96.0	84.5	11.7	33.6

[a]1-g Florisil or silica gel solid-phase extraction cartridges were used. The amount of pesticides loaded onto the cartridges was 0.5μg per compound. Fraction 1 was eluted with 3 mL hexane, Fraction 2 with 5 mL 26 percent methylene chloride in hexane, and Fraction 3 with 5 mL 10 percent acetone in hexane. The number of replicates is 4.

[b]Under these conditions, all PCBs elute quantitatively in Fraction 1 from both cartridges.

[c]F stands for Florisil cartridge, SG for silica gel cartridge.

[d]Data not available.

[e]These compounds coelute on the 30-m x 0.25-mm ID DB-5 fused-silica capillary column.

For the Florisil and silica gel cartridge evaluation, the charged cartridges were eluted with 3 mL hexane (Fraction 1), followed by 5 mL 26-percent methylene chloride in hexane (Fraction 2) and 5 mL 10-percent acetone in hexane (Fraction 3). Under these conditions, silica gel proved to be superior to Florisil because it allowed complete separation of the PCBs from all but four organochlorine pesticides, quantitative recovery of all compounds, and almost complete separation of the Method 8080/8081 organochlorine pesticides from 16 phthalate esters. The four organochlorine pesticides that eluted with the 16 phthalate esters could be identified and quantified without any difficulty because they were resolved from the phthalate esters on a 30-m x 0.25-mm ID DB-5 fused-silica capillary column. The

Table 5b. Elution Patterns and Average Percent Recoveries of the Organochlorine Pesticides Using Florisil and Silica Gel Cartridges[a,b]

	5 μg/compound					
	Fract. 1		Fract. 2		Fract. 3	
Compound	F	SG	F	SG	F	SG
alpha-BHC	79.0		34.1	111.0		
beta-BHC			104.0	111.0		
gamma-BHC (Lindane)	32.0		82.8	111.0		
delta-BHC			103.0	110.0		
Heptachlor	94.8	105.0				
Aldrin	94.6	108.0				
Heptachlor epoxide			104.0	112.0		
gamma-Chlordane	44.2		73.2	108.0		
Endosulfan I			104.0	114.0		
4,4'-DDE	96.8	109.0				
Dieldrin			105.0	114.0		
Endrin			d	d		
Endosulfan II			60.3		58.3	110.0
4,4'-DDD	37.7		77.3	114.0		
Endrin aldehyde			47.5	52.4	72.7	41.9
Endosulfan sulfate[e]						
4,4'-DDT[e]	51.6	44.6	12.2	19.4	56.8	63.3
4,4'-Methoxychlor			105.0	94.3	11.2	26.1

[a]1-g Florisil or silica gel solid-phase extraction cartridges were used. The amount of pesticides loaded onto the cartridges was 5.0 μg per compound. Fraction 1 was eluted with 3 mL hexane, Fraction 2 with 5 mL 26 percent methylene chloride in hexane, and Fraction 3 with 5 mL 10 percent acetone in hexane. The number of replicates is 4.

[b]Under these conditions, all PCBs elute quantitatively in Fraction 1 from both cartridges.

[c]F stands for Florisil cartridge, SG for silica gel cartridge.

[d]Data not available.

[e]These compounds coelute on the 30-m x 0.25-mm ID DB-5 fused-silica capillary column.

* the dual-column approach, whenever feasible, results in shorter analysis times and reduces operator error;

* the use of disposable solid-phase extraction cartridges for sample extract cleanup often results in better separation. Furthermore, it reduces solvent and adsorbent consumption and labor cost in sample preparation. In addition, a significant error resulting from operator and material variables that may affect the quality of the results can be eliminated.

procedure was tested at 2 concentrations in quadruplicate. The results in Table 5 show elution patterns and compound recoveries for 18 organochlorine pesticides.

Seven Aroclor mixtures were tested individually on the silica gel and the Florisil SPE cartridges. In each case the cartridge was loaded with 10 μg of the corresponding Aroclor and eluted with 3 mL hexane. The recovery data indicate that Aroclors are recovered quantitatively from either cartridge with 3 mL hexane. An additional fraction was collected by eluting the cartridge with an additional 3 mL hexane to verify that indeed the PCBs are removed completely from the cartridge with 3 mL hexane. Larger cartridges may require additional solvent to ensure complete removal of the PCBs.

CHLORINATED HYDROCARBONS

Cleanup of the sample extracts containing the 22 chlorinated hydrocarbons was performed initially according to the procedure given in EPA Method 3620 [1]. In this procedure, Florisil (60/80 mesh), activated at 130°C for at least 16 hours, is used, and the test compounds are eluted with 200 mL petroleum ether. Under these conditions, most of the chlorinated hydrocarbons were recovered quantitatively (recovery $>$ 78 percent), except for the BHC isomers, benzal chloride, and benzotrichloride. When a second fraction was collected by eluting the Florisil column with 200 mL of diethyl ether/petroleum ether (1:1), the BHC isomers were recovered quantitatively; however, benzal chloride and benzotrichloride were still not recovered. When Florisil cartridges were used with hexane/acetone (9:1) as eluant, the recoveries were greater than 90 percent (except for hexachlorobenzene at 78 percent) (Table 6), and the relative standard deviations of five determinations ranged between 0.4 and 6.5 percent, with most values less than 2 percent. The benzal chloride and benzotrichloride recoveries were 99 and 90 percent, respectively. This procedure was tested with several solid matrices; the results are summarized in Table 6. More details are provided in Reference 5.

CONCLUSIONS

The experimental results presented in this paper clearly demonstrate that:

* replacement of packed GC columns by open tubular or capillary fused-silica columns results in improved resolution;

Table 6. Average Percent Recoveries of Chlorinated Hydrocarbons from Florisil Cartridge Cleanup

Compound	Standards in hexane[a] Fraction 1	2	Sandy Loam Soil	Sedi- ment	Environmental Materials NBS SRM -1572	-1632a	-1633a
Hexachloroethane	95	0	88	79	78	70	76
1,3-Dichlorobenzene	101	0	86	71	67	58	69
1,4-Dichlorobenzene	100	0	93	78	120	59	39
1,2-Dichlorobenzene	102	0	98	84	80	73	83
Benzyl chloride	101	0	96	96	112	85	96
1,3,5-Trichlorobenzene	98	0	92	75	46	59	68
Hexachlorobutadiene	95	0	83	65	57	54	65
Benzal chloride[b]	99	0	d	102	96	94	103
1,2,4-Trichlorobenzene[b]	99	0	d	102	96	94	103
Benzotrichloride	90	0	88	84	90	75	82
1,2,3-Trichlorobenzene	97	0	103	89	70	73	91
Hexachlorocyclopentadiene	103	0	36	31	123	28	35
1,2,4,5-Tetrachlorobenzene[c]	98	0	86	70	50	63	61
1,2,3,5-Tetrachlorobenzene[c]	97	0	86	70	50	63	61
1,2,3,4-Tetrachlorobenzene	99	0	104	80	69	75	83
2-Chloronaphthalene	95	0	90	78	84	80	78
Pentachlorobenzene	104	0	90	76	58	74	71
Hexachlorobenzene	78	0	99	76	61	67	78
alpha-BHC	100	2.0	102	98	99	98	101
gamma-BHC	99	2.0	105	98	101	100	102
beta-BHC	95	1.2	95	85	85	90	97
delta-BHC	97	1.2	110	114	98	75	100

[a]Fraction 1 was eluted with 5 mL hexane/acetone (9:1), Fraction 2 with an additional 5 mL hexane/acetone. The number of determinations is 5.

[b,c]These pairs cannot be resolved on the DB-210 column.

[d]Data not available.

REFERENCES

[1] Test Methods for Evaluating Solid Waste - Physical/Chemical Methods, Vol. 1B. U.S. Environmental Protection Agency, Washington, DC, 1986.

[2] Majors, R. E., "New Devices and Instrumentations for Sample Preparation in Chromatography", LC.GC 7(2), 92-96, 1989.

[3] Beckert, W. F., and Lopez-Avila, V., "Evaluation of SW-846 Method 8060 for Phthalate Esters", presented at the Fifth Annual Waste Testing and Quality Assurance Symposium, July 24-28, 1989, Washington, DC, USA.

[4] Lopez-Avila, V., Schoen, S., Milanes, J., and Beckert, W. F., "Single-Laboratory Evaluation of EPA Method 8080 for Determination of Chlorinated Pesticides and Polychlorinated Biphenyls in Hazardous Wastes", J. Assoc. Off. Anal. Chem. 71(2), 375-387, 1988.

[5] Beckert, W. F., and Lopez-Avila, V., "Evaluation of SW-846 Method 8120 for Chlorinated Hydrocarbons", Proc. Symp. on Waste Testing and Quality Assurance, July 11-15, 1988, Vol. II, G-102 - G-120.

[6] Lopez-Avila, V., Milanes, J., Dodhiwala, N. S., and Beckert, W. F., "Cleanup of Environmental Sample Extracts Using Florisil Solid-Phase Extraction Cartridges", J. Chromatogr. Sci. 27(5), 209-215, 1989.

THE RATIONALE FOR GEOLOGIC DISPOSAL OF HIGH-LEVEL RADIOACTIVE WASTE IN THE UNITED STATES

J. J. DLUGOSZ

Formerly with U.S. Department of Energy, Las Vegas, Nevada, now with U.S. Environmental Protection Agency, Environmental Monitoring Systems Laboratory, Las Vegas, Nevada 89119

M. S. BEDINGER

University of Nevada Las Vegas, Environmental Research Center, Las Vegas, Nevada 89154

ABSTRACT

The program for disposal of high-level and transuranic radioactive waste in the United States calls for establishment of a mined repository in geologic formations that will provide for retrievability of the waste for 50 years. The goal of radioactive waste isolation in mined repositories is to prevent unacceptable concentrations of radionuclides from migrating to the accessible environment. Early concepts of high-level radioactive waste disposal assumed that containment of the waste by the enclosing rock would be virtually complete. With the application of broader scientific and engineering disciplines, specifically geology and hydrology to the study of prospective environments for waste disposal, the realization came that total isolation of the waste in the immediate vicinity of the repository will be difficult to ensure, and this includes geological and hydrological methods used for

Chemistry for the Protection of the Environment
Edited by L. Pawlowski *et al.*, Plenum Press, New York, 1991

predicting conditions or events that may prevail during the extremely long time required for waste isolation. Some of the earth processes that are of major concern during the waste-isolation storage time include: (1) rates of radionuclide transport in the ground-water flow system which, in turn, reflect chemical reactions of radionuclides with ground water and earth materials; (2) climatic changes; and (3) tectonic and associated erosional events. To compensate for the limitations in our knowledge, the current rationale for waste isolation emphasizes the need for a series of independent barriers to radionuclide migrations. Multiplicity of these barriers, both engineered and natural, will compensate for uncertainties in predicting natural or man-induced conditions and events that may occur during the time required for waste isolation.

The site selected for intense study for suitability of high-level and transuranic radioactive waste isolation is Yucca Mountain in the desert of the southern part of the State of Nevada. The target burial zone is in welded tuff at a depth below the surface of about 300 meters. The burial zone is in the unsaturated zone about 260 meters above the water table. The climate is arid; the average precipitation is about 150 mm/yr. Studies are currently ongoing to determine if the site will provide long term isolation of radioactive waste.

INTRODUCTION

The geologic disposal of high-level radioactive waste was recommended by the U.S. National Academy of Sciences in 1957 and has proven to be considerably more complex, both technically and politically, than was originally conceived. Aside from physical exhumation by exhumation by such processes as erosion, meteorite impact, gaseous transport, or igneous activity, ground-water flow is the only natural mechanism by which radionuclides could be transported to the accessible environment from an underground repository. By the end of this century, the United States plans to begin operating the first geologic repository for the permanent disposal of commercial spent nuclear fuel and high-level radioactive waste. Public Law 97-425, the Nuclear Waste Policy Act of 1982 (the Act), specifies the process for selecting a repository site, and constructing, operating, closing, and decommissioning the repository. Congress approved geologic disposal by declaring that one of the key purposes of the Act is "to establish a schedule for the siting, construction, and operation of repositories that will provide reasonable assurance that the public and the environment will be adequately protected from the hazards posed by high-level radioactive waste and such spent nuclear fuel as may be disposed of in a repository" [Section 111(b)(l)].

In February 1983, the U.S. Department of Energy (DOE) carried out the first requirement of the Act by formally identifying potentially acceptable sites in the following locations (the host rock of each is shown in parentheses):

1. Vacherie Dome, Louisiana (salt dome)

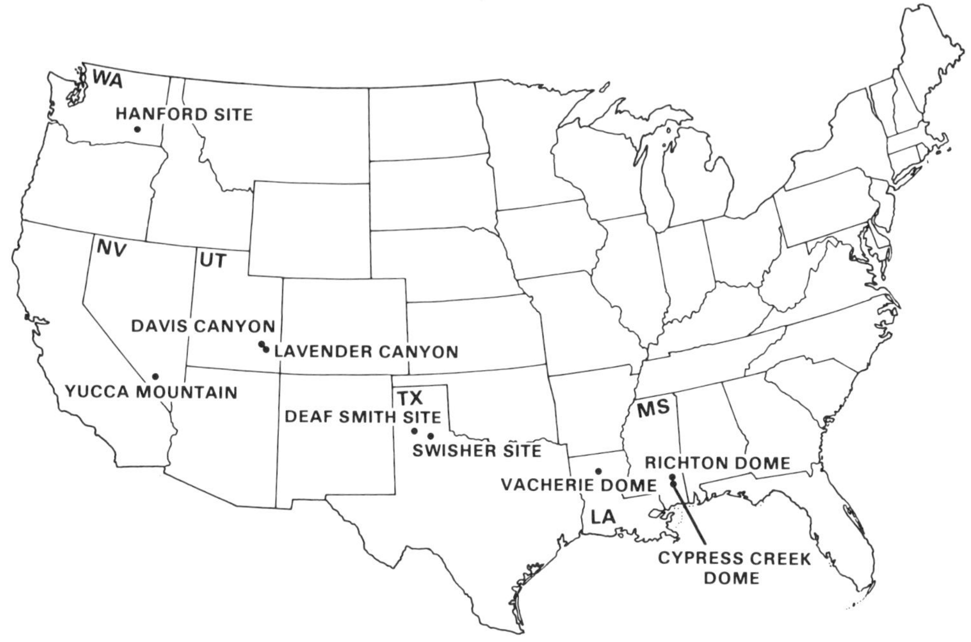

Fig. 1. Potentially Acceptable Sites for the First Repository [4].

2. Cypress Creek Dome, Mississippi (salt dome)
3. Richton Dome, Mississippi (salt dome)
4. Yucca Mountain, Nevada (welded tuff)
5. Deaf Smith County, Texas (bedded salt)
6. Swisher County, Texas (bedded salt)
7. Davis Canyon, Utah (bedded salt)
8. Lavender Canyon, Utah (bedded salt)
9. Hanford Site, Washington (basalt flows)

The locations of these sites are shown in Fig. 1.

The Act further required the DOE to issue general guidelines to be used in determining the suitability of sites. In February 1983, the DOE published draft guidelines for the recommendation of sites for nuclear waste repositories [1]. The DOE revised the guidelines after receiving extensive comments from the Nuclear Regulatory Commission (NRC), the states, Indian tribes, other Federal agencies, and the public. The NRC concurred with the revised guidelines in June 1984, and the final guidelines were promulgated in December 1984 [2].

The Act required that, after the guidelines are issued, the DOE must nominate at least five sites as suitable for site characterization. The act specified that during site-characterization, the DOE will construct exploratory shafts for underground testing to determine whether geologic conditions will allow the construction of a repository that will safely isolate radioactive waste. The Act required the DOE to prepare site-characterization plans for review by the NRC, states, Indian tribes, and the public. After site characterization and an environmental impact statements are completed, the DOE is to recommend one of the characterized sites for development as the first repository.

Although the authors have relied heavily upon the Environmental Assessment of the Yucca Mountain site [3] this report does not necessarily reflect the views or current plans of the U.S. Department of Energy.

YUCCA MOUNTAIN SITE

The Yucca Mountain site is in Nye County, Nevada, on and adjacent to the southwest portion of the Nevada Test Site, about 137 kilometers northwest of Las Vegas (Fig. 2). The Yucca Mountain site is on three adjacent parcels of Federal land, each under separate control of the DOE, the U.S. Air Force, and the U.S. Bureau of Land Management.

The Yucca Mountain site in Nevada was selected for site characterization from among the nine prospective sites for intensive characterization. The Yucca Mountain Site is in the Basin and Range Province, which is characterized by complex structure and stratigraphy. Tertiary and Quaternary tectonic events created the characteristic northward-trending mountain blocks and basins; these events are a continuation of a long geologic history of faulting, volcanism, and plutonism in the Province. Although they share a geologic history of earth movement and igneous activity, numerous sections of the Basin and Range Province have distinctly different aspects in age and nature of terrain, rocks, and tectonic events.

Nye County which is largely rural has a population density of 0.5 person per square mile. The three unincorporated towns closest to the proposed site are Amargosa Valley, Beatty, and Pahrump. The total population of Nye County in 1980 was 9,048. The 1980 population of Clark County, which adjoins Nye County on the east, was 463,087, with a density of 58.8 persons per square mile. Approximately 96 percent of this population resides in the Las Vegas valley. Incorporated cities in the Las Vegas valley include Henderson, Las Vegas, and North Las Vegas. Unincorporated towns and communities in the Las Vegas valley are East Las Vegas, Enterprise, Grandview, Lone Mountain, Paradise, Spring Valley, Sunrise Manor, and Winchester.

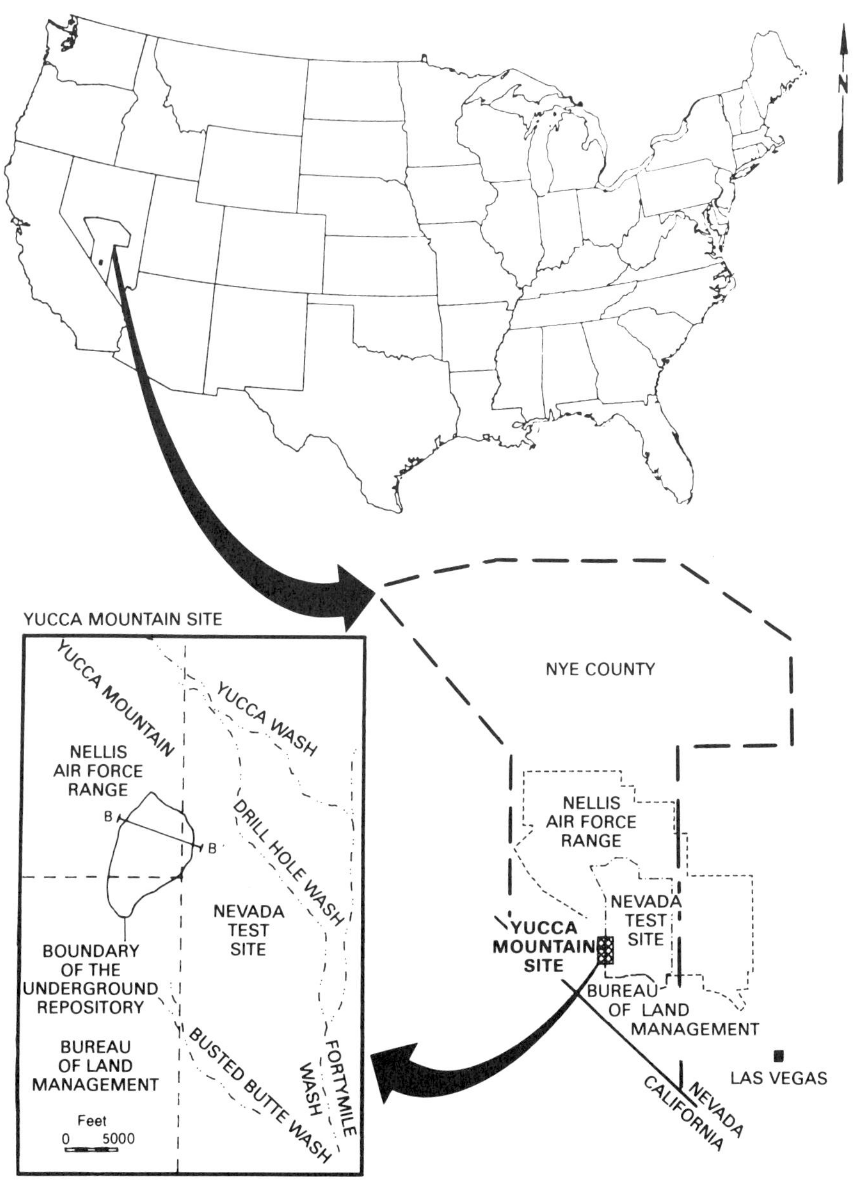

Fig. 2. Location of the Yucca Mountain Site in Southern Nevada (from DOE, 1988).

GEOLOGIC CONDITIONS

This section summarizes the stratigraphy, structure, seismicity, and mineral-resource potential of the Yucca Mountain site and nearby areas. Descriptions of stratigraphy and structure of the Nevada Test Site and vicinity are found in Lipman and others [5], Eckel [6], Byers and others [7], Stewart [8], and Maldonado and Koether [9].

The regional stratigraphic setting of Yucca Mountain is characterized by the four major rock groups shown in Figure 3. The first and oldest of these groups, the Precambrian crystalline rocks, are not exposed in the vicinity of Yucca Mountain but may occur at great depths beneath portions of the site. The second group, Upper Precambrian and Paleozoic sedimentary rocks, is present at the surface about 15 kilometers east of Yucca Mountain at Calico Hills, where it is composed of Devonian and Mississippian argillite and carbonates. This group is also observed 30 to 40 kilometers southeast of Yucca Mountain in the Specter Range and Skeleton Hills, where predominantly Cambrian and Ordovician carbonates and some quartzite are exposed. Carbonates and quartzite of similar age are also present at Bare Mountain about 14 kilometers west of Yucca Mountain. Silurian carbonates have been encountered at a depth of 1,250 meters in one drill hole 2 kilometers east of Yucca Mountain.

The third major group, Tertiary volcanic rocks, occurs at Yucca Mountain and comprises approximately 3,000 meters of the total stratigraphic section. These rocks are composed chiefly of high silica rhyolitic ash-flow tuffs, with smaller amounts of dacitic lava flows and flow breccia and minor amounts of tuffaceous sedimentary rocks and bedded tuffs.

These rocks form the southern end of the southwestern Nevada volcanic field, a large plateau segmented by faults and built chiefly of rhyolitic ash flows and related volcanic material. The ash flows that formed this plateau were erupted between about 8 and 16 million years ago from a complex of overlapping, nearly circular volcanic depressions called calderas. Collectively, the calderas comprise an area of about 1,800 square kilometers. Outcrops throughout the region indicate that the volcanic rocks extruded from this caldera complex once covered an area of more that 6,500 square kilometers.

Quaternary and Tertiary deposits compose the fourth group, which is represented at Yucca Mountain by alluvium and unsorted debris-flow deposits in channels that are cut into the uppermost layers of volcanic rocks and by alluvial-fan deposits that form aprons along the east and west sides of the mountain. Alluvium more than 200 meters in thickness blankets the volcanic rocks beneath Crater Flat to the west and Jackass Flats to the east of Yucca Mountain. Eolian (windblown) sands, caliche, and soil zones also occur in these thicker Quaternary sections. In Crater Flat, basalt flows and cinder cones of Quaternary age are present at the surface, and flows are also present in the alluvium at depth.

Basin fill began to accumulate with origin of the basin and range structure in the middle Tertiary, and accumulation has continued in the Quaternary. The composition of basin fill ranges from silts and clays to sands and gravels. Evaporites include carbonates and simple to complex salts in the playas. Permeability of basin fill differs with grain size and with the degree of consolidation and cementation. In some basins, deeper deposits are more consolidated and less permeable than the overlying, less consolidated deposits, and the fine-grained deposits are in the central parts of the basins.

Faulting and volcanism about 10 to 40 million years ago produced the early features of the Basin and Range Province. In the vicinity of Yucca Mountain, tectonic activity has steadily decreased over the past 10 million years. Minor volcanic activity has continued during basin filling and, most recently, has produced thin, areally restricted flows and cones of basaltic material on Crater Flat, west of Yucca Mountain. Some faults in the vicinity of Yucca Mountain show evidence of continued movement during the past 2 million years. Investigations to date, covering a 1,100-square-kilometer area around the site, have identified 32 faults that offset or fracture Quaternary deposits. These Quaternary faults have been divided into three broad age groups: 5 faults last moved between 270,000 and 40,000 years ago; 4 faults last moved about 1 million years ago; and 23 faults last moved probably between 2 million and 1.2 million years ago. Recently available but unevaluated thermoluminescence data may indicate on the order of 1 to 10 centimeters of fault displacement in eastern Crater Flat less than 6,000 years ago. Yucca Mountain and areas to the west and south have had a relatively low level of seismicity throughout the historical record.

There is no evidence that the Yucca Mountain site contains commercially attractive geothermal, gold, uranium, hydrocarbon, oil shale, or coal resources, although geothermal resources are found in the general area of the site. Under foreseeable economic conditions and in spite of the many small, predominantly gold, mining operations in the area, there is thought to be little potential at the site for extracting the mineral resources because of a lack of rock alteration commonly associated with these economic mineral deposits.

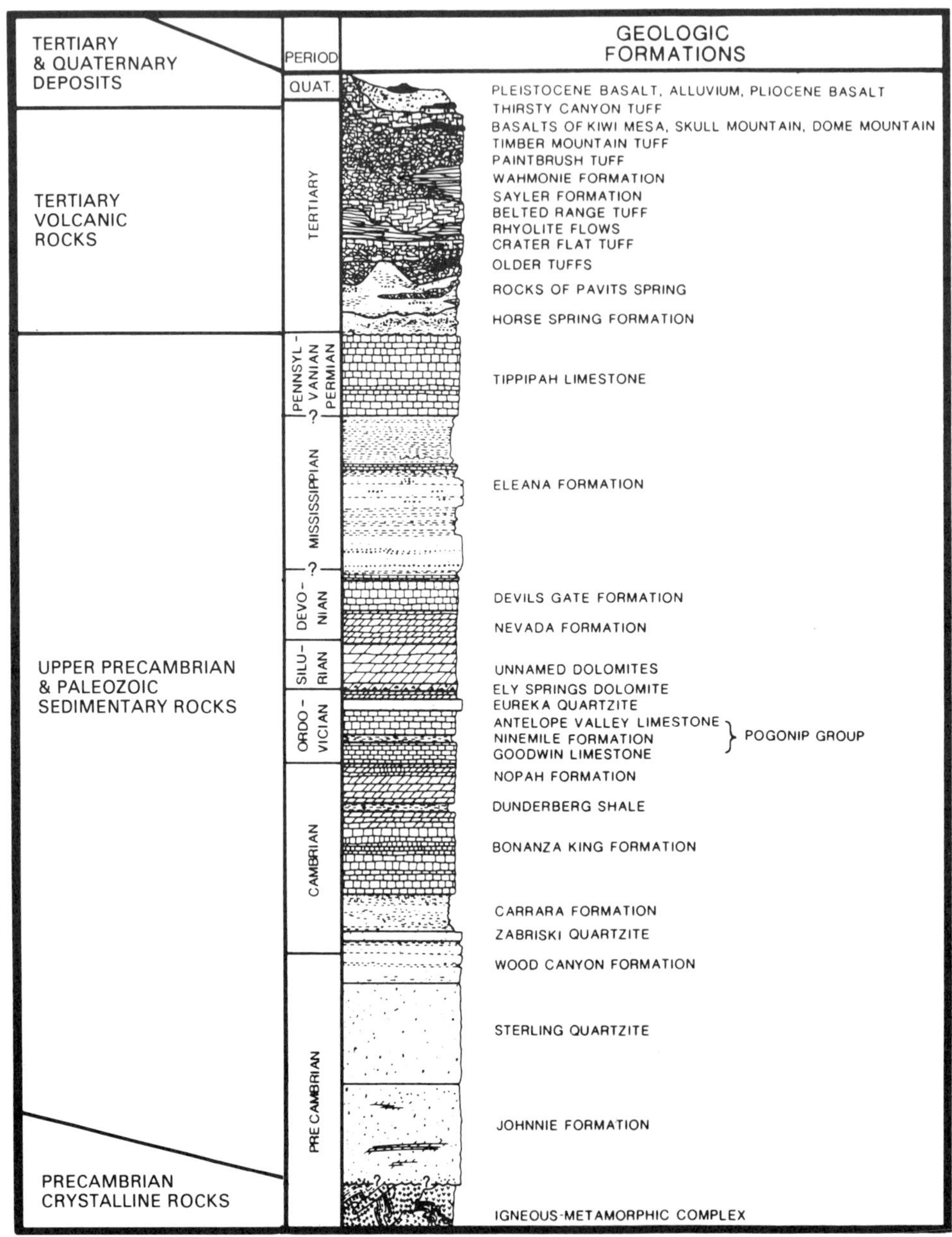

Fig. 3. Geologic Formations in the Vicinity of Yucca Mountain (after DOE, 1986).

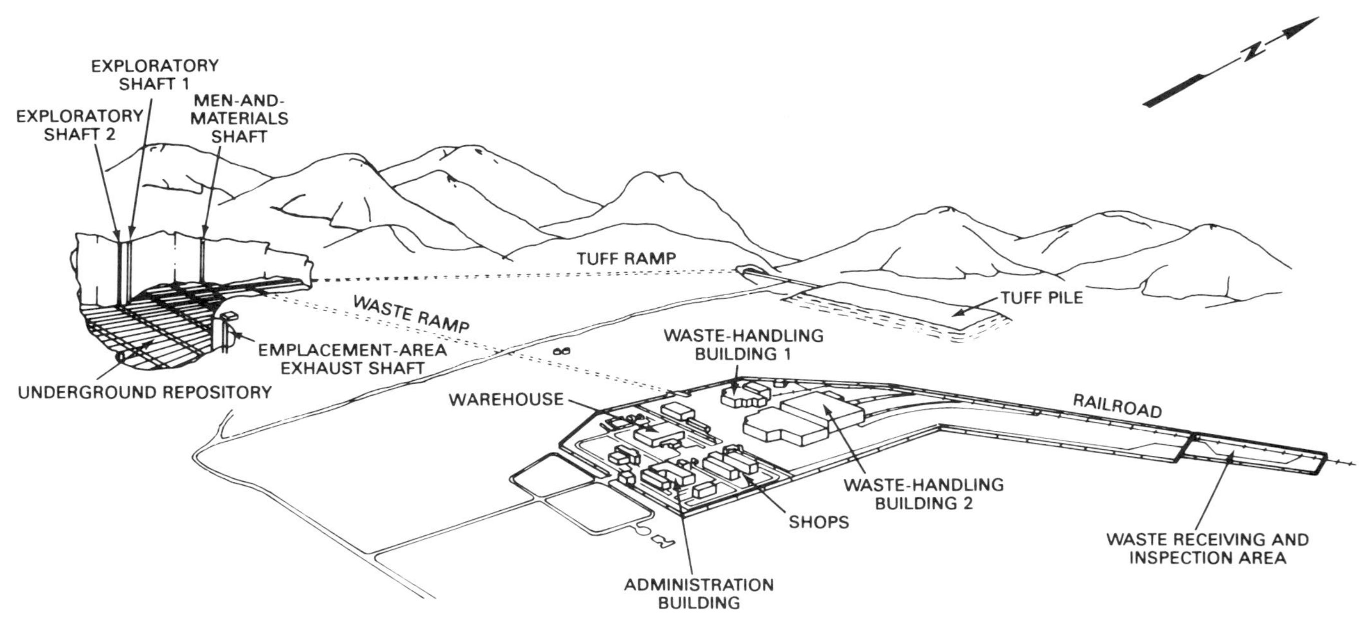

Fig. 4. Perspective of the Proposed Repository at Yucca Mountain (from DOE, 1988).

HYDROLOGIC CONDITIONS

Climate in the Basin and Range Province is arid to semiarid; annual precipitation averages about 28 centimeters. Variability in areal distribution is directly related to topographic relief between the ranges and the intervening basins. Precipitation varies from less than 10 centimeters in the basins to as much as 30 to 40 centimeters at higher elevations in many of the ranges. The mean annual free-water-surface evaporation ranges from 90 to 250 centimeters.

The large difference between precipitation and potential rate of water loss by evaporation and transpiration results in little recharge to ground water, little runoff, and few perennial streams and lakes. Average annual runoff from the Basin and Range Province generally is less than 50 millimeters. The few perennial streams that originate within the Basin and Range Province have their sources at higher elevations in the ranges within the Province, or in the bounding highlands and ranges, principally the Sierra Nevada, west of the province, or the Wasatch Mountains, east of the Province. Two perennial streams flowing through the Province, the Colorado River and the Rio Grande, originate outside the Province.

Water deficiency of the Basin and Range Province, measured by the excess of potential evapotranspiration compared to precipitation, is a significant characteristic with respect to waste isolation. As a consequence of the climate, the ground-water recharge is low and depth to ground water is great in some areas. The unsaturated zone, where rates of vertical movement of water are slow, is a prospective zone for disposal of high-level radioactive waste.

The hydrologic system encompassing Yucca Mountain exhibits low precipitation, thick unsaturated zone, deep water tables, and closed topographic and ground-water basins [10]. Ground water is recharged by the slow infiltration and downward percolation of precipitation and surface water through intergranular pores and perhaps through fractures in the rocks overlying the water table. At Yucca Mountain, most of the precipitation, averaging 150 millimeters per year, is returned to the atmosphere through evaporation and plant transpiration before it can infiltrate deep enough to become ground-water recharge. Only a small fraction (probably 3 percent or less) of the annual precipitation reaches the depth proposed for the repository. The ground-water flow system encompassing Yucca Mountain has been delineated by Bedinger and others [11]. Czarnecki [12] suggested that the ground water flow from the Yucca Mountain area discharges at Alkali Flats in the Franklin Lake Playa.

At Yucca Mountain, a repository would be constructed in the unsaturated zone 140 to 380 meters above the water table. The movement of ground water in the unsaturated zone is typified by a very low flux of water moving downward, primarily through the intergranular pores of the tuff layers [13]. In the saturated zone below, water moves laterally through fractures and pores in the tuffs and in the underlying carbonate-rock aquifers. Flow of ground water in the region has been modeled by Waddell [14]; Czarnecki [15] employed a digital model of the flow system to examine the effects of changes in climate on the ground-water flow system.

GEOLOGIC REPOSITORY

A geologic repository will be developed much like a large mine. An inclined adit will be constructed to allow for the removal of excavated material and to permit the construction of tunnels and disposal rooms at depths about 300 meters below the surface. Surface facilities will be provided for receiving and preparing the waste for emplacement underground (Fig. 4). The surface and underground facilities will occupy about 400 and 2,000 acres of land, respectively. When the repository has been filled to capacity and its performance has been shown to be satisfactory, the surface facilities will be decommissioned and all shafts and boreholes will be permanently sealed.

A repository can be viewed as a system of multiple barriers, both natural and engineered, that act together to contain and safely isolate the waste. The engineered barriers will include the waste package and the underground facility. The waste package will consist of the waste (either spent nuclear fuel or solidified high-level waste) a metal container; an air pocket will separate the waste container from the host rock. The waste container and airpocket will enhance long-term isolation by delaying eventual contact between the waste and the geologic environment. These barriers will further limit any ground-water circulation around the waste packages and will impede the subsequent transport of radionuclides to the accessible environment (about 5 kilometers from the repository).

The geologic, hydrologic, and geochemical features of the site constitute natural barriers to the long-term movement of radionuclides into the accessible environment. These natural barriers will provide waste isolation by impeding radionuclide transport through the saturated and unsaturated flow system and will possess characteristics that reduce the potential for human interference in the future.

Although the DOE plans to use engineered barriers–as required by both NRC in 10 CFR Part 60 and the U.S. Environmental Protection Agency (EPA) in 40 CFR Part 191–the DOE places primary reliance on the natural barriers for waste isolation. Therefore, in evaluating the suitability of sites, the use of an engineered-barrier system will be considered to the extent necessary to meet the performance requirements specified by the NRC and the EPA but have not been relied on to compensate for deficiencies in the natural barriers.

REFERENCES

[1] DOE (U. S. Department of Energy), 1983, Proposed general guidelines for recommendations of sites for nuclear waste repositories: Federal Register, vol. 48, p. 5670, February 7, 1983.

[2] DOE (U. S. Department of Energy), 1984, General guidelines for recommendations of sites for nuclear waste repositories, final siting guidelines: 10CFR Part 60, Federal Register, vol. 49, p.47714, December, 1984.

[3] DOE (U. S. Department of Energy), 1986, Environmental assessment: Yucca Mountain Site, Nevada Research and Development Area, Nevada, DOE/RW-0073, vol. 1, 449 p.

[4] DOE (U. S. Department of Energy), 1988, Overview, site characterization, Yucca Mountain site, Nevada, DOE/RW-0198, 164 p.

[5] Lipman, P. W., Christiansen, R. L., and O'Connor, J. T., 1966, A Compositionally zoned ash-flow sheet in Southern Nevada: U.S. Geological Survey Report.

[6] Eckel, E. B. (ed.), 1968, Nevada Test Site: Geological Society of America Memoir 110, 288 p.

[7] Byers, F. M., Jr., Carr, W. J., Orkild, P. P., Quinlivan, W. D., and Sargent, K.A., 1976, Volcanic suites and related cauldrons of the Timber Mountain-Oasis Valley Caldera Complex, Southern Nevada: U.S. Geological Survey Professional Paper 919.

[8] Stewart, J. H., 1980, Geology of Nevada, A discussion to accompany the geologic map of Nevada: Nevada Bureau of Mines and Geology, Special Publication No. 4, University of Nevada, Reno.

[9] Maldonado, F., and Koether, S. L., 1983, Stratigraphy, structure, and some petrographic features of Tertiary volcanic rocks at the USW G-2 drill hole, Yucca Mountain, Nye County, Nevada: U.S. Geological Survey Open-File Report 83-732.

[10] Winograd, I. J., and Thordarson, W., 1975, Hydrogeologic and hydrochemical framework, south-central Great Basin, Nevada-California, with special reference to the Nevada Test U.S. Geological Survey Professional Paper 712-C.

[11] Bedinger, M. S., Langer, W. H., and Reed, J. E., 1989, Groundwater hydrology in Bedinger, M. S., Sargent, K. A., and Langer, W. H. (eds.), Characterization of the Death Valley Region, Nevada and California: U.S. Geological Survey Professional Paper 1370-F, in press.

[12] Czarnecki, J. B., 1987, Should the Furnace Creek Ranch-Franklin Lake Playa Ground-water Subbasin simply be the Franklin Lake Playa Ground-water Subbasin?: EOS, vol. 68, n0. 44, p. 1292.

[13] Montazer, P., and Wilson, W. E., 1984, Conceptual hydrologic model of flow in the unsaturated zone, Yucca Mountain, Nevada: U.S. Geological Survey Water Resources Investigations Report 84-4345.

[14] Waddell, R. K., 1982, Two-dimensional, steady-state model of ground-water flow, Nevada Test Site and vicinity, Nevada-California: U.S. Geological Survey Water Resources Investigations Report 82-4085.

[15] Czarnecki, J. B., 1985, Simulated effects of increased recharge on the ground-water flow system of Yucca Mountain and vicinity, Nevada-California: U.S. Geological Survey Water Resources Investigations Report 84-4344.

OPTICAL MONITORING OF NATURAL ORGANIC MATTER IN THE AQUATIC ENVIRONMENT

V. PENNANEN

Department of Limnology, University of Helsinki
Viikki, SF-00710 Helsinki, Finland

Naturally occurring organic substances that can be characterized as ranging from yellow to black in colour, of high molecular weight, and refractory, are regarded as humic substances [1]. For the global carbon cycle, humic substances serve as a major reservoir of organic carbon [2]. Humic substances interact with pollutants and a more detailed knowledge of these interactions is needed to a better understanding of the behaviour of pollutants in the environment [3].

Owing to the lack of definite structural properties and biochemical functions, humic substances are defined operationally, according to the solubility as humin, humin and fulvic acids [1]. The water-extractable fraction (viz. fulvic acids) of soil hums is regarded the main source of humus in aquatic environments [4].

Beside the pedogenic organic matter leached from soils into water bodies, aquogenic organic compounds are formed in the water column by degradation of plankton and other organisms found inside the system [5]. The degree of these two components (aquogenic and pedogenic, or autochtonous and allochtonous, respectively) of organic matter varies according to the humosity and productivity of the lake.

Chemistry for the Protection of the Environment
Edited by L. Pawlowski *et al.*, Plenum Press, New York, 1991

Optical monitoring of the concentration of aquatic organic matter is based on the fact that strong statistical correlation has been found between several optical parameters and the observed concentration of organic carbon at different locations. Although local and temporal variations are recognizable, it is evident that a general relationship of some sort should exist [6]. It is presented that a combination of fluorescence and absorption spectroscopy might form a basis for local and even global monitoring of humus resources. Some examples of the application of optical measurement of humic matter in lakes of different degree of humosity in Finland are presented and interferences of different types are discussed.

REFERENCES

[1] G.R. Aiken, D.M. McKnight, R.L. Wershaw and P. Macarthy, in G.R. Aiken, D.M. Mc Knight, R.L. Wershaw and P. MacCarthy (Eds.). Humic Substances in Soil, Sediment, and Water. Geochemistry, isolation and characterization, John Wiley & Sons, New York, Ch. 1, pp. 1-9, 1985.

[2] B. Bolin, The carbon cycle, Sci. Am., 223 125-132, 1970.

[3] R. Saint-Fort and S.A. Visser, Study of the interactions between atrazine, diazinon and lindane with humic acids of various molecular weights, J. Environ. Sci. Health, A23 613-624, 1988.

[4] E.T. Gjessing, Physical and Chemical Characteristics of Aquatic Humus. Ann Arbor Science, Ann Arbor Mi, 1976.

[5] J. Buffle, O. Zali, J. Zumstein and R. De Vitre, Analytical methods for the direct determination of inorganic and species: seasonal changes of iron, sulfur, and pedogenic and aquogenic organic constituents in the eutrophic lake Bret, Switzerland. Sci. Total Environ. 64 41-59, 1987.

[6] V. Pennanen, Humic fractions in dimictic lakes in Finland, Thesis, University of Helsinki, Finland, Ch. 4, p. 25, 1988.

MONITORED BIODEGRADATION OF CONCENTRATED WASTE WATERS BY MEANS OF TRADITIONAL AND MASS SPECTROMETRIC METHODS

V. BALICE[1], C. CARRIERI[2], B. RINDONE[2], and A. ROZZI[3]

[1] Istituto Ricerca sulle Acque, Via De Blasio 5, 70123 Bari, Italy

[2] Dipartimento di Chimica Organica ed Industriale Università di Milano, Via Venezian 21, 20133 Milano Italy

[3] Istituto Ingegneria Sanitaria, Politecnico di Milano via F. lli Gorlini 1, 20100 Milano, Italy

ABSTRACT

Tests have been conducted on olive mill effluents (OME) stored in open basins to evaluate organic degradation as a function of gross parameters such as chemical oxygen demand (COD), total organic carbon (TOC) and tannins. Organic degradation is almost 15% at the tested OME concentration (120 g/l) and degradation was found to increase with decreasing of initial OME concentration. More sophisticated analytical determinations using capillary gas chromatography-mass spectrometry showed that some organic compounds (i.e., polyphenols, long-chain fatty acids, etc.) considered scarcely biodegradable, are to some degree and at times completely degraded or converted into intermediate products.

Chemistry for the Protection of the Environment
Edited by L. Pawlowski *et al.*, Plenum Press, New York, 1991

INTRODUCTION

Studies carried out in the last decade on concentrated wastes, and particularly on olive oil wastewater stored in open basins, have shown a variable degree of organic substance degradation as biological oxygen demand (BOD_5 or COD ranging from 30-35% to 60-75%) [1-4].

The Water Reclaim Plan of the Apulia Region [5] entails the treatment of olive mill effluents (OME) mixed with sewage on twenty-eight depuration platforms by thermal and biological processes.

To avoid organic overloading or oversizing of biological plants OME storage basins have been proposed to equalize the organic load accumulated during the two-three months of the milling campaign over the whole year. Furthermore the presence in the OME of not easily degradable organic compounds (i.e., phenol derivatives, tannins, fatty acids, etc.) makes it difficult to remain within the effluent discharge limits (in Italy 40 mg/l BOD_5, 160 mg/l COD) which makes it hard to identify the single compounds which contribute to BOD_5 and COD by means of more sophisticated analytical techniques to show the fate of these compounds during the storage and eventually to select and optimize the treatment process.

With this aim, studies were carried out on OME in order to:

- determine the degradation of organic matter by measuring gross parameters such as COD, TOC, tannins during a six month period of lagooning, which is the average time of OME storage on platforms. This reduces the polluting load of stored wastes, and thus has an effect on the sizing of treatment plants;

- determine the relation between degradation of organic matter and initial OME COD concentration ranging from 50 to 150 g/l;

- use gas chromatographic and mass spectrometric methods at the end of the storage period to ascertain the fate of some components considered scarcely degradable.

EXPERIMENTAL

OME from press mills were stored in basins of different sizes (see Table 1). Basin A was equipped with a pump which recycled OME before sampling. In reservoir B_1, equipped with a stirrer, lime was added to neutralize OME after pH control, while reservoir B_2 was considered a blank.

The COD, organic acids and tannins were determined by Standard Methods procedures [6]. Organic carbon determinations were carried out with a TOC 915 Analyzer (Beckman), and alcohol analyses with a Sigma 1 gas chromatograph (Perkin Elmer) equipped with a FID detector. Injection, column, and detector temperatures were, respectively, 130, 110 and 130 °C. The chromatographic column was a 2 mm x 200 cm glass column equipped with 0.2% Carbowax 1500 on Carbopack C 80/100. The nitrogen carrier gas flow rate was 20 cc/min.

The following extraction procedure was adopted for gas chromatographic-mass spectrometric determinations OME extraction with ethyl acetate at pH=2.

Capillary gas chromatographic determinations were performed with arian Vista 6000 gas chromatograph equipped with a 4% SE column (Length=25 m). Split splitless injection and detector temperatures were respectively 230 and 270 °C. The following oven-programmed temperature was used: 1 minute at 120 °C followed by a temperature gradient of 6 °C/min to 250 °C and final hold of 10 minutes at this temperature.

Derivatization of ethyl acetate was done by methylation.

The same derivatized extracts were analyzed by a coupled gas chromatograph-mass spectrometer type VG 7070 E R, operating in electron impact at 70 eV or in chemical ionization (with isobutane as the ionizing gas).

RESULTS AND DISCUSSION

The concentration values found for all samples were corrected to the initial volume for each basin, taking into account evaporation and rain effects.

A first series of degradation tests was performed during a period of six months.

In Fig. 1, 2 and 3 the trends of load are reported respectively as COD (g/l O_2), TOC (g/l C) and tannins (g/l tannic acid) while the percentage degradation of organic compounds is reported in Table 1.

Organic compound degradation ranges from 15% to 25% in basin B_1 (lime addition), for TOC and COD parameters.

Tannin degradation in basin B_1 is around 45.5%, which is probably due to calcium-tannin precipitation.

The lower organic degradation values (15-25%) compared with those previously found (35-75%) [2, 4] could be related to the higher concentration of the OME used in this study (120 g/l COD) against 60-70 g/l COD of the OME used on those references.

There is an upper concentration limit of inorganic and organic compounds above which biological reactions are partly inhibited.

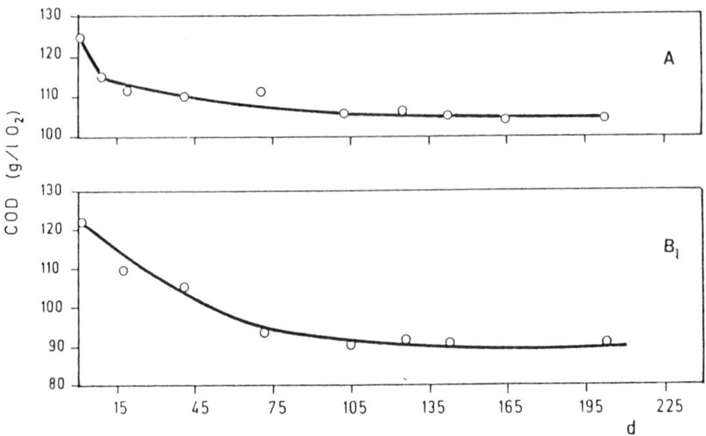

Fig. 1. Organic load variations for different basins as a function of COD.

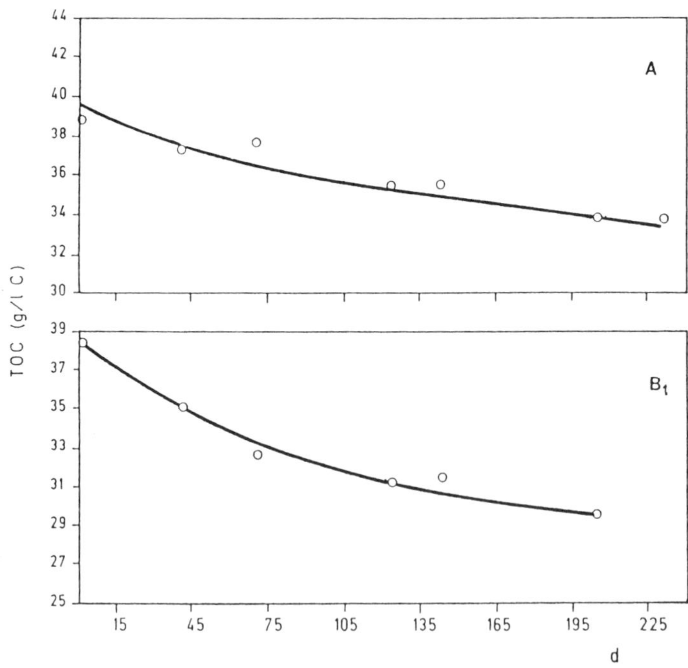

Fig. 2. Organic load variations for different basins as a function of TOC.

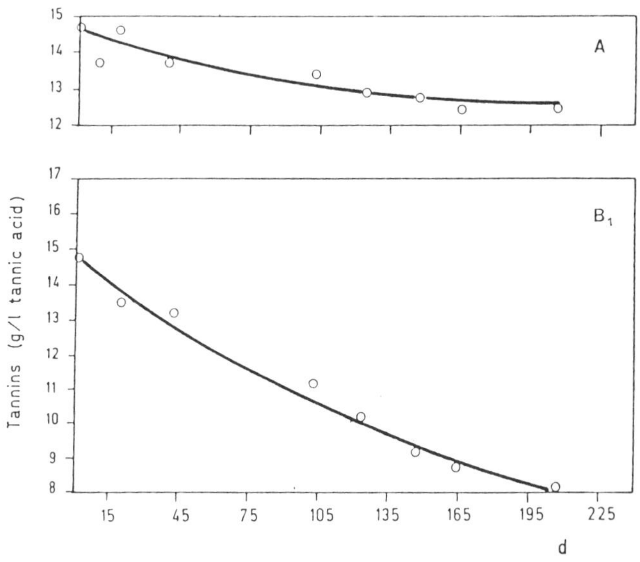

Fig. 3. Trend of tannins for different basins.

Table 1. Dimensions of basins and organic compound degradation

Basin	Degradation percentage			S	V	h
	as COD	as TOC	as tannins	(m^2)	(m^3)	(m)
A	14.6	12.9	14.6	1.696	5.11	3.01
B_1	26.8	24.6	45.5	1.265	1.221	0.96
B_2	14.8	13.1	14.3	1.265	1.221	0.96

A second series of tests was carried out to evaluate the correlation between organic degradation and OME initial concentration. Percentage degradation related to COD, TOC and tannins after a six month period, is reported in Fig. 4.

There is clearly a threshold concentration ($\sim$ 110 g/l as COD, 36 g/l as TOC, and 13.5 g/l as tannins) after which the degradation reaches a constant minimum value, which indicates an inhibition of biological degradation due either to concentration of organic compounds or to the presence of refractory organic substances.

In fact previous tests made on biological plants [7] have shown that some organic compounds such as phenol derivatives, tannins and long-chain fatty acids are not easily biodegradable, but can easily lead to serious problems due to the stringent limits of the Italian law.

This comes out the necessity of investigation with more sophisticated techniques the fate of these organic compounds after a long period of storage

of OME. By applying the OME gas chromatographic mass spectrometric technique the following compounds were found:
- aromatic compounds of phenolic structure;
- non-ionic detergents of aliphatic structure with various ethoxyl groups; and
- phthalic acid derivatives

whose molecular compositions are reported in Fig. 5.

The monitoring efficiency of this method was tested by investigating the behaviour of fresh OME and after long-term storage in open basins. Both types of samples were diluted with sewage in the ratio of 1:7 to simulate influent to an upflow anaerobic sludge blanket plant.

The capillary gas chromatographic profiles are shown in Fig. 6 and 7. It can easily be seen that after storage mono- and diunsaturated C_{18} fatty acids are almost completely degraded (peaks 16 and 17) while C_6 saturated acid (peak 14) is still present. Also the amount of phthalic acid derivatives is greatly reduced (peak 1) while the dihydroxy derivative of phenolic compounds has been completely removed peak 9).

This monitoring methodology is now being applied to biological treatment plants.

CONCLUSIONS

The following conclusions can be drawn concerning the fate of OME stored in open basins:

- Organic degradation of OME as COD, TOC and tannins is about 15% of the initial organic load (120 g/l COD).

Lime neutralization of OME has a twofold function: to increase the degradation rate and to precipitate as calcium salts some compounds determined as tannins.

Organic degradation is dependent on initial OME concentration. There is a concentration limit (110 g/l COD) beyond which the degradation reaches a plateau at a minimum level. Degradation increases as OME initial concentration decreases.

- More sophisticated analytical techniques show that some organic compounds considered scarcely biodegradable in biological plants such as phenol derivatives, long-chain fatty acids, etc. are either partially or completely degraded after long-term storage in open basins.

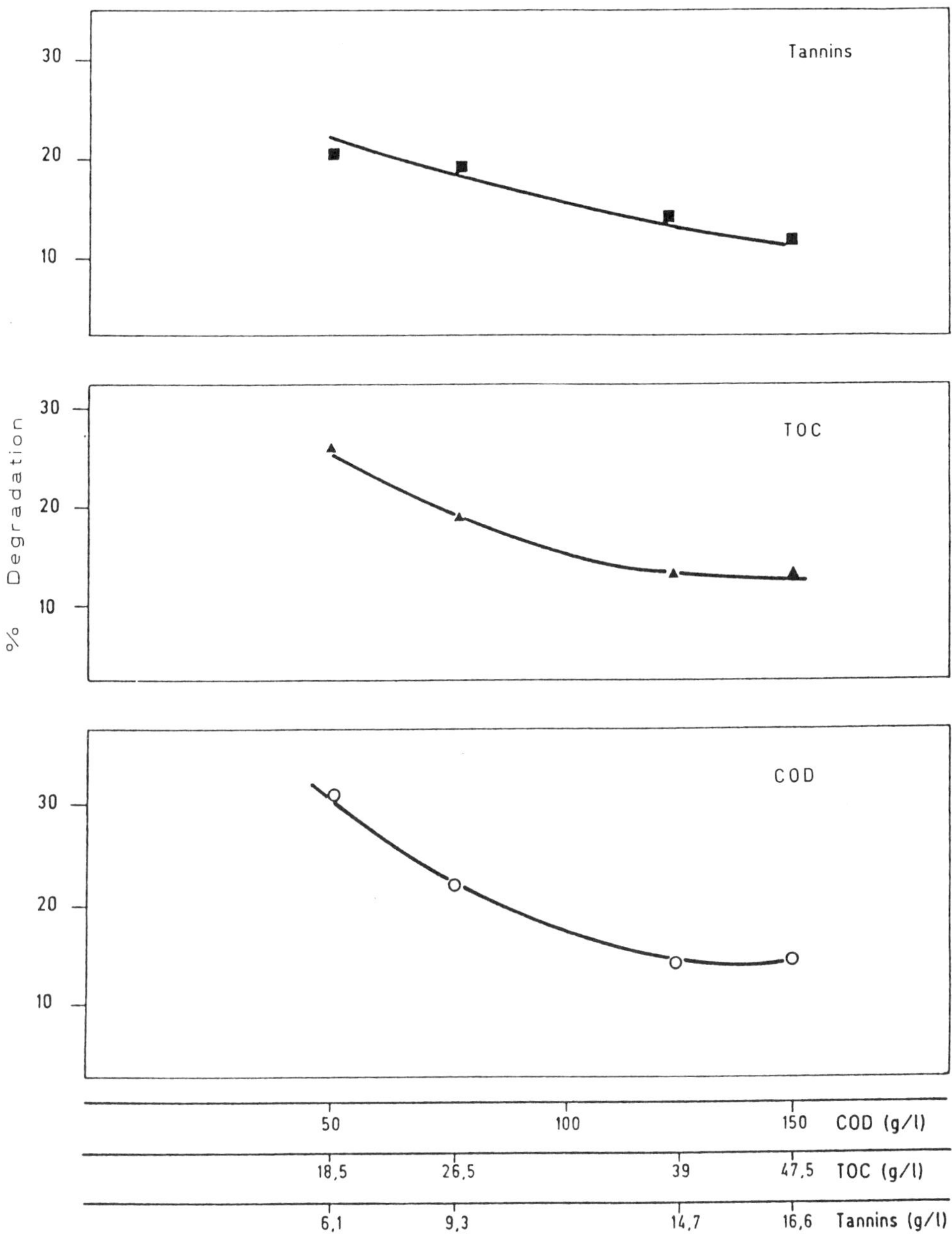

Fig. 4. COD, TOC and tannin degradations versus initial OME concentration.

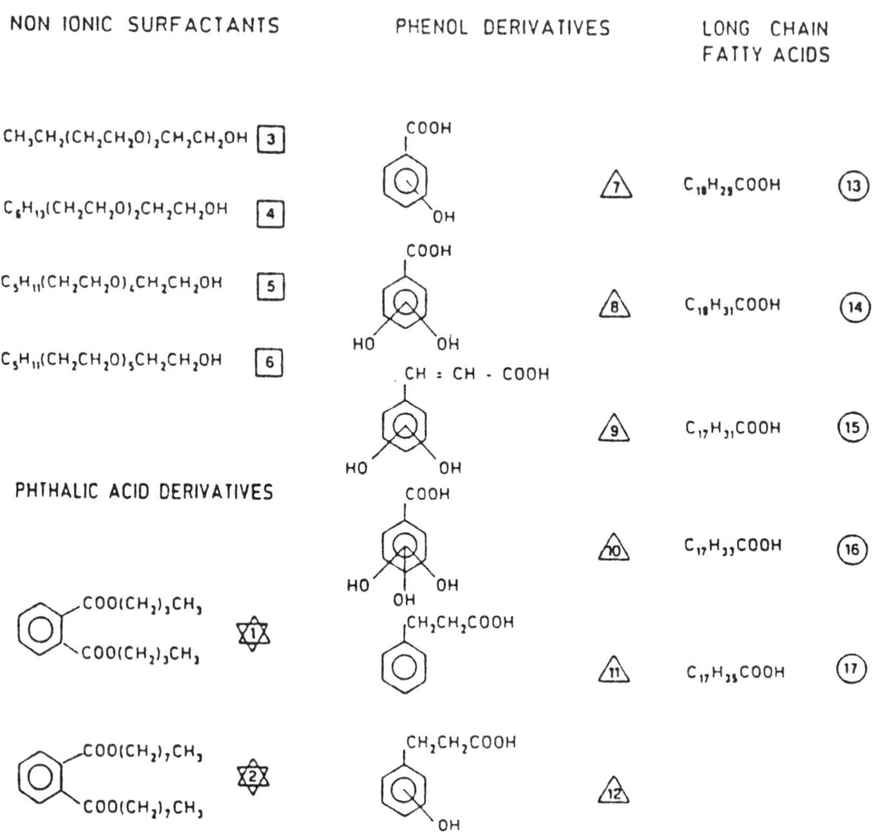

Fig. 5. Chemical composition of some organic compounds of OME.

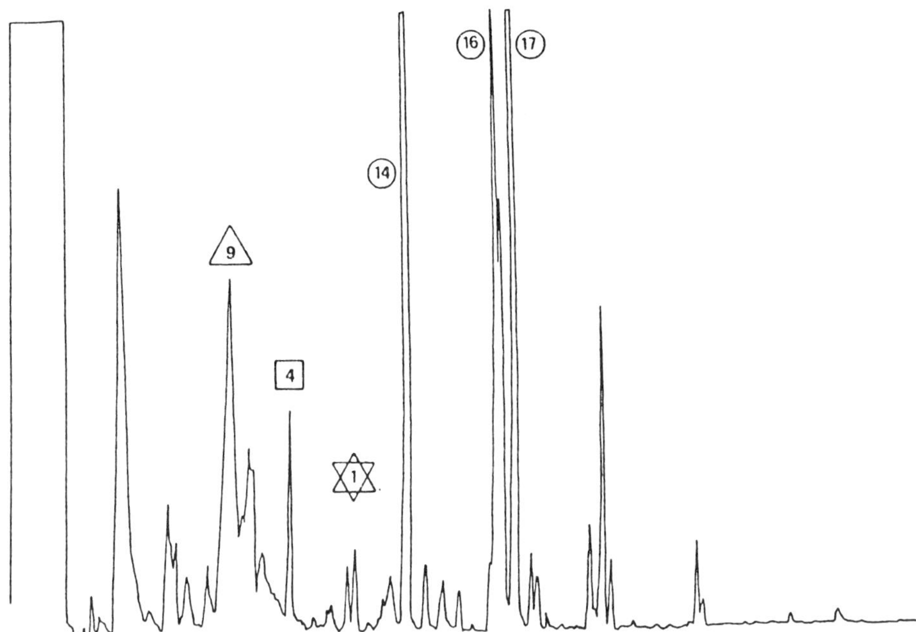

Fig. 6. Freshly diluted OME extracted at pH 2 and derivatized by methylation.

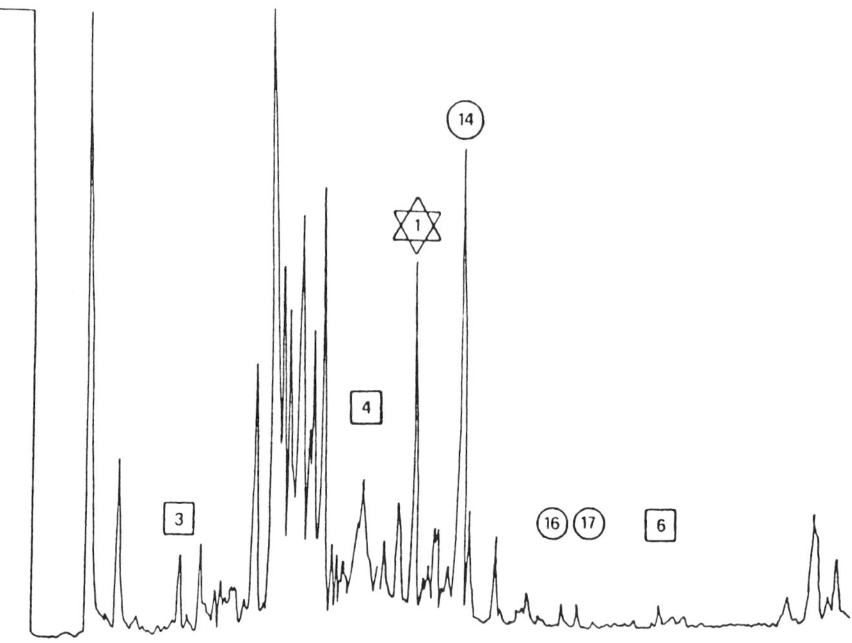

Fig. 7. Final stored OME extracted at pH 2 and derivatized by methylation.

REFERENCES

[1] Bradley R.M., L. Baruchello, "Sul trattamento degli scarichi liquidi nell'industria dell'olio di oliva" AES, 4, 63-66, 1980

[2] Fiestas J.A. Ros de Ursinos, "Depuration de aquas residuales en las industrias de aceitunas y aceite de oliva" Grasas y aceites 28, 2, 113-121, 1977

[3] Cabello R.L., J. A. Fiestas Ros de Ursinos, "Evaluacion de Alpechines en balsas de evaporacion de aquas residuales", Madrid 18-20 Nov, 1981

[4] Di Fazio A., S. Troisi, "Dati sulla qualità dell'acqua sotterranea in una zona della Penisola Salantina" Istituto di Ricerca sulle Acque - Rapporto tecnico n. 103, Maggio, 1984

[5] Regione Puglia "Piano Regionale di Risanamento delle Acque" Maggio, 1983.

[6] "Standard Methods for the Examination of Water and Wastewater" 15th Edition, AWWA-WPCF, 1980

[7] Balice V., C. Carrieri, O. Cera and B. Rindone, "The fate of tannin-like compounds from olive mill effluents in biological treatments". Proceedings of the Fifth International Symposium on Anaerobic Digestion, Bologna 22-26 May, 275-279, 1988

FAST ANALYSIS OF HEAVY METALS IN CONTAMINATED SOILS USING FIELD-PORTABLE X-RAY FLUORESCENCE TECHNOLOGY AND GEOSTATISTICS [1]

G. A. RAAB, R. E. ENWALL, W. H. COLE,
C. A.KUHARIC and J. S.DUGGAN

Lockheed Engineering & Sciences Company
1050 East Flamingo Road, Suite 208
Las Vegas, Nevada 89119, USA

ABSTRACT

There are many sites within the United States which are contaminated with heavy metals as well as other chemicals. The contaminants on hazardous waste sites must be determined before the remedial investigation/feasibility studies can be completed. In general, samples under investigation must be sent to an approved laboratory for analysis at a cost of approximately US$300 per sample and 21 days turnaround. Field-portable X-ray fluorescence technology provides the remediator with a fast, economical way of screening 100 samples per day on hazardous waste sites for inorganic constituents. If we assume 200 samples for one site are analyzed by traditional methods, the cost is US$60,000. This would easily pay for the field-portable X-ray fluorescence instrument, which can be used for all subsequent sites at substantial savings with realtime data.

Some field-portable X-ray fluorescence equipment employs energy dispersive technology to simultaneously analyze for elements. Energy dispersive X-ray fluorescence technology used in conjunction with software employing fundamental parameter equations allows for "standardless" calibration. Other equipment uses

[2]Presented at the 95th Annual Northwest Mining Association Convention, Spokane, Washington, December 6–8, 1989. Reprinted by permission of the NW Mining Association.

calibration curves based on sitespecific standards and bases its software on modified fundamental parameter calculations.

Used in conjunction with a microcomputer and related peripheral devices such as printers and plotters, field-portable X-ray fluorescence equipment allows immediate processing of data to support timely, on-site decisions. Geostatistical software permits application of regionalized variable theory for optimizing sampling design, and for generation of minimum error isopleth maps of contaminant concentration to guide future sampling or remedial activities.

NOTICE: Although this research was funded in part by the U.S. EPA through Contract 68-03-3249 to Lockheed Engineering & Sciences Company, it has not undergone Agency review and does not necessarily reflect Agency policy.

INTRODUCTION

In 1980 the US Environmental Protection Agency (EPA) decided to provide standardized and specialized analytical services through the Contract Laboratory Program (CLP) to support a variety of government regulated sampling activities, from those associated with the smallest preliminary site investigation to those of largescale, complex remedial, monitoring and enforcement actions. The central and overriding assumption governing the structure and function of the CLP is the basic requirement to provide legally-defensible analytical results for use in supporting governmental enforcement actions.

The CLP maintains a high level of quality assurance and documentation in all aspects of program activities. Commercial laboratories must qualify on a quarterly basis. Companies contracted by the EPA to conduct on-site work send their samples to these laboratories within the CLP. As a result, the costs of conducting the analyses under strict requirements and maintaining the documentation are much higher than the unregulated analyses by other laboratories. The cost for 24 inorganic analytes per sample is about US$300 and a turn around time of 21 days.

The use of field-portable X-ray fluorescence (FPXRF) provides a way to reduce costs, increase productivity and produce data in real time. With the instrument cost at about US$46,000, the initial investment seems high. When taken in perspective of a US$300,000 sampling program, and the chance to have data concerning the site before sampling, the cost becomes reasonable. With each subsequent site the FPXRF analysis cost per sample goes down.

EQUIPMENT

The instrument we use is a field-portable, energy dispersive X-ray fluorescence spectrometer. The unit is self-contained, battery powered, microprocessor based and weighs 8.5 kg. The surface analysis probe is specifically designed for field use. The FPXRF unit is also hermetically sealed and can be decontaminated with soap and water. The probe includes a radioisotope of 244-curium, a Xe-gas proportional counter and the associated electronics. The source is protected by a government approved safety shutter.

The electronic unit has 32 calibration memories called 'models'. Each model can be independently calibrated for as many as six elements. These can be used to measure elements from silicon to uranium assuming the proper isotope source is available. The unknown sample intensities are regressed against the calibration curves to yield concentrations. The routine data are recorded by hand as this unit has no means for storing multiple analyses.

METHODOLOGY

-Historical Investigations and Aerial Photography

Before we travel to a new site, we first conduct an historical investigation for published documents. This includes obtaining remedial investigations/feasibility studies, special reports, and aerial photographic investigations. When no aerial photographs exist, we fly our own. We include the latest aerial photograph in our site-screening report (SSR) to aid the on-site coordinator in developing ground truth in relationship to the concentration isopleth maps.

Our photographic interpreters use stereoscopic pairs of historical and/or current aerial photographs to perform the analysis. Stereo viewing enhances the interpretation because it allows the analyst to observe the vertical as well as horizontal spatial relationships of natural and cultural features. We use stereoscopy to distinguish between various shapes, tones, textures, and colors that can be found within the study area, which aid in determining the extent and nature of contamination.

-Fast Screening

The portability of the FPXRF unit allows field personnel to conduct analyses on site. The screening in situ (SIS) measurement allows the field analyst to make a direct measurement on the surface of the soil without collecting a physical sample or preparing it. With an analysis time of 30 seconds and no sample preparation, we can conduct hundreds of analyses in several days.

The data we generate with the FPXRF must be used with discretion.

The SIS measurement represents the first several millimeters of soil. Hence, it reflects the surface conditions only. For measurements representing depth the most effective method is to take core samples and analyze them.

-Geostatistics

In most cases, contaminant concentration values obtained for a set of samples are spatially autocorrelated within the sampling domain, invalidating the assumption of data independence demanded by classical sampling theory. Regionalized variable theory was developed and formalized in the early 1960s to fill the gap left by classical statistics, and has led to the development of an array of statistical tools, known collectively as "geostatistics", for analyzing spatially autocorrelated data. The structure of spatial autocorrelation is represented by the semivariogram which mathematically defines the average change in concentration as a function of distance in a given direction for a particular contaminant. Semivariograms also provide the means for calculating errors inherent in spatial estimation (or interpolation); thus they can be used to evaluate different sampling patterns and to design spatially optimum schemes in advance of subsequent sampling phases.

Clearly the most important inferential steps in the progression of a site from initial screening to final evaluation of remediation involve development of spatial models describing contaminant concentrations at given locations throughout the site. Spatial models are usually derived by interpolating concentration values from sampled locations to unsampled locations occurring at the nodes of predefined regular grids. Such models provide the basis for such graphical decision-making tools as concentration isopleth (or contour) maps and isometric plots. Many procedures exist for interpolating concentration values at unsampled locations, but nearly all require arbitrary, a priori assumptions about the spatial structure of data variability. Unfortunately, the correspondence between the assumed and actual structure is nearly always a matter of chance. Geostatistics, on the other hand, provides an interpolation tool known as "kriging" that utilizes the definition of spatial variability provided by semivariograms. Provided the concentration data meet certain assumptions, the results of kriging are optimized to be unbiased and to incur minimum possible interpolation errors. Kriging also provides an estimate of the error incurred at each interpolated location. Maps of kriging errors provide site decision-makers with the bases for evaluating the reliability of their sampling data and to identify areas requiring further sampling.

Current computer technology allows facilities for application of geostatistical data analysis to accompany the FPXRF system to field sites, thereby providing rapid results for decision-making. Small, transportable personal computers (PCs) are available which can be configured with large work space memory and disk storage. Several geostatistical software packages have been

Table 1. Pb Concentration of the Site-Specific Calibration Standards

standard #	wt% Pb
1	0.000
2	0.041
3	0.412
4	0.656
5	0.841
6	1.444

developed for PCs, and include both commercial and public domain sources. Small, high-speed and relatively inexpensive multi-pen plotters can be selected for field operations to produce publishing quality maps and diagrams on-site. A case study is presented later in this paper to provide an example of the detailed geostatistical data analysis and graphical output achieved for an actual field site.

SITE SPECIFIC CALIBRATION STANDARDS

Without the use of a full fundamental parameters program to calculate elemental concentrations, XRF uses a calibration curve. In the case of the SIS measurement, the problems of heterogeneity, bulk density, and particle size distribution in the soil matrix can strongly affect the analysis. Since we cannot eliminate these problems in the routine analysis, we introduce the same error in the calibration curve. We accomplish this by using site-specific calibration (SSC) standards.

We collect the SSC standards on a reconnaissance visit to the site. The FPXRF unit is calibrated with soil standards containing the analytes of interest on the proposed site but not necessarily matching the on-site soil matrix. By using the site typical standards for a calibration curve in the FPXRF unit, the reconnaissance team can then screen the proposed site for the relative ranges of analyte concentrations and collect samples representative of that relative range. Once those samples are analyzed in the laboratory, then we can use the suite of site-specific soils to calibrate the FPXRF unit with a site specific curve (see Table 1 and Figure 1) for use in a detailed and much more accurate screening of the reconnoitered site.

We have the SSC standards prepared according to Buckley and Cranston [1] as follows: our chemists digest one split totally with aqua regia and hydrofluoric acid; they treat the second split with a nitric acid extrac-

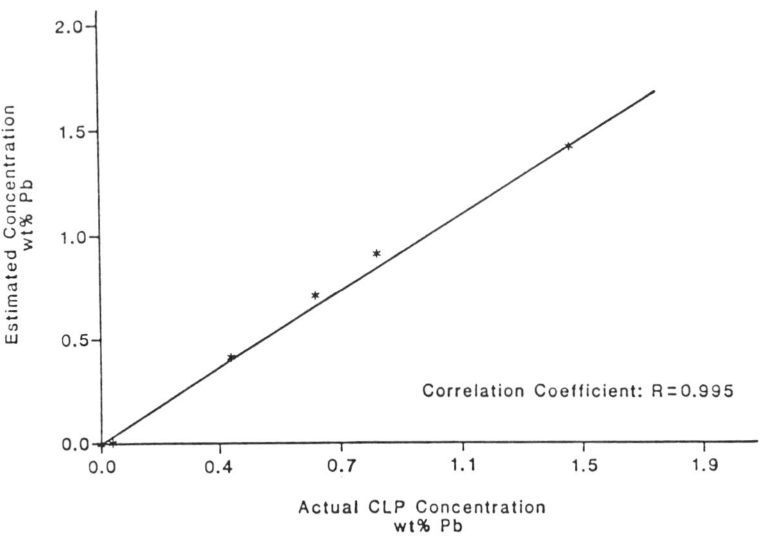

Fig. 1. Pb calibration curve for SSC standards.

tion according to CLP protocol and the third split we leave untreated. The chemists analyze the first and second digestates on the inductively coupled argon plasma spectrometer (ICP) or atomic absorption spectrometer (AAS) under the strict instrument quality control constraints as required by the CLP. They analyze the third untreated split with a laboratory-grade XRF. Upon completion of the analyses we compile the data into a table as we have done in Table 2.

Table 2 shows a cross comparison of data obtained from the different preparation and analytical methods. As XRF and the corresponding sample preparation (or lack thereof) are not yet accepted CLP protocol, tables such as this will enable us to show comparability between XRF and ICP, as well as allow us to demonstrate traceability to standards certified by the National Institute of Standard Testing (NIST).

For the calibration programming of the field-portable instrument we use the XRF results. Once calibrated, we now have standards with the same bulk density, heterogeneity, and particle size distribution as the routine samples we will measure on the site. These standards are now our certified reference materials.

Table 2. Site-Specific Standards a Cr Site

ELEMENT	ICP		X-Ray Fluorescence
	CLP Extraction	Total Digestion	Laboratory-grade wave-length dispersive*
	All values in mg/kg		
Ba	110	642	716
Co	9	45	34
Cr	100	264	232
Cu	29.4	48	54
Mn	434	1,461	1,704
Ni	7.5	41	42
Pb	216	515.1	326
V	63.7	220	196
Zn	102	195	177

* Analyses performed by P. Dalheim, Mineralab, Denver, Colorado

QUALITY ASSURANCE

Keeping in mind that this is a screening process and not under strict laboratory control, the matrix determines our limits for precision. The detection limits are also dependent upon the matrix and are not set a priori.

All the samples we send for analysis are traceable to the standard reference materials certified by the NIST, by including blind audits of the standard reference materials and in the qualification of the analytical equipment.

To assure the quality of data obtained in any environmental sampling program, we must address precision, accuracy, representativeness, completeness, comparability.

We determine precision of the FPXRF instrument by repeated measurements of a low-level sample (i.e., near the detection limit) and a midcalibration-range sample (e.g., after every tenth routine analysis). The low-level sample should be a near-background-level sample taken on the site. We set the precision objective at $\pm$ 20 percent relative standard deviation for sample results above a quantifiable concentration for the analyte of interest.

We determine accuracy of the FPXRF instrument by repeatedly analyzing a reference sample which can also double as the midcalibration range sample used to determine precision.

An adequate sampling scheme (i.e., the sample grid must delineate the hot spots) will satisfy the requirements of representativeness. For optimizing a sampling grid we have found geostatistics to be the most effective tool.

We address completeness on a site-specific basis: If samples or sample data are lost or misplaced, the completeness is less than 100 percent.

If we use a single FPXRF instrument on a site, then data comparability is not in question. Comparability of XRF data to data from a CLP laboratory, or to data obtained from wet chemistry methods in general, is more complex. Because the CLP employs leach or partial extraction methods for the analytes of interest, results obtained by using these methods are not directly comparable to the total elemental analysis provided by XRF. Complete sample digestion, which is a total elemental analysis, will provide a more comparable method to XRF. The CLP extractions often show lower results than does XRF, whereas complete sample digestion and XRF will yield tantamount results. Field technicians must be aware this complication in comparability can arise by using the CLP digest numbers rather than the total elemental values for the calibration standards.

Much of the quality assurance and quality control protocol published in the past applies to analysis of aqueous samples and may not apply to XRF analysis of soils [3]. By running low-level and midcalibration-range samples at regular intervals, it can be shown that gain change and baseline drift do not occur in XRF analysis with the regularity that they do in AAS and ICP analysis. Matrix spikes are unnecessary in XRF; enhancement and adsorption effects are well understood and corrected for by the microprocessor on field-portable instruments and by fundamental parameters software on the more powerful instruments.

- Realistic Application of Quality Assurance in the Field

In XRF analysis of soils, the most pronounced sources of error are introduced by heterogeneity, where the error is either a positive or a negative bias, and by a wide range of particle size fractions, where the error is a negative bias in the specimen to be analyzed. To insure against these sources of error in the laboratory, the sample can be dried, sieved, pulverized, thoroughly homogenized and then aliquoted and pressed to a pellet which becomes the specimen for analysis. These sorts of sample preparation and analysis steps are generally written into laboratory standard operating procedures which insure that every sample coming into the laboratory is handled in a uniform manner.

The crew operating out of a field laboratory generally does not have the luxury of being able to operate under standard operating procedures as they encounter a plethora of sample handling situations from one site to another. On one site, the project manager may only want in situ analyses for trend evaluation, while on another site, samples might await the field crew, requiring analysis in or out of the sample container with no sample preparation. A third situation might require extensive sample preparation prior to specimen analysis. Below we will discuss the individual situations.

The simplest situation, with regard to sample preparation, is in situ analysis; the reason being that there is no physical sample taken, thus no sample handling error. Screening in situ (SIS) analyses are a quick and cost effective way to determine trends of contaminants on a site. This situation may be the simplest but also has the highest degree of error in that there has been no effort to reduce the heterogeneity or the potentially wide range of particle size fractions. Data obtained from the SIS mode of analysis are best suited for trend analysis in determining the extent and distribution of the contaminants on a site. When data are used in this manner, the emphasis is not on the accuracy and precision of the individual sample but rather on the statistical distribution of many relative sample concentrations.

The aforementioned trend analysis can further be used to guide the decision of where to collect corroboratory samples to be sent to a laboratory. The purpose is to reduce the number of samples and more wisely select those that must be sent to the laboratory, thus cutting overall costs and data turnaround time as well as increasing the quality of the sampling strategy.

An extension of the SIS mode of analysis is the situation where numerous samples have been taken prior to the arrival of the field laboratory, and the project manager requires rapid screening of the samples to decrease the number of samples to be delivered to a laboratory. Where the sample container is large enough (e.g., a large sealable plastic bag) the XRF probe can be inserted into the container and a direct measurement can be made. Otherwise the sample must be poured out onto a clean surface for direct measurement. The degree of heterogeneity without further sample preparation was fixed at the moment of placing the sample in the sample container. Several measurements (with the sample probe moved between each) reported as a mean value will begin to define the variability (i.e. the heterogeneity) of the sample.

If data of the level of quality suitable for trend analysis is unacceptable and the goal is to approach laboratory analytical results, then thorough sample preparation to reduce heterogeneity, bulk density, and particle size distribution is necessary prior to analysis.

As in all data acquisition programs, the end use of the data should guide the initial data quality objectives (DQOs). If trend analysis of the contaminants on a site is the end goal, then SIS analysis will suit the DQOs. If, for some reason, low-level samples are of extreme importance, as may be the case in looking for analyte migration from soil horizons to a water table, then thorough sample preparation prior to specimen analysis would be required to meet the DQOs. In either case, corroboratory samples should be sent to an off-site laboratory to assure that the chosen mode of sampling and analysis has met the DQOs.

CASE STUDY

-Site Description

An early application of the field XRF system took place at a contaminated waste site consisting of a former metal plating concern. Waste electrolytic plating solutions containing high concentrations of lead and chromium with lesser concentrations of other heavy metals were occasionally released on the ground surface and migrated toward a topographic depression on the eastern margin of the site. Surficial deposits of heavy metals eventually migrated downward leading to contamination of subsurface formations and an underlying unconfined aquifer.

A remedial investigation/feasibility study had been completed, and final remedial design was underway. A request was made by the EPA Site Remediation Manager for rapid XRF screening of drill core and ground surface to assist in selection of bulk samples for bench leach and metal stabilization tests, and for detailed spatial definition of surface contamination in a fairly small but important portion of the contaminated area. This case study describes the XRF sampling and spatial definition of surficial lead concentrations within the detailed area. Chromium has been excluded from discussion for the sake of brevity.

-Sampling Design and Procedure

The detailed area bounds several point sources of surficial contamination and measures 200 feet in the east-west direction by 120 feet in the north-south direction. A square sampling grid was surveyed over the area using a tape and tripod-mounted pocket transit. Grid lines were oriented parallel to the boundaries of the area. An internodal distance of ten feet was used to achieve the spatial resolution required by the site manager. This distance was halved along one east-west grid line near the middle of the area in order to provide extended resolution through a known zone of exceptionally high contaminant variability. Each point was marked to facilitate physical location and bulk sampling of contaminated zones.

SIS sampling procedures followed for this survey were very simple. The X-ray probe was placed in direct contact with the ground surface at the sampling location following removal of any loose foreign material such as construction debris or dead vegetation. Brief notes were taken concerning the composition and condition of the ground surface. Along most of the grid lines, gamma exposure and X-ray counting continued for an interval of fifteen seconds at each sample location. An exception was the grid line with five-foot spacings along which every third location was counted for five consecutive intervals without moving the probe. These replicates were used to evaluate the minimum variability that can be expected with the SIS mode of sampling.

RESULTS

The on-site coordinator had planned a total of 245 sampling points, but the presence of construction materials and other obstructions limited the survey to 212 points actually sampled. Figure 2 presents a map showing the resulting sample locations and lead concentrations expressed as milligrams per kilogram.

Descriptive statistics for the lead values are as follows:

arithmetic average:	371 mg/kg
standard deviation:	761 mg/kg
coefficient of variation (CV):	205%

The high CV is consistent with a very strong positive skewness shown by the histogram. Transformation of the lead values to natural logarithms yields a probability plot that roughly approximates a lognormal distribution.

We generated directional semivariograms for the log-transformed data and these appear in Figure 3. All four appear to be transitional structures that approach a sill of about 1.4. Three of the semivariograms approach the sill at an effective range of about 30 feet. The north-south semivariogram appears to be slightly different, but the difference is judged to be insignificant for the purposes of this study. Lack of anisotropy displayed by the semivariograms is consistent with the fact that the site is used for commercial operations involving heavy surface disturbance and transport of lead-bearing soil and dust by vehicles. Additional surface disturbance is caused by site safety procedures that require hosing the ground with water to minimize dust generation. All of the foregoing activities would tend to increase the spatial dispersion of the lead and diminish any anisotropy that could be expected from topographic control of lead transport along the east-west direction.

An isotropic exponential model was fitted to the logarithmic semivariograms shown in Figure 3 and converted to a general relative model using the procedure described in David [4]. Ordinary kriging using the relative semivariogram model was performed on untransformed lead values to obtain a spatially unbiased estimate of the global mean concentration in the detailed area. A discrete spatial model of lead concentrations was interpolated from the untransformed data by simple point kriging using the global mean of 336 ppm and the relative semivariogram model. The spatial model consists of a square grid with an internodal distance of 5 feet, and appears in Figure 4.

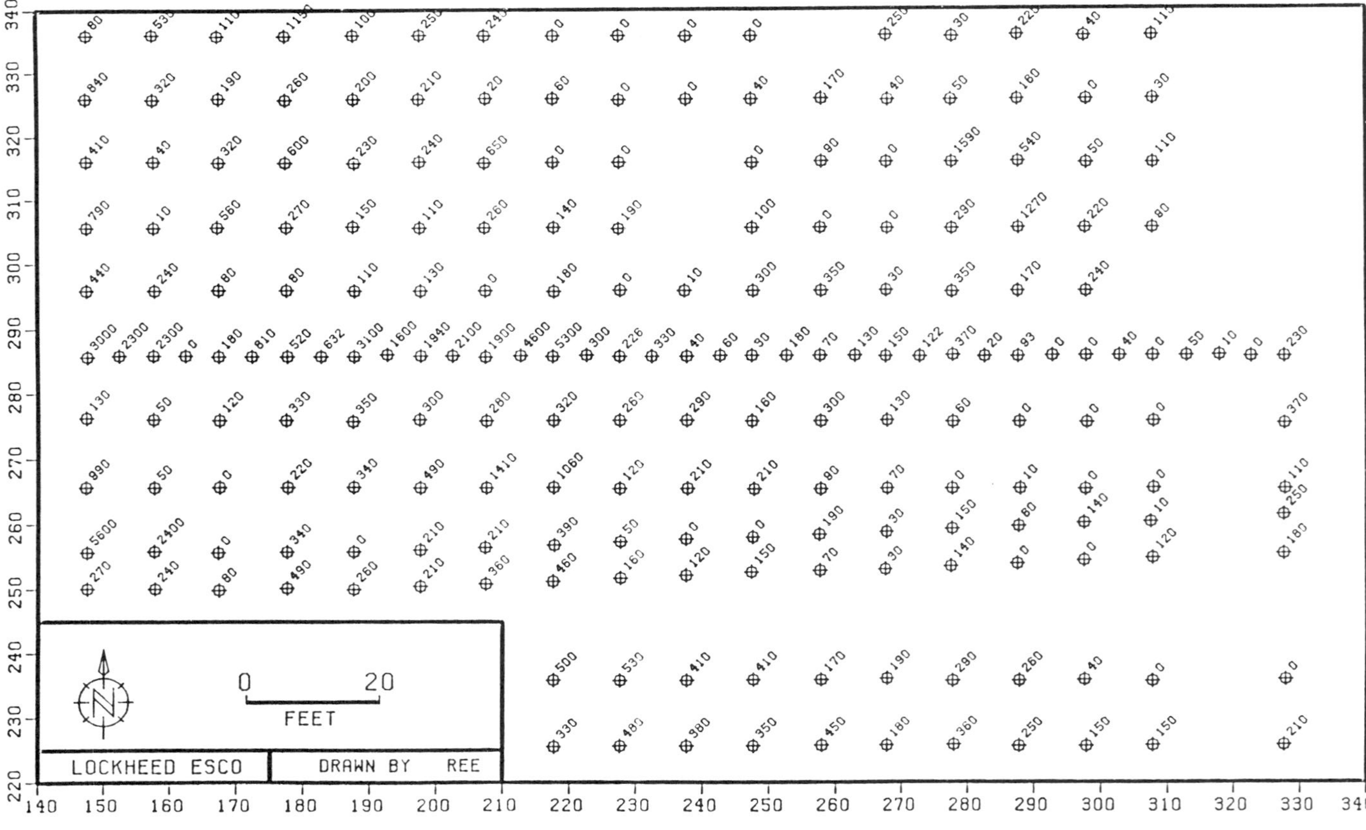

Fig. 2. Case study site with sample locations. lead concentrations in mg/kg.

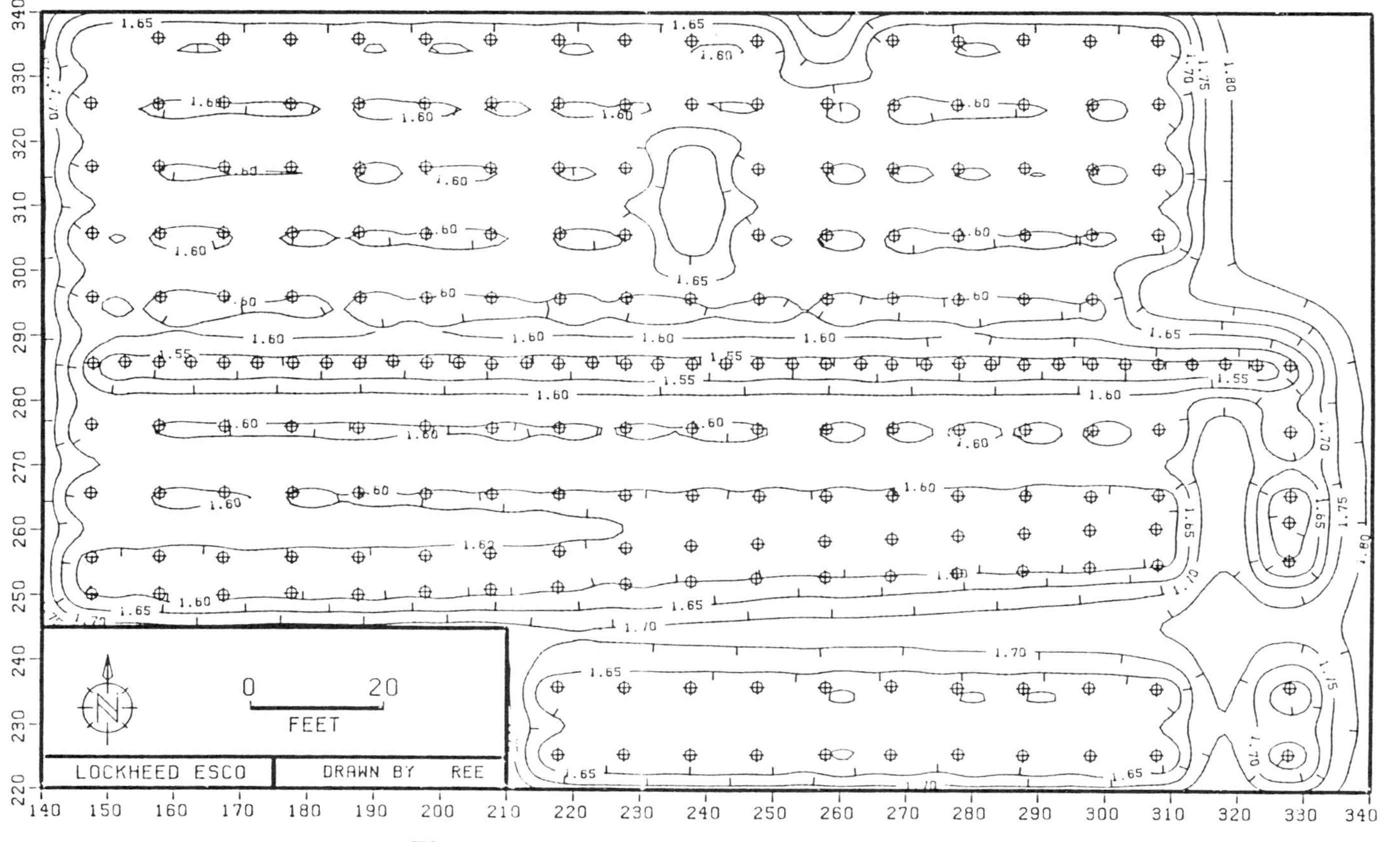

Fig. 3. Directional logarithmic semivariograms.

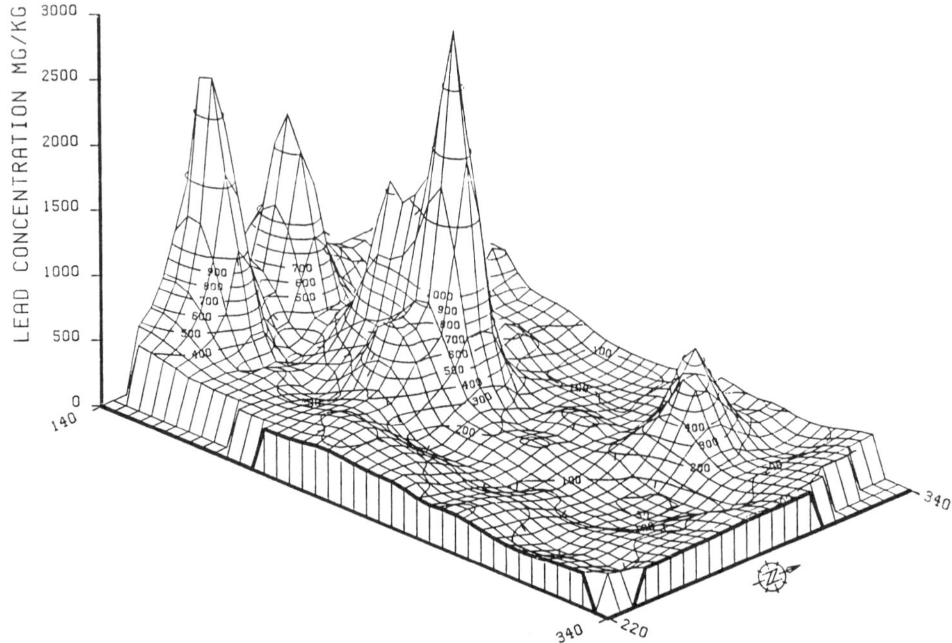

Fig. 4. Spatial lead model with isopleth projections. View from the southeast to northwest.

Lead isopleths were interpolated through the grid model as shown in Figure 4, and are graphically displayed as contours in the map of Figure 5. The contours clearly delineate three fairly small zones with elevated lead concentrations in the western portion of the area. One small zone with slightly elevated concentrations occurs toward the northeast corner. Kriging errors, expressed as the ratio of kriging standard deviation divided by the local mean concentration (represented by the kriged estimate), were calculated for each grid node. Isopleths of these errors appear as contours in Figure 6. As expected, errors increase away from sampling locations. In the case of a regular sampling pattern such as the one utilized in this case study, the pattern of errors is also very regular and predictable. However, if the sampling pattern is very irregular, the pattern of errors may be complex and necessitate a map to facilitate evaluation of the reliability of spatial interpolation in different portions of the site. Such a map can also be used to optimize the design of subsequent sampling. The most cost-effective future samples should be taken at locations with large kriging errors since these locations

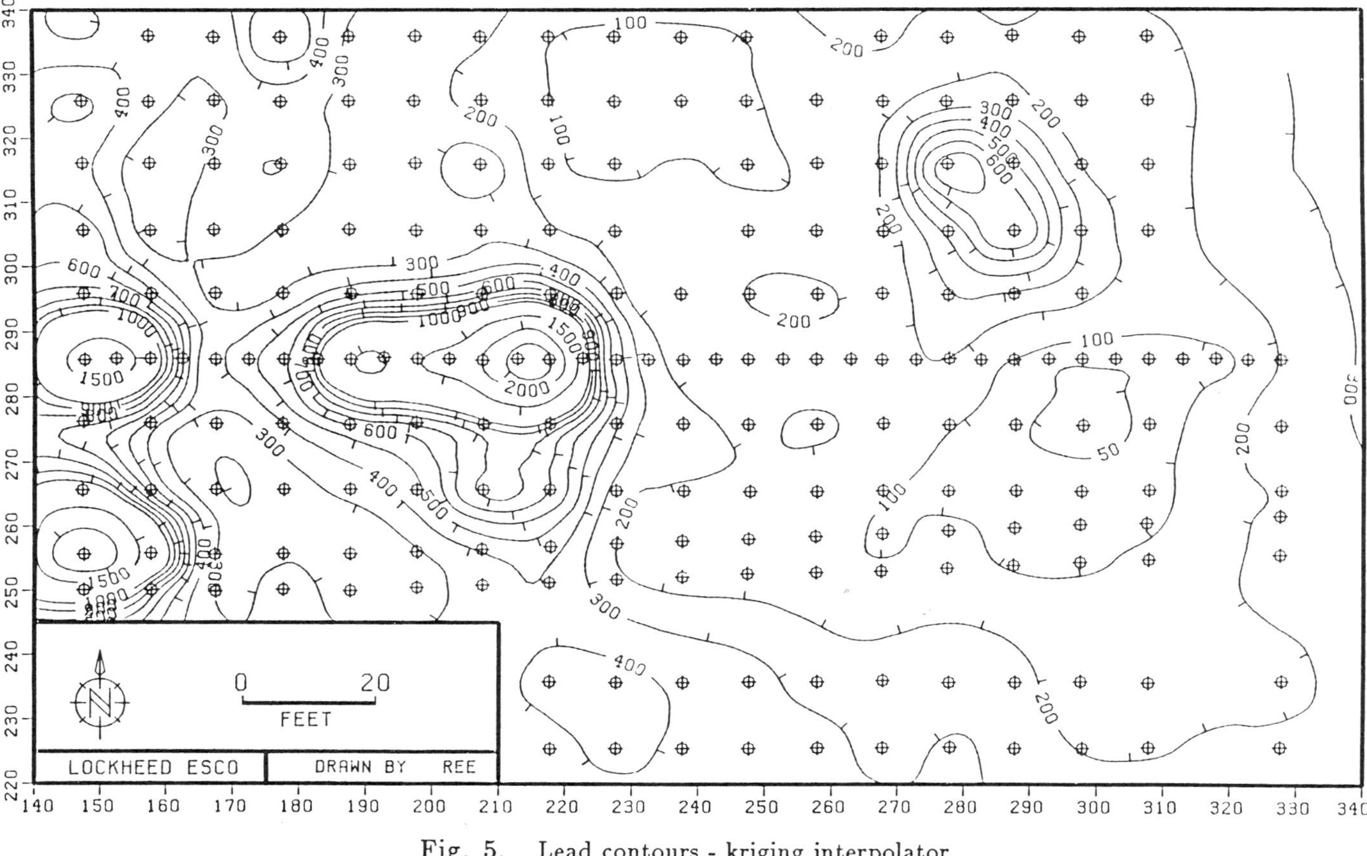

Fig. 5. Lead contours - kriging interpolator.

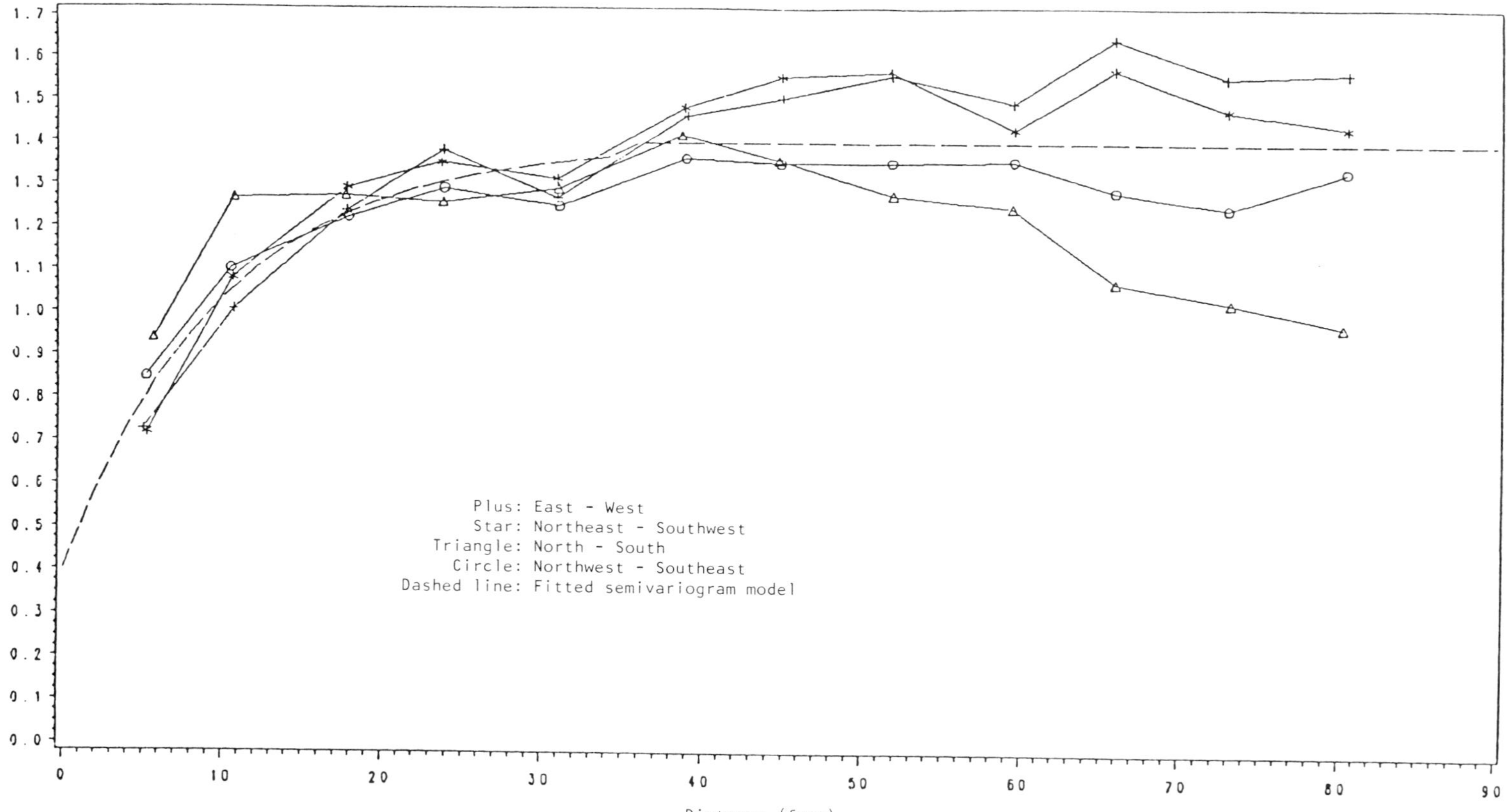

Fig. 6. Interpolation errors. Kriging relative standard deviation.

Table 3. FPXRF Detection Limits for Several Sites (Raab et al., in press)

	standard deviation	Pb	As	Cu	Cr
		All values in mg/kg			
1	sigma	25	8	50	15
3	sigma = IDL	75	24	150	45
10	sigma = LOQ	250	80	500	150

will result in maximum reduction of kriging errors. The reader should be cautioned against attaching probability levels to kriging standard deviations because estimation errors incurred in linear kriging are autocorrelated and do not necessarily follow a normal distribution.

DETECTION LIMITS AND LIMITS OF QUANTIFICATION

The American Chemical Society (ACS) defines a limit of detection as "...the lowest concentration level that can be determined to be statistically different from a blank" [2]. This is further defined as three times the standard deviation of a series of blanks (3 sigma).

The ACS defines a limit of quantitation (LOQ) as "...the level above which quantitative results may be obtained with a specified degree of confidence" [2]. The recommended value for the limit of quantitation is 10 sigma. This is said to correspond "...to an uncertainty of ± 30% in the measured value (10 sigma ± 3 sigma) at the 99% confidence level" [2].

A value falling below 3 sigma is considered to be "not detected"; this is not to be interpreted as a zero level of concentration but merely below the sensitivity of the instrument. Values between 3 sigma and 10 sigma are considered to be in the "region of less-certain quantitation". Values greater than 10 sigma are considered to be in the region of quantitation. The ACS states quite emphatically that "...quantitative interpretation, decision making and regulatory actions should be limited to data at or above the limit of quantitation." See Table 3 for examples of detection limits on four metals we attained on several of our sites [5].

The above discussion is supplied as a source of information. More important, though, is that this program is designed to be a screening technique. In that light, the major emphasis should not be placed on individual samples but rather on the trend of analyte concentration over the area of interest.

Table 4. Routine Samples for an Fe Site. (Raab et al., in press)

SAMPLE	Laboratory-grade	EDXRF	FPXRF
	Pellets	Loose soil in cup	Loose soil in cup
	All values in wt.% Fe		
#1	0.56	0.53	0.54
#2	0.73	0.68	0.75
#3	0.76	0.70	0.93
#4	0.71	0.60	0.67
#5	0.87	0.70	0.99

ROUTINE SAMPLING

On the case study site we conducted 750 measurements in a three-day time period. The most time-consuming portion of our on-site work is the surveying of the sample locations. In reality, the actual analysis goes very fast.

In order to further optimize a field sampling plan, we have developed a new approach to sampling contaminated sites: we collect physical samples for only those routine samples which fall below the LOQ and send to a laboratory for corraboratory analyses. Table 4 shows the comparability of the same samples analyzed with the XRF and FPXRF units. Technicians split the samples, and grind one to approximately 0.053 mm. They place the pulverized samples into a pellet die and press it for 5 minutes at 30,000 pounds per square inch. The other sample is untreated. The pellet theoretically gives us the greatest accuracy. We see that the FPXRF measurement yields very close values to both sets of analyses in the laboratory. The FPXRF measurements reflect the average of three separate measurements.

Table 4 shows that the FPXRF analyses of high concentrations demonstrate a high degree of correlation to the laboratory results [5]. Therefore, we need not collect samples with field values of concentrations higher than ten sigma (= LOQ). The questionable data are those field values below ten sigma. The advantage in using the FPXRF develops on those sites where the majority of concentrations are higher than ten sigma.

CONCLUSIONS

We can effectively use SIS measurements to routinely analyze soils on hazardous waste sites providing valuable information in real time with defensible quality assurance procedures.

We can successfully measure high concentrations without corroboratory analyses. This allows field personnel to examine real-time data before collecting physical samples.

We can measure hundreds of locations in a short period of time and never collect a physical sample.

We can use geostatistics to optimize sampling plans.

We can provide a document to the regional project manager with calibration curves, routine and quality control data, aerial photographs, and concentration isopleth maps.

ACKNOWLEDGEMENTS

The authors would like to thank L. A. Eccles, J. J. D'Lugosz, and S. J. Simon for their advice and help in reviewing this document. We are indebted to Peggy Dalheim from Mineralab in Denver, Colorado, for providing analyses at a critical time.

REFERENCES

[1] Buckley, D.E. and Cranston, R.E., 1971. Atomic Adsorption of 18 Elements from a Single Decomposition. Chem Ged. 7: 273-284.

[2] ACS Committee on Environmental Quality, "Principles of Environmental Analysis" Anal. Chem. 1983, 55, 2210-2218.

[3] U.S. EPA. 1987. Data Quality Objectives for Remedial Response Activities: Development Process. EPA/540/G-87/004. U.S. Environmental Protection Agency, Washington, D.C., 154pp.

[4] David, M. 1988. Handbook of Applied Advanced Geostatistical Ore Reserve Estimation. Elsevier Scientific Publishing Co. New York. 216pp.

[5] Raab, G.A., C.A. Kuharic, W.H. Cole II, R.E. Enwall, and J.S. Duggan, 1990. In: Advances in X-Ray Analysis, Volume 33, Proceedings of the Thirty-eighth Annual Conference on Application of X-Ray Analysis, held July 31-August 4, 1989 in Denver, Colorado. Plenum Press.

tion of samples but to their analyses as well. In the subsequent laboratory analysis of a sample which may have been very well collected, preserved, and transported to the laboratory, each of the above factors plays an analogous role: (1) sorption or degradation in the sampling container and analytical transference device, e.g., syringe, pipet, or beaker; (2) changes in temperature or pressure (particularly applicable to the extreme pressure/temperature changes occurring in the syringe and then the GC itself); (3) cross-contamination of syringes, purge and trap devices, sample lines, injectors, columns, and detectors.

The problem of sorption of organics in sampling devices and in the passage of samples through analytical tubing was addressed in an article by Barcelona et al. [6]. In that discussion, the sorption of various organic liquids in different organic materials is well documented [1]. This problem is seemingly one of a particularly Sisyphean nature; i.e., the containment of a substance within a like substance is akin to rolling a stone up a hill only to have it fall immediately down the other side. Certainly the problem encountered by the industry in attempting to contain petroleum products in unlined fiberglass tanks attests to this dilemma. The development of permeation tube calibration systems is based on the phenomenon [4].

The present method addresses all of the problems mentioned in that, to put it most succinctly, it combines immediacy and simplicity of analysis. That is, it is easily transportable to the field, which eliminates problems of sample transfer and storage, and it provides an immediate analysis of a large volume of sample, which speaks generally to the problem of representativeness.

THE EXTRACTION/COLORIMETRIC TECHNIQUE FOR AROMATICS

The Hanby Field Test Method for aromatics in water described here comes in the form of a kit complete with necessary reagents and apparatus to perform immediate analyses at the groundwater well site. It is contained in a rugged plastic case with enough reagents to perform thirty field analyses. Within the case are contained: a 500-ml separatory funnel, a tripod ring stand, a 10-ml graduated cylinder, 2 reagent (liquid) bottles, one desiccant jar with 30 reagent (powder) vials, a color chart depicting test results for eleven typical aromatics, plastic safety glasses and 12 pairs of gloves. Upon arrival at the site the kit is opened and the tripod ring stand is assembled. A 500-ml water sample is introduced into the separatory funnel which is placed in the ring stand. Next, 5 ml of the extraction reagent is poured into the separatory funnel using the 10-ml graduated cylinder. The sample is vigorously extracted for two minutes with occasional release of the slight pressure buildup which occurs. The funnel is placed back in the ring stand and the extraction phase is allowed to separate to the bottom for five minutes. After phase separation is complete the lower extraction layer is drained into a test tube, allowing a small amount of the extraction solvent to remain in the separatory funnel. Then one of the reagent vials is opened and the contents immediately poured into the test tube. The tube is shaken for two minutes allowing the catalyst to be dispersed well throughout the extraction reagent so that color development, which is concentrated in the powder, will be uniform. Hue and intensity of the color of the catalyst which has settled in the tube is now compared to the standard aromatics pictured in the color chart.

The wide range of intense colors produced in Friedel-Crafts reactions has been observed since the discovery of this reaction. A brief description of the reaction, as well as the color involved, is given by Shriner et al. [5]. In this novel adaptation of Friedel-Crafts alkylation chemistry, one of the reactants, the alkyl halide, is used as the extractant. The alkyl halide extractant plus the aromatic compound present in the water sample are caused to form electrophilic aromatic substitution products by the Lewis acid catalyst, which is added in great enough amount to also act as the necessary dehydrant to allow the Friedel-Crafts reaction to proceed. These products are generally very large molecules, i.e., phenyl groups clustered around the alkyl moiety, which have a high degree of electron delocalization. These two factors are the principal reasons for the extreme sensitivity of this procedure. That is, large molecules are produced which are very intensely colored.

In the field conditions where this procedure is by and large carried out, the reaction is exposed to sunlight. This means that there will be a "window" in which to observe the color that is produced. This is because of the general instability of the reaction products to photochemical oxidation. Strong sunlight will cause most of the colors produced to fade to various shades of brown within just a minute or two; therefore, it is advisable to perform the test in a shaded area.

PURGE AND TRAP GC COMPARISON STUDIES

Comparison of the Field Test Kit method to analyses performed with a purge and trap GC were made using standard solutions of benzene, toluene, ethyl benzene and o-xylene (BTEX) the purge and trap/GC used for this study was a Tekmar LSC-3 and a Hewlett-Packard 5890, Supelcowax widebore capillary 60-meter column, programmed from 45°C (for 3 min.) to 120°C @ 8°C/min. A five ml. aliquot of the standard concentrations, 0.2,

1.0, and 10.0 ppm, was injected into the purge vessel of the Tekmar. The sample was purged with helium for 10 minutes at 180°C. Purge flow was twenty ml/min helium. Two separate preparations of the BTEX standards were analyzed by the Field Test Kit and by purge and trap GC. Peak areas were compared for all chromatograms and these results were normalized with regard to the theoretical response ratio of 0.2, 1.0 and 10.0. This simple average error calculation showed a typical purge and trap GC variation in analysis of ±9.1%.

The results of the Field Test Kit method judging by comparison to the color chart are, of course, somewhat subjective. Rigorous evaluation of color intensities are being conducted using exact quantities of reagent and further UV/VIS spectrophotometric reflectance readings. These studies are being continued on a Varian DMS 300 as well as on a Cary 2200 UV/VIS/NIR spectrophotometer.

APPLICATIONS OF THE METHOD

Obviously, this method will have a wide variety of applications in field investigations. Recent regulations for the monitoring of underground storage tanks require that soil/groundwater investigations be carried out regularly to insure that no leakage has occurred. It can be seen that the use of this technique which is easily learned and can be performed at an extremely low cost will provide an immediate and definitive answer to these requirements.

OHIO RIVER STUDY

In the evening of January 2nd, 1988, the collapse of a tank containing approximately 3.5 million gallons of diesel fuel precipitated one of the worst inland oil spills in the country's history. Approximately one million gallons of the oil washed in a huge wave over the containing dikes around the tanks at the Ashland Oil plant at West Elizabeth, Pennsylvania, and into the Monongahela River.

Monday morning, two days after the spill, I contacted Mike Burns of the Western Pennsylvania Water Company in regard to using the Hanby Field Test Kit at the company's water treatment facility on the Monongahela south of Pittsburgh. Mike asked me to bring one of the kits to the plant. The next day I flew into Pittsburgh and was met by Mike at the West Penn Water Works Treatment facility where I demonstrated the use of the kit for

Table 1. Ohio River Sampling for Diesel Oil, January 7, 1988

Ohio River Mile Point	Fluorometer Reading	Hanby Field Test (ppm)
85.5	6	0.20
85.0	8	0.20
85.0	11	0.20
84.5	6	0.20
84.5	10-15	0.50
84.5	10	0.20
82.0	20	1.00
81.0	25	1.00
80.0	30	1.50
79.0	33	2.50
77.0	57	10.00
76.0	43	8.00
75.0	35	7.00
74.0	48	5.00
70.0	29	3.00

the personnel at the plant. Then Mike suggested I call John Potter, the chief chemist at the Water Treatment Plant in Wheeling, West Virginia, which was the next major facility taking water from the Ohio River. John said the kit sounded like it would fill a real need for a rapid analysis of the river water at the plant's intakes. The next day, Wednesday, I was demonstrating use of the kit to the personnel at Wheeling. It was immediately put into use on a round-the-clock basis when they realized that in just a few minutes they could get visual indication down to 100 ppb of the diesel aromatic components.

The next morning I met with the West Virginia Department of Natural Resources personnel who were in Wheeling to monitor the oil spill. That afternoon I was invited by the office of the EPA in Wheeling to join EPA chemist Bob Donaghy, West Virginia Department of Natural Resources Inspectors Sam Perris and Brad Swiger, and the Ohio River Valley Water Sanitation Commission Coordinator of Field Operations, Jerry Schulte, on the river tugboat Debbie Sue to make a run up the Ohio River from Wheeling to try to locate the front of the spill. We continued approximately eighteen miles up the Virginia side, mid-channel, and the Ohio side. As Table 1 indicates, the results from the EPA fluorometer and the Hanby Field Test Kit tracked each other fairly consistently at each point.

VALDEZ OIL SPILL

At 12:04 a.m., March 24, 1989, the oil tanker Exxon Valdez ran onto Bligh Reef in Prince William Sound. Of the 1.26 million barrels of Prudhoe Bay crude oil the ship was carrying, approximately 10.1 million gallons immediately poured out of the ruptured tanks into the clear blue water of the sound. Of the series of unfortunate circumstances involved in this event, such as the unpreparedness of crew, port and pipeline officials to immediately begin containment efforts, perhaps none was more critical than the imminent release of hundreds of millions of salmon fry which had been hatched and raised in the half-dozen fish hatcheries in and around Prince William Sound.

On Monday, March 27, I contacted the Alaska Department of Environmental Conservation in regard to use of the test kit for on-site monitoring of aromatic contamination of the water caused by the crude oil. Prior analyses of Prudhoe Bay crude had revealed that in comparison to other sources of oil it was particularly high in aromatic content (25%) with the majority components of this fraction being: naphthalenes (9.9%), phenanthrenes (3.1%), and pyrenes (1.5%).[1] Mr. David Kanuth of the Alaska D.E.C. had obtained one of the test kits in January and had subsequently used it at a smaller oil release several weeks later. It was decided that the kit would be helpful at the Valdez incident and commissioner Dennis Kelso requested that I fly a kit to three of the hatcheries which would probably be most impacted by the spill. On Thursday, March 30, at 8:30 a.m., Dick Fanell of the D.E.C., Mark Kuwada of the Alaska Fish and Game Department, and I took off in a pontoon-equipped Cessna 206 piloted by Ken Lobe and flew SSW from the Valdez harbor to the three fish hatcheries we were scheduled to visit. Within minutes of being airborne we were able to see the huge fingers of oil spreading southerly from the area where the tanker was still impaled on the reef. Thirty minutes later we landed in the bay at the first hatchery, Esther.

After demonstrating the use of the test kit to the hatchery manager two samples were obtained from the water near the fish pens. These samples were analyzed with the test kit, and the indications using the color chart results for gasoline as a guide were 0.2 ppm and 1.5 ppm. These results had not been expected as no visible oil had been apparent from the air as we had flown into the hatchery.

The next hatchery visited was the Main Bay Hatchery. A test kit was also left there following demonstration of its usage. One sample was taken near the fish pens which indicated a concentration of 0.5 ppm aromatics. The third landing was made in the water of the Port San Juan hatchery. Two samples were taken indicating concentrations of 0.5 and 0.6 ppm.

---

[1]Information supplied by the Alaska Department of Environmental Conservation.

The following day a return to Houston was made in order to prepare a comprehensive series of standards from 125 ppb to 20 ppm of the Prudhoe Bay crude in seawater. It was necessary for this work to be performed at the Houston laboratory as a large assortment of glassware and appropriate reagents were required to simultaneously prepare the eleven different standards which were utilized in the study.

The procedure for the eleven concentrations employed called for first preparing a solution of the Prudhoe Bay crude in hexane (5:100), then making a 100 to 1 dilution of the hexane stock in acetone. This resulted in a 500-ppm stock which was then used to prepare 500-ml standards of: Blk 125, 250, 500, 750, 1000, 2000, 3000, 4000, 5000, 10000, and 20000 ppb. When the 500-ml standards were all prepared (in Galveston Bay saltwater) each was extracted according to the test kit method with 5-ml extractant reagent for two minutes. After all extractions had been collected in 16×100-ml test tubes, the color development catalyst was added and the tubes were shaken for two minutes in a darkened room. Upon completion each tube was placed in a rack previously prepared and labeled, then the rack containing the twelve tubes was taken into a sunlit room for the photographs (f16. 125th sec). This whole procedure was repeated twice more as a quality control measure.

A remarkable difference in hue is observed as the concentration of the crude oil in the water increases. At approximately 1 ppm the blue colors of the polynuclear aromatics begin to neutralize the more orange colors of the single-ring aromatics. This effect is particularly noticeable in the extractant (liquid) phase of the test tube so that the liquid actually appears colorless above the gray color of the catalyst. In the photographs taken with the liquid phase masked an evenly progressive increase in intensity of color or saturation is evidenced.

Enlarged copies of these photographs were sent to the various entities utilizing the Field Test Kit including the Alaska Department of Environmental Conservation, the Department of Fish and Wildlife and eight of the fish hatcheries themselves. Due to the extent of the spread of the oil and its contamination of the surrounding shoreline it must be assumed that the partitioning of these aromatic components into this prolific marine area will continue for quite a long period.

SOIL STUDIES

Site investigations at leaking underground soil tanks, hazardous waste sites, surface spills, etc., normally involve at least preliminary evaluations of the soil prior to the installation of monitoring wells. The use of the Hanby

Field Test Kit for soils has found wide application in this regard and has proven to be far superior to methods such as soilgas techniques involving the use of OVA's or field-portable chromatographies to sniff the headspace of samples.

The technique developed at Hanby Analytical Laboratories for use of the kit is as follows:

1. Place 100 grams well-crumbled soil and one packet of the flocculant salt in a quart jar.

2. Add 500 ml of distilled water and shake well for 15-20 minutes so that water/soil is very well mixed.

3. Pour mixture into a 1-liter Imhoff cone. Allow mixture to clarify for thirty minutes.*

4. Decant 250 ml of the clarified portion of the waterwash into the Field Test Kit separatory funnel and perform the FTK extraction/colorimetric test.

* Place a piece of Parafilm or aluminum foil over each Imhoff cone to keep volatilization losses minimized.

UV/VIS SPECTROPHOTOMETER STUDIES

Determinations of principal wavelengths and reflectance data were made in correspondence with aromatic compounds depicted on the color chart. These investigations were conducted by preparing a range of concentrations of selected aromatic compounds, performing the Hanby extraction/colorimetric procedure and then immediately measuring the reflectance of the catalyst.

METHOD

Ten parts-per-million (vol/vol) solutions of benzene, toluene, o-xylene, special unleaded gasoline, naphthalene and diesel were prepared by injecting 20-microliter amounts of each compound into 2.0 liters of de-ionized water at 20°-21°C and stirring for one hour. Dilutions from the stock solutions were prepared to 0.01, 0.02, 1.0 and 5.0 ppm. The extraction/colorimetric procedure employed with the Field Test Kit was modified to fit the requirements of the UV/VIS reflectance apparatus. Four ml of the extraction reagent were used to extract the water samples for two minutes. The extraction solvent was then drained into a cuvette. Two grams of the catalyst material was added to the cuvette which was covered with its teflon cap, and the mixture was shaken vigorously for three minutes. The cuvette was placed in the spectrophotometer and scanned over a range of 350 nm to 600 nm.

INSTRUMENTAL PARAMETERS

For this study a Varian DMS 300 UV/VIS spectrophotometer was utilized. Instrument settings were: Slit width 2 nm, tungsten source, scan rate 50 nm/min. All of the scans were corrected to 100% transmittance baseline using a blank sample which was scanned in reference to a barium sulfate reflectance disk. The sample compartment was fitted with a diffuse reflectance accessory which was modified by blocking out the top portion of the light path so that only the catalyst in the bottom half of the cuvette would be scanned.

CONCLUSIONS

The development of a field method for the analysis of organic contaminants at sub-part-per-million levels in water has been proven to be a valuable tool in the establishment and the sampling of groundwater monitoring wells. The accuracy of the method has been proven to far exceed that of direct injection gas chromatography. A rapid soil-wash method has also been developed employing the Hanby Field Test Kit technique which has proven to be effective on top and deep soils over a range of 5 mg/kg to 10,000 mg/kg gasoline in soil. Development of instrumental spectrophotometric techniques will allow even greater sensitivity and qualitative analysis of aromatic contaminants in soil and groundwater.

A variation of the procedure involving the extraction of a sample with an aromatic solvent and then addition of the Lewis acid catalyst allows for the determination of the presence of alkyl halides, e.g., trichloroethylene. In this version of the test a reflectance adapter for the spectrophotometer is not necessary since the color is not concentrated in the catalyst but is developed in the extractant solvent.

REFERENCES

[1] TRC Environmental Consultants, Inc. Laboratory Study on Solubilities of Petroleum Hydrocarbons in Groundwater, American Petroleum Institute, Washington, DC, 1985.

[2] Keith, L.H., Identification and Analysis of Organic Pollutants in Water, Ann Arbor Science Publishers, Inc., Ann Arbor, 1981.

[3] Johnson, R.L., Pankov, J.F., Cherry, J.A., Design of a Ground-Water Sampler for

Collecting Volatile Organics and Dissolved Gases in Small-Diameter Wells. Ground Water, V. 25, pp 448-454, 1987.

[4] O'Keefe, A.E., Primary Standards for Trace Gas Analysis. Analytical Chemistry, V. 30, pp 760-768, 1986.

[5] Shriner, R.L., Fuson, R.C., Curtin, D.Y., Morrill, T.C., The Systematic Identification of Organic Compounds, John Wiley & Sons, New York, 1980.

[6] Barcelona, J.J., Sample Tubing Effects on Ground-Water Samples. Analytical Chemistry, V. 30, pp 460-465, 1985.

THE ROLE OF LUMINESCENCE AND SPECTRAL PATTERN RECOGNITION IN ENVIRONMENTAL PROGRAMS

D. EASTWOOD and R. L. LIDBERG

Lockheed Engineering and Sciences Company
Suite 301, 1050 E. Flamingo Road, Las Vegas
Nevada 89119

K. J. SIDDIQUI

Creighton University,
Dept. of Math and Computer Sciences
California at 24th Street, Omaha
Nebraska 68178

Fluorescence/luminescence spectroscopy has been used successfully for a number of years for such environmental applications as analysis of polyaromatic hydrocarbons (PAHs) and forensic oil identification. Use of a variety of spectral pattern recognition techniques has improved the applicability of luminescence for classificationand identification purposes in addition to its uses for quantitation. Other spectral techniques such as infrared have been more extensively investigated by pattern recognition techniques but differences in spectral structure and information content make direct studies on luminescence spectral libraries of hazardous chemicals important.

The present paper reviews and summarizes the current state-of-the-

Chemistry for the Protection of the Environment
Edited by L. Pawlowski *et al.*, Plenum Press, New York, 1991

art of luminescence as applied to environmental programs in the United States. Recent work performed by the authors and other researchers in forensic identification, classification and quantitation of petroleum oils and hazardous chemicals, and related pattern recognition techniques are described. Current studies by the authors of real-world samples involving petroleum oil samples and extension of such identification techniques to other hazardous chemicals such as polychlorinated biphenyl mixtures are discussed. Suggestions are made for future objectives and directions for luminescence research to aid in environmental analysis for hazardous chemicals.

Oil spill identification and classification for forensic purposes by a variety of techniques including fluorescence were important areas of research in the United States in the 1970s. Much of this work was funded by the United States Coast Guard who found that fluorescence was the simplest, fastest and most economical method of the multi-method approach tested. Luminescence applied to oil identification was reviewed in an article by Eastwood [1]. Pattern recognition approaches for fluorescence spectra of petroleum oils were discussed in articles by Killeen et al. [2] and Chien and Killeen [3]. More recently interest in oil identification and classification has revived in part due to North Sea oil explorations [4] and in part due to renewed funding of the Superfund program to clean up hazardous waste sites in the United States. Remote sensing has also been applied with considerable success to classification of oil spills [5].

Instrumental and analytical techniques which have been widely used to improve luminescence sensitivity and selectivity for environmental applications include: synchronous, variable synchronous (and constant energy separation) and derivative spectroscopy, laser excitation (often combined with matrix isolation techniques with inert gases or Shpol'skii matrices), fluorescence lifetime determinations, room temperature phosphorescence in controlled media (including heavy metal sensitized micelles, cyclodextrins, colloids, and soil substrates), low-temperature measurements between 4°K and 77°K, and generation of excitation-emission matrices to aid in resolving mixtures. Often, these techniques are combined and are supplemented by increasingly sophisticated computer algorithms to aid in determination of fluorescent components of mixtures by a variety of pattern recognition and chemometric techniques. These methods have been used on real-world samples either directly or as hyphenated techniques in combination with separatory techniques such as high performance liquid chromatography (HPLC) or thin layer chromatography (TLC), often with photodiode arrays or multichannel optical analyzers as detectors. These publications and authors are too numerous to mention but are documented in the cited review articles and publications [6-11].

Chemicals of environmental interest to which these techniques have been applied include, but are not limited to, the following: petroleum oils, PAHs, metal complexes, humic and fulvic acids, polychlorinated and polybrominated biphenyls (PCBs and PBBs) and other halogenated aromatics, and phenols.

Applications for comparison with the work of the authors would include Shpol'skii, quasilinear or matrix isolation spectra of petroleum oils by scientists such as Wehry [12] and room temperature fluorescence synchronous spectra of coal liquids by Vo-Dinh [13]. Specific applications for polychlorinated biphenyls would include earlier low-temperature luminescence studies by Brownrigg and Hornig [14] and room temperature phosphorescence studies using micelles and cyclodextrins by Cline-Love [15,16].

Data compression by Fourier transformation of fluorescence spectral files was performed by Faulkner and co-workers in 1977 [17] (unfortunately on uncorrected spectra). In 1985, Rossi and Warner [18] reported an algorithm for fluorescence spectral matching of excitation emission matrices (EEM) which evaluated spectral matching by correlation and intervector distances between Fourier spectra of EEMs.

Earlier, Eastwood and co-workers at the Coast Guard R&D Center had studied many weathered and unweathered oils by fluorescence/low-temperature luminescence (LTL) and had documented oil emission, excitation and synchronous spectra [1,19]. Other techniques which were also investigated included derivative spectroscopy, contour or total luminescence spectroscopy, polarization and lifetime studies, and use of special solvents. Killeen [2] used a vector analysis approach to interpolate fluorescence spectra and to correct for weathering.

Sogliero and Eastwood [20] also applied pattern recognition considerations to the fluorescence/luminescence spectra of hazardous chemicals. A computer library of digitized LTL and fluorescence spectra was generated and the effect on spectral quality of instrumental parameters was assessed using similarity measures such as the normalized area under the spectral curve and the power (sum of squared intensities of the spectrum). They [21] found that, for a library of LTL spectra of 60 hazardous chemicals, a successful feature set consisted of only 6 components (the first 4 noncentral sample components of the spectrum, the normalized area under the spectral envelope, and the wavelength corresponding to maximum emission intensity). This feature set performed well in a test using cluster analysis involving 2000 pairwise comparisons of the spectral features selected. Mulkerrin and Wampler [22] also used moments of fluorescence spectra for pattern recognition.

In more recent work Eastwood and Lidberg [23] applied fluorescence spectroscopy to screening, classifying and semiquantitating real-world hazardous waste samples containing oil from sites in Alaska. Fluorescence spec-

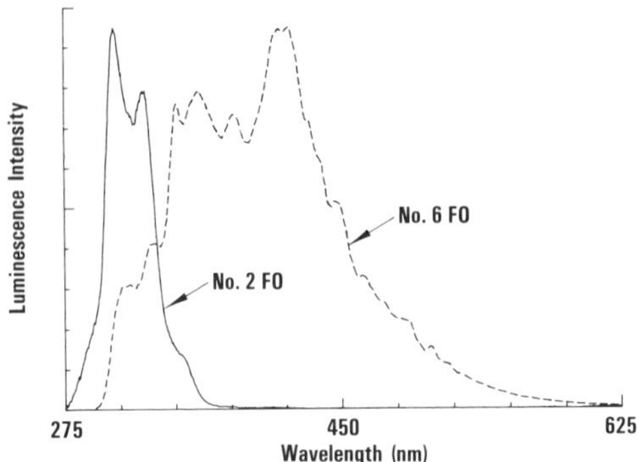

Fig. 1. Synchronous emission spectra of a No. 2 and No. 6 fuel oil. 20 μg/mL in cyclohexane, $\Delta\lambda = 25$ nm, bandpass - 4.5, 0.9 nm, room temperature.

tral libraries were developed in-house for reference petroleum oils. The pattern recognition approach was similar to that of Killeen et al. [2]. It was found that the additional structure of the synchronous spectra was especially useful for distinguishing between No. 6 fuel oils and heavier crudes.

Figure 1 demonstrates the use of synchronous luminescence spectroscopy in the classification of petroleum oils. There is a clear distinction between the No. 2 and No. 6 fuel oils. The use of synchronous emission spectroscopy, $\Delta\lambda = 25$ nm, leads to increased structure when compared to standard emission spectra exciting at 254 nm [23]. Low-temperature luminescence of PCB mixtures and chlorinated biphenyl isomers in a heptane (Shpol'skii) matrix, Figures 2 and 3, provides increased detection limits because of increased phosphorescence emission yield. The Shpol'skii matrix also provides enhanced spectral structure which improves pattern recognition capabilities. The pattern recognition of each structure is made by partitioning the feature space [24,25].

These spectral pattern recognition methods can be grouped into two main categories:

1. Waveform (template) matching and correlations (including the vector analysis approach). This technique takes as features the amplitude of the waveform at equal intervals and simply measures the similarity between the unknown waveform and the stored references (templates) by matching and correlating points at different intervals. The problem, arising from using the points themselves as features, is high dimensionality of the resulting feature-vector, implying higher computational cost and storage.

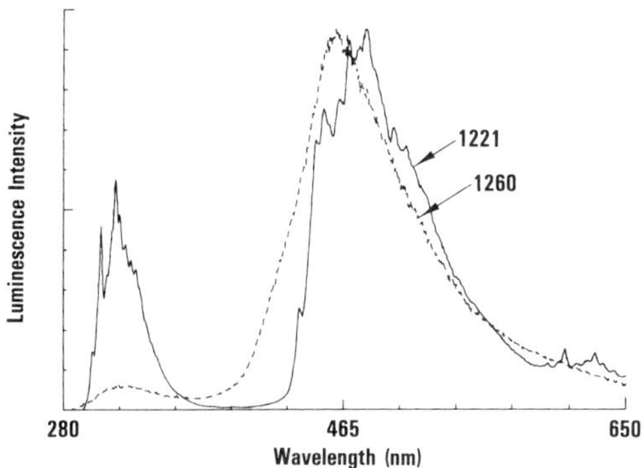

Fig. 2. Low temperature fluorescence and phosphorescence of commercial PCB mixtures, 1221 and 1260. 15 μg/mL in heptane, temperature = 77 K, ex = 270 nm, bandpass = 4.5, 0.9 nm, cutoff filter = 305 nm.

2. Transformation and series expansion. Attempts have been made to reduce the dimensionality and extract features invariant to some noise by transformations and series expansions. Although, in the theory of pattern recognition a number of transformations such as Fourier, Walsh, Haar and Hadamard, as well as Karhunen-Loeve series expansions, have been explored [24-26], in luminescence spectroscopy only the Fourier transform and the Kalman filter have been investigated [28]. Both of the above methods are characterized by their high sensitivity to noise. From a practical point of view, it is relatively easy to build masks for these techniques, even though the matching process which includes normalization and proper positioning of the waveform is somewhat complex. Such algorithms are slow and often rely on complementary techniques or extra features for higher performance.

In this paper, both these approaches were investigated and a pattern recognition system for the classification of luminescence spectra was designed. Complete details of this system will be presented elsewhere.

In both the methods discussed below, a set of luminescence data was created by digitizing the low-temperature luminescence spectra of six commercial PCB mixtures.

As an example of the first approach the vector analysis method was investigated on PCB mixtures and selected isomers, both by phosphorescence

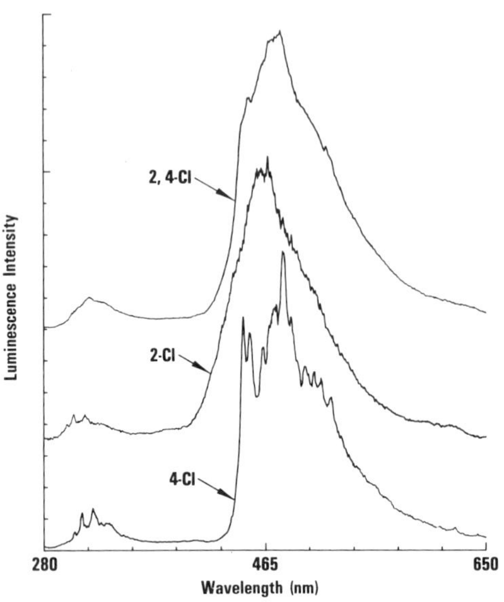

Fig. 3. Low temperature fluorescence and phosphorescence of chlorinated biphenyl isomers. 16 μg/mL in heptane, temperature = 77K, ex = 254 nm, bandpass = 4.5, 0.9 nm, cutoff filter = 280 nm.

only and by fluorescence and phosphorescence.

Each spectrum, digitized into n intensity values, was considered a vector in n-dimensional space. Two spectra with identical intensity values over the entire spectral range would have an angular distance in n-dimensional space of zero. Thus, as the difference in spectral structure increases, the calculated angle between the vectors also increases. The angle between two vectors $\overline{X}$ and $\overline{Z}$ is defined as

$$\theta_{x,z} = \arccos \frac{\overline{X} \cdot \overline{Z}}{|X|\,|Z|}$$

Table 1 shows the results of a vector analysis search of PCB mixture 1254 against the reference library. Angular distances are given for both fluorescence and phosphorescence (280–650 nm) and for phosphorescence only (370–650 nm).

Another approach in the second category of methods involved three phases: transformation, pattern analysis, and classification. These phases are described below.

All luminescence spectra of six commercial PCB mixtures were sub-

Table 1. Vector Analysis of Commercial PCB Mixture 1254 against PCB Reference Library

PCB Mixture	Angular Distance (Radians)	
	Phosphorescence[1]	Phosphorescence[2]
1254	0.0008[3]	0.0009[3]
1248	0.0965	0.1005
1260	0.1133	0.1137
1242	0.1541	0.1731
1016	0.1679	0.1760
1211	0.2179	0.3303

1) 140 Dimensional Vector
2) 185 Dimensional Vector
3) Note Computer rounding error

Table 2. Training Matrix Used in the Hierarchical Euclidean Distance Classifier

Mixtures	Features																
	1	2	3	4	5	6	7	8	9	10	11	12	13	14	15	16	17
16	1890	24	20	15	10	13	21	571	2655	11	0	0	0	0	1	1	2
21	2065	94	31	25	9	16	75	730	3152	141	5	1	0	1	1	2	8
42	2045	52	16	12	19	46	587	5707	119	5	0	0	1	2	2	4	
48	1767	25	38	20	25	34	24	475	2548	110	1	2	1	2	2	3	2
54	1679	30	30	24	21	28	30	404	2351	100	2	3	2	1	2	3	3
60	1573	34	24	26	20	17	31	346	2214	59	2	1	1	3	2	1	2

1) Entries are multipled by 1000
2) Features 1-8: represent areas of 8 zones with 4 zones in quartiles 1 and 4 each,
9: total area (sum area in 4 quartiles),
10-17: represent maximum intensities in the zones listed in 1-8

jected to Discrete Hartley Transform (DHT). The transformation is usually required to convert the high-dimensional data into a 2-dimensional form, to reduce the processing time without losing discriminatory information, and to make certain information more readily accessible in a particular mapping space.

The advantage of choosing DHT was that the Fast Fourier Transform (FFT) can be easily extracted and that it keeps all of the properties of the FFT. The computer program for DHT which we adopted from the work of Bracewell [24] ran significantly faster than FFT. Moreover, the simplicity of the transform is an added advantage and the transform does not involve complex multiplication. Usually, the selection of features is done on an ad hoc basis. The primary objective of the feature selection is the reduction of dimensionality of the input pattern without losing important discriminatory information.

Considering this objective, the features that we found discriminatory were area: under 4 zones of size 32 in the first and fourth quartiles; intensity: maximum intensity in each zone; and total area. (see Table 2).

To extract these features the DHT was divided into 4 quartiles, each with 4 zones of size 32. Quartiles 2 and 3 were ignored since they contain very little discriminatory information.

A feature vector was then extracted for each transformed and normalized spectrum so that each spectrum was represented by its feature-vector. These data were then arbitrarily divided into a training set and a testing set using standard methods [25, 26]. The discriminatory characteristics of the learned system, which are summarized in Table 2, were organized in a two-dimensional matrix called the training matrix.

To use a smaller feature set efficiently to discriminate six pattern classes, as opposed to a single-layer classifier, a simple hierarchical classifier was used for identification. Euclidean distance was used to determine the dissimilarity of a feature for two patterns at a given node. Using the hierarchical classifier based on a dissimilarity function, the unknown group of feature-vectors were classified, identifying one class at each level of the hierarchy until only two feature-vectors remained, which were identified at the next level. The classifier was able to identify all PCB mixtures correctly. This method is far superior to the vector analysis approach and is efficient and less complex computationally.

What does the future hold for environmental luminescence? Better instrumentation and optics and a wider choice of detectors (together with increased use of high-energy lasers combined with the techniques previously discussed) will increase the sensitivity (already down to a few molecules per unit volume) and selectivity of luminescence techniques. Use of fluorescence immunoassay with monoclonal antibodies, chemical derivatization, chemical sensors or optrodes, chemiluminescence, fluorescence quenching, and room temperature phosphorescence will increase the number of chemicals to be studied by molecular luminescence and, hence, the general utility of such techniques. Real-time and in situ measurements using fiber-optic chemical sensor (optrode) techniques (especially for monitoring ground water [9] or leachates) and improved techniques for remote detection of chemical spills by fluorescence will eliminate many current sampling and identification problems.

Better fluorescence/luminescence standards, interlaboratory studies leading to standard methods, and more accurately corrected spectrofluorometers and detectors will increase the accuracy and reproducibility of the techniques. Finally, the use of intelligent systems such as pattern recognition systems, knowledge-based systems and expert systems [28] will not only allow the spectroscopist to more easily select the appropriate chemometric techniques but also to perform them with increased accuracy and reliability. Multivariate factor analysis, cluster analysis, Fourier and other transforms, adaptive techniques and filters such as the Kalman filter, fractal theory, and many other computer algorithms will combine with better computer hardware and software and with better corrected fluorescence/luminescence spectral libraries of hazardous chemicals to simplify the task of the environmental analyst.

REFERENCES

[1] Eastwood, D. in *Modern Fluorescence Spectroscopy 4* (E.L. Wehry, Ed.), Plenum, NY, 251-275, 1981.

[2] Killeen, T.J., Eastwood, D.,, Hendrick, M.S., Talanta, 1981, 28, 1-6.

[3] Chien, Y.T., Killeen, J.T., in *Handbook of Statistics 2* (P.R. Krishnaiah and L.N. Kanal, Eds.), 1982, North Holland, 651-671.

[4] Urdal, K., Vogt, N.B., Sporstal, S.P., Lichtenthaler, R.G., Mostad, H., Kolset, K., Nordenson, S., Esbensen, K., Marine Pollution Bulletin, 1986, 17, 366-393.

[5] Sandness, G.A., Ailes, S.B., *Proceedings of the Eleventh International Symposium on Remote Sensing of Environment*, 1977, 2, 1445.

[6] Vo-Dinh, T., *Room Temperature Phosphorimetry for Chemical Analysis*, 1984, John Wiley and Sons, New York.

[7] Brownrigg, J.T., Busch, D.A., Giering, L.D., *A Luminescence Survey of Hazardous Materials*, 1979, Baird Corp DOT-CG-91-78-1888.

[8] Seitz, W.R., Anal Chem, 1984, 56, 16A-34A.

[9] Chudyk, W.A., Carraba, M.M., Kenny, J.E., Anal. Chem, 1985, 57, 1237-1242.

[10] Schulman, S.G., *Molecular Luminescence Spectroscopy, Methods and Applications*, Part 1, 1985, 77, John Wiley and Sons, NY.

[11] Hurtubise, R.J., "Trace Analysis by Luminescence Spectroscopy" in *Trace Analysis, Spectroscopic Methods for Molecules*, C.D. Christian and J.B.Callis, Eds., 1986, John Wiley & Sons, NY, Ch. 2.

[12] Conrad, V.B., Gore, R.R., Jammons, J.L., Maple, J.R., Perry, M.B., Wehry, E.L., in *New Directions in Molecular Luminescence*, ASTM STP 822, D. Eastwood, Ed, ASTM, Philadelphia, 1983, 17-31.

[13] Vo-Dinh, T. in *New Directions in Molecular Luminescence*, ASTM STP 822, D. Eastwood, Ed., 1983, ASTM Philadelphia, 5-16.

[14] Brownrigg, J.T., Hornig, A.W., *Archives of Environmental Contamination and Toxicology*, 1976, 4, 175.

[15] Weinberger, R., Rembish, K., Cline-Love, L.J. in *Advances in Luminescence Spectroscopy*, ASTM STP 863, L.J. Cline-Love and D. Eastwood, Eds., ASTM, Philadelphia, 1983, 40-51.

[16] Femia, R.A., Scypinski, Cline-Love, L.J., Environ. Sci. Technol., 1985, 19, 155-159.

[17] Yim, K.W.K., Miller, R.C., Faulkner, L.R., Anal. Chem. 1977, 49, 2069-2074.

[18] Rossi, T.M., Warner, I.M., Appl. Spectrosc. 1985, 39, 949-959.

[19] Chemistry Branch, U.S. Coast Guard R & D Center, "Oil Spill Identification System," Report No. DOT-CG-D-52-77, June 1977.

[20] Sogliero, G., Eastwood, D., Gilbert, J., *Advances in Luminescence Spectroscopy*, ASTM STP 863, L.J. Cline-Love and D. Eastwood, Eds., ASTM, Philadelphia, 1985, 95-115.

[21] Sogliero, G., Eastwood, D., Ehmer, E., Appl. Spectrosc, 1982, 36, 110-116.

[22] Mulkerrin, M.G., Wampler, J.E., Anal. Chem 1982, 54, 1778-1782.

[23] Eastwood, D., Lidberg, R.L., in *Management of Uncontrolled Hazardous Waste Sites, Proceedings of 7th Natl. Conf.*, Hazardous Materials Control Research Institute, Silver Spring, MD, 1986, 370-379.

[24] Bracewell, R.N., *The Fourier Transform and Its Applications*, 1986, McGraw-Hill, NY, Chapter 19, 20.

[25] Fu, K.W. *Sequential Methods in Pattern Recognition and Machine Learning*, 1986, Academic, NY.

[26] Tou, J.T., Gonzales, R.C., *Pattern Recognition Principles*, 1974, Addison-Wesley, Reading, MA.

[27] Rutan, S.C., Brown, S.D., Anal. Chem Acta. 160, 99-119.

[28] Liebman, S.A., in *Artificial Intelligence Applications in Chemistry* (Pierce, T.H., Hohne, B.A., Eds.), American Chemical Society, Washington, D.C., 1986, 306.

HEAVY METALS CONTENT IN THE PLANT BIOMASS OF LAWNS IN CITY RESIDENTIAL DISTRICTS

H. ZIMNY, CZ. WYSOCKI,
and E. KORZENIEWSKA

Department of Environment Protection, Warsaw Agricultural University, Nowoursynowska 166, 02-766 Warsaw, Poland

ABSTRACT

Air pollution in the urban environment causes, among other things, accumulation of heavy metals in soil and plants. The study was undertaken to determine the extent to which conditions in urban areas influence heavy metal accumulation in the biomass of lawn vegetation in residential districts. The investigations showed that the content of elements such as Mn, Zn, Cr, Pb, Cu, Ni, Cd and Co does not exceed values dangerous to this type of environment. Only the iron content is increased to values close to those in more degraded habitats such as, for instance, along streets.

INTRODUCTION

Human impact on the environment increases with the development of industry and urbanization. Dust and gases emitted by municipal installations such as hydroelectric and gas works, boiler houses and various industrial establishments are a major source of air pollution in cities. Moreover, motor vehicles contribute considerably to air pollution since their exhaust

Chemistry for the Protection of the Environment
Edited by L. Pawlowski *et al.*, Plenum Press, New York, 1991

Table 1. Spatial Structure of Chosen Residential Districts (after Lipińska [13])

No. of plot	Name of study area Residential Districts (RD) Park (P)	Surface ha	Buildings %	Greenery %	Other surfaces (streets, squares, etc. %)
1.	S lużew nad Dolink a(RD)	13.3	12.2	57.7	30.1
2.	Wierzbno (RD)	7.8	20.1	53.4	26.5
3.	Batorego (RD)	4.7	14.1	65.4	20.5
4.	Sadyba (RD)	7.3	13.5	57.7	28.8
5.	Ursynów SGGW-AR (P)	4.5	8.0	80.0	12.0

gases contain large quantities of toxic substances such as sulphur oxides, carbon, nitrogen, various hydrocarbons, and heavy metals [1, 2]. The impact of urban conditions on the environment results in more and more frequent disturbances in soil and vegetation, especially in the reduction of biomass, and above all, in the greenery along the streets [3, 4].

The aim of the present study is to determine to what extent pollution of the urban environment has affected the presence of heavy metals in the vegetation of lawns in residential districts. Investigations performed so far have been concerned mainly with the metal content in roadside lawns and in parks [3, 5].

MATERIAL AND METHODS

Determinations of the heavy metals content in the plant biomass of lawns were performed in four selected residential districts: S lużew nad Dolink a, Wierzbno, Batorego, and Sadyba, each characterized by a large proportion of greenery its spatial structure (Table 1). Ursynów park was used as a control area.

For analysis of biomass, material was collected in the central part of the above-named residential districts. The investigations were performed in 1987 and 1988 during the spring, summer, and early autumn. The plants were not washed before analysis. The plant material was dried at 84°C and, after grinding, was incinerated in a muffle furnace at 480°C and finally, was wetted with concentrated nitric and perchloric acids.

The content of such elements as iron, zinc, chromium, lead, copper,

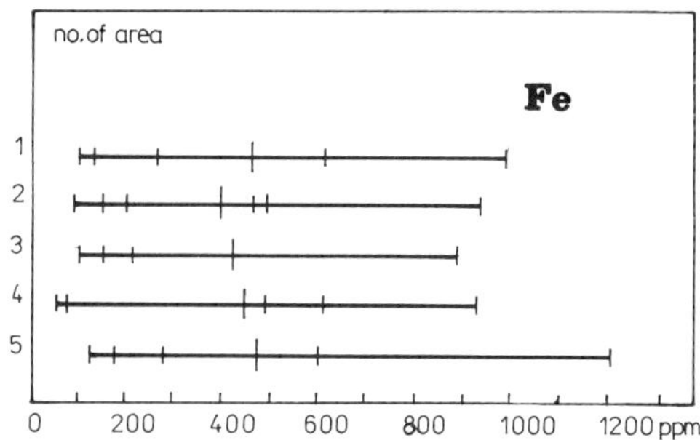

Fig. 1. Iron content in the biomass of lawns.

nickel, cadmium, and cobalt was determined by atomic absorption spectrophotometry (AAS) in a Perkin-Elmer 300 apparatus. The results are presented in graphic form as mean values.

RESULTS AND DISCUSSION

The iron content in the lawn biomass areas varied from 51 to 924 ppm, while the control surface varied from 119 to 1185 ppm (Fig. 1). These values exceed those given by Czarnowska [5] for parks (90-220 ppm), but they were comparable to those given for roadside lawns (500-1200 ppm).

The mean manganese content in the examined districts is 34.2 to 48.4 ppm which is somewhat lower than the content of the control surface (62.5 ppm, Fig. 2). When compared with the results of other investigations, these values may be considered the mean or even lower than the value reported by Walczyna et al. [6], where in meadow grasses 150 ppm of manganese was recorded. Results of investigations conducted by Czarnowska [3] in some Warsaw residential districts indicate a value of 18 to 117 ppm of manganese. The results of Ylaranta and Sillanpaa [7] investigations on the manganese content in the biomass of various plants, including grasses, indicate a timothy (*Phleum pratense*) content, of around 80 to 160 ppm. Results of the analysis to determine the level of zinc in the vegetation indicate that in most districts, the zinc value was higher than that of the control surface (Fig. 2). The mean value of zinc in the residential districts is 47.4 to 78.8 ppm and 54.8 ppm in the control surface (*Phleum pratense*). These values indicate a higher zinc level as compared with that in timothy (*Phleum pratense*) occurring in meadow communities where about 30 ppm was noted [7].

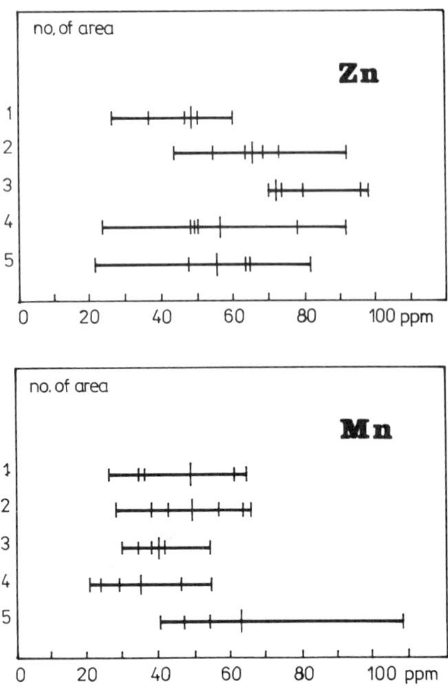

Fig. 2. Manganese and zinc contents in the biomass of lawns.

The Czarnowska paper [5] reports that lawn vegetation in residential districts in Warsaw contains from 24 to 63 ppm of zinc. The mean chromium content in the plant material of residential districts of Warsaw varies from 2.20 to 3.93 and does not differ widely from the control value (Fig. 3). When compared with the results of other authors, the Cr content in these districts is very low. Czarnowska [5] noted that about 12.5 ppm of chromium was detected in the lawn material. Recent results may be compared with the mean results previously obtained by the author on park lawns. In meadow communities, which are similar in composition to lawns, a mean chromium content in the biomass of 2 ppm was found, thus previous values are not much lower than present values [8].

Lead is one of the most dangerous elements threatening the environment. Motor vehicles are the major source of lead contamination in cities. The lead content in the plant material from residential districts does not differ from the values (a mean of 5.3 ppm) given by Czarnowska [5]. Lead is mainly cumulated in street zones, where its value in the biomass reaches more than 20 ppm [9]. Among the examined plots, the lowest value was found in the control surface ($\overline{x} = 3.46$ ppm, Fig. 3). The mean copper content in biomass of lawns of residential districts varies from 7.3 to 10.9 ppm,

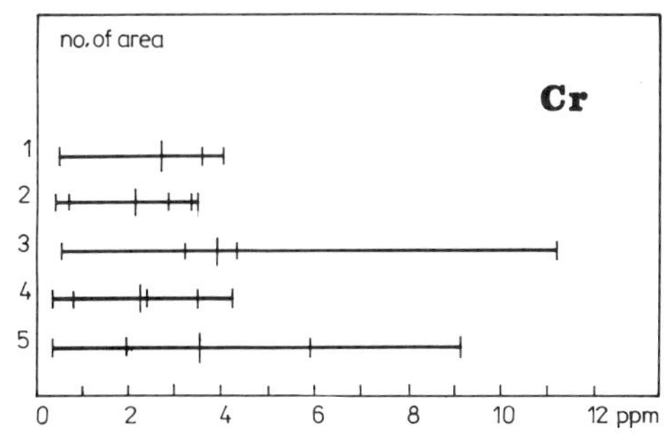

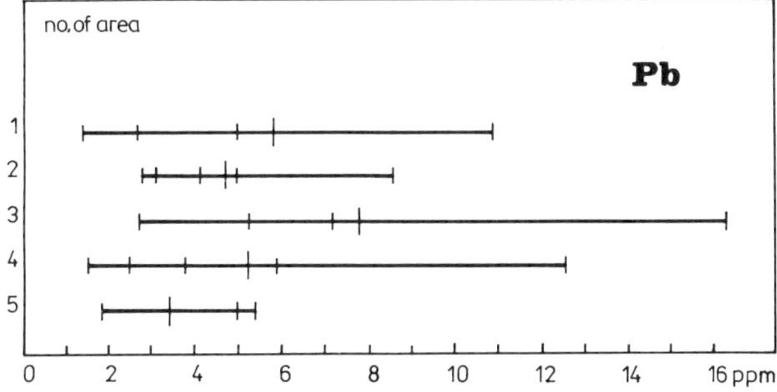

Fig. 3. Chromium and lead contents in the biomass of lawns.

and on the control surface it is 7.7 ppm (Fig. 4). These values are close to those obtained by Czarnowska [5], which are 5 to 15 ppm.

Walczyna et al. [6] report that grasses in meadow communities contain a mean of 6 ppm of copper. Ylaranta and Sillanpaa [7] mention in their paper values of 4-8 ppm for timothy (*Phleum pratense*). Results obtained in this study indicate copper is not at a level dangerous to the environment.

Nickel is another heavy metal that does not pose a major threat to the environment. Its content varies from 1.02 to 2.63 ppm in the studied districts, and 1.17 ppm on the control plots (Fig. 4). These values are much higher than those found in meadow grassess where the mean nickel value was 0.58 [10]. Considerably higher amounts of nickel were noted in the lawn vegetation along highways in the U.S.A., namely 3.8 to 5 ppm [11].

Another element potentially toxic to the environment is cadmium. The mean content of the later in the residential districts varies from 0.28 to 0.76 ppm, and is 0.49 ppm on the control plot (Fig. 5). Compared with the level of cadmium typically found in plants (0.05-5 ppm), the present values need not be considered toxic [12].

A similar relation was observed in regards to cobalt, of which the mean value in the lawn biomass is 0.15 to 0.25 ppm, and 0.14 ppm in the

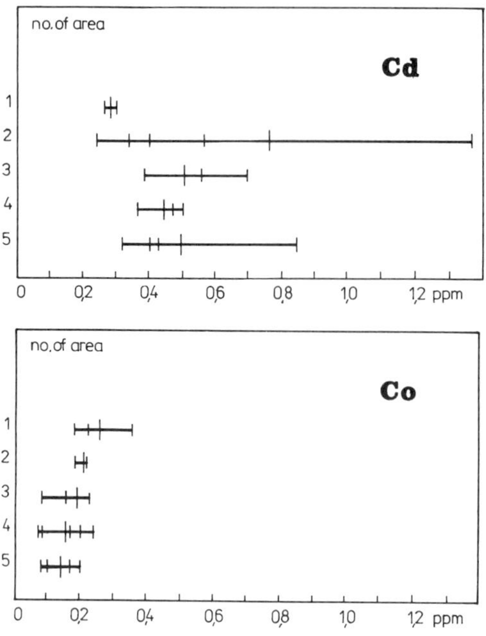

Fig. 4. Copper and nickel contents in the biomass of lawns.

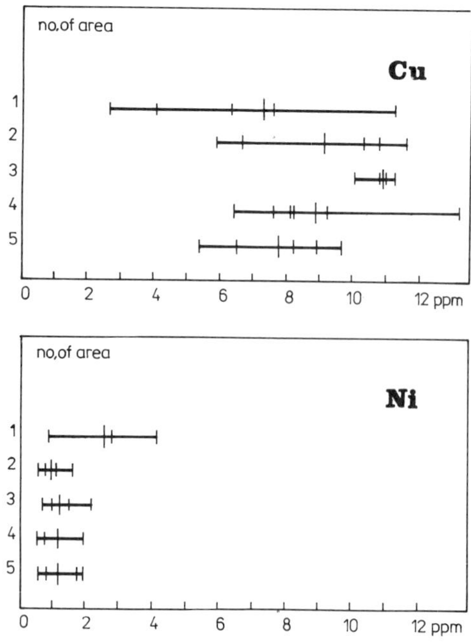

Fig. 5. Cadmium and cobalt contents in the biomass of lawns.

control plot (Fig. 5). The most typical value for cobalt found in plants is 0.02-1 ppm.

The results of this study indicate that most heavy metals, with the exception of iron, are found in the biomass of the studied areas in amounts that are not dangerous to the natural environment.

REFERENCES

[1] J. Lisicka, A. Ozięblo, Atmosphere Conservation, 3: 69-72, 1977.

[2] W. Skorupski, The vegetation influence on the urban habitat management, IKS issued by PWN, Warsaw, 95-108, 1984.

[3] K. Czarnowska, Warsaw Agricultural University Science Fascule, No. 106, 1978.

[4] Cz. Wysocki, H. Zimny, Pol. Ecol. Stud., 9, 1-2: 207-223, 1983.

[5] K. Czarnowska, Pol. J. Soil. Sci., vol. VII: 117-122, 1974.

[6] J. Walczyna et al., Agricultural Science Development Problems, 175: 49-63, 1975.

[7] T. Ylaranta, M. Sillanpaa, Ann. Agricul. Fenniae, Agrogeologica et Chimica, vol. 23: 158-170, 1984.

[8] M. Andrzejewski et al., Agricultural and Forestry Science Commitee Works, PTPN 31: 11-19, 1971.

[9] A. Strusiński, PZH Annuals, vol. 29, T.4: 411-417, 1978.

[10] K. Boratyński et al., Soil Scien. Annual, 22: 205-264, 1971.

[11] J. V. Langerwerff, A. W. Specht, Env. Sci. Technol., 4: 583 1970.

[12] A. Kabata-Pendias, M. Piotrowska, Environment Protection and Water Administration, Department, Typescript, 1984.

[13] A. Lipińska, PWN, Warsaw, 1977.

INCIDENCE AND LEVEL OF AFLATOXIN M_1 IN LIQUID AND IMPORTED POWDERED MILK IN JORDAN WITH SPECIAL REFERENCE TO AFLATOXIN B_1 IN CORRESPONDING FEEDS

R. M. NATOUR[1], J. A. RABBA [2], M. S. NOWAR[3], A. SALHAB[4], and A. MAHASNEH[5]

1 Prof. of Biology, Fac. Sci. The University of Jordan, Amman
2 Postgraduate Student, Fac. Sci. The University of Jordan, Amman
3 Prof. of Animal Nutrition, Fac. Agr. The University of Jordan, Amman
4 Prof. of Toxicology, Fac. Med. The University of Jordan, Amman
5 Prof. of Microbiology, Fac. Sci. The University of Jordan, Amman

ABSTRACT

The purpose of this study was to investigate the incidence of aflatoxins in cows milk and corresponding feedstuffs in Jordan. Detection of aflatoxins in imported powdered milk was also considered. A total of 133 liquid milk samples and 137 samples of corresponding feedstuffs collected from 5 local dairy farms in Jordan were tested. In addition, 41 imported powdered milk samples obtained from 3 dairy processing plants were tested. The collected samples were assayed for the level of aflatoxin M_1 in milk and aflatoxin B_1 in feedstuffs. Results of the analysis revealed the presence of aflatoxin M_1 in 5 liquid milk samples with an incidence rate of 3.8% and with toxin levels ranging from traces to 18.7 ppb. Aflatoxin B_1 was found in 3 feed samples with an incidence rate of 2.2%. Toxin levels were 66.0 and 85.2 ppb in two dairy ration samples and 17 ppm in the corn sample. On the other hand, the powdered milk samples were free from any detectable levels of aflatoxin M_1.

Tabuca or cassava, one of the ingredients of the dairy ration in Farm C, was contaminated with an unknown metabolite. This metabolite was not detected in the corresponding liquid milk. The characteristics of the unknown metabolite, including its UV absorption spectrum, indicated that it does not belong to the aflatoxins group.

INTRODUCTION

Contamination of human food and animal feed with mycotoxins and toxigenic fungi is a problem all over the world. When consumed, these contaminants can induce unnatural biological changes in man known as mycotoxicosis.

Aflatoxins are groups of mycotoxins, some of which are very toxic and able to induce three forms of genetic damage, carcinogenicity, mutagenicity, and teratogenicity [1,2,3]. Aflatoxins may also be transferred from feed to the tissues of livestock raised for human consumption. Oil seed meals and corn are primary sources of the toxins found in dairy rations [6].

Allcroft and Carnaghan [4] discovered that a small proportion of ingested aflatoxin B appears in cows milk as a metabolite called aflatoxin M [5]. Thus, milk is a secondary source of aflatoxin to man [6]. Aflatoxin B_1 is the most potent of the different aflatoxins, followed by aflatoxin M_1 which had 72% and 3.2% of toxicity (on one day old ducklings) and mutagenicity (on bacteria), respectively [7 and 8].

The concentrations of aflatoxin M_1 in milk is directly proportional to the total daily amount of aflatoxin B_1 ingested by the animal [9,10,11]. Aflatoxin M_1 is secreted into milk at the rate of 1-3% of the aflatoxin consumed [12], and it can be detected in milk when the level of aflatoxin B_1 in concentrate feed reaches 46 ppb [13]. Stoloff [14] reports that for each part of aflatoxin M_1 to appear in a cow's milk, approximately 300 parts of aflatoxin B_1 must have been ingested by that cow.

Aflatoxin M_1 appears in a cow's milk within two days of ingestion of aflatoxin B_1 contaminated feed, or within five hours of an orally administered dose of pure aflatoxin B_1 [15]. However, when a significant amount of aflatoxin is consumed, aflatoxin M_1 will appear in the milk at least 12 hours after ingestion [9]. Aflatoxin M_1 disappeared from milk in 3 to 4 days after eliminating aflatoxin contaminated feed from the cow's diet. The level of aflatoxin M_1 decreased with time in naturally contaminated raw milk [16]. When stored at zero ^oC, the level of aflatoxin decreased approximately 40% after 4 days and 60% after 6 days. Ciegler [17] found that out of the 1000 microorganisms, only Flavobacterium aurantiacum NRRL B_1-184 removed aflatoxin B_1 irreversibly from a nutrient solution. This organism could remove aflatoxin M_1 from contaminated milk [18]. Aflatoxin M_1 in milk is transformed to the manufactured products [19].

In Jordan, there are about 90 commercial dairy farms spread throughout the regions of Dhlail, Zarqa, Mafraq, and Irbid. These farms contain about 4500 cows, of which 52% belong to the cooperative society of Cattle Breeders. A substantial portion of the relatively large quantities of milk produced annually in Jordan finds its way to various dairy plants for processing into pasteurized milk, yogurt, and *labaneh*. Therefore, this study intended to partially evaluate the incidences and levels of aflatoxin M_1 in fresh milk and aflatoxin B_1 in corresponding feedstuffs in Jordan. The incidence of M_1 in imported powdered milk was also investigated.

MATERIALS AND METHODS

Samples of fresh liquid milk and feed were collected from the following farms:

1- Dairy farm 1, in Dhlail
2- Dairy farm 2, in Dhlail
3- Dairy farm A, in Deir Alla
4- Dairy farm B, in Al-Khaldieh
5- Dairy farm C, in Zarqa

Dairy farms 1 and 2 belong to the local Cooperative Society for Dairy Production.

A total of 41 samples of imported dried (powdered) milk were collected from 3 dairy processing plants: Jordan Dairy Company, Al-Raie Dairy Company, and Jordan University Dairy Pilot Plant.

A representative liquid whole milk sample was taken from each farm. The sample was collected in a 200–ml glass bottle containing 10 ml of 0.4% sodium azide solution as a preservative. Samples were stored in a refrigerator at 4°C and analyzed within 1-4 days.

A sample of concentrate and roughage feed from each farm was collected in a sterile polyethylene bag and taken to the laboratory and mixed to form a composite sample representative of the complete dairy rations prepared under the conditions of each farm. Samples were taken one to four days prior to collecting milk samples. These samples were used for aflatoxin analysis. One wheat bran sample and 26 concentrate samples were taken from the milk of the Cooperative Society for toxin analysis. A corn sample from farm C was also analyzed for toxins.

Liquid and powdered milk samples were assayed for aflatoxin M_1. Liquid milk samples were extracted using the method developed by Roberts and Allcroft [20]. Powdered milk samples were analyzed according to AOAC [21].

Extraction of aflatoxins from feed samples was done according to the BF method (Method II) in AOAC [21]. In some cases, the SM1 method [22] was also used.

The TLC plates coated with 0.25 to 50 nm silica gel (Adsorbosil1) were spotted with the chloroform extract of toxins and developed first with diethylether and then with chloroform /acetone mixture (9:1 v/v) or with chloroform /acetone /propan-2-OL (80:15:5 v/v/v) which was used to separate M_1 from M_2 [23].

Aflatoxins were identified on the basis of migration with aflatoxin standard and by their characteristic fluoresence color under short and long ultraviolet light.

Aflatoxin concentrations were determined on the basis of absorption of their methanol extract against methanol in the reference cell at 360 nm wavelength, and by using an equation for calculating aflatoxin concentration in the volume of methanol added and consequently in the analyzed samples of milk or feed [21,24].

$$\mu g\ aflatoxin/ml = A \cdot MW \cdot 1000 \cdot CF/E$$

where: A = absorbance
MW = molecular weight of aflatoxin:

$B_1 = 312$ $G_1 = 328$ $M = 328$
$B_1 = 314$ $G_2 = 330$

E = molar absorption:
$B_1 = 19{,}000$ $G_1 = 17{,}100$ $M = 19{,}950$
$B_1 = 20{,}900$ $G_2 = 18{,}200$
CF = correction factor.

Aflatoxin was confirmed by the method of Stoloff [25] using trifluroacetic acid, an acetic acid /thionylchloride mixture and a formic acid / thionylchloride mixture. When trace amounts of aflatoxins were present, the plates were sprayed with H_2SO_4 (25 or 50%) [26].

RESULTS AND DISCUSSION

The incidence and level of aflatoxin M_1 in cows milk and aflatoxin B_1 in the corresponding dairy ration samples, are given in Table 1. Out of 133 liquid milk samples, only 5 samples (3.8%) were found to be contaminated with aflatoxin M_1 at a level ranging from trace to 18.7 ppb. The incidence of

Table 1. Aflatoxins M_1 and B_1 in Liquid Milk of Cows and Corresponding Dairy Ration Samples Collected from Some Dairy Farms in Jordan

Source Samples	Milk Samples				Feed Samples				
	Tested	Contaminated			Type	Tested	Contaminated		
	No.	No.	%	Aflatoxin M_1 (ppb)		No.	No.	%	Alfatoxin B_1 (ppb)
1- The Coop	Society								
The mill					C	26	0	0	0
					WB	2	0	0	0
Farm 1	21	1	4.8	7.9	DR	21	0	0	0
Farm 2	21	1	4.8	14.8	DR	21	0	0	0
D plant	24	1	4.2	18.7					
2- D Farm A	19	0	0	0	DR	19	0	0	0
D Farm B	24	0	0	0	DR	24	0	0	0
D Farm C	24	2	8.3	Tr	DR	24	2	8.3	66 & 85.2
					Corn	1	1	100	17000

(1) Formulated from their ingredients in laboratory. (2) For dairy production (Cooperative Society of Cattle Breeders).
C = Concentrate; WB = Wheat Bran; DR = Dairy Ration; Tr = Trace, D = Dairy.

aflatoxin M_1 was 4.8, 4.8, 4.2, and 8.3% in the milk samples collected from Farm 1, Farm 2, Dairy plant, and Farm C, respectively. The corresponding levels of aflatoxin M_1 were successively 7.9 ppb, 14.8 ppb, 18.7 ppb, and trace.

The milk sample from Farm C, showed the highest incidence (8.3%) and the lowest level (trace) of aflatoxin M_1. In contrast, the sample from the Dairy plant showed an opposite trend being the lowest in incidence (4.2%) and having the highest level (18.7 ppb) of aflatoxin M_1.

The same trend was noticed in the selected survey [6] for the incidence and the level of aflatoxin M_1 in cow's milk (Table 2) , where in E. Germany [27], the incidence of aflatoxin M_1 was the lowest (11.1%), but the level was the highest (1.7-6.5 ppb) when compared with the corresponding figures of milk samples from Belgium, W. Germany, Netherlands, U.K., and the U.S.

The incidence rate and level of aflatoxin M_1 in milk samples from the aforementioned developed countries (Belgium, E. Germany, W. Germany, Netherlands, U.K., and the U.S.) ranged from 11.1 to 79.6% and from trace to 6.5 ppb, respectively (Table 2). This indicates that the incidence rate of aflatoxin M_1 in cow's milk in Jordan was low (4.2 to 8.3%) but the level was high (traces to 18.7 ppb). This level is considered to be hazardous, since aflatoxin M_1, as its parent compound aflatoxin B_1, induces tumor formation when fed to rainbow trout at a level of 0.5 to 2.0 $\mu g/kg$ body weight [28].

Analysis of 137 feed samples for aflatoxin (Table 2) indicates the pres-

Table 2. Selected Surveys for the Incidence and Level of Aflatoxin M_1 in Cows Milk (Paterson, 1983 with, minor modifications)

Country	Total No. samples analyzed	Samples Containing Aflatoxin No.	% incidence	Range μg/L or kg	Reference No.
			Liquid (fresh) milk		
Belgium	68	42	61.8	0.02 - 0.2	29
E. Germany	36	4	11.1	1.7 - 6.5	27
W. Germany	61	28	45.9	0.01 - 0.25	30
W. Germany	419	79	18.9	Trace - 0.54	9
W. Germany	260	118	45.4	0.05 - 0.33	31
India	21	3	14.3	up to 13.3	32
Netherlands	93	74	79.6	0.09 - 0.5	33
U.K.	278	85	30.6	0.03 - 0.52	34
U.S.	302	177	58.6	0.1 - more than 0.5	35
Range			11.1 - 79.6	Trace- 6.5 - 13.3	
			Dried (powdered) milk		
E. Germany	18	1	5.6	6.4 30	
W. Germany	55	36	65.5	Trace - 4.0	36
W. Germany	120	7	5.8	0.05 - 0.13	37
S. Africa	56	0	0	—	38
Range			0 - 65.5	Trace - 6.4	

ence of aflatoxin B_1 in only 3 samples (2.2%). These samples were taken from Farm C and included 2 samples of the complete dairy ration and one sample of corn. The levels of aflatoxin B_1 were 66.0 and 85.2 ppb in the dairy ration samples and 17 ppm in the corn sample.

However, the previous studies in Jordan [39] show a lower incidence (0.16%) and level of aflatoxins (B_1: 33 ppb and B_2: 12 ppb) in feed than those found in the present study; the rate of incidence of aflatoxin B_1 in feed is still lower than in other developed countries. The incidence rate and level of the aflatoxin B_1 in feed were: 41.8% and 140 ppb in Australia [40], 27% and 5-1500 ppb in South Africa [41] and 5.3% and 5.6 ppb in New Zealand [42]. On the other hand, in Poland, aflatoxins were not detected in any of the feed samples examined [43]. In Kuwait, Natour et al. [44] indicated that 63.9% of 58 examined feed samples were found to contain variable amounts of one or more aflatoxin. The results are as follows: 6samples contained 201 ppb of B_1, 8 contained 335 ppb of B_2, 12 contained 220 ppb of G_1 and 12 contained 220 ppb of G_2.

A considerably high level of aflatoxin B_1 in the corn used in Farm C (Table 1) agreed with the levels reported by many workers. Corn is a suitable substrate for natural production of aflatoxin [45 and 46].

Table 2. Selected Surveys for the Incidence and Level of Aflatoxin M_1 in Cows Milk (Paterson, 1983 with, minor modifications)

Country	Total No. samples analyzed	Samples Containing Aflatoxin No.	% incidence	Range μg/L or kg	Reference No.
			Liquid (fresh) milk		
Belgium	68	42	61.8	0.02 - 0.2	29
E. Germany	36	4	11.1	1.7 - 6.5	27
W. Germany	61	28	45.9	0.01 - 0.25	30
W. Germany	419	79	18.9	Trace - 0.54	9
W. Germany	260	118	45.4	0.05 - 0.33	31
India	21	3	14.3	up to 13.3	32
Netherlands	93	74	79.6	0.09 - 0.5	33
U.K.	278	85	30.6	0.03 - 0.52	34
U.S.	302	177	58.6	0.1 - more than 0.5	35
Range			11.1 - 79.6	Trace- 6.5 - 13.3	
		Dried (powdered) milk			
E. Germany	18	1	5.6	6.4 30	
W. Germany	55	36	65.5	Trace - 4.0	36
W. Germany	120	7	5.8	0.05 - 0.13	37
S. Africa	56	0	0	—-	38
Range			0 - 65.5	Trace - 6.4	

ence of aflatoxin B_1 in only 3 samples (2.2%). These samples were taken from Farm C and included 2 samples of the complete dairy ration and one sample of corn. The levels of aflatoxin B_1 were 66.0 and 85.2 ppb in the dairy ration samples and 17 ppm in the corn sample.

However, the previous studies in Jordan [39] show a lower incidence (0.16%) and level of aflatoxins (B_1: 33 ppb and B_2: 12 ppb) in feed than those found in the present study; the rate of incidence of aflatoxin B_1 in feed is still lower than in other developed countries. The incidence rate and level of the aflatoxin B_1 in feed were: 41.8% and 140 ppb in Australia [40], 27% and 5-1500 ppb in South Africa [41] and 5.3% and 5.6 ppb in New Zealand [42]. On the other hand, in Poland, aflatoxins were not detected in any of the feed samples examined [43]. In Kuwait, Natour et al. [44] indicated that 63.9% of 58 examined feed samples were found to contain variable amounts of one or more aflatoxin. The results are as follows: 6samples contained 201 ppb of B_1, 8 contained 335 ppb of B_2, 12 contained 220 ppb of G_1 and 12 contained 220 ppb of G_2.

A considerably high level of aflatoxin B_1 in the corn used in Farm C (Table 1) agreed with the levels reported by many workers. Corn is a suitable substrate for natural production of aflatoxin [45 and 46].

The relationship between the incidence of aflatoxin M_1 in liquid cows milk and incidence of aflatoxin B_1 in feed revealed a positive relationship only in Farm C (Table 1). The two contaminated dairy ration samples obtained from this Farm contained 66.0 and 85.2 ppb aflatoxin B_1, while only traces of M_1 were found in the produced milk. This result agrees with the trend reported by Roberts and Allcroft [20], that aflatoxin M_1 may appear in the milk of cows fed on a complete ration containing more than 46 ppb aflatoxin B_1.

On the other hand, the three milk samples from Farm 1, Farm 2, and the dairy plants (all belong to the Cooperative Society) were contaminated with aflatoxin M_1 at levels of 7.9, 14.8, and 18.7 ppb, respectively, while the corresponding feeds were free from aflatoxins. In this case, either the analyzed samples are not representative of the feed and/or the cows ate contaminated feed that was not sampled 1–4 days prior to the collection of milk samples.

One of the most crucial points made in this study is that a high level of aflatoxin M_1 (18.7 ppb) exists in the milk sample taken at the dairy plant, which is a composite of the milk samples taken at various dairy farms belonging to the Cooperative Society. Milk samples obtained from farms 1 and 2 contained aflatoxin M_1 at levels of 7.9 and 14.8 ppb, respectively. This would indicate that the milk produced in some or all other dairy farms of the Cooperative Society was highly contaminated with aflatoxin M_1 (more than 18 ppb) to the extent the average that aflatoxin M_1 in the pooled samples in the dairy processing plant was 18.7 ppb.

Stoloff [14] reported that for each part of aflatoxin M_1 to appear in a cows milk, approximately 300 parts of aflatoxin B_1 must have been ingested by that cow. Thus, the presence of aflatoxin M_1 in liquid milk indicates the presence of aflatoxin B_1 in the animal feed used at the various farms. The corn ingredient in Farm C was highly contaminated with aflatoxin B_1 (17 ppm). Aside from this, improper mixing of ingredients during formation and preparation of rations may result in a high aflatoxin B_1 level in the ration. When a cow ingests feed ration containing a high level of aflatoxin B_1, an increased level of aflatoxin M_1 will be found in the milk it produces.

Analysis to determine the presence of aflatoxins in ration samples revealed the presence of an interesting substance in 20 of the 24 samples taken from Farm C. These samples were analyzed and the results indicated the presence of this substance in *tabuca*. Column chromatographic purification of the sample extract indicated that the interfering agent was not removed by hexane or the anhydrous diethylether fractions, which are commonly used to remove interfering substances. However, this interfering substance appeared in the chloroform-methanol (97:3 v/v) fraction which was used to elute the aflatoxin from the chromatographic column.

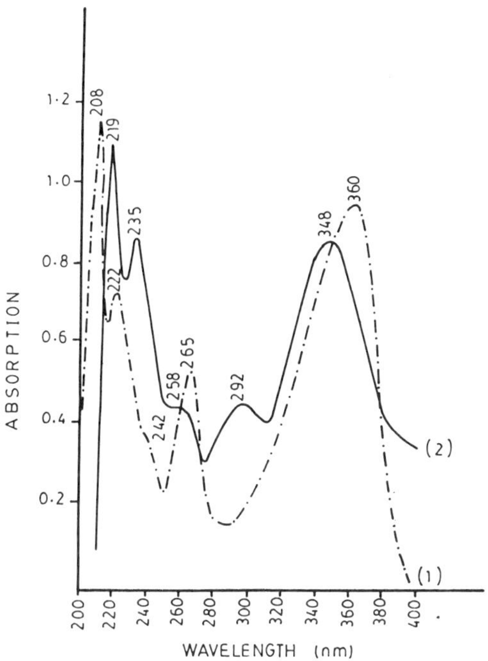

Fig. 1. Ultraviolet absorption spectrum of aflatoxin B_1 (1) and that of the unknown metabolite in methanol (2). (This metabolite extracted from the poultry pellet "Tabuca".)

The unknown metabolite was characterized using blue violet florescence under UV light; RF (in developing system chloroform acetone 9:1 v/v) was lower than that for aflatoxin B_1 (0.68) and B_2 (0.61), being 0.47.

Many tests were run to identify this unknown metabolite. The UV-absorption spectrum of this metabolite (in methanol) showed only a minor difference from that of aflatoxin B_1 (Figure 1). Contrary to aflatoxin B_1 the RF value of this unknown metabolite did not change when treated with trifluoroacetic acid or formic acid-thionylchloride. The unknown metabolite was somewhat unstable on TLC plates and tended to decompose when exposed to air or light. Accordingly, this evidence indicates that this unknown metabolite is not related to the aflatoxin group.

Seven cows in Farm C showed signs of illness, such as a loss of appetite and swelling of the abdominal cavity. These signs, especially the loss of appetitem were noticed in various types of livestock including cows that had been poisoned with aflatoxin [47 and 29]. Affected cows may have been exposed to a relatively high levels of aflatoxin B_1 by either consuming a small

Table 3. Incidence of Aflatoxin M_1 in Powdered Milk Samples

Sample source	Type of milk	Place of origin	No. of tested samples	No. of contaminated samples
Jordan Dairy Company	Skim	Different origin	12	none
Al-Raie Company	Full cream	Holland	12	none
Jordan University	Full cream	Ireland	6	none
Dairy Pilot Plant	Skim	Denmark	11	none

portion of highly contaminated corn or consuming rations that contained a high level of aflatoxin B_1 as a result of improper mixing of the concentrate ingredients.

Table 3 shows that aflatoxin M_1 was not present in 41 powdered milk samples collected from three processing plants in Jordan. Patterson [6] surveyed the incidence and level of aflatoxin M_1 in dried milk samples taken from East and West Germany and South Africa. Aflatoxin M_1 was present in the East and West Germany samples but was not found in the South African sample (Table 2).

ACKNOWLEDGEMENT

The authors wish to express their gratitude and sincere thanks to the Faculty for Scientific Research, University of Jordan, Amman, for the grant offered by this Faculty to finance this research.

REFERENCES

[1] Wogan, G. N., Aflatoxins and their relationship to hepatocellular carcinomas. In "Hepatocellular Carcinoma", Okieda, K. and Peters, R. L. (Eds.), John Wiley & Sons. New York, 25, 1976.

[2] Legator, M. S., Magnetic effect of aflatoxin, J.A.V.M.A., **155**, 2080, 1969.

[3] Nowar, M. S., Sherif, R. M., Gbreel, M. G., and El-Darawany, A. M., Teratogenicity of aflatoxins (B_1 and G_1) mixture in rats and chick embryos. The Egyptian Journal of Environmental Mutagenesis, Tetratogenesis and Carcinogenesis "In press", 1988.

[4] Allcroft, R. and Carnaghan, R. B. A., Groundnut toxicity. An examination for toxin in human food products from animals fed toxic groundnut meal. Vet. Rec. **75**; 257-263, 1963.

[5] Allcroft, R., Rogers, H., Lewis, G., Nabney, J. and Best P. E., Metabolism of aflatoxin in sheep. Excretion of the milk toxin. Nature **209**;154-155, 1966.

[6] Patterson, D. S. P., Foodborne diseases: Aflatoxicosis. In "Handbook of Foodborne Diseases of Biological Origin", Rechcigl, M. (Ed) CRC Press, Inc., Boca Raton, Florida, pp. 323-351, 1983.

[7] Halzapfel, C. W., Steyn, P. S. and Purchase, I. F. H., Isolation and Structure of Aflatoxin M. Tetrahedron Letters **26**: 2799-2803, 1966.

[8] Hayes, A. W., Mycotoxins: a review of biological effects and their role in human disease, Clim. Toxicol., 17, 45, 1980.

[9] Kiermeier, F., The Significance of Aflatoxins in the Dairy Industry, Annu. Bull. Int. Dairy Fed., No. 98, 1, 1977.

[10] Patterson, D. S. P., Aflatoxin and related compounds, In "Mycotoxic Fungi, Mycotoxins, Mycotoxicoses" an Encyclopedia Handbook, Vol. 1, Wyllie, T. D. and Morehouse, L. G., (Eds.), Marcel Dekker, New York, 1977.

[11] Purchase, I. F. H., Aflatoxin residues in food and animal origin. Food Cosmet. Toxicol., 10, 531, 1977.

[12] Applebaum, S. R., Bracket, E. R., Wiseman, W. D. and Marth, H. E., Aflatoxin: Toxicity to dairy cattle and occurrence in milk and milk products. A review. J. Food Prof. **45**: 752-777, 1982.

[13] Polan, C. E., Hayes, J. R. and Campbell, T. C. Consumption and fate of aflatoxin B_1 by lactating cows. J. Agr. Food Chem., **22**: 635-638, 1977.

[14] Stoloff, L. Incidence, distribution and disposition of products containing aflatoxin. Proceedings of the American Phytopathological Society, **3**: 156-178, 1976.

[15] Lynch, G. P. Mycotoxins in feedstuffs and their effect on dairy cattle. J. Dairy Sci., **55**: 1243-1255, 1972.

[16] Mckinney, J. D., Cavanagh, G. C., Bell, J. T., Hoverslund, A. S., Nelson, D. M., Pearson, J. and Selkirk, R. J. Effects of ammoniation of aflatoxin in ration fed lactating cows. J. Am. Oil Chem. Soc., **50**: 79-84, 1973.

[17] Ciegler, A., Lillehoj, E. B., Patterson, R. E. and Hall, H. H. Microbial detoxification of aflatoxin. Appl. Microbiol. **14**: 934-937, 1966.

[18] Lillehoj, E. B., Stubblefield, R. D., Shannon, G. M. and Shotwell, O. L. Aflatoxin M_1 from aqueous solutions by Flavobacterium aurantiacum. Mycopatho. Mycol. App. **45**: 259- 266, 1971.

[19] Wiseman, D. W. and Marth, E. H. Stability of Aflatoxin M_1 during manufacture and storage of butter-like spread, nonfat dried milk and dried buttermilk. J. Food. Prot., **47**: 633-636, 1983.

[20] Robert, B. A. and Allcroft, R. A note on the semiquantitative estimation of aflatoxin M_1 in liquid milk by thin layer chromatography. Food Cosmet. Toxicol. **6**: 339-340, 1968.

[21] Association of Official Analytical Chemists. Natural poisons. In "William, H. (Ed.) Official Method, 26.001-26.090, 1975.

[22] Seitz, L. M. and Mohr, H. E. New method for quantitation of aflatoxin in corn. Cereal Chem., **54**: 179-183, 1977.

[23] Patterson, D. S. P., Glancy, E. M. and Robers, B. A. The estimation of aflatoxin M_1 in milk using a two-dimensional thin-layer chromatographic method suitable for survey work. Food Cosmet. Toxicol., **16**: 49-50, 1978.

[24] Masri, M. S., Page, J. R. and Garcia, V. C. Analysis for aflatoxin in milk. J.A.C.O.C., **51**:594-600, 1968.

[25] Stoloff, L. Collaborative study of a method for identification of aflatoxin B_1 by derivative formation. J.A.O.A.C., **50**: 354- 360, 1967.

[26] Pryzybylski, W. Formation of aflatoxin derivatives on thin layer chromatographic plates. J.A.O.A.C., **58**: 163, 1975.

[27] Fritz, W., Donath, R. and Engst, R. Determination and occurrence of aflatoxin M_1 and B_1 in milk and dairy produce, Nahrung, **21**: 79, 1977.

[28] Sinnhuber, R. O., Lee, D. J., Wales, J. H., Landers, M. K. and Keyl, A. C. Aflatoxin M_1 : A potent liver carcinogen for rainbow trout, Fed. Proc., **29**: 268, 1970.

[29] Van Pe'e, W. van Brabant, J. and Joostens, J. The detection and concentration of aflatoxin M_1 in milk and milk products, Rev. Agr. (Bruxelles), **30**: 403, 1977.

[30] Kiermeier, F. Aflatoxin residues in fluid milk, Pure Appl. Chem., **35**: 271-273, 1973.

[31] Polzhofer, K. Determination of aflatoxins in milk and milk products, Z. Lebensm. Unters. Forsch., **163**: 175-177, 1977.

[32] Paul, R. A. J., Kalra, M. S. and Sign, A. Incidence of aflatoxin in milk products. Indian J. Dairy Sci., **29**: 318- 321, 1976.

[33] Schuller, P. L., Verhulsdonk, C. A. and Paulsch, W. E. Aflatoxin M_1 in liquid and powdered milk, Zesz. Probl. Postepow. Nauk. Roln., **189**: 225,1977.

[34] Paterson, D. S. P. Unpublished data (see Patterson, D. S. P., 1983), 1977.

[35] Anon. FDA is preparing to set action level for aflatoxin in fluid milk, Food, Chem. New, **19(37)**: 38, 1977.

[36] Hanssen, E. and Jung, M. Control of aflatoxins in the food industry, Pure Appl. Chem., **35**: 239, 1973.

[37] Jung, M. and Hassen, E. The occurrence of aflatoxin M_1 in dried milk products, Food, Cosmet. Toxicol., **12**: 131, 1974.

[38] Luck, H., Steyn, M. and Wehner, F. C. A survey of milk powder for aflatoxin content, S. Afr. J. Dairy Technol., **8**: 85, 1977.

[39] Masri, Z. Contamination of animal feeds used in Jordan in aflatoxins and aflatoxin producing fungi. M.Sc. Thesis, University of Jordan, Amman-Jordan, 1981.

[40] Bryden, W. L., Lloyed, A. B. and Comming, R. B. Aflatoxin contamination of Australian animal feeds and suspected cases of mycotoxicosis. Aust. Vet. J., **56**: 176-180, 1980.

[41] Dutton, M. F. and Westlake, K. Occurrence of mycotoxins in cereals and animal feedstuffs in Natal, South Africa, J. Assoc. Off. Anal. Chem., **68**: 839-842, 1985.

[42] Staton, D. W. A survey of some foods for aflatoxins. Food Technol. Anst., **12**: 25, 1977.

[43] Chelkowski, J. and Golinski, P. Mycotoxins and toxigenic fungi in mixed feeds and their cereal components. In "Proc. 5th Int. IUPAC Symp. on Mycotoxins and Phycotoxins, Vienna, Austria, Sept., 1-3: 68-71, 1971.

[44] Natour, R. M., Al-Awadi, A., Illian, M. and Salman, A. J. Aflatoxin and aflatoxigenic fungi in poultry feed in Kuwait. Dirasat, **XII** 43-52, 1985.

[45] Goldblatt, L. A. Aflatoxin: Scientific background, control and implications, Academic Press, New York, 1969.

[46] Heathcote, J. G. and Hibbert, J. R. Aflatoxins: Chemical and Biological Aspects. Elsevier Scientific Publication Company, New York, 1978.

[47] Alcroft, R. Aflatoxicosis in farm animals. In (Goldblat, L. A. Ed.). Aflatoxins: Scientific Background, Control and Implication. Academic Press, New York, 237-264, 1969.

FLOW ANALYSIS FOR DETERMINATION OF HYDROGEN PEROXIDE IN ENVIRONMENTAL WATER BY THE USE OF BIOMIMETRIC FUNCTIONAL RESIN

Y. SAITO[1], J. ODO[1], M. MIFUNE[1], M. CHIKUMA[2] and H. TANAKA[3]

[1] Faculty of Pharmaceutical Sciences, Okayama University, Tsushima-Naka Okayama, 700 JAPAN

[2] Osaka University of Pharmaceutical Sciences, Matsubara City, Osaka, 580 JAPAN

[3] Faculty of Pharmaceutical Sciences, Kyoto University, Sakyo-Ku, Kyoto, 606 JAPAN

ABSTRACT

We have developed a novel method of flow analysis for hydrogen peroxide, which is applicable to environmental water samples, including rain. The present method is based on the measurement of the absorbance of dye formed by the following reaction using an anion-exchange resin functionalized with manganese (III)-tetrakis(sulfophenyl)porphine (Mn-$TPPS_r$) as a catalyst instead of horseradish peroxidase.

$$\text{4-Aminoantipyrine} + \text{Phenol (or 4-Chlorophenol)} + 2H_2O_2 \xrightarrow{\text{Mn-TPPSr or peroxidase}} \text{Quinoid dye} + 4H_2O$$

The use of Mn-$TPPS_r$ is more advantageous than that of enzyme in stability of the catalyst. The present method is satisfactory in reproducibility, of procedure, sensitivity, and significant for the investigation of the rain problem.

Chemistry for the Protection of the Environment
Edited by L. Pawlowski *et al.*, Plenum Press, New York, 1991

INTRODUCTION

The amount of hydrogen peroxide in rain is keenly related to the acid-rain problem. Horseradish peroxidase has been widely used as a catalyst for the determination of hydrogen peroxide in both environmental and clinical analyses. Peroxidase acts specifically on hydrogen peroxide, but is not very stable. In an attempt to develop a stable artificial mimesis of this enzyme, we have investigated the catalytic activities of ion-exchange resins loaded with some metal-complexes of porphyrins, and found that some resins bear high enzyme-like activities. The peroxidase-like activity of the resin which enhances the dye formation reaction has been effectively applied to the determination of hydrogen peroxide [7,8].

This paper deals with a new flow method for continuous and rapid determination of hydrogen peroxide in water by the use of the ion exchange resin loaded with manganese(III)-tetrakis(sulfophenyl)-porphine (Mn-$TPPS_r$) instead of peroxidase. Since Mn-$TPPS_r$ is stable and its activity does not decrease at room temperature, the method presented here may be applicable to various water samples instead of the conventional peroxidase method.

EXPERIMENTAL

Materials

Manganese(III)-tetrakis(sulfophenyl)-porphine (Mn-$TPPS_r$) solution was prepared from manganese chloride and tetrakis(sulfophenyl)porphine purchased from Tokyo Kasei Co., Ltd., by the method described previously[4,5]. The resin loaded with Mn-TPPS (25 μmole Mn-TPPS/1 g dry resin, Mn-$TPPS_r$) was prepared from the aqueous solution of Mn-TPPS and anion-exchange resins, Amberlite IRA 900 and CG-400, and Dowex MSA-1 and I-X8, as reported previously [1]. No elution of Mn-TPPS was observed even in 1_M solution sodium chloride. Mn-$TPPS_r$ was stable when stored for at least one year at room temperature in a dark place. No decrease of its peroxidase-like activity was observed. Other reagents used were of analytical or reagent grade.

Apparatus

The flow system shown in Figure 1 is simply constructed of a dual pump (model KHD-W-52, Kyowa Seimitu, Japan), a sample injector (20 μl,

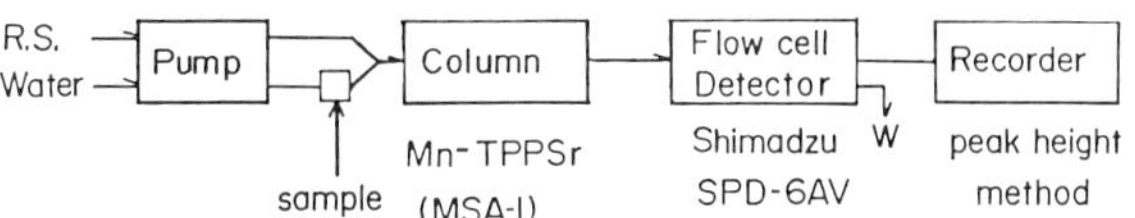

Fig. 1. Scheme of flow system.

model 7125, Rheodyne, USA), a catalytic column packed with Mn-$TPPS_r$, and a UV-VIS spectrometric detector equipped with a recorder (model SPD-AV, Shimadzu, Japan). These instruments were connected with stainless steel tubes (ϕ=1 mm).

Recommended Procedure

1) 4-Aminoantipyrine (4-AAP)/4-chlorophenol system: Reagent solution A and water are applied to flow through the dual pump (flow rate: 2.0 ml/min). Catalytic column packed with Mn-$TPPS_r$ is ϕ 2-mm x 50-mm stainless column. Sample solution containing hydrogen peroxide (20 μl) is injected into the injector. Hydrogen peroxide is determined by peak height at 505 nm.

The reagent solution A is a 1:1:4 mixture of 4-AAP (1 mg/ml), 4-chlorophenol (10 mg/ml) and phosphate buffer solution (pH 8.0).

2) 4-AAP/phenol system: Reagent solution B and water are applied to flow through the dual pump (flow rate: 1.0 ml/min). Catalytic column is ϕ 1-mm x 400-mm Teflon column. Sample solution (50 μl) is injected into the injector. Hydrogen peroxide is determined by the peak height at 505 nm.

The reagent solution B is a 1:1:4 mixture of 4-AAP (1 mg/ml), phenol (10 mg/ml) and borate buffer solution (pH 8.4).

RESULTS AND DISCUSSION

Coupling Reagent to 4-AAP

Phenol and 4-chlorophenol were examined as a coupling reagent to 4-AAP. Sensivibity of phenol was lower than that of 4-chlorophenol, but its reaction time for the formation of the dye was shorter than that of 4-chlorophenol.

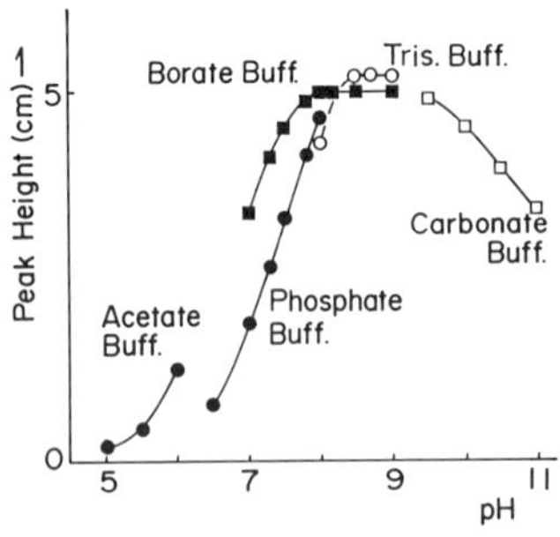

Fig. 2. Effects of buffer and pH on peak height.

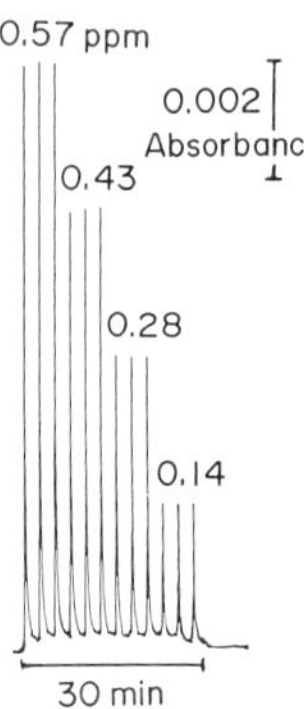

Fig. 3. Peaks of hydrogen peroxide.

Selection of the Resin Loaded with Metal-Complex of Porphyrin

As a resin loaded with metal complex of porphyrin, we have chosen Mn-TPPS$_r$, which showed the highest peroxidase-like activity among the resins tested in both the 4-AAP/phenol and the 4-AAP/4-chlorophenol systems [7].

Anion-exchange resins, Dowex MSA-1 and I-X8, and Amberlite IRA 900 and CG 400, were examined in the 4-AAP/4-chlorophenol system for the selection of the anion-exchange resin for the preparation of Mn-TPPS$_r$. MSA-1 (100-200 mesh) loaded with Mn-TPPS was found to be the best. The

2x50-mm or 1x400-mm columns were used in the 4-AAP/4-chlorophenol and the 4-AAP/phenol system, respectively, in consideration of the capacity of peroxidase-like activity and the adsorption of the resulting quinoid dye on Mn-$TPPS_r$.

Selection of Conditions

1) Buffer and pH: The effect of pH on the peak height in the 4-AAP/phenol system was examined between pH buffer solutions. As shown in Figure 2, a constant and nearly maximum peak height was obtained between pH 7.0 and 9.0 (borate buffer) in the 4-AAP/phenol system. We selected borate buffer (pH 8.4). In 4-AAP/4-chlorophenol system, the maximum peak height was obtained at pH 8.0 of phosphate buffer. Borate buffer (pH 8.0) gave only 75% of the maximum peak height. Phosphate buffer (pH 8.0) was selected in the 4-AAP/4-chlorophenol system.

2) Temperature: In both the systems, increasing temperatures between 25 and 50°C at the catalytic column did not show, a significant increase in the peak height. We chose 35°C as column temperature.

3) Concentration of reagent: Since the constant and maximum height was obtained by using concentrations higher than 7 mg/ml phenol or 4-chlorophenol, and 0.8 mg/ml 4-aminoantipyrine, we prepared the reagent solution by adding 10 mg/ml phenol (or 4-chlorophenol) and 1 mg/ml 4-AAP to the buffer solution.

Peaks, Calibration Curves, Sensibility and Reproducibility

Sharp peaks were obtained by following the recommended procedure, as shown in Figure 3. Calibration curves satisfy Beer's law in the range between 10 ng/ml (ppb) and 1μg/ml (ppm) by using 50 μl of the sample solution and 1x400-mm column, and up to at least 5 μg/ml by using 20 μl sample and 4x50-mm column. The coefficients of variations (n=10) were less than 1% in both the systems.

Repeated Uses (Durability)

In order to examine durability of the column-packed Mn-$TPPS_r$, the effect of repeated uses was investigated. No decrease of peak height was observed in a minimum of 300 times of use. Satisfactory durability of the column was thus confirmed.

Interference

In the presented method, interference caused by various ions was ex-

Table 1. Recoveries by the Present Method and by the Chemiluminescent Method

sample	present method				chemiluminescent method		
	added ppb	found ppb	recov. %	C.V.* &(n=5)	found ppb	recov. %	C.V.* %(n=3)
A	0	0	-		0	-	
	944	925	98	0.65	809	86	
	472	481	102	0.23	370	78	
	236	236	100	0.72	191	81	
	118	120	102	0.72	98	83	
B	0	0	-		0	-	
	915	915	100	0.24	897	98	
	458	444	97	0.49	449	98	
	229	215	94	0.12	218	95	5.45
	115	112	97	0.80	115	100	4.64
	57	54	95	1.20	55	96	8.54

* C.V.: Coefficient of variation.

amined. Change of peak height less than 4.3% was observed when K^+, Na^+, Mg^{2+}, Br^- and CO_3^{2-} were present in 100 times their amount of hydrogen peroxide. Ca^{2+} and Fe^{3+} caused an 8.7% increase and a 33% decrease of peak height, respectively, in the presence of 100 times amount of hydrogen peroxide. However, in their presence of 10 times their amount, increase and decrease of peak height were less than 3%.

APPLICATION TO ENVIRONMENTAL WATERS

The present method by 4-AAP/phenol system was applied to the determination of hydrogen peroxide in two samples of rain water (samples A and B, taken in Soka-city Saitama in the vicinity of Tokyo, Japan in September, 1987). No hydrogen peroxide was detected in these samples, which was added 180 ml of 0.5% sodium pyrophosphate solution to avoid the decomposition of hydrogen peroxide. Recoveries of hydrogen peroxide were made from samples A and B when spiked with hydrogen peroxide (57-944 ppb). The results are presented in Table 1 together with recoveries obtained by the chemiluminescent method [9]. Recoveries by the present method, 94-105%, were better than those obtained by the chemiluminescent method. In addition, reproducibility of the present method was found to be excellent, as seen in the coefficients of variation.

In conclusion, the flow analysis method using Mn-$TPPS_r$ presented in this paper is effective for the determination of hydrogen peroxide in a rain sample.

ACKNOWLEDGEMENTS

The authors thank Mr. Katuhito Sato and Mr. Kunihiko Murata, Kanto Kagaku Co., Ltd., and Mr. Takaaki Tai, Miss Kumiko Maturo and Miss Misato Tani, Okayama University, for their technical assistance.

REFERENCES

[1] Y. Saito, M. Mifune, T. Kawaguchi, J. Odo, Y. Tanaka, M. Chikuma and H. Tanaka, Chem. Pharm. Bull., **34** 2885, 1986.

[2] Y. Saito, M. Mifune, J. Odo, Y. Tanaka, M. Chikuma and H. Tanaka, Reactive Polymers, **4**, 243, 1986.

[3] Y. Saito, M. Satouchi, M. Mifune, T. Tai, J. Odo, Y. Tanaka, M. Chikuma and H. Tanaka, Bull. Chem. Soc., Jpn., **60**, 2227, 1987

[4] Y. Saito, M. Mifune, S. Nakashima, Y. Tanaka, M. Chikuma and H. Tanaka, Chem. Phar. Bull., **34**, 5016, 1986.

[5] Y. Saito, M. Mifune, S. Nakashima, H. Nakayama, J. Odo, Y. Tanaka, M. Chikuma and H. Tanaka, Chem. Pharm. Bull., **35**, 869, 1987.

[6] Y. Saito, M. Mifune, S. Nakashima, J. Odo, Y. Tanaka, M. Chikuma and H. Tanaka, Anal. Sci., **1987**, 171, 1987.

[7] Y. Saito, M. Mifune, S. Nakashima, J. Odo, Y. Tanaka, M. Chikuma and H. Tanaka, Talanta, **34**, 667, 1987.

[8] Y. Saito, M. Mifune, S. Nakashima, J. Odo, Y. Tanaka, M. Chikuma and H. Tanaka, J. Pharmacobio-Dyn., **10**, s-16, 1987.

[9] K. Yoshizumi, K. Aoki, I. Nouchi, T. Okita, T. Kobayashi, Kakamura and M. Tajima, Atmos. Environ., **18**, 395, 1984.

NONIONIC DETERGENTS AS TRACERS OF GROUNDWATER POLLUTION CAUSED BY MUNICIPAL SEWAGE

U. ZOLLER, E. ASHASH, G. AYALI and S. SHAFIR

Division of Chemical Studies, Department of Biology
Haifa University-Oranim, P.O. Kiryat Tivon 36910, Israel

B. AZMON

The Hydrological Service, Ministry of Agriculture – The Water Commission, Northern District P.O. Box 140, Haifa, Israel

ABSTRACT

This preliminary study to aimed at determining the extent which Israel's groundwaters is in the vicinity of streams polluted by municipal sewage, have been contaminated by synthetic detergents, the locally used nonbiodegradable alkylphenol-based nonionics in particular. Streams polluted by sewage (domestic and/or industrial) and six adjacent water wells were sampled simultaneously, and their nonionic detergent concentration determined.

It was found that groundwaters are being contaminated by nonionic synthetic detergents (0.1-0.8 mg/liter), the origin of which are the adjacent surface water bodies polluted by sewage.

The data collected indicate an inverse relationship between the distance of a given well from the polluted stream and the concentration of the nonionic detergent

Chemistry for the Protection of the Environment
Edited by L. Pawlowski *et al.*, Plenum Press, New York, 1991

in the well. No obvious, simple pattern appears to emerge, concerning the detergent concentration-hydraulic head relationships. However, the data appear to suggest that the deeper the well, the higher the concentration of the nonionic detergents in the well's water, assuming all other involved factors are similar. The mean velocity of movement of the detergent pollutant front in the soil was found to be about 0.4 meter/day. This means that the detergent pollutants are exposed to natural biochemical degradation and physiochemical adsorption processes within water/soil systems for more than a month before they reach the groundwaters.

Clearly, neither the naturally occurring biodegradation nor physical soil adsorption processes in the subsurface water/soil systems are capable of avoiding groundwater contamination by nonbiodegradable ("hard") detergents. In view of the increasing demand for potable water and the ever-increasing reuse of water resources, their contamination by detergents constitutes a real environmental threat.

INTRODUCTION

Groundwater contamination is a major environmental problem the total extent of which has not been fully identified nor is it thoroughly understood. Such contamination can result from a wide variety of both point and nonpoint sources, and has become an issue of major concern and public attention in recent years [1]. Of particular issue has been the entry of anthorpogenic ("man–made") organic materials into subsurface soil and/or groundwater by leachate from all kinds of wastewater including municipal sewage [2].

Currently, more than 30 million metric tons of synthetic detergents and soaps are produced worldwide [3], most of which are discharged into municipal wastewaters. Consequently, they constitute a significant part of the organic loadings in the wastewaters and a factor in determining the quality profile of municipal sewage [4].

Alkylphenol-based nonionic surfactants of the APEO type and their metabolites are known to be major refractory constituents of treated wastewater effluents [5]. The persistent existence of this class of low biodegradable ("hard") nonionic surfactants in aquatic environments requires the study of their transfer, fate, and chemical effects both in surface and subsurface soil/water systems.

The issue referred to in our study is illustrated in Fig. 1. It can be formulated in general terms as follows: What are the biochemical-hydrological relationships between a well and the river which flows near it, particularly as far as water soluble anthropogenic organic pollutants in the river are concerned.

The biochemical aspects involve the factors which affect the survival

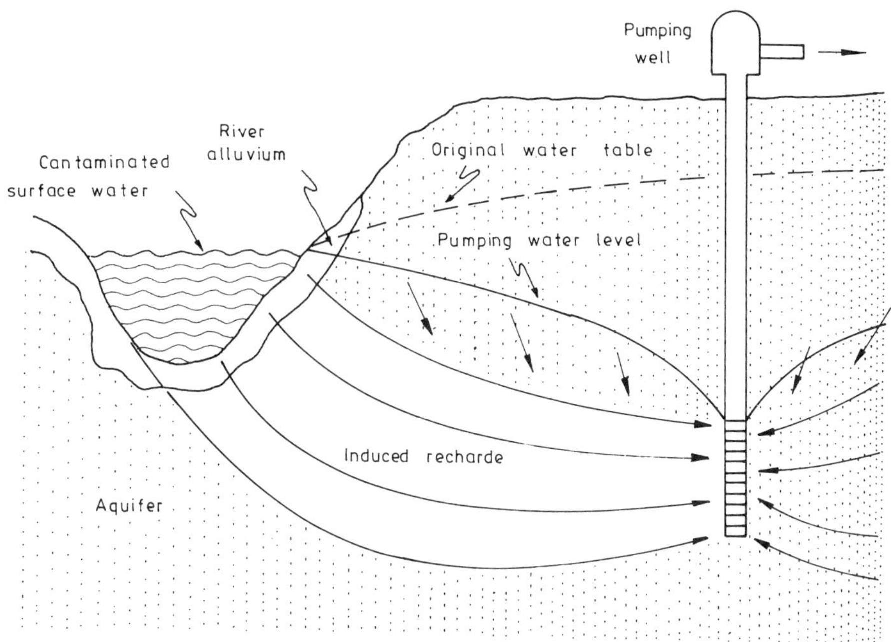

Fig. 1. Induced infiltration from stream. Stream pollution can result in groundwater pollution.

of such pollutants (synthetic, "hard" nonionic surfactants in our case) in both surface and subsurface water systems, the factors which determine the adsorption-desorption properties of these substances on solid surfaces, and their fate on bio- and/or chemicaldegradation, under both aerobic and anaerobic conditions. The hydrological aspects involve the permeabilities of different types of soil to water solutions substantially loaded with organic substances and the capacity of the soils either to transfer or to adsorb these chemicals.

A crucial analytical aspect is the reliable determination of low concentrations of nonionic detergents in the water in the presence of all other contaminants (uncontrolled "in vivo" conditions) [6]. Since the nonbiodegradable (hard) nonionic detergents are expected to survive for long periods of time in wastewaters as well as in surface and subsurface natural waters, our study has been primarily confined to the determination of nonionic trace concentrations in well-water systems and in streams carrying municipal sewage nearby. The wells were located in the vicinity of the streams (but were not equidistant from them) and pumped water from different depths.

Major factors which should be considered as determining the extent of well pollution from municipal sewage draining into nearby surface streams (rivers) include:

1. The extent and composition of stream pollution.
2. The distance of the well from the contaminated stream [7].
3. The geological crosssection between the well and the stream including the types of soil.
4. The existence of a hydraulic head (HH) between the stream and the well.
5. The well's depth (i.e. the depth from which the water of the well is being pumped).

PURPOSE

Since: (a) several municipal sewage systems in Israel (as well as in many other countries) are channeled into surface water streams after either a brief treatment or no treatment at all [4]; (b) hard nonionic surfactants of the nonylphenol-ethoxylate type (APEO) comprise about 10 percent of the synthetic detergents found in Israel municipal wastewaters [4]; and (c) even though groundwater/subsurface systems are not biologically inert, the rate of degradation of hard nonionic surfactants has been shown to be substantially slower in them than in surface water systems [8]; a study has been undertaken in order to determine: (a) the extent to which our groundwaters – particularly in the vicinity of streams polluted by municipal sewage – have been contaminated by synthetic detergents (specially: nonionic); (b) the extent to which nonionic detergents found in water wells may be reliably used as tracers for the monitoring of aquifer pollution by municipal wastewater; and (c) the nature of the geological-hydrological-physicochemical relationships as far as synthetic detergents in subsurface water/soil systems are concerned.

METHOD AND PROCEDURE

Two groups of wells – three each within the basin of a different stream constituted the set of wells sampled. The wells are not located equidistantly in the vicinity of the streams (each pumps water from a different depth). The first group (designated D.I., I.B., and M.C.) are located near Hadera stream and their water is pumped from a Pleistocene sandy aquifer. The second group (designated A.N., S.T., and Z.A.) is located near Hadas stream and their water are pumped from a Cenomanian-Turonian aquifer. The existence of HH facilitates potential pollution of the surveyed wells surveyed by the adjacent streams which contain municipal sewage wastewater.

Water samples were collected simultaneously from the wells and the corresponding polluted streams at the shortest distances from the former during 1985-1986. The nonionic surfactants content of the water samples has been determined by the use of the colorimetric cobalt thiocyanate-nonionic detergent complex (modified SDA-CTAS) method [6,9]. The following specially prepared calibration curve (absorption versus concentration) was used:

$$Abs. = 0.0644 \cdot [conc.] + 1.8 \cdot 10^{-3} \text{ (corr. coef. } \rho = 0.92)$$

The wells and the streams were sampled several times during one year, and the tabulated concentrations data represent a weighted average value.

The concentration of the nonionic detergents determined in the wells and in the corresponding stream samples were delineated and analyzed in terms of the type and extent of stream pollution, the distance of the well from the stream, the HH and the depth of the well, as well as the long-term environmental consequences.

RESULTS AND DISCUSSION

Data concerning the characteristics of each well (distance from stream, depth, and HH between the stream and the well) as well as representative concentrations of nonionic detergents found both in the wells and the corresponding streams are given in Table 1.

The nonionic detergent concentrations in the streams (right-handed column in Table 1) are in accord with what was found in recent studies [4], and thus represent typical concentrations in Israel surface waters as of 1985–1986. The concentrations of nonionic detergents found in the adjacent water wells are alarming and should become a matter of deep concern even though the presence of the hard surfactants in groundwaters could be expected [4]. Not only do substantial amounts of the nonionic detergents which reach surface waters find their way into groundwaters, but one of their metabolites - the nonylphenol is known for its increasing concentrations and long-term persistence in wastewaters particularly under anaerobic conditions [11, 12].

Traditionally, the detergent content of groundwater was determined using the classical MBAS method which would account only for the anionic detergents and overlook the presence of nonionic detergents. One suspects that this may be the reason that detergents were not considered as being a major threat to groundwater and its reuse. In those countries in which

Table 1. Nonionic Detergent Concentrations in Water Wells Adjacent to Streams Polluted by Municipal Sewage

Stream/Well	Well Characteristics			Nonionic Concentrations[a,b]	
	Distance	Depth	Hydraulic Head	Well	Stream[d]
		(meters)			(mg/liter)
Hadera/D.I.	40	61	12.9	0.78	1.7
I.B.	40	50	11.7	0.77	2.6
M.C.	160	50	12.5	0.12	1.6
Hadas/A.N.	150	91	-	0.44	2.4
S.T.	150	156	21.0	0.48	2.1
Z.A.	300	80	20.6	0.27	2.0

a. Expressed in terms of a weighted average mixture of ethoxylated nonylphenol-based nonionic detergents (NPnEO)[6,10].
b. Determined during 1985–1986. d. Refers to concentration found in the nearest spot to the corresponding well.

hard nonionic detergents are extensively used in the household, and where institutional-industrial detergent formulations are used [13], the need for groundwater monitoring for nonionic detergents is clearly an imperative.

Although data concerning the particular "dilution-survival" ratio for the systems studied are not available as yet, the average 23% dilution of nonionic detergents found in the water wells compared with the initial concentrations in the adjacent streams (average: 0.48/2.07 mg/liter) suggests that this probably represents the lower limit of nonbiodegradable detergent pollutants reaching groundwaters under similar conditions.

Inspection of the data given in Table 1 indicates an inverse relationship between the distance of a given well from the polluted stream and the concentration of the nonionic detergent in the well: the closer the well the higher its contents of nonionic detergents. This pattern appears to be general for each group of local wells separately or for the entire set of wells surveyed as a system. In view of the complexity of the multivariant systems involved (i.e., soil-water-detergents-microorganisms-groundwater) [14], and given the lack of sufficient data as yet, the establishment of a meaningful correlation curve (concentration versus distance) at this stage is premature. No obvious, simple pattern appears to emerge from the data collected as far as the detergent concentration-hydraulic head relationships are concerned. The data appear to suggest that the deeper the well the higher the concentration of the nonionic detergents in the well's water within each set of wells, assuming all other involved factors are similar, and taking into consideration the initial concentration of the detergents in the closest spot of the corresponding adjacent stream. More data are needed, however, if any existing relationships are to be established.

The mean velocity of movement of the detergent pollutant front (S) may be given at first approximation – by the following equation: $S = \frac{C \cdot V}{R}$

R - The retention capacity per unit gross volume of soil for the pollutant (micrograms of pollutant per cubic centimeter of soil).

C - Detergent (the pollutant) concentration of the feed wastewater solution at the source (mg/liter).

V - Average velocity of water transporting in the given soil (meters per day).

R may be estimated from the difference between the concentration of the nonionic detergents in the stream and the adjacent well, and V may be obtained from:

$$V = KI$$

where:
K -Permeability (meter/day)
I - Hydraulic Gradient $= \frac{\Delta H}{L}$; L = distance of well from the stream.

Assuming K=5 meter/day^{wf}, one obtains for the most distant wells M.C. ("Hadera" stream) and Z.A. ("Hadas" stream) the values of ca. S=0.42 and 0.4 m/day, respectively. This means that it should take the nonionic detergents about 75 and 38 days, respectively, to reach these wells from the corresponding polluted streams. It thus turns out that even such long periods of time are not sufficient for the complete primary chemical and/or biological degradation of the nonionic detergents and that their penetration into groundwater and aquifers is unavoidable.

SUMMARY AND CONCLUSION

Clearly, groundwaters are being contaminated by surfactants, the origin of which are surface streams polluted by municipal sewage. Neither the naturally occurring biodegradation in the subsurface water/soil systems nor physical soil adsorption processes are capable of avoiding this contamination.

Based on the preliminary results presented above, our approach of using the nonbiodegradable nonionic detergents of the NP_nEO type as tracers of groundwater and aquifer pollution caused by municipal sewage is valid, practical, easy and inexpensive to perform, and can be reliably used routinely in groundwater monitoring. Obviously, this methodology is only applicable when 'hard' NP_nEO nonionic detergents (or, alternatively, any other type of nonbiodegradable surfactants) constitute a substantial component of household cleaning formulations. Under such circumstances, the suggested method can be used even if municipal wastewaters are treated before being disposed into surface water bodies in the environment, since the resistance of these organic surfactants toward various sewage treatment processes and their persistence in the environment particularly under anaerobic conditions are well-known and documented [4, 15, 16].

It is obvious that in view of the increasing demand for portable water and the ever-increasing extensive use and reuse of our precious groundwater resources, the continuation of uncontrolled use of *replaceable* 'hard' nonionic detergents in household and institutional formulations constitutes a real environmental threat with all the consequences involved.

ACKNOWLEDGEMENTS

The technical help of the Hydrological Service, The Water Commission, Northern District, Israel and the financial assistance of "Oranium" – the School of Education of The Kibbutz Movement in Israel are highly appreciated.

REFERENCES

[1] Burmaster, D.E., and R.H. Harris, *Tech. Rev.*, **85**, 50-62, 1982.

[2] Pye, V.I., R. Patrick, and J. Quarles, *Groundwater Contamination in the United States*, University of Pennsylvania Press, Philadelphia, Pa. 1983.

[3] Schneider, R. *Tenside/Surfactants/Detergents*, **21**, 212-215, 1984.

[4] Zoller, U. *Journal Amer. Oil Chem. Soc.*, **62**, 1006-1008, 1985.

[5] Schneider, R., et al. *Water Sci & Tech.*, **19**, 449-460, 1987.

[6] Zoller, U. and R. Romano, *Environmental International*, **9**, 55-61, 1983.

[7] Flynn, J.M., A. Andreoli, and A.A. Guerrera, *Journal AWWA*, 1551-1662, 1958.

[8] Larson, R.J., and R.M. Ventullo, Proceedings of the 3rd National Symposium on Aquifer Restoration and Groundwater Monitoring, U.S.A., pp. 402-409, 1983.

[9] Boyer, S.L., et al. *Environ. Sci. & Tech.*, **11**, (13), 1167-1171, 1977.

[10] Zoller, U. In L. Pawlowski, et al. (Eds). *Chemistry for Protection of the Environment*, Elsevier, Amsterdam, pp. 161-166, 1984.

[11] Giger, W., P.H. Bunner, and C. Schaffner, *Science*, **225**, 623-626, 1984.

[12] Kravetz, L., H. Chung, K.F. Guin, W.T. Shebs, and L.S. Smith, *Tenside/Surfactants/Detergents*, **2**, 1-6, 1984.

[13] Haupt, D.H., *J. Amer. Chem. Soc.*, **60**, 1914-1918, 1983.

[14] a. Wilson, J.T., J.F. McNabb, D.L. Balkwill, and W.C. Ghiorse, *Ground Water*, **21**, 125-132, 1983.
b. Ewing, B.B., L.W. Lepke, and S.K. Banerji, A publication of the University of Illinois, Urbana, Illinois, 1960.
c. Howells, W.G., et al. *Proceedings of the World Surfactants Congress*, München, Germany, pp. 437-452, 1984.

[15] Gerike, P., and W. Jasiak, *Tenside/Surfactants/Detergents*, **23**, 299-304, 1986.

[16] Giger, W., M., Ahel, M. Koch, H.U. Laubscher, C. Schaffner, and J. Schneider, *Wat. Sci. Tech.*, **19**, 449-460, 1987.

CHANGES IN CHEMICAL COMPOSITION OF SOILS AS A RESULT OF IRRIGATION WITH POTATO STARCH WASTE WATER

H. MARZEC

University of Technology and Agriculture
85-029 Bydgoszcz, Poland

ABSTRACT

Systematic studies of physicochemical composition of potato starch waste water (PSWW) used to irrigate grassland have been performed. The application of PSWW results in changes in chemical composition of the soil. Shortly after irrigation, (i.e., 1 month), an increase in content of nitrogen, potassium, magnesium, phosphorus and organic carbon took place. After the next 8 months, a decrease in the content of these elements was noticed. Application of PSWW caused a decrease in the content of calcium and in soil pH as well as an increase in hydrolytic acidity of soils especially when a high PSWW dose was used.

INTRODUCTION

PSWW contains considerable amounts of organic compounds and fertilizing constituents (N, P, K). To prevent the migration of these constituents to water basins, it is necessary to retain them in the soil in order to utilize their fertilizing properties [1–8]. A two-stage method of PSWW purification has commonly been used: initial mechanical treatment and biological treatment which proceeds in the soil environment. For agricultural utilization

Chemistry for the Protection of the Environment
Edited by L. Pawlowski *et al.*, Plenum Press, New York, 1991

Table 1. Physicochemical Composition of PSWW Originated in Trzemeszno Potato Starch Factory after Autumn Campaign 1987

Parameters		Date of sampling						Mean values	Polish mean
		22.09	02.10	19.10	28.10	06.11	23.11		
pH		6.2	6.1	5.7	5.6	5.8	4.6	5.7	5.8
Total suspension	mg/l	472	790	991	731	1116	498	767	495
Mineral suspension	mg/l	159	142	277	166	282	199	204	90
Organic suspension	mg/l	313	648	714	566	834	299	562	375
BOD_5	mgO_2/l	1450	1636	1507	1751	3848	3520	2285	1220
Nitrogen	mg/l	94	165	160	194	277	185	179	142
Calcium	mg/l	173	142	137	136	92	96	129	76
Magnesium	mg/l	24	18	35	46	24	26	29	-
Phosphorus	mg/l	42	18	25	27	27	57	33	24
Potassium	mg/l	306	334	364	386	526	440	393	300

of PSWW, light, less fertile and insufficiently wet soils should be chosen. The literature on the effect of PSWW on soil fertility shows that sewage utilization causes a number of changes to the physicochemical properties of soils; these changes can be beneficial on detrimental. The magnitude and direction of these changes depend mainly on the chemical composition of the PSWW used, as well as climatic and soil conditions.

EXPERIMENTAL

Materials and Methods

In the year 1986, the field investigation was started on the experimental area "Koz lowo". The investigation covered a region of 0.5 ha. It was sowed with suitable mixtures of grasses. The effect of two doses of PSWW on the chemical properties of soils was studied in the years 1987 and 1988. The whole area of investigation was divided into three parts separated from each other by protective zones. One part was not irrigated while two others were sprinkled with 180 mm (low dose) and 360 mm (high dose) of PSWW.

During the period of experiment, samples of soils were collected three times:

- in autumn 1987, before irrigation,
- in winter 1988, after irrigation,

Table 2a. Results of Chemical Analyses of Soils Before and After Irrigation with PSWW in the Years 1987 and 1988

Dose 0 mm				
		Data		
		25.09.87	21.01.88	23.09.88
Parameters	Unit			
N	mg/100g	73.0	69.5	63.7
P	soil	8.0	7.9	4.6
K	"	10.2	11.1	9.7
Ca	"	43.5	39.0	26.7
Mg	"	2.5	2.9	2.4
C	"	914.0	721.0	681.0
pH in H_2O	pH	5.4	5.5	5.0
pH in KCl	pH	5.1	5.1	4.6
Hh	me/100g	1.5	1.5	1.8
C/N		12.5	10.4	10.7

- in autumn 1988, before irrigation.

The following determinations were carried out:
- organic carbon by Tiurin method,
- total nitrogen by Kjeldahl method,
- available phosphorus by Egner-Riem procedure,
- exchangeable calcium and potassium by flame photometry,
- available Mg by Schachtschabel method,
- pH of soil (in water and in 1m KCl) potentiometrically,
- hydrolytic acidity (Hh) by Kappen method.

Simultaneously, during the period of autumn 1987 in the Trzemeszno Potato Starch Factory, the chemical composition of mixed PSWW used in this experiment was studied.

RESULTS AND DISCUSSION

Characterization of PSWW

The chemical composition of PSWW analyzed during autumn 1987 varied considerably from an average composition of mixed waste water described by Kutera [3]. In the present work, the values of all determinations were considerably higher than those assumed as Polish means. Especially large values have been found for suspensions, BOD_5 and calcium. In the final

Table 2b. Results of Chemical Analyses of Soils Before and After Irrigation with PSWW in the Years 1987 and 1988

Dose 180 mm				
		Data		
		Before	After	Before
Parameters	Unit	25.09.87	21.01.88	23.09.88
N	mg/100g	73.5	94.5	65.9
P	soil	8.4	11.8	5.5
K	"	11.6	21.5	18.3
Ca	"	79.5	76.0	55.4
Mg	"	2.2	3.8	2.8
C	"	926.0	957.0	823.0
pH in H_2O	pH	4.8	4.8	5.0
pH in KCl	pH	4.4	4.3	4.7
Hh	me/100g	0.9	1.5	2.2
C/N		12.6	10.1	12.5

stage of the studies, the highest values for BOD_5 were noticed (Table 1).

Effect of PSWW Irrigation on Chemical Properties of Soils

Results obtained from the analyses show (Table 2) that the content of total nitrogen in soil was at its highest immediately after irrigation with PSWW. This rapid increase by 30–50% was caused by the large amount of nitrogen contained in PSWW. A significiant decrease in nitrogen content was observed in autumn 1988.

Irrigation with PSWW also resulted in increased potassium and phosphorus content. In relation to potassium, it was shown that its content increased 2 and 2.3 times after irrigation with low PSWW and high PSWW doses, respectively. On the other hand, a slightly lesser increase in the content of available phophorus in soil has been observed. However, after eight months, the contents of both elements decreased.

Similar behavior (i.e., increase and then decrease in content) has been observed for magnesium and organic carbon after irrigation. Studies of Kutera and Czyżyk [2] showed a beneficial effect of PSWW application on an increase in organic carbon content of soil humus layer.

Over a period of a year, hydrolytic acidity increased generally two times on average. Soil reaction did not change after irrigation with a 180–mm dose of PSWW, while the use of a 360–mm dose caused a decrease in soil pH by 1.0. Analysis of the soil chemical composition done in January

Table 2c. Results of Chemical Analyses of Soils Before and After Irrigation with PSWW in the Years 1987 and 1988

Dose 360 mm		Data		
		Before	After	Before
Parameters	Unit	25.01.87	21.01.88	23.09.88
N	mg/100g	64.0	96.0	53.6
P	soil	5.7	11.0	4.1
K	"	6.4	15.0	7.7
Ca	"	76.0	55.4	37.9
Mg	"	2.4	2.7	2.5
C	"	711.0	840.0	705.0
pH in H_2O	pH	5.6	4.6	5.2
pH in KCl	pH	5.2	4.2	4.9
Hh	me/100g	1.0	1.9	2.1
C/N		11.1	8.8	13.2

showed that irrigation with PSWW caused a decrease in calcium content, particularly in the case of the high dose of PSWW application (by $\sim$ 28%).

Further decreases of this important element by 27 and 32% for 180- and 360-mm doses (respectively) occurred during the next eight months (from February to September, 1988).

Generally, a beneficial effect of PSWW irrigation on content of N, P, K and Mg in soil has been noticed. These constituents can be taken up by plants during growth. It must be stressed, however that PSWW irrigation, particularly in the case of the high dose, also caused a certain detrimental effects like a decrease in exchangeable calcium content and increase in hydrolytic acidity. Similar observations were noted by other authors [8].

CONCLUSIONS

- PSWW from Trzemeszno Potato Starch Factory is more polluted with mineral and organic compounds than those effluents produced by other potato starch factories in Poland.
- High contents of these compounds in sewage clearly indicate that there is insufficient initial purification conducted in the potato starch factory.
- Chemical analyses of untreated soils indicated that after the period of one year, the contents of N, P, K, Ca, Mg and organic carbon decreased due to uptake by plants and/or leaching by rainfall.

- Simultaneously an increase in hydrolytic acidity and decrease in pH of soil have been observed.

- The irrigation of soils with a 180–mm dose of PSWW had an effect on the increase in N, P, K, Mg and organic carbon contents of the soils. Soil pH was kept at the same level.

- The use of high PSWW (i.e., 360 mm) caused the increase in N, P, K, Mg contents of the soils, decrease in Ca content and drop in soil pH. Such high dose, can be used simultaneously with calcium supplemental fertilization.

REFERENCES

[1] Förster, L., Teichardt, R., Arch. Acker u.Pflanzenbau u.Bodenkd., 31, 725, 1987.

[2] Kutera, J., Czyżyk, W., Biblioteka Wiad. IMUZ, 27, 1, 1968.

[3] Kutera, J., Wiad. Mel. i Lak., 10, 267, 1983.

[4] Multan, H., Nazaruk, M., Rytel, Z., Zawada, E., Wiad. Mel. i Lak., 8 - 9, 213, 1977.

[5] Stehlik, K., Musil, J., Rostl. Vyroba, 28, 1285, 1982.

[6] Török, L., Vermes, L., Wiad. Mel. i Lak., 2, 33, 1972.

[7] Wiewiórka, A., Wiad. Mel. i Lak., 8 - 9, 236, 1984.

[8] Wiśniewski, W., Gonet, S., Majcherczyk, Z., Szczygielska, T., Zesz. Nauk. ATR, 8, 119, 1979.

WASTE CHARACTERIZATION AND MONITORING METHODOLOGY IN THE UNITED STATES

D. FRIEDMAN

Office of Solid Waste
United States Environmental Protection Agency
Washington, DC 20460, USA

INTRODUCTION

In 1976, as a consequence of the discovery of several instances of serious ground water contamination from industrial chemical waste, the Congress of the United States enacted the Resource Conservation and Recovery Act (RCRA). This law expanded the environmental pollution prevention and control authority to include solids, liquids, sludges, and other wastes. The RCRA established a two-part system for waste management–Subtitle C addresses hazardous waste, and Subtitle D addresses solid wastes not designated as hazardous (e.g., municipal wastes). Prior to the enactment of RCRA, the management of these wastes was not subject to Federal regulatory control.

Under the authority of RCRA, the United States Environmental Protection Agency (EPA) has developed a comprehensive waste management program designed to protect human health and the environment from exposure to hazardous wastes. EPA issued the initial hazardous waste regulations in 1980 establishing a comprehensive Federally mandated and enforced system of "cradle to grave" waste management and control. The management

Chemistry for the Protection of the Environment
Edited by L. Pawlowski *et al.*, Plenum Press, New York, 1991

of nonhazardous wastes is the responsibility of the individual states.

An essential element of the regulatory framework for managing hazardous wastes is testing. Areas in which analytical and property testing plays a major role include hazardous waste identification, proper management method selection, facility operational monitoring, and site investigation and hazard assessment. EPA has developed and evaluated appropriate testing methodologies to aid the regulators and the regulated community in conducting these various types of testing and assessments necessary for RCRA compliance. All the approved or required sampling and testing methods are contained in the compendium: Test Methods for Evaluating Solid Waste (SW-846) [1].

HAZARDOUS WASTE IDENTIFICATION

The RCRA directed the U.S. EPA to identify wastes requiring controlled management though specific listing and on the basis of specific quantifiable properties called "characteristics". The characteristics developed to date include leachability, reactivity, corrosivity, and ignitability. A key concept in the hazardous waste definition system is that of mismanagement. The test methods that address these hazardous waste characteristics are an important part of the EPA's regulatory framework.

The Toxicity Characteristic

During the regulatory development period, EPA collected information about waste generation and management to establish major exposure pathways that affect human health and the environment. Because of the overwhelming complexity of the task at hand, the Agency concentrated its initial efforts on limiting human exposure to toxic chemicals resulting from waste storage, transportation, treatment, and disposal. Data available, at the time, indicated that a major pathway for human exposure to hazardous wastes was through ground water. When toxic wastes are placed in a nonsecure landfill, mobile components may leach from the waste into the subsurface environment beneath the disposal site. Ultimately, ground water contamination can result from the leaching. The codisposal of industrial chemical wastes with municipal wastes in an unlined landfill was identified as a reasonable mismanagement scenario to employ in defining the toxicity characteristic for hazardous wastes based on leaching.

Using as a basis, the sanitary landfill codisposal scenario, the Agency developed the first of four characteristics of a hazardous waste. This characteristic is called Extraction Procedure Toxicity. Extraction Procedure Toxicity is concerned with the potential for contaminating ground water a waste poses. It employs a potential for leachability estimation procedure known as the Extraction Procedure, or Method 1310, combined with regulatory thresholds based on drinking water standards and considerations of subsurface transport changes in contaminant plume concentrations.

Since its initial promulgation in 1980, EPA has been working to improve the characteristics in several ways. These include: development of a more accurate and precise test procedure for estimating the changes that will occur as the contaminated plume of ground water moves through the subsurface environment; expanding the number of chemicals for which hazardous waste identification regulatory thresholds have been developed; and refining the health thresholds to reflect new data. As a result of these activities, in 1986, the Agency proposed changes to the characteristics and, in 1990, expects to promulgate these changes [2].

EPA is, however, continuing to work toward improving and expanding the scope of the characteristics. Studies are now underway to further refine the leaching tests so that they more accurately reflect the ground water impact potential both of viscous, oily materials and of solidified or other large particle size materials; refine the subsurface fate and transport models with respect to metals, anions (e.g., CN); and biodegradable and hydrolytially unstable species; and continue to expand the list of chemicals for which thresholds have been established.

The Reactivity Characteristic

While the toxicity characteristic focus is on the effects of long-term exposures, the reactivity characteristic deals with acute, short-term issues. These include such diverse concerns as whether the waste is likely to explode when subjected to initiation stresses normally encountered during waste management, and whether the waste might release toxic gases (e.g., HCN, H_2S, H_3P) in high enough concentrations to cause asphyxiation when comingled with non-hazardous wastes.

While the Agency has developed several tests and regulatory criteria for identifying such wastes, the tests have not yet been fully validated and incorporated into our regulatory program.

Among the test procedures that are presently employed for defining reactive wastes are: the Card Gap Test for explosivity, and a Releasable Toxic Gas Test for determining a waste's hydrogen sulfide and hydrogen cyanide generation potential. Using the gas generation potential test, wastes that release more than 250 mg/Kg of hydrogen cyanide or 500 mg/Kg of hydrogen sulfide are classified as hazardous [1].

The Corrosivity Characteristic

The corrosivity characteristic is another of the hazardous waste identification criteria which identifies wastes that pose several very different hazards. One of these is the acute hazard posed by wastes that, upon contact, attack skin and other living tissue. A second, and very different concern is with wastes that require special containers to prevent leaks and resulting uncontrolled introduction of toxic materials into the environment. A final concern addressed by the corrosivity characteristic is a waste's potential to cause leaching of otherwise immobile toxicants from land disposal units.

At the present time the corrosivity characteristic applies only to liquid wastes. Liquid wastes are deemed to be hazardous if they are substantially corrosive to mild steel (i.e., >0.6 cm/yr) or are aqueous and either strongly acidic or strongly alkaline (i.e., pH less than 2.0 or greater than 12.5). Future consideration is being given to expanding the corrosivity characteristic to include physical solids which form highly corrosive solutions when in contact with water.

The Ignitability Characteristic

Fire is an ever-present danger when handling certain wastes. The ignitability characteristic identifies those wastes whose composition and properties make them particularly dangerous. Wastes are characterized as hazardous if they are particularly sensitive to ignition and, when ignited, burn vigorously and persistently. An example of wastes that are classified as hazardous are liquids that have a flash point below 60 degrees C, and certain finely divided metals.

MANAGEMENT METHOD SELECTION

Solid wastes that are classified as hazardous wastes must be managed in accordance with RCRA regulations pertaining to storage, transportation, treatment, and disposal. These regulations include a permitting requirement. Only RCRA-permitted facilities may accept hazardous wastes. Furthermore, facilities may only accept, for management, those wastes for which they have a permit. In addition, Congress has directed EPA, in the 1984 amendments to RCRA, to promulgate land disposal restrictions regulations which place stringent limits on the properties of materials that can be placed into or on the land (e.g., landfills, surface impoundments, land treatment units).

Selection of appropriate management methods is linked closely to the characteristics and properties of the waste. Appropriate characterization depends on the management options being considered. Waste management options cover a broad range including storage in drums, tanks, and surface impoundments; chemical, thermal, or physical treatment; and disposal in landfills. A few examples should better illustrate how waste management options affect the degree of required waste characterization.

In order to minimize leachate formation in landfills, the RCRA regulations prohibit landfilling of drummed waste that contains "free liquid". To identify if a waste contains "free liquid" the EPA has developed method 9095. The method employs a disposable, conical filter screen and release of any liquid through the screen indicates that the waste contains "free liquid". To perform the test takes only 10 minutes and is an example of a simple, quick screening test to classify waste materials for management purposes.

A more complicated testing scheme is needed, however, to determine whether a waste complies with the land disposal restrictions regulations. Land disposal standards have been, and still are being, set for all hazardous wastes. The RCRA bans the landfilling of hazardous wastes until and unless they have been treated to Best Demonstrated Available Technology (BDAT) standards. EPA is required to establish BDAT standards for hazardous wastes. For organic wastes, EPA has determined that BDAT for such wastes is incineration. Since incineration destroys the toxic organics, the BDAT thresholds are based on the residual content of the toxic organics in the treatment residue – the ash. Testing protocols for such wastes then become conventional compositional analysis.

For inorganic wastes the problem is more complicated. While some of the species, such as the anion cyanide, can be destroyed by chemical or biological treatment, the toxic metal species can never be destroyed. To render such wastes less hazardous, solidification and vitrification are among the most effective treatment methods. These processes render the metals less leachable and therefore less likely to contaminate ground water when land disposed. For such wastes, BDAT thresholds are based on potential leachability. To quantify the extent to which a waste's potential leachability has been decreased by the treatment process, the EPA employs method 1311 [3]. Method 1311 is a relatively aggressive extraction procedure that serves as a measure of treatment process effectiveness. It is not meant to accurately predict the leachate quality resulting from land disposal of a solidified waste.

We have seen how qualitative waste management standards and appropriate testing methods have been established to protect ground water resources. For my last example, I will like to switch to a different concern, namely protection of air quality. Studies in the United States have shown that industrial waste management is a significant source of organic chemical contamination of the air and a major contributor to ozone formation. To lessen such emissions, the EPA is developing fugitive emission standards for hazardous waste management facilities. The standards are designed to minimize the release of ozone precursor compounds from waste management operations. To determine which wastes need this special handling, the EPA has developed a new test for measuring the volatile organic chemical content of wastes. The method, and the thresholds that are soon to be proposed, will identify those wastes that require special handling to prevent the release of ozone-forming chemicals to the air.

FACILITY OPERATIONAL MONITORING

We have seen how waste characterization plays a substantial role both in determining the appropriate level of regulatory control for a waste and selecting suitable waste treatment and disposal techniques. At this time, I would like to briefly discuss another critical role monitoring methodology plays in the waste management program.

Once a facility is treating and disposing of hazardous wastes, it is critical that monitoring be carried out to insure that the facility is operating correctly and not endangering human health or the environment. This monitoring can take many forms. In the short time we have available, I will touch on only a few examples to illustrate the diverse nature of both the monitoring and the methodology that is available and under development.

Harm to ground water resources has been and remains one of our major concerns with respect to land disposal. While the technology of landfill design and operation has undergone a significant advancement over the past decade, the containment techniques (e.g., liners, caps, leachate collection) have not been used long enough to withstand the test of time. Therefore, prudent waste management requires that we monitor land disposal facilities in order to identify failure of the containment system early enough to take remedial action before human or environmental exposure occurs.

Design of landfill monitoring systems poses several problems. These include: where to place monitoring wells, what species to monitor for to serve as an early warning system, what methods to employ for detection of these indicator parameters, and finally how to evaluate the monitoring data to know when containment system failure occurs. While well placement and statistical data evaluation are critically important to the design and operation of a monitoring network the subject is too complex for discussion in this paper. Instead, I will concentrate on selection of indicator parameters and monitoring methods to use for detecting their presence.

The objective of ground water monitoring is to give as early an indicator of containment system failure as possible. For this reason, the parameters that are selected for monitoring should be those that are the most mobile and the ones that can be detected at very low levels. While initially EPA selected pH, specific conductivity, total organic carbon, and total organic halogen as appropriate indicators, we soon found that, with the exception of total organic halogen, these were not sensitive, reliable parameters. The ground water properties of pH, specific conductance and total organic carbon vary too much as a natural consequence of the change in seasons for them to serve as reliable indicators. Studies aimed at selecting better indicators led to the selection of a different approach to facility monitoring, namely, to monitor using a suite of sensitive, broad spectrum analytical methods chosen for their cost-effectiveness and ability to detect a wide variety of potential contaminants. These methods include purge and trap gas chromatography using a mass spectrometer detector (methods 5030 + 8240/8260) with liquid/liquid extraction followed by gas chromatography using a mass spectrometer detector (methods 3510/3520 + 8270) for the non-purgeable gas chromatographable organics, combined with plasma spectroscopy (method 6010) for metals. Such a combination has the advantage of being able to detect the presence of a wide variety of organic and metal species at very low levels. It would be hard to imagine a site leaking without at least one of these contaminants reaching the ground water. In fact, for sites that have been shown to be leaking, and causing environmental contamination, the purgeable organics are the first organic contaminants usually found.

I turn now to a totally different problem, namely, how to monitor an incinerator for proper operation. This is an area where United States testing methodology is undergoing rapid change. At the present time, we rely primarily on measuring parameters such as CO, combustion chamber temperature, and unburned hydrocarbons in the stack gas to insure proper incineration conditions. Studies have shown, however, that these parameters are not sufficient indicators of toxic organic compound destruction. Therefore, the EPA and the operators of hazardous waste incinerators are working to develop and evaluate new continuous measurement techniques for use in monitoring thermal treatment units. Parameters to be measured using these techniques include: specific organic chemicals that are present in the waste stream and difficult to combust or that are expected products of incomplete combustion, hydrogen halides present in the stack downstream of the scrubber, and specific toxic metals of concern such as hexavalent chromium.

Promising techniques that may be suitable include: ion trap mass spectroscopy for organic compounds, electrolytic conductivity based continuous monitors for hydrogen halides, long path length infra-red spectroscopy and laser techniques for monitoring stack emission for organic compounds and other species of concern, and special filters and detectors for trapping and identifying the presence and concentration of chromium VI.

While my discussion has focused primarily on containment and treatment effectiveness monitoring, another important aspect of monitoring relates to protection of workers at a contaminated waste management site undergoing remediation. Here one of our major concerns is with workers breathing organic compounds released to the air during earth moving or other remediation activities. To protect against such exposures portable monitoring techniques have been developed. For the most part the monitoring problem revolves around sample collection. Two very different approaches to sample collection that have been developed are the Summa canisters and the absorbent "badges".

The Summa canisters are highly polished and cleaned aluminum pressure vessels that are evacuated and used to collect samples of the air at the site for later analysis by conventional gas chromatography. The absorbent badges are small containers that are filled with a polymeric material, such as Tenax resin, and are worn by the worker. Organic compounds present in the air around the individual are adsorbed onto the resin and trapped for later analysis. The organic compounds absorbed onto the resin are easily desorbed and analyzed in a manner similiar to that with the canisters. The badges are rugged and can be used to develop an integrated exposure over a selected time period, a manner similar to that used with the film badges common in X-ray facilities and the nuclear industry.

SITE ASSESSMENT

The final testing application I will very briefly touch on is that of site assessment. Assessing the extent and hazard posed by contamination at an operating or abandoned waste management site is a very important, expensive, time consuming, and difficult problem. In the United States, tens of millions of dollars are spent annually by EPA and industry on this type of testing.

While the techniques employed are generally those of conventional sampling and analytical chemistry, the field is one of the most exciting in terms of new approaches and techniques. These new techniques will permit real-time screening and analysis of both wastes and environmental media right at the site. Savings, by not having to transport the samples to a distant laboratory, and not having to wait for the results before continuing the sampling or cleanup operation, are potentially enormous. At the same time, with these new techniques, the accuracy of our assessments will increase.

QUALITY ASSURANCE

I do not want to stop until I have, at least briefly, raised one critical aspect of testing and monitoring. That is quality assurance. Environmental decision making relies on data. Good decisions require good data. The data that we use must be of known and acceptable quality. A good quality assurance program, that includes all aspects of sampling, analysis and data reduction, is integral to obtaining good data. The quality assurance program is the cornerstone of the monitoring program. It should not be considered as a luxury but rather as an integral component of all data gathering efforts. Doing it right the first time is almost always cheaper and better than having to repeat the study.

The public demands protection from environmental pollution. The costs in terms of money and our credibility as good scientists and responsible regulators, mandate that the data we gather be accurate and sufficient to make good decisions.

CONCLUSION

As you can see, testing plays a critical role in all aspects of the waste management process. This is an exciting area of science and one that is still undergoing great changes. While I could go on talking for the rest of the morning, will stop and let our other speakers and poster presenters expand on many of the subjects that I have briefly touched on.

REFERENCES

[1] Test Methods for Evaluating Solid Waste. Volumes IA, IB, IC, and II. Office of Solid Waste. U. S. Environmental Protection Agency, Washington, DC 20460. SW-846, Third Edition, November 1986 as amended in 1989. Available from the U. S. Government Printing Office, Washington, DC 20402 as document number 955-001-00000-1.

[2] *Federal Register,* Volume 51, Number 114, June 13, 1986; pp. 21648-21693. Available from U. S. Environmental Protection Agency, Office of Solid Waste, Methods Section (OS-331), Washington, DC 20460.

A NEW INSTRUMENT FOR REAL-TIME ASSESSMENT OF WASTEWATER TOXICITY

M. I. BEACH

N–CON Systems Co., Inc. Larchmont NY 10538, USA

F. CADENA

Civil Engineering Departament
New Mexico State University, Las Cruces, NM 88003
USA

ABSTRACT

Simplified, rapid, real-time methods for assessing toxicity of wastewater are essential for controlling discharges to both surface waters and wastewater treatment plants.

This presentation describes a new instrument that combines the action of living micro-organisms, representative of the treatment system or receiving water, with a method of determining and recording their oxygen uptake when exposed to a continuous wastestream. The high speed of the instrument's reaction to toxic or inhibitory conditions, combined with a number of user selected alarm points gives operating personnel adequate warning so that remedial action can be taken. A personal computer records the data in both numeric and graphic form, and performs routine self–checks and daily calibration of the dissolved oxygen sensor. Stored records may be recalled for on screen review, printout or transfer to spreadsheets or reports.

Chemistry for the Protection of the Environment
Edited by L. Pawlowski *et al.*, Plenum Press, New York, 1991

INTRODUCTION

Throughout the world there is a rising concern about the effect of toxic effluents being discharged to receiving waters. This concern has focused increased attention on both the types and quantities of materials coming into wastewater treatment plants. When incoming wastewater contains materials that disrupt the normal activities of the micro-organisms that break down the organic materials in the treatment, the system is unable to function properly and discharges both untreated organic and toxic material to the receiving waters. In order to protect the micro-organisms, methods must be found to detect and prevent the introduction of toxic chemicals, or major pH changes into the system. Frequent analysis of wastewater is both costly and time consuming. Delay in data availability makes traditional tests such as Biochemical Oxygen Demand (BOD_5) and Chemical Oxygen Demand (COD) unsuitable for process control. In addition, these tests do not necessarily indicate toxicity. Simple, rapid, real-time methods for assessing toxicity of wastewater are essential for controlling discharges to both surface waters and to wastewater treatment plants.

Many wastewater treatment plants have systems which are used to detect the changes in the dissolved oxygen content of the mixed liquor to determine how the plant is operating. However, once the toxic or inhibitory substances have reached the biological system, very little can be done to reduce their effect on the biomass and prevent their discharge.

The role of the toxicity monitor is to determine the potential effect of incoming wastewater on the biomass in the treatment system in time to prevent a biological process upset. It must therefore be installed as early in the influent system as practical. Where an industrial facility is the source of toxic wastes, a toxicity monitor should be installed before the effluent enters the collection system, to allow maximum time to take remedial action.

To fill the requirements of a true toxicity monitor the instrument must:

1. show the effect of influent on the plant's own biomass
2. be truly continuous and on-line
3. have rapid response time
4. provide alarms when inhibitory or toxic conditions begin
5. require minimum operator attention

Because of the critical nature of the information it is providing, the monitor must also be able to:

1. check its own operation and calibration with suitable alarms in the event of a problem
2. provide a permanent record of "non–toxic", as well as problem conditions.
3. provide data in a form that can be easily compared and used in reports.

To meet this need a new instrument called a Bioscan has been developed. The Bioscan combines the power of a personal computer and a dissolved oxygen sensor with a submerged biological filter composed of the same microorganisms as in the treatment system. When installed in a wastewater treatment plant, it can be used for continuous, real-time monitoring of inhibitory or toxic materials in the influent. Effluents being discharged to receiving waters are monitored with a biological filter composed of highly sensitive micro–organisms that will not acclimate.

OPERATING PRINCIPLE

The operating principle of the Bioscan is simple. Aerobic micro–organisms require oxygen as they break down organic matter. If their activity is normal, the dissolved oxygen in the water is consumed. Toxic or inhibitory materials impair biomass respiration and little or no dissolved oxygen is used. Therefore, the quantity of dissolved oxygen and the rate at which it is consumed are indications that the material flowing through the biological filter is biodegradable, inhibitory or toxic.

DESCRIPTION OF THE INSTRUMENT

The key to the Bioscan's operation is the design of the submerged biological filter that supports a microbial population similar to the wastewater treatment plant activated sludge process. The biological filter is designed to maintain the layer of micro–organisms at a constant thickness. The wastewater influent or industrial waste is mixed with a readily biodegradable substrate and is aerated to saturation (7–8 mg/L). The substrate assures that the biomass will have sufficient food for vigorous oxygen demand. The ratio of wastewater to biodegradable substrate can be set by adjusting the rate of the individual feed pumps. The usual ratio is approximately 1 to 10. The mixture flows through the biological filter at a rate of 40 ml per minute. This rate allows the healthy biomass to almost deplete the available food and dissolved oxygen. If the wastewater contains toxic materials that inhibit or kill the biomass in the filter, the micro-organisms consume little or no oxygen. Thus, the quantity of dissolved oxygen (DO) passing out of the filter is an indication of inhibition or toxicity. The dissolved oxygen remaining in the effluent of the filter is monitored by the system's computer through a dissolved oxygen sensor. The results are continuously recorded by the computer and shown both numerically and as a graph on the computer display.

For quick and simple observation, the computer display is divided horizontally into three levels. If the graph is in the GREEN area of the screen, operation is normal. If the graph moves into the Green/Red stripe area it indicates inhibitory conditions. Entry of the graph into the RED area indicates a toxic condition. Alarms may be set through the computer to provide a local or a remote signal, such as on the central control panel.

The high speed of the biological filter's reaction to inhibitory or toxic conditions, combined with the user selected alarm points, gives operating personnel adequate warning so that remedial action may be taken.

After a toxic condition alarm, the computer switches off the sample stream and delivers only aerated biodegradable substrate to the filter to allow it to regenerate. When the filter has recovered, sample delivery is automatically restarted by the computer. If the filter is completely poisoned, it can be regenerated in 3 to 6 hours by flushing it out and reseeding it with filtered activated sludge. If continued monitoring is essential following a toxic event that requires the regeneration of the biological filter, a second or stand-by biological filter can be supplied. The stand-by biological filter is made up of the same biomass and maintained on the aerated biodegradable substrate. The stand-by filter can be put on line automatically by the computer, after a predetermined delay, or by the operator while the primary biological filter is being regenerated. The ratio of wastewater to biodegradable substrate should be reduced following a toxic incident to prevent poisoning the filter, but should still be high enough to monitor the inhibitory effects.

Self testing of the dissolved oxygen system is provided by the computer on a daily basis. An oxygen saturated sample (approx 7.5 mg/l) bypasses the filter and is sent directly to the oxygen measuring cell. After about 30 seconds to stabilize, the computer determines the deviation, if any, from the previous day's calibration check. An alarm is provided if the deviation exceeds 5%. The operator would then manually clean and recalibrate the DO cell as necessary. Sample pump delivery adjustments and dissolved oxygen system calibration can be done manually by plant operators without interfering with the computer settings.

Completed daily records are stored in the computer memory. The computer automatically assigns the correct date to each chart. User may select American, European or Julian dating as desired. Graphic and numeric records are easily recalled for on screen review, print out or transfer to spread sheets or reports.

CONCLUSION

Biomonitoring using this new technology combining the representative biomass and the power of a personal computer will help to reduce the incidence of damage to the treatment system. This will help to protect receiving waters from discharges of untreated organics along with the toxic materials. In addition, this new monitoring system provides a permanent record of compliance or non–compliance with toxic discharge regulations.

PATHWAYS OF CHEMICALS IN THE ENVIRONMENT

ENVIRONMENTAL PHOTOTRANSFORMATION OF THE HERBICIDE BROMOXYNIL (3,5-DIBROMO-4-HYDROXYBENZONITRILE) IN AQUATIC SYSTEMS CONTAINING SODIUM CHLORIDE

J. KOCHANY*, G. G. CHOUDHRY** and G. R. B. WEBSTER*

*Pesticide Research Laboratory Department of Soil Science University of Manitoba Winnipeg, Manitoba, Canada R3T

** The Great Lakes Institute and Department of Civil and Environmental Engineering University of Windsor, Windsor Ontario, Canada N9B 3P4

ABSTRACT

Aqueous solution phase photochemistry of the herbicide bromoxynil (3,5-dibromo-4-hydroxybenzonitrile) (1) in the presence of various concentrations of sodium chloride was extensively investigated with ultraviolet radiations near 313 nm. In the presence of 0.5 to 25.0 X 10^{-3} M NaCl, the quantum yields for the phototransformation of the herbicide bromoxynil (1) amounted to 0.045 ± 0.005 to 0.017 ± 0.007, whereas such data was 0.052 ± 0.004 in the absence of sodium chloride. These quantum yield data for the photolysis of 1 in the absence and presence of sodium chloride followed the Stern-Volmer equation, thereby indicating that NaCl apparently exerted quenching effects on the photolysis of the compound 1. The rates of photodegradation of the herbicide 1 in water were slower with than without the presence of sodium chloride.

Chemistry for the Protection of the Environment
Edited by L. Pawlowski *et al.*, Plenum Press, New York, 1991

For instance, the photolysis of the 7.8 X 10^{-6} M aqueous solution of the herbicide 1 in the presence of NaCl (10.0 X 10^{-3} M) resulted in the formation of 3-bromo-4-hydroxybenzonitrile (2), 3-bromo-5-chloro-4-hydroxybenzonitrile (3), 3-chloro-4-hydroxybenzonitrile (4), and 4-hydroxybenzonitrile (4A). In the case of this photoreaction of the chemical 1, the percentages of maximum concentrations of the photoproducts 3, 2, 4, and 4A were obtained after 10.5, 20.0, 30.0, and 44.0-min exposures to UV light, respectively, the percentage disappearance of the starting herbicide 1 being ca. 90% for 44-min photolysis. Photoproducts 2, 3, 4, and 4A were identified by interpreting the GC-MS data.

The formation of 4-hydroxybenzonitrile (4A) decreased with the rise in concentration of NaCl. For example, when a mixture of aqueous solutions of 3.0 mL bromoxynil (2.0 X 10^{-4} M) plus 0.5 mL NaCl (0.5 M) was exposed to UV light for up to 3 h, the photoproduct 4A could not be observed; however, other products, namely, phenols 2-4 were produced.

INTRODUCTION

Bromoxynil (3,5-dibromo-4-hydroxybenzonitrile) (1) as its potassium salt or its octanoate or butyrate esters is an active ingredient of several commercially available herbicides, viz., Buctril M, Pardner, and Sabre, which are commonly used for weed control in Canadian prairie agriculture [1]. Bromoxynil esters hydrolyze to provide phenol (1) in non-sterile water [2]. Thus, in natural waters, 3,5-dibromo-4-hydroxybenzonitrile (1) is anticipated to be formed easily from any of the formulations utilized.

Because of frequent applications of bromoxynil (1) and its formulations, there is a growing concern that it may contaminate the soil and water systems. The occurrence of sodium chloride in almost every soil and aquatic environment is well-known, thus natural waters contaminated with the herbicide 1 will also contain NaCl.

Pesticides present in aquatic systems can undergo both biotic and abiotic transformations. The latter reactions may include chemical or photochemical processes. Among these environmental pathways, photoalterations of pesticides in water by sunlight are of particular importance [3]. Sunlight phototransformations may proceed via direct and/or indirect photoreactions [3-8].

Environmental photochemistry of several herbicides was recently reviewed by Marcheterre et al. [9]. To the best of our knowledge, nothing has been published in the context of photolytic fate of bromoxynil (1). However, we recently reported the environmental phototransformation of 1 dissolved in water in the absence [10] and in the presence [11, 12] of soil fulvic acids. The purpose of this study was to determine the aquatic photolytic fate of this herbicide in the presence of NaCl as well as to identify its photoproducts.

MATERIALS AND METHODS

Reagents and Chemicals: HPLC-grade methanol, chloroform, and water were supplied by Caledon Laboratories Ltd., Georgetown, ON, Canada. Bromoxynil (1) (98% purity) was obtained from Chem. Service, West Chester, PA, U.S.A. Other chemicals such as sodium chloride, pyridine, p-nitroacetophenone (PNAP), phosphoric acid, sodium phosphate, sodium sulphate, potassium chromate, sodium carbonate, and 4-hydroxybenzonitrile (4A) were supplied by Aldrich Chemical Company Inc., Milwaukee, WI, U.S.A.

Ultraviolet (UV) Spectroscopy: UV spectra of solutions of the herbicide 1, both with and without the presence of NaCl, were recorded on a Hewlett-Packard model 8452 A Diode Array Spectrophotometer equipped with an IBM PC/AT microcomputer and ThinkJet Printer.

Photochemical Experiments: Solutions of bromoxynil (1) (7.8×10^{-6} M) in water together with those in aqueous 0.5-25.0 $\times 10^{-3}$ M NaCl were irradiated at around 313 nm utilizing a Rayonet Photochemical Reactor under previously described conditions [13]. Radiations around wavelength (λ) 313 nm were isolated with an aqueous chemical filter solution consisting of potassium chromate (0.270 g/L) and sodium carbonate (1.000 g/L) [14]. For monitoring the light intensity (I_λ), a chemical actinometer, namely, aqueous p-nitrocatophenone (PNAP) (1.00×10^{-5} M)/pyridine (1.41×10^{-3} M) system reported by Dulin and Mill [15] was used. The I_λ range was 3.0-4.0 $\times 10^{-6}$ E L^{-1} sec^{-1}.

For the identification of the photoproducts of bromoxynil (1), irradiations of 1 were performed on synthetic (preparative) scale. For such experiments, a 2.0×10^{-4} M solution of 1 plus NaCl (0.5 M) both in water were used in the ratio of 3.0:0.5 (v/v). Photolysis time was extended to 3.0 h in order to obtain sufficient concentration of chlorophenols as photoproducts (see below). The 10.0 mL photolyzate of 1 in 7.143×10^{-2} M NaCl was acidified with 0.1 mL phosphoric acid (1.0 M) and was subsequently extracted with three 2.0-mL portions of chloroform. The chloroform extract was dried over sodium sulphate for 30.0 min and was analyzed by GC-MS.

HPLC Analyses: Quantitative analyses of bromoxynil (1) and its photoproducts were carried out on a Waters Associates HPLC system (Model 6000A pump and M440 UV detector set at 254 nm) with AUFS = 0.01. A 25 cm X 3.2 mm (i.d.) C_{18} Bondapack column was used. Samples were injected through Rheodyne Model 7125 injector having 100 L loop. The column temperature was ca. 20°C. The mobile phase consisted of a mixture of methanol and 0.055 M phosphate buffer of 2.6 pH (2:3 v/v) [16]. The mixed solvents were membrane filtered (FH; 0.5 μm, Millipore Corp.) and degassed with helium (Linde-Union Carbide, U.S.A.). The solvent peak was used as the reference for dead volume determination. For 2.1 mL/min flow rate of the mobile phase, the retention times of bromoxynil (1) and its photoproducts in the presence of NaCl were as follows:

bromoxynil (1) = 7.2 min
3-bromo-4-hydroxybenzonitrile (2) = 5.2 min
3-bromo-5-chloro-4-hydroxybenzonitrile (3) = 6.8 min
3-chloro-4-hydroxybenzonitrile (4) = 4.7 min
4-hydroxybenzonitrile (4A) = 2.8 min.

Gas Chromatography - Mass Spectrometry: A Hewlett-Packard 5890 gas chromatograph coupled with a VG 7070E - HF mass spectrometer was used for the identification of photoproducts. The gas chromatograph was equipped with a 40 m capillary DB-5 column supplied by Chromatography Specialties Inc., Brockville, ON, Canada, the conditions for the operation being as follows:

injector temperature: 230°C
column temperature T_1: 130° for 2 min
rate: 5°C/min to T_2 = 220°C kept for 5 min.

Both chromatograms and mass spectra were recorded and analyzed using a VG-11-250T data system.

Identification of Photoproducts: Due to the unavailability of authentic standards, structures of the photoproducts 2-4 formed from the herbicide 1 in water containing sodium chloride (see Reaction 1) were determined by interpretation of GC-MS data. However, the photoproduct 4-hydroxybenzonitrile (4A) was identified by comparing the HPLC retention time and the GC-MS data of the product 4A with those of its authentic standard.

RESULTS AND DISCUSSION

UV Light Absorption Spectra: Fig. 1a includes the UV light absorption spectrum of a mixture of bromoxynil (1) (7.8 μM) plus sodium chloride (10.0 mM) dissolved in water. The UV spectra obtained after the irradiation of this reaction mixture for 10.0, 20.0, and 30.0 min are recorded in Fig. 1b, c, and d, respectively. It is evident from Fig. 1 that as the photolysis time increases, both broad absorption maxima seen around 284 and 225 nm disappear, while the value of absorbance in the 200-313 nm range increases. This observation indicates that the concentration of the aromatic moiety of the starting material 1 is lost as photodegradation proceeds. In the absence of NaCl, the broad absorption maximum at about 290 nm in the UV spectrum of aqueous solution of 1 is accompanied by a shoulder at 300 nm [10].

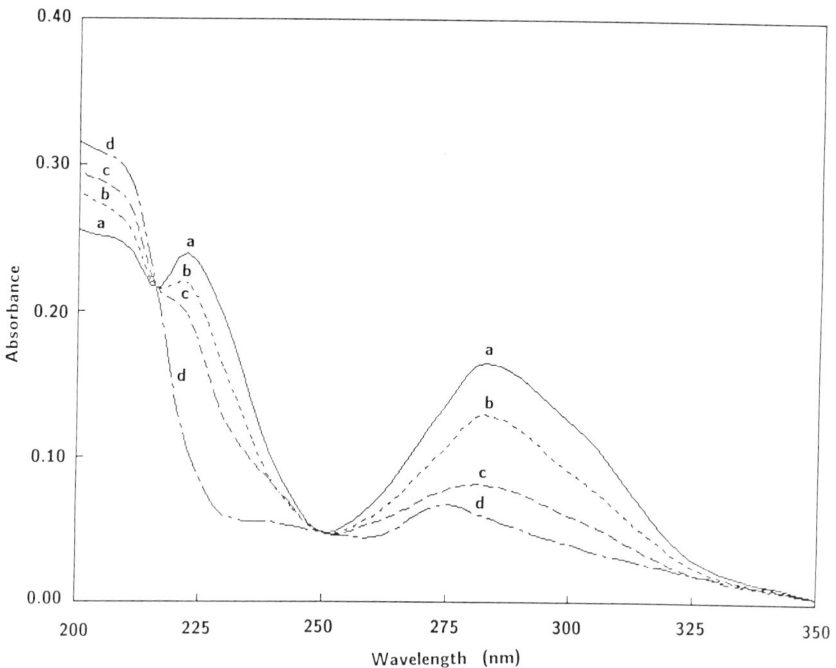

Fig. 1. Changes in the UV light absorption spectra of 7.8 X 10^{-6} M aqueous solution of bromoxynil (1) containing 10.0 X 10^{-3} M NaCl caused by 0.0 (a), 10.0 (b), 20.0 (c), and 30.0 min (d) exposures to radiations around 313 nm.

Quantum Yields (ϕ) for phototransformation: For the determination of quantum yields (ϕ) for phototransformation of the herbicide 1 in water containing NaCl, the theory and methods reported by Choudhry and Webster [3,14] were used. The ϕ values for compound 1 at around 313 nm are documented in Table 1. When the concentration of NaCl is raised from 0.5 to 25.0 mM, the quantum yield for the photodegradation of 1 decreases from 0.045 $\pm$ 0.005 to 0.017 $\pm$ 0.007, thereby showing that the rate of photolysis of 1 is reduced in the presence of sodium chloride.

Monitoring and Identification of Photoproducts: When aqueous solution of bromoxynil (1) (7.8 μM) with NaCl (10.0 mM) was subjected to photolysis for up to 60.0 min, four photoproducts, 3-bromo-4-hydroxybenzonitrile (2), 3-bromo-5-chloro-4-hydroxybenzonitrile (3), 3-chloro-4-hydroxybenzonitrile (4), and 4-hydroxybenzonitrile (4A), were observed (Reaction 1).

$$\mathbf{1} \xrightarrow[H_2O,\ NaCl]{h\nu} \mathbf{2} + \mathbf{3} + \mathbf{4} + \mathbf{4A} \qquad (1)$$

Table 1. Quantum Yields[a] (ϕ) for the Phototransformation of Aqueous Solutions[b] of Bromoxynil (1)[c] Containing Various Amounts of Sodium Chloride at around 313 nm.

NaCl (10^{-3} M)	Quantum yields
0.0	0.052 ± 0.004
0.5	0.045 ± 0.005
1.0	0.040 ± 0.005
2.5	0.034 ± 0.004
5.0	0.029 ± 0.007
10.0	0.024 ± 0.005
15.0	0.021 ± 0.007
25.0	0.017 ± 0.007

[a] Previously described procedures by Choudhry and Webster [14], were utilized.

[b] The pH of the sample solutions was in the range of 6.5 to 7.0.

[c] The concentration of the candidate herbicide used was 7.8 X 10^{-6} M.

The chlorophenols 3 and 4 seen amongst the photoproducts of 1 appear to be yielded via the photoincorporation of chloride ions into the substrate 1 as well as into the primary photoproduct 2 as discussed elsewhere [17]. The phenols 2 and 4A are also produced in the case of photolysis of 1 without the presence of sodium chloride. The monitoring of the concentrations of bromoxynil (1) along with its products 2-4A during the Reaction 1 are described in Fig. 2. It is apparent that under the present circumstances the maximum concentrations of the benzonitriles 3, 2, 4, and 4A are achieved after irradiation of 1 for 10.5, 20.0 30.0, and 44.0 min, respectively.

In the case of photoreaction of bromoxynil (1) (7.8 μM) in water for 15.0 min, plots of the ratios of concentrations of 3-bromo-5-chloro-4-hydroxybenzonitrile (3) to 1 together with those of 3-chloro-4-hydroxybenzonitrile (4) to 3-bromo-4-hydroxybenzonitrile (2) versus concentration of sodium chloride in the range of 0.0-25.0 mM are given in Fig. 3. Similar plots for the ratios of concentrations of benzonitriles 4 to 3 and for those of 4 to 3 plus 2 are recorded in Fig. 4. Positive correlations between [NaCl] and [3]/[1] as well as [4]/[2] ratios (see Fig. 3) clearly show

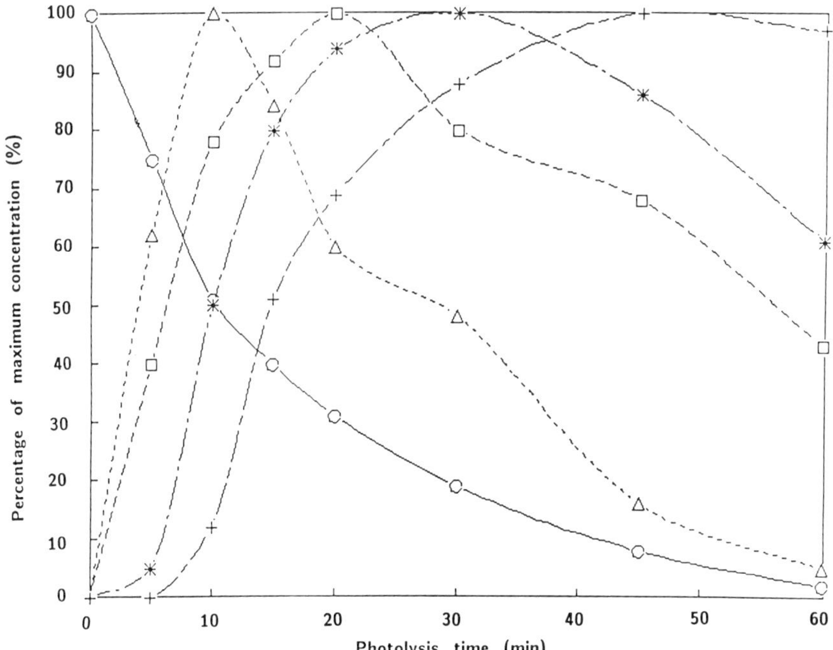

Fig. 2. Plots of changes in the concentrations of aqueous bromoxynil (1) (7.8 X 10^{-6} M) (– – ○ – – ○ ––) and its photoproducts, namely, 3-bromo-4-hydroxybenzonitrile (2) (—□—□—), 3-bromo-5-chloro-4-hydroxybenzonitrile (3) (—△-△—), 3-chloro-4-hydroxybenzonitrile (4) (—*—*—), and 4-hydroxybenzonitrile (4A (—+—+—) in the presence of 10 X 10^{-3} M NaCl versus irradiation time at around 313 nm.

that the photoproducts, viz., chlorophenols 3 and 4, are formed through the photoincorporation of Cl^- ions into the starting material 1 and into its primary photoproduct 2, respectively. Likewise, failure of an achievement of a positive correlation between [NaCl] and [4]/[3] ratios (Fig. 4) reveals the fact that the chlorobenzonitrile 4 is not exclusively generated via the photodegradation of the bromochlorobenzonitrile 3. However, plots of [NaCl] versus [4]/[2] (Fig. 3) and [4]/([3]+[2]) ratios (Fig. 4), both having positive correlations, indicate that the product 3 acts as a possible minor source of 4 in this photoreaction.

Total ion chromatograms of the chloroform extracts of aqueous solutions of 1.710 X 10^{-4} M bromoxynil (1) containing 7.143 X 10^{-2} M sodium chloride photolyzed for 1 h and 3 h are given in Figs. 5a and b. The con-

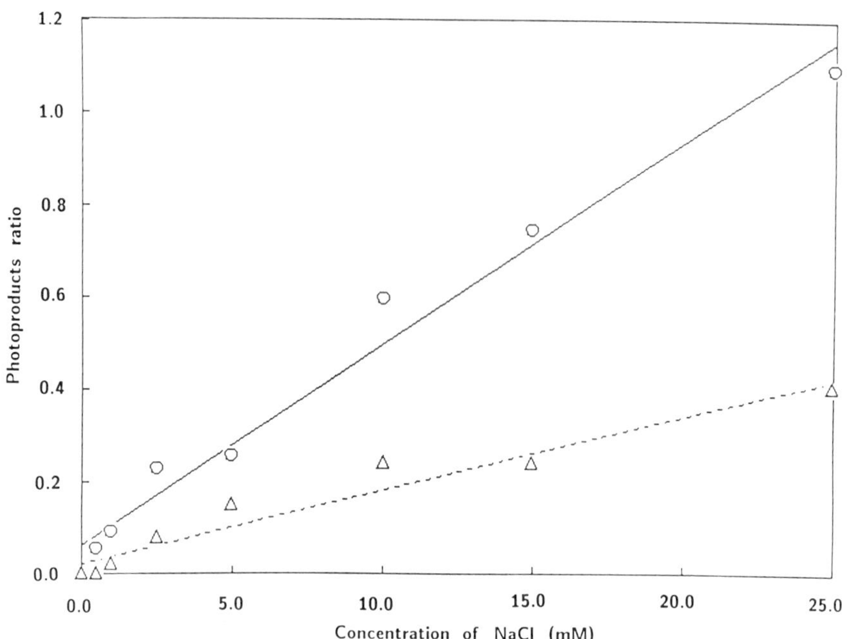

Fig. 3. Effects of various concentrations of NaCl on the distribution of products of 7.8 X 10.0^{-6} bromoxynil (1) dissolved in water: (– – o – – o ––) ratio of concentrations of 3-bromo-5-chloro-4-hydroxybenzonitrile (3)/bromoxynil (1) and (-△-△-) ratio of concentrations of 3-chloro-4-hydroxybenzonitrile (4)/3-bromo-4-hydroxybenzonitrile (2). The photoreaction time was 15.0 min.

centrations of halogenated benzene derivative products 2-4 increase as the photolysis time of 1 is increased. It is noteworthy that contrary to the above cited results (see Reaction 1 and the text pertaining to it), under the present conditions, 4-hydroxybenzonitrile (4A) does not appear as a photoproduct of 1 (Figs. 5a and b) (see also below).

Effects of Sodium Chloride on the Phototransformation of Bromoxynil (1) and its Photoproduct 4-Hydroxybenzonitrile (4A): Fig. 6 includes a plot of amounts of phototransformed herbicide 1 versus sodium chloride concentration, photoreaction time being 15 min. This figure again shows that the rate of the photodestruction of 1 is suppressed as the amount of NaCl is raised.

A time plot for the photoformation of 4-hydroxybenzonitrile (4A) from 7.8 μM bromoxynil (1) in water in the presence of 1.0, 10.0, and 50.0 mM sodium chloride is depicted in Fig. 7. Up to ca. 50.0 min, the rate

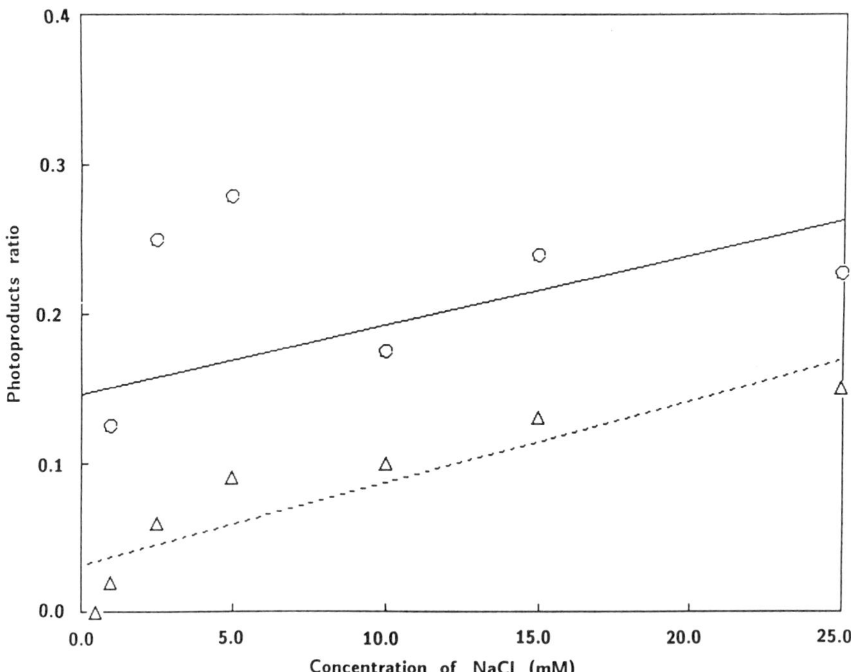

Fig. 4. Effects of various concentrations of NaCl on the distribution of the products of 7.8 X 10^{-6} M bromoxynil (1) dissolved in water: (– – – o – – – o – –) ratio of concentrations of 3-chloro-4-hydroxybenzonitrile (4)/3-bromo-5-chloro-4-hydroxybenzonitrile (3) and (–△–△–) ratio of concentrations of 3-chloro-4-hydroxybenzonitrile (4)/3-bromo-5-chloro-4-hydroxybenzonitrile (3) plus 3-bromo-4-hydroxybenzonitrile (2). The photoreaction time was 15.0 min.

of production of phenol 4A is reduced with the rise in the concentration of NaCl. NaCl in general appears to enhance the further photodecomposition of the product 4A (see also below).

In the case of photoreaction of the herbicide 1 (7.8 μM) in water for 15.0 min, the effects of different concentrations of NaCl on the photogeneration of phenol 4A from 1 are mentioned in Fig. 8. Initially when the concentration of NaCl increases from 0.0 to 5.0 mM, the amount of the photoproduct 4A increases; whereas, further rise in the concentration of this inorganic salt reduces the chemical yield of 4A.

Stern-Volmer Treatment: A Stern-Volmer treatment of the quantum yield (ϕ) data recorded in Table 1 for the photoconversion of the candidate

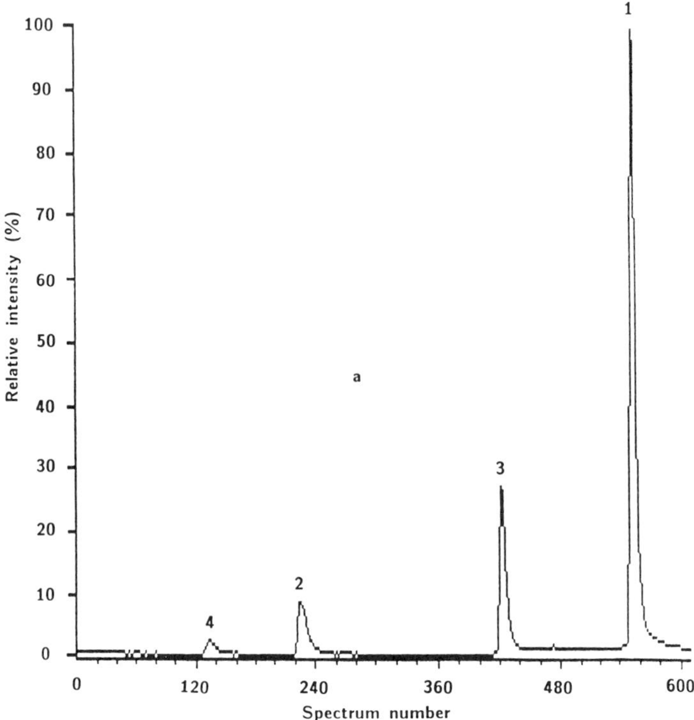

Fig. 5a. Plot of the total ion chromatograms (TIC) of the chloroform extracts of the 1.0 h photolyzates of a mixture of aqueous solutions of 3.0 mL bromoxynil (1) (2.0×10^{-4} M) plus 0.5 mL NaCl (0.5 M). The peaks labelled with numbers represent the compounds which have been identified and structures are described in Reaction (1).

compound 1 as aqueous solutions containing varying amounts of NaCl is given in Fig. 9. The plot shown in this figure was obtained utilizing equation (1) [18, 19]:

$$\phi_o/\phi_q = 1 + k_q r([NaCl]) \tag{1}$$

where ϕ_o and ϕ_q are the quantum yields for the phototransformation of 1 without and with the presence of quencher (e.g., NaCl), respectively; while k_q, r, and [NaCl] are the rate constant of quenching of the excited state molecules of 1, the lifetime of the excited state of the molecules of 1 acting as source of photoreaction, and the concentration of the quencher, i.e., sodium chloride, respectively. The straight line obtained by plotting ϕ_o/ϕ_q values versus concentrations of sodium chloride (Fig. 9) seemingly shows

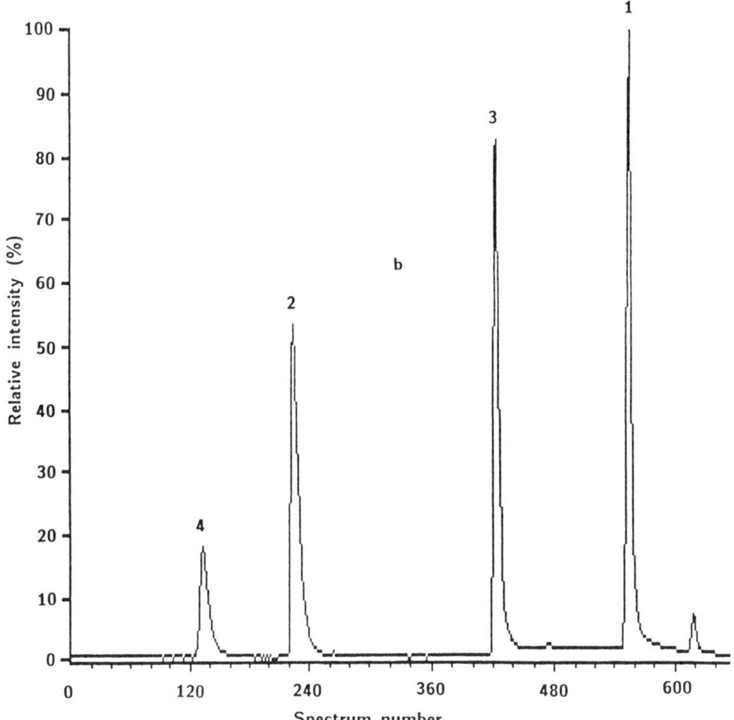

Fig. 5b. Plot of the total ion chromatograms (TIC) of the chloroform extracts of the 3.0 h photolyzates of a mixture of aqueous solutions of 3.0 mL bromoxynil (1) (2.0×10^{-4} M) plus 0.5 mL NaCl (0.5 M). The peaks labelled with numbers represent the compounds which have been identified and structures are described in Reaction (1).

that, in these reactions, NaCl behaves as photoquencher. Strictly speaking equation (1) can only be applied to those photoreactions which involve photoquenchers which will accept electronic excitational energy from the excited molecules of the candidate substrate, i.e., 1. However, NaCl molecules do not to our knowledge possess such photochemical properties. Further research work is required to explain the apparent photoquenching behavior of NaCl observed in the photolysis of bromoxynil (1).

Mass Spectral Data: Mass spectral data of bromoxynil (1) as well as those of two photoproducts, namely, phenols 2 and 4A, have been previously reported elsewhere [10]. Figs. 10a and b include mass spectra of 3-bromo-5-chloro-4-hydroxybenzonitrile (3) and 3-chloro-4-hydroxybenzonitrile (4) pro-

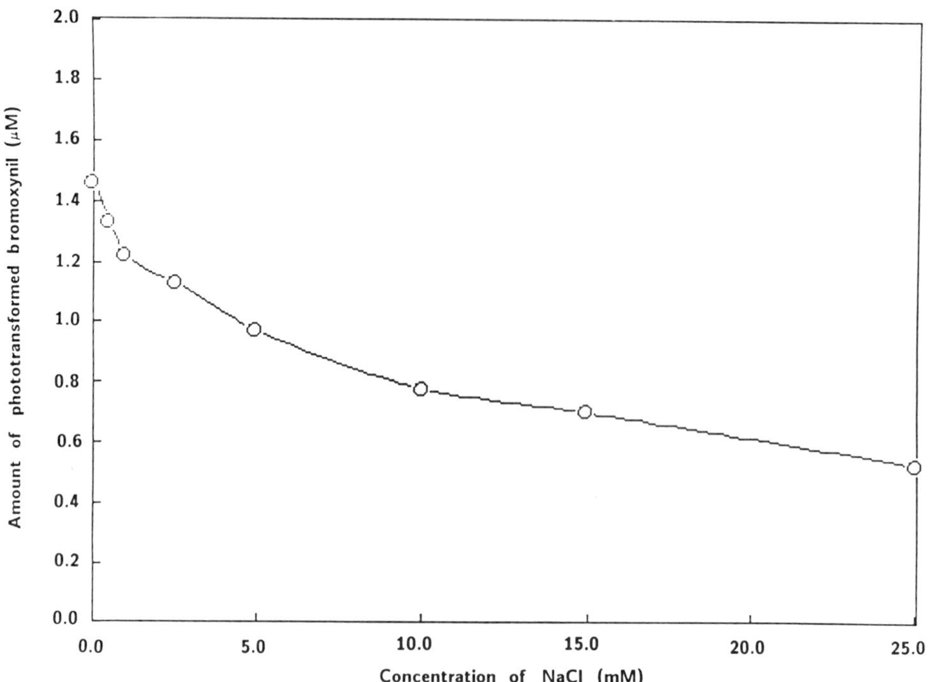

Fig. 6. Effects of various concentrations of sodium chloride on the photodestruction of 7.8 X 10^{-6} M bromoxynil (1) dissolved in water. The photoreaction time was 15.0 min.

duced via the photoincorporation of chloride ions into the compounds 1 and 2, respectively, during the photolysis of 1 in the presence of NaCl. These spectra were obtained based upon fragments containing ^{35}Cl and ^{79}Br isotopes solely. In Fig. 10a, the pattern of the parent molecular ion ($M^{+\cdot}$) clusters provides evidence for the presence of one Cl and one Br atom in 3 [19]. The m/e value of the parent molecular ion of 3 is 231 which is an odd number, indicating that this compound may contain an odd number of N atoms. The peaks seen at 124, 88, and 62 m/e in the mass spectrum of 3 (see Fig. 10a) are attributable to the loss of COBr, HCl, and CN groups from species of 231 (i.e., $M^{+\cdot}$), 124, and 88 m/e value, respectively. Similarly, the structure 3-chloro-4-hydroxybenzonitrile (4) can be ascribed to the photoproduct whose mass spectrum is given in Fig. 10b. The possible sources of the peaks occurring at 117, 89, and 62 m/e are $M^{+\cdot}$ - HCl (i.e., 153 - HCl), 117 - CO, and 89 - HCN, respectively.

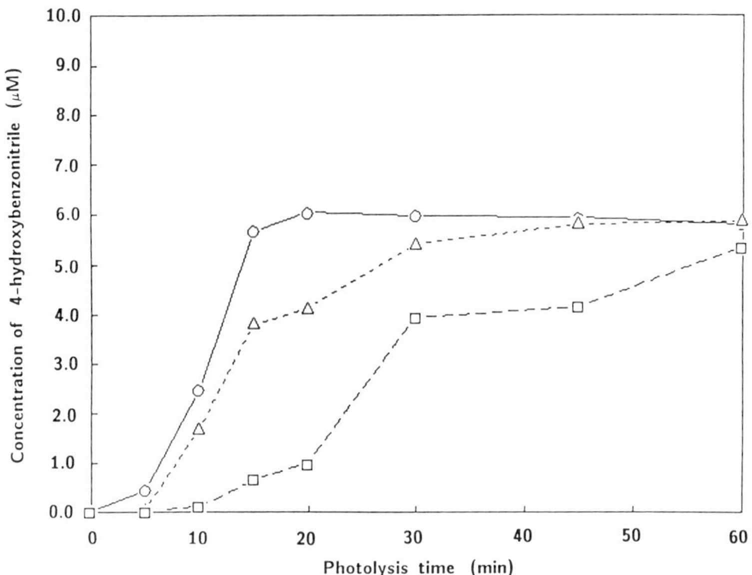

Fig. 7. Effects of 1.0 (– – ∘ – – ∘ – –), 10.0 (—△-△—), and 50.0 X 10^{-3} M (—□—□—) sodium chloride on the formation of the photoproduct 4-hydroxybenzonitrile (4A) from 7.8 X 10^{-6} M aqueous solution of bromoxynil (1).

CONCLUSIONS

The present laboratory findings concerning the environmental aquatic photolytic fate of the herbicide bromoxynil (3,5-dibromo-4-hydroxybenzonitrile) (1) in the presence of sodium chloride at around 313 nm lead to several environmentally significant conclusions. For instance, when the concentration of NaCl present in the natural water bodies is raised from 0.0 to 25.0 mM, the quantum yield for the photoconversions of the herbicide 1 decreases from 0.052 ± 0.004 to 0.017 ± 0.007. In other words, the rate of photodegradation of 1 will in general decrease with the rise of the amount of NaCl present in the aquatic systems. However, the rate of the photodecomposition of the secondary photoproduct 4-hydroxybenzonitrile (4A) in water will increase as the NaCl concentration increases. More importantly, in the presence of sodium chloride, in addition to usual reductively debrominated products, viz., 3-bromo-4-hydroxybenzonitrile (2) and phenol 4A, the herbicide 1 present in aquatic environments will also yield chloride-photoincorporated

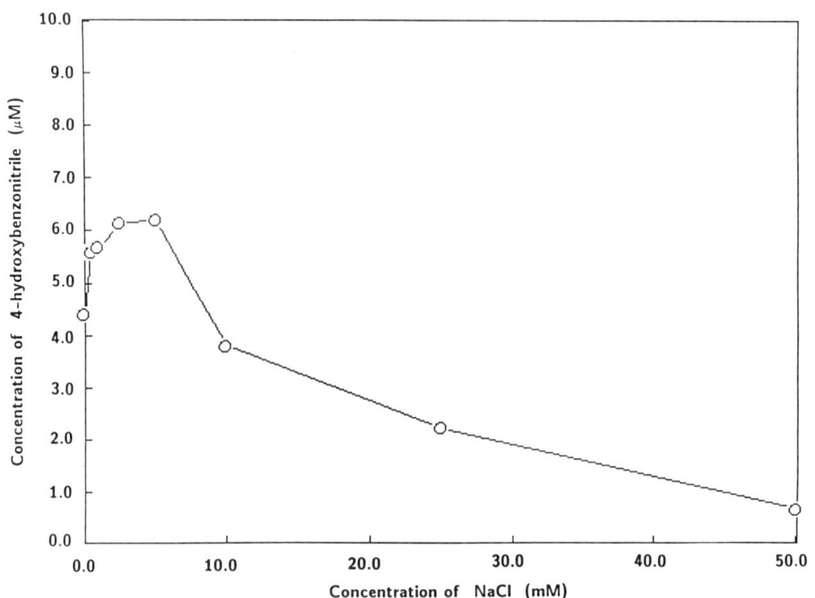

Fig. 8. Changes in the concentrations of the photoproduct 4-hydroxybenzonitrile (4A) during the photolysis of 7.8 X 10^{-6} M bromoxynil (1) in water containing various amounts of sodium chloride. Photoreaction time was 15.0 min.

products, namely, 3-bromo-5-chloro-4-hydroxybenzonitrile (3) and 3-chloro-4-hydroxybenzonitrile (4) on exposures to sunlight.

ACKNOWLEDGEMENT

The authors gratefully acknowledge the support of the Natural Sciences and Engineering Research Council of Canada (NSERC) through an NSERC International Scientific Exchange Award (Code: ISETR 106) to G.G.C. to support Dr. J. Kochany's visit and work at the University of Manitoba. In addition, an Agriculture Canada Operating Grant (No. 86003) to G.G.C. for the partial support of this Research Project is acknowledged. Furthermore, the authors are grateful to Mr. W. Buchannon, Department of Chemistry, University of Manitoba, Winnipeg, for running the GC-MS spectra and for his expert scientific advice.

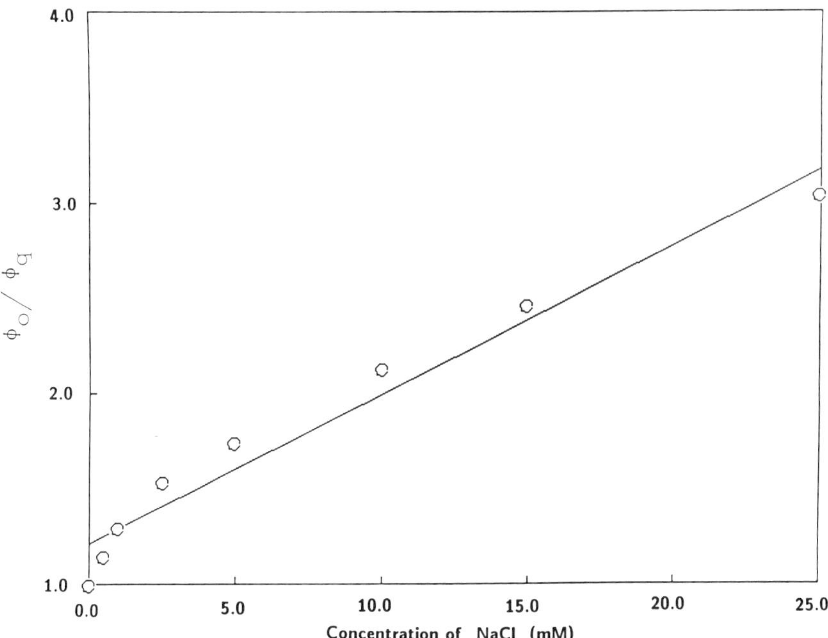

Fig. 9. Stern-Volmer [eq. (1)] plot for the phototransformation of bromoxynil (1) dissolved in water in the absence and presence of sodium chloride. The parameters ϕ_o/ϕ_q are ratios of photochemical quantum yields without and with the presence of NaCl. Photoreaction time was 15.0 min.

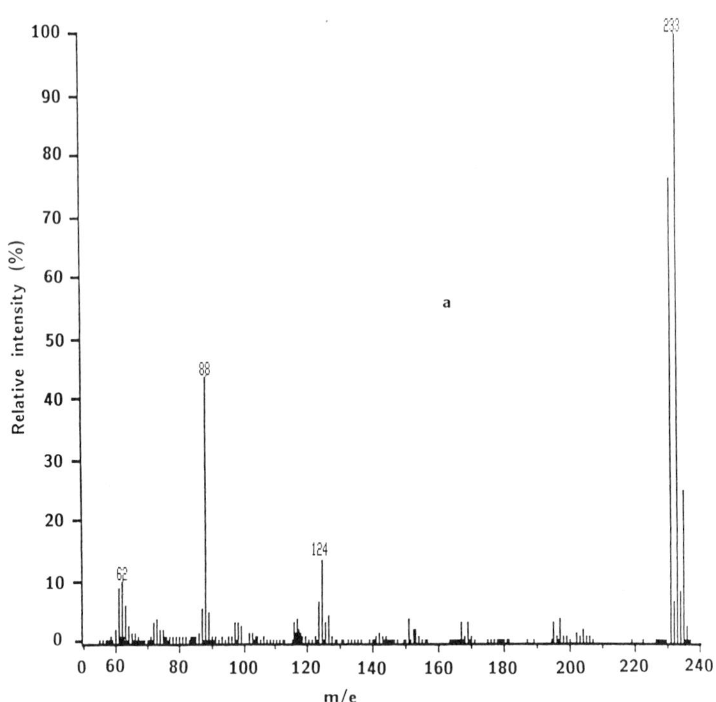

Fig. 10a. Mass spectrum of the photoproduct 3-bromo-5-chloro-4-hydroxybenzonitrile (3) obtained by the photolysis of a mixture of 3.0 mL of bromoxynil (1) (2.0×10^{-4}M) plus 0.5 mL NaCl (0.5 M) for 3.0 h.

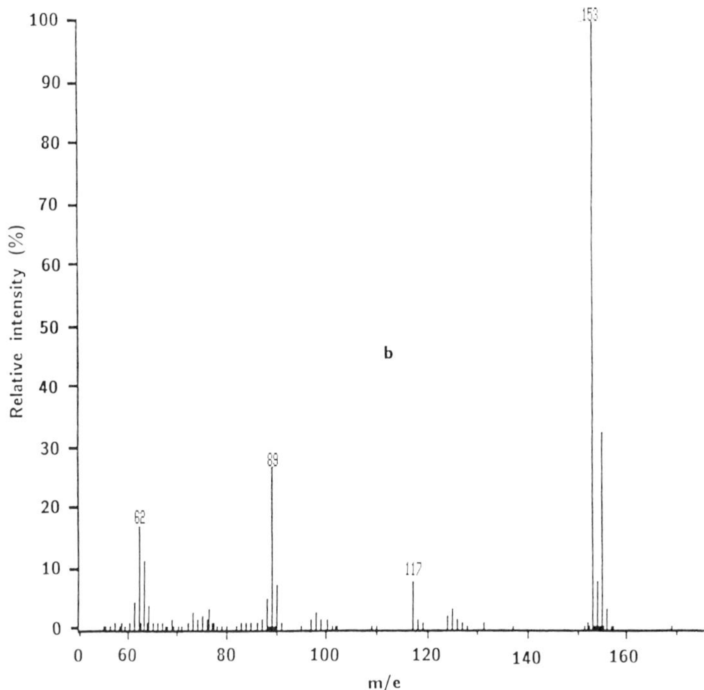

Fig. 10b. Mass spectr of the photoproduct 3-chloro-4-hydroxybenzonitrile (4) obtained by the photolysis of a mixture of 3.0 mL of bromoxynil (1) (2.0×10^{-4}M) plus 0.5 mL NaCl (0.5 M) for 3.0 h.

REFERENCES

[1] Manitoba Agriculture, Guide to Chemical Weed Control, Winnipeg, Government of Manitoba, Canada., 1-139, 1988.

[2] Brown D.F., McDonough L.M., McCool D.K. and Papendick R.J., J. Agric. Food Chem., 195, **32**, 1984.

[3] Choudhry G.G. and Webster G.R.B., Residue Rev., 79, **96**, 1985.

[4] Zepp R.G., Experimental approaches to environmental photochemistry, in The Handbook of Environmental Chemistry, O. Hutzinger, ed., Springer-Verlag, Berlin, 19-41, Vol. 2 / Part B, 1982.

[5] Zafiriou O.C., Joussot-Dubien J., Zepp R.G. and Zika R.G., Environ. Sci. Technol., 358A, **18**, 1984.

[6] Choudhry G.G., Humic Substances: Structural, Photophysical, Photochemical and Free Radical Aspects and Interactions with Environmental Chemicals, Gordon and Breach, New York, 1-185, 1984.

[7] Choudhry G.G., Webster G.R.B. and Hutzinger O., Toxicol. Environ. Chem., 27, **13**, 1986.

[8] Zika R.G. and Cooper W.J., Eds., Photochemistry of Environmental Aquatic Systems, American Chemical Society, Washington, D.C., 1-228, 1987.

[9] Marcheterre L., Choudhry G.G. and Webster G.R.B., Rev. Environ. Contam. Toxicol., 61, 103, 1988.

[10] Kochany J., Choudhry G.G. and Webster G.R.B., Pesticide Sci., in press, 1989.

[11] Kochany J., Choudhry G.G. and Webster G.R.B., Intertnat. J. Environ. Analytical Chem., in press, 1989.

[12] Kochany J., Choudhry G.G. and Webster G.R.B., Sci. Total Environ., in press, 1989.

[13] Choudhry G.G., Roof A.A.M. and Hutzinger O., J. Chem. Soc. Perkin Trans. I, 2957, 1982.

[14] Choudhry G.G. and Webster G.R.B., Chemosphere, 9, **14**, 1985.

[15] Dulin D. and Mill T., Environ. Sci. Technol., 815, **18**, 1982.

[16] Kochany J. and Lipczynska-Kochany E., Sci. Total Environ., 69, **67**, 1987.

[17] Kochany J., Choudhry G.G. and Webster G.R.B., Arch. Environ. Contam. Toxicol., in press, 1989.

[18] Calvert J.G. and Pitts J.N., Photochemistry, John Wiley, New York, 1-830, 1966.

[19] Barltrop J.A. and Coycle J.D., Excited States in Organic Chemistry, John Wiley, New York, 1-375, 1975.

[20] McLafferty F.W., Interpretation of Mass Spectra, W.A. Benjamin, New York, 16-30, 1973.

PESTICIDE LOSSES IN SURFACE RUNOFF FROM IRRIGATED FIELDS

W. F. SPENCER and M. M. CLIATH

USDA–Agricultural Research Service U.S. Salinity Laboratory Department of Soil and Environmental Sciences University of California Riverside, CA 92521

ABSTRACT

Pesticide concentrations were determined in surface irrigation runoff water following the application of pesticides to large fields of cotton, sugar beets, lettuce, alfalfa, onions, or melons in Imperial Valley, CA. The concentrations and total amounts of pesticides in runoff water were dependent upon the characteristics of the pesticides and methods and rates of application. The percentages of the applied pesticides lost in runoff water were generally low with seasonal totals for insecticides in runoff below 1% of the amounts applied. Seasonal losses of the soil-applied herbicides were usually 1 to 2% of the amounts applied. Time elapsed between pesticide application and the irrigation event was inversely related to log concentration, indicating an approximate first-order rate of decrease in concentration with time. The concentration of pesticides in the 0- to 1-cm depth of soil was a good indicator of concentrations and amounts to be expected in runoff water.

Chemistry for the Protection of the Environment
Edited by L. Pawlowski *et al.*, Plenum Press, New York, 1991

INTRODUCTION

The use of pesticides has resulted in enormous benefits to mankind, chiefly in the areas of public health and agricultural production. However, their use is not without attendant risk. One problem is their mobility. They do not always stay where they are applied. One of the many pathways by which pesticides leave treated fields is in runoff water. Agriculture has received considerable blame for the presence of pesticides in surface and ground waters. Many investigators have reported pesticide losses in runoff water from rainfall in humid areas and much effort has gone into modeling such runoff [1-4]. Since very little published data were available on pesticides in runoff from irrigated fields, we undertook to establish the importance of irrigation runoff water in moving pesticides from treated areas. This paper presents some of the results of the studies on runoff of pesticides from irrigated fields, examines some of the processes affecting concentrations and amounts of pesticides in runoff water, and discusses some of the relationships which may be useful in modeling.

METHODS

Field Procedures

The field work was done in Imperial Valley, CA on farmers' fields varying in size from about 28 to 60 ha. The farmers determined which crops were to be grown and which pesticides were to be applied. Each field was instrumented with a propeller water meter for measuring volumes of incoming irrigation water and a Parshall flume for measuring outgoing runoff water. The experimental sites were characterized by soil type descriptions from soil maps of the individual fields. Most of the soils were fine textured and slowly permeable to water. Pesticides were determined in irrigation runoff water from fields planted to cotton, sugar beets, lettuce, alfalfa, melons, and onions irrigated with furrow or border surface irrigation systems over a 4-year period. Insecticides were aerially applied; herbicides, except EPTC, were applied with a ground rig directly to the soil. EPTC was applied in surface irrigation water at about 2 mg/L EPTC.

Water samples were obtained from the throat of the runoff flume by hand sampling with a beaker and transferring the water into a 1-liter Pyrex glass bottle. The samples were immediately placed into an ice-water bath in an ice chest. They remained in the ice chest during transport and were then stored in a cold room at 4°C prior to processing in the laboratory. Based on the relationship between pesticide concentration and the runoff hydrograph, we sampled runoff water most frequently during the first irrigation after a pesticide application. For the second irrigation, samples were taken at approximately 1/2-hour intervals for the first two or three hours after start of runoff and then at 2- to 4-hour intervals. For subsequent irrigations, samples were taken four to six times per event.

Soil samples to the 1-cm depth were collected before some irrigations, from the irrigation furrow, or the area to be wetted, to relate soil concentrations to pesticide runoff. Samples were obtained with a 3-cm diameter stainless steel tube with a metal plug installed in the tube to prevent sampling from depths below 1 cm.

Laboratory Procedures

In the laboratory, each water sample was analyzed for recently applied pesticides of interest and the sediment load was determined. Water samples were analyzed for a total of 20 pesticides (14 insecticides and 6 herbicides). Whole water samples (water and sediment) were extracted four times by liquid/liquid extraction with 50 mL of organic solvent to 500 mL of water by shaking in a separatory funnel. Pesticide concentrations in the extracts were determined by gas liquid chromatography (GLC) using appropriate detectors for the pesticides of interest. Soil samples from the 0- to 1-cm depth from treated fields were screened through a 4-mm screen and mixed thoroughly, then 50-g soil samples were Soxhlet extracted for 4 hr with 250 mL of a 1:1 mixture of acetone and hexane. The extracts were concentrated to the appropriate volume and analyzed by GLC.

Data Calculations

Runoff water volumes were calculated from hydrographs, obtained with water-stage recorders, by integrating the area under the hydrograph using the trapezoidal rule. Most of the irrigations resulted in multi-peak runoff hydrographs. The total amounts of pesticides removed from treated fields in runoff water were calculated from flow volumes and pesticide concentrations. This was done by integrating the flux of the pesticide calculated from the product of the pesticide concentration and the water flow at the time discreet samples were taken.

The relationship between pesticides in runoff and several independent variables was determined by simple linear regression analysis. Pesticide concentrations were correlated with the following five independent variables: 1) time elapsed since the last pesticide application, 2) accumulative water applied since the last pesticide application, 3) runoff volume, 4) sediment in the runoff water, and 5) soil pesticide content in the 0- to 1-cm depth.

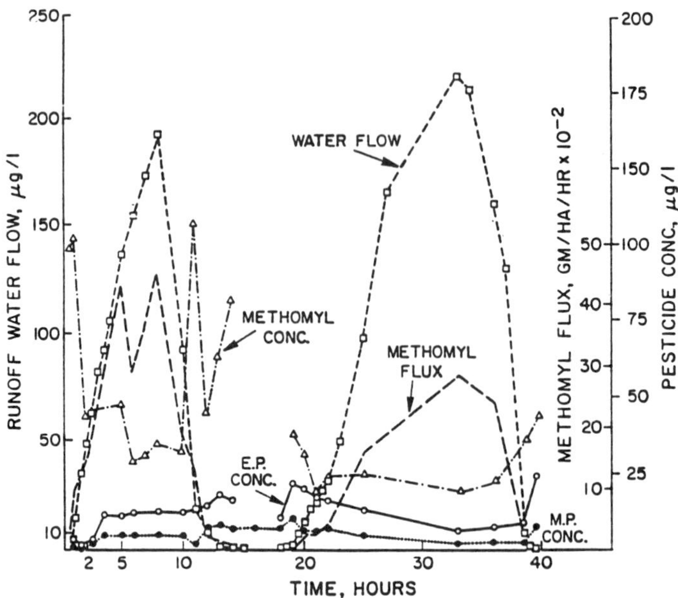

Fig. 1. Water runoff hydrograph; concentrations of methomyl, ethyl parathion (E.P.), and methyl parathion (M.P.); and flux of methomyl at various times after initiation of runoff from Field 4, 7 days after the last of three applications of the insecticides to sugar beets at rates of 0.85, 0.53, and 0.26 kg/ha, respectively.

Additional details of methods and procedures for the runoff study were reported by Spencer et al. [5].

RESULTS AND DISCUSSION

The basic data obtained were the water runoff hydrograph and the concentrations of pesticides in the runoff water with time after start of runoff as illustrated in Fig. 1. Fig. 1 shows the water runoff hydrograph; concentrations of methomyl, ethyl parathion, and methyl parathion; and flux of methomyl at various times after initiation of runoff.

Table 1 illustrates the type of information obtained from each runoff event and each pesticide application during a crop season. This is for prometryn in irrigation runoff water from a 28 ha cotton field irrigated 12 times during the growing season. In this case, 1.6 kg/ha prometryn was applied in two applications during the growing season, and 27.15 g/ha prometryn was lost in runoff water equivalent to 1.7% of the amount applied. The data illus-

Table 1. Prometryn in Irrigation Runoff Water from a 28-ha Cottonfield Irrigated 12 Times with a Furrow Irrigation System

Application		Irrigation runoff			Pesticides in runoff				
		Water applied L/ha x 10^5	Runoff volume L/ha x 10^5	Days since pesticide application	Concentration[1]			.	Percent of total applied
Date	Rate				Minimum	Maximum	Mean	Am't	
	kg/ha	cm	cm		μg/L	μg/L	μg/L	g/ha	.
3/25-27	0.3[2]	18.3	5.3	2	N.D.	24.	10.9	5.8	1.9
-	-	12.5	3.0	33	N.D.	60.	6.3	1.9	.6
-	-	12.2	(3.0)[3]	56	1.1	2.1	1.5	(.44)	.15
-	-	12.9	3.8	78	N.D.	3.6	1.1	.4	.13
6/25	1.3[4]	12.2	4.0	1	24.	77.	33.	13.3	.83
-	-	10.8	.9	12	N.D.	30.	20.5	1.8	.11
-	-	10.7	1.8	21	2.3	16.	6.2	1.12	.07
-	-	10.4	1.8	32	4.4	9.3	6.1	1.07	.07
-	-	9.7	1.8	40	N.D.	8.9	3.3	.61	.04
-	-	4.7	1.8	54	N.D.	13.5	3.9	.71	.04
-	-	11.0	.7	67	.-	N.D.<0.8	.-	.-	.-
-	-	10.0	2.6	82	.-	N.D.<0.8	.-	.-	.-
Totals	1.6	.	.		.	.	.	27.15	1.70

[1] Mean concentration equals amount of pesticide divided by volume of runoff water. N.D. = not detectable at the levels indicated.

[2] Herbicides applied to top of beds.

[3] Runoff values in parentheses were estimated from partial hydrographs and/or other irrigations because of malfunction of water-stage recorder.

[4] Herbicides applied by spraying over the entire ground surface with a Lilliston rig.

trate how concentrations change with time after pesticide applications. The amounts in runoff were highest during the first irrigation after applications, with amounts generally decreasing with time.

Tables 2, 3, and 4 summarize total seasonal losses of herbicides and insecticides applied to various crops. Each figure represents the runoff from several individual irrigations during the growth of the crop. The amounts of pesticides in runoff water varied considerably, depending upon the pesticide, its method of application, and the irrigation runoff water volume or irrigation efficiency. The percentages of the applied pesticides lost in runoff were generally very low with the exception of: 1) EPTC applied directly in irrigation water, 2) insecticides applied by air while irrigation water was on the field, and 3) one herbicide, prometryn, applied at midseason to cotton as a directed spray to the furrow and bed surfaces.

Table 2. Seasonal Runoff Losses of Herbicides and Insecticides Applied to Cotton

Pesticide	Year	Total amount applied kg/ha	Pesticides in runoff: Total amount g/ha	Pesticides in runoff: Percent of total applied
		herbicides		
DCPA	3	7.6	92.6	1.22
"	4	3.4	47.7	1.40
"	4	3.6	49.2	1.37
Dinitramine	2	1.3	17.3	1.32
Prometryn	2	3.1	157.	5.0
"	3	1.6	23.5	1.5
"	4	1.3	12.4	0.95
Trifluralin	3	1.1	3.21	0.29
"	2	0.96	1.38	0.14
		organophosphate insecticides		
Chlorpyrifos	2	1.68	0.29	0.02
"	3	0.70	0.61	0.09
"	4	2.44	5.96	0.24[1]
"	3	0.96	0.23	0.02
"	4	2.88	2.00	0.07
Methidathion	3	1.12	2.3	0.21
"	1	1.12	1.76	0.16
"	2	1.02	20.5	2.0[1]
		pyrethroid insecticides		
Fenvalerate	2	0.57	0.58	0.10[2]
"	3	0.27	0.15	0.05
"	4	0.49	1.02	0.21[2]
"	3	1.23	1.06	0.09[2]
"	2	0.54	0.50	0.09[2]
Permethrin	2	0.71	1.12	0.16[2]
"	3	0.51	N.D.	N.D.
"	4	1.22	0.34	0.028
"	3	0.10	N.D.	N.D.
"	1	0.51	N.D.	N.D.
"	2	0.60	0.54	0.09

[1] 69% of chlorpyrifos and 99% of methidathion seasonal loss followed an aerial application during irrigation.

[2] Includes an aerial application during irrigation.

Table 3. Seasonal Runoff Losses of Insecticides Applied to Various Crops

Insecticide	Year	Crop	Total amount applied kg/ha	Pesticides in runoff: Total amount g/ha	Pesticides in runoff: Percent of total applied
E. Parathion	4	Lettuce	0.83	0.18	0.02
"	3	Lettuce	4.20	18.4	0.44
"	1	Sugar Beets	1.11	5.71	0.51
"	2	Sugar Beets	2.07	6.52	0.31
"	2	Sugar Beets	2.51	8.50	0.34
"	3	Alfalfa	0.28	0.18	0.06
M. Parathion	4	Lettuce	0.42	0.013	0.003
"	3	Lettuce	2.1	6.70	0.32
"	1	Sugar Beets	0.56	1.71	0.31
"	2	Sugar Beets	1.02	1.67	0.16
"	2	Sugar Beets	1.25	2.10	0.17
"	3	Alfalfa	0.14	0.02	0.01
Diazinon	2	Sugar Beets	1.12	0.75	0.07
"	2	Sugar Beets	0.48	0.19	0.04
"	4	Melons	2.69	1.47	0.05
Malathion	4	Cotton	1.31	N.D.	N.D.
"	1	Cotton	1.12	0.004	0.0003
"	3	Sugar Beets	1.96	1.8	0.09
"	2	Alfalfa	4.46	3.93	0.09
"	3	Alfalfa	1.70	N.D.	N.D.
"	1	Lettuce	3.20	1.67	0.05
Methomyl	3	Cotton	0.4	6.9	1.73
"	4	Lettuce	3.41	16.3	0.48
"	3	Lettuce	3.16	21.1	0.67
"	1	Sugar Beets	3.49	8.78	0.25
"	2	Sugar Beets	4.01	11.7	0.29
"	4	Alfalfa	1.93	5.3	0.27
"	2	Sugar Beets	5.82	20.4	0.35
"	3	Alfalfa	0.79	1.00	0.13
"	3-4	Alfalfa	5.54	39.6	0.71

Table 4. Seasonal Losses of EPTC When Applied in Irrigation Water

Year	Crop	Total amount applied kg/ha	Pesticides in runoff	
			Total amount g/ha	Percent of total applied
2	Sugar beets	2.8	202	7.2
4	Alfalfa	6.6	420	6.4
3-4	Alfalfa	13.8	934	6.8

On cotton, most of the soil-applied herbicides were lost to the extent of about 1-2% of the application, except for trifluralin which was lost in lesser amounts equivalent to about 0.2% of that applied (Tab. 2). Runoff of trifluralin was lower probably because of its higher volatility compared with its solubility, thus it tends to volatilize from the soil surface as it moves up in evaporating water. A 5% loss of prometryn in year 2 was due to an exceedingly high loss in the first runoff event after a midseason application as a directed spray which resulted in high concentrations in the irrigation furrows.

In general, lower runoff rates occurred with aerially applied insecticides, than with soil-applied herbicides, and the herbicides persisted longer than did the insecticides. Concentrations were highest in the first irrigation following pesticide application, and the amounts in runoff were dependent upon the relative persistence of the pesticide at the soil surface. For example, ethyl parathion is known to be more persistent than methyl parathion. They were always applied together in a 2:1 mixture of ethyl to methyl parathion, but appeared in runoff in a 4:1 ratio of ethyl to methyl parathion. This resulted in about double the percent loss of applied ethyl- compared to methyl-parathion. However, an average of only 0.28% of ethyl parathion and 0.16% of methyl parathion was lost in the runoff water (Tab. 3). With other less persistent organophosphates, the percent losses were even lower than with parathion, except for irrigation events where the pesticides were applied during an irrigation (Tab. 2).

Table 4 illustrates what happened when pesticides were applied in irrigation water. When EPTC was applied at approximately 2 mg/L in surface irrigation water, relatively high losses occurred. On alfalfa, the runoff water contained EPTC concentrations as high as 1.6 mg/L with total amounts lost during individual irrigations varying from 2.6% to 13% of that applied. It was apparent that runoff of EPTC was directly related to the volume of runoff water, and the quantities lost can be greatly decreased by preventing excess water loss.

Amounts and concentrations of pesticides in runoff were most highly correlated with the time elapsed since the last pesticide application and with the soil pesticide content. Pesticides in runoff were not highly correlated with the amount of sediment in runoff, the runoff water volume, or the accumulative water applied since the last pesticide application. Time elapsed between pesticide application and the irrigation event was inversely related to log pesticide concentration indicating a first-order rate of decrease in runoff concentration with time (Tab. 5). For example, the linear regression model indicated a correlation coefficient, r, of -0.872 and -0.932, with half-lives for the runoff water concentration decrease of 12 and 7.7 days for methyl and ethyl parathion, respectively.

A highly significant correlation was observed between concentrations of pesticides in the 0- to 1-cm soil depth and concentrations of pesticides in runoff water with the exception of the pyrethroids (Tab. 6). Even when data for ethyl- and methyl- parathion were combined, log concentration in runoff versus log soil concentration had a correlation coefficient of 0.954. The lack of correlation between pyrethroid concentrations in soil and runoff appears to be due to the erratic runoff behavior of the pyrethroids, which caused the linear regressions to be influenced by one or two abnormally high pyrethroid concentrations in runoff water when soil concentrations were low.

The persistence of pesticides in the 0- to 1-cm soil depth was indicated by simple linear regressions relating time since the last pesticide application to log concentration of pesticides in soil. With most pesticides, the decrease in concentration followed a first-order rate with highly significant correlations between time and log concentrations or log amounts remaining in the soil. For example, ethyl parathion disappeared from the 0- to 1-cm soil depth with a half-life of 13 days and methyl parathion with a half-life of 7.8 days (data not shown).

The minimum sampling frequency for estimating total pesticide losses was examined for several runoff events. For this purpose, total pesticide runoff was calculated using various combinations of sampling times with respect to the water runoff hydrograph. Most of the irrigations resulted in multi-peak hydrographs due to the fact that the fields were irrigated in thirds or halves. A comparison of pesticide losses in several irrigation runoff events calculated using average concentrations during peak flows versus the full integration method, as previously described, is shown in Table 7. The calculations indicate that in most cases, total pesticide runoff could have been accurately estimated using pesticide concentration data from only the peak flows.

For example, runoff of methomyl from Field 4 was estimated to be 6.2 g/ha using 24 data points taken over the entire irrigation period. The irrigation runoff event had two periods of peak flow as indicated in Fig. 1,

Table 5. Simple Correlation (r) and Linear Regression Coefficients (a,b)[1] for Log_{10} Pesticide Concentration in Runoff (C_R, μg/L) vs Time Elapsed since Last Pesticide Application (t, days) for All Irrigations

Pesticide	n[2]	r	a	b
ORGANOPHOSPHATES				
Diazinon	11	-0.234	-0.006	-0.262
Chlorpyrifos	17	-0.449	-0.029	0.006
Malathion	13	-0.620*	-0.069	0.842
Methidathion	10	-0.867**	-0.062	0.990
Sulprofos	3	0.263	0.270	-0.585
E. Parathion	21	-0.872**	-0.025	0.889
M. Parathion	19	-0.932**	-0.039	0.594
PYRETHROIDS				
Permethrin	23	-0.130	-0.015	-0.784
Fenvalerate	13	-0.519*	-0.059	-0.606
CARBAMATE				
Methomyl	44	-0.574**	-0.014	1.236
ORGANOCHLORINES				
Endosulfan	7	-0.882**	-0.018	1.49
Ethylan	3	-0.992**	-0.139	2.270
HERBICIDES				
Prometryn	35	-0.641**	-0.0193	1.551
DCPA	35	-0.680**	-0.0095	1.638
Trifluralin	12	-0.601*	-0.022	0.708
Dinitramine	10	-0.339	-0.0067	0.615
EPTC	18	-0.580**	-0.024	0.499

[1] $\text{Log } C_R = at + b$.

[2] n = numbers of observations.

[3] Asterisks indicate statistical significance at the 5% (*) or 1% (**) level.

Table 6. Simple Correlation (r) and Linear Regression Coefficients for Log Runoff Concentration (C_R, μg/L) vs Log Soil Concentration (C_S, μg/g) in to 0- to 1-cm Depth Zone

Pesticides	n	log C_R vs log C_S [1] r	a	b
ORGANOPHOSPHATES				
Methidathion	3	0.711	1.524	1.711
E. Parathion	6	0.996**[2]	1.425	0.727
M. Parathion	6	0.763*	0.817	0.329
PYRETHROIDS				
Permethrin	7	-0.299	-0.486	-0.806
Fenvalerate	4	-0.724	- 1.892	-3.252
CARBAMATE				
Methomyl	10	0.954**	1.163	1.164
HERBICIDES				
Prometryn	12	0.845**	0.572	1.355
DCPA	8	0.921**	1.113	1.049
Trifluralin[3]	4	0.999**	0.715	0.702
Dinitramine	3	0.991**	1.226	1.035

[1] $\log C_R = a \log C_S + b$

[2] Asterisks indicate statistical significance at the 5% (*) or 1% (*) level.

[3] Trifluralin on cotton only.

Table 7. Comparison of Pesticide Losses in Individual Runoff Events Calculated Using the Full Integration Method with Values Calculated Using Average Concentrations during Peak Flows Only

Field	Chemical	Pesticide amounts in runoff		
		Integration	Peak Flow Averaging[1]	
		g/ha	g/ha	difference(%)
4	Methomyl	6.2	6.1	-2
	Ethyl parathion	3.1	2.6	-16
	Methyl parathion	1.2	1.0	-17
5	Cycloate	.82	.86	+5
	Ethyl parathion	8.5	7.3	-14
	Methyl parathion	2.1	1.9	-10
5	Diazinon	.17	.17	0
1	Prometryn	9.3	9.7	+4
	Chlorpyrifos	.18	.16	-11
1	Prometryn	7.9	7.5	-5
1	Prometryn	12.8	12.9	+1
1	Prometryn	72	38	-47

[1] Amounts $= C_1V_1 + C_2V_2 + - - - C_nV_n$, where C_1, C_2, C_n and V_1, V_2, V_n are average concentrations and runoff volumes for flow peaks 1, 2, and n, respectively.

and using the average of three concentrations of methomyl from each peak flow and the flow volume under the peaks resulted in a calculated runoff of 6.1 g/ha methomyl, which was within $\pm$ 2% of the integrated estimate using 24 data points.

Other runoff events provided similar comparisons when both the pesticide concentration and the water volume for each peak of water flow was used. Over all, the data indicate that the total pesticide runoff can be quite accurately estimated using only pesticide concentrations from the peak flows. The estimate from the last event in Table 7 is an exception. In this case, prometryn had been recently applied as a post-emergent herbicide with much of the pesticide remaining on the soil surface in the irrigation furrows. In similar situations, when the first runoff water may contain very high concentrations of the pesticide, sampling only peak flows may miss a significant part of the runoff flux and result in underestimates of the total pesticide loss by a considerable amount, such as the -47% in this case.

CONCLUSIONS

The results indicate that the amounts of pesticides lost in irrigation runoff water are quite low. However, they can be considerably lessened by choice of pesticide and its method of application, extending the waiting period between application and irrigation, and by increasing irrigation efficiency through preventing excessive loss of irrigation water in runoff. When pesticides are applied directly in irrigation water, irrigation runoff losses should be prevented or held to a minimum. Pesticides should not be applied when irrigation water is on the field. Tile drain effluents contribute insignificant amounts of presently used pesticides to irrigation return flows compared with the amounts from irrigation runoff water.

REFERENCES

[1] Caro J. H., In Control of water pollution from cropland: Vol II, An Overview, U.S. Department of Agriculture Report ARS-H-5-2, 91-119, 1976.

[2] Knisel W. G., CREAMS: A field-scale model for chemicals, runoff, and erosion from agricultural management systems. USDA-Cons. Research Rpt. No. 26, 640 pp, 1980.

[3] Wauchope R. D., J. Environ. Qual. 7:459-472, 1978.

[4] Willis L. H. and McDowell L. L., Environ. Toxicol. Chem. 1:267-280, 1972.

[5] Spencer W. F., Cliath M. M., Blair J. W., and LeMert R. A., USDA-Cons. Research Rpt. No. 31. 72 pp, 1985.

STORMWATER CONTAMINATION IN AN URBANIZING WATERSHED

C. G. UCHRIN, T. MALDONATO,
Y. H. PANG, and R. YU

Department of Environmental Science, Cook College
New Jersey Agricultural Experiment Station Rutgers
University, New Brunswick, NJ 08903, USA

ABSTRACT

Results are presented from a study examining the effects of urbanization on a watershed with respect to four metals: chromium, copper, lead and zinc. Zinc was found to be ubiquitous to the basin. Copper and chromium were found to increase instream as a function of stormwater runoff during low flow conditions. Chromium was not detected in the stream. Copper, lead and chromium were found in the runoff from four specific sites: forest, farm, urban and suburban. A correlation was developed relating normalized instream lead and copper loadings to land use type.

INTRODUCTION

As a watershed urbanizes, it is exposed to growing environmental pressures such as increases in water demand, increases in impervious areas, and increases in various pollutant loadings to local receiving waters. Urban runoff is considered one of the most important nonpoint sources of metal input to adjacent waters [1] and urbanization has been found to result in higher metals loadings in stormwater runoff [2].

Chemistry for the Protection of the Environment
Edited by L. Pawlowski *et al.*, Plenum Press, New York, 1991

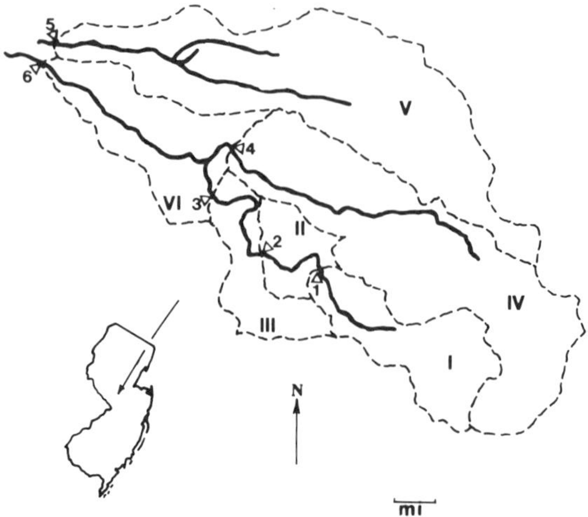

Fig. 1. Upper Millstone River basin; dashed lines delineate drainage basin boundaries, Roman numerals denote drainage basins, solid lines show waterways, and Arabic numerals and triangles denote stream sampling points.

This paper presents research results from a study conducted examining the current state of nonpoint source metals (copper, chromium, lead and zinc) in the Upper Millstone River in Central New Jersey. The Upper Millstone discharges to Carnegie Lake, Princeton, and its proximity to U.S. Highway No. 1, a major artery in the New York–Philadelphia corridor, and the state capital, Trenton, makes the basin a prime location for industrial, commercial and residential growth. A 300% increase in population has been projected for the 1984–1999 period. A map of the area is displayed in Fig. 1 and current land use is given in Table 1.

METHODOLOGY

Sampling and analysis were performed on the Upper Millstone River and stormwater inputs between June 1987 and December 1988. A variety of parameters were analyzed; however, only the metals (copper, chromium, lead, and zinc) are discussed here. Two types of stations were selected for this study: instream stations and stormwater discharges.

Table 1. Land Use (km^2) in the Upper Millstone River Basin

Basin	Forest	Farm	Residential	Water	Industrial	Urban
I	18.3	9.4	6.9	1.2	0.1	0.0
II	0.5	3.3	1.7	0.8	0.6	0.6
III	3.1	4.6	0.4	0.3	0.0	3.4
IV	23.7	23.7	2.7	1.9	0.6	0.8
V	19.0	25.8	8.2	2.4	0.6	1.2
VI	15.6	28.5	7.0	5.4	1.9	6.4

Baseline Studies

Baseline studies were conducted sampling the six stream stations during dry weather periods throughout the entire sampling period. Baseline samples were collected approximately every two weeks only when more than five consecutive dry days preceded the sampling. All baseline samples were collected between 9 and 10 a.m. to ensure consistency within the diurnal period. Stream sampling stations are noted on Fig. 1.

In-Stream Stormwater Studies

Four stormwater sampling surveys were conducted examining the response of the six stream stations to stormwater inputs. The storms occured on November 5, 1988, March 26-27, 1989, August 23-26, 1989, and November 1-5, 1989. The stream sampling points used in the baseline studies were also used.

Site-Specific Stormwater Studies

A stormwater sampling survey was conducted examining the response of four different drainage basin types: farm, forest, suburban, and urban. This study was conducted in conjunction with the final in-stream stormwater sampling survey.

Metals Analysis

All metals samples were digested using nitric acid as outlined in APHA [3], section 302D. Samples were analyzed for copper, total chromium, and lead using a Perkin Elmer Model 503 graphite furnace atomic absorption spectrophotometer (AAS). Zinc analyses were performed using a Perkin Elmer Model 603 flame photometric AAS.

Table 2. Baseline Heavy Metal Concentrations (mg/l) in the Upper Millstone River Basin (n=16)

Station	Chromium		Copper		Lead		Zinc	
	Mean	S.D.	Mean	S.D.	Mean	S.D.	Mean	S.D.
1	0.0	0.1	9.6	10.9	1.7	1.9	78.6	51.0
2	0.1	0.3	7.1	4.2	3.0	3.4	90.2	87.4
3	0.2	0.6	12.0	9.0	4.0	2.6	114.0	103.0
4	0.3	0.6	8.4	8.4	1.7	2.2	87.4	95.1
5	0.3	0.6	8.2	6.2	2.5	2.5	113.0	150.0
6	0.2	0.1	7.8	8.3	2.7	3.0	89.3	67.1

RESULTS

Baseline Studies

Results from the baseline studies are summarized in Table 2. As can be seen, chromium was virtually undetected, while low levels of lead and copper were found. Higher levels of zinc were found.

In-Stream Stormwater Studies

Data are presented in Tables 3–8 for the August, 1989 survey and in Tables 9–14 for the November, 1989 survey. The August event occured during a period of particulary low flow, while the November event was during relatively high flow, as can be seen from the gage heights. It should be noted that these are not stream depths and the U.S. Geological Survey is in the process of developing stream rating curves to convert these into stream flow. It is interesting to observe that the November data do not significantly deviate from the dry weather baseline data, while the low flow August data show order–of–magnitude increases as a function of stream gage height for copper at stations 2, 4 and 5 and lead at station 3. Zinc appears ubiquitous and chromium, which would be associated with heavy industry, is not present. Station 3 has the highest urban land use which may explain the higher lead concentrations. Thus in this basin, copper and lead appear to be the most telltale indicators of urbanization effects on the river.

Site-Specific Stormwater Studies

Results from the site-specific stormwater studies are displayed in Fig-

Table 3. Heavy Metals (μg/l) at Station 1 during August 23-26, 1989 Rain Event

Date	Time	Gage-m	Chromium	Copper	Lead	Zinc
8/23	21:20	0.34	ND	45.0	4.0	43
8/24	00:35	0.35	ND	40.0	1.0	25
8/24	03:00	0.36	ND	3.0	1.0	5
8/24	04:50	0.47	ND	9.0	5.0	43
8/24	06:12	0.49	ND	9.0	3.0	25
8/24	08:40	0.49	ND	4.0	1.0	85
8/24	13:10	0.49	ND	4.0	2.0	5
8/24	19:45	0.50	ND	5.0	2.0	73
8/25	09:30	0.59	ND	26.0	1.0	113
8/25	19:50	0.54	ND	26.0	2.0	165
8/26	11:30	0.45	ND	4.0	1.0	155
8/27	11:40	0.38	ND	7.0	2.0	113

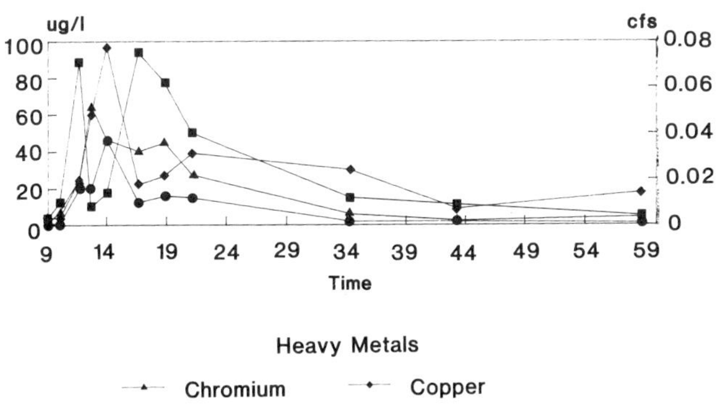

Fig. 2. Site-specific farm drainage area response to 11/1/89 storm, metals concentration and flow versus time-hours (0 = 000 hrs, 11/1/89).

ures 2, 3, 4 and 5 for the farm, forest, urban and suburban drainage basins, respectively. As is evident, there is a pronounced relationship between flow and heavy metal concentration. It is interesting to note that while chromium was not detectable in the receiving stream, it is present in the runoff from all of the basins. Again, zinc data were high and did not display any trend.

Table 4. Heavy Metals ($\mu g/l$) at Station 2 during August 23-26, 1989 Rain Event

Date	Time	Gage-m	Chromium	Copper	Lead	Zinc
8/23	21:55	0.41	ND	3.0	8.0	35
8/24	00:55	0.62	ND	3.0	8.0	35
8/24	03:15	0.66	ND	28.0	11.0	63
8/24	04:20	0.67	ND	32.0	61.0	590
8/24	06:25	0.67	ND	22.0	12.0	73
8/24	08:30	0.42	ND	12.0	13.0	85
8/24	13:25	0.43	ND	9.0	15.0	405
8/24	19:25	0.52	ND	6.0	9.0	43
8/25	09:05	0.63	ND	4.0	9.0	43
8/25	20:05	0.72	ND	12.0	19.0	145
8/26	11:45	0.62	ND	10.0	22.0	73
8/27	11:55	0.55	ND	6.0	11.0	65

Table 5. Heavy Metals ($\mu g/l$) at Station 3 during August 23-26, 1989 Rain Event

Date	Time	Gage-m	Chromium	Copper	Lead	Zinc
8/23	20:55	0.02	ND	8.0	4.0	25
8/24	00:55	0.05	ND	3.0	8.0	15
8/24	01:35	0.06	ND	5.0	3.0	25
8/24	03:28	0.15	ND	5.0	13.0	65
8/24	04:05	0.35	ND	5.0	7.0	35
8/24	05:15	0.70	ND	7.0	25.0	55
8/24	05:55	0.68	ND	4.0	19.0	25
8/24	06:50	0.50	ND	10.0	8.0	15
8/24	08:15	0.56	ND	4.0	4.0	65
8/24	13:47	0.48	ND	2.0	4.0	105
8/24	19:30	0.18	ND	3.0	4.0	85
8/25	10:08	0.33	ND	1.0	2.0	55
8/25	20:20	0.30	ND	3.0	4.0	135
8/26	11:55	0.23	ND	2.0	14.0	105
8/27	12:08	0.14	ND	3.0	3.0	43

Table 6. Heavy Metals (μg/l) at Station 4 during August 23-26, 1989 Rain Event

Date	Time	Gage-m	Chromium	Copper	Lead	Zinc
8/23	21:10	-0.03	ND	1.0	1.0	4
8/24	01:20	-0.02	ND	21.0	1.0	25
8/24	03:50	-0.01	ND	6.0	5.0	73
8/24	05:28	0.14	ND	10.0	5.0	55
8/24	07:00	0.16	ND	10.0	8.0	73
8/24	09:15	0.15	ND	36.0	6.0	95
8/24	14:06	0.17	ND	13.0	3.0	95
8/24	20:12	0.18	ND	34.0	3.0	85
8/25	10:20	0.21	ND	2.0	ND	65
8/25	20:30	0.22	ND	11.0	2.0	95
8/26	12:07	0.22	ND	2.0	1.0	43
8/27	12:13	0.10	ND	16.0	2.0	73

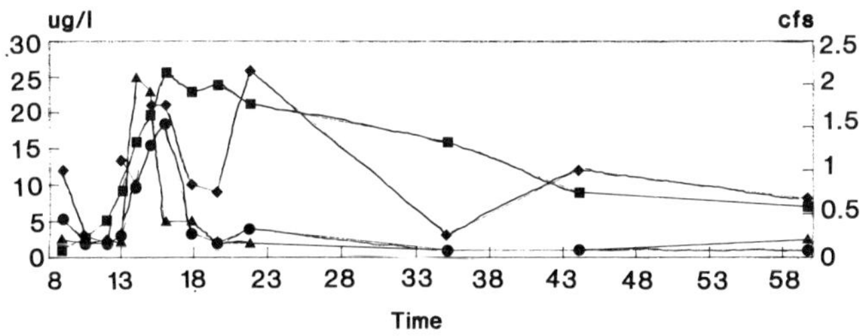

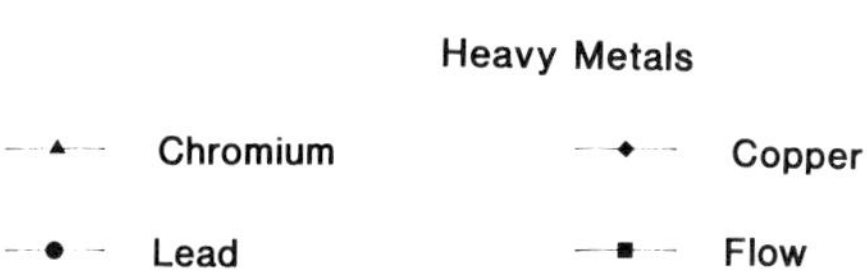

Fig. 3. Site-specific forest drainage area response to 11/1/89 storm, metals concentration and flow versus time-hours (0 = 000 hrs, 11/1/89).

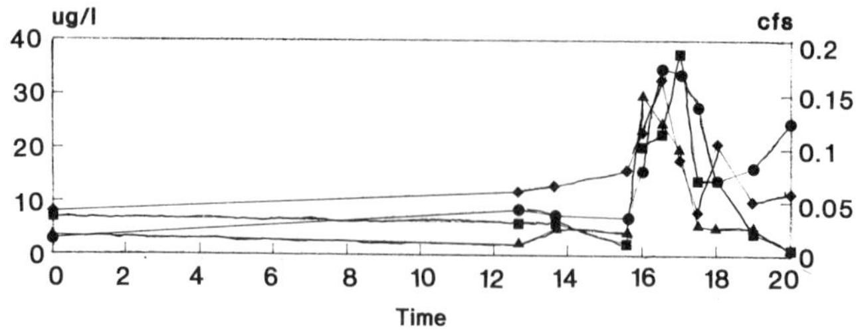

Fig. 4. Site-specific urban drainage area response to 11/5/89 storm, metals concentration and flow versus time-hours (0 = 000 hrs, 11/5/89).

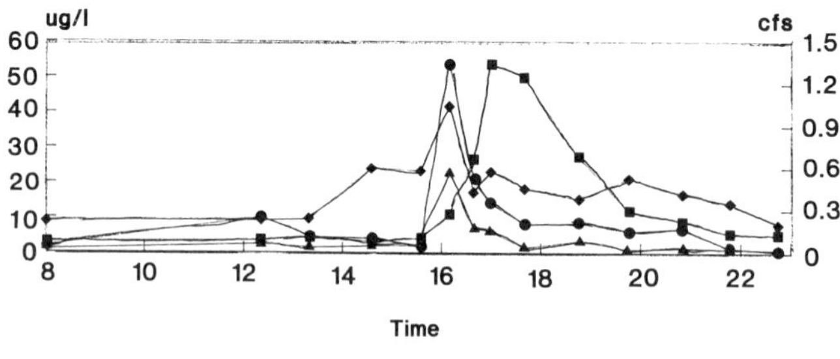

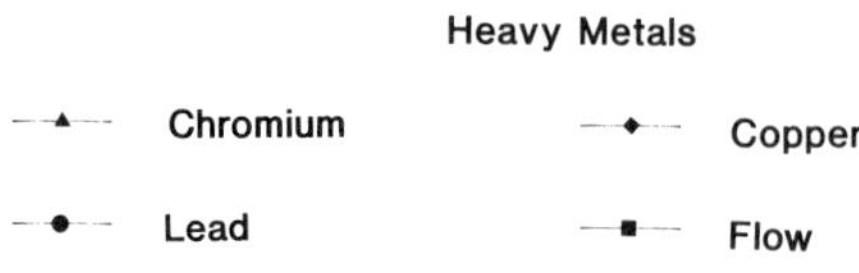

Fig. 5. Site-specific suburban drainage area response to 11/5/89 storm, metals concentration and flow versus time-hours (0 = 000 hrs, 11/5/89).

Table 7. Heavy Metals (μg/l) at Station 5 during August 23-26, 1989 Rain Event

Date	Time	Gage-m	Chromium	Copper	Lead	Zinc
8/23	21:20	0.34	ND	2.0	3.0	65
8/24	03:20	0.34	ND	8.0	1.0	125
8/24	05:20	0.38	ND	22.0	2.0	195
8/24	07:35	0.40	ND	60.0	2.0	750
8/24	09:35	0.41	ND	25.0	1.0	85
8/24	14:24	0.45	ND	62.0	1.0	165
8/24	20:27	0.43	ND	95.0	1.0	65
8/25	10:35	0.38	ND	37.0	ND	125
8/25	20:45	0.37	ND	6.0	1.0	85
8/26	12:20	0.38	ND	70.0	2.0	85
8/27	12:25	0.36	ND	16.0	1.0	35

Table 8. Heavy Metals (μg/l) at Station 6 during August 23-26, 1989 Rain Event

Date	Time	Gage-m	Chromium	Copper	Lead	Zinc
8/23	21:40	1.01	ND	51.0	2.0	105
8/24	03:40	1.01	ND	26.0	1.0	65
8/24	05:30	1.02	ND	21.0	ND	95
8/24	07:50	1.04	ND	10.0	ND	5
8/24	09:45	1.08	ND	14.0	4.0	95
8/24	14:35	1.16	ND	5.0	1.0	113
8/24	20:35	1.21	ND	6.0	4.0	105
8/25	10:45	1.22	ND	6.0	3.0	105
8/25	20:55	1.18	ND	6.0	ND	490
8/26	12:30	1.15	ND	2.0	ND	73
8/27	12:35	1.08	ND	28.0	1.0	135

Table 9. Heavy Metals (μg/l) at Station 1 during November 5, 1989 Rain Event

Date	Time	Gage-m	Chromium	Copper	Lead	Zinc
11/4	23:58	0.54	ND	2.0	ND	20
11/5	12:26	0.54	<1.5	3.0	2.5	125
11/5	16:05	0.54	ND	2.0	<1.0	30
11/5	18:43	0.56	ND	3.0	1.0	40
11/5	22:36	0.55	ND	1.0	ND	25
11/6	10:10	0.57	ND	4.0	ND	58

Table 10. Heavy Metals (μg/l) at Station 2 during November 5, 1989 Rain Event

Date	Time	Gage-m	Chromium	Copper	Lead	Zinc
11/5	00:29	0.56	1.0	1.0	<1.0	30
11/5	12:43	0.56	ND	3.0	1.5	30
11/5	16:23	0.58	1.0	8.0	3.5	45
11/5	19:00	0.58	ND	3.0	1.0	25
11/5	22:51	0.59	ND	<1.0	<1.0	30
11/6	10:23	0.56	ND	3.5	ND	45

Table 11. Heavy metals (μg/l) at Station 3 during November 5, 1989 Rain Event

Date	Time	Gage-m	Chromium	Copper	Lead	Zinc
11/5	01:10	0.24	1.5	3.0	3.0	30
11/5	12:05	0.26	1.5	3.5	2.5	40
11/5	13:33	0.27	1.0	3.0	3.5	30
11/5	15:43	0.27	1.0	3.0	3.5	45
11/5	16:45	0.30	3.0	3.0	4.5	40
11/5	17:32	0.35	1.0	5.0	<1.0	45
11/5	18:17	0.37	4.5	10.0	3.5	165
11/5	20:03	0.37	3.0	4.0	2.5	95
11/5	23:05	0.35	<1.0	2.0	<1.0	35
11/6	09:55	0.24	<1.0	3.0	ND	45

Table 12. Heavy Metals ($\mu g/l$) at Station 4 during November 5, 1989 Rain Event

Date	Time	Gage-m	Chromium	Copper	Lead	Zinc
11/5	00:50	0.30	<1.0	2.0	1.0	30
11/5	13:22	0.28	<1.0	1.5	<1.0	45
11/5	16:57	0.30	1.0	3.0	2.5	60
11/5	19:36	0.30	1.5	5.0	3.0	20
11/5	23:18	0.30	ND	<1.0	<1.0	30
11/6	11:05	0.30	ND	<1.0	<1.0	45

Table 13. Heavy Metals ($\mu g/l$) at Station 5 during November 5, 1989 Rain Event

Date	Time	Gage-m	Chromium	Copper	Lead	Zinc
11/5	07:30	0.36	ND	5.0	ND	65
11/5	12:55	0.39	ND	3.5	<1.0	100
11/5	17:20	0.39	<1.0	2.5	<1.0	30
11/5	20:30	0.38	ND	<1.0	<1.0	70
11/5	23:55	0.37	<1.0	3.5	<1.0	58
11/6	11:25	0.37	<1.0	1.5	<1.0	75

Table 14. Heavy Metals ($\mu g/l$) at Station 6 during November 5, 1989 Rain Event

Date	Time	Gage-m	Chromium	Copper	Lead	Zinc
11/5	07:45	1.04	ND	2.0	ND	40
11/5	13:10	1.03	2.5	3.0	1.0	25
11/5	17:30	1.02	ND	5.0	3.0	45
11/5	20:45	1.02	1.0	3.5	3.0	45
11/5	23:55	1.03	ND	3.5	<1.0	10
11/6	11:53	1.03	ND	ND	ND	30

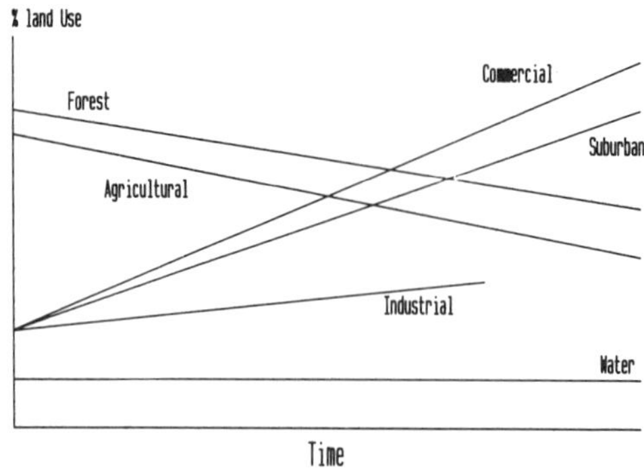

Fig. 6. Conceptual description of land use change due to urbanization.

DISCUSSION

A conceptual scenario of land use shifting due to urbanization is displayed in Fig. 6. It is hypothesized that this shifting of land use will alter the characteristics of runoff, in particular, heavy metals loadings to receiving streams. This study attempted to correlate the mass instream loadings of the two most sensitive metals, copper and lead, to land use type. An empirical correlation of this type was attempted since a mass balance approach may not work, since significant amounts of metal can be input to the water column by bottom scour. Results examining filtered and unfiltered samples demonstrated that almost all of the metals were found to be associated with the particulate fraction.

A simple multiple linear regression model was used as follows:

$$M = b_0 + b_1(for) + b_2(frm) + b_3(res) + b_4(ind) + b_5(urb) + b_6(wat) \quad (1)$$

where M is the mass loading of the appropriate metal, normalized to rainfall and drainage area (lb-metal/sq mi-D.A./in-rain), for, frm, res, wat, ind and urb are the respective percentages of forest, farm, residental, water, industrial and urban areas, and the subscripted b terms are the regression parameters. Since this equation is a linear mixture model in which the sum of the independent variables is unity, one of the independent variables must be deleted [4]. There are several ways to perform this statistically [5], however, it is reasonable to presume that the surface water area of a basin will not appreciably change; therefore this term was selected for deletion.

Results are displayed in Table 16. The best correlations (r^2) were found for natural logarithm of lead and square root of copper as M, but these were only fair (0.55 and 0.50, respectively); however, the variation

Table 15. **Precipitation Data (inches)**

Date	Time	Accum.	Date	Time	Accum.	Date	Time	Accum.
8/23/89	19:30	0.00	11/1/89	05:00	0.00	11/5/89	11:00	0.00
8/23/89	22:00	0.10	11/1/89	09:00	0.05	11/5/89	12:15	0.06
8/24/89	00:00	0.20	11/1/89	09:30	0.13	11/5/89	13:40	0.07
8/24/89	02:30	0.21	11/1/89	10:00	0.22	11/5/89	15:30	0.07
8/24/89	03:20	0.31	11/1/89	10:30	0.35	11/5/89	15:50	0.12
8/24/89	03:40	0.47	11/1/89	11:00	0.48	11/5/89	16:00	0.15
8/24/89	04:00	0.69	11/1/89	11:30	0.60	11/5/89	16:20	0.16
8/24/89	04:20	0.93	11/1/89	12:00	0.69	11/5/89	16:30	0.24
8/24/89	04:40	1.15	11/1/89	12:30	0.75	11/5/89	16:50	0.26
8/24/89	05:30	1.26	11/1/89	13:00	0.81	11/5/89	17:15	0.27
8/24/89	06:30	1.27	11/1/89	14:00	0.96	11/5/89	17:50	0.27
8/24/89	07:50	1.29	11/1/89	15:00	0.98	11/5/89	23:45	0.27
8/24/89	08:50	1.29	11/1/89	16:00	1.00			
8/24/89	21:00	1.29	11/1/89	20:00	1.00			
8/25/89	00:30	1.51						
8/25/89	11:30	1.51						
8/26/89	00:30	1.52						

Table 16. Regression Model Parameters

M	b_0	b_1	b_2	b_3	b_4	b_5	r^2
$Cu^{(1/2)}$	-17.6	0.242	0.098	0.299	1.170	0.166	0.50
ln(Pb)	-53.2	0.682	0.288	0.780	2.586	0.624	0.55

Table 17. Land Use Percentages

Station	Forest	Farm	Residential	Water	Industrial	Urban
1	51.1	37.6	6.7	4.2	0.4	0.0
2	38.6	39.7	11.1	6.0	2.1	2.5
3	34.7	39.4	8.7	4.9	1.6	10.7
4	44.4	44.3	4.9	3.6	1.2	1.6
5	33.2	45.1	14.3	4.3	1.1	2.1
6	33.6	42.9	8.4	5.9	2.0	7.2

in the independent variables is not great. Table 17 displays the drainage area percentages used for the appropriate stations. These models can thus be modified and recalibrated as development takes place; this study can be used as the baseline. It is anticipated that as urbanization increases, the metals loading in the streams will also increase.

ACKNOWLEDGEMENTS

The work described herein was supported in part by a grant from the New Jersey Department of Environmental Protection, Division of Water Resources (Dr. T. R. Jeng, Project Director) and the New Jersey Agricultural Experiment Station, Publication No. D-27525-4-89.

REFERENCES

[1] Forstner U., and Wittmann G.T.W., Metal Pollution in the Aquatic Environment, Springer-Verlag, New York, 1983.

[2] Wilber W.G., and Hunter J.V., *Wat. Res. Bullet.*, 13, 721-734, 1977.

[3] APHA, Standard Methods for the Examination of Water and Wastewater, 16th Edition, American Public Health Association, Washington, D.C.,1985.

[4] Scheffe H., *J. Royal Stat. Soc., Series B,* 20, 344-360, 1958.

[5] Snee R.D., and Marquardt D.W., *Technometrics*, 18, 19-30, 1976.

AN ASSESSMENT OF HAZARD FROM SOLID INDUSTRIAL WASTES: A CASE STUDY

I. TWARDOWSKA

Polish Academy of Sciences
Institute of Environmental Engineering
M. Curie-Skłodowska st. 34, 41-819 Zabrze Poland

ABSTRACT

To assess possibility of pollution from a dump, an identification of mineralogical and chemical composition of waste material, leach experiments and field studies have been carried out. It has been found that the base layer of the dump, constructed of unstabilized slag, is an effective insulation shield preventing leaching of contaminants to the aquatic environment through the development of a slag layer impermeable to air and water and reduction conditions within the slag layer. It has also been ascertained that this material, after being mechanically crushed, becomes highly permeable, which promotes intensive leaching of toxic contaminants such as Cr^{6+}. Therefore, destroying the dump, instead of eliminating a hazard, could cause the formation of at least two active sources of water contamination by chromium at an old and new site.

INTRODUCTION

Solid industrial waste dumps are one of the major sources of contamination for the aquatic environment from pollutants leached from waste material by percolating atmospheric precipitation. The adequate assessment of the extent and seriousness of the potential hazards to water resources from designed or already existing waste dumps, as well as the selection of the most effective and economical protection measures, requires conversance with composition and properties of deposited wastes and their geochemical transformations due to weathering.

This statement can be exemplified by the case of the slag dump of the only Polish ferrochrome production plant "Siechnice", at present in liquidation for ecological reasons. The dump is to be moved to another site because of the possibility of serious contamination of water in the combined ground- and bank-filtered water supply for Wroc law city.

Simultaneously, investigations have been carried out to define the extent of hazard due to leaching and transport of contaminants to the aquatic environment from this dump.

OBJECT, MATERIAL AND METHODS

At the waste dump covering 8.5 hectares, highly alkaline slags from the traditional process of ferrochrome smelting in open electric furnaces using partially dolomitized burnt lime as a slag–forming additive have been deposited for 29 years. As a result of allotropic transformations of 2 $CaO{\cdot}SiO_2$ from α to γ phase crystals in the course of solidification, slag turns into powder ($0 < 4\ \mu m \approx 90\%$). Addition of boron-containing calcite material during smelting stabilizes 2 $CaO{\cdot}SiO_2$ crystals in phase β, which keeps slags in the form of hard, coarse-grained material.

The investigated dump has been constructed in two layers: the base layer, 10-12 m thick, comprised of instabilized slag deposited in the period 1961-1978, and the upper layer, 15 m thick, where stabilized slag has been placed since 1978.

Investigations consisted of an identification of mineralogical and chemical composition of slag and its transformations due to weathering; leach experiments on the representative sample of freshly produced slag; an estimation of the natural moisture content and water retention capacity quoted on a dry-solids basis, as well as the chemical composition of pore water in vertical profile of the dump. The ground water quality in the vicinity of the dump has also been examined.

For the experiments, a representative sample of freshly produced slag was taken as an average of every shift throughout one month. Because of an inexact stabilization process, the slag was partially unstable. The unstabilized material accounted for about 10% of the total slag discharge. Wastes deposited at the upper layer of the dump therefore formed a mixture of coarse-grained stabilized material and unstabilized powdered slag.

To determine the dynamics of contaminants leaching from the stabilized slag layer, leach experiments simulating infiltration of water from atmospheric precipitation through the dump under conditions of unsaturated (va dose) zone were carried out. To slag 2, 3 and 10 m thick - modeled as a series of five connected sections 2 m high a flow rate of leaching water corresponding to the ten-year average daily precipitation of 3.59 mm was applied.

Field studies carried out in the investigated area (Fig. 1) included the sampling of slag at the waste dump and ground water at the dump forehead. Samples of slag of natural moisture content were taken every 1 m of the vertical profile from 4 boreholes 5 to 13 m deep, drilled at an open part of the top and at a terrace of the unstabilized base layer. The uppermost (0-20 cm) layer of the unstabilized part of the dump was also sampled at points. The samples of wastes from the top of the stabilized layer of the dump were taken by scraping some 10 m^2 of the uppermost (0-20 cm) layer at 3 points. Ground water has been periodically sampled since 1987 from the 16 piezometric boreholes and from 8 wells in the vicinity of the dump.

Pore solutions from the slag samples of natural moisture content taken from the boreholes were extracted by a pressure method [1]. Water extracts, leachates and ground water samples were examined by standard methods, using atomic absorption spectroscopy (AAS Perkin-Elmer Mod. 1100 B) for determination of heavy metals.

RESULTS AND DISCUSSION

The major components of slag are spinels, silicate minerals, metallic chromium and Si-rich external glass matrices enclosing crystalline phases. Metallic chromium occurs in the form of drops of inexactly separated residual alloy and as spinels, $MgO \cdot Cr_2O_3 \cdot Al_2O_3$, but the bulk of chromium concentrates in glass matrices as small irregular inclusions.

The chemical composition of slag (Table 1) reflects the mineralogical composition of raw material and slag. The predominant constituents in average sample are calcium, silica, chromium, magnesium and aluminium. Chromium concentrations ranged from 6.0 to some 10.5% as Cr_2O_3, the content of water–extractable species in freshly produced material being much lower and ranging from 0.006 to 0.008%. Cr_t.

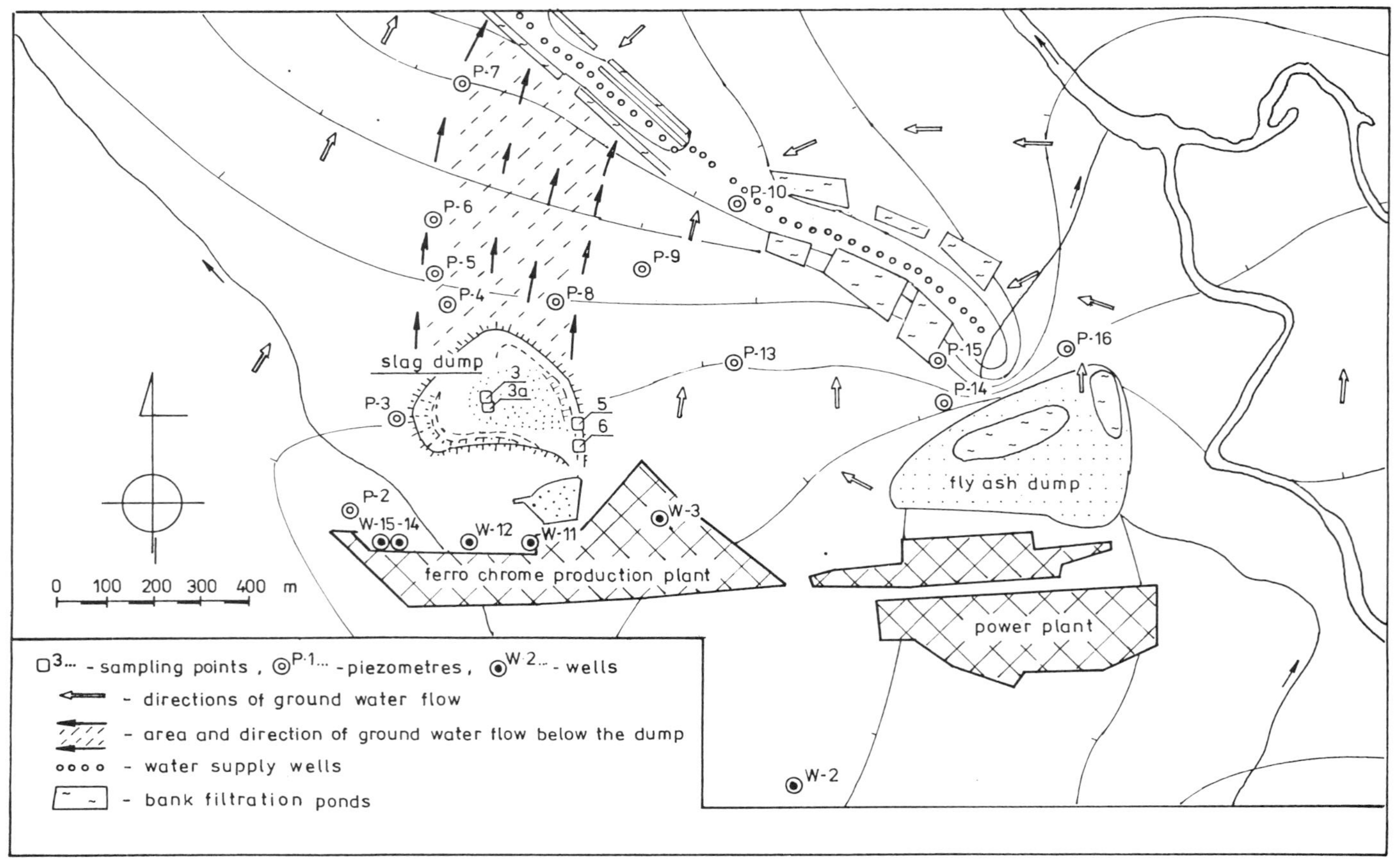

Fig. 1. Scheme of the hydrodynamic field and localisation of sampling points in the investigated area.

Table 1. Chemical Composition of the Freshly Deposited Slag From Ferro Chrome Production Plant (wt%, dry basic)

Constituents	Stabilized slag		Instabilized slag
	Average sample	Fine-grained sample	Average sample
Loss-on-ignition	0.50	0.90	2.20
SiO_2	29.88	26.02	35.29
TiO_2	0.25	0.20	0.23
Al_2O_3	7.37	5.67	7.40
Cr_2O_4	8.69	10.52	6.02
Fe_2O_3	2.07	1.18	0.86
FeO	0.63	1.22	1.15
CaO	42.14	45.00	46.01
MgO	8.36	9.00	0.44
SO_3	0.05	trace	trace
$Cr_{soluble}$	0.008	0.009	0.006
Ag_2O_3	0.0004	0.0007	0.0004
B_2O_3	1.30	1.03	0.80
CoO	0.00068	0.00064	0.00065
CuO	<0.012	<0.015	<0.012
MnO	0.18	0.065	0.25
NiO	0.0030	0.0044	0.0030
PbO	0.13	0.054	0.024
SrO	0.08	0.095	0.07
Tl_2O	<0.02	<0.02	<0.02
V_2O_5	0.0060	0.0034	0.0058
Compounds extractable in H_2O:			
Cl^-	0.0770		0.0030-0.0700
SO_4^{2-}	0.0148		trace-0.330
Cr_t	0.00035		0.0060
Cr^{3+}	0.00020		
Cr^{6+}	0.00015		0.0060

Of the trace heavy metals, boron, manganese–zinc, lead, strontium, copper, vanadium, nickel, cobalt and silver occurred in the average sample in concentrations from 1.30 to 0.0004 wt% in descending sequence. No germanium and mercury was detected. Beryllium, arsenic and tin were present in almost undetectable concentrations. Tallium content was below the detection limit.

The chemical composition of stabilized and unstabilized material is similar. The total chromium content in the investigated sample of unstabilized slag was somewhat lower than in the stabilized one, but water extractability of Cr_t was as much as 40 times higher in unstabilized stag, undoubtedly because of the substantially larger surface area of this highly dispersed material.

The dynamics of constituents leaching from the layers of stabilized slag under conditions simulating infiltration of precipitation water displayed, on the one hand, a high leachability of soluble constituents, and on the other hand, a strong dependence on the geochemical constraints imposed by equilibrium conditions, in conformity with the general theory of solid solutions and mineral reactions [2,3,4].

The mineralogical and chemical composition of slag ($CaO:SiO_2 = 1.3$ to 1.7) determines the permanent highly alkaline character of slags and leachates, with pH values ranging from 8 to above 12. The major anions in leachate are components of alkalinity, i.e., carbonates, bicarbonates and hydroxides balanced mainly by sodium and potassium cations. Minor ionic components are sulphate and chloride anions, as well as calcium and magnesium cations occurring in considerably lesser quantities. Total dissolved solids and ionic strength of leachate show a declining tendency with time and water exchange rate, which means scarcity of macro-components generation within a slag layer due to weathering (Fig. 2).

Of the toxic constituents, the leaching of chromium is of particular importance, as Cr_t species are known carcinogens and occupy fairly large Eh- pH fields [2]. The dynamics of Cr_t and Cr^{6+} leaching from the layer of stabilized slag may be described by an exponential function or also expressed as two simple regressions in semi-log scale, distinguishing two phases - phase (I) of intensive leaching and phase (II) of extensive leaching under stable conditions (Figs. 3,4).

Maximum determined concentrations of chromium at the start of phase (I) leaching from the 2– to 10– m thick layers of stabilized slag ranged from 15.5 to 110 mg/dm^3 Cr_t and from 4.75 to 38.96 mg/dm^3 Cr^{6+} and many times exceeded the standard limit of 0.05 mg/dm^3 Cr^{6+} permitted for drinking water. The Cr^{6+} concentrations accounted for 23 to 62%, average 44%, of the total chromium content in leachates (Cr_t). The rest of leachable chromium was present as Cr^{3+} cationic species.

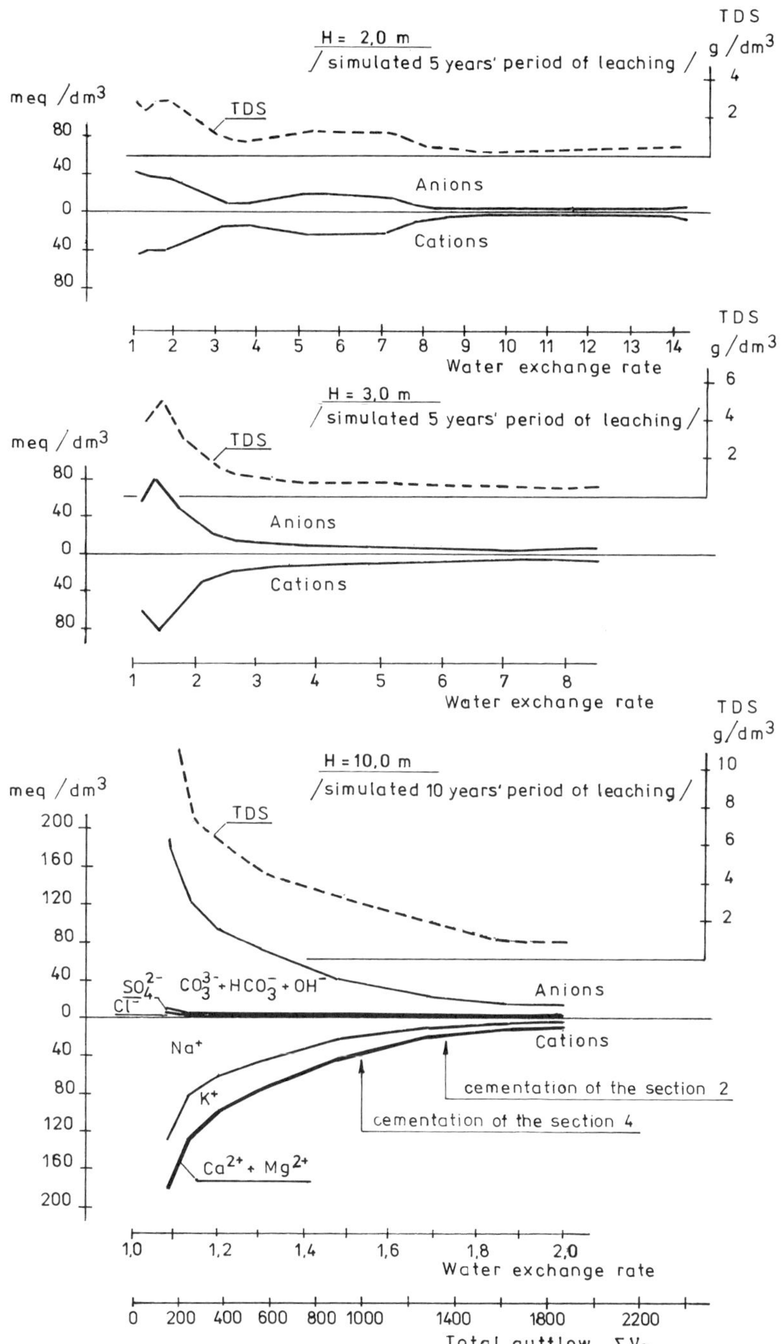

Fig. 2. Chemical composition of leachate from the layers of stabilized slag during the simulated leaching by precipitation water

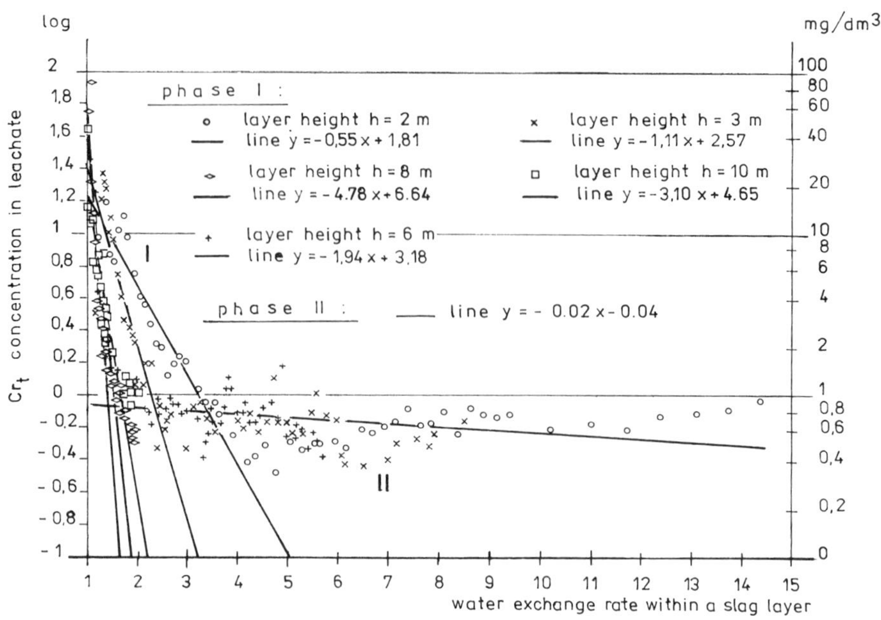

Fig. 3. Dynamics of total chromium leaching from 2–, 3–, 6–, 8– and 10–m–thick layers of freshly produced stabilized slag.

The character of chromium migration during the vertical flow of infiltrating water through the stabilized slag layer in phase (I) (see Fig. 5) was found to be similar to the pattern of the vertical redistribution of highly soluble constituents such as chlorides in course of the gradual filling of a retention capacity of material up to the start of free percolation of water through the full thickness of the layer [5]. The average retention capacity of stabilized slag is not high and amounted to 6.45 wt% of dry basis.

Gradual decrease with time of chromium concentrations in leachate occurred in accordance with the decline of chemical oxygen demand (COD) and change of oxidation conditions (Eh>0) to reduction ones (Eh<0) due to partial compaction and cementation of slag as a result of weathering processes, particularly active in the unstabilized fraction of slag. Stabilized material is much more resistant to decomposition.

The boundary point between two phases for Cr_t and Cr^{6+} corresponds with the range of water exchange rate from 1.4 to 3.2, adversely depending on the thickness of the leached slag layer.

In phase (II), as a consequence of formation of reduction conditions (Eh ranging from -65 to -84 mV), the concentration of chromium in leachate

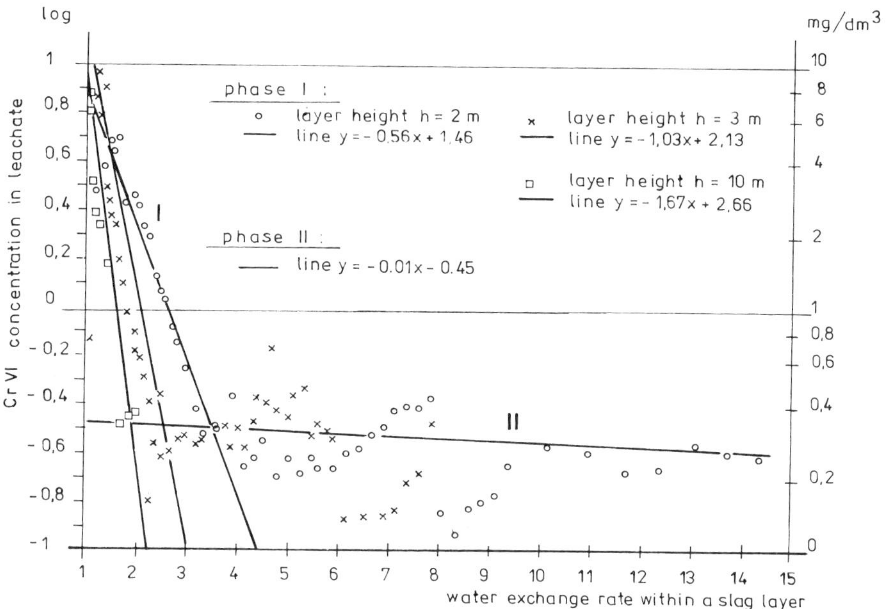

Fig. 4. Dynamics of Cr^{6+} leaching from 2–, 3– and 10–m–thick layers of freshly produced stabilized slag.

was relatively low, from 0.34 to 1.12 mg/dm^3 Cr_t (Fig. 3) and from 0.12 to 0.47 mg/dm^3 Cr^{6+} (Fig. 4). The limiting factor for Cr^{3+} dissolution is high pH values ranging from 8 to 12 pH, and for the formation of soluble Cr^{6+} species, the range of Eh potential (Fig. 6).

The basic weathering transformations in slag were devitrification of glass matrices resulting in the formation of amorphous Si-rich calcareous material and a two-stage process of hydrolysis (Equation 1) and carbonization (Equation 2) of CaO owing to water and air penetration through the slag layer:

$$CaO + H_2O \rightarrow Ca(OH)_2 + 6 \text{ Kcal} \tag{1}$$

$$Ca(OH)_2 + CO_2 \rightarrow CaCO_3 + H_2O \tag{2}$$

An other secondary mineral was goethite FeO(OH), partly filling veins and pores of weathered material. Silicate minerals such as melilite, larnite and diopside were also susceptible to decomposition, forming micro-crystals, while spinels and metallic chromium were resistant to weathering.

It was ascertained that the permeability to water of waste material

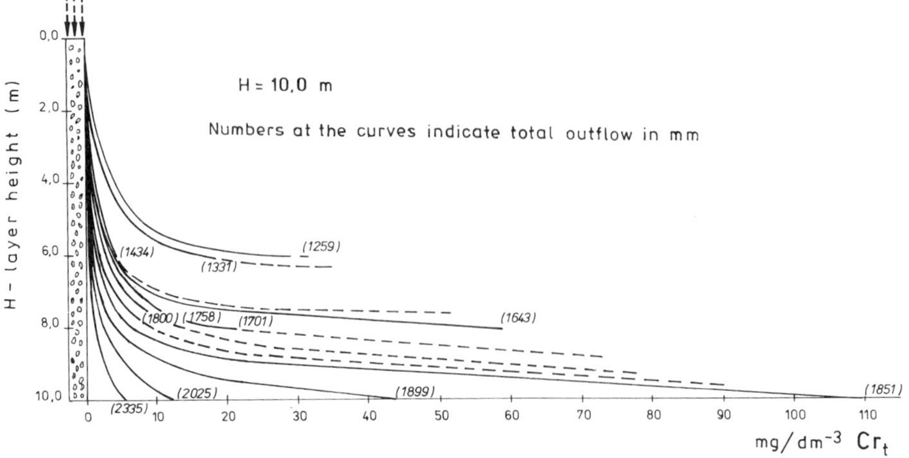

Fig. 5. Re-distribution of total chromium loads along a vertical profile of 10 m thick slag layer during the simulated leaching by precipitation water.

depended on the proportion of unstabilized slag to stabilized slag, which was much more resistant to decomposition. The layers 2 and 3 m thick of the stabilized coarse-grained slag remained highly permeable to water ($k=10^{-4}$ m/s) throughout the leach cycle. In 10–m–long series connection, utter decay of water flow in two sections occurred 493 and 720 days after start of leaching process in consequence of sealing by cemented instabilized slag concentrated in 20– to 50–cm–long parts of these columns.

The water pollution potential from heavy metals other than chromium leached from slag is low (Table 2) because of their stability at permanently high pH values [2,3,4]. Dynamics of leaching of these metals displayed no correlation with time and layer thickness and showed high dispersion of concentrations within the range generally much below the maximum standard concentrations for drinking water, up to complete decline of occurrence.

The leach experiments on the stabilized slag have hence ascertained a clear leachability of chromium, including Cr^{6+}, in the initial stage of leaching, and also a high permeability to water, provided that the proportion of instabilized material is not substantial. Therefore, the pollution potential of stabilized slag deposited on a waste dump is high.

Considering a consistently better water extractability of chromium from instabilized slag, the chromium concentrations in pore solutions and leachate from an instabilized slag layer are expected to be adequately higher than those from stabilized material, should water infiltration throughout this

Table 2. Concentrations of Trace Heavy Metals in Leachates from Slag Layers (mg/dm^3)

Heavy metals	Thickness of slag layer (m)										Standard limits for drinking water
	2 m		3 m		6 m		8 m		10 m		
	mean	min / max	mean	min / max	mean	min / max	mean	min / max	mean	min / max	
Boron, B	0.495	0.250 / 3.000	1.275	0.250 / 3.800		—		—	1.229	0.250 / 3.000	-
Cadmium, Cd	0.009	0.005 / 0.093	0.012	0.005 / 0.103	0.010	0.005 / 0.064	0.002	0.001 / 0.005	0.003	0.005 / 0.007	0.05
Cobalt, Co	0.052	0.005 / 0.197	0.052	0.005 / 0.691	0.009	0.005 / 0.103	0.014	0.005 / 0.075	0.015	0.005 / 0.095	-
Copper, Cu	0.153	0.005 / 0.620	0.065	0.005 / 0.398	0.106	0.005 / 0.859	0.196	0.005 / 1.498	0.175	0.005 / 0.955	0.50
Iron, Fe_t	0.239	0.005 / 1.200	0.260	0.005 / 1.500	0.222	0.005 / 1.630	0.123	0.005 / 0.304	0.180	0.005 / 1.040	0.50
Manganese, Mn	0.013	0.005 / 0.270	0.012	0.005 / 0.090	0.022	0.005 / 0.273	0.012	0.005 / 0.145	0.019	0.005 / 0.174	0.10
Nickel, Ni	0.189	0.005 / 2.430	0.090	0.005 / 1.046	0.011	0.005 / 0.134	0.006	0.005 / 0.052	0.0023	0.005 / 0.080	-
Lead, Pb	0.109	0.012 / 0.653	0.115	0.012 / 0.567	0.137	0.012 / 0.363	0.104	0.012 / 0.600	0.109	0.012 / 0.400	0.10
Zinc, Zn	0.042	0.005 / 0.208	0.044	0.005 / 0.240	0.057	0.005 / 0.524	0.218	0.005 / 1.133	0.141	0.005 / 0.956	5.00

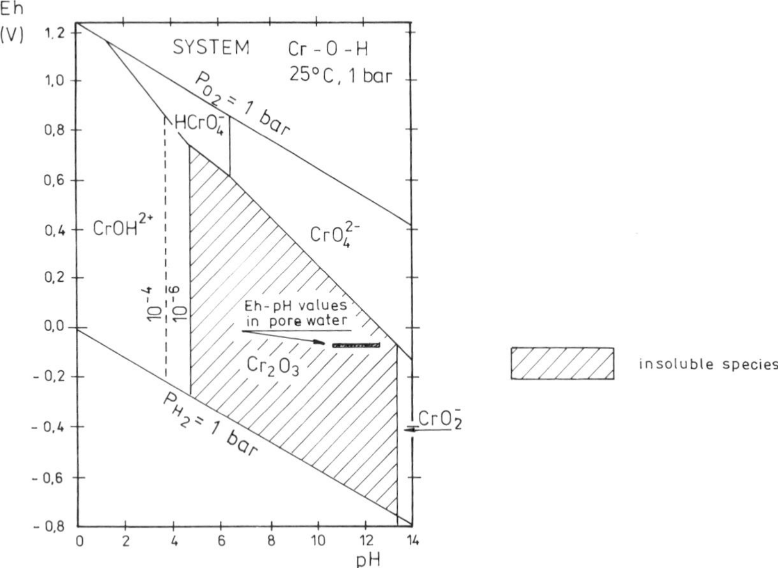

Fig. 6. The range of Eh-pH values in pore solutions within the layer of unstabilized slag at the background of Eh-pH diagram for part of the system Cr-O-H.

layer under oxidation conditions occur. The investigations on pore water extracted from slag samples taken along the four vertical profiles of the instabilized base layer of the dump (Fig. 1) displayed, however, rather slight vertical redistribution of macro-components and chromium (Fig. 7). The pattern of distribution of concentrations along such profile suggests that the process of vertical transport of macro-components stopped at the initial stage corresponding with the I phase of leaching and no further load transport occurs, while the chromium concentrations in pore solutions conformed to the II phase of leaching, Cr^{6+} being present in the range from 0.29 to 1.20 mg/dm^3, with maximum at the depth 2 to 3 m below the surface (Fig. 7,8).

Results of the mineralogical investigations on the slag samples taken from the instabilized base layer of the dump afford a clear explanation of this phenomenon. The complete two-stage transformation of CaO into $CaCO_3$ according to the equations 1 and 2 was detected only in some 2–cm–thick crust at the exposed surface of the layer. In the entire profile of the dump below the crust calcium oxide appeared to undergo only hydrolysis forming amorphous calcium hydroxide (Equation 1). Considering that the sole source of carbon dioxide for carbonatization according to the equation 2 is air,

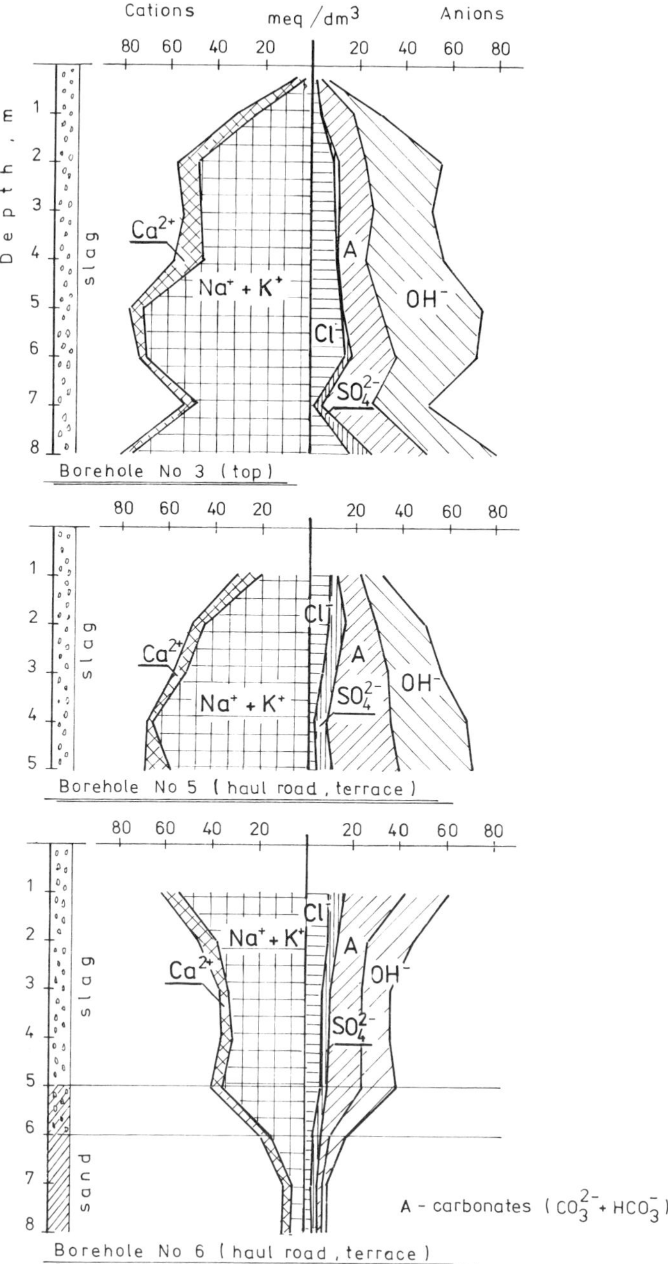

Fig. 7. Distribution of macro-components in pore solutions along vertical profiles of the unstabilized slag layer.

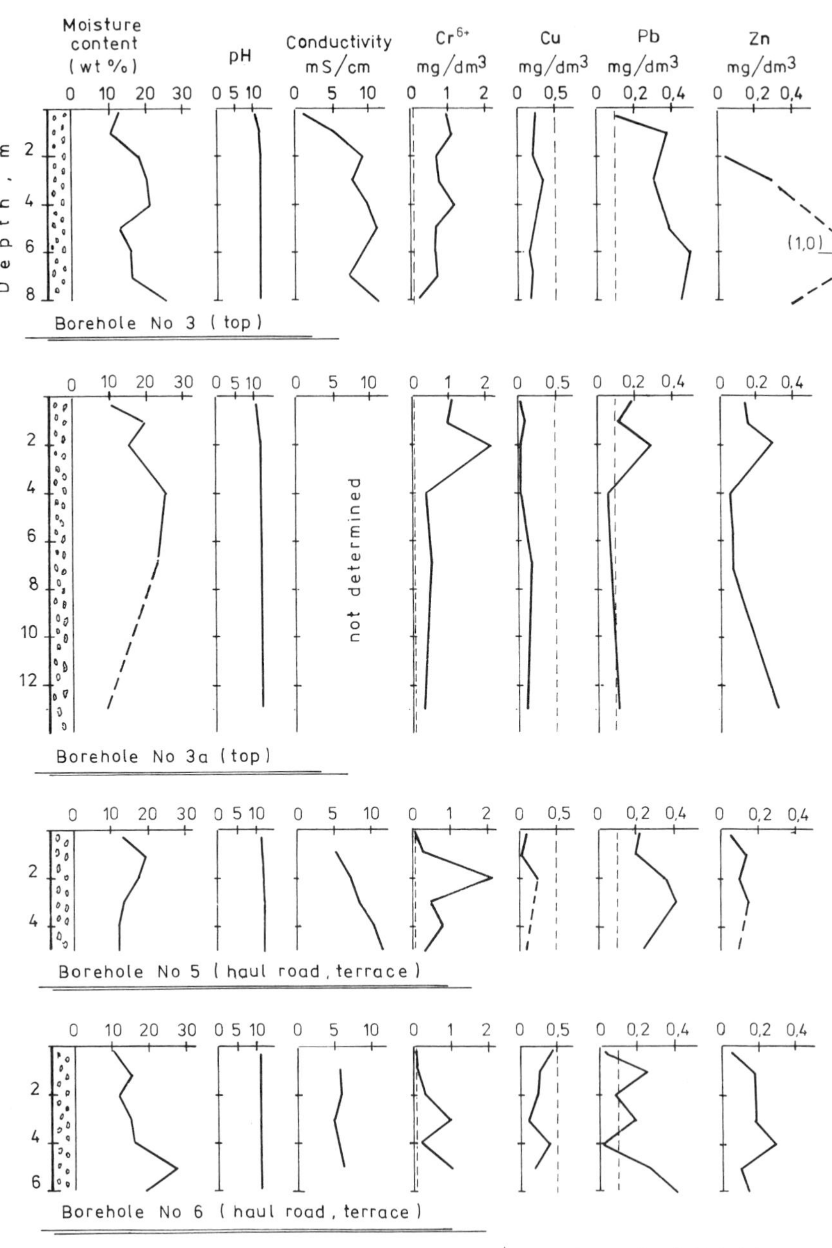

Fig. 8. Moisture content, pH and concentration of heavy metals in pore solutions along vertical profiles of the unstabilized slag layer.

these observations support the contention that the permeability to air of the unstabilized weathered slag layer is effectively nil. Consequently, the reduction conditions within this layer have been developed what resulted in stability of Cr^{6+} in a highly alkaline environment (Fig. 6).

Relating to the impermeability to air, in the unstabilized base layer the impermeability to water has been also developed. Excavations carried-out on the dump displayed that the unstabilized base layer had transformed into homogeneous cemented material. The cementation effect followed by practically complete decay of water flow ($k<10^{-8}$ m/s) in the partly unstabilized slag layer during leach experiments has been mentioned above. The phenomenon of cementation of unstabilized slag has also been reported by Magosz et al. [6] in column leach experiments. The cementation and cease of outflow occurred in these experiments in 60-70 days, which is much faster than in the experiments presented here, where only partly unstabilized slag was used and from 4 to 12 times doses of leaching water were applied.

The mineralogical examinations of the cemented unstabilized slag taken from the dump and cemented parts of the 10–m–long series of columns have ascertained that binding material consists of isotropic highly dispersed products of slag weathering that are comprised of amorphous decomposed glass matrices. The particles of silica and calcium hydroxide have been cemented together through the agency of adsorbed water which promotes the mobility of the ions near the surface of the solid [7]. Cementation causes the development of an internal porosity trapping the infiltrating water, whereas the open porosity becomes, in practice, negligible. This way the base layer of unstabilized slag has formed an effective shield against the infiltration of contaminants from the upper stabilized slag layer.

The periodical survey of ground water quality in the examined area carried out in the years 1986-1988 did not display any visible impact of infiltration from the dump on water quality in the direction of flow below the dump (Table 3), the occasional larger concentrations of Cr^{3+} and detectable contents of Cr^{6+} in the entire area being attributed to spreading of fly dust, mainly from electric furnaces and to some extent from the unprotected top of the slag dump and ore heaps. In one point above the dump site (W-11), the constant high contamination of water by chromium has been blamed on siting of a well in the buried slag. Generally, the results of these surveys support the conclusion that the base layer of cemented unstabilized slag successfully protects against contaminant infiltration and leaching.

On the other hand, it has been ascertained that this material, once being mechanically crushed, becomes highly permeable ($k=10^{-4}$ m/s) because of the formation of an open porosity. The application of mechanical pressure does not substantially suppress its filtration properties. The laboratory experiments showed that the filtration coefficients of the crushed slag

Table 3. Concentration of Contaminants in Ground Water in the Vicinity of the Slag Dump in the Years 1986-1988 (mg/dm^3)

Sampling points		pH range	Cr^{3+} (-)		Cr^{6+} (0.05)		Cu (0.50)		Pb (0.10)		Zn (5.0)	
			mean	min max	mean	min max	mean	min max	mean	min max	mean	min max
Above the dump		7.10		0.000		0.000		0.005		0.012		0.009
	W - 1	7.40	0.018	0.080	0.0004	0.0023	0.010	0.020	0.019	0.030	0.083	0.248
		6.90		0.000		0.000		0.006		0.005		0.024
	W - 2	7.40	0.009	0.012	0.000	0.000	0.009	0.012	0.024	0.050	0.282	0.094
		7.20		0.040		0.000		0.002		0.005		0.024
	W - 3	7.90	0.158	0.410	0.0253	0.0835	0.004	0.008	0.025	0.025	0.049	0.094
		7.00		0.000		0.230		0.004		0.000		0.036
	W - 11	9.00	0.417	1.400	0.5480	1.241	0.008	0.010	0.009	0.025	0.048	0.068
		6.90		0.000		0.000		0.019		0.000		1.790
	W - 12	7.40	0.006	0.010	0.000	0.000	0.028	0.037	0.031	0.073	2.463	3.000
		6.60		0.000		0.000		0.010		0.010		0.179
	W - 14	7.40	0.007	0.012	0.0025	0.010	0.019	0.034	0.021	0.030	0.691	1.800
		7.00		0.000		0.000		0.006		0.000		0.154
	W - 15	7.40	0.020	0.060	0.000	0.000	0.011	0.014	0.010	0.021	0.367	0.843
		6.86		0.000		0.000		0.005		0.016		0.005
	P - 2	7.70	0.017	0.080	0.0009	0.004	0.024	0.045	0.047	0.140	0.112	0.226
		6.80		0.000		0.000		0.003		0.010		0.005
	P - 1	8.00	0.027	0.050	0.0202	0.0428	0.010	0.012	0.039	0.070		5.455
Below the dump		7.20		<0.0023		0.000		0.005		0.000		0.005
	P - 4	7.35	0.008	0.020	0.0012	0.005	0.013	0.022	0.043	0.150	0.074	0.186
		7.00		0.000		0.000		0.020		0.012		0.005
	P - 5	7.30	0.015	0.030	0.0013	0.0067	0.038	0.046	0.042	0.090	0.516	1.420
		7.00		0.008		0.000		0.004		0.000		0.070
	P - 6	7.20	0.025	0.056	0.0008	0.0023	0.019	0.030	0.052	0.180	0.195	0.530
		6.79		0.008		0.000		0.020		0.010		0.110
	P - 7	7.10	0.020	0.060	0.000	0.000	0.057	0.164	0.051	0.200	0.218	0.476
		6.70		0.000		0.000		0.005		0.000		0.005
	P - 8	8.30	0.030	0.140	0.000	0.000	0.018	0.026	0.022	0.090	0.064	0.156
Aloof the dump		7.10		0.000		0.000		0.005		0.000		0.042
	P - 3	7.40	0.016	0.066	0.000	0.000	0.037	0.116	0.028	0.150	0.121	0.302
		7.10		0.000		0.000		0.012		0.000		0.064
	P - 9	7.20	0.023	0.048	0.000	<0.0023	0.029	0.066	0.060	0.150	0.417	1.690
		7.10		0.000		0.000		0.005		0.000		0.010
	P - 10	7.68	0.030	0.150	0.0014	0.0068	0.0265	0.066	0.023	0.090	0.162	0.500
		6.90		0.000		0.000		0.004		0.000		0.005
	P - 11	7.68	0.022	0.082	0.0004	0.0023	0.034	0.076	0.027	0.070	0.092	0.192
		7.00		0.000		0.000		0.003		0.000		0.005
	P - 12	7.20	0.013	0.036	0.000	<0.0023	0.030	0.072	0.049	0.180	0.120	0.236
		7.02		0.000		0.000		0.003		0.000		0.005
	P - 13	7.30	0.011	0.036	0.000	<0.0023	0.024	0.036	0.030	0.036	0.154	0.640
		8.80		0.000		0.000		0.000		0.000		0.005
	P - 14	10.42	0.018	0.090	0.0014	0.0135	0.017	0.034	0.025	0.053	0.117	0.400
		6.22		0.000		0.000		0.006		0.000		0.005
	P - 15	8.00	0.029	0.150	0.000	<0.0023	0.043	0.106	0.040	0.120	0.136	0.115
		6.46		0.000		0.000		0.005		0.000		0.005
	P - 16	8.40	0.065	0.400	0.0013	0.0090	0.026	0.076	0.022	0.130	0.127	0.500

Numbers in parenthesis - maximum standard level permitted in drinking water extracted from the boreholes at the dump ranged from $0.93 \cdot 10^{-4}$-$2.26 \cdot 10^{-4}$ m/s at 12.5 KPa to $0.33 \cdot 10^{-4}$-$0.76 \cdot 10^{-4}$ m/s at 200 KPa, mean values accounting for $0.86 \cdot 10^{-4}$-$1.29 \cdot 10^{-4}$ m/s for this range of pressure applied. Under conditions of high permeability to air and water, the intensive leaching according to phase (I) is being promoted and contaminants are easily generated and released from lumpy slag. Therefore, the desruction of the dump and replacement of disturbed and admixed material, instead of eliminating hazard, could cause the formation of at least two active sources of water contamination by chromium at an old and new site.

At the same time, simple and relatively inexpensive additional measures such as application of top drainage system on the clay bed layer to prevent unnecessary entry of precipitation water into the dump, covering of a the top surface and slopes with soil and biological reclamation of the dump to promote evapotranspiration and protect against raising dust, would effectively eliminate the potential hazard of ground water contamination.

CONCLUSIONS

1. The adequate assessment of the pollution potential from designed and already existing waste dumps requires the detailed knowledge of geochemical transformations of deposited rejects and their interaction with the aquatic environment.

2. The most economical and effective protection measures against contamination of the aquatic environment in the vicinity of a dump should support positive natural processes and arrest negative ones, as the mere replacement of disturbed wastes from existing dumps could cause an opposite effect instead of an intended hazard elimination.

3. In the particular case of the slag dump of the ferro–chrome production plant "Siechnice", the base layer of unstabilized slag forms an insulation shield against the generation and infiltration of contaminants from the dump to the ground water. Simple additional measures would, hence, successfully eliminate the potential contamination hazard from this dump.

ACKNOWLEDGEMENTS

This study was supported financially by the Central Research Programme 03.11. Mineralogical and chemical composition of slags was examined at the Institute of Geology of the Silesian Technical University in

Gliwice, Poland. Pore solutions were extracted and partially examined at the University of Mining and Metallurgy, Institute of Hydrogeology and Engineering Geology, Cracow, Poland.

REFERENCES

[1] Kriukow P.A.: Gornyje, poczvennyje i ilovyje rastvory. (Monograph, in Russian), Izd. Nauka, Novosybirsk, 1971.

[2] Brookins D.G.: Eh-pH diagrams for geochemistry. Springer Verlag, Berlin-Heidelberg-New York-London-Paris-Tokyo, p. 176, 1988.

[3] Garrels R.M. and C.L. Christ: Solutions, minerals and equilibria. Harper and Rowley, New York, p. 453, 1965.

[4] Ganguly J., S.K. Saxena: Mixtures and mineral reactions. Springer Verlag, Berlin-Heidelberg-New York-London-Paris-Tokyo, p. 291, 1988.

[5] Herzig J., J. Szczepańska, S. Witczak, I. Twardowska: Chlorides in the Carboniferous rocks of the Upper Silesian Coal Basin. Fuel **65**, pp. 1134-1141, 1986.

[6] Magosz S., K. Kaczmarczyk, A. Lysek, Z. Mazur, J. Paluch: Badania wplywu żużla chromowego na wody powierzchniowe i gruntowe na terenie oddzialywania huty "Siechnice". Sci. Report C $_2$-34/I/73 (in Polish), p. 62, 1974.

[7] Gregg S.J., K.S.W. Sing: Adsorption, surface area and porosity. Academic Press, London and New York, p. 317, 1967.

LEAD CONCENTRATIONS IN DRINKING WATER AT THE U.S. COAST GUARD ACADEMY: A CASE STUDY

T. R. REILLY and G. B. DUERSTOCK

ABSTRACT

Drinking water samples taken from various locations in the dormitory facilities and water supply piping at the U.S. Coast Guard Academy (CGA) in New London, CT, were analyzed for lead content using a Perkin-Elmer model 3030 atomic absorption (aa) spectrophotometer. Results indicate that CGA drinking water meets the current Environmental Protection Agency lead maximum contaminant level (MCL) of 50 PPB. However, in one area it appears that the lead concentrations might exceed the newly proposed lead MCL of PPB scheduled for implementation in 1990. The accuracy of the results was verified by reanalyzing samples via AA spectrophotometer and inductively coupled plasma (ICP) techniques at the Coast Guard Research and Development Center. Lead–based solder and fittings are shown to be the most likely causes of elevated lead concentrations. Flushing the lines of standing water at the sites showing elevated lead concentrations proves to be a viable short term solution for reducing lead concentrations to levels beneath the proposed MCL.

INTRODUCTION

Concern over elevated lead (Pb) concentrations in drinking water has become an increasingly debated environmental issue for the past decade.

Chemistry for the Protection of the Environment
Edited by L. Pawlowski *et al.*, Plenum Press, New York, 1991

Studies conducted in the northeastern U.S., and specifically in Connecticut, have documented Pb concentrations in excess of the current .05 mg/l (50 PPB) maximum contaminent level (MCL) found in 40 CFR 141.11(b) [1]. In response to recent epidemiological studies concerning adverse health effects at relatively low Pb concentrations in blood, EPA published proposed regulations on 18 August 1988 reducing the Pb MCL level to 5 PPB for water leaving the water suppliers' tanks and an average of 10 PPB for water at the receiver's faucet [2].

Many New England buildings and homes are aging and have both Pb–containing solder and piping materials. Because construction of the U.S. Coast Guard Academy commenced in 1932, the potential for elevated Pb levels in the drinking water was a concern. Current building codes prohibit use of Pb–containing piping materials, but only in 1985 did EPA prohibit the use of Pb–based solders in drinking water systems. Two of the more commonly used solder types were a 50-50 tin-lead mix and "liquid" solder, which contains Pb in excess of 50%.

Pb leaches into the water supply in many forms; The total Pb (II) solubility (S) equation below defines some of the Pb constituents [3]:

$$S = [Pb^{2+}] + [PbOH^{+}] + [Pb(OH)_2] + [Pb(OH)_3^{-}] + 2[Pb_2OH^{3+}] + 3[Pb(OH)_4^{2+}] + 4[Pb_4(OH)_4^{4+}] + 6[Pb_6(OH)_8^{4+}] + [PbCO_3] + [Pb(CO_3)_2^{2-}]$$

The Pb leaching phenomena is affected by numerous factors including [1]:

1. The pH of the water; Pb concentrations are higher in acidic water.
2. Dissolved oxygen; a higher oxygen content tends to increase the Pb concentration in the water.
3. The water/pipe contact time; more Pb will leach into standing water.

The purpose of this study was to survey the U.S. Coast Guard Academy drinking water for Pb content and then determine the source of any excessive levels of Pb contamination.

MATERIALS AND METHODS

All water samples from sinks and drinking fountains were drawn in 250 ml polypropylene bottles between 4:00 AM and 5:00 AM to allow for the maximum achievable standing time. EPA protocols require standing times from 8 to 18 hours [4]. Water samples drawn directly from the water

Table 1. Initial Coast Guard Academy Survey; μg/liter of Pb; Standard Deviations in Parentheses

Location	P–E 3030 12/88	P–E 3030 1/89	P–E 5100 1/89
Chase A1 FT	4(0.1)	6(2.1)	5.8
Chase D1 FT	9(0.1)	6(0.2)	8.2
Ward FT	60(1)	55(0.9)	49.8
Billard FT	12(1)	8(0.2)	8.1
Smith FT	8(0.1)	11(2.1)	9.8

main or from hydrants were also taken between 9:00 AM and 10:00 AM for comparison purposes. All water samples were treated with 2% ultrapure, atomic absorption quality nitric acid, and stored at room temperature.

Lead analysis was effected through use of a Perkin-Elmer model 3030 atomic absorption (AA) spectrophotometer; a graphite furnace atomizer was employed. The data represent the average of at least three readings per sample. The following temperature program was used [5]:

Step	Temp(^{0}C)	Ramp(^{0}C/sec)	Hold(sec)	Flow Rate(ml/min)
1	110	10	25	300
2	500	10	20	30
3	2300	1	5	0

For data comparison purposes, selected data were reanalyzed using U.S. Coast Guard Research and Development (R&D) Perkin-Elmer model 5100 spectrophotometr AA. Additionally, the R&D Perkin-Elmer Plasma 40 model ICP unit was used on selected samples to analyze for potential interferences as well as for data comparison purposes.

RESULTS AND DISCUSSION

Water was initially sampled from five water fountains located on the U.S. Coast Guard Academy grounds (Table 1). To assure the validity of the results, these samples were analyzed by two different AA spectrophotometers, Perkin-Elmer model numbers 3030 and 5100.

Note that the data correlation between instruments is excellent. Because one 60 PPB sample was discovered in the cafeteria food–processing area of the cadet (student) dormitory, the emphasis for the duration of this study focused on drinking water supplied to and drawn from the cadet dormitory, a four–story complex housing approximately 1000 cadets. Water samples were drawn from three faucets and one fountain in the food prepa-

Table 2. Cadet Dormitory Survey; μg/liter of Pb; Standard Deviations in Parentheses

Location	Nov 88	Dec 88	Feb 89	Apr 89	Jul 89
Ward FT	60(1)	7(0.5)	1(0.1)	2(0.1)	–
Ward S1	–	10(0.1)	4(0.1)	5(0.2)	22(1.0)
Ward S2	–	26(0.2)	12(0.1)	11(1.2)	42(2.5)
Ward S3	–	23(0.3)	21(1.0)	3(1.0)	9(0.9)
A1 FT	4(0.1)	–	9(0.2)	6(0.2)	14(1.2)
B2 FT	–	–	10(1.2)	6(0.2)	17(0.2)
C1 Ft	–	–	3(0.2)	2(0.2)	6(0.1)
D1 FT	9(3.1)	–	4(2.1)	2(0.1)	4(0.1)
A3 FT	–	–	2(0.2)	2(0.1)	4(0.3)
D3 FT	–	–	1(0.2)	3(1.0)	5(0.1)
B2 S	–	–	1(0.2)	3(1.0)	5(0.1)
D1 S	–	–	3(0.1)	2(0.2)	–

Notes:

1. Ward denotes food processing area.
2. FT denotes drinking fountain.
3. S denotes sink faucet.

ration area, along with six randomly selected fountains and two sinks from the living portion of the dormitory (Table 2). Drinking water from fountains and lavatory sinks in the cadet dormitory averaged Pb concentrations of 5.3 PPB which is in compliance with the proposed 10 PPB Pb standard. Water in the food–processing area, however, averaged Pb concentrations of 16.1 PPB which exceeds the proposed 10 PPB MCL for Pb.

Initial efforts to characterize the New London, CT city source water were unsuccessful because the sampled water represented long stagnant water in deadend segments of the water/fire main that had not been flushed for approximately one year (Table 3). Of interest with this data set are the high Pb readings that are achieved under long–stagnant conditions. After allowing the incoming city source water, located at the first connection inboard of the academy water main (labeled MAIN GATE in Table 3) to flow freely for two hours, additional water samples were drawn which better characterized the city source water. The city source water averaged Pb concentrations of 10 PPB and a pH of 6.8. Our observation is that the city source water meets the 5 PPB MCL proposed for suppliers as long as normal flow rates within the water main are maintained. The samples drawn from faucets and sinks in the living portion of the cadet dormitory corroborate this ob-

Table 3. Water Main (Stagnant Water); μg/liter of Pb; Standard Deviations in Parentheses

Location	P–E 3030 (CGA)	P–E 5100 (R&D)	ICP (R&D)
25 January 1989 (9.30 AM)			
Main Gate	Offscale	3120	–
Ch # 1	395(1)	485	–
Ch # 2	400(1)	395	–
Smith	160 (1.3)	168	–
Johnson	82(1)	72.7	87
Leamy	96(1.3)	87.5	63
26 January 1989 (4.30 AM)			
Main Gate	43(2.1)	96.1	59
Ch # 1	25(2.6)	35.6	26
Ch # 2	100(2.6)	106.7	82
Smith	12(1)	29.1	37
IeAMY	38(1.2)	45.3	44

servation (Table 2). Our assessment is that the city source water does not contribute to the elevated Pb levels found in the cadet dormitory. Hence, by deduction, the contaminating Pb source must originate in the academy's piping materials.

Having deduced that the source of the Pb contamination originates in the piping materials, determining whether or not the Pb source was from pipes, fittings or solder became the next priority. A review of various U.S. Coast Guard Academy piping plants dating back to the 1930's indicates that service drinking water pipes are type K copper tubing complying with U.S. Federal specification WW-T-799. Perusal of this specification shows a demand for 99.5% copper purity, which eliminated the drinking water service piping as a possible lead leaching source [6]. The specifications for the original water main are, unfortunately, lost. Although the water main cannot be completely ruled out as a potential Pb source due to the elevated readings (Table 3), it must be considered to be a minor Pb contributor under normal water consumption conditions based on the low Pb concentrations displayed in the living area portion of the dormitory (Table 2). Our assessment is that, under normal water consumption rates, the pipes contribute negligible quantities of Pb to the drinking water. Shifting focus to solder and fitting, the blueprints describe the underground copper water lines as, "...assembled to red brass of the flared type for pipe 2 inches in diameter and smaller, and of the solder type for pipe of 2.5 inches in diameter and larger." Solder, to a

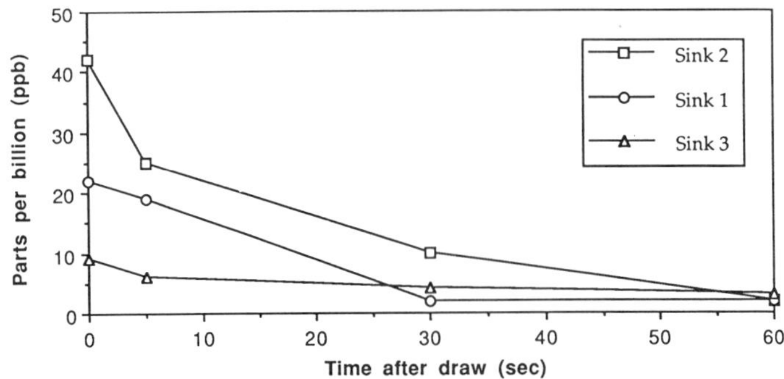

Fig. 1. Lead concentration (ppb) vs. time (sec) after draw for three sinks in cadet food preparation area.

significant degree, and red brass, to a lesser degree, contain Pb and became the primary Pb contamination suspects. We next attempted to isolate a Pb source within the piping located between the faucets and fountains and the water main. A dynamic test was devised wherein four samples were drawn from the three food-handing area faucets previously tested (Table 2). The first sample was drawn in accordance with the EPA protocols; the water was allowed to continue flowing and the subsequent three samples were taken after 5 seconds, 30 seconds and 60 seconds had elapsed (Fig. 1.). Water flow rates were kept in the laminar regime to reduce the possibility of mixing. Since the Pb concentrations decrease significantly with time, we deduce that the solder found in the plumbing near the faucets is the primary source of Pb contamination in the food preparation area of the cadet dormitory.

CONCLUSION

The short term solution to remedy the elevated Pb concentrations found in the food-handing portion of the cadet dormitory is simply to instruct all food handlers to run the water for one minute prior to using water for consumption purposes. Further study will be conducted to pinpoint actual Pb source positions and to determine the extent of the materials that will eventually need to be replaced.

REFERENCES

[1] Birden H., Calabrese E., Stoddard A. *Lead dissolution from soldered joints,* Journal AWWA, 66–70, November, 1985.

[2] Federal Register vol 53 n 160, 31516-31578, 1988.

[3] Schock M. *Response of lead solubility to dissolved carbonate in drinking water*, Journal AWWA, 695–704, December, 1980.

[4] EPA, *Suggested sampling procedures to determine lead in drinking water in buildings other than single family homes,* 1988.

[5] Perkin-Elmer model 3030 AA technical manual, 1985.

[6] Federal specification WW–T–799, copper, 1962.

INVESTIGATION OF LEAD CONTENT IN SOILS, VEGETABLES AND FRUITS CULTIVATED CLOSE TO THE LEAD EMITTING IRENA GLASSWORKS AT INOWROCLAW DURING THE YEARS 1982–1983 AND 1987

S. ZOMMER-URBAŃSKA, M. KUKLIŃSKI and P. W. TOPOLEWSKI

Medical Center of Postgraduate Education Faculty of Pharmacy, 85-626, Bydgoszcz, Poland

ABSTRACT

Lead contents measured in soils in 1982 and in the edible tissues of fruits and vegetables during 1982–1983 were compared with results determined for 1987. The analyses were carried out using atomic absorption techniques. Samples of soil and plant material were collected from plots situated in the vicinity of the Irena Glassworks at Inowroclaw. It was found that the lead content in investigated samples of both soil and plants, on may occasions, exceeded accepted boundary values. We found in 1987 that in spite of decreased lead emission, plant contamination had not diminished. High lead concentrations may result from both industrial and traffic emissions.

INTRODUCTION

Lead is a toxic element which accumulates both in surface layers of soils [1] and in humans [2, 3]. Absorption of lead by plants depends markedly on soil acidity. In comparison with acidic soils, metal uptake (excluding molybdenum and selenium) by plants from neutral and weakly alkaline soils is more difficult. In the last decades, lead content in the atmosphere, soils, plants and food-stuffs has increased in Poland [4, 5]. Only a few papers have been published in Poland which consider either the content of lead in surface soils and in edible plant tissues or the effect lead has upon an area which receives high input from regional factories such as works producing crystal glass [6, 7, 8, 9].

The purpose of this work was to investigate the effects lead emissions from the Irena Glassworks at Inowroclaw have upon the content of this element in soils, vegetables and fruits taken from neighboring gardens.

MATERIALS AND METHODS

These studies were carried out originally during 1982–1983 and were then repeated in 1987. Samples of plant material were collected from sites located up to 650 m away from a lead emitter, on the grounds of three Workers Allotments. These allotments are called Irena, Waryński and Transportowiec and are used mostly by area residents to grow produce for home consumption. They are situated north and northwest from the main lead emitter. Meteorological data showed that winds blowing in these directions make up 32% of all yearly winds.

Lead was determined by the means of atomic absorption spectrophotometry (AAS) analysis as reported in the literature [10, 11]. Samples of soils or plant material were mineralized using a mixture of perchloric and nitric acid in a ratio of 4 : 1. Ammonium-pirrolidindithio-carbamat (APDC) was used to form a lead complex. The complex was then extracted into an organic phase using methyl-isobutyl-ketone saturated with deionized water as an extraction solvent. The determination was performed using AAS spectrophotometer type Unicam SP 98A.

RESULTS

The average concentration of lead in surface soil layers at a distance of 100 meters from the emitter amounted to 104 mg/kg. The lead content in the samples thus exceeded five-fold the accepted value of 20 mg/kg [6]. However, lead contamination of soils depends upon the distance from the emission source. At distances of 250, 500, 750, from 1000 to 1500 and then from 3000 to 6000 meters, lead contents amount to 76, 54, 39, 30 and 24 mg/kg, respectively. Determination of soil pH values showed that light soil acidification occurred at some sites. This acidification can favor lead uptake by plants.

Table 1. Average Lad Content in Selected Vegetables and Fruits Growing in the Emission Area of the Irena Glassworks at Inowroclaw in 1982-83 and 1987 (mg/kg of dry product)

		1982		1983		1987
		Sampling sites(*)				
	Products	1	2	1	2	3
	I. Vegetables					
1.	Bean	1.81	1.17	—	1.23	2.00
2.	Bean (shelled)	—	1.57	—	—	0.75
3.	Cabbage (white)	2.45	2.72	2.56	3.49	5.80
4.	Celery (leaves)	5.07	5.42	5.80	7.76	6.10
5.	Celery (roots)	—	—	4.08	3.21	3.50
6.	Chive	—	5.00	3.54	6.14	17.50
7.	Dill	4.75	4.90	—	4.61	16.00
8.	Leek	6.52	4.15	6.51	5.15	4.00
9.	Lettuce	8.48	9.82	—	6.95	15.80
10.	Marjoram	—	9.77	10.80	11.30	11.80
11.	Parsley (leaves)	12.10	9.06	9.80	17.40	12.00
12.	Parsley (roots)	3.30	2.95	3.19	3.70	2.70
13.	Tomatoes	3.31	3.16	3.66	3.73	8.80
14.	Red beet (young leaves)	8.55	7.88	4.35	8.33	15.80
15.	Red beet (roots)	4.20	8.40	4.84	5.70	1.80
16.	Rhubarb	3.20	5.32	3.61	3.99	14.50
17.	Turnip-cabbage	—	2.55	—	3.56	1.80
	II. Fruits					
18.	Apples (early)	1.91	—	1.50	2.04	5.00
19.	Apples (late)	3.70	2.17	2.67	4.02	10.20
20.	Gooseberry	8.12	8.33	—	7.21	18.50
21.	Currents (red)	5.49	—	—	5.66	26.50

(*) 1 – the Irena Workers Allotments.
2 – the Waryński Workers Allotments.
3 – the Transportowiec Workers Allotments.

In 1982 and 1983, lead was determined in selected plant species gathered from the Irena and Waryński Workers Allotments (Table 1). It was found that in about 80% of the studied plant species, the lead content was higher than Polish standards set in 1980, i.e. 2 mg/kg of dry product. Analysis of the data obtained indicates that nearly 25% of the plants samples showed a four-fold excess of this permissible lead value. However, analyses of lead content in vegetables and fruits collected in 1987 from the Transportowiec Workers Allotments showed, in spite of a relatively long distance from the lead emitter, an excess of the Polish standards in more than 80 of the cases. 55% of the cases attained a five-fold excess and 40%, a seven-fold excess (Table 1). Other authors have reported that, in some regions of Poland, vegetables and most fruits contain metals in quantities up to seven-fold higher than the metal contents in plants cultivated in unpolluted areas [12].

On the basis of the presented data, it can be concluded that the major cause of contamination in plants cultivated in the Irena and Waryński Workers Allotments is lead emission from the Irena Glassworks at Inowrocław. Contamination of plants gather in 1987 from the Transportowiec Workers Allotments may be caused by industrial emission but there may also be a contribution from nearby traffic arteries.

CONCLUSION

1. Aboveground parts of plants both subjected to prolonged contact with a contaminated atmosphere and characterized by a large surface area (for example, red beets, parsley and marjoram) contain large amounts of lead. This may indicate that lead is absorbed mainly from the atmosphere.

2. Some fruits such as gooseberries and currents, when compared to others, are also characterized by a considerable ability to absorb lead.

3. Plants with pods such as broad beans contain, after being shelled, very small amounts of lead. Their cultivation should be encouraged in lead polluted areas.

4. It is unadvisable to found allotments in either neighborhoods near sources of significant lead emission such as crystal glassworks or near busy traffic arteries.

REFERENCES

[1] Czarnowska, K., Accumulation of heavy metals in soil and plants on the terrain of Warsaw, Part II, Conferences of Polish Academy of Science, Dept. of Agricultural and Forest Sciences, Wrocław, 1974.

[2] Somers, E., The toxic potential of trace metals in food, J. of Food Sci., 39, 215, 1974.

[3] WHO Technical Report Series, No. 532, 1973.

[4] Nikonorow, M., Chemical and biological contamination of food, WNT, Warsaw, 1976.

[5] Nikonorow, M., Current problems in connection with contamination of environment, Part I: Contamination of soil, water, atmospheric air as well as plants and feeding stuffs, Roczn. PZH, 32, 393, 1981.

[6] Zommer-Urbańska, S., Kuklińska, M., Determination of lead content in the soil in the range of lead emission from the "Irena" Glass Factory in Inowrocław, Roczn. PZH, 38, 37 1987.

[7] Bukliński, R., Bloniarz, J., Influence of emission from Glass Work "Krosno" on some trace elements content in selected vegetables and fruits, Roczn. PZH, 20, 222, 1987.

[8] Zommer-Urbańska, S., Kuklińska, M., Lead level in certain vegetables and fruits raised in the range of emission of load compounds by the "Irena" Glass Works in Inowroclaw, Roczn. PZH, 36, 196, 1985.

[9] Zommer-Urbańska, S., Topolewski, P., Kuczyńska, J., Wojciech, P., Investigation of fluorine and lead content in certain vegetables and fruits cultivated in the range of emission of these elements by The Glassworks "Irena" at Inowroclaw, Roczn. PZH (in print).

[10] Pinta, M., Atomic Absorption Spectrophotometry, PWN, Warsaw, 1977.

[11] Sapek, A., Methods of chemical analysis of meadow vegetation, soil and water, IMUZ, Falenty, 1972.

[12] Bukliński, R., Bloniarz, J., Influence of emission in the region of Steelworks and Power Station "Stalowa Wola" on content of trace elements in selected vegetables and fruit cultivated in the range of emission of these elements, Roczn PZH, 33, 121, 1982.

AN INVESTIGATION OF THE INTERACTION OF TRIALKYLTIN COMPOUNDS WITH HUMIC ACIDS

JERZY RADECKI

Department of Chemistry
Academy of Agriculture and Technology
10-728 Olsztyn-Kortowo, Poland

ABSTRACT

As previous work has shown, the trialkyltin compounds are more toxic than their inorganic homologs [1,2] and thus, their behaviour in the environment should be investigated. Humic acids are widely distributed in soils and have diverse roles within geochemistry [3]. They combine with toxic metals and thus change the mobility and bioavailability of these metals in soil.

From an environmental standpoint, complexing of heavy metals by humic acids is of importance in determining the fate of these pollutants in the environment.

This paper describes the application of paper chromatography and polarography to compare the nature and strength of the binding of Sn^{+4}, $(CH_3)_3Sn^+$, $(C_2H_5)_3Sn^+$, $(C_3H_7)_3Sn^+$.

MATERIALS AND METHODS

Humic acids samples were taken from forest soils at a depth of 20-25cm. The extraction and purification of humic acids and analyses of total acidity and carboxyl groups were carried out in accordance with the method

Chemistry for the Protection of the Environment
Edited by L. Pawlowski *et al.*, Plenum Press, New York, 1991

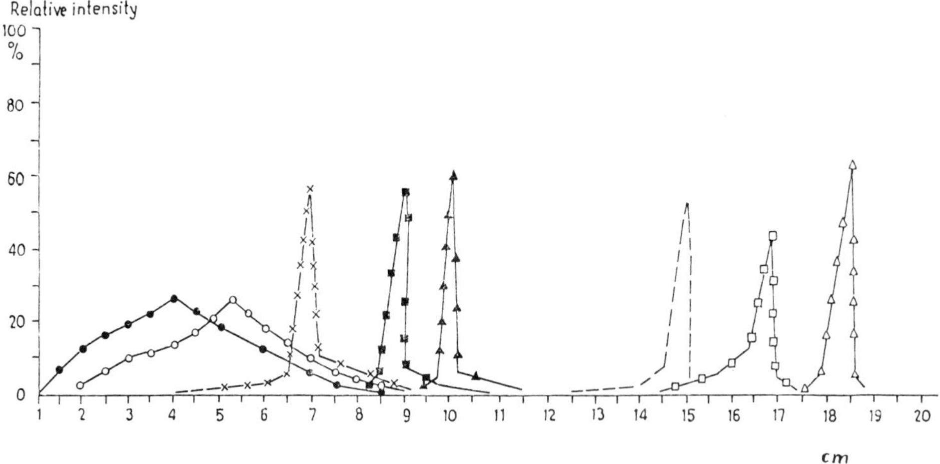

Fig. 1. Typical chromatogram of tin compounds on blank paper; Sn^{+4}-o- $(CH_3)_3Sn^+$-Δ,$(C_2H_5)_3Sn^+$-□- $(C_3H_7)_3Sn^+$—, and on the paper impregnated with humic acid Sn^{+4}-•- $(CH_3)_3Sn^+$-Δ,$(C_2H_5)_3Sn^+$-□- $(C_3H_7)_3Sn^+$-x-x-

reported by Schnitzer and Khan [4]. Samples of soil and humic acids were analyzed by an elemental analyzer for C, H and O contents.

One gram of humic acid sample was dissolved in an appropriate volume (ca. 50ml) of 0.1 M NaOH and the solution was adjusted to pH 5.0 with 1M CH_3COOH and diluted with water to 100ml.

A piece of filter paper (Whatman NO 3 M.m. Chr.) for paper chromatography was dipped in the prepared solution, taken out, and then air-dried. Standard stock solutions (10000 $\mu g^* ml^{-1}$) were prepared by dissolving the appropriate amount of the relevant tin compound in methanol.

One microlitre of each tin compound solution was separately spotted on the starting line of a piece of humic acid impregnated paper by means of microsyringe and then developed for 7 to 10h (25cm) with a solution made by adding 0,05M $Cd^{+2}$to a 0,1M acetate buffer solution (pH 5.0: 80% ethanol). Thereafter the paper was dried and cut into chromatographic section of 0,5cm each. Analysis of Sn content was performed for each section using the polarographic method already reported [5]. Samples were mineralized with a mixture of $HNO_3/HClO_4$. In addition, filter paper not impregnated with humic acid was run as a control. The Rf values for each tin compound were obtained from the ratios between the spot fronts in both actual and blank tests. The spot front or peak position was obtained by a simple proportional calculation using the concentration of the section in which the maximum concentration of tested metal was found and those of the sections on both sides of the maximum section.

Table 1. Some chemical properties of soil humic acid samples

Sample	Soil		Humic acid			Functional groups		
	C %	N %	C %	O %	H %	total acidity	carbo-xyl meq/g	pheno-lic
1	1,09	0,06	53,2	34,1	4,2	5,94	3,30	2,63
2	0,84	0,04	55,4	31,2	3,8	5,61	3,10	2,51
3	0,64	0,08	57,2	33,7	3,2	6,10	3,70	2,40
4	1,62	0,22	54,8	36,4	4,8	6,70	4,25	2,45
5	0.97	0,09	56,3	37,2	3,9	6,42	4,15	2,27
6	1,20	0,15	52,8	38,5	5,1	6,30	4,10	2,20

Thus:

$$x = \frac{x_m + 0.5(c_{-1} - c_{+1})}{c_{+1} - c_m} \quad \text{for } c_{-1} \pm c_{+1}$$

and:

$$x = \frac{x_m + 0.5(c_{+1} - c_{-1})}{c_m - c_{+1}} \quad \text{for } c_{+1} \pm c_{-1}$$

where x is the position of the peak, x_m is the position of the section where a metal exhibits its maximum concentration and c_m, c_{-1}, c_{+1} are the concentration of the metal in x_m and in the sections on both sides of x_m.

RESULTS AND DISCUSSION

Typical chromatographs are shown in Figure 1. Table 1 presents some chemical properties of the soil and humic acid samples. Table 2 summarizes the Rf values obtained showing that the values for the different tin derivatives varied considerably depending on the type of humic acid. However it was found that the strength of the interaction (which is inversely proportional to the value of Rf) between various tin compounds and humic acids is in the following order:

$S^{+4} \geq (C\,H_3)_3Sn^+ > (C_2H_5)_3Sn^+ > (C_3H_7)_3Sn^+$

independent of which acid was used (Table 2).

To estimate the relationship between the chemical structure of the humic acids and the strength of interaction between the investigated compounds, correlation analyses were carried out on the Rf values for all the compounds under consideration and the amount of carboxyl and phenolic groups in the humic acids (Table 3). These results indicated that carboxyl groups are the principal binding sites for the studied tin derivatives. The

Table 2. Rf values for lead compounds

Sample No	m	e	p	sn
1	0,521	0,481	0,425	0,251
2	0,621	0,573	0,500	0,312
3	0,372	0,310	0,261	0,372
4	0,321	0,250	0,192	0,137
5	0,481	0,410	0,320	0,212
6	0,472	0,390	0,310	0,200
Average	0.4645	0.402	0.342	0.216

LSD at p=0,01 for tin compounds - 0,032

m = $(CH_3)_3Sn^+$, e = $(C_2H_5)_3Sn^+$, p = $(C_3H_7)_3Sn^+$,
sn = Sn^{+4}

Table 3. Correlation coefficients between Rf and functional groups

Funtional group	m	e	p	sn
total acidity	-0.807**	-0.846**	-0.934**	-0.911**
COOH	-0,719**	-0,785**	-0,857**	-0,914**
OH	0,385	0,325	0,440	-0,629*

* - significant at p = 0,05 ** - significant at p = 0,01

m = $(CH_3)_3Sn^+$, e = $(C_2H_5)_3Sn^+$, p = $(C_3H_7)_3Sn^+$,
sn = Sn^{+4}

obtained results suggest that alkyltin derivatives have lower binding ability in the soil environment than inorganic tin compounds and thus will be more easily leached from the soil to the water reservoir.

REFERENCES

[1] N.J. Snoeij, A.H. Perninks, Seinen. Biological Activity of Organotin Compounds. Environ. Res. 44, 335-353, 1987.

[2] P.J. Smith. Toxicological Date on Organotin Compounds. International Tin Research Institute Greenford Middlesex England. 1978.

[3] M. Schnitzer, P.A. Papst. Organo-metallic interactions in soils. Soil Sci. 109, 333-340, 1970.

[4] M. Schnitzer. S.U. Khan. Humic substances in the environment. Dekker, New York, 29-54, 1972.

[5] A.M. Bond. Direct Current, Alternating Current, Rapid and Inverse Polarographic Methods for Determination of Tin. Anal. Chem. 42, 1165-1168, 1970.

COPPER COMPLEXATION CAPACITY

OF RIVER YAMUNA IN DELHI

D. K. BANERJEE and E. P. JAGADEESH

School of Environmental Sciences
Jawaharlal Nehru University
New Delhi-110067, India

ABSTRACT

River Yamuna flows through the urban territory of Delhi for a stretch of about 50 km. During its course, 17 major and minor drains discharge domestic and industrial wastewaters directly into the river. It also receives leachates and sluicing waters from ash ponds of thermal power plants. The extent to which the river gets polluted in the process needs accurate assessment. Towards this end, copper complexation capacity (CC_{Cu}) and apparent stability constants of copper complexes (K_s) were determined for water samples collected from six different sites along the river, using ion selective electrode potentiometry. The total dissolved organics content and also humic and fulvic acid fractions were ascertained by absorbance measurements at 254, 465 and 665 nm, respectively. The absorbance ratios showed a positive correlation with CC_{Cu} at all the sites. However, this correlation was stronger at upstream sites, before the river enters the urban territory, than at downstream sites, where anthropogenic perturbations are liable to be greater. Apparent (conditional) stability constants also showed a similar pattern of correlation with absorbance ratios. CC_{Cu} and K_s values were also determined for the treated effluent discharged from a sewage treatment works before it enters into the river.

The study revealed that at the upstream sites where the anthropogenic perturbances are limited, the naturally occurring organic ligands represented by humic and fulvic acid fractions determine the complexing capacity of the river and the stability constants of complexes, whereas at sites within the urban territory, influ-

Chemistry for the Protection of the Environment
Edited by L. Pawlowski *et al.*, Plenum Press, New York, 1991

ence of other strong complexing agents, possibly of anthropogenic origin, is very apparent. The order of K_s values obtained was quite comparable to such values cited in literature for other riverine environments.

Such investigations are important as data along these lines can be helpful in quality assessment of river waters.

INTRODUCTION

It has been increasingly realized in the realm of environmental studies of trace metals in aquatic systems that their total concentration is not so important as the forms in which they occur. This, in fact, determines their bioavailability, toxicity and transport in an aquatic system [1,2,3].

The speciation of metals, in turn, is influenced by factors like pH, Eh, temperature, dissolved oxygen, suspended particulate matter, etc., and also by organic and inorganic ligands present in the water bodies. Natural waters contain a variety of metal complexing agents like humic, fulvic and tannic acids and also colloidal particles like oxides of manganese, iron and aluminum. Polluted waters contain various ligands of anthropogenic origin, in addition to the natural ones. The identification and quantification of the individual ligands and their complexation characteristics are quite difficult. Also, their extraction procedures themselves may bring about changes in the structure of ligands and thereby in their complexation characteristics. However, the complexation capacity of a natural water body and the apparent stability constant of the metal complexes formed give a reasonably valid representation of its buffering capacity towards toxic metals.

Complexation capacity has thus become an important water quality parameter, as it assesses the concentration of heavy metals that can be added on or discharged into a waterbody before free metals start appearing in it. Since copper is the most toxic of the common metal ions, the complexing capacity of a water body is measured in terms of that of Cu^{2+} (CC_{Cu}). A variety of methods varying from ion exchange to electro-analytical potentiometry are available for complexation capacity and stability constants determinations.

River Yamuna, which originates from the Himalayan glaciers, flows through Delhi, a major urban territory (Fig.1), for a stretch of about 50 km. During its course, 17 major and minor drains discharge wastewater directly into the river through its western bank [4]. At present, the total wastewater discharge of Delhi is estimated at 1480 MLD out of which only half, 745 MLD, is collected for treatment [5]. The rest finds its way into the river and at places, the river looks like an open sewer. The extent to which the river gets polluted needs accurate assessment for planning pollution control measures and the present study is aimed towards this end.

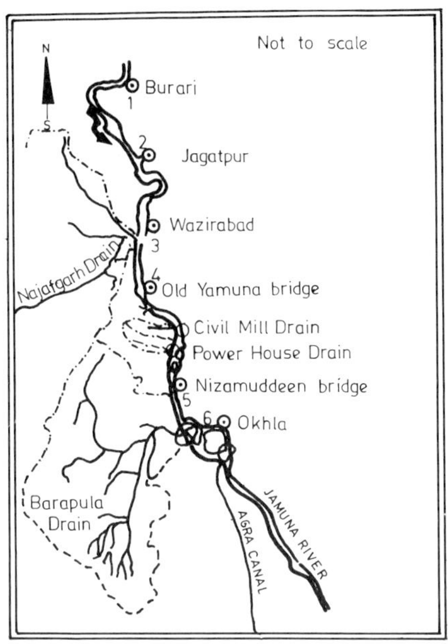

Fig. 1. Yamuna river around Delhi.

METHODOLOGY

Six sampling sites were chosen on the river, three upstream before the river enters the urban territory of Delhi and three within it. The sites were selected on the basis of their special characteristics like heavy algal growth (site 1), shallow and turbid waters (site 2), placid waters near a barrage (site 3), vicinity of an important effluent–carrying drain (site 4), entry of ash pond leachates from a thermal power plant (site 5), and immense aquatic weed growth (site 6). Site 6 is also at the extreme end of the urban territory, the river leaves Delhi thereafter. In addition, samples of treated effluents from a sewage treatment works (near site 6) were also collected.

Water samples were collected from three points on each site, and each in replicate, in pre-cleaned polyethylene containers [6] of 2.5 l capacity, after rinsing them in river water. The same procedure was followed for treated effluent also. pH and dissolved oxygen were measured at the site itself, and the samples were brought immediately to the laboratory to be preserved at 4°C. Total alkalinity and chloride estimations were done within 24 hours of collecting the samples. All physico-chemical parameters were determined using the standard methods [7].

Total dissolved organics, humic and fulvic acid fractions in the sam-

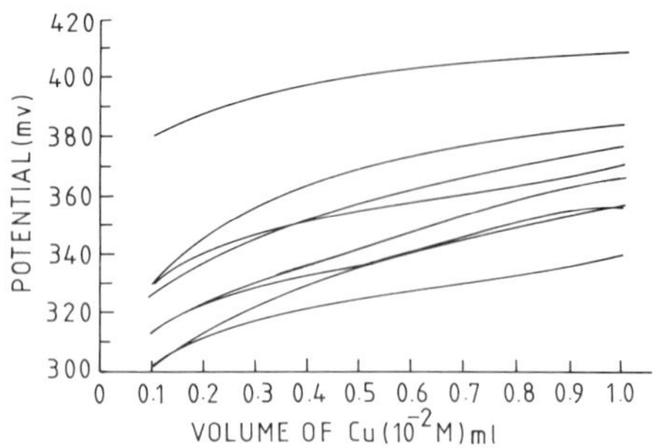

Fig. 2. Potential vs volume of Cu added.

ples were ascertained by measuring their absorbances at 254 nm, 465 nm and 665 nm, respectively [8], by using a Bausch and Lomb Spectronic 1001 spectrophotometer. Unfiltered samples were used in these experiments, as the usual millipore filtration removes many of the humic substances associated with colloidal particle size range [9,10].

For ISE potentiometry, 50 ml of each sample was pipetted out in pre-cleaned polypropylene beakers and 1 ml of ionic strength adjustor (5M $NaNO_3$) added to each sample. Cupric ion activity was measured using an Orion Ionalyser - 901 with single junction reference electrode and Cu^{2+} selective ion electrode.

Initially, a trial titration was conducted with addition of 1 ml of cupric nitrate solution (suprapure) of concentrations ranging from 10^{-6} to 10^{-1} M and the resulting electrode potentials were measured in mv. These were plotted against concentrations of added copper. The region of a sharp change in slope indicated the approximate range of complexation capacities. From these preliminary investigations, the concentration of cupric nitrate for addition in microliter quantities to the samples was found to be 10^{-2} M. A Teflon stirrer was used at constant speed in all titrations and relative potentials in mv were noted. The Cu electrode was washed with distilled water, dried and kept soaking for several minutes in a 10^{-3} M EDTA solution to facilitate cleaning of the membrane sensing surface [11]. Also, the measurements were made under constant light intensity. 50 ml samples of distilled water were also spiked with the same quantities of cupric nitrate solution at an adjusted pH of $\sim$ 8, and the relative potentials were measured under the same conditions as before (Fig. 2). From the data thus obtained, copper complexation capacity and conditional stability constants were ascertained by Vandenberg plots (Figs. 3-9) [12,13]. From the slopes and intercepts of

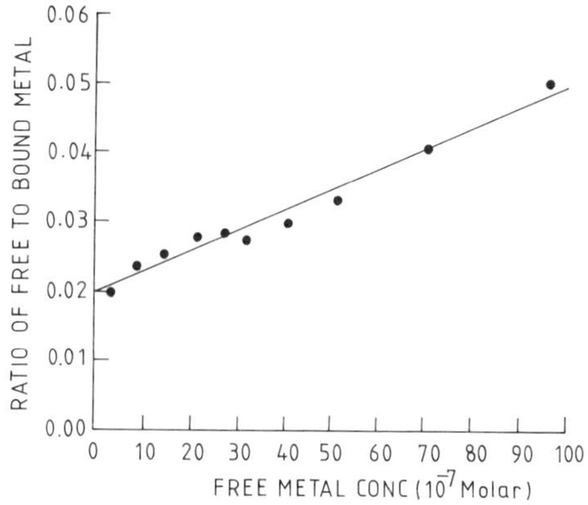

Fig. 3. Correlation between free and bound metal, Site 1.

Table 1. Copper Complexation Capacity, Conditional Stability Constants and Absorbance of Samples

Sample No. & site	CC_{Cu} (10^{-4}M)	K_sx10^{-5}	log K_s	abs_{254}	abs_{465}	abs_{665}	abs_{465}/abs_{254}	abs_{665}/abs_{254}
1.BURARI	2.187	2.507	5.39	0.011	0.004	0.002	0.364	0.182
2.JAGATPUR	3.925	1.550	5.19	0.016	0.005	0.002	0.313	0.125
3.WAZIRABAD	1.924	1.320	5.17	0.012	0.004	0.002	0.333	0.167
4.OLD YAMUNA BRIDGE	2.640	7.542	5.87	0.026	0.004	0.001	0.153	0.038
5.NIZAMUDDIN BRIDGE	2.205	7.742	5.89	0.041	0.008	0.003	0.195	0.073
6.OKHLA	5.008	8.399	5.92	0.070	0.014	0.004	0.200	0.057
7.TREATED EFFLUENT	2.220	15.26	6.18	0.072	0.011	0.002	0.153	0.028

these plots, complexation capacity and stability constants were calculated by the equation

$$\frac{M}{C_M - M} = \frac{1}{CC_{Cu}}M + \frac{1}{KCC_{Cu}}$$

where the terms have their usual significance.

RESULTS AND DISCUSSION

Table 1 shows the values of copper complexing capacity (CC_{Cu}), conditional stability-constants (K_s) and absorbance ratios at various sampling sites. Some secondary parameters like pH, D.O., chloride and alkalinity are given in Table 2. Complexation capacity varied from 192 to 500 μm/l. It

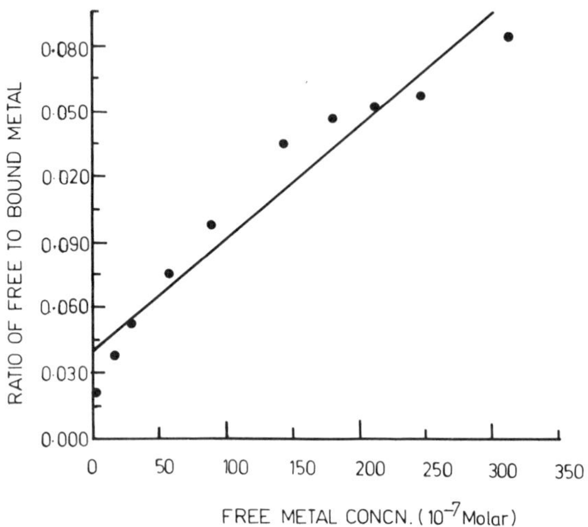

Fig. 4. Correlation between free and bound metal, Site 2.

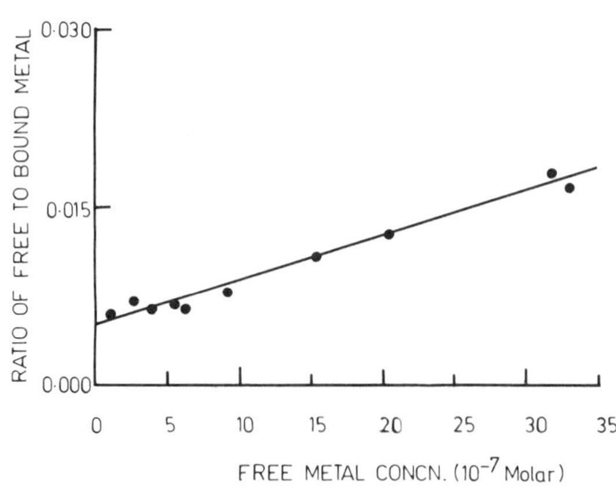

Fig. 5. Correlation between free and bound metal, Site 3.

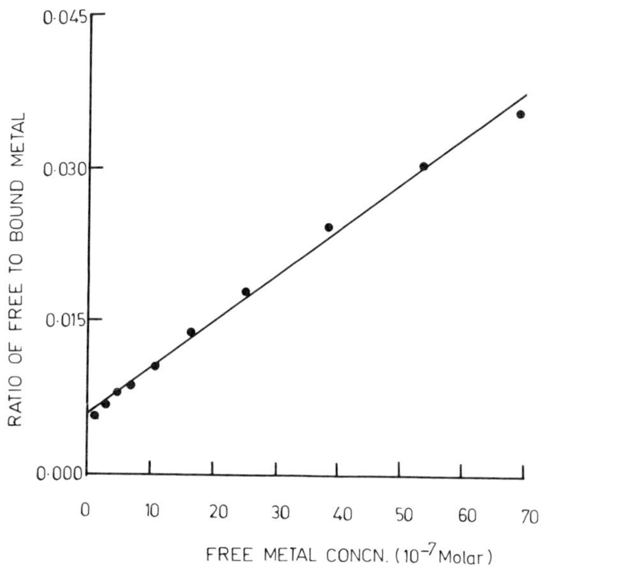

Fig. 6. Correlation between free and bound metal, Site 4.

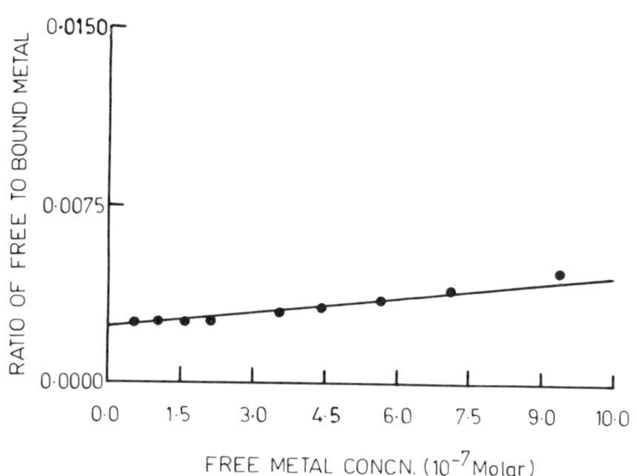

Fig. 7. Correlation between free and bound metal, Site 5.

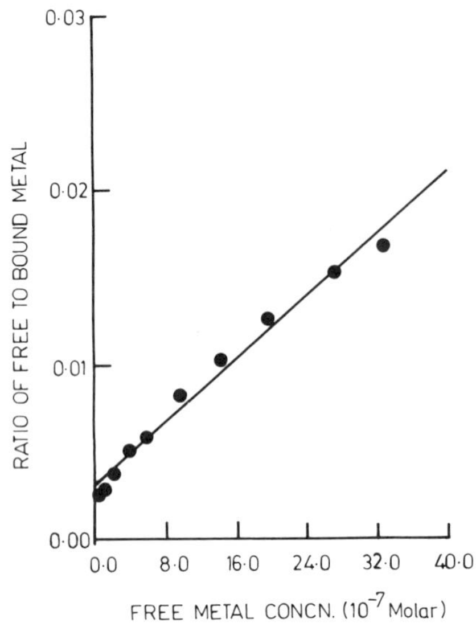

Fig. 8. Correlation between free and bound metal, Site 6.

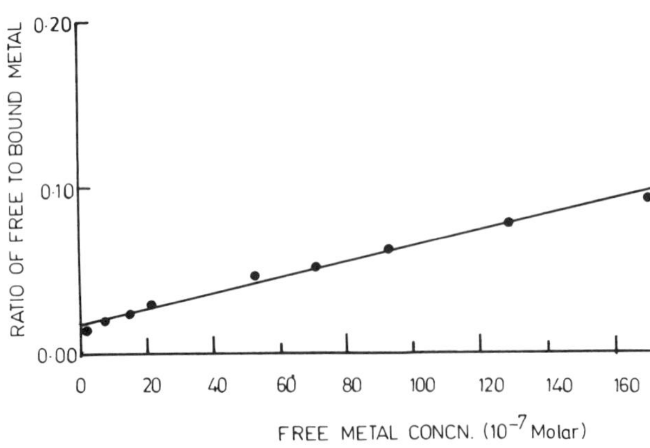

Fig. 9. Correlation between free and bound metal, treated effluent.

Table 2. Analytical Characteristics of Samples

Sample No. & site	pH	D.O. (mg/l)	Total alkalinity (mg/l $CaCO_3$)	Chloride (mg/l Cl)
1. BURARI	8.1	9.8	128.1	2.98
2. JAGATPUR	8.3	9.8	129.3	2.98
3. WAZIRABAD	8.2	9.6	126.4	4.97
4. OLD YAMUNA BRIDGE	8.3	1.10	170.8	12.92
5. NIZAMUDDIN BRIDGE	8.0	1.65	168.3	12.92
6. OKHLA	8.2	0.28	247.6	18.89
7. TREATED EFFLUENT	8.4	-	170.8	-

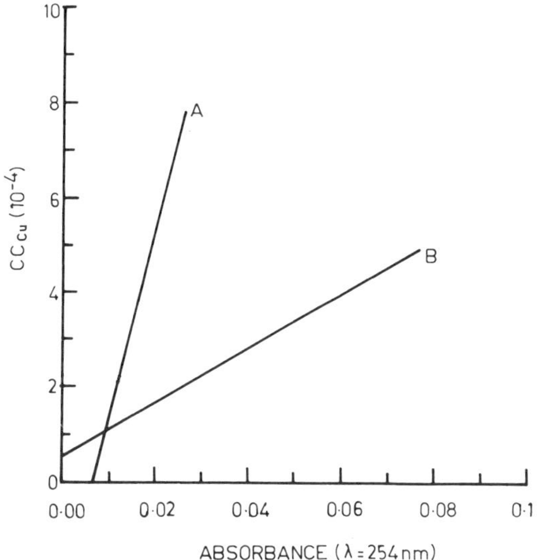

Fig. 10. Correlation between copper complexation capacity and absorbance (254 nm); A-upstream of Delhi, B-river in Delhi.

showed, uniformly at all sites, a positive correlation with total dissolved organics, as ascertained by absorption at 254 nm (Fig. 10). However, this correlation was more prominent at upstream sites than downstream, as evidenced by the linear regression equations which worked out to be:

$$CC_{Cu} \text{ (upstream)} = -2.40 + 391.2 \text{ (abs}_{254}\text{)}; \ r = 0.95$$
$$CC_{Cu} \text{ (downstream)} = 0.562 + 59.2 \text{ (abs}_{254}\text{)}; \ r = 0.88$$

This can be ascribed to the fact that at these sites, the amount of anthropogenic interference is less (upstream) and so the dissolved organics contain higher fractions of naturally occurring organic ligands. The ratios of ab-

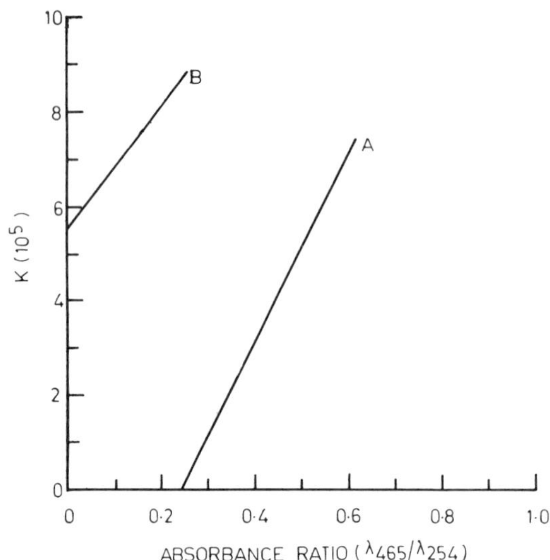

Fig. 11. Correlation between conditional stability-constants and absorbance rations (465 nm/254 nm); A-upstream of Delhi, B-river in Delhi

sorbances at different wave lengths (abs_{465}/abs_{254} and abs_{665}/abs_{254}) which are functions of the humic acid-fulvic acid fractions of the dissolved organics [14], were found to be significantly higher at upstream sites, supporting the aforesaid assumption. Analysis of bottom sediments from the river Yamuna has shown the clay content to vary from 1.28 to 2.40% [15]. The comparatively higher order of complexation capacity of this river, as found by the present study, can be ascribed to this high clay content. Also, alkalinity, or the carbonate/bicarbonate content of the river, is high (Table 2). A multiple regression analysis with LOTUS-123 software package taking complexation capacity as a dependent parameter and dissolved oxygen, pH, chloride and total alkalinity as independent parameters showed a good linear correlation with R^2= 0.8887. Conditional stability constants calculated from Vandenberg plots (Figs. 3-9) ranged from 5.10 to 5.92 for the different sites. These values are of the same order as reported for other river waters by different workers [16,17]. The stability constant values also showed greater positive correlation with absorbance ratios (abs_{465}/abs_{254} and abs_{665}/abs_{254}) at upstream sites as compared to downstream ones (Figs. 11-12).The linear regression equations are as follows:

$$K_s \text{ (upstream)} = -4.99 + 20.14\ \text{abs}_{465}/\text{abs}_{254};\ r = 0.839$$
$$K_s \text{ (downstream)} = 5.52 + 12.99\ \text{abs}_{465}/\text{abs}_{254};\ r = 0.74$$

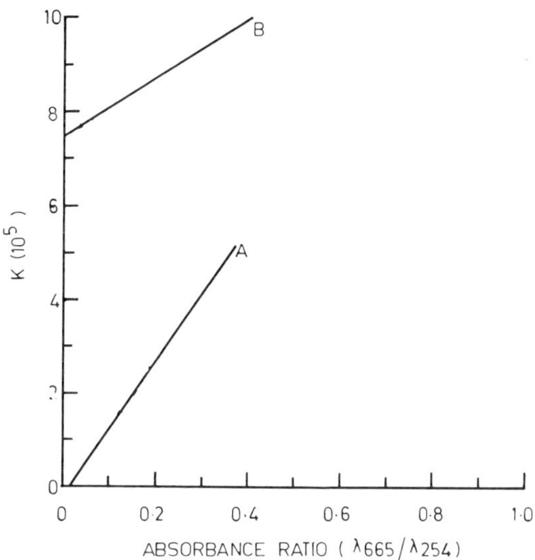

Fig. 12. Correlation between conditional stability-constants and absorbance rations (665 nm/254 nm); A-upstream of Delhi, B-river in Delhi.

$$K_s \text{ (upstream)} = -0.10 + 11.96 \text{ abs}_{665}/\text{abs}_{254};\ r = 0.56$$
$$K_s \text{ (downstream)} = 7.50 + 6.93 \text{ abs}_{665}/\text{abs}_{254};\ r = 0.27$$

The stability constants are governed by the strength of the complexing ligands. The domestic and industrial effluents which are discharged predominantly at the downstream (within the urban territory) sites are liable to have strong complexing agents, which increase the overall conditional stability constants at these sites. The treated effluent (near site 6), however, showed a decrease in complexing capacity. This can be ascribed to the removal of a major portion of natural organic ligands and colloidal organic matter in the aeration process involved in wastewater treatment. Other inorganic complexing agents in the effluent also are liable to agglomerate and settle down in settling tanks. The stability constants, however, showed a slight increase. This may be explained by the fact that some recalcitrant complexing agents of anthropogenic origin may remain unaffected by the effluent treatment. Also, processes such as hydrolysis and aeration convert organic pollutants into strong complexing agents for metal ions [18]. The treatment process could result in modified ligand structure and thereby alter stability of complexes, resulting in an overall increase in conditional stability constants. These facts can be further examined be chromatographic separation of such ligands using resins like XAD-2, Amberlite, Sephadex, etc., and determining stability constants separately before and after effluent treatment.

Site 2 had shallow waters which were turbid in appearance. Under these circumstances the possibility of the presence of inorganic ligands like the hydroxides and oxyhydroxides of Al, Fe and Mn is greater. Therefore, there is an increase in the complexation capacity. A decrease at site 3 is in consonance with the absorbance ratios, showing a proportionate decrease in naturally occurring ligands. At this site, there is a barrage; the waters are placid, deep and clear. The suspended colloidal–sized inorganics are liable to be agglomerated and settled. At site 4, the biggest effluent–carrying drain of Delhi (Najafgarh drain discharging $\sim$ 300 MLD) joins the river. It must be contributing a large share of the ligands of anthropogenic origin. The resulting increase in the complexation capacity is clearly seen. At site 5 a slight decrease in CC_{Cu} is noticed. At this site, leachates from the fly ash ponds of a thermal power station (Indraprastha) contribute a considerable amount of metal ions [19], thereby saturating some of the binding sites of ligands. Site 6 shows a sharp increase in complexation capacity. The speciality of this site is that some effluent–carrying drains open into the river just before this site. This site is also marked with immense growth of aquatic weeds. These features increase the dissolved organic matter to almost double that at site 5 (Table 1) and this accounts for increased complexation capacity.

The conditional stability constant at site 1 is distinctly different from those at other upstream sites. This could be ascribed to the phytoplankton blooms which result in the release of algal exudates. Release of strong copper complexing agents, and conditional stability constants in the order of 10^8 to 10^{12}, have been reported earlier [20,21]. The presence of two classes of ligands has also been observed in natural waters with phytoplankton growth [22], one present at low concentrations ($3.7 \cdot 10^{-8}$M), but having relatively high complex formation constant (log K= 10.1-10.7) while the other has a higher concentration (1.8 - $2.6 \cdot 10^{-6}$M) and a much lower complex formation constant (log K = 7.4). A similar effect of an organic complexing exudate from the phytoplankton bloom at site 1 must have increased the overall conditional stabilityconstant, as compared to other sites. It would be informative to further investigate this aspect by culturing the algal species exclusively found in this area and isolating their exudates for complexation studies.

The K_s values clearly demarcate the quality of upstream and urban zone waters. There is an increase in the K_s values for the latter, indicating that the relatively strong complexing agents of anthropogenic origin are present in these waters as compared to more naturally occurring ligands in upstream waters. This is in agreement with the higher value of K_s for the treated effluent, the reasons for which have been explained earlier.

REFERENCES

[1] Florence, T.M. and Batley, G.E., Chemical speciation in natural waters, CRC Critical Reviews in Analytical Chemistry (1980), 9, 219.

[2] Harrison, R.M. and Laxen, D.P.H., Metals in the environment 1 - Chemistry, Chemistry in Britain (1980), 16, 316.

[3] Leppard, G.G. (ed), Trace element speciation in surface waters and its ecological implications, NATO Conference Series 1 - Ecology, vol.6, Plenum Press, New York (1983).

[4] Central Board for the Prevention and Control of Water Pollution, Union Territory of Delhi: CUPS/2/1978-79; CBPCWP, New Delhi (1979).

[5] Central Pollution Control Board, News letter, vol.2, No.2, CPCB, New Delhi (1989).

[6] Laxen, D.P.H. and Harrison, R.M., Cleaning methods for polythene containers prior to the determination of trace metals in freshwater samples, Analytical Chemistry (1981), 53, 345.

[7] American Public Health Association, Standard Methods for Examination of Water and Wastewater, 16th Edn, APHA, Washington, D.C. (1985).

[8] Dobbs, R.A., Wise, R.H. and Dean, R.B., The use of ultraviolet absorbance for monitoring the total organic carbon content of water and wastewater, Water Research (1972), 6, 1173.

[9] Florence, T.M., The speciation of trace elements in waters, Talanta (1982), 29, 345.

[10] Power, J.F. and Langford, C.H., Optical absorbance of dissolved organic matter in natural water studies using the thermal lens effect, Analytical Chemistry (1988), 60, 842.

[11] McCrady, J.K. and Chapman, G.A., Determination of copper complexing capacity of natural river water, well water and artificially reconstituted water, Water Research (1979), 13, 143.

[12] Krammer, C.J.M., On the copper complexation capacity in the marine environment, Ph.D. thesis, University of Groningen, Netherlands (1985).

[13] Neubecker, T.A. and Allen, H.E., The measurement of complexation capacity and conditional stability constants for ligands in natural waters, Water Research (1983), 17, 1

[14] Chen, Y., Senesi N. and Schnitzer, M., Information provided on humic substances by E_4/E_6 ratios, Soil Science Society of America Journal (1977), 41, 352.

[15] Kurian, P., Studies on characterization and complexation of humic acids from Yamuna sediments, M.Phil dissertation, Jawaharlal Nehru University, New Delhi (1987).

[16] Hart, B.T., Complexing capacity in natural waters, Environmental Technology Letters (1981), 2, 75.

[17] Shuman, M.S. and Woodward, Jr., G.P., Stability constants of copper organic chelates in aquatic samples, Environmental Science & Technology (1977), 11, 809.

[18] Mertell A.E., Motekaitis, R.J. and Smith, R.M., Structure stability relationships of metal complexes and metal speciation in environmental aqueous solutions, Environmental Toxicology and Chemistry (1988), 7, 417.

[19] Sreenivasan, S., Impact of coal ash effluents from a thermal power plant on water quality of a receiving river in Delhi, Ph.D. thesis, Jawaharlal Nehru University, New Delhi (1986).

[20] McKnight, D.M. and Morel, F.M., Release of weak and strong copper complexing agents by algae, Limnology and Oceanography, (1979), 24, 823.

[21] Gachter, R., Davis, J.S. and Mares, A., Regulation of copper availability to phytoplankton by macromolecules in lake waters, Environmental Science & Technology (1978), 12, 1416.

[22] Jardin, W.F. and Pearson, H.W., A study of the copper complexing compounds released by some species of cyanobacteria, Water Research (1984), 18, 985.

PARTITIONING OF ELEMENTS BETWEEN WATER AND SUSPENDED MATTER: KINETIC APPROACH

M. CAMBIAGHI[o], G. CICERI[o]
W. MARTINOTTI[*] and A. TOPPETTI[o]

o CISE-Tecnologie Innovative, Via Reggio Emilia, 39
20090 Segrate (MI), Italy

* ENEL-Thermal and Nuclear Research Center
Via Rubattino, 54, 20134 Milano, Italy

ABSTRACT

Results concerning the sorption and desorption kinetics of some radionuclides (^{137}Cs, ^{58}Co, ^{51}Cr, ^{54}Mn, ^{65}Zn, ^{85}Sr, ^{59}Fe) on suspended matter from the Po river water are reported and discussed.

The different mathematical models for the interpolation of the experimental data include the following mechanisms:

$$X \underset{k_{-1}}{\overset{k_1}{\rightleftharpoons}} Y \quad ; \quad X \underset{k_{-1}}{\overset{k_1}{\rightleftharpoons}} Y \underset{k_{-2}}{\overset{k_2}{\rightleftharpoons}} Z \quad ; \quad X \underset{k_{-1}}{\overset{k_1}{\rightleftharpoons}} Y \overset{k_2}{\longrightarrow} Z$$

mechanism (a) mechanism (b) mechanism (c)

where X is the element in the liquid phase, Y and Z are the same element sorbed on the solid phase.

Mechanism (a), usually the worst one in the interpolation of the experimental data, was rather inappropriate to describe very fast kinetic sorption processes. On the other hand, it is very useful for its simple application to radiologial assessment models.

Chemistry for the Protection of the Environment
Edited by L. Pawlowski *et al.*, Plenum Press, New York, 1991

Mechanism (b) generally appeared the more appropriate for describing the sorption process, while mechanism (c) was sometimes useful for Mn and Co, considering the irreversible sorption on solid phase.

A correlation analysis was used to elucidate the dependence of the kinetic constant values on some solution and suspended–matter properties (e.g., pH, particulate load, stable element content in the two phases).

Moreover, the kinetic constant values were statistically treated for a possible use in stochastic models.

INTRODUCTION

The fate of trace elements and radionuclides in aquatic ecosystems depends on their interaction with both biotic and abiotic compartments. Among the latter, water and suspended matter are important pathways of transport.

The interaction of an element at the solid-liquid interface can be determined using either a thermodynamic [1, 3] or a kinetic [4, 5] approach.

As far as the thermodynamic approach is concerned (equilibrium conditions in the system), the solid-liquid interactions are evaluated by the distribution coefficient (K_d), defined as follows:

$$K_d = \frac{\text{element concentration in the solid phase}}{\text{element concentration in the liquid phase}}$$

In open systems the thermodynamic approach can be used only for long term description of relatively fast processes, where steady-state conditions are reached.

However, when the time scale of the process is the same as the period of interest, the kinetic approach must be used to control the system evolution. Unfortunately, the kinetic approach results in a more difficult application than the thermodynamic one, requiring a deep knowledge of the chemical reactions of the element between the liquid and the the solid phase, as well as of its speciation in water and suspended matter. A necessary macroscopic and phenomenological simplification must be done, studying the kinetics of the sorption and desorption reactions under defined natural conditions (e.g., pH, salinity, temperature) and searching for simple mathematical models to describe them.

In this paper radiotracers were used to study the sorption of Cs, Co, Cr, Mn, Zn, Sr and Fe on suspended matter from Po river water samples in order to compute the sorption and desorption kinetic constants for use in radiological assessment models.

Three different mathematical models for the interpolation of the experimental data were tested. The obtained forward and backward constant

Fig. 1. Sites and dates of collections of Po river water samples. Year 1986.

values are presented and correlated with some properties of the solution and the suspended matter, such as pH, particulate load, stable element content in the two phases, etc., for their use in deterministic models.

The kinetic constant values were also statistically elaborated for possible use in stochastic models.

EXPERIMENTAL

Sample Collection and Treatment

Seven water samples were collected at different times from two stations of the Po river (Northern Italy)(Fig. 1).

Radiotracers (Table 1) were added to the unfiltered samples (about 1 l). The mixture was shaken and subsamples (about 50 ml) taken at fixed time intervals (from 30-60 minutes to 2 weeks). The subsamples were filtered through a 0.45 μm Millipore filter and 25 ml aliquots of the solutions were analyzed by gamma-ray spectrometry against similar standards, using a Ge(Li) detector coupled by a multichannel analyzer (Silena-Cato)(Fig. 2). Carrier-free radionuclides were used when available to minimize sample alteration.

Table 1. Characteristics of the Radionuclides Added to the Water Samples

Radionuclide	Half-life (days)	Chemical form
^{137}Cs	10950	Cs^{+}
^{54}Mn	278	Mn^{2+}
^{51}Cr	27.8	Cr^{3+}
^{58}Co	71.3	Co^{2+}
^{59}Fe	45.1	Fe^{3+}
^{65}Zn	244	Zn^{2+}
^{85}Sr	64	Sr^{2+}

The obtained results, corrected for radionuclide decay, were used to compute K_d values as a function of time.

In a similar way, other aliquots of the seven Po river water samples, filtered through 0.4 μm Nuclepore polycarbonate filter just after collection, were submitted to the same procedure for studying radionuclides, insolubilization processes.

The average Po river water quality parameters, the characteristics and the element content (total and leachable) of the suspended matter are presented in Tables 2a, 2b and 2c, respectively.

Mathematical Models Description

The following mechanisms were considered to interpolate the experimental data:

$$X \underset{k_{-1}}{\overset{k_1}{\rightleftharpoons}} Y \quad (a); \quad X \underset{k_{-1}}{\overset{k_1}{\rightleftharpoons}} Y \underset{k_{-2}}{\overset{k_2}{\rightleftharpoons}} Z \quad (b); \quad X \underset{k_{-1}}{\overset{k_1}{\rightleftharpoons}} Y \overset{k_2}{\longrightarrow} Z \quad (c)$$

where X is the element in the liquid phase, Y and Z are the same element sorbed on the solid phase.

A detailed mathematical description of the three mechanisms is reported in the Annex.

The forward (k_1, k_2) and backward (k_{-1}, k_{-2}) kinetic constants were computed from the solution of the differential equations, using a nonlinear regression method [6] and a proper computer subroutine [7] to interpolate the experimental values (K_d as a function of time).

The mechanism considering an irreversible first–order reaction only was also taken into account but not reported here because it did not yield any useful results. The mechanism including two parallel, reversible first–

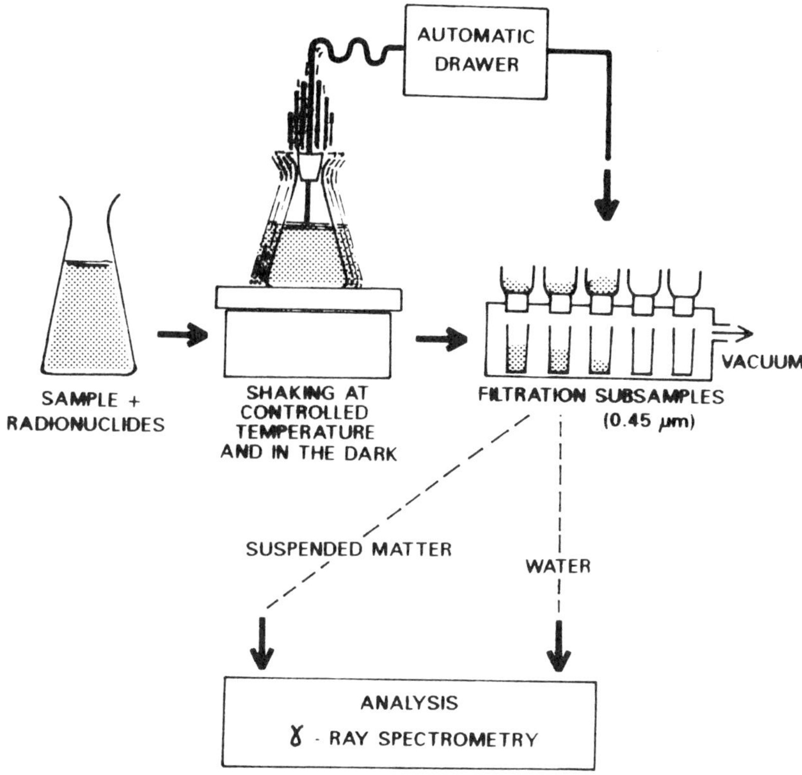

Fig. 2. Experimental approach for kinetic tests

order reactions (not reported here) produced results very similar to those obtained by mechanism (b).

RESULTS AND DISCUSSION

Average minimum and maximum values of forward and backward kinetic constants as well as particulate load and K_d values obtained from the application of the three different mathematical models are reported in Table 3 for each radionuclide.

As an example, Fig. 3 shows one of the forty interpolation profiles obtained by applying the three different mathematical models to the same sample and to the same radionuclide.

The mathematical model producing, for each experimental run, minimum standard deviation values between the computed and expected (exper-

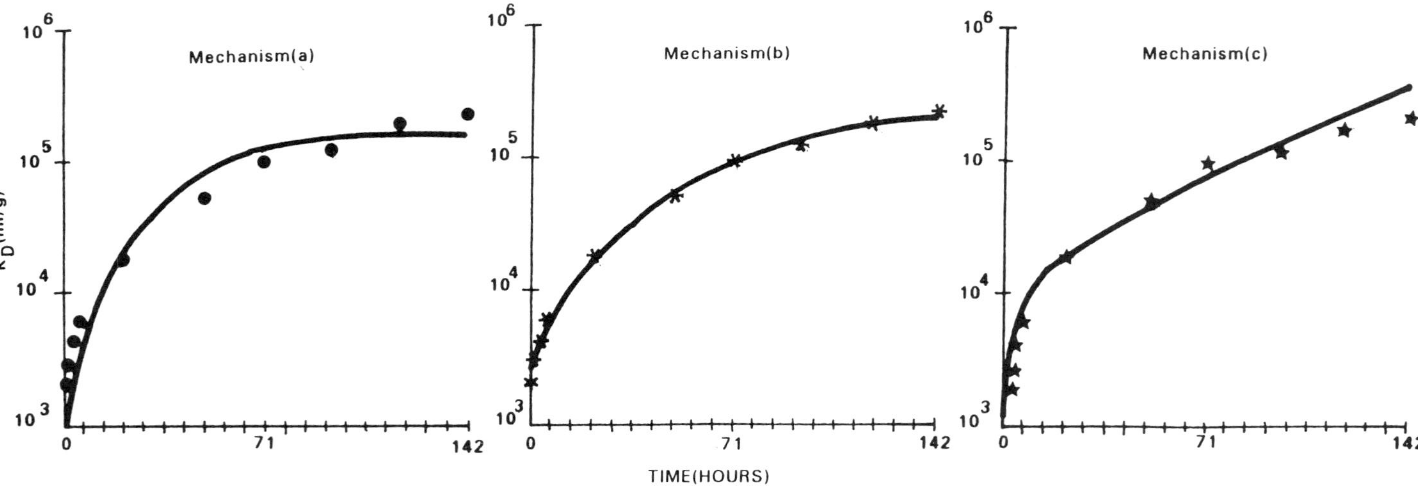

Fig. 3. ^{54}MN: interpolation profiles obtained applying the three different mathematical models to experimental data (k_d's as a function of time).

Table 4. Ability of the Three Different Mechanisms in the Interpolation of the Experimental Data (bulk samples)

Element	mechanism			Sample size
	(a)	(b)	(c)	
^{137}Cs	0	7	0	7
^{54}Mn	0	6	1	1
^{51}Cr	0	3	0	1
^{58}Co	0	6	1	7
^{59}Fe	0	7	0	7
^{65}Zn	0	7	0	7
^{85}Sr	0	3	0	3

Table 5. Ability of the Three Different Mechanisms in the Interpolation of the Experimental Data (filtered samples)

Element	mechanism			Sample size
	(a)	(b)	(c)	
^{137}Cs	0	5	0	5
^{54}Mn	2	0	3	5
^{51}Cr	1	1	0	2
^{58}Co	1	3	1	5
^{59}Fe	0	5	0	5
^{65}Zn	0	4	1	5
^{85}Sr	0	1	0	1

imental) data was considered as the best one in the considered case. Tables 4 and 5 summarize the obtained results for the bulk samples (water and suspended matter) and for the filtered samples, respectively.

These tables present data for Cr(III) only. For Cr(VI) the sorption profiles were similar to those of Cr(III), showing that the radionuclide is likely to be quickly reduced to Cr(III) by the organic matter before sorption on the solid phase.

Among the three considered, mechanism (b) was the best one for each unfiltered sample and for each radionuclide, except for Co and Mn in one sample only, for which mechanism (c) appeared more appropriate. The latter is related to autocatalytic oxidation of Mn(II) to Mn(IV) [4,5,8], and to the great affinity of Co towards δ-MnO_2 [9]. For Co and Mn is likely that longer term laboratory runs (3 to 4 weeks) would lead to experimental data for which mechanism (c) might be more appropriate, pointing out a continuous K_d increase as a function of time, as reported for the marine environment [4].

The insolubilization of Mn in filtered samples (Tab. 5) was well described by mechanisms (a) and (c). A comparison between these results and those regarding unfiltered samples (Tab. 4) shows the important role of suspended matter in the whole mechanism of Mn insolubilization.

The role of the suspended matter in the insolubilization process can also be observed for Cr, Zn and Co, but less distinctly.

The knowledge of the best mathematical model enabling us to explain the interaction mechanism of a given radionuclide between water and suspended matter is important for a better understanding of the chemico-physical processes occurring in aquatic ecosystems. Moreover, in radiological assessment models the use of the simplest mechanism may be necessary in relation to the mathematical code complexity and adequate for other introduced simplifications. From this point of view, mechanism (a) may be a good approximation of the experimental data.

As an example, in Fig. 4, the K_d values obtained by applying mechanisms (a) and (c) to interpolate the experimental data, as a function of time, are compared to those obtained by mechanism (b) (the best one), in terms of percentage deviation from the reference mechanism.

The use of the kinetic constants in deterministic models requires the knowledge of the physical laws elucidating their dependence on the most important chemico-physical environmental parameters, such as particulate load and grain-size, pH, composition of the solid and liquid phases, etc. Since the three considered exchanging mechanisms are a combination of first-order reactions only, the computed k_1, k_2, k_{-1} and k_{-2} may be apparent constants, the variability of which, for the same element, may depend on an undefined number of environmental parameters.

Table 6 presents the results of the correlation analysis between the kinetic constants obtained by applying mechanism (b), and the following water and suspended matter properties of the seven considered samples: pH; particulate load; total stable element, total Fe, total Mn, silt (4-63 μm fraction), clay (0.45-4 μm fraction) and organic carbon (P.O.C.) concentrations in the solid phase; dissolved organic carbon (D.O.C.), stable element, Mn and Fe concentrations in water, Na and K concentrations in water (for Cs only). The positive (+) and negative (-) correlation ($r \geq 0.7$) is indicated with "*" or "**" if the first species error is less than 5% or 1%, respectively.

The constants k_1, k_{-1} and k_2 of Fe were negatively correlated to the silt concentration in the suspended matter.

The constants k_2 and k_{-2} of Mn were negatively correlated to pH, while k_{-2} was positively correlated also to Fe concentration in the solution.

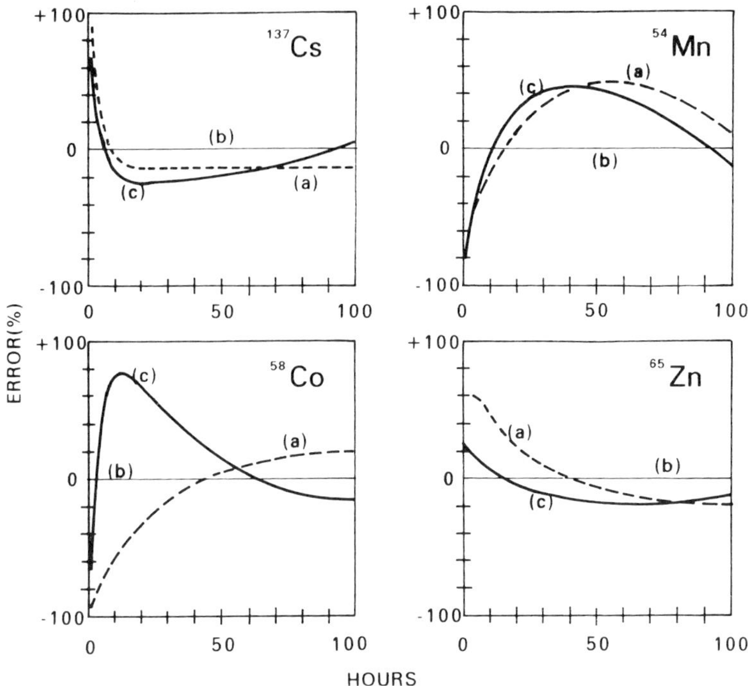

Fig. 4. Percentage deviations from the reference mechanism (b) of K_d values, as a function of time, obtained applying mechanisms (a) and (c).

For Zn positive correlations were found between k_{-1} and P.O.C., and between both k_2 and k_{-2} and both stable Zn (Fig. 5d) and Mn concentrations in water, while only k_{-2} was correlated to the stable Fe concentration in water.

A lot of correlations were found for Cs: k_2 was positively correlated to the concentration of stable Mn, Na (Fig. 5a), K, and organic carbon in water, and to the total stable Cs concentration in the suspended matter (negatively), whereas k_{-2} was negatively correlated to the concentration of total stable Cs in the suspended matter (Fig. 5c), and positively to both P.O.C. and D.O.C. (Fig. 5b).

No correlation was found for Co. For Cr and Sr the correlation analysis was not applied because of a poor number of kinetic constants available.

Some of these results can hardly be explained in terms of element affinity towards the water and suspended-matter characteristics or in terms of the peculiarity of the sorption mechanism as reported by Buchholtz and

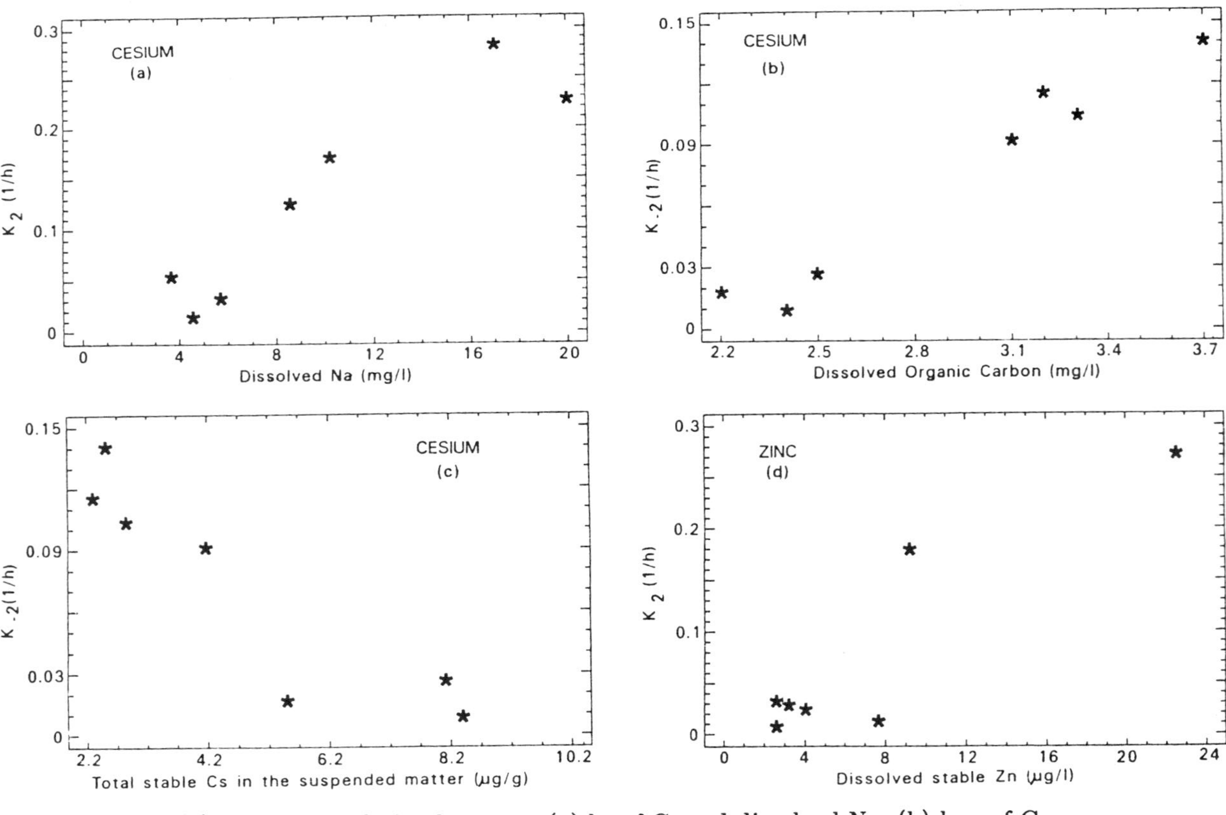

Fig. 5. Correlation between: (a) k_2 of Cs and dissolved Na; (b) k_{-2} of Cs and dissolved organic carbon; (c) k_{-2} of Cs and total stable Cs in the suspended matter; (d) k_{-2} of Zn and dissolved stable Zn.

coworkers for the marine environment [5]. Moreover, some expected correlations, such as those between particulate load (which presents a wide range in the considered samples) and forward kinetic constants, or between Cs sorption rate and clay content in the suspended matter, were not found.

On the other hand, the found correlations are probably not able to explain the total variance of the kinetic constants. This fact makes the use of the kinetic constants in deterministic models very difficult.

A second possible approach to the problem of the kinetic constants dispersion is used in stochastic models that consider the data variability as "natural", especially if the goal is their use in previsional models rather than the need to understand the chemico-physical mechanisms controlling element sorption on suspended matter. Then the problem is dealt with searching for a distribution (normal, log-normal, etc.) allowing us to describe the data dispersion. As an example, in Figs. 6 and 7 the probability plotting technique was used to represent in a probabilistic scale the cumulative frequency, and in a logarithmic scale the kinetic constant values. According to the Kolmogorov-Smirnov test [10] a straight line gives a good interpolation of the experimental data, proving the ability of the log-normal statistical model in describing the kinetic constant distribution, as already verified for K_d values [3,11].

CONCLUSIONS

The interaction of an element at the solid-liquid interface must be treated using the kinetic approach, when the time scale of the process is the same as the period of interest.

The complexity of this kind of approach necessitates studying the kinetics of the sorption and desorption reactions under defined natural conditions (e.g., pH, salinity, temperature), searching for simple mathematical models able to interpolate the experimental data.

Among the three considered mathematical models, mechanism (b) appeared the more appropriate for describing the sorption process, while mechanism (c) was useful for Mn and Co only, considering the irreversible sorption on solid phase, related to the autocatalytic oxidation of Mn(II) to Mn(IV) and to the great affinity of Co towards δ-MnO_2. Mechanism (a) was usually the worst one for data interpolation, appearing not quite appropriate to describe very fast kinetics. However, since the observed deviations in the sorption prediction (in terms of K_d values at different times) were very small in comparison with the reference mechanism (mechanism (b)), this mechanism may be very useful in radiological assessment models because of its simple applicability.

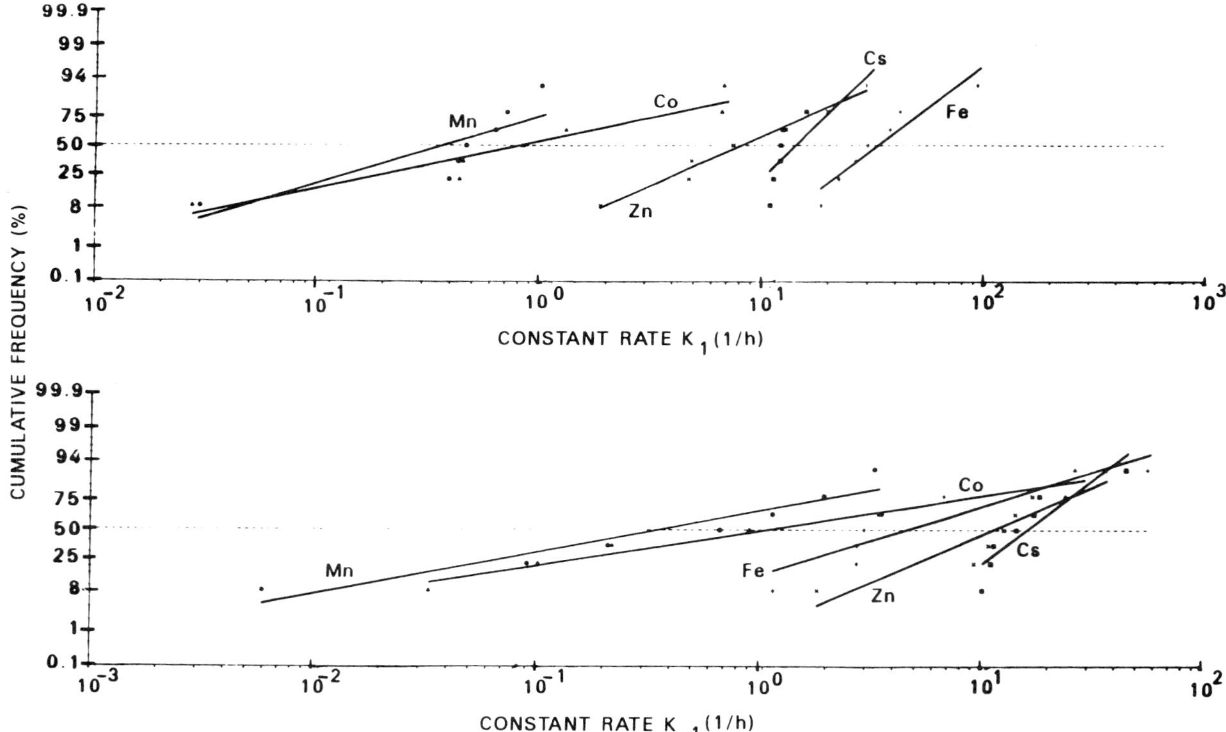

Fig. 6. Probability plotting of the kinetic constants (k_1 and k_{-1}) using mechanism (b).

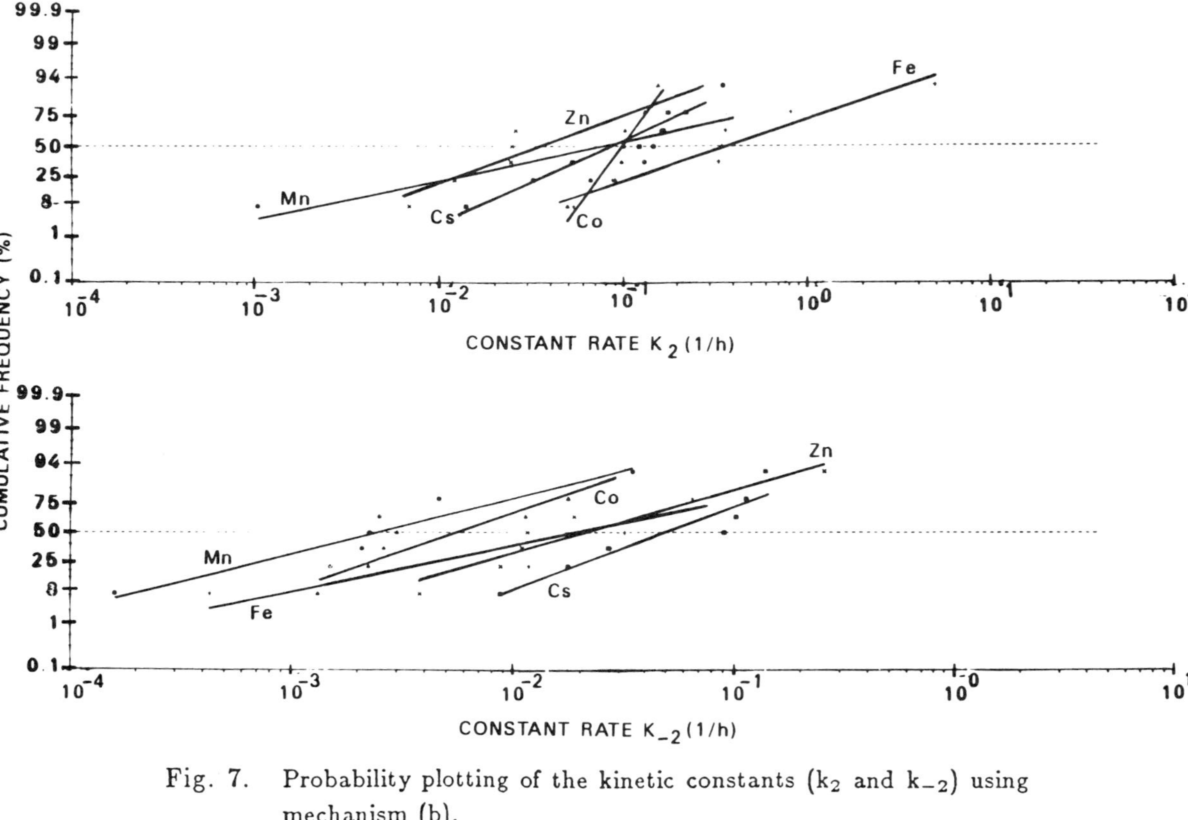

Fig. 7. Probability plotting of the kinetic constants (k_2 and k_{-2}) using mechanism (b).

The obtained forward and backward constant values (referring to mechanism (b)) were correlated to some solutions and suspended–matter properties for use in deterministic models.

Significant correlations were found between k_1, k_{-1} and k_2 of Fe and silt concentration in the suspended matter; k_2 and k_{-2} of Mn and pH; k_{-2} of Mn and Fe concentration in solution; k_{-1} of Zn and P.O.C.; k_2 and k_{-2} of Zn and both stable Zn and stable Mn concentrations in water; k_{-2} of Zn and stable Fe concentration in water; k_2 of Cs and stable Mn, Na, K, and organic carbon concentrations in water, and the total concentration of stable Cs in the suspended matter; k_{-2} of Cs and the total concentration of stable Cs in the suspended matter and both P.O.C. and D.O.C. No correlation was found for Co.

Alternatively, for use in stochastic previsional models, the kinetic constant values were statistically treated to find a distribution (log-normal) able to describe the data dispersion.

REFERENCES

[1] Y. Onishi, R.J. Serne, E.M. Arnold, C.E. Cowan, F.L. Thompson, Critical review: Radionuclide transport, sediment transport, and water quality mathematical modelling; and radionuclide adsorption/desorption mechanisms, NUREG/CR-1322, DNL. 2901/RE U.S. NCR Washington, D.C. (1980).

[2] Y.H. Li, L. Burkhardt, M.R. Buchholtz, P. O'Hara, P.H. Santschi, Partitioning of radiotracers between suspended particles and seawater, Geochim. Cosmochim. Acta, 48, 2011, 1984.

[3] G. Ciceri, A.L. Traversi, W. Martinotti, G. Queirazza, Radionuclide partitioning between water and suspended matter: comparison of different methodologies, in: Chemistry for Protection of the Environment 1987, Studies in Environm. Sci. n. 34, L. Pawlowski, E. Mentasti, W.J. Lacy, C. Sarzanini Eds., Elsevier, Amsterdam-Oxford-New York-Tokyo, 353, 1988.

[4] U.P. Nyffeler, Y.H. Li, P.H. Santschi, A kinetic approach to describe trace-element distribution between particles and solution in natural aquatic systems, Geochim. Cosmochim. Acta, 48, 1513, 1984.

[5] M.R. Buchholtz, P.H. Santschi, W.S. Broecker, Comparison of radiotracer Kd values from batch equilibration experiments with in situ determination in the deep sea using the Manop-Lander: the importance of geochemical mechanisms in controlling ion uptake and migration, in: Application of distribution coefficients to radiological assessment models, T.H. Sibley and C. Myttenaere Eds., Elsevier App. Sci. Pub., London and New York, 192, 1986.

[6] R. Fletcher, A modified Marquard subroutine nonlinear least squares, Atomic Energy Research Establishment (AERE), Harwell (U.K.), AERE-R 6799, 1971.

[7] M.P. Hopper, A catalogue of subroutine (1978), Atomic Energy Research Establishment (AERE), Harwell (U.K.), AERE-R 9185, 1981.

[8] J. Murray, J. Dillard, G. Giovanoli, H. Moers, W. Stumm, Oxidation of Mn(II): initial mineralogy, oxidation state and ageing, Geochim. Cosmochim. Acta, 49, 463, 1985.

[9] J. Murray, The interaction of metal ion at the manganese dioxide-solution interface, Geochim. Cosmochim. Acta, 39, 505, 1975.

[10] J.D. Gibbons, Nonparametric methods for quantitative analysis, Holt, Rinehart and Winston Eds., New York, 1976.

[11] M. Cambiaghi, G. Ciceri, W. Martinotti, A. Toppetti, Partitioning of elements between water and suspended matter: Thermodynamic approach, (in preparation).

ANNEX

Description of program (a)

The program is used to check the fitness to experimental data of a first order reaction mechanism that can be expressed as follows:

$$X \underset{k_{-1}}{\overset{k_1}{\rightleftharpoons}} Y \tag{1}$$

where X and Y are the trace elements present in the liquid and on the solid phase respectively; k_1 and k_{-1} are the kinetic constants.

The time (t) variation of trace element content in each phases is given by:

$$\frac{dX}{dt} = -k_1 X + k_{-1} Y$$
$$\frac{dY}{dt} = k_1 X - k_{-1} Y \tag{2}$$

The first order ordinary linear differential equation system (2), with the following initial conditions:

$$X(0) = X_o$$

$$Y(0) = 0$$

and the mass conservation equation $X + Y = X_o$, can be solved as follows:

$$X = X_o \frac{k_{-1}}{k_1 + k_{-1}} + \left(1 - \frac{k_{-1}}{k_1 + k_{-1}}\right) \exp[-(k_1 + k_{-1})\, t]$$
$$Y = Xo - X \tag{3}$$

After finding the kinetic constants (k_1 and k_{-1}), and being C_p the particulate load, it is possible to calculate K (thermodynamic constant) and the equilibrium distribution coefficient (K_d):

$$K_d = \lim_{t->\infty} \frac{Y(t)}{X(t)} = K\frac{1}{C_p} \tag{4}$$

which, if a due account is given to eq. (3), becomes:

$$K = \frac{k_1}{k_{-1}} \tag{5}$$

Description of program (b)

In the second mechanism under consideration the trace element sorbed on the particle surface, is assumed as interactable not only with the solute but also with the element bound in the lattice held position of the particle. Such a mechanism can be outlined as follows:

$$X \underset{k_{-1}}{\overset{k_1}{\rightleftharpoons}} Y \underset{k_{-2}}{\overset{k_2}{\rightleftharpoons}} Z \tag{6}$$

where X, Y and Z denote the solute element, that sorbed by particle surface structure and that sorbed in a reversible way in the inner structure.

The time (t) variation of the trace element content in each phase is given by the following system of differential equations:

$$\frac{dX}{dt} = -k_1X + k_{-1}Y$$

$$\frac{dY}{dt} = k_1X - k_{-2}Z - (k_{-1} + k_2)Y \tag{7}$$

$$\frac{dZ}{dt} = k_2Y - k_{-2}Z$$

The differential equation system (7) with the following initial conditions:

$$X(0) = Xo$$

$$Y(0) = 0$$

$$Z(0) = 0$$

and the mass conservation equation X + Y + Z = Xo, can be solved as follows:

$$Y = \alpha_1 + \frac{1}{l_1 - l_2}[\alpha_2 \exp(l_1 t) - \alpha_3 \exp(l_2 t)]$$

$$Z = \beta_1 + \frac{1}{l_1 - l_2}[\beta_2 exp(l_1 t) - \beta_3 exp(l_2 t)] \tag{8}$$

$$X = Xo - Y - Z$$

Expressions of coefficients l, α and β are:

$$\begin{aligned} l_1 = \ & 1/2\{-(k_1 + k_{-1} + k_2 + k_{-2}) + \\ & +[(k_1 + k_{-1} + k_2 - k_{-2})^2 + 4(k_{-2} - k_1)k_2]^{1/2}\} \end{aligned}$$

$$\begin{aligned} l_2 = \ & 1/2\{-(k_1 + k_{-1} + k_2 + k_{-2}) + \\ & -[(k_1 + k_{-1} + k_2 - k_{-2})^2 + 4(k_{-2} - k_1)k_2]^{1/2}\} \end{aligned}$$

$$\alpha_1 = A/C$$

$$\alpha_2 = (l_1 + k_{-2})(-A/C) + (k_{-2} - k_1)(-B/C)$$

$$\alpha_3 = (l_2 + k_{-2})(-A/C) + (k_{-2} - k_1)(-B/C)$$

$$\beta_1 = B/C$$

$$\beta_2 = k_2(-A/C) - (l_2 + k_{-2})(-B/C)$$

$$\beta_3 = k_2(-A/C) - (l_1 + k_{-2})(-B/C)$$

with:

$$\begin{aligned} A = \ & k_1 k_{-2} Xo; \\ B = \ & k_1 k_2 Xo; \\ C = \ & (k_1 + k_{-1} + k_2)k_{-2} - (k_{-2} - k_1)k_2. \end{aligned}$$

After finding the kinetic constants (k_1, k_{-1}, k_2 and k_{-2}) it is possible to calculate K (thermodynamic constant) and, through the particulate load (Cp), the equilibrium distribution coefficient (K_d):

$$K_d = \lim_{t \to \infty} \frac{Y(t) + Z(t)}{X(t)} = K \frac{1}{Cp} \tag{9}$$

which, if due account is given to eq. (8) becomes:

$$K = \frac{k_1 k_{-2} + k_1 k_2}{k_{-1} k_{-2}} \tag{10}$$

Description of program (c)

The third mechanism under consideration is different from mechanism (b) because the second reaction step is assumed as irreversible:

$$X \underset{k_{-1}}{\overset{k_1}{\rightleftharpoons}} Y \overset{k_2}{\rightarrow} Z \tag{11}$$

The differential equations that give the time element content variation in each single phase become:

$$\frac{dX}{dt} = -k_1 X + k_{-1} Y$$

$$\frac{dY}{dt} = k_1 X - k_{-1} Y - k_2 Y \tag{12}$$

$$\frac{dZ}{dt} = k_2 Y$$

which, with the following initial conditions:

$$X(0) = Xo$$

$$Y(0) = 0$$

$$Z(0) = 0$$

and with the mass conservation equation X + Y + Z = Xo, can be solved as follows:

$$X = \frac{1}{l_1 - l_2}[\alpha_1 \exp(l_1 t) - \alpha_2 \exp(l_2 t)]$$

$$Y = \frac{1}{l_1 - l_2}[\beta_1 \exp(l_1 t) - \beta_2 \exp(l_2 t)] \tag{13}$$

$$Z = Xo - X - Y$$

Expression of coefficients l, α and β are:

$$l_1 = 1/2\{-(k_1 + k_{-1} + k_2) + \\ +[(k_1 - k_{-1} - k_2)^2 + 4k_1k_{-1}]^{1/2}\}$$

$$l_2 = 1/2\{-(k_1 + k_{-1} + k_2) + \\ -[(k_1 - k_{-1} - k_2)^2 + 4k_1k_{-1}]^{1/2}\}$$

$$\alpha_1 = (l_1 + k_{-1} + k_2)Xo$$

$$\alpha_2 = (l_2 + k_{-1} + k_2)Xo$$

$$\beta_1 = k_1Xo$$

For $t \to \infty$, for example when the system is in equilibrium condition, eqs. (13) state that the element is completely sorbed in the inner structure of the suspended matter which is in agreement with the irreversible reaction assumption.

The computer codes are available in FORTRAN IV and run on IBM 4341.

ORGANICALLY BOUND CHLORINE IN MARINE ORGANISMS: CHEMICAL PROPERTIES AND POSSIBLE BIOCHEMICAL ORIGIN

A. JERNELOV

Swedish Environmental Research Institute
S-100 31 Stockholm, Sweden

INTRODUCTION

During the last year a number of mystifying features relating to organically bound chlorine in aquatic organisms have been highlighted in lectures and articles.

Fish from the Baltic Sea are obviously a health hazard for fish-eating mammals, such as seals and otters.

About 90% of the content of organically bound chlorine found in the fat of Baltic fish is made up of unidentified substances.

Chemical indications suggest that a large fraction of the organically bound chlorine in fish fat consists of high molecular weight substances despite the fact that such large compounds should not be able to penetrate cell membranes.

Fish in lakes on the islands around the Baltic and fish from mountain lakes far away from known sources of organo-chlorine compounds contain

Chemistry for the Protection of the Environment
Edited by L. Pawlowski *et al.*, Plenum Press, New York, 1991

concentrations of organically bound chlorine comparable with those found in fish from the Baltic.

Thus, in summary:

We know that fish from the Baltic and oligotrophic lakes contain high concentrations of organo-chlorine compounds.

We do not know what compounds or group of compounds it is, or from where they come.

We suspect that these unidentified substances contribute to the catastrophic situation for Baltic fish-eating mammals.

Another complicating factor is that the analyses of extractable organically bound chlorine (EOCl) in fat from fish and other aquatic organisms often give results with a very high variability. Apparently, small differences in analytical methods give rise to large differences in results.

In this paper I will present some new results and a hypothesis aimed at explaining some of the "mystifying" facts with regard to geographical distribution and chemical form of occurrence of organically bound chlorine in aquatic organisms. Before doing so, however, I would like to remind the reader of some important lessons from the field of research on mercury in the environment:

- Methylmercury can be biologically formed from inorganic mercury. Thus, organisms cannot only degrade but may also build organic compounds.
- A process that for one organism is functionally a detoxification, may for other organisms in the ecosystem mean that a lethal toxin is produced.
- The concentration of a substance in an organism (e.g., mercury in fish) is not only a function of the amount of the substance present in the environment but also of the pattern of turnover and accumulation of the substance in a given ecosystem (high concentrations in fish in acid and oligotrophic lakes).

SOME NEW RESULTS

About twenty years ago, IVL started work with the fate and effects of EDC-tar (by-products from PVC manufacturing) after dumping in the marine environment. The original results were presented at an FAO conference in Rome on marine pollution. As a result IVL became involved in a number of international studies relating to EDC-tar and vinyl chloride in the marine environment. One of the latest of these was in relation to a ship with a cargo of vinyl chloride that sunk in the Adriatic Sea.

In a study of vinyl chloride accumulation in sea bass, a quick build up to a concentration of 70ppm was measured as chlorine was found, followed by a rapid decrease in concentration after transfer of fish to clean water, so that after 3 weeks vinyl chloride could not be detected in the sea bass fat. These results were in agreement with those from previous studies including the original work referred to above.

This time, however, we also analyzed for EOCl. From a base content of 30ppm at the start of vinyl chloride exposure, the EOCl concentration increased as was expected to 100ppm. After 3 weeks it had gone down to 60 ppm where it remained throughout the experimental period (6 weeks).

This indicates that the vinyl chloride was transformed to some other organo-chlorine compound with a very slow rate of degradation/excretion.

An important finding in earlier studies of accumulation and excretion of organo-chlorine compounds in fish is that, PCB's as well as DDT and several of its metabolites are in equilibrium solution between the fat of the fish and the surrounding water.

In a recent experiment we compared PCB and Σ DDT with EOCl to see if the same processes regulate excretion of the well known organo-chlorine compounds and the unknown ones which constitute 90% of EOCl. The results show that eels which spent three weeks in aquaria water which is filtered over frequently re-newed activated carbon have 20% lower concentrations of PCB's and Σ DDT than eels which spent the same time in aquaria without activated carbon in the filters. The EOCl concentrations, however, remain the same in both groups of eels.

The result indicates that while the excretion of identified compounds like PCB's and DDT is regulated by the equilibrium solution between fish fat and surrounding water, the main part of the organically bound chlorine is more tightly bound to the fat in the cells of the fish.

PRINCIPLE FOR SYNTHESES OF FATTY ACIDS

The syntheses of fatty acids in an organism starts with the binding of an acetyl or butyroyl group to coenzyme A. From this base the long carbon chains of the fatty acids are then formed through the addition of building blocks with two carbon atoms, with reduction, dehydration and further reduction following each addition of a two-carbon building block. When the fat reserves are mobilized the breakdown of the fatty acids occurs in a similar fashion, though naturally in the reverse order.

HYPOTHESES

Based on the above the following chain of hypotheses can be formulated:

- Organically bound chlorine in fish fat is to a large extent in the form of chlorinated fatty acids.
- These are formed in the fish itself.
- Building blocks are low molecular weight chlorinated organic compounds which may have an atmospheric distribution.
- A deficit of fat and a shortage of building blocks for fatty acids make fish in oligotrophic waters like the Baltic more prone to utilize low molecular weight chlorinated compounds as building blocks for fatty acids than are fish in eutropic water like the North Sea. The result is a higher apparent accumulation factor.

POSSIBLE EFFECTS OF SALINITY ON TOTAL PHOSPHORUS CONCENTRATION IN LAKE KINNERET

Y. AVNIMELECH

Technion-Israel Institute of Technology
Haifa 32000, Israel

ABSTRACT

An increase in total phosphorus found to occur in Lake Kinneret during the last 2 decades may be related to a concomitant decrease in salt concentration in the Lake.

The effects of sodium and calcium addition on sediments suspended in H_2O and in filtered lake water was evaluated. It was found that the addition of 1 meq $CaCl_2$/l Lake Kinneret water led to a 5–fold increase in the median settling velocity of the suspended sediments. It is concluded that the salt concentration of the Lake is close to the flocculation value of the sediment particles.

The changes in salt concentrations during the last decades, and the seasonal changes of calcium concentration may explain an increased resuspension of sediments and thus an increase in the supply of particulate phosphorus to the Lake.

INTRODUCTION

Total phosphorus concentrations (i.e., phosphorus in non-filtered water) in Lake Kinneret have increased during the last decade [6]. No concomitant increase in total phosphorus inputs from the watershed was found [2]. Yet, a significant decrease in the salinity of the lake has occurred during the

Chemistry for the Protection of the Environment
Edited by L. Pawlowski *et al.*, Plenum Press, New York, 1991

last two decades, mainly due to the construction and operation of the saline water diversion canal [5]. A possible relationship between those two changes is tested in this work.

Solids suspended in the water may carry relatively large amounts of phosphorus. The average phosphorus content of Lake Kinneret sediments is on the order of 0.2% [5]. Thus, even a very low suspended solid concentration of 10 mg/l will contain a load of 0.02 mg P/l, which is approximately the total phosphorus concentration in Lake Kinneret. Any change in the suspended solid contents in the lake will, accordingly, imply a marked change in phosphorus.

The effects of salts on the flocculation and the resultant stability of colloidal clay or other particles were studied extensively [7]. A sigmoidal flocculation curve is obtained (Figure 1) and is defined by three zones. In the low salt range, where the induction of flocculation by the salts is negligible, the particles are in a state of dispersion and the suspension is stable. With an increase in salt concentration, a range of flocculation inducement is reached. The inflection point of the exponential flocculation curve is defined as the flocculation vale. A further increase of salinity no longer affects the suspension stability. According to the double layer theory, the flocculation is induced by the divalent cations at a lower salt concentration as compared with the monovalent cations. The theoretical ratio between the two values is 64 [7].

It can be expected that even minor changes in salinity of the lake water will have a marked effect on total phosphorus, if the flocculation value of Lake Kinneret sediments is in the range of the salt concentration in the lake water. It is not anticipated that at a lower or higher concentration range such effects will be significant.

The hypothesis that salinity changes in Lake Kinneret may have induced changes in the phosphorus content of Lake Kinneret is tested in the present work.

MATERIALS AND METHODS

Settling velocities of suspended particles were measured continually in a settling column. A premixed suspension in a glass cylinder was placed in a tube containing 4 collimated light sources and four light-sensitive elements across it. Scanning of turbidity along a sedimentation column, as a function of time, yields data that can be used to calculate the distribution of settling velocities in the particle population [1]. The optical densities at

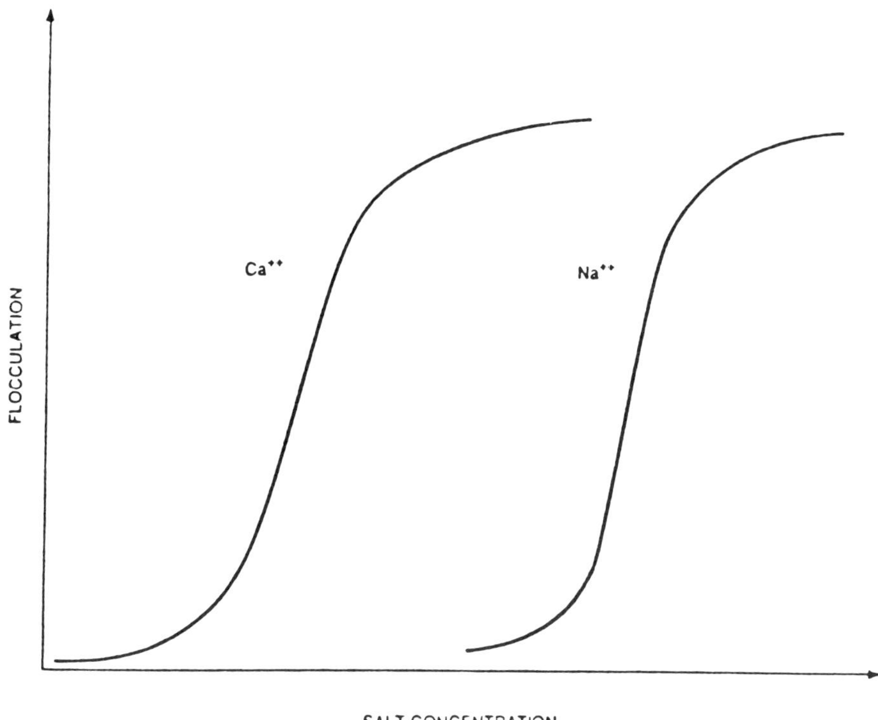

Fig. 1. Effects of NaCl and $CaCl_2$ on the flocculation of colloids (schematic).

the 4 ports along the sedimentation column were continually scanned, controlled and recorded by a computer. Measurements were conducted along a 30–600-minute time span. Minimal detectable settling velocity was about 5-10 mm/hr.

Sediments were sampled from the bottom of Lake Kinneret and stored at 40°C. Suspension for analysis was made by resuspending 20 g wet sediment per liter of distilled water or filtered lake water. The suspension was left still for one hour for sedimentation of the coarse particles. The decanted suspension used for analysis contained about 1500 mg/l suspended solids. The distilled water extract contained 0.6 meq Ca++/l and 0.15 meq Na+/l.

Filtered Lake Kinneret water had an electrical conductivity of 1.07 mmho/cm, and Ca++, Na+ and Cl- concentrations of 5.1, 2.4 and 6.2 meq/l, respectively.

RESULTS AND DISCUSSION

The distributions of settling velocities for the sediment particle population suspended in distilled water or in salt solutions are given in Fig.2. The curve for the particles suspended in distilled water indicates very slow settling velocities of the particle population, where only 10-20% of the particles settled at a rate of 20 mm/hr or above, while 80% of the particles settled at a rate of 10 mm/hr or less. The equivalent radius can be calculated according to Stokes law

$$V = \frac{r^2 g(\Delta\rho)}{\eta} \quad (1)$$

where V is the settling velocity, r is the radius of the particle, g the gravity, η the water viscosity and $\Delta\rho$ the difference between the particles' and fluid's densities. The density of clay particles is usually considered to be 2.6-2.7 g/cm^3 [8]. The density of flocs can be appreciably lower. Thus, the values calculated here, using the assumption that the density is 2.65 g/cm^3, can be underestimates of the floc radii.

The equivalent radii of particles in the H_2O suspension is, according to Stokes Law, in the range of 1.7-2.4 μm, which is the range of clay particles [8]. Settling velocity is shown to increase with addition of salts. Addition of 100 meq NaCl/l leads to a different settling velocity distribution. Through addition of 100 meq NaCl/l, 30% of the population had a settling velocity of 70 mm/h or more ($r \geq 4.5$ μm) and 20% of the population sank at rates of above 300 mm/h ($r \geq 9.2$ μm). Doubling the added NaCl concentration led to a steeper response. The median settling velocity rose to 270 mm/hr ($r \geq 8.8$ μm) and about 25% of the particle population sank at rates higher than 1000 mm/hr ($r \geq 17$ μm). The sensitivity of settling velocity increase to addition of salts was, as expected, higher for the addition of the divalent calcium. With addition of 2 meq $CaCl_2$/l the medium settling velocity rose to about 100 mm/hr ($r \geq 5.3$ μm) and by adding 3 meq/l the median settling velocity rose to 220 mm/hr ($r \geq 7.9$ μm). It can be seen that the theoretical exponential increase in the sinking rates with addition of electrolytes is obtained, and that the difference between the effects of Na^+ and Ca^{++} is close to the theoretically expected one. The effects of the addition of $CaCl_2$ to filtered Lake Kinneret water were also tested, are given in Fig. 3. The median settling rate of particles in the raw filtered water was 160 mm/hr ($r \geq 6.8$ μm). With addition of 0.5 meq $CaCl_2$/l the median velocity rose to 250 mm/hr ($r \geq 8.5$ μm), and by adding 1.0 meq/l the median velocity rose steeply to 850 mm/hr ($r \geq 15.6$ μm). These results indicate that at present Lake Kinneret water has an electrolyte concentration and composition just below that of the flocculation value of its sediment's colloidal fraction.

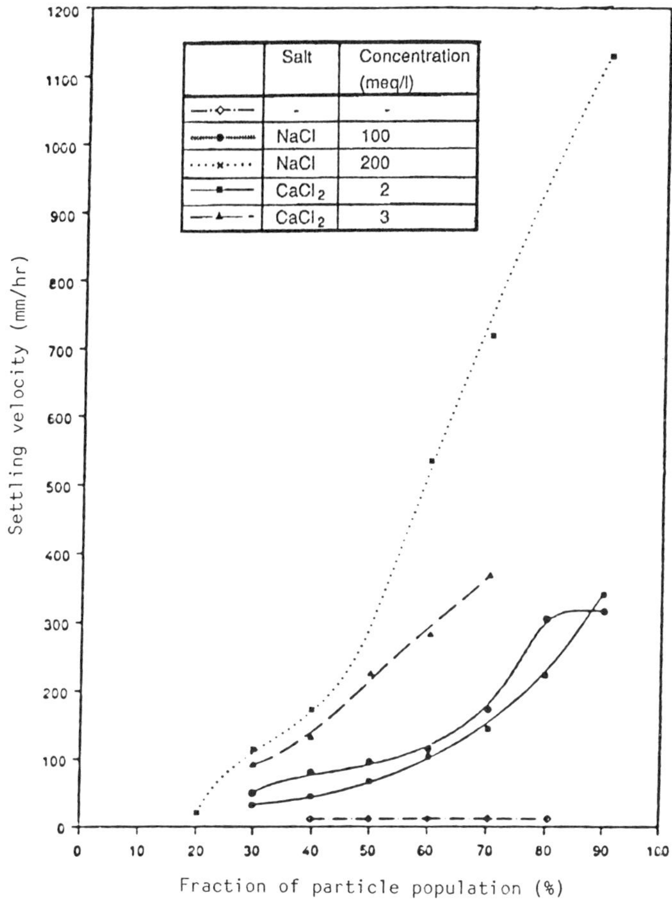

Fig. 2. Effects of NaCl and $CaCl_2$ on the settling velocity distribution of particle population from Lake Kinneret sediments.

Lake Kinneret salt composition in the 1950's [4] and for the period 1968-1976 [5] is given in Table 1. A significant reduction in salinity has occurred during this period. Chlorides, being the main anion, dropped from 9.9 meq/l to 7.0 meq/l (present concentration is 6.2 meq/l) [3]. Sodium concentration dropped by about 3 meq/l, calcium by about 0.4 meq/l, while calcium + magnesium concentrations dropped by 1.7 meq/l. These changes are certainly within the range that could have a significant effect on the stability of sediment suspension and thus on phosphorus concentration.

An additional relevant process is the seasonal variation in calcium

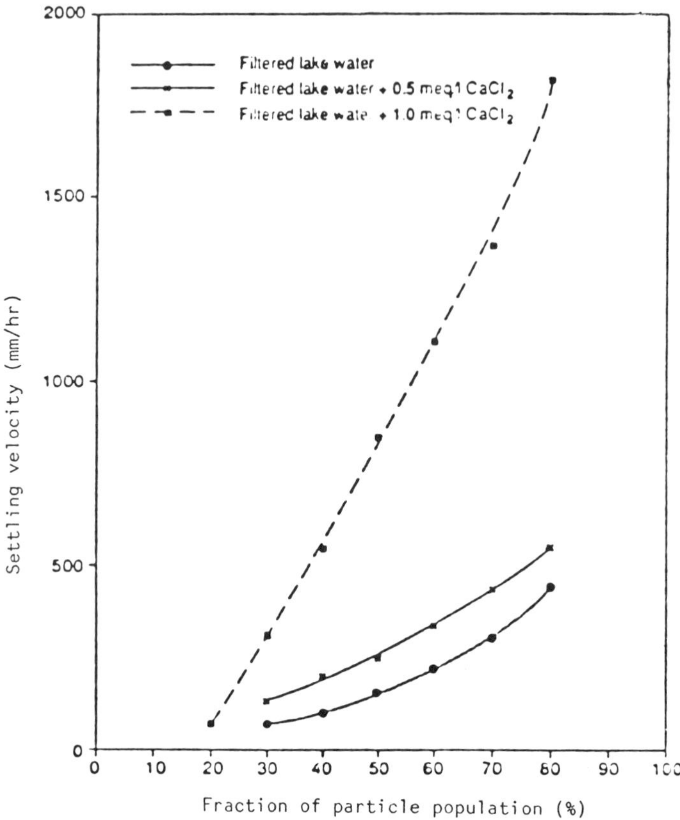

Fig. 3. Effects of $CaCl_2$ on the settling velocities of sediment particles suspended in filtered Lake Kinneret water. Percentage of particles having a settling velocity equal to or lower than the value indicated.

concentration in the lake hypolimnion. Calcium concentration in the hypolimnion is highest in the winter mixing period (Dec.-March) and is then about 2.5 meq/l. During the summer, calcium concentration is reduced to concentrations of around 2 meq/l. It was found that during that period there is a net release of phosphorus from the lake sediments into the water. It is possible that the reduction in calcium concentrations, both long-term and seasonal, in combination with the wind pattern of the summer and the fall, both induce sediment resuspension and a resultant phosphorus enrichment in the lake.

Table 1. Major Ion Concentration in Lake Kinneret (meq/l)

Author	Period	Na	Ca	Ca+Mg	Cl
Oren (1957)	1951,1957	7.0	3.0	6.2	9.9
Serruya (1978)	1968-1976	5.1	2.65	4.5	7.0
Gophen (1984)	1983		2.35		6.2
Nishri (1988) [9]	1988	4.26	2.45	4.8	5.8

CONCLUSIONS

The hypothesis tested in this work is that it is possible that the increase in total phosphorus found recently in Lake Kinneret is due to the salinity reduction that occurred in the same period.

It was shown that even a slight addition of suspended sediments may significantly increase phosphorus concentration in the lake. Total phosphorus concentration in the lake, being on the average 0.017 mg/l, is equivalent to the phosphorus content of 8.5 mg/l of suspended sediment.

It was found in laboratory experiments that the salt composition of the lake is within the range of the flocculation value for the lake sediments, thus even a slight change in salinity will have a significant effect on the size and settling velocities of the sediment particles. An addition of 1 meq/l of $CaCl_2$ to sediments suspended in lake water led to a 5-fold increase in the median settling velocity of the particles.

Comparing Lake Kinneret water composition during the last 3 decades, it is found that the change in concentration of the divalent ions calcium and magnesium is certainly large enough to have a significant effect on the stability of the sediment suspension. During the period 1960-1980, the divalent cation concentration in the lake was reduced by about 1.7 meq/l. Calcium concentration, also changing along a seasonal pattern, decreases from 2.5 meq/l in the winter to less than 2.0 meq/l in the summer. Internal phosphorus flux from the sediments to the water was found at the summer period and may be related to the lowering of calcium as well as the summer wind pattern.

The flux of resuspended phosphorus–containing particles may also affect the level of soluble phosphorus through the effect on dissolution kinetics.

More work is needed for a direct measurement of the plausible mechanism discussed here.

REFERENCES

[1] Avnimelech, Y. and G. Miller, A fluorometric scanning technique for measuring phytoplankton sinking rates. J. Freshwater Ecology 1: 629-636, 1982.

[2] Gaifman, Y. and H. Dexter, Lake Kinneret watershed monitoring report, Mekorot Water Co., Nazareth, Israel (Mimeogr.) 1984.

[3] Gophen, M., Lake Kinneret limnological laboratory annual report (Mimeogr.), 1984.

[4] Oren, O.H., Physical and chemical characteristics of Lake Tiberias Water Planning of Israel Ltd., No. 1445/58, 1957.

[5] Serruya, C., The lacustrine environment, pp.123-216. In: Serruya C. (Ed.) Lake Kinneret, Dr. W. Junk Publ. The Hague, pp. 501, 1978.

[6] Smith, S.V., Serruya, S. Geifman, Y. and Berman, T., Annual mass balance of water, P,N, Ca and Cl in Lake Kinneret (Israel). Limnol. and Oceanogr. In Press.

[7] Verwey, E.J.W. and Overbeek, J.T.G., Theories of the stability of lyophobic colloids. p.119. Elsevier N.Y. 1948.

[8] Grim, R.E., Clay nineralogy. p. 596., McGraw Hill Pub. N. Y., 1968.

[9] Nishiri, unpublished data, 1988.

ENVIRONMENTAL POLLUTION IN THE VICINITY OF A WASTE-GYPSUM LANDFILL

R. SZPADT

Wrocław Technical University
50-370 Wrocław, Wybrzeże Wyspiańskiego 27, Poland

Z. AUGUSTYN

Wrocław Geological Enterprise
Kołłątaja 24, 50-007 Wrocław, Poland

ABSTRACT

The manufacture of wet phosphoric acid from apatite gives rise to solid and semi-solid wastes (waste gypsum, sodium fluosilicate and wastewater sludges) which should be classified as hazardous wastes and disposed of accordingly.

Both groups of wastes contain calcium sulphate hemihydrate and dihydrate, sodium fluosilicate, sodium sulphate, calcium sulphate, calcium fluoride, sodium fluoride, phosphoric acid and sulphuric acid. Rare elements (yttrium and lanthanide series) are the most valuable compounds of waste gypsum.

In Poland, the landfills where those wastes are disposed of meet neither legislative regulations nor technological and environmental requirements. This practice brings about serious soil and groundwater contamination in the landfill vicinity.

Thus remedial and preventive measures should be taken immediately to abate present and future damage. Two ways to go are suggested. One of these

Chemistry for the Protection of the Environment
Edited by L. Pawlowski *et al.*, Plenum Press, New York, 1991

- a long-term option - consists in the processing of waste gypsum toward recovery of rare elements and production of building gypsum. The other one - a short-term option - aims at modifying the technology of landfill operation and taking current preventive measures (compacting the waste materials, covering the landfill slopes with an inert material) so as to minimize top dust blowing, as well as surface water runoff or infiltration.

INTRODUCTION

Solid wastes and sludges generated during manufacture of wet phosphoric acid are amongst the most harmful waste materials of the inorganic industry. They should be classified as hazardous wastes and disposed of accordingly.

Actually, most of those wastes generated in Poland have been landfilled without taking any safety measures. The landfill itself accounts for serious environmental damage in the immediate vicinity. This paper gives the results of a case study of landfill behavior and related environmental pollution.

CHARACTERIZATION OF THE WASTES

The manufacture of wet phosphoric acid (and its derivatives) from apatite gives rise to three major groups of wastes (included are also those produced during lime treatment of process wastewater) :

- waste gypsum (in specialized literature also referred to as phosphogypsum), i.e. calcium sulphate, unit load of 2.50 kg $CaSO_4$/kg H_3PO_4 (2.67 kg $CaSO_4 \cdot 1/2H_2O$ or 3.17 kg $CaSO_4 \cdot 2H_2O$),
- sodium fluosilicate (in sludge form), unit load of 0.08 kg Na_2SiF_6/kg H_3PO_4,
- sludges from lime treatment.

Before 1984, only calcium sulphate dihydrate was landfilled. Since 1984, as a result of some modifications in the manufacturing technology, the landfill has received calcium sulphate hemihydrate with a small admixture of dihydrate. Both sludges have been mixed and removed to one of the two lagoons for temporary storage.

The chemical composition and physical properties of waste gypsum as well as of the sludge mixture are presented in Table 1. As shown by these data, calcium sulphate is the main component of waste gypsum. The sludge mixture consists primarily of sodium fluosilicate, sodium phosphate as well as calcium sulphate and calcium phosphate [1,2]. Both groups of the

Table 1. Typical Chemical Composition of Wastes and of their Water Extracts [1,2]

Parameter	Unit	Waste gypsum from current production	Waste gypsum from landfill	Mixed sludges
Water content at 50° C	wet wt. %	20.2	11.2	-
Water content at 105° C	wet wt. %	25.3	23.3	61.2
Loss on ignition at 600° C	dry wt. %	6.6	6.1	12.8
Calcium	dry wt. %	22.5	21.2	9.4
Iron	dry wt. %	0.04	0.10	1.8
Sulphates	dry wt. %	58.8	53.5	-
Water-soluble phosphates, PO_4^{-3}	dry wt. %	1.15	1.39	3.73
Water-soluble fluorides, F^-	dry wt. %	0.08	0.21	0.19
RE_2O_3	dry wt. %	0.62	0.62	-
Total water holding capacity (Field capacity)	wet wt. %	54.0	40.0	-
Water retention capacity	wet wt. %	28.7	13.8	-
Water extracts				
pH		2.6	1.7-4.7	4.1-5.1
Conductivity	mS/cm	3.8	2.8	2.3
Sulphates	g SO_4^{-2}/m^3	2328	1649	824
Fluorides	g F^-/m^3	57	39	67
Phosphates	g PO_4^{-3}/m^3	860	974	1460
Calcium	g Ca/m^3	800	577	88

REMARKS: All data expressed in % of dry wt. are related to the drying residue at 105° C. Water extracts were prepared by shaking 100 g of raw wastes with 1 liter of distilled water.

waste materials contain CaF_2, SiO_2, NaF, H_3PO_4, H_2SO_4 and some rare elements (RE), i.e., yttrium and the lanthanide series (La, Ce, Pr, Nd, Sm, Eu, Gd, Tb, Dy, Ho, Er, Tm, Yb, Lu) in an average amount of 0.62 % of dry wt., expressed as the sum of oxides (RE_2O_3). Figure 1 shows the average concentrations of particular RE in wastes and in groundwater.

It is worth pointing out that waste gypsum from the processing of apatite excavated on the Kola Peninsula (Soviet Union) displays a very low radioactivity level (comparable with that of natural gypsum) [3].

The pHs of the wastes and of their water extracts are strongly acidic due to the presence of sulphuric and phosphoric acid which are leached first from the wastes.

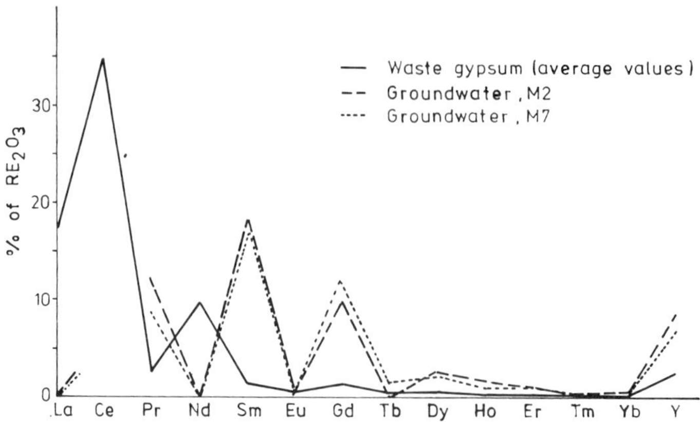

Fig. 1. Comparison of RE contents in waste gypsum and groundwater.

Waste material from the actual production contains mainly calcium sulphate hemihydrate and has a moisture content of ca. 20 wt.% (at 50° C).

Water holding capacity amounts to 54% of the initial wet weight and comprises:

- actual moisture content (ca. 20 wt.%),
- hydration ability (max. 14 wt.%),
- water retention capacity (ca. 20 wt.%).

CHARACTERIZATION OF THE LANDFILL SITE

The landfill under study covers an area of ca. 11 hectares. Figure 2 shows the landfill site and its nearest vicinity. Till 1970, only lime sludges and clinker wastes were disposed of in the landfill. They now create the bottom layer of the higher part of the landfill. Since 1970, waste gypsum has also been deposited on the landfill, thus covering the alkaline bottom wastes. Since 1984, mixed sludges (partly dewatered and removed from a filled lagoon) have been disposed of in the lower part of the landfill.

Actually, the total weight and the total volume of the deposited material is determined as amounting to ca. 2.0 million tons and ca. 1.6 million cubic meters, respectively.

The landfill subsoil is a glacial outwash aquifer with an average thickness of ca. 7 m (Figure 3). It consists of beds and lenses of fine to coarse sand and gravel with thin lenses and beds of fine to medium sand and silt interbedded with the coarser material. The glacial outwash aquifer is underlain by a glacial till.

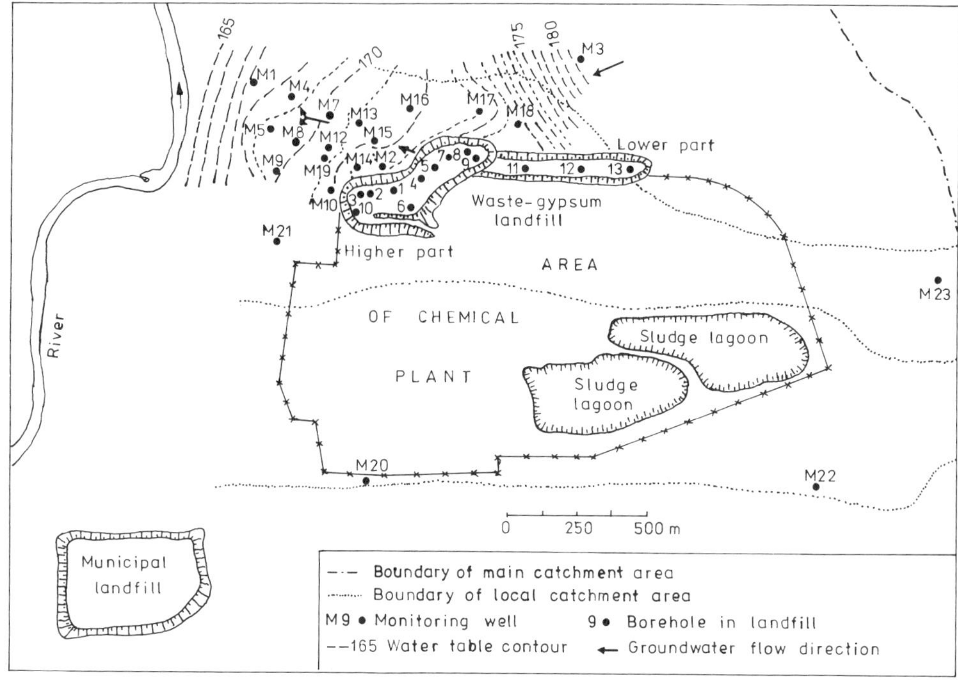

Fig. 2. Landfill site and its vicinity.

The material deposited in the landfill was sampled in 13 boreholes at every 0.5 m of depth. 21 monitoring wells were located in the vicinity of the landfill for groundwater sampling. Soil samples were collected near the monitoring wells.

ENVIRONMENTAL IMPACT OF THE LANDFILL

Investigations in the vicinity of the landfill have shown that soil and groundwater are highly polluted in terms of acidity and increased contents of sulphates, phosphates and fluorides. The highest concentrations were measured in soil samples taken predominantly at the monitoring wells M2, M10, M14, and also M12, M16 and M17, viz., in the nearest vicinity of the landfill. This suggests that slope runoff and dust blowing and deposition are the main sources of soil contamination. Water-soluble contaminants are leached with rainwater and migrate to the deeper zones of the soil profile and to groundwater. Some pollutants, as for example fluorides and sulphates, are

LANDFILL

Waste gypsum – white, dusty, slightly wet
Waste gypsum – white, dusty, wet
Waste gypsum – white-grey, wet
Deposited alkaline material

SUBSOIL

Cultivated soil
Fine and medium sand with gravel
Coarse sand and gravel with cobbles
Grey-yellow sandy loam
Organic dusts and aggradate muds
Grey-brown glacial till
Water bearing zone
0.05 Concentration of RE_2O_3 in soil, dry weight %.

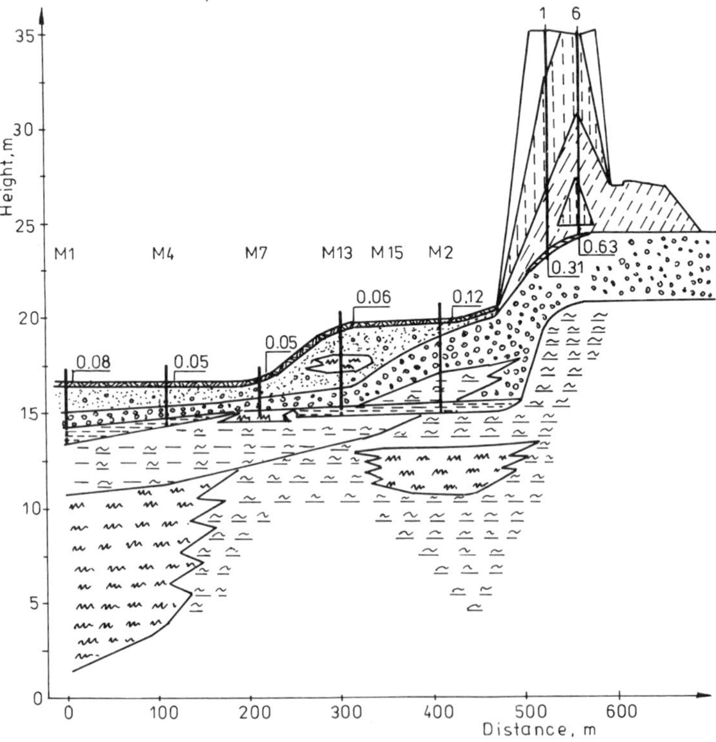

Fig. 3. Hydrogeological cross-section of the waste-gypsum landfill and of its subsoil.

Table 2. Typical Chemical Composition of Groundwater and Soil in the Landfill Vicinity

Monitoring well	Groundwater				Soil (mg/kg dry wt.)		
	pH	Sulphates g/m^3	Fluorides g/m^3	RE_2O_3 g/m^3	Sulphates	Fluorides	RE_2O_3
M 1	6.5-7.3	727	0.09	1.10	2800	9	800
M 2	5.6-6.8	1271	0.13	0.48	3000	12.4	1200
M 3	6.5-6.8	1031	0.11	0.51	-	-	400
M 4	6.4-7.2	767	0.13	0.64	-	-	500
M 5	6.1-7.0	216	0.15	1.01	-	-	-
M 7	6.0-6.3	926	0.02	0.59	trace	10	500
M 8	6.3-7.1	845	0.18	0.40	-	-	-
M 9	5.8-6.7	1213	0.10	0.46	3200	15	-
M 10	5.4-5.9	1400	0.06	0.91	4200	28	-
M 12	5.8-7.0	883	0.14	0.47	3300	7.4	600
M 13	5.9-7.1	976	0.17	0.68	-	-	-
M 14	6.2-6.6	855	0.07	0.38	4400	9.7	-
M 15	5.5-6.5	1186	0.10	0.67	-	-	-
M 16	5.7-6.3	1174	0.42	0.91	2900	11.2	-
M 17	3.9-4.3	2022	2.07	0.38	-	12.6	-
M 18	6.0-7.1	1162	0.10	0.46	-	-	900
M 19	4.9-6.5	949	0.07	0.42	-	-	-
M 20	6.6-7.6	525	0.12	0.30	-	3.6	600
M 21	7.3	915	n.d.	-	-	-	-
M 22	6.7	425	n.d.	-	-	-	-
M 23	6.6	224	n.d.	-	-	-	-

n.d. - not detected

retained by the clay minerals. Analysis of water extracts from soil materials (sampled at various depths at the monitoring well M10) has confirmed these observations (Fig. 4).

The highest concentrations of soluble pollutants were measured in water samples from the monitoring well M17, located at the lower part of the landfill (1903 - 2131 g SO_4^{-2}/m^3, 0.26 - 2.79 g F^-/m^3, pH of 3.92 - 4.31). Water samples from the monitoring wells located at the higher part of the landfill also showed considerable concentrations of pollutants (513 - 1540 g SO_4^{-2}/m^3, 0 - 0.3 g F^-/m^3, pH of 5.45 - 7.20). Natural groundwater of the landfill area is characterized by relatively high concentrations of calcium sulphate (225 - 425 g SO_4^{-2}/m^3, 130 - 175 g Ca/m^3). Thus, only a certain portion of calcium sulphate downstream of the landfill should be attributed to its impact.

The general groundwater pollution in the vicinity of the landfill comes from the following major sources [2,4]:

- leachates and excess water from the sludges deposited in the lower

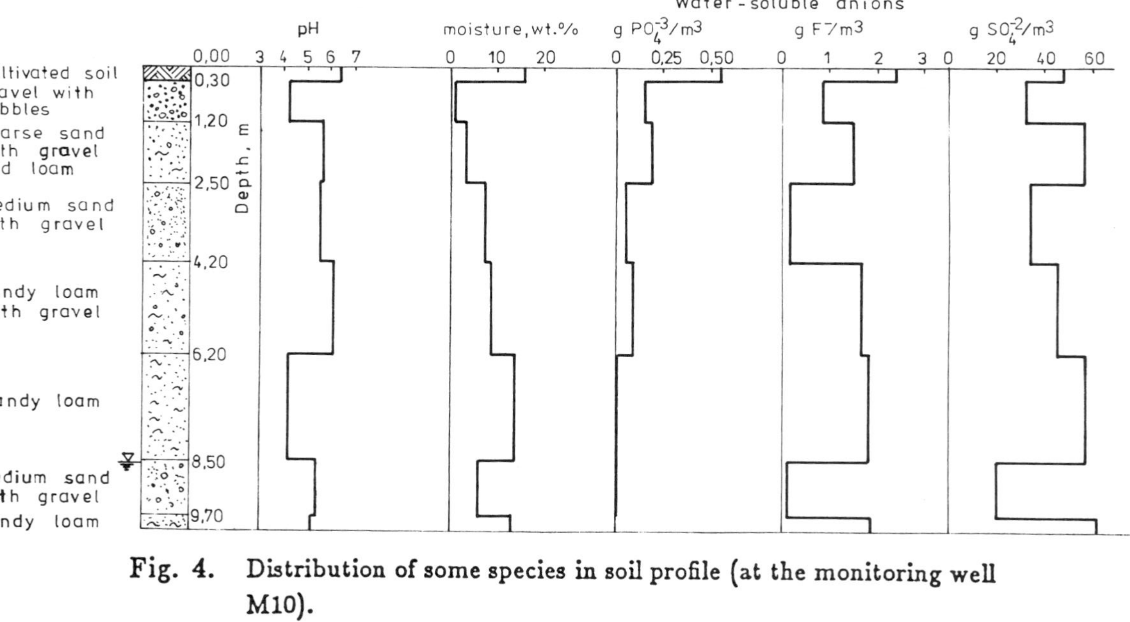

Fig. 4. Distribution of some species in soil profile (at the monitoring well M10).

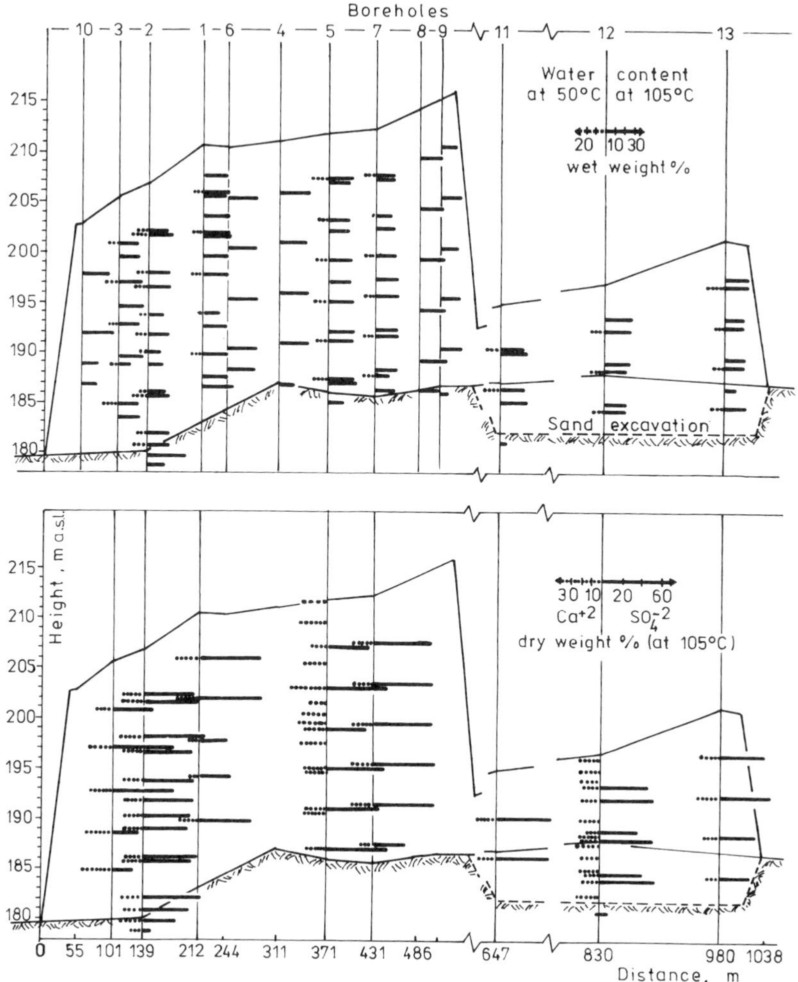

Fig. 5. Distribution of calcium, sulphates and water content in landfill profile.

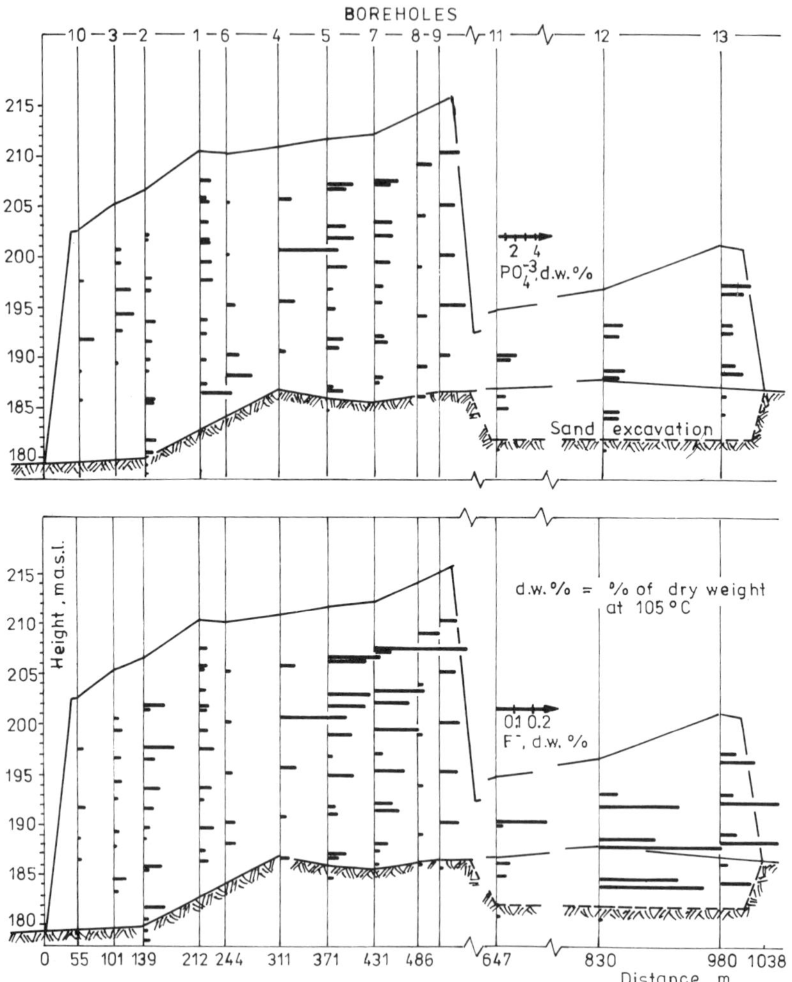

Fig. 6. Distribution of water-soluble fluorides and phosphates in landfill profile.

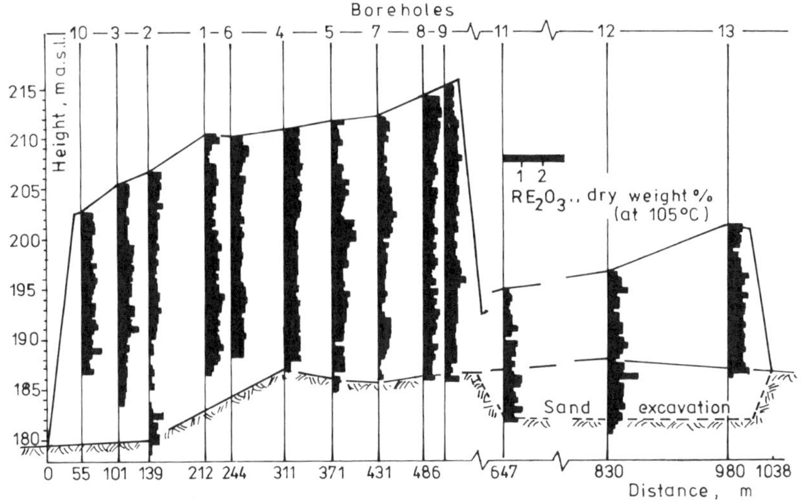

Fig. 7. Distribution of RE in landfill profile.

part of the landfill located predominantly in a sand excavation,

- direct contact of waste gypsum with the groundwater on a bottom of the sand excavation,
- runoff from landfill slopes,
- runoff from roads and from the area of the nearby chemical plant,
- leaching of dust deposited on the soil surface in the landfill vicinity,
- infiltration of excess water from sludge lagoons.

Direct infiltration of rainwater into the waste gypsum deposit of the higher landfill part, or the escape of pollutants by leaching, has only a slight effect on the total groundwater pollution.

The infiltration process has been considerably limited by the disposal of calcium sulphate hemihydrate which formed a cover on the landfill top and on the slopes. The hydration of hemihydrate to dihydrate yielded a solid crust, thus promoting more intensive runoff. Other factors limiting the infiltration effect are the low permeability, the high water holding capacity and the high capillary rise of the well compacted waste gypsum.

Owing to these advantageous properties of the waste gypsum the total annual precipitation of 0.655 m can be stored in the 1.6 m thick compacted gypsum layer, irrespective of water evaporation and runoff. Total water content of waste calcium sulphate hemihydrate at its field capacity amounts to 43 wt.% (the actual water content of these waste materials approaches 25 wt.%). Waste material sampled from the boreholes (dihydrate) showed a total water content of 20 to 33 wt.% (Figure 5), with field capacity at 36 to 39 wt.%. In both cases, the waste materials still have an available water retention capacity for the storage of infiltrating water. Well compacted waste gypsum is characterized by a very high capillary rise (with a velocity of ca. 0.6 m/h) which promotes water transport from the landfill interior and evaporation from the landfill surface.

Analysis of borehole samples showed no differences in the content of main pollutants between deposited and actually generated waste materials. No regularity was found to occur in the vertical distribution of moisture, water-soluble phosphates and fluorides, as well as the RE in deposited waste gypsum (Figures 5, 6 and 7). This should be attributed to the very low rate of rainwater infiltration into the deposited material and to the poor vertical transport to the landfill bottom.

PREVENTIVE AND REMEDIAL MEASURES

Deposited waste material is considered as a prospective source of RE and building gypsum. The stripping and processing of waste gypsum is the best method of landfill remediation. A joint concept of the processing of waste gypsum has been developed, but its implementation requires extremely high investment efforts [3].

Taking this fact into account, short-term preventive and remedial measures have been proposed.

They can be itemized as follows:

- construction of a ditch all along the landfill slopes to collect runoff water and to treat it jointly with the effluents from the nearby chemical plant,
- strong compacting of waste gypsum during its incorporation into the landfill mass in order to minimize top dust blowing and maximize the rainwater retention and evaporation from the landfill top layer,
- covering the sludges deposited on the lower part of the landfill with waste hemihydrate to reduce leaching of easily soluble pollutants,
- covering the landfill slopes, using earth and stabilized waste material from municipal landfill and planting natural vegetation (grass).

All these measures should prevent further environmental degradation in the landfill vicinity.

CONCLUSIONS

1. In terms of acidity, water solubility and concentrations of harmful species (mainly fluorides), waste gypsum and sludges from manufacture of

wet phosphoric acid should be classified as hazardous wastes.

2. Environmental pollution in the landfill vicinity comes predominantly from top dust blowing (and deposition on the soil), from slope and near-slope runoff, as well as from the leaching of pollutants as a result of direct contact with groundwater in the lower part of the landfill.

3. Water infiltration and percolation through landfill has no significant effect on the level of groundwater pollution.

4. The best method of landfill remediation is stripping of the deposited material toward recovery of rare elements and building gypsum.

5. Short-term remedial measures must be undertaken to prevent further environmental damage in the landfill vicinity.

ACKNOWLEDGMENTS

This study was part of the Central Research Programs No. 03.08 and 03.11 sponsored by the Polish Government.

The authors are greatly indebted to their co-workers affiliated with the cooperating institutions for valuable assistance.

REFERENCES

[1] Augustyn Z., Rare-Elements Resources in the Phosphogypsum Landfill. Report of the Wrocław Geological Enterprise, 1988, (In Polish).

[2] Szpadt R., Environmental Impact of the Phosphogypsum Landfill. Report of Inst. of Env. Prot. Eng. Tech. Univ. of Wrocław, 1988, (In Polish).

[3] Kijkowska R., Kowalczyk J., Mazanek C., Pawlowska-Lozińska D., Apatite Phosphogypsum - Material for Obtaining Rare Elements Gypsum. Wydawnictwo Geologiczne, Warszawa, 1988, (In Polish).

[4] Kozerski B., Burcz T., Wojtkiewicz P., Hydrogeological Aspects of Phosphogypsum Landfill in Wiślinka. Proc. of the 4. Conf. on Actual Problems of Hydrogeology. Gdańsk, pp. 69-80, 1988, (In Polish).

PHYSICOCHEMICAL TREATMENT: ION EXCHANGE

MEETING NEUTRAL EFFLUENT REQUIREMENTS IN MODERN ION EXCHANGE DEMINERALISERS

J. FARRAR

The Permutit Company Limited London, England

ABSTRACT

This paper discusses the latest techniques used to meet neutral effluent requirements when producing demineralized water by ion exchange. A brief introduction is included to give some background to those unfamiliar to the techniques used in producing demineralized water. A short cycle ion exchange (Scion^R) design is discussed and examples used to demonstrate the advantages of reducing the cycle times of water demineralizers. Variations of effluent pH with different residence times are given so that effluent handling equipment size may be optimized.

INTRODUCTION

Most demineralized water used today is produced from two stage ion exchange processes. The first stage generally uses a strong acid cation ion exchange resin which is in the hydrogen form. This results in the exchange of

Chemistry for the Protection of the Environment
Edited by L. Pawlowski *et al.*, Plenum Press, New York, 1991

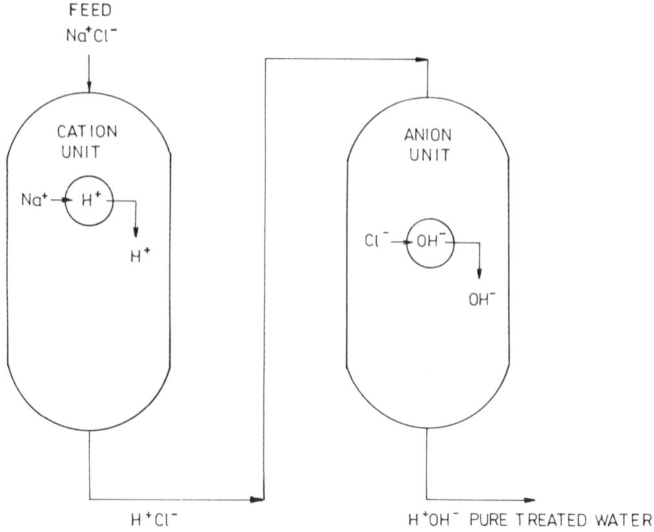

Fig. 1. Flowsheet of typical ion exchange system.

all the cations in the feed producing the corresponding acid. This acid is then passed through the second stage anion exchanger where the anions associated with the acid are exchanged for hydroxide ions. These combine with the hydrogen ions from the acid to give pure water. This is best illustrated in Fig. 1.

The resins eventually become "exhausted" with all the available exchange sites being "used up" and the plant has to be regenerated. This is usually carried out using hydrochloric or sulphuric acid on the cation unit with sodium hydroxide being used on the anion. Originally the regenerant was passed through the resin in the same direction as the "service" (or exhaustion flow); however, although this is simple it is inherently inefficient. The more efficient way is to pass the regenerant in the opposite direction to the service flow. This method of regeneration is termed "counter current" or "counter flow." These counter current regeneration (CCR) techniques were developed through the 1960's and 1970's and after initial problems these are now well established. Permutit offers five different counter flow regeneration systems. Each has its own particular optimum application but all produce similar treated water qualities at comparable chemical efficiency. For completeness a brief description of the various systems is now given.

WATER HOLD DOWN SYSTEM

If the regenerant is passed upflow through the resin bed and then out through a connection in the top of the unit the bed will expand and mix in the same way as it does during backwashing. This is prevented from happening in the water hold down system by positioning a header and lateral collecting system, similar to that used in mixed beds, just below the top of the resin bed. During the regenerant injection a blocking flow of water is introduced into the top of the unit, the upward flowing regenerant and the blocking flow leaving the column via the sub-surface collector.

Providing that an adequate downflow is used, the pressure drop the through resin above the collector plus the bridging effect around the collector generates sufficient counter pressure to overcome the tendency of the resin bed to fluidize.

The sub-surface collector is usually covered with a depth of 100–150 mm of resin which is not regenerated and hence not utilized in the exchange cycle. In some installations this resin is replaced by inert resin or polyethylene chips to reduce costs.

A reverse flush through the sub-surface collector is included before the chemical regeneration stage to remove any particulate matter filtered out on the bed surface during the previous service run.

Occasionally a whole bed backwash is required to alleviate pressure drop build up and a double regeneration would be required to restore the treated water quality and operating capacity of the unit.

The major drawback with this system is the large volume of additional waste water arising from the counterflow and the additional costs of providing larger effluent sumps, pumps, etc. Nowadays its use tends to be restricted to situations where a separate weak acid cation unit is located upstream of the CCR strong acid cation unit, such that the diluted waste sulphuric acid from the water hold down system can be thoroughfared directly to the weak cation unit at the optimum concentration.

AIR HOLD DOWN SYSTEMS

This system is similar in many respects to the water hold down system except that the counter pressure required to prevent bed fluidisation during regeneration is provided by a flow of low pressure air. In this instance the unit has to be drained down to the level of the sub-surface collector prior to starting the regenerant upflow (and subsequently refilled at the conclusion of the rinse stage).

Generally speaking this process has replaced the water hold down

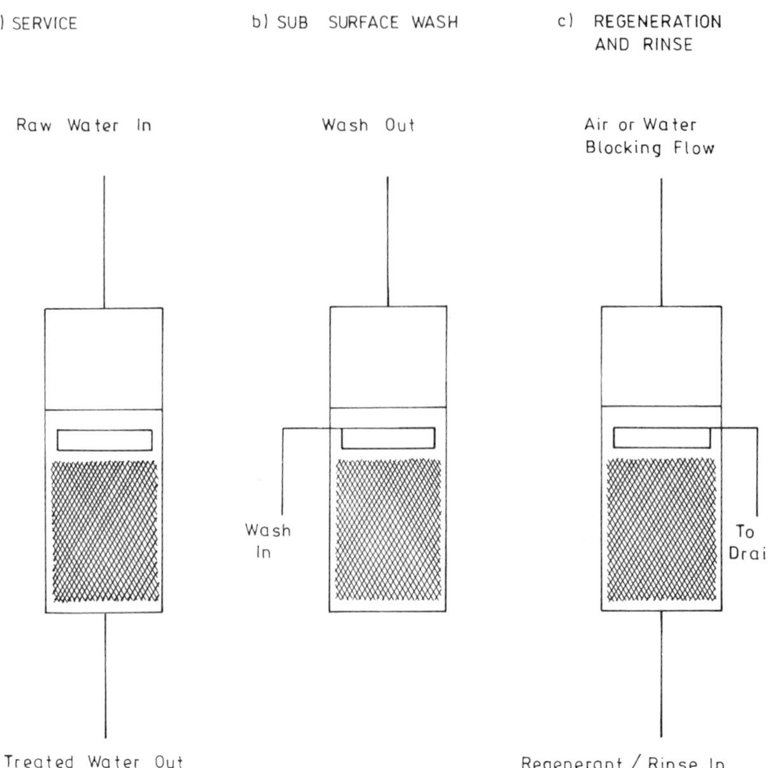

Fig. 2. Air or water hold down CCR techniques.

system particularly for layered bed designs using sulphuric acid regeneration. This is because of the very large waste water volumes that water hold down systems produce (see Fig. 2).

SPLIT CCR SYSTEM

This is a simple development from the water hold down system in that the bed hold down pressure is provided by part of the regenerant flow instead of a separate supply of water. The technique involves burying the spent regenerant collector deeper in the resin bed (usually one quarter of the way down the bed). The regenerant flow is divided into two streams. One part flows upwards through the lower part of the bed, the other part simultaneously flows downwards through the upper part of the bed and acts as the blocking flow. Both regenerant streams leave the unit via the buried collector.

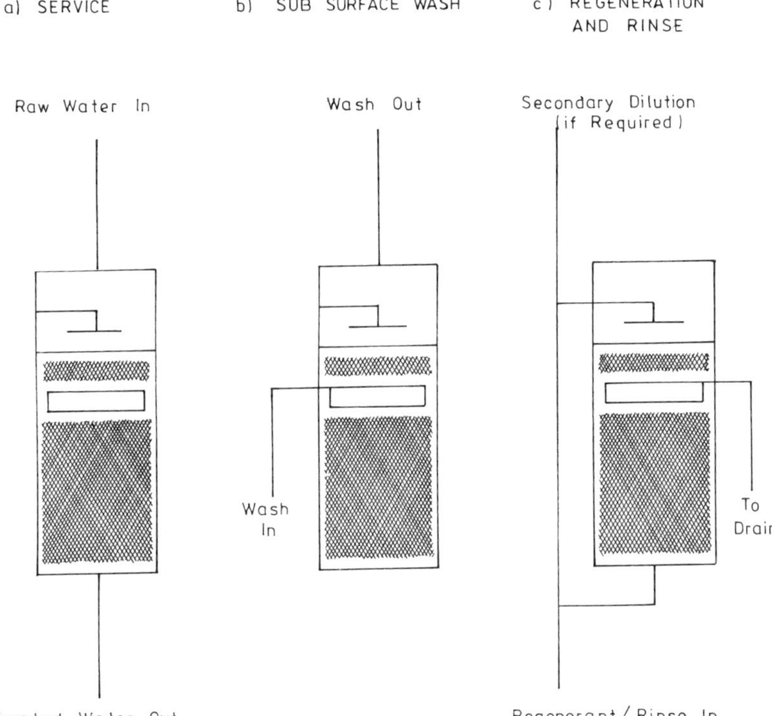

Fig. 3. Split CCR technique.

The bottom three quarters of the resin bed use this regenerated counter current and the top one quarter of the bed above the collector is regenerated coflow with respect to the downward service flow. All the resin in the unit is now utilized in the exchange cycle and the waste water consumption is substantially reduced compared to the water hold down system (see Fig. 3).

There is usually very little difference in installed costs, running costs or treated water quality between the air hold down and split CCR systems. Quite often the choice is dictated by customer preference.

COMPACT-ION SYSTEMS

This is our name for a downflow regeneration, upflow service packed bed column designed to maximize use of the column volume and minimize

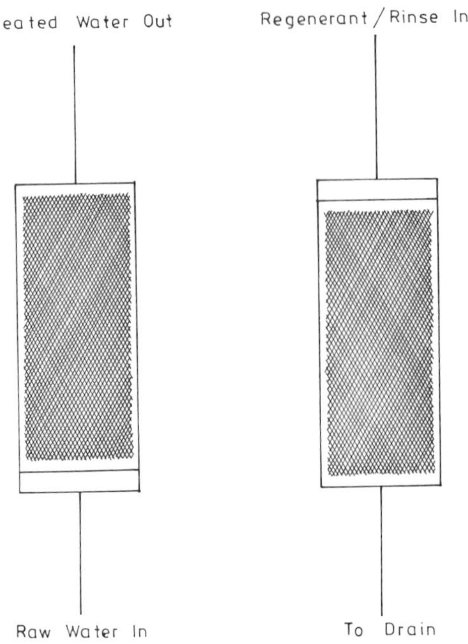

Fig. 4. Compact-ion CCR system.

equipment costs. Developed originally as a water softener its application was extended to two stage demineralising in the mid 1970's.

We also offer a system using the reverse of the compact ion, i.e., downflow service, upflow regeneration.

As the units are full of resin, backwashing and cleaning of the bed is impossible so the feedwater must be free of suspended solids. Alternatively a separate resin backwashing tank can be provided with facilities for transferring some or all the resin out of the operator vessel for occasional cleaning.

When these counter flow regenerated systems were first developed they gave about 30% savings in regenerant costs and effluent volumes over the old "coflow" system and over the intervening years a few more percent has been trimmed off. However, there was very little change in the overall design from the mid 1970's onwards. Most designs have typical exhaustion periods of 8–24 hours with standby units which are brought into use when a unit is regenerating. This means that the plants tend to be large with big resin volumes and that the effluent is produced in large volumes two or three times a day. This in turn results in large effluent handling tanks and associated equipment.

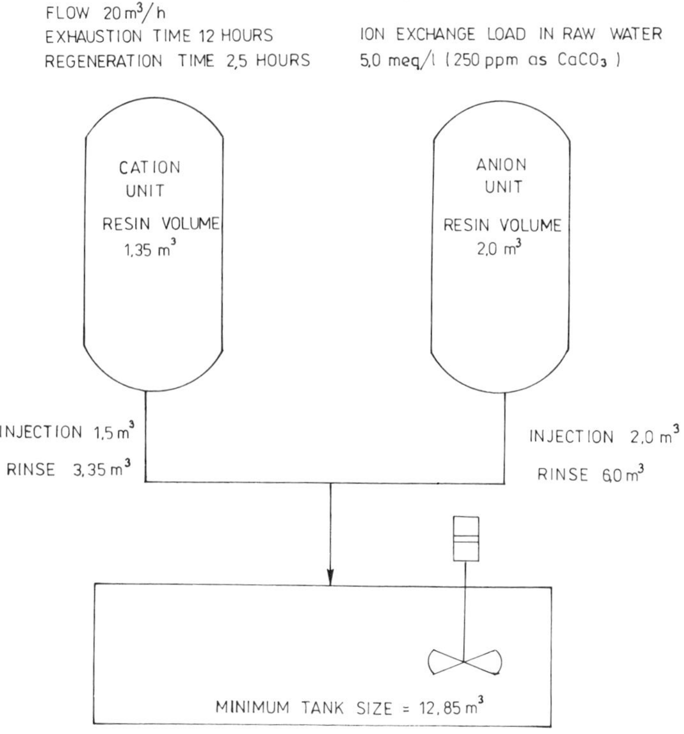

Fig. 5. Typical conventional plant design.

The discharge of effluents from demineralizers is normally only limited by pH. Various limits are used, anything from pH 5 – 11 to pH 6.0 to 7.0 may be encountered. Most modern plants are designed with effluent pH limits in mind. This usually means that the excess chemicals arising from the regeneration of the resins are chemically balanced. If this is so and the whole of the regeneration effluent is mixed, very little (if any) additional chemical will have to be added to achieve the desired pH. It is probably best to illustrate the volumes involved by siting an example. (See Fig. 5)

This then is a typical design for a two-stage demineralizer. It could have been designed in 1972, 1982 and even 1989.

SCION (SHORT CYCLE ION EXCHANGE)

We had arrived at a mature stagnant design. It was time to take a fresh look at demineralization. We already had established from our work on continuous ion exchange process that fixed bed units offered the most economic route to producing demineralized water. We undertook extensive lab trials to push ion exchange to its limits. This work was prompted by the development of the Hipol^R [1] and Tripol^R [2] processes. These operated at high flows and taught us much about the kinetics of modern ion exchange resins [3].

During the lab work it was found that by optimizing resin type and bead size it was possible to cut the average run exhaustion time from the usual 8 to 16 hours to between 1 and 3 hours to less than 30 mins. This meant that a drastic cut in the size of demineralization plants was possible.

After proving the initial process would work using small diameter glass columns we built a 23 m^3/h prototype and ran it in our research and development labs for 12 months before installing it in a brewery. We subsequently built 10 further preproduction prototype units. These plants worked well and Scions^R of various sizes are now one of our main product lines, with over 70 plants sold in the last 18 months.

One of the major benefits of reducing the cycle time was to ease the problems associated with effluent disposal. If you can reduce the cycle times from say 12 hours to 2 hours then assuming all other things are equal this results in a reduction of active resin volume per regeneration by the same amount (although the total effluent per day remains the same). Thus one gets a small, regular quantity of effluent rather than a large, difficult to handle, irregular quantity. But if you combine the short cycle with the following:

packed bed counter flow regeneration

equal resin volumes in the cation and anion units

equal flows during regeneration

stoichiometrically equal regenerant quantities

equal chemical injection and rinse times

combined effluent discharge pipework

optimized resin type/grade selection

solid state control

then you have the basis of being able to mix the effluents and produce a dischargeable waste with no additional treatment. The small quantity of effluent that a Scion^R plant produces is best illustrated by referring to the previous example of a conventional counterflow regenerated plant and comparing this to the equivalent Scion^R unit.

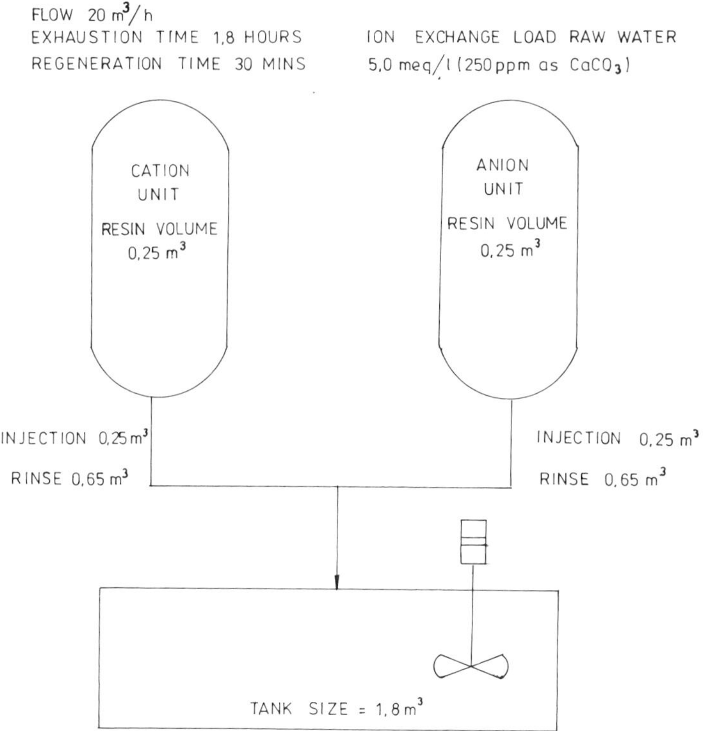

Fig. 6. Typical Scion plant design.

You can clearly see that the Scion^R is a fraction of the size of the conventional unit. The effluent volume is similarly reduced to only 1.8 m^3 from the conventional plants, 12.85 m^3.

In some cases even this size of effluent tank is an embarrassment. In order to reduce the size of the tank even further we have conducted extensive trials using production Scion^R units.

Originally we had believed that as we had "equalized" the cation and anion regeneration flows, resin volumes, regenerant levels, etc., then we would get an approximately balanced effluent at the point where the two waste streams met, i.e., that there would be no need to mix the effluent in a tank in order to meet the desired pH limits. Fig. 7 is a pH trace obtained from just such a trial.

You will note that the pH starts low, levels out then rises at the end of the regeneration period. It was decided to try some simple mixing tanks

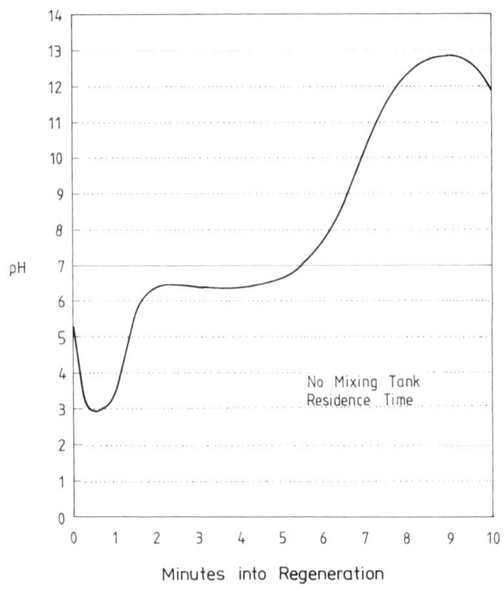

Fig. 7. Scion effluent pH vs time.

without using additional stirrers. These were successful and as Fig. 8 shows, the hold up time in the tank determines the final pH range.

This allows us to tailor the effluent treatment system to suit any particular pH band. Our normal effluent pH requirement is pH 5 to 10. This means that a 4 min. hold up tank is used. This saves considerable space even when compared to the already small Scion effluent volume. If you refer to Fig. 6 you will see that the total Scion effluent volume is 1.8 m^3. A 4 min. hold up tank would reduce that to something like 0.4 m^3, a saving of over 75%. If you compare the size of this tank to the 12.85 m^3 required by the conventional plant (see Fig. 5) the enormous saving that can be made is obvious. All one has to do to realize this startling advantage is to switch to short cycle exchange and then optimize the effluent tank size.

This paper highlights the importance of considering all aspects of running an ion exchange demineralized before choosing a particular design. Effluent requirements are becoming of increasing importance. The short cycle ion exchange process clearly offers great advantages in this important area.

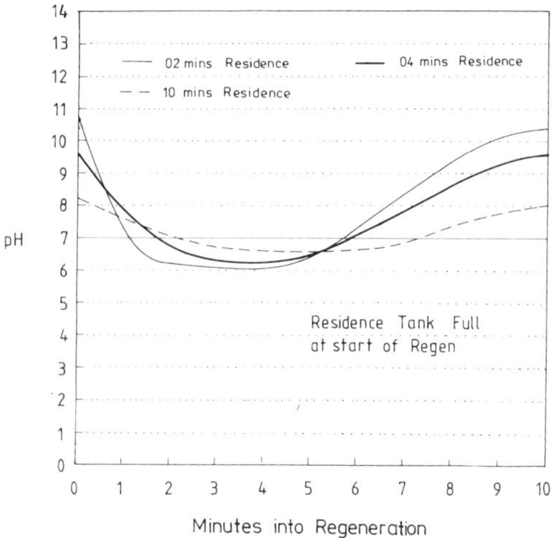

Fig. 8. Scion effluent pH vs time.

REFERENCES

[1] Jackson E.W., and Smith J.H., Make-Up Treatment Counter Current Regeneration Experience in the United Kingdom, 38th International Water Conference, Pittsburgh, USA 1977.

[2] Smith J.H., and Peploe T.A., The Tripol Process - a new approach to ammonia cycle condensate polishing, B.N.E.S. 2nd International Conference on Water Chemistry of Nuclear Reactor Systems, 1980.

[3] Smith J.H., Mixed vs Single Bed Ion Exchange Kinetics, 43rd International Water Conference, Pittsburgh, USA, 1982.

SELF-HEATING HAZARD IN AMMONIUM NITRATE REMOVAL BY ION EXCHANGE

A. BISKUPSKI, A. KOŁACZKOWSKI,
Z. MEISSNER, M. MALINOWSKI
and B. SORICH*

Institute of Inorganic Technology
and Mineral Fertilizers
Technical University of Wrocław
ul. Smoluchowskiego 25 Pl-50-372 Wrocław, Poland
* Nitrogen Plant Kędzierzyn Pl-47-220 Kędzierzyn-Koźle, Poland

ABSTRACT

The possibility of a spontaneous temperature increase in an ion exchange bed in ammonium nitrate recovery from waste water is discussed. The results of laboratory experiments confirmed in full-scale installations are considered to be guidelines for fire and explosion prevention.

INTRODUCTION

The NH_3 and NH_4NO_3 concentration in a steam condensate generated in an ammonium nitrate installation can rise to 5 and 10 g/dm^3, respectively, sometimes even higher. From this condensate it is possible to obtain two important products: demineralized water and an NH_4NO_3 water solution of higher concentration (up to 30% by mass), by means of the ion exchange process [1].

This process consists of reactions described by the following equations for the treatment stage:

(1) $R\text{-}H + NH_4NO_3 = R\text{-}NH_4 + HNO_3$,
(2) $R + HNO_3 = RH\text{-}NO_3$

and for the regeneration stage

(3) $R\text{-}NH_4 + HNO_3 = R\text{-}H + NH_4NO_3$,
(4) $RH\text{-}NO_3 + NH_4OH = R + NH_4NO_3 + H_2O$

where R-H, R and $R\text{-}NH_4$, $RH\text{-}NO_3$ stand for the cation or anion exchanger before and after the treatment stage, respectively. As the quality of the demineralized water obtained in reaction (2) meets the technical requirements of a high pressure boiler, it may be introduced into the water circuit. The ammonium nitrate solution, being the product of reactions (3) and (4), shows a quite significant concentration and is usually recycled into the main production process.

The recovery of ammonium nitrate and demineralized water by means of the ion exchange process may be risky. The danger may be caused by the possibility of an uncontrolled, spontaneous temperature rise in the ion exchange bed, which may lead to the destruction of the ion exchange column and even to fire and explosion. Such occurrences were recorded by Calmon [2], who considered the ammonium nitrate decomposition, as well as the interaction between nitric acid (oxidizer) and ion exchanger (reducer), to be the possible main heat sources in this system.

Laboratory investigations on the self-heating phenomenon [3,4] in ammonium nitrate removal by ion exchange have shown that the greatest temperature increase takes place during the regeneration of the ion exchange bed. The temperature maximum moves along the column according to the direction of flow and increases continuously. A simplified mathematical model of this heat accumulation in an ion exchange bed was proposed by Meissner and Kolaczkowski [5].

The thermal stability of the system H_2O - NH_4NO_3 - HNO_3 - resin, investigated by Biskupski et al. [6,7], is influenced by different components. The decomposition rate of the particular components depends mainly on the nitric acid concentration.The influence of chloride, ferric, and chromate ions is less important and becomes noticeable only in the presence of nitric acid.

In spite of these laboratory–scale experiments it remained unknown whether the parameters of the process in a full-scale installation coincided with really safe ones. Hence the aim of the present investigations was to verify the technical parameters of the process in a full-scale installation.

EXPERIMENTAL

In the laboratory–scale investigations the maximum temperature increase (ΔT) in the ion exchange column did not exceed the value of 6.2 deg, even at high nitric acid or ammonia concentrations, 25% by mass. Similar values of temperature increase were usually noticed in the full-scale installation as well. However, in one case a temperature increase of more than 60°C was reached for some minutes, which is far over the assumed limit. All parameters of the process were checked, especially NH_4NO_3, HNO_3 or NH_3 concentrations in the solutions, and none of them exceeded the limited value. So it was necessary to take into consideration all the factors which may stimulate the temperature increase in the ion exchange bed.

It was assumed that, besides the NH_4NO_3, HNO_3, and NH_3 concentrations, the following are very important factors influencing the self-heating phenomenon: 1. the concentration of Cl^-, NO_x, and Fe ions in the reagents, 2. the working life of the ion exchange resin, 3. the accumulation of impurities in the ion exchange resin.

To establish the influence of these factors on the self-heating phenomenon, not only were the impurities concentration levels determined, but also a special test was performed. It consisted in determining the temperature profile in the ion exchange column during regeneration. The possible trend of changes in such a temperature profile would be regarded as the evidence of the influence of impurities or ion exchanger destruction on the decomposition of ammonium nitrate. As the temperature profile determined in the full-scale ion exchange column was not precise enough, the self-heating phenomenon was also followed on the basis of a temperature profile determined in adiabatic conditions in a laboratory ion exchange column of about 26 mm in diameter and about 1.8 m in height. Thermistors and conductivity sensors arranged every 300 mm along the ion exchange bed made it possible to determine the temperature and composition of reagents.

RESULTS

In a full-scale installation the condensate and effluent from the ion exchange columns during their treatment and regeneration stage were analyzed. The concentrations of ammonium nitrate, nitric acid, ammonia, nitrogen oxides (NOx), organic substances as well as chloride, ferric and ferrous ions are shown in Tables 1 and 2. The concentrations of nitric acid and ammonia were determined by acid-base titration, that of ammonium nitrate by means of the formalin method and that of nitrogen compounds in 3+ and 4+ oxidation states by Saltzmann's method modified by Liskow. For the al-

Table 1. Full-Scale Installation Performance Characteristics for KPS Resin Regeneration

Cy-cle	Concentration maximum									
	Inlet				Outlet					
	HNO_3	NO_2	Σ Fe	Cl^-	HNO_3	NH_4NO_3	NO_2	Fe^{3+}	Cl^-	PV
	[g/dm^3]	[mg/dm^3]			[g/dm^3]		[mg/dm^3]			[mgO_2/dm^3]
4	168	970	3	7	99	147	609	24	7	50
20	191	1020	2	5	190	175	735	4	5	19
46	188	375	7	16	12	153	71	5	9	3
100	189	609	2	37	109	139	441	2	29	7

Table 2. Full-Scale Installation Performance Characteristics for AD-41 Resin Regeneration

Cy-cle	Concentration maximum								
	Inlet*			Outlet					
	NH_3	Σ Fe	Cl^-	NH_3	NH_4NO_3	NO_2^-	Fe^{3+}	Cl^-	PV
	[g/dm^3]	[mg/dm^3]		[g/dm^3]		[mg/dm^3]			[mgO_2/dm^3]
4	158	10	0	15	101	200	3	16	32
19	127	1	0	19	157	190	1	48	20
55	141	2	0	45	90	103	2	34	38
105	161	1	1	38	133	777	1	18	121

* NO_2^- concentration at inlet equal to 0.

kaline solutions the nitrogen concentration was denoted as nitrite and for the acidic solutions as NO_2. The organic substance content was determined on the basis of the permanganate value (PV). Chloride ions were determined by Volhard's method, ferric and ferrous ions as rhodanate using the colorimetric analysis. The ferrous compound content in the ion exchange resin was determined according to the manufacturer's (VEB Chemiekombinat Bitterfeld) recommendation.

The results of some experiments presented in Figs. 1 and 2 show the dependence of effluent component content on the time of regeneration.

The results of laboratory-scale investigations are shown in Tables 3 and 4. There the following main parameters can be found: treatment (W) or regeneration (R) stage, number of cycles (treatment + regeneration stages), the concentration of different components in the reagents before and after the ion exchange process, the velocity of liquid flow, the mole equivalent capacity (MEC) of 6· 300 ion exchanger, and the maximum temperature increase. The temperature changes versus effluent volume for different distances from the bed surface (2 to 6·300 mm, 7 - effluent temperature) are given in Figs. 3 and 4.

Table 3. Results of Laboratory-Scale Experiments for KPS Resin

Cycle	Concentration [mole/dm^3]		Flow rate [m/s]·10^3	ΔT [deg]	MEC [mole/dm^3]	NO_2 [mg/dm^3]	PV [mgO_2/dm^3]
	NH_4NO_3	HNO_3					
0 W	0.138		3.50	0.22	1.83		
0 R		5.64	1.07	1.70			
5 W	0.135		3.50	0.20	1.90	0.0	0.9
5 R		5.64	1.05	1.46		2.9	20.8
20 W	0.130		3.50	0.18	1.84	0.0	0.6
20 R		5.72	1.09	1.53		2.2	16.5
46 W	0.121		3.60	0.18	1.89	0.0	0.7
46 R		5.57	1.07	1.34		6.5	11.8
100 W	0.122		3.50	0.11	1.78	0.0	0.4
100 R		5.78	1.15	1.85		2.2	10.0

Table 4. Results of Laboratory-Scale Experiments for AD-41 Resin

Cycle	Concentration [mole/dm^3]		Flow rate [m/s]·10^3	ΔT [deg]	MEC [mol/dm^3]	NO_2^- [mg/dm^3]	PV [mgO_2/dm^3]
	HNO_3	NH_3					
6 W	0.139		3.60	2.00	1.26	0.1	1.7
6 R		14.3	1.00	5.12		178.8	225.0
20 W	0.137		3.60	2.00	1.22	0.1	2.7
20 R		14.3	1.10	3.90		160.0	49.0
55 W	0.121		4.60	2.00	1.57	1.5	0.9
55 R		14.2	0.96	5.77		1.7	50.0
102 W	0.122		3.50	1.95	1.22	3.6	1.5
102 R		14.3	0.84	6.22		165.0	49.0

DISCUSSION

The investigations reported by Meissner and Kolaczkowski [5] have shown that the greatest temperature increase in the ion exchange bed takes place during the regeneration stage. The temperature maximum moves along the column according to the direction of flow of regeneration liquors and increases continuously (Figs. 3 and 4). The temperature increase in the shifting reaction zone depends on the heat generation in the ion exchange bed volume unit, conditioned by nitric acid or ammonia concentration in the regeneration liquors used. Heat generation may rise abruptly when a significant nitric acid or ammonium nitrate decomposition takes place. It is stimulated by a higher concentration of both these compounds, and also by the impurity content in the reagents or in the ion exchange resin.

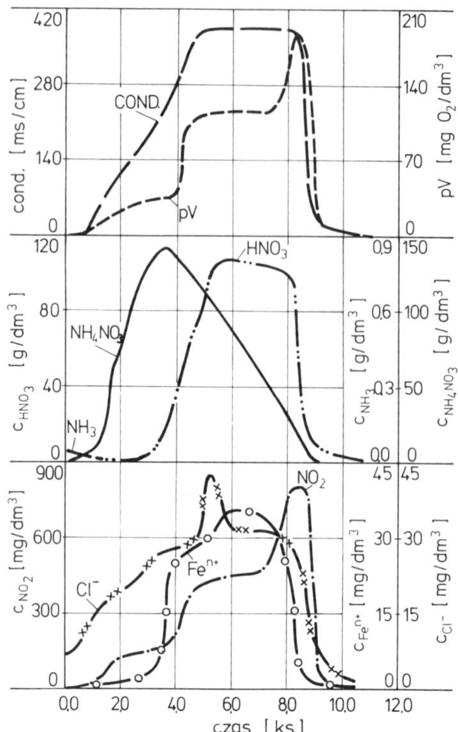

Fig. 1. The dependence of the different component content in the effluent on the time of regeneration for KPS resin.

It is well known that nitric acid, chloride, nitrite and ferric ions, as well as organic substances stimulate the decomposition of ammonium nitrate in solid state and in solution. From Fig. 1 and 2 it follows that for a certain period of time the effluent contains these substances simultaneously in their highest concentration, although the average concentration of these components in the reagents is relatively low. The additional heat generated in the ion exchange bed volume unit as a consequence of ammonium nitrate and nitric acid decomposition caused by high impurities concentrations, contributes to the temperature increase in the shifting reaction zone. Under disadvantageous conditions the heat accumulation in the shifting reaction zone may result in an abrupt temperature rise and initiate a violent decomposition in the ion exchange bed. Therefore it is of great importance for the technical safety of the process to reduce the impurities concentration level.

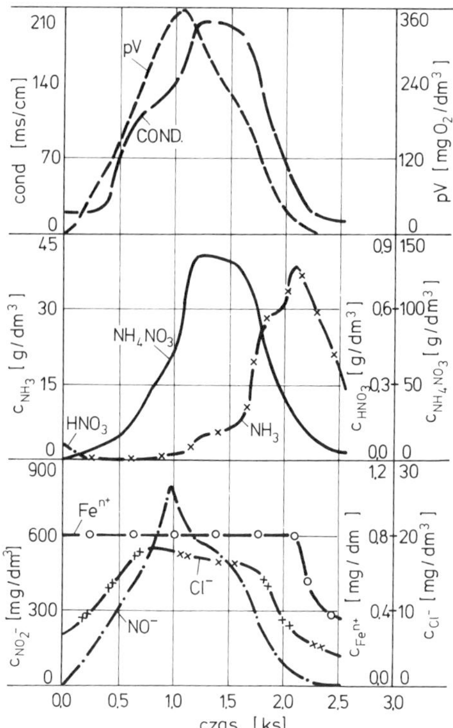

Fig. 2. The dependence of the different component content in the effluent on the time of regeneration for AD-41 resin.

It should be noticed (Table 1) that the main source of those impurities may be nitric acid which contains up to about 1000 mg/dm^3 NO_2, 40 mg/dm^3 Cl^-, and 7 mg/dm^3 Fe^{3+}. The steam condensate (Table 5) also contains a significant amount of impurities (2.5 mg/dm^3 Cl^-, 13.6 mg/dm^3 NO_2, 6.8 mg/dm^3 NO_2^- and 3.7 mg/dm^3 O_2 permanganate value). So the first step to reduce the nitrite and chloride ion concentration level is to improve the quality of the nitric acid used for regeneration.

The laboratory investigations have confirmed that the ion exchanger work life (number of cycles) has no influence on the temperature increase under standard conditions. So it can be assumed that by this time the ammonium nitrate decomposition has not been affected by the ion exchanger breakdown or a possible impurity accumulation.

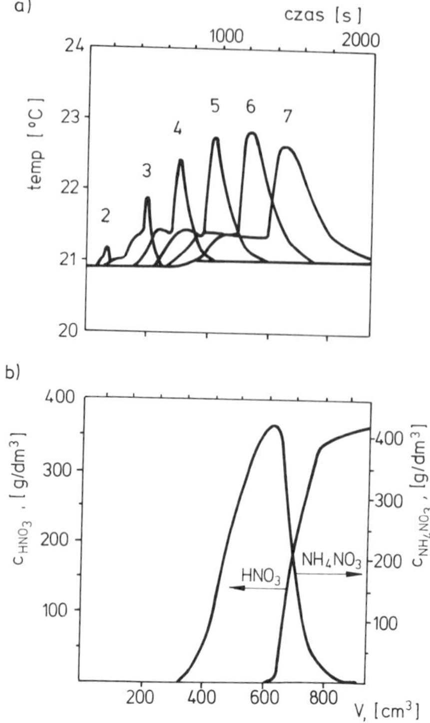

Fig. 3. The temperature changes versus effluent volume for different distances from the bed surface and component concentration at outlet during KPS resin regeneration.

CONCLUDING REMARKS

On the basis of these investigations the short-time temperature increase, noted only once during one year of the ion exchange operation, cannot be exactly explained. However, it is very likely to have been caused by an unusual joint increase of chloride, nitrite, nitric acid and organic substances concentration in the ion exchange bed. It stimulated the exothermic decomposition of ammonium nitrate and resulted in a greater heat accumulation and temperature rise in the shifting reaction zone.

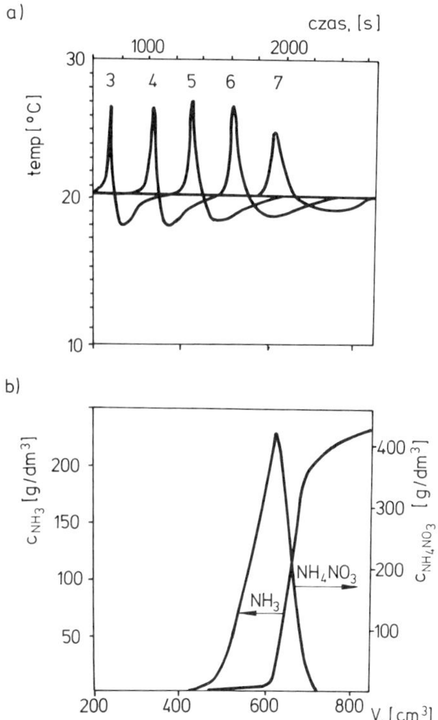

Fig. 4. The temperature changes versus effluent volume for different distances from the bed surface and component concentration at outlet during AD-41 resin regeneration.

Table 5. Some Typical Impurities Concentration in Condensate [mg/dm^3]

NH_3	NH_4NO_3	Σ Fe	Cl^-	NO_x	PV
680	3260	1.0	0.7	4.2	3.7
460	1590	0.3	0.4	7.4	1.8
200	2700	0.3	0.2	7.4	1.9
630	1060	0.4	2.1	11.7	
690	1020	0.4	1.4	12.3	
760	930	0.4	1.8	13.6	
1000	1040	0.4	2.5	13.5	

REFERENCES

[1] Pawlowski, L., Recirculation of the waste components in the form of concentrated post-regeneration solutions of the ion exchange separation. *Uniwersytet im. Marii Curie-Sklodowskiej*, Lublin, 1980.

[2] Calmon, C., Explosion hazards of using nitric acid in ion-exchange equipment. *Chem. Eng.* **17**, 271-274, 1980.

[3] Meissner, Z., Kolaczkowski, A. and Sorich, B., Safety problems in the ion exchange process admitted to the purification of waste waters bearing ammonium nitrate, *Pr. Nauk. PWr. I-26* **31** *Ser. Konf.* **16**, 77-84, 1986.

[4] Meissner, Z., Kolaczkowski, A., Adamczyk, R., An attempt to mathematical modelling of heat phenomena in the ion exchange process, *Pr. Nauk. PWr. I-26* **34** *Ser. Konf.* **18**, 51-55, 1987.

[5] Meissner, Z., Kolaczkowski, A., Safety aspects of the purification of waste water containing ammonium nitrate. *Chem. Eng. Sci.* 43 (8), 2119-2124, 1988.

[6] Biskupski, A., Tatarek, M., Wöjtowicz, K., Chemical composition changes of the solutions in purification of ammonium nitrate containing waste waters by means of ion exchange method, *Pr. Nauk. PWr. I-26* **31** *Ser. Konf.* **16**, 85-94, 1986.

[7] Biskupski, A., Sorich, B., Tatarek, M., Thermal decomposition of chosen ion exchangers and aqueous solutions of ammonium nitrate and nitric acid, *Pr. Nauk. PWr. I-26* **31** *Ser. Konf.* **16**, 95-102 1986.

PHYSICOCHEMICAL TREATMENT: COAGULATION, FLOCCULATION AND SORPTION

THE THERMAL REGENERATION OF EXHAUSTED ACTIVATED CARBON: THE BALANCE BETWEEN WEIGHT LOSS AND REGENERATION EFFICIENCY

R. J. MARTIN[1] and W. J. NG[2]

[1]Department of Civil Engineering, University of Birmingham, England

[2]Department of Civil Engineering, National University of Singapore

ABSTRACT

Following experimental research studies carried out at the University of Birmingham, UK, this paper presents data for different gaseous atmospheres and different temperatures in the thermal regeneration of exhausted activated carbon, and their effects on the balance between minimization of carbon weight loss and maximization of restoration of adsorption capacity to the carbon.

Thermal regeneration is essentially a destructive process. Different gaseous atmospheres were shown to possess different threshold temperatures, below which adsorbates could not properly be volatilized away from the carbon pores. The threshold temperature for a particular gaseous atmosphere must be exceeded for any real success in regeneration of the carbon. Exceeding the threshold temperature, however, is accompanied by carbon weight loss. It was found that reactivation conditions could be selected to increase adsorption capacity per unit weight of activated carbon at the expense of losing some carbon. Repeated cycles of exhaustion and regeneration showed that an emphasis on quality does not yield lasting profit; progressively more make-up carbon must be added to ensure a satisfactory adsorption capacity of the carbon bed in total.

Because carbon losses progressively accelerated with each subsequent cycle, the mildest conditions necessary for the removal of the adsorbate and reactivation of the carbon should be used.

Chemistry for the Protection of the Environment
Edited by L. Pawlowski *et al.*, Plenum Press, New York, 1991

INTRODUCTION

Wherever the concentration of population in urban centres has resulted in pollution and water resource problems, activated carbon adsorption has developed as an important unit process for the removal of a wide range of organic pollutants from waters and wastewaters. Adsorption by granular activated carbon (GAC) beds is by far the most effective technology for the broad spectrum of organics of concern.

Spent carbon must be subjected to regeneration if it is to be used again. The most common technique practiced in regeneration is thermal volatization in which adsorbed impurities within the pores of the exhausted carbon are desorbed by volatilization and oxidation at high temperature. The process of thermal volatilization is characterized by carbon weight loss (perhaps 5-10%) due to oxidation and attrition, and by the cost of energy in heating the carbon to around 800-850°C [1]. Recent studies have shown thermal regeneration unit costs to range from about $1.54 per kg for a very small treatment unit, to about $0.48 per kg for a large reactivator [2]. Reactivation of spent carbon and addition of make-up carbon can account for 50% or more of the total cost (amortized capital expenditure plus operating costs) of a GAC system [2-4].

It is clear that the cost of thermal regeneration must be reduced if activated carbon technology is to develop the wider circle of applications that it could, and should do, as we approach the 21st century. One approach is that of regional regeneration wherein a number of GAC users share regeneration costs using one regeneration facility [2-4]. An alternative approach is to scientifically examine the thermal regeneration process with a view to making it more cost-effective, quite apart from the economies of scale.

Thermal regeneration is, effectively, a balance between minimization of carbon weight loss and maximization of restoration of adsorption capacity to the carbon. Following experimental research studies carried out at the University of Birmingham, UK, this paper presents data on different gaseous atmospheres and different temperatures and their effects on the balance between weight loss and regeneration efficiency.

EXPERIMENTAL PROCEDURE

Before any experimental work was begun, it was resolved that the furnace used for the studies should ideally have the following features.

(a) The oven should be capable of achieving temperatures of at least 1000°C.

(b) The oven should be susceptible to accurate temperature control with uniformity of temperature throughout the oven interior.

(c) The oven should have facilities for the introduction and control of the gaseous atmosphere within the reaction chamber.

(d) The structure of the furnace, and especially that of the reaction chamber, should be tolerant of rapid temperature changes.

(e) The reaction chamber should be inert to the gaseous atmospheres introduced to it.

(f) The reaction chamber should not contaminate the carbon samples introduced to the chamber.

With these features in mind, a number of commercially available laboratory furnaces was investigated. Not only were they too expensive for the limited budget of the authors, they were also not satisfactory for the planned experiments. Consequently, it was concluded that it would be better to design and build a furnace specific to the work on thermal regeneration of exhausted carbon.

The choice of the reaction chamber material was the first step. Metals were ruled out as they would scale at temperatures around 1000°C in the gaseous atmospheres to be used; contamination of the small carbon samples would have made any weight analysis interpretation invalid. There would also be problems of metal expansion and contraction and possible chemical reactivity with organic adsorbates in the exhausted carbon. Vitreous silica glass was eventually chosen as a suitable material. It is chemically inert (or as inert as it is possible to be), and has a very small coefficient of expansion. Silica glass is, however, very fragile and difficult to shape without specialist equipment and these factors influenced the ultimate design of the reaction chamber and, in turn, that of the furnace.

The furnace was designed as a tube furnace. Some features of the furnace are noted here in summary form; a full account of the design and construction of the furnace has been recorded by Ng [5].

A 1 m long silica glass tube (OD 37 mm) was inserted in a 0.75 m long alumina-porcelain tube (ID 39 mm); the latter tube gave the chamber some structural strength. It is worth noting that the alumina-porcelain could not itself be used as the reaction chamber material because it would have been attacked by steam at high temperatures.

The silica glass tube consisted of a central 820 mm length of clear glass. A 40 mm length of translucent silica glass separated this central portion from the outer 50 mm length of clear glass at each end. The translucent sections were put in to act as heat diffusers to maintain comparatively cool ends thereby protecting the end fittings from damage. Six heating elements made of crusilite were arranged around the alumina-porcelain tube to provide a central hot zone of length 230 mm. The power supply to the heating elements was protected with both voltage and current cut-out circuits. A platinum-rhodium thermocouple was used to monitor the temperature of the gaseous atmospheres in the reaction chamber hot zone; this particular thermocouple was chosen because it could operate in oxidizing atmospheres. A chromel-alumel thermocouple was used to monitor the temperature of the carbon. (The gas temperature, T_g, was always lower than the carbon temperature, T_c.) At the ends of the silica tube, silicon rubber O-rings were used to maintain seals between the tube and aluminium flanges.

The flange at the gas entry end was designed so that the gaseous atmosphere could be altered during an experiment. The flange at the gas exit end was fitted with a plate with a 5 mm hole to permit the exit of the gaseous atmosphere.

The tube furnace was surrounded by insulating bricks with a minimum of 75 mm insulation surrounding the tube. The completed furnace was further insulated by boxing in the bricks with 12.5 mm insulation board on all sides. The whole assembly was held together with angle-iron.

Gases were drawn from cylinders and flows measured with in-line flow-meters before entering the furnace.

Filtrasorb 400 granular activated carbon supplied by Chemviron Limited was used throughout the experimental research. It has been extensively used in previous adsorption and desorption research studies by the authors, and has been found to be effective for the adsorption of a wide range of pollutants of small and large molecular weight.

Carbon samples were exhausted with Rhodamine B or humic acid using established techniques [6-9]. Spectrophotometry was used to determine the equilibrium concentrations of the two compounds (Rhodamine B 550 nm/visible range; humic acid 300 nm/UV range). Exhausted carbon samples were stored in a desiccator prior to thermal regeneration and subsequent re-exhaustion. Because the thermal regeneration procedure is being reported here for the first time, a reasonable degree of detail is used to describe the procedure.

An exhausted carbon sample, generally around 150 mg, was put into a fireclay boat of dimensions 50 x 10 x 10 mm. (All fireclay boats were fired to a temperature of 1000°C to remove any contaminants before use.) The boat was quickly introduced to the center of the hot zone, and the exit flange plate fitted in place. The boat and sample were left in the hot zone for 20 minutes after which the boat was gradually withdrawn to allow cooling. This gradual withdrawal followed a set pattern. As soon as the 20 minute heating phase had finished, the boat was was withdrawn to a position 250 mm from the exit end and left there for 5 minutes. The boat was then withdrawn a further 50 mm after 5 minutes; this was repeated such that when the 20 minute cooling phase had finished, the boat was approximately 100 mm from the exit flange plate. From this point, and after a total of 40 minutes residence in the tube, the boat was carefully removed from the furnace and placed in a desiccator. The carbon sample was subsequently weighed when cool.

The regenerated sample was then subjected to re-exhaustion to assess the effects of the thermal regeneration for the specific conditions of temperature and gas composition used.

CALCULATIONS OF WEIGHT CHANGE AND REGENERATION EFFICIENCY

The following weight measurements were taken.

W_o = original weight (150 mg) of virgin carbon or weight of carbon before start of each exhaustion-regeneration cycle.

W_d = weight of dried exhausted carbon before regeneration.

W_r = weight of regenerated carbon.

Let W_1 be the difference in weight between W_o and W_r . Let W_2 be the difference in weight between W_d and W_r.

$$W_d > W_o > W_r \quad \text{in general}$$

W_1 is the loss in carbon weight following regeneration whereas W_2 is the loss in weight of both carbon and adsorbate. Both W_1 and W_2 are usually presented as percentage weight losses.

$$W_1(\%) = \frac{W_o - W_r}{W_o} \times 100$$

$$W_2(\%) = \frac{W_d - W_r}{W_d} \times 100$$

The following method of calculation was employed to quantify regeneration efficiency. The original capacity (A_o) of the carbon for a particular adsorbate was deemed to be that quantity of solute adsorbed from solution per unit

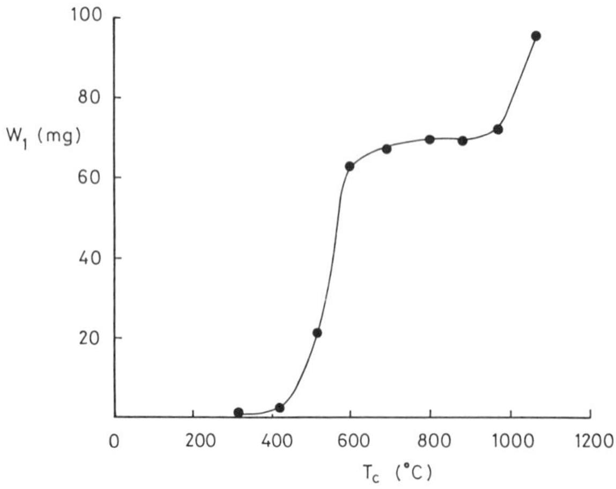

Fig. 1. Weight loss v. temperature.

weight of carbon (experimental conditions chosen to ensure equilibration and exhaustion). The capacity of the regenerated carbon (A_r) was deemed to be that quantity of the same solute adsorbed from solution per unit weight of regenerated carbon (again ensuring both equilibration and exhaustion).

$$\text{Regeneration efficiency } (RE\%) = \frac{A_r}{A_o} \times 100$$

This method of calculation has been successfully used by the authors in previous studies [6-9]. It is simple and readily shows up anomalous values.

RESULTS AND DISCUSSION

A series of preliminary tests was carried out in order to test the furnace and operating procedures.

In test 1, virgin carbon samples were heated over a range of temperatures in an open-system, i.e., no inlet gases and with the exit flange plate not fitted. The most immediate fact to emerge was that the platinum-rhodium thermocouple, positioned just inside the hot zone, registered a lower temperature than did the chromel-alumel thermocouple. Weight loss data (W_1) confirmed the expectation that weight loss would increase with temperature.

However, as Figure 1 shows, this relationship was not at all straightforward. The results for test 1 are presented in Table 1; in addition to temperature and weight loss data, Table 1 also includes data for A_r for Rho-

Table 1. Tests on Virgin Carbon

W_o (mg)	T_g (°C)	T_c (°C)	W_r (mg)	W_l (mg)	W_l (%)	A_r (mg/mg)
150	225	315	149.2	0.8	0.5	0.0370
150	325	420	148.2	1.8	1.2	0.0373
150	410	513	128.9	21.1	14.1	0.0433
150	510	598	87.1	62.9	41.9	0.0640
150	610	692	83.4	66.6	44.4	0.0662
150	710	793	81.0	69.0	46.0	0.0651
150	805	880	81.5	68.5	45.7	0.0647
150	905	969	78.0	72.0	48.0	0.0715
150	1000	1065	54.6	95.4	63.6	0.1016

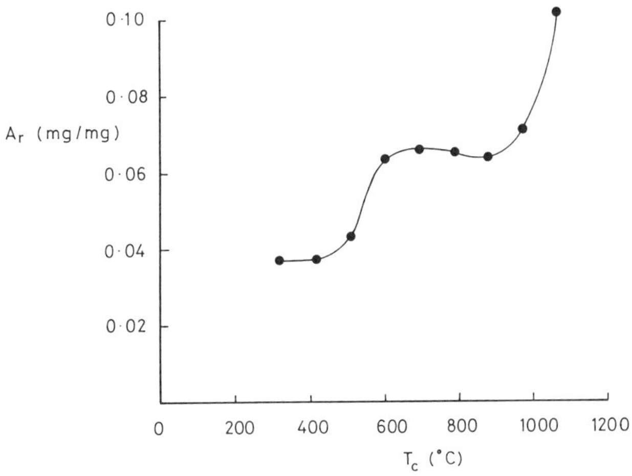

Fig. 2. Adsorption capacity v. temperature.

damine B. Again as expected, the adsorption capacity of the heat-treated carbon for Rhodamine B generally increased with increasing temperature.

However, as Figure 2 shows, this relationship too was not uniform.

Figure 1 appears to have 4 phases: approximately (i) up to 400°C, (ii) 400-600°C, (iii) 600-900°C and (iv) beyond 900°C. Figure 2 also appears to have the same 4 phases. Thus, it is interesting to note that weight loss and adsorption capacity behaved in an approximately parallel manner with change in temperature. New adsorption sites were obviously created by the burning of the carbon.

In test 2, virgin carbon samples were heated over a range of tem-

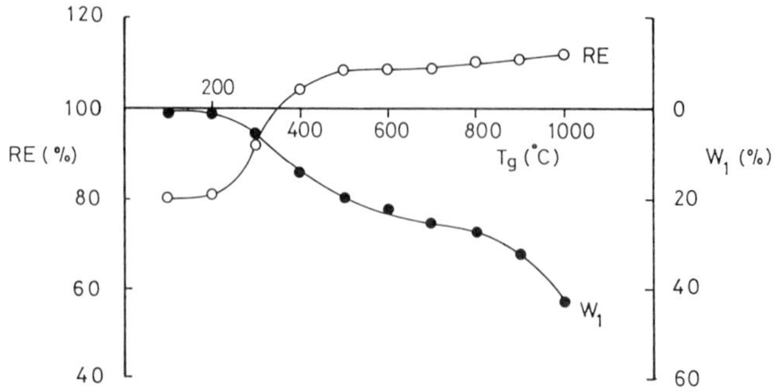

Fig. 3. The cross-over point.

peratures in a gas flow of 100 ml/min of nitrogen and with the exit flange plate still not fitted. The results and graphs of the results showed that the 4 phase structure was essentially maintained, though in a much less obvious manner. The effect of the nitrogen was to suppress weight losses, particularly in phases (ii) and (iv). The suppression of weight loss was accompanied by lower A_r values for Rhodamine B in subsequent adsorption experiments on the heat-treated carbon samples. It is of interest to note that although lower A_r values were recorded than in test 1 (A_r being the adsorptive capacity on a per unit carbon weight basis), the adsorptive capacities of the carbon samples on a complete batch sample basis were higher then in test 1. Thus, on a batch sample basis, equilibrium concentrations of Rhodamine B left in solution during exhaustion were lower in test 2 than in test 1.

Thus, tests 1 and 2 confirmed very early in the research programme that there is a delicate balance between weight loss and regeneration efficiency. Put another way, there is a delicate balance between employing reactivation conditions which increase the adsorption capacity of the carbon per unit weight of carbon and employing those conditions which increase the adsorption capacity of the carbon for the entire batch of carbon being regenerated. The effect of the nitrogen was to boost total activity of the carbon solely because of considerably lower carbon destruction; the effect of air was to enhance carbon destruction and thereby enhance activity per unit weight of carbon.

Tests were continued, but for carbon exhausted with Rhodamine B before thermal treatment. Initial tests were conducted in an open-system. Results for weight loss and regeneration efficiency are presented in Figure 3.

Because of the parallel nature of the graphs in Figure 1 and Figure 2, Figure 3 is presented with $RE(\%)$ and $W_1(\%)$ scales running in opposite directions so that the temperature cross-over point could be ascertained. Thus, Figure 3 shows a cross-over point at T_g approx. 300°C. The significance of this cross-over point is that below this temperature, carbon losses were small and RE values were somewhat low; above the cross-over point temperature, the reverse is true with increasing carbon losses and increasing RE values. (This graphical procedure was useful used for many other series of tests.) Thus, the cross-over point highlights the balance between weight loss and regeneration efficiency.

It will be observed that in Figure 3, RE values of greater than 100% were recorded. Clearly, the heat-treatment yielded a carbon of higher adsorption capacity per unit weight of carbon compared with the virgin carbon exhausted prior to heat-treatment. The cross-over point of approx. 300°C confirmed earlier studies by the authors on chemical regeneration and subsequent volatilization of exhausted activated carbon which showed that for a conventional laboratory furnace, carbon losses could be expected from around 275°C onwards [9].

Weight losses and RE values for Rhodamine B were plotted against temperature for 100, 150, 200, 250, 300 and 400 ml/min of nitrogen, all with the exit flange plate not fitted. (The significance of the plate is that when not fitted, there is some scope for air getting into the carbon during heating. When the plate is fitted and with a flow of gas being passed through the tube furnace, the scope for air getting in is as near zero as it can possibly be.) As before, the introduction of nitrogen significantly reduced carbon losses. As the gas flow was increased, there was an upward drift of the cross-over point. At 100 ml/min of nitrogen, the cross-over point of T_c approx. 600°C was easy to distinguish and although the graphs were much less obviously S-shaped, the basic 4 phase structure could still be discerned. Beyond 150 ml/min nitrogen flow, the cross-over point stabilized within the 700-800°C region (T_c) and was considerably less easy to distinguish as the S-shape of the graphs disappeared. Whereas Figure 3 shows a significant divergence of RE and W_1 beyond the cross-over point, the presence of nitrogen at a rate of 200 ml/min and over led to RE/W_1 divergences that were negligible; below the cross-over point temperatures, RE values were low. Thus, maintaining a high temperature and minimizing carbon weight loss does not in itself guarantee good thermal regeneration conditions.

To test the efficacy of the regeneration technique over a number of exhaustion-regeneration cycles, carbon was heated in different combinations of gases over 3 cycles at $T_g = 510^oC$ ($T_c = 620^oC$ approx.). Results are presented in Table 2. The results shown are averages of two runs. Rhodamine B was used as the adsorbate for the six carbon samples. The exit flange plate was fitted for these tests.

Using oxygen-free nitrogen, it was observed that the first regeneration

Table 2. Repeated Exhaustion-Regeneration

Gas flow (ml/min)	Cycle no.	W_1 (%)	W_2 (%)	RE (%)	A_r (mg/mg)
N_2 400	1	0.9	3.8	94.6	0.0369
	2	-1.8	2.2	85.3	0.0326
	3	-1.6	2.4	78.2	0.0295
N_2 375 CO_2 25	1	0.6	3.6	93.7	0.0358
	2	-1.5	2.5	83.1	0.0316
	3	-1.7	2.3	76.1	0.0282
N_2 375 Air 25	1	11.3	14.0	103.7	0.0449
	2	15.8	19.2	105.2	0.0542
	3	23.5	25.9	105.4	0.0709

was reasonably successful; a small carbon weight loss was recorded. Beyond the 1st cycle, however, the carbon showed weight gains. Previous studies by the authors have shown that weight gains may arise from residual adsorbate and from the formation of pyrolysis products in the carbon pores [9]. (The weight gains in Table 2 are signified by the presence of a minus sign.) It is clear that the blockage of pores has an adverse effect on subsequent adsorption and thus, RE values are seen to fall.

Very similar results were obtained for a mixture of oxygen-free nitrogen and carbon dioxide. Thus, the temperature used in these studies was too low for regeneration to be effective for N_2 and N_2/CO_2.

The use of air in a nitrogen/air gas mixture resulted in significant carbon weight losses. It is especially important that the W_1 data show increasing weight loss with each subsequent cycle. Very successful RE values may be observed, clearly as a result of the burning of carbon and the creation of new adsorptive sites and increased porosity of the remaining carbon. It is obvious that repeated cycles for N_2/air at the temperatures used would very rapidly destroy the carbon; after 3 cycles, the cumulative carbon weight loss was around 50%.

It was observed that in the nitrogen/air tests, the carbon granules visibly decreased in size after each thermal regeneration. (This was also true in other tests where temperatures and gaseous conditions were such that carbon destruction was significant.) Reduction in granule size facilitates further destruction.

CONCLUSIONS

This research paper is the first of what is intended to be a series of papers following experimental work by the authors on the thermal regeneration of exhausted activated carbon. Further reports will deal with a wider variety of gaseous atmospheres and operating temperatures and greater repetitions of the exhaustion-regeneration cycle.

The initial results may be summarized as follows:

1. Thermal regeneration is a balance between minimization of carbon weight loss and maximization of restoration of adsorption capacity to the carbon.

2. Thermal regeneration is essentially a destructive process. A small degree of destruction is desirable in order to create new adsorption sites. A high degree of destruction is not desirable as excessive carbon weight loss is encountered.

3. Different gaseous atmospheres were shown to possess different threshold temperatures, below which adsorbates could not properly be volatilized away from the carbon pores. The threshold temperature for a particular gaseous atmosphere must therefore be exceeded for any real success in regeneration of the carbon. Exceeding the threshold temperature, however, was accompanied by carbon weight loss.

4. The threshold temperature could be determined by plots of regeneration efficiency and carbon weight loss against operating temperatures; cross-over points from the plots could be used to define these threshold temperatures.

5. It was observed that reactivation conditions could be selected to increase adsorption capacity per unit weight of activated carbon at the expense of losing some carbon. Repeated cycles of exhaustion and regeneration showed that an emphasis on quality of carbon and a disregard for quantity does not yield lasting profit; progressively more make-up carbon would have to be added to ensure a satisfactory adsorption capacity of the carbon bed in total.

6. Because carbon losses progressively accelerated with each subsequent cycle, the mildest conditions necessary for the removal of the adsorbate and reactivation of the carbon should be used.

REFERENCES

[1] Guymont, F.J., The effect of capital and operating costs on GAC adsorption system design, "Activated Carbon Adsorption of Organics from the Aqueous Phase", volume 2, edited by M.J. McGuire and I.H. Suffet, Ann Arbor Science, Michigan, chapter 23, 531 - 538, 1980.

[2] Adams, J.Q. and Clark, R.M., Cost estimates for GAC treatment systems, Journal

of the American Waterworks Association, 81, No. 1, 35 -42, 1989.

[3] Adams, J.Q. and Clark, R.M., Development of cost equations for GAC treatment systems, Journal of Environmental Engineering (ASCE), 114, 672 - 688, 1988.

[4] Adams, J.Q. Clark, R.M., Lykins, B.W., DeMarco, J. and Kittredge, D., GAC treatment cost experience at two drinking water utilities, Journal of Environmental Engineering (ASCE), 114, 944 - 961, 1988.

[5] Ng, W.J., The regeneration of activated carbon, PhD thesis, Department of Civil Engineering, University of Birmingham, 1980.

[6] Martin, R.J. and Ng, W.J., Chemical regeneration of exhausted activated carbon-I, Water Research, 18, 59 - 73, 1984.

[7] Martin, R.J. and Ng, W.J., Chemical regeneration of exhausted activated carbon-II, Water Research, 19, 1527 - 1535, 1985.

[8] Martin, R.J. and Ng, W.J., The repeated exhaustion and chemical regeneration of activated carbon, Water Research, 21, 961 - 965, 1987.

[9] Martin, R.J. and Ng, W.J., The chemical regeneration and subsequent volatilization of exhausted activated carbon, "Chemistry for Protection of the Environment 1987", edited by L. Pawlowski, E. Mentasti, W.J. Lacy and C. Sarzanini, Elsevier, Amsterdam, 189 - 198, 1988.

DIFFERENTIAL SCANNING CALORIMETRY (DSC) OF EXHAUSTED AND NON-EXHAUSTED ACTIVATED CARBON

M. BAUDU, P. LE CLOIREC, and G. MARTIN

Laboratoire Chimie des Nuisances et Génie de l'Environment
E.N.S. Chimie, Avenue Général Leclerc 35700 Rennes, France

ABSTRACT

The authors studied the desorption of water and organic materials from exhausted or non-exhausted activated carbons by using differential scanning calorimetry (DSC). Water is found to be always present in the inner structure of activated carbon. The desorption enthalpies have been determined and compared with the literature data for several carbons loaded with organic compounds. This energy of solid-solute interactions has been related to a physical and chemical parameter of the solute (molecular refraction). A fair correlation is found for aromatic and aliphatic molecules.

INTRODUCTION

A large number of publications report the removal of organic compounds from aqueous solutions by adsorption onto activated carbons. Generally, only the solutions and the variations of solute concentrations are studied; models using Arrhenius's equation allow an approach to calculate the adsorption energy [1, 2]. Some studies give adsorption energy data on the

Chemistry for the Protection of the Environment
Edited by L. Pawlowski *et al.*, Plenum Press, New York, 1991

Table 1. Characteristcs of Activated Carbon

Parameters	
Origin	coconut
Size (mm)	1 - 1.5
Porosity	microporous
Specific area (m^2/g)	1200
Density	0.52

carbon before and after adsorption of dissolved organic compounds. Thus, Beljokova [3] and Metiri [4] used microcalorimetry analysis. Bonsal et al. [5] reported several works on calorimetry. Gas chromatography is one of the most recent methods to calculate adsorption energies [6, 7]. However, differential scanning calorimetry (DSC) has been employed to study virgin carbon or the regeneration of activated carbon loaded with halogenated or aliphatic compounds [9, 10].

The objective of the work was to obtain quantitative information on adsorption energies using the differential scanning calorimetrie analysis. Several aromatic and aliphatic compounds loaded onto activated carbon will be analysed. A correlation between molar refraction and energy will be proposed. This could give us a better knowledge of the activated carbon and help to approach the adsorption mechanism.

MATERIAL AND METHODS

Activated Carbon

The activated carbon used for this study was Picactif NC60 (Pica Company, Levallois, France). Its characteristics are summarized in Table 1.

Sample Preparation

The virgin carbon was left in a oven at 1000°C under a nitrogen flow during 4 hours in order to remove water and oxygen adsorbed onto the carbon surface. Then, the grains were transferred into a dry box swept with nitrogen. Samples were loaded with different organic compound vapors or with water. The adsorbed molecules have been chosen to present a large range of aromatic compounds (Table 2). Numerous works have already shown a good adsorption of these solutes, or similar ones, onto activated carbon [11].

Table 2. Desorption Energy Data for Various Aliphatic and Aromatic Compounds

n°	compound	q_d	ΔH_v	E_{ad}	MR
0	Water	40	39.5		
1	Benzene	39	30.7	8.3	26.5
2	Chlorobenzene	48	36.51	11.5	31.4
3	Toluene	45	33.43	11.6	31.1
4	Benzaldehyde	57.5	38.36	19.1	32.2
5	Phenol	50.5	40.20	10.3	28.0
6	Naphthalene	64	40.45	23.55	43.0
7	Methanol	36.25	35.20	1.05	8.24
8	n-Butanol	50.5	43.76	6.74	22.0
9	Dichloromethane	33	27.16	5.84	16.43

Methods

DSC analysis was performed on a Mettler TA 3000 apparatus. The activated carbon sample was put in an aluminium crucible (180 μl) and was weighed in a dry box to avoid water or oxygen readsorption. The furnace of the DSC machine was under a helium flow (4 l/h) to avoid carbon combustion and to get a good baseline. The temperature gradient was 10°C/minute from room temperature to 500°C. The reference crucible was filled with dry virgin activated carbon. Previous works have shown that a better baseline was obtained when using this kind of reference [12].

RESULTS AND DISCUSSION

The Adsorption Energies

The desorption energy is obtained by integrating the DSC signal $d\Delta H/dT = f(T)$, ΔH being the endo or exothermal enthalpy and T the temperature. The graphs given in Figures 1 and 2 are for carbons loaded with some organic molecules and compared to the virgin carbon (the starting material). Large peaks are generally obtained when aromatic compounds are present on the support. The shape of these curves is discussed in terms of energy distribution. For the same molecule, different adsorption energies can take place depending on the molecular structures and on the sites where the solid-solute interactions occur. For methanol or dichloromethane, a narrow peak seems to indicate that no site is preferred; the mechanism could thus

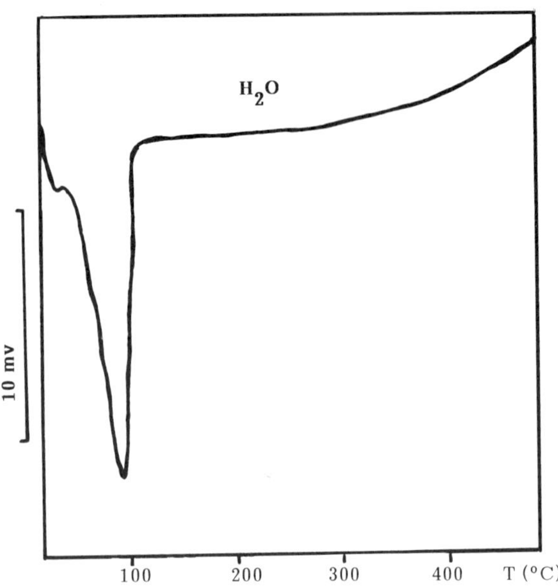

Fig. 1. DSC signal for water desorption from a virgin activated carbon sample.

be physisorption. For aromatic compounds with a hydrophilic group, the DSC signal shows a large energy distribution. Both chemical and physical adsorption mechanisms are suggested. The adsorption energy data are in fact an average value of a large range of energy.

This desorption energy value (q_d) obtained by DSC analysis, could be written as the sum of following energy values:

$$q_d = E_{ad} + \Delta H_v + E_d$$

where:

q_d = desorption energy,

E_{ad} = adsorption energy,

ΔH_v = vaporization enthalpy,

E_d = diffusion energy.

The energy E_d is very small compared with the other values and is nearly zero. Since the vaporization enthalpy is known by determining the

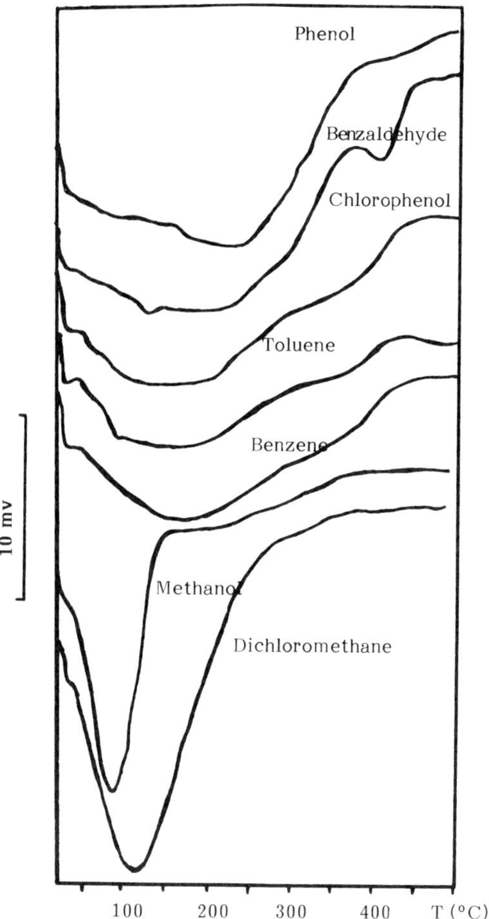

Fig. 2. DSC signal for organic molecule desorption from loaded activated carbon samples.

desorption energy, the adsorption energy can be approximated. The data for aromatic and aliphatic compounds are given in Table 2. The water desorption energy of the starting material sample (virgin carbon) is found to be very close to the vaporization enthalpy. A large part of water present in the inner pores can be assumed to be free water. This assumption has been confirmed by previous work on carbon with solid state NMR [13]. The values of adsorption energy (Table 2) are close to the data found in the literature, but where other analytical techniques, like chromatographic or calorimetric analyses, were used [6, 7].

Correlation Between Adsorption Energy and Solute Structure

The shapes of the DSC graphs allow the proposal of a mechanism of physical and/or chemical adsorption. These solid-solute interactions depend on the structure of the adsorbates. Abe [14] and Le Cloirec [15] proposed a correlation between the adsorption equilibrium constants and the molar refractions and the hydrophilic groups of the molecules. Referring to these previous works, we propose to correlate the adsorption energy to the molar refraction. Table 2 shows the data used to find the following equation:

$$E = 0.632\, MR - 5.89$$

correlation coefficient = 0.930

with:

E_{ad} = adsorption energy (kJ/mole of adsorbed molecules),

MR = molar refraction.

We can thus deduce a linear relationship between the adsorption energy of a molecule on the activated carbon surface and its molar refraction. For the tested molecules, a higher molar refraction gives a better adsorption energy for this kind of support. However, at this time no relation is found between interaction energy and adsorption capacity. In order to compare the experimental and calculated values of energies, Figure 3 was plotted. A good correlation was found.

CONCLUSION

The following conclusion may be proposed:

- Differential scanning calorimetry (DSC) is a simple and quick method to determine the adsorption energy of aromatic and aliphatic compounds.

- Some adsorption mechanisms can be deduced from the DSC signal shape. An adsorption energy distribution is shown for aromatic molecules containing hydrophilic groups in their structure.

- A linear correlation between adsorption energy determined by DSC and molar refraction is observed.

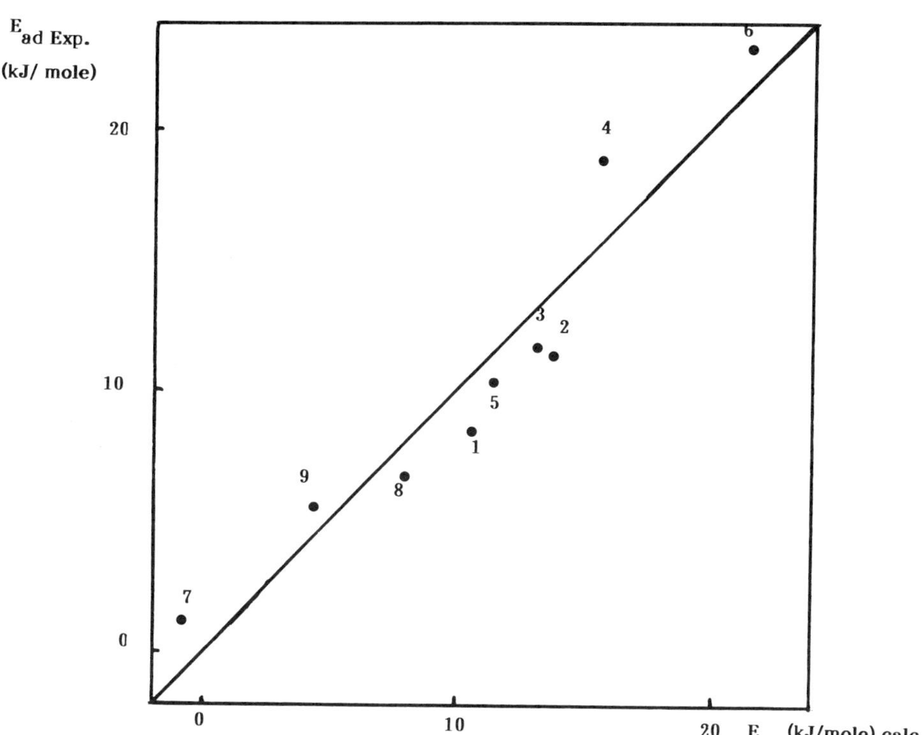

Fig. 3. Correlation between experimental and calculated adsorption energies.

REFERENCES

[1] Gaid, K., Thèse Doctorat es Sciences Physiques, Université de Rennes I, Rennes, France, B, 344, 200, 1981.

[2] Vidal-Madjar, C., Guiochon, G., Bull. Soc. Chim., 9, 3110, 1971.

[3] Beljokova, L.D., Kiselev, A.B., Kovaleva, N.V., Bull. Soc. Chim., 1, 285, 1967.

[4] Metiri, J., Spartyka, G., Brun, B., Calorim. Anal. Therm., 15, 155, 1984.

[5] Bansal, R.C., Donnet, J.B., Stoekli, F., Active Carbon, Marcel Dekker Inc., New York, 1988.

[6] Vukov, A.J., Gray, D.G., Langmuir, 4, 743, 1988.

[7] Jakoda, A., Kavazoe, K., Suzuki, M., J. Chem. Eng. Jap., 20, 199, 1987.

[8] Rosenwald, R.J., Dubow, J.B., Rajeshwar, K., Thermochimica Acta, 53, 321, 1982.

[9] Amicarelli, V., Baldassare, G., Liberti, L., Thermochimica Acta, 30, 247, 1979.

[10] Amicarelli, V., Baldassare, G., Liberti, L., Thermochimica Acta, 30, 255, 1979.

[11] Le Cloirec, P., Thèse Doctorat es Sciences Physiques, Université de Rennes I, Rennes, France, B, 423, 325, 1985.

[12] Baudu, M., Le Cloirec, P., Martin, G., proposed for publication to Carbon; (1989).

[13] Le Cloirec, P., Martin, G., Gallier, J., Carbon, 26, 275, 1988.

[14] Abe, I., Hayashik, K., Kitagawa, M., Hiroshima, T., Bull. Chem. Soc. Japan, 56, 1002, 1983.

[15] Le Cloirec, P., Guernion, C., Benbarka, B., Martin, G., Rev. Sci. Eau, 5, 259, 1986.

AN APPLICATION OF THE FRACTAL THEORY: THE DETERMINATION OF FRACTAL STRUCTURE OF FLOCS IN COAGULATION-FLOCCULATION

M. FRANCESCHI, A. GIROU, A. VERDIER[1]
R. BURLOT, G. GENTY and L. HUMBERT[2]

[1] Institut National des Sciences Appliqueés Toulouse, France
[2] Institut de Geodynamique de Bordeaux, Bordeaux, France

ABSTRACT

We present a floc formation study in order to develop a new approach for the automation of the coagulation-flocculation process.

The structure of "flocculi" formed during the rapid mixing step determines their aggregation and their elimination during mixing and decantation steps.

The characterization of their structure is necessary for a better understanding of the different mechanisms involved.

The fractal dimension of "flocculi" was determined after the measuring of their morphological parameters. These morphological measurements were realized by coupling a photographic method with digital image processing.

The influence of the nature and concentration of flocculants was investigated and experiments were also conducted with different clay concentrations.

According to the results it would seem that the fractal dimension of "flocculi" depends on:

- the nature and concentration of flocculants,
- the clay concentration initially in the solution,
- the mechanisms involved during the "flocculi" formation step.

Chemistry for the Protection of the Environment
Edited by L. Pawlowski *et al.*, Plenum Press, New York, 1991

NOTATIONS

PDME : Polycondensate of dimethylamine and epichlorhydrin
$FeCl_3$: Iron chloride
SA : Aluminium sulfate $Al_2(SO_4)_3 \cdot 18H_2O$
S : 2 dimensional area of floc
P : Perimeter of floc
D : Fractal dimension
ofc : Optimal flocculation concentration
a : Coefficient of linear regression
b : Coefficient of linear regression
Tur : Residual turbidity

INTRODUCTION

The study of the aggregation process is of both practical and fundamental interest in water science.

In the coagulation-flocculation process the elimination of colloidal clay particles is due to the formation of aggregates, called flocs, by using polymers or mineral coagulants like aluminium sulfate or iron chloride. The size and structure of these aggregates seem to be very important parameters in industrial application.

The elimination of flocs by sedimentation leads to the formation of sludge. For industrial application it which is necessary to produce a small volume of sludge, which is directly correlated with the structure and nature of formed flocs. Flocs frequently have a complex random structure with low average density. Such systems were often described by qualitative terms such as wispy, ramified or tenous. Recently it has been shown experimentally [1-6] and through computer simulation that the structure of colloidal aggregates can be characterized by the concept of fractal geometry. It is the purpose of this publication to show that the fractal characterization of flocs can be correlated with other results obtained in experimental coagulation-flocculation tests.

BIBLIOGRAPHY ON FRACTAL GEOMETRY AND ITS APPLICATION TO COLLOIDS

The structure of some very familiar objects (clouds, coastlines, vascular systems in biological structures, structures obtained in turbulent flow, aggregates etc.) cannot be described in terms of Euclidean geometry.

During the past 2-3 decades, mathematicians and in particular B. B.

Fig. 1. Simple representation of a fractal structure.

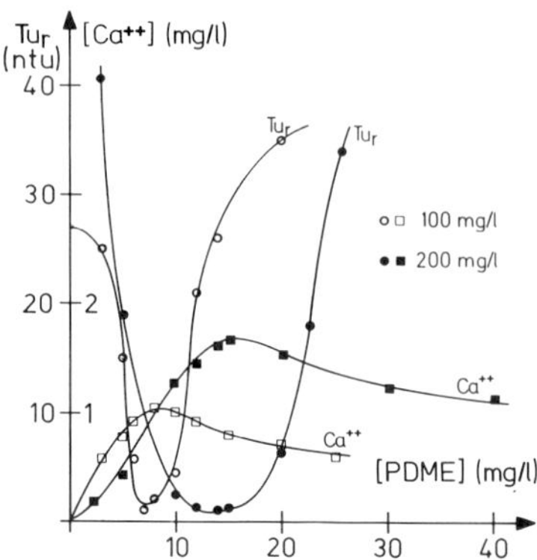

Fig. 2. Flocculation results for PDME.

Mandelbrot [7-9] developed a new geometry which provides a basis for describing and understanding many of these structures and processes. In recent years the fractal geometry has been found to be relevant to an extremely broad range of systems and phenomena of scientific interest.

Most of the familiar objects mentioned are invariant to a change of length scale: this means that these objects look the same under different magnifications.

For example, Figure 1 gives a very simple representation of a fractal structure: In Figure 1a, five particles have been joined together to form a cross. In Figure 1b, five crosses have been joined to form a larger cross of 25 particles. Figures 1c and 1d show the third and fourth stages of the hierarchical generation process. Each time the mass increases by a factor of 5. The overall size of the aggregate increases by a factor of 3.

The fractal dimensionality is given by:

$$\begin{aligned} D &= \log(M)/\log(l) \\ &= \log(5)/\log(3) \\ &= 1.465 \end{aligned}$$

For colloidal aggregates and other random fractals in real space it is convenient to think of the fractal dimensionality in terms of the scaling relationship between mass M and length l:

$$M \sim l^D$$

This type of relation is frequently employed to measure the fractal dimensionality D. For example, if the radius of gyration Rg is measured as a function of the aggregate mass M:

$$Rg \sim M^{\beta}$$

This power law relation can be characterized by a fractal dimensionality $D = 1/\beta$.

Such relations have been used by different authors for the determination of D for colloid aggregates.

For example: Tuel [2] determined the fractal dimensionality of precipitated silica by using the relation

$$S \sim R^{D}$$

S : surface of aggregates
R : radius of aggregates

More recently Aratani [9] determined the fractal dimension of flocs by using a relation between the surface and the perimeter of flocs formed by flocculation of kaolinite clay:

$$S^{1/2} \sim P^{1/D}$$

Aratani proposes to use D as a parameter to describe the ruggedness of the flocs.

So, even if the fractal geometry seems to be a very interesting method to a better understanding of floc structures we can note that authors have used different definitions of the fractal dimension D. And there is not enough experimental data to develop a new theory of the floc formation phase.

MATERIALS AND METHODS

All the flocculation tests were made with bentonite suspensions (Prolabo n°21793364). The suspensions were prepared by mixing a given quantity of clay (100 or 200 mg/l) in distilled water with conductivity less than 1.5μS/cm. These concentrations of clays are representative of the suspended matters in natural water.

The flocculation experiments were made possible by the use of a hydrocure FLH 6 Jar Test apparatus. The procedure employed in all the tests was:

- rapid mixing phase : 1 min at 200 r.p.m
- slow mixing phase : 10 min at 40 r.p.m
- settling phase : 10 min

The coagulants and flocculant used were:

- Iron chloride $FeCl_3$, in solution at 10 g/l

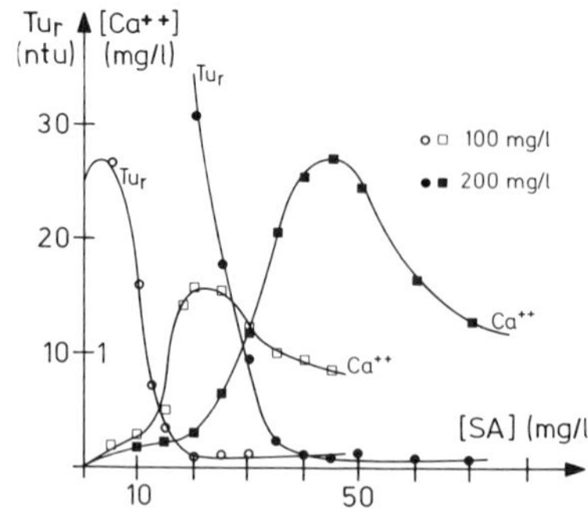

Fig. 3. Coagulation results for SA.

- Aluminium sulfate $Al_2(SO_4)_3$ in solution at 10 g/l
- An organic polycondensate synthetized at the organic chemistry laboratory at INSA. This polymer is obtained by polycondensation of dimethylamine and epichlorohydrin.

The structure of the macromolecule is:

$$\left[- \overset{\displaystyle R}{\underset{\displaystyle R}{\overset{|\oplus}{N}}} - CH_2 - \overset{Cl^-}{CHOH} - CH_2 - \right]_n \qquad R = R' = CH_3$$

This polymer is not a commercial product but it has a great efficiency in flocculation and its action mechanisms have been studied in detail [10].

After rapid mixing, slow mixing and settling phase, samplings were made for different analyses (pH, adsorption of flocculant or coagulants, photographies for image analysis). pH measurements were made with a TACUSSEL pH meter (minisis 8000). The concentrations of aluminium, iron and calcium were determined by atomic absorption (PERKIN ELMER 3030) and residual flocculant was analyzed by a total carbon analyzer (IONICS 1254).

The photographs of flocs were analysed at the Geodynamic Institute of Bordeaux III (France). All the different steps of the image analysis were developed in a previous paper [11].

For the determination of the fractal dimension of floc D we have applied the method developed by Aratani [9]: the 2 dimensional area of flocs depends on the fractal dimension by the relation

$$S^{1/2} \sim P^{1/D}$$

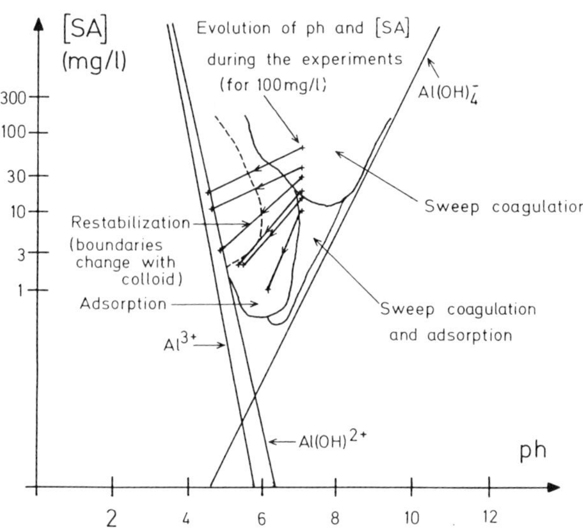

Fig. 4. Operation diagram for aluminium sulfate coagulation.

S : 2 dimensional area of floc
P : perimeter of floc
D : fractal dimension

RESULTS AND DISCUSSION

Figures 2, 3 and 4 show the results obtained in flocculation tests for the three products used. In the same figures we have represented the evolution of residual turbidity, and the evolution of calcium concentration in the solution.

For PDME (Fig. 2) the optimum flocculation concentration (ofc) is 7 mg/l for a bentonite concentration equal to 100 mg/l and 14 mg/l for 200 mg/l. After the ofc there is an increase in the residual turbidity due to a restabilization of the suspension by charge inversion [10]. Evolution of Ca^{2+} concentration in solution shows that the adsorption mechanism of PDME is due to ion exchange between ammonium groups of PDME and calcium ions of bentonite clay.

The results obtained with aluminium sulfate (SA) are represented in Fig. 3. For a bentonite concentration of 100 mg/l the ofc is 20 mg/l and 40 mg/l when the clay concentration is 200 mg/l. The action of SA

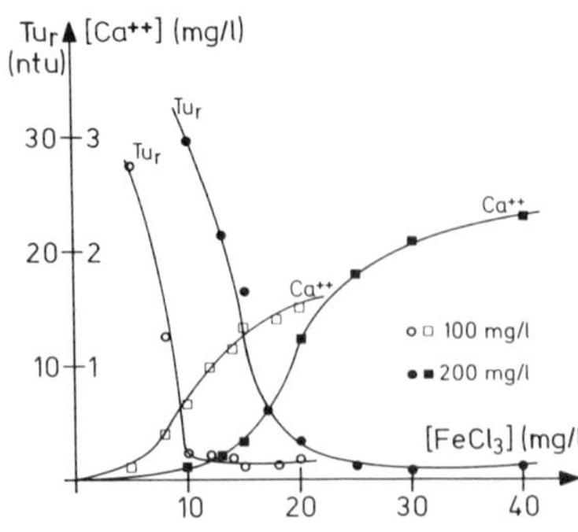

Fig. 5. Coagulation results for $FeCl_3$.

as a coagulant is due to the formation of hydrolyzed species like $Al(OH)^{2+}$, $Al(OH)_2^+$ and to other polymerized species which can adsorb on clay particles and to the precipitation of $Al(OH)_3$ [11] which can eliminate colloid by sweep coagulation.

An augmentation in SA concentration provokes a change in the repartitioning of the different species and the flocculation mechanisms change as the SA concentration increases (Fig. 4). In our study, changes in flocculation mechanisms are shown by the change in the calcium concentration (Fig. 3). For the two clay concentrations, before the ofc, there is an increase of the Ca^{2+} exchanged by the clay. After the ofc, Ca^{2+} concentration diminishes with no change in the residual turbidity.

Figure 4 represents a coagulation diagram established by Amirtharajah [11]. We have included on it the evolution of pH and the value of SA concentration during the coagulation experiments. Before the ofc, coagulation is due to the adsorption of hydrolyzed species by ion exchange with Ca^{2+}. After the ofc, coagulation is due to adsorption and sweep coagulation. This change in the action mechanisms can explain the diminution in Ca^{2+} concentration in solution when sweep coagulation becomes the predominant mechanism.

Fig. 5 shows the results obtained with $FeCl_3$; as for SA the action of $FeCl_3$ is due to hydrolyzed species and to the precipitation of $Fe(OH)_3$ [11].

Evolution of Ca^{2+} concentration shows that the coagulation effect is due to ion exchange of the hydrolyzed ions, but the amount of Ca^{2+} exchanged is less important than with SA. These differences can be explained with the coagulation diagram (Fig. 6). On this diagram established by Amirtharajah [11] we have reported the evolutions of $FeCl_3$ concentration and pH in solution during the experiments for bentonite concentration equal

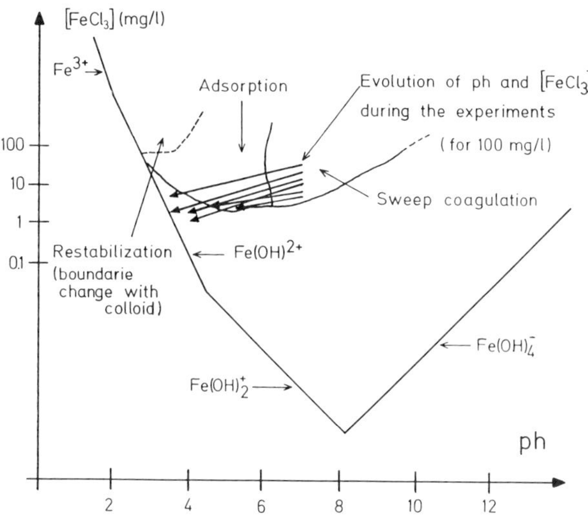

Fig. 6. Operation diagram for iron chloride coagulation.

to 100 mg/l. As for SA, the evolution of $FeCl_3$ concentration and pH during the experiment shows that we have succesful sweep coagulation and adsorption.

For SA sweep coagulation becomes more and more important when SA concentration increases. This is not true for $FeCl_3$ (Fig.6): the coagulation of clay particle is always due to sweep coagulation and adsorption of hydrolyzed species. The same analysis can be made when bentonite concentration is 200 mg/l.

For each coagulant and flocculant we have determined the fractal dimensions D of flocs for different concentrations.

Fig. 7 and 8 show the results obtained with PDME. For the two clay concentrations D value is more important at the ofc than before. This augmentation on D value is due to change in floc structure when the floc size increases. The value of D obtained for PDME concentration greater than the ofc is lower; it seems that there is similitude between floc structure before the ofc and after the ofc (Fig.8).

For SA the results show an increase of D with the efficiency of coagulation. But for a clay concentration equal to 100 mg/l all the fractal dimensions were determined, for SA concentration gives a great efficiency of coagulation (15 mg/l : Tur = 3 ntu; 30 mg/l : Tur = 1.2 ntu etc.). So even if the change in D is not very important (1.42 for 15 mg/l to 1.56 for 60 mg/l), this change can be attributed to the modification of the flocculation

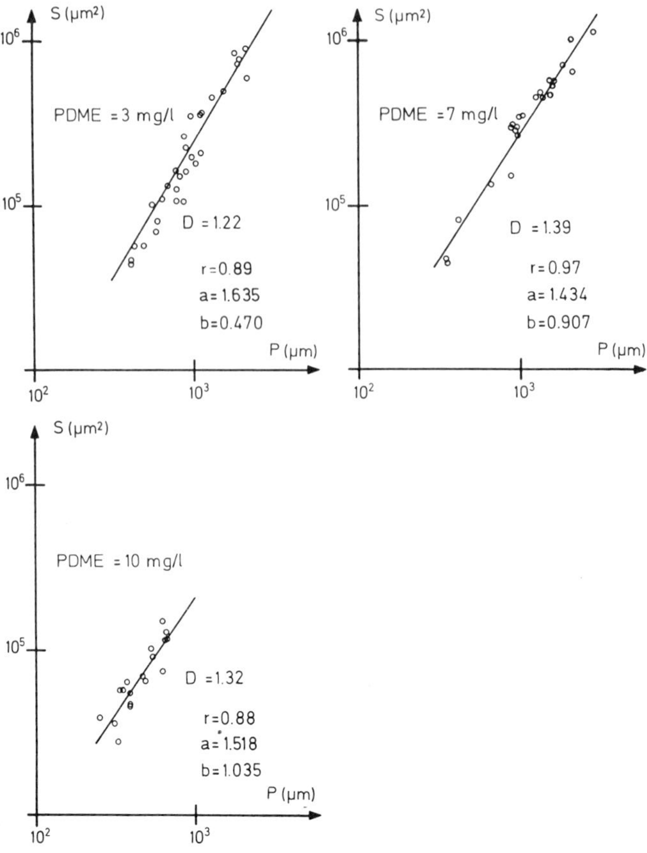

Fig. 7. Determination of the fractal dimension D for PDME.

mechanism. When SA increases, sweep coagulation becomes more and more important and induces a change in the structure of the formed flocks.

For 200 mg/l the evolution of D is the same: we note an increase of D with SA concentration. The fractal dimension obtained for 20 mg/l is low because for this concentration the coagulation efficiency is not very high (Tur = 18 ntu). For SA concentration near the ofc and greater than the ofc we have the same evolution of D as for 100 mg/l. This result confirms the hypothesis of change in D when the coagulation mechanisms change for the same efficiency. Indeed the values of D are lower than for 100 mg/l (Fig. 14).

Figures 11 and 12 show the results obtained with $FeCl_3$, for bentonite concentrations of 100 and 200 mg/l. Results show that there is not a significant change in the fractal dimensions D. Figure 6 shows that in

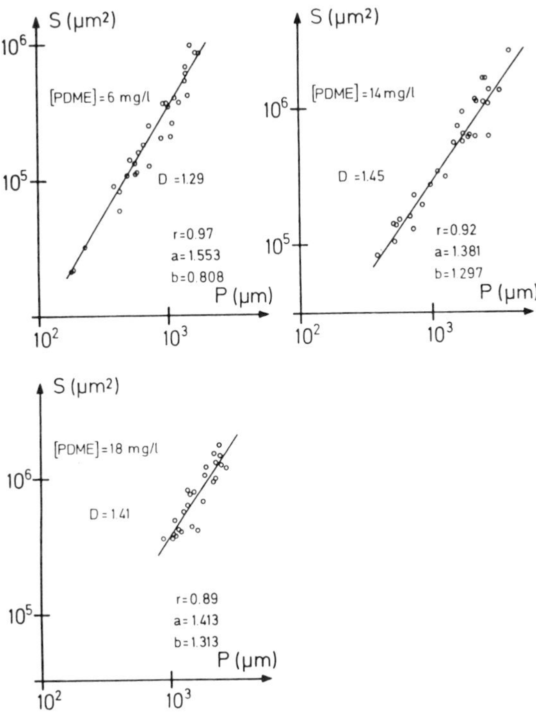

Fig. 8. Determination of the fractal dimension D for PDME. Bentonite concentration: 200 mg/l.

our experiments we are always in a zone where the coagulation is made by adsorption-destabilization and this is confirmed by Fig. 5: coagulation is due to the adsorption of hydrolyzed species with ion exchange with calcium. So the structure and the fractal dimension of flocs are similar in all the experiments.

For 200 mg/l of bentonite the fractal dimensions of floc are greater than for 100 mg/l (Fig. 12). This result is opposite of results obtained with SA (Fig. 15).

CONCLUSIONS

This study has shown that the structure and form of flocs formed by coagulation-flocculation can be described by the concept of fractal geometry.

Change in the fractal dimension of flocs can be correlated with the nature of the coagulant or flocculant used.

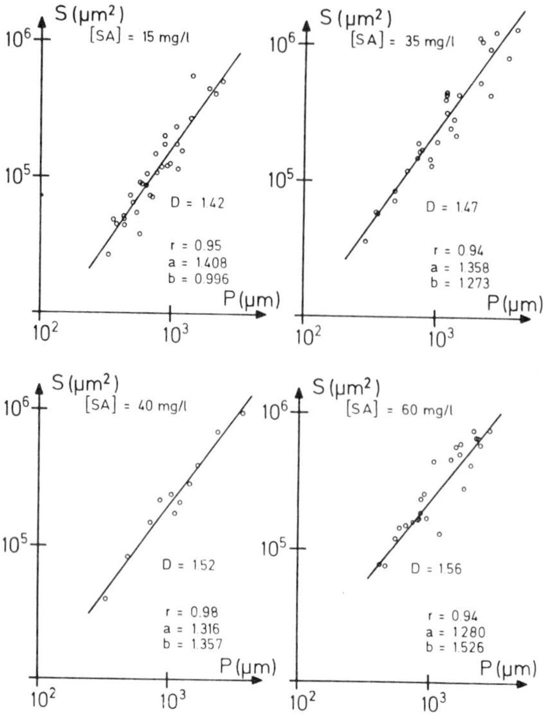

Fig. 9. Determination of the fractal dimension D for SA. Bentonite concentration: 100 mg/l.

The action mechanisms of coagulants, like SA or $FeCl_3$, are due to hydrolyzed species and to $Al(OH)_3$ or $Fe(OH)_3$. When the coagulant concentration increases, the species in solution change and the fractal dimension of flocs changes too. In this study, fractal geometry was applied to flocs formed during the rapid mixing step. If these concepts are applied to the slow mixing step it will be possible to obtain, for bigger flocs, the same value of fractal dimension D as for the example in Fig. 1. This case presents a very good argument for the theory of flocs growing by agglomeration of microflocs (of same structure) to form big flocs with a conservation of the fractal dimension. If the fractal dimension D is not conserved, other mechanisms like "coalescence" may be responsible for floc groth.

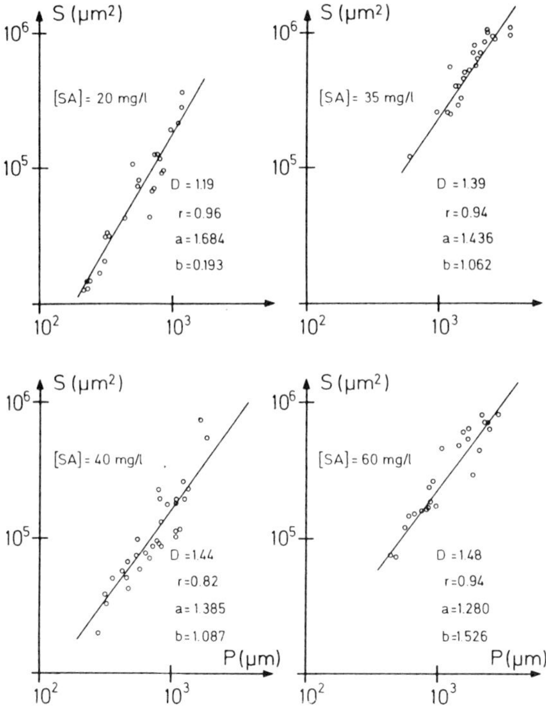

Fig. 10. Determination of the fractal dimension D for SA. Bentonite concentration: 200 mg/l.

But before we draw conclusions, more experiments must investigate the influence of the different parameters:

- **nature of coagulants and flocculants,**
- **nature and concentration of colloids,**
- **ionic strength, etc.**

After that, we think that fractal geometry can help us to provide an application of the formation mechanisms, and it gives us a basis for better understanding of the properties of these aggregates.

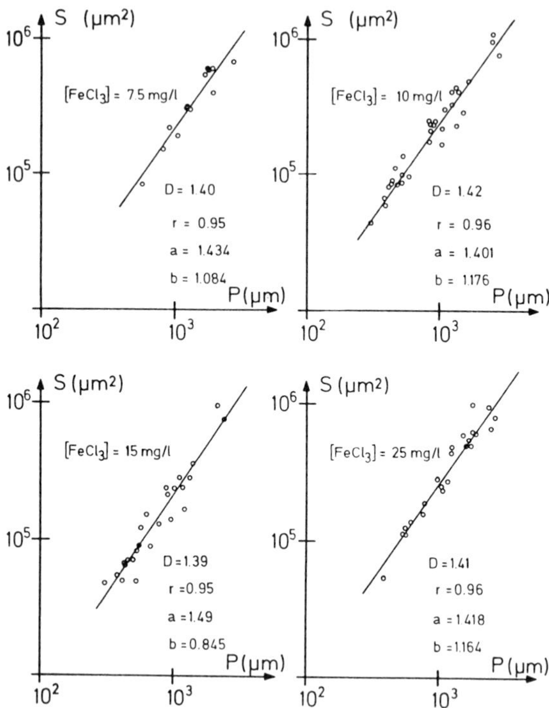

Fig. 11. Determination of the fractal dimension D for $FeCl_3$. Bentonite concentration: 100 mg/l.

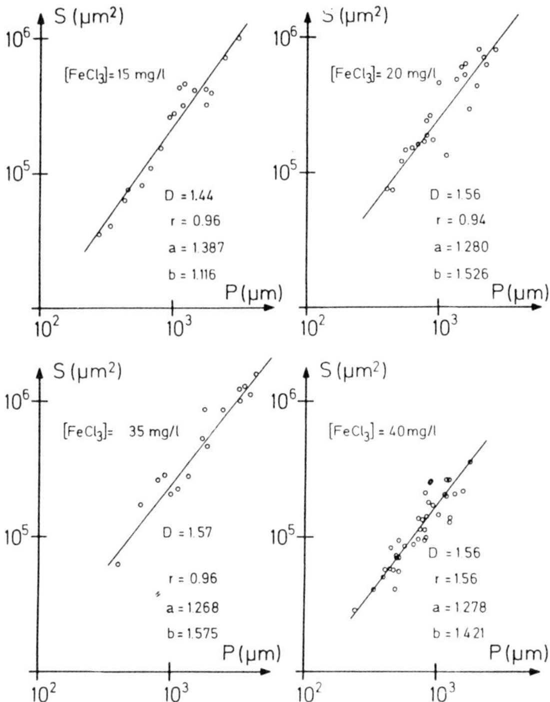

Fig. 12. Determination of the fractal dimension D for $FeCl_3$. Bentonite concentration: 200 mg/l.

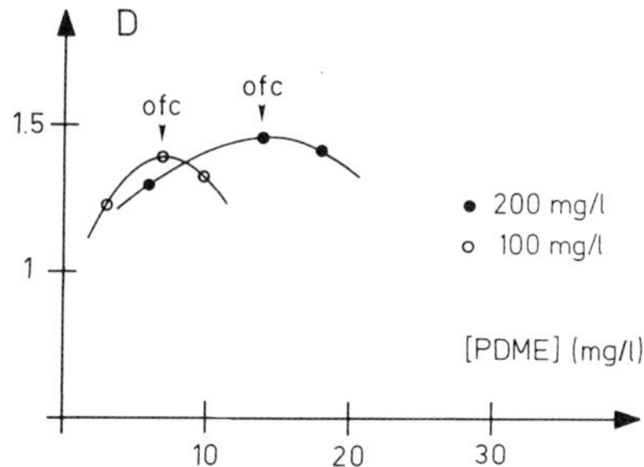

Fig. 13. Evolution of the fractal dimension D with the concentration of PDME.

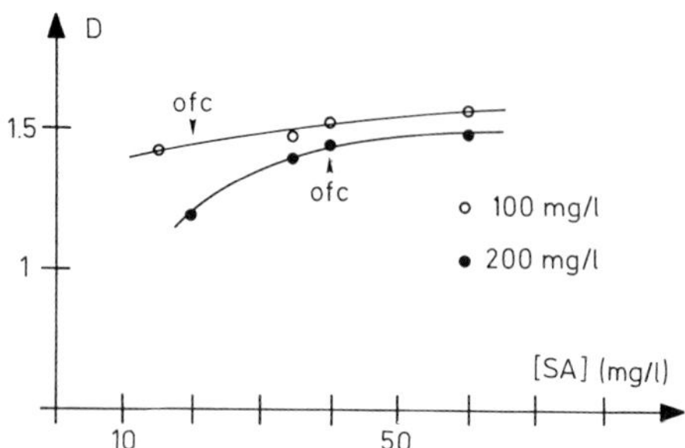

Fig. 14. Evolution of the fractal dimension D with the concentration of SA.

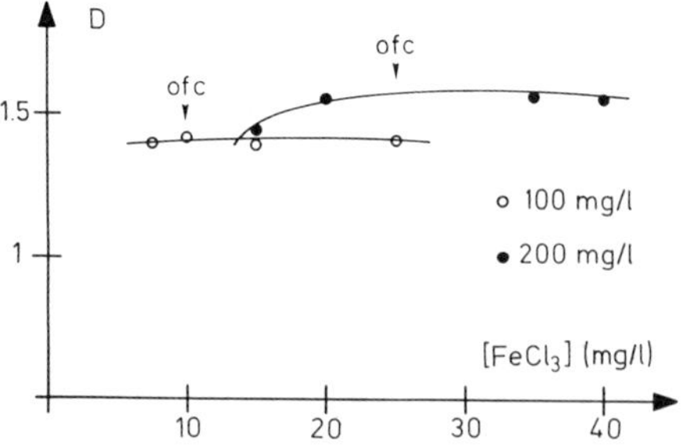

Fig. 15. Evolution of the fractal dimension D with the concentration of $FeCl_3$.

REFERENCES

[1] Wong H., Cabane B., ,masundaran P., "Highly ordered microstructure of flocculated agg ates." C ids and surfaces, 30. p. 355-360, 1988.

[2] Tuel A., Dautry P., Ho mel M., Legrand A. P., Morawski J. C., "Image analysis of fractal aggregates of precipitated silica." Progress in colloid and polymer science. n°76 p. 32-36, 1988.

[3] Schaefer D. W., Martin J. E., "Fractal geometry of colloidal aggregates" Physical reviews letters. vol. 52, n°26 p. 2371-2374. June 1984.

[4] Meakin P. "Fractal aggregates" Advances in colloid and interface sciences, 28. p. 249-331, 1988.

[5] Meakin P. "Diffusion-limited aggregation in three dimensions: result of a new cluster-cluster aggregation model." Journal of Colloid and Interface Sciences. Vol. 102. n°2 December 1984.

[6] Jullien R., Botet R., Mors P. M., "Computer simulations of cluster-cluster aggregations." Faraday Discuss. Chem. Soc., 83, p. 123-137. 1987.

[7] Mandelbrot B. B. "The fractal geometry of nature." W. H. Freeman and Company. New-York. 1982.

[8] Mandelbrot B. B. "Les objets fractals: forme et dimension." Eds. Flammarion. Paris 1975.

[9] Aratani T., Fujii T. M. T., Miyanani K., "Evaluation of floc form by fractal dimension." Kagaku Kogaku Ronbunshu, Vol. 14 n°3 p. 395-400, 1988.

[10] Kim H. S. "Contribution à l'étude des interactions entre une suspension argileuse et des polyelectrolytes cationiques." Thèse de doctorat de 3ème Cycle. Université Paul-Sabatier. Toulouse France.

[11] Amirtharajah A., Johnson P. N., "Ferric chloride and alum as single and dual coagulants." J.A.W.W.A. p. 232-239, 1983.

[12] Franceschi M., Girou A., Verdier A., Genty D., Humbert L., "Possibilities of the numerical treatment of image to assess quantitatively the morphological characters of flocs." 4th International conference on water quality and treatment. Siofok-Lake Balaton. Hungary 25-27 May 1988.

RADIAL CLARIFIERS AND THICKENERS WITH LAMELLA MODULES

J. GEGA, W. KOWALSKI, and J. WROBEL

Academy of Mining and Metallurgy
Institute of Metallurgical Machines and Automatics
Cracow, Poland

ABSTRACT

This paper presents the main technological, designing and economical problems of radial clarifiers and thickeners, working on the basis of Lamella modules. To accelerate sedimentation ferromagnetic particles of electromagnetic coagulation was applied. The results of the research have found practical use in the steel and mining industries. Due to modernization radial clarifiers and thickeners, an increase of their efficiencies, or a considerable in particle contamination drop in outlet water, was obtained. The cost of modernization is equal to 10-20% of demanded cost for the same results realized by traditional facilities.

INTRODUCTION

To achieve a higher cleaning efficiency in a clarifier, the rate of particle sedimentation or sedimentation area should be increased. In fact, it is possible and advisable to simultaneously apply both ways of intensification of the sedimentation process.

In practice, the rate of sedimentation is increased by coagulation of

Chemistry for the Protection of the Environment
Edited by L. Pawlowski *et al.*, Plenum Press, New York, 1991

particles. In the case of suspensions from metallurgical plants which contain considerable amounts of ferromagnetic components, the electromagnetic coagulation method is a very cheap and highly effective one [1].

The practical way of increasing the sedimentation area is the realization of the so-called shallow-depth sedimentation concept. It consists in distribution of a deep stream of suspension into a number of narrow parallel streams. The clarifiers, to which this concept is applied, are called Lamella module clarifiers.

Research on construction of highly efficient clarifiers with application to the electromagnetic coagulation method has been carried out for several years at the Institute of Metallurgical Machines and Automatics, Academy of Mining and Metallurgy, Cracow. These clarifiers have found practical use in the steel industry.

Present practical solutions afforded by Lamella clarifiers include constructions in which plates have been replaced by light packs of plastic ducts. The characteristic parameter of a given type of Lamella module is the specific surface factor. The value of the specific surface factor referred to the layer of packets is, theoretically, equal to a multiple of sedimentation area increase in the clarifier filled in with packets of modules, in comparison with the traditional clarifier. For the Kary modules this factor is 6; it is 12 for the ENVISED modules produced in Poland [2].

In rectangular clarifiers, the modules are placed along the clarifier symmetrically towards the chamber inside, which is destined for elements of the sediment scraping system (Fig. 1). In radial clarifiers the modules make a layer in the space between the filling chamber and the walls of the clarifier. Dimensions of the filling chamber are larger than in the traditional clarifiers, which results from placement of the drive and the supporting structure of the scraping system inside it. When electromagnetic coagulation is applied the dimensions of the chamber should be selected so as to fulfill the function of the reaction chamber.

MATHEMATICAL MODELLING OF THE SHALLOW SEDIMENTATION PROCESS

The research on construction of Lamella clarifiers has been carried out along with the research on designing. A mathematical modelling of shallow sedimentation processes constitutes a basis for the methods of clarifier design. In the Institute of Metallurgical Machines and Automatics, the work is going on in three main directions, linked with the methods based on mechanics of continuous media [3], probabilistic modelling of physical phenomena [4], and consisting in searching for precise analytical solutions of

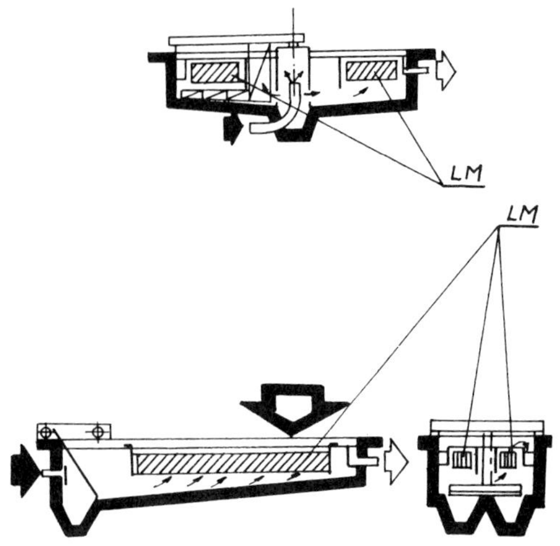

Fig. 1. Lamella module (LM) arrangement.

the classical model of mechanics for the particles sedimentation of the polydisperse solid phases of suspensions. Mathematical models tend to search for the relationship between sedimentation efficiency and main dimensions of Lamella packets, geometrical and hydromechanical clarifiers, properties of liquid and solid phases of a suspension. Positive verifications of elaborate designing methods allowed with their application to designing of highly efficient Lamella clarifiers equipped with the electromagnetic coagulation system.

AN ANALYTICAL-EXPERIMENTAL DETERMINATION METHOD OF EFFECTIVENESS FOR THE CLARIFIERS WITH LAMELLA MODULES CLARIFIERS

As opposed to the mathematical models of the shallow sedimentation process used so far, the model proposed by the authors is far more precise. The fundamental thing is the description of behavior of a real suspension under its stationary flow conditions through a Lamella module element. The basis of the model is the set of two differential equations, one of which is the balanced equation of the flowing mass with the exchange to surroundings, while the latter is the kinetics equation for the process of suspension sedimentation. Behavior of a suspension is defined by the sedimentation activity coefficients as a function of place x, the average speed of liquid flow v and condition is to determine the coefficient α experimentally. The set of the balanced equation for exchanged masses (1) and the kinetics equation can be presented as follows:

$$\frac{\partial C(x,t)}{\partial x} + \frac{\partial M(x,t)}{\partial t} + v \cdot \frac{\partial C(x,t)}{\partial x} = 0 \qquad (1)$$

$$\frac{\partial M(x,t)}{\partial t} = \alpha(x,v,h) \cdot v(X,t) \qquad (2)$$

where:
$C(x,t)$ - concentration of particles suspended in a liquid medium,
$M(x,t)$ - concentration of solid particles deposited at the bottom of the element.
The solution to the above set of equations (1) and (2) enables the determination of the function C(x,t) in the following form:

$$C(x,t) = \frac{\alpha(0,v,h)}{\alpha(x,v,h)} \cdot \exp[-\int_0^x W(s)ds] \qquad (3)$$

and the function $M(x,t)$ in the form:

$$M(x,t) = \alpha(0,v,h) \cdot v \cdot C \cdot t \cdot \exp[-\int_0^x W(s)ds] \qquad (4)$$

where:
$C_0 = C(x=0,t)$
$W(x) = \alpha(x,v,h) - \frac{\alpha'(x,v,h)}{\alpha(x,v,h)}$

After transformations of (3), the relationship describing the sedimentation activity coefficient $\alpha(x,v,h)$ is obtained in the following form:

$$\alpha(x,v,h) = -\frac{\partial[\ln C_E(x)]}{\partial x} \qquad (5)$$

where:
$C_E(x)$ - function of concentration distribution for suspension, which flows along the way x, from experiment.

The basic criterion for the evaluation of the usefulness of the tube clarifier is effectiveness of solid particles sedimentation. For the applied model, the effectiveness coefficient ψ has been introduced, as the ratio of the mass that sedimented in a tube element of length l to ∞ and the mass that could settle over a whole length of the element from 0 to ∞. According to this definition, the effectiveness coefficient ψ is the relation:

$$\psi = \frac{\int_l^\infty \exp[-\int_0^x W(s)ds]dx}{\int_0^\infty \exp[-\int_0^x W(s)ds]dx} \qquad (6)$$

The relationship between the effectiveness coefficient ψ and the conventional effectiveness η is expressed by:

$$\eta = 1 - \psi \tag{7}$$

A PROBABILISTIC MODEL OF THE SEDIMENTATION PROCESS

The model was developed on the assumption that grain-size distribution is of logarithmic-normal character. The grain-size distribution function for particle sediment along an element [4] has been determined:

$$k(x) = -\frac{1}{2\sqrt{2\pi}(x + tg\theta)^{5/2}} \cdot \exp\left\{-\frac{1}{2}\left[\frac{ln(x + tg\theta) - m_x}{\sigma}\right]^2 + \frac{3}{2}m_x - \frac{9}{2}\sigma^2\right\} \tag{8}$$

while:

$$m_x = ln\frac{9\mu_o \cdot S \cdot v}{2(\rho - \rho_o)g \cdot cos\theta} - 2m_r \tag{9}$$

where:
v - rate of suspension flow,
μ_o, ρ, ρ_o - viscosity and density of solid phase and liquid,
θ - angle of element's slope,
S - value of parameter Yao.

The final formula defining the length of the element has a form:

$$\lambda = \frac{99}{16} \cdot \frac{\mu_0 \cdot v}{(\rho - \rho_o)g \cdot cos\theta} \cdot \exp[2(p\sigma - \sigma - m_r)] - tg\theta \tag{10}$$

where:
p - value of the normal distribution distribuant equal to sedimentation efficiency
m_r, σ - parameters of distribution of grain radii-sizes.

MODELLING OF THE SEDIMENTATION PROCESS ON THE BASIS OF THE CLASSICAL SEDIMENTATION THEORY

On the basis of the generalized Hazen sedimentation theory for Lamella sedimentation, the mathematical model and designing method for clarifiers with modules have been developed. Efficiency of sedimentation η is expressed from the equation [5]:

$$\eta = 1 - \int_o^{dg} f(d)dd + \int_o^{dg} (\frac{d}{dg})^2 \cdot f(d)dd \tag{11}$$

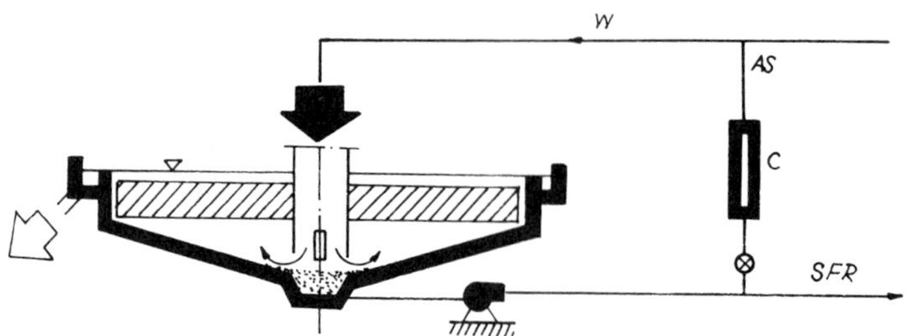

Fig. 2. The scheme suspension circulation in a highly efficient clarifier. CL - cleaned water; SFR - sludge for removal; AS - activated sludge; C - coagulator; W - waste.

where:
$f(d)$ - function of particle-size distributions,
d_g - size of boundary particle, calculated from the Stokes formula with consideration of surface load q referred to the real sedimentation area:

$$d_g = \sqrt{\frac{18\mu_0 q}{(\rho - \rho_0)g}} \tag{12}$$

INDUSTRIAL APPLICATIONS OF LAMELLA CLARIFIERS WITH ELECTROMAGNETIC COAGULATION

On the basis of theoretical research and experiments carried out in the Institute, an industrial Lamella clarifier has been designed and constructed for wastewater circulation in the gas cleaning plant from steel vacuum degasifing in electric steel mill of the VOD-VAD system. The clarifier is equipped with the electromagnetic coagulation system (Fig. 2) [6]. Concentration of solid phase particles entering the clarifier is ca 0.3 kg/m^3. About 80% of grains are less than 5 μm. The chamber of circular clarifier, which is filled in with Lamella packets of the Kary form 800 m^3/h, undergo cleaning. The surface load capacity referred to liquid surface is 5 $m^3/m^2/h$. Despite the remarkable fineness of particles and high surface load capacity, the concentration in cleaned water ranges from 0.03-0.04 kg/m^3. Another example of application of highly efficient sedimentation devices in metallurgy is a clarifier for water cleaning plant from wet dust collectors in converter plant [7]. The clarifier has been equipped with the Lamella modules ENVISED. The intensity of waste water flow is about 700 m^3/h and the surface load capacity exceeds 5 $m^3/m^2/h$. The average content of solid particles in overflow is 0.080 kg/m^3, while concentration in the input suspension is about 3.5 kg/m^3. The clarifier's diameter is 15 m. A traditional clarifier, without the Lamella modules and electromagnetic coagulation system, which guarantees the same remarkable effect of sedimentation, would be 45 m in diameter.

CLARIFIERS WITH A LAMELLA MODULE RING

Modernization of radial clarifiers with diameters of up to 15 m consists in filling up almost all space of the clarifiers with the Lamella modules, which results in a multiple rise in efficiency or in a considerable improvement in quality of cleaned water. In many cases, however, intensification of operation of large-diameter clarifiers is desirable. This especially regards the thickeners, which are destined for simultaneous highly effective densification and clarification processes. For that reason the concept has been worked out to use module rings in thickeners, which will cause an increase in efficiency or improvement in cleaned water quality. Module rings can be used in clarifiers of around 30 and 45 m in diameter.

In radial clarifiers with diameters above 15 m, the scraping system usually has peripheral drive. The elements of a scraping system construction move to cross the axis section areas of the clarifier, which excludes the possibility of filling total space of the clarifier with Lamella modules. There is no objection, however, to the clarifier being partially filled with the Lamella modules. In practice this idea can be realized in the following way. The the module rings ara placed inside the clarifier, and are shielded by clarifier walls and by a special system of partitions forcing an appropriate flow of suspension. The Lamella modules and partitions should be placed either on the supporting construction fixed in the clarifier wall or on the self-supporting construction from the clarifier's wall to its base. The Lamella rings along with limiting walls can be regarded as an extra "secondary" clarifier, placed inside a primary clarifier. The scheme of such a clarifier with the Lamella modules in the form of a ring is presented in Fig. 3. It can be said that this clarifier is, in fact, a system of two clarifiers in series.

Placement of the Lamella ring in the clarifier must follow the changes in construction of the scraping system for sediment which enables the installation of the modules along with the supporting construction without a fall in effectiveness and reliability of the scraping system. In practice this is accomplished by removing these elements of grating which are close to overflow and within the part of the clarifier destined for the module rings, and also by reinforcing other constructional elements so as to maintain the current capacity for sediment transportation.

Rough estimation of modernization costs show that they will consti-

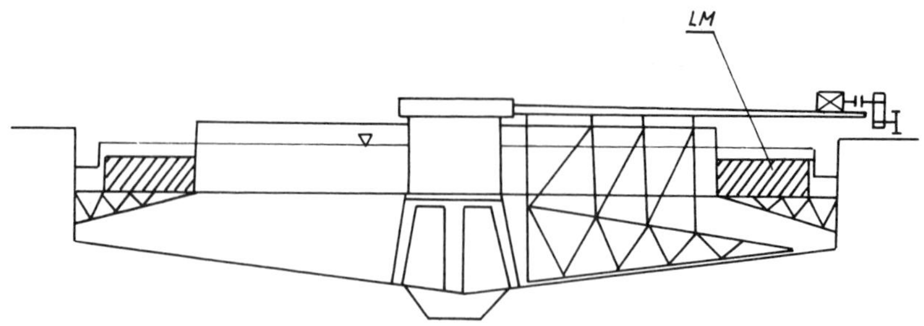

Fig. 3. The diagram of the constructional design of the clarifier with a Lamella module (LM) ring.

Rough estimation of modernization costs show that they will constitute 10-20% of the capital cost necessary to gain the same effects by building new clarification plant.

CONCLUSION

Based on experiments, theoretical analyses and results of industrial exploitation of highly efficient clarifiers and thickeners, the following conclusions may be formulated:

- Construction of new objects of this type and modernization of existing objects should be regarded as the main directions in introducing highly efficient Lamella modules into industrial practice.

- Newlybuilt highly efficient clarifiers usually take up an area five times less than that of traditional clarifiers.

- The application of the electromagnetic coagulation method for iron-bearing suspensions additionally increases the efficiency of the cleaning process.

- Modernization of existing clarifiers, i.e., supplying them with the Lamella modules and with the electromagnetic coagulation system, enables an increase in the effectiveness of their operation with low financial outlay.

REFERENCES

[1] G ega J., Wróbel J., Kowalski W.: Intensyfikacja procesu oczyszczania ścieków hutniczych w obiegach zamkni etych z zastosowaniem koagulacji elektromagnetycznej, Mat. konf. nt. "BEZODPADOWE TECHNOLOGIE'86," Budapeszt, 1986.

[2] Kowalski W.: Ocena teoretyczna wk ladu wielostrumieniowego ENVISED, Mat.

XVII Krak. Konf. Nauk.-Techn. Przeróbki. Kopalin, Kraków-Zakopane, 261-271, 1983.

[3] Gega J.: Badania procesu płytkiej sedymentacji w celu określania głównych wymiarów elementów wielostrumieniowych w osadnikach hutniczych, Zesz. Nauk. AGH, s. Elektr. i Mech. Górn. i Hutn., 94, Prace Inst. Masz. Hutn. i Aut., 6, Kraków 1979.

[4] Czajkowski M.: Metoda projektowania osadników wielostrumieniowych na podstawie probabilistycznego modelu matematycznego procesu płytkiej sedymentacji, Zeszyty Naukowe AGH, w druku.

[5] Kowalski W.: Aspekty teoretyczne projektowania osadników Dorra z pierścieniem wkładów wielostrumieniowych, Mat. VIII Gliwickiego Sympozjum nt. "Teoria i praktyka procesów przeróbczych," Gliwice, 1989.

[6] Gega J., Wróbel J., Kowalski W., Czajkowski M.: Projektowanie i eksploatacja wysokosprawnych osadników w zamkniętych obiegach wodnych w hutnictwie, Mat. konf. "BEZODPADOWE TECHNOLOGIE '86," Budapeszt, 1986.

[7] Czajkowski M., Gega J., Kowalski W.: Badania procesu oczyszczania zawiesiny ze stalowni konwertorowo-tlenowej w wysokosprawnym osadniku wielostrumieniowym, Mat. II Ogólnopol. Konf. Nauk. Inst. Inż. San. Pol. Częst., Ustronie Morskie, 1987.

USE OF GRANULAR ACTIVATED CARBON FOR THE REMOVAL OF BIOHAZARDOUS COMPOUNDS FROM DRINKING WATER

C. H. NADOLNEY

Office of Toxic Substances
U.S. Environmental Protection Agency[1]
Washington, D.C. 20460, USA

ABSTRACT

Treatment plant procedures are being developed for the removal of biohazardous organic compounds from source waters. Many of these are unknown biologically active compounds of low volatility that are present at low concentrations; others are known mutagens and carcinogens. Bioassays performed on subfractions of extracted organics have demonstrated the presence of additional and as yet unidentified mutagens. The toxic organic contaminants probably arise from two main sources: man-made industrial chemicals entering the source waters from point discharges, spills, and agricultural runoff, and as conversion bioproducts of disinfection during the treatment processes used in the preparation of drinking water. An effective method for the removal of many bioreactive water-borne contaminants

[1]The ideas and options are the author's; they do not represent the views and positions of the agency.

Chemistry for the Protection of the Environment
Edited by L. Pawlowski *et al.*, Plenum Press, New York, 1991

is through adsorption by granular activated carbon (GAC). The role, effectiveness and limitations of GAC in the removal of residues of organic compounds in drinking water will be assessed.

INTRODUCTION

One of the most important environmental problems facing the United States and other industrially developed nations is the extensive contamination of drinking water resources by toxic substances. The presence of trace levels of organic compounds in groundwater and surface waters has heightened concern among the scientific community and throughout the public sectors as to the human health impact of chronic exposure to these organic contaminants. The full dimensions of this problem are not known because of the costs of sampling and chemical analysis for toxic contaminants. The toxic contaminants probably arise from two main sources: man-made industrial chemicals entering the source waters from point discharges, spills, and agricultural runoff [1], and as conversion bioproducts of disinfection during the treatment process used in the preparation of drinking water [2]. Bioassays have identified trace levels of several known potent mutagens and carcinogens of the organic contaminants in finished drinking water prepared by conventional methods [3, 4]; however, these are reduced significantly in the effluents of raw water treated by GAC [5, 6].

In the United States, a complex set of laws and institutions attempt to protect drinking water resources from such natural and man-made contamination. Chief among these on the federal level are the Safe Drinking Water Act (SDWA), the Clean Water Act (CWA), the Resource Conservation and Recovery Act (RCRA), and the Comprehensive Environmental Response, Compensation, and Liability Act (CERCLA or Superfund), laws implemented by the U.S. Environmental Protection Agency (USEPA). State regulatory efforts cover an even larger patchwork of laws and agencies. The two basic characteristics of all the regulations and programs are cleanup and prevention.

In recent years, the increased regulatory pressures have caused operators of several water systems that are vulnerable to contamination of their water supplies by synthetic (SOCs) and volatile (VOCs) organic chemicals to modify their treatment plants to include granular activated charcoal (GAC) adsorption [7]. The use of GAC in the preparation of safe drinking water in the United States had been proposed by the National Academy of Sciences [8].

METHODS

The use of carbon materials for the removal of impurities from drinking water dates to ancient times [9]. Now it is used in municipal and industrial processes, including wineries and breweries, paper and pulp, pharmaceuticals, food, petroleum, and petrochemical applications. GAC has been used widely in water treatment systems to remove dissolved "nuisance" organics which contribute to odor and taste problems. It is also a filtration medium for the removal of suspended particulates and colloidal matter. GAC effectively removes VOCs (e.g., alkanes, alkenes, and aromatics) as well as pesticides and herbicides from groundwater [10].

The GAC adsorption systems used in drinking water treatments usually are installed in a stationary bed having a depth of 0.76 to 3.5 m (2.5 to 10 ft). Water under treatment is passed downward through the adsorbent. The adsorbed material accumulates at the top of the bed until the amount adsorbed reaches a maximum. Flow rates vary from 6.2 to 12.2 $m^3/m^2/h$ (2.5 to 5 gpm/sq ft). The "breakthrough" condition occurs when the carbon reaches saturation and the contaminant level begins to rise in the effluent. When the level reaches a predetermined point, regeneration of the carbon is required. Small systems typically dispose of spent carbon or regenerate it off-site. Systems above 3 to 5 mg/d usually provide on-site regeneration for economic reasons, as do systems where carbon usage exceeds 1000 lb/d [10]. Contact time is the single factor affecting carbon performance. Generally, an empty bed contact time (EBCT) of 5 to 15 minutes provides optimal adsorption of the organics found commonly in raw water. The annual cost for carbon replacement is a function of volume and the frequency of carbon replacement [11].

RESULTS AND DISCUSSION

The treatment of groundwater and surface water sources in plant systems featuring GAC adsorption demonstrates the efficient and preferential removal of water-borne mutagens relative to other chemicals that make up the total organic carbon (TOC). The same finding applies to the potential for mutagen formation following rechlorination of effluent water [5, 13]. GAC has an excellent performance efficiency (rated as 70-100%) for the removal of VOCs, pesticides and herbicides, and polychlorinated biphenyl compounds (PCBs) [10]. Among the VOCs, the following three classes of organic compounds have been rated excellent for removal efficiency by GAC technology: a) alkanes, including carbon tetrachloride, 1,2-dichloroethane, 1,1,1-trichloroethane, 1,2-dichloropropane, ethylene dibromide, and dibromodichloropropane; b) alkenes, including 1,1-dichloroethylene, cis-1,2-dichloroethylene, trans-1,2-dichloroethylene, and trichloroethylene; vinyl chloride has an average removal efficiency (rated as 30-69%); and c) pesticides, including pentachlorophenol, 2,4-D, Alachlor, carbofuran, lindane, toxaphene, heptachlor, chlordane, 2,4,5-TP, and methoxychlor.

The ultimate design of a GAC water treatment facility should be based on an economic study as well as on the plant's operation and maintenance characteristics, ease of implementation, and flexibility of the alternatives to meet future upgrade treatment criteria standards for drinking water established by the USEPA. A recent state-of-the-art study on cost estimating functions and equations for GAC treatment systems has been reported [12]. Factors affecting cost are the size of the system, carbon exhaustion frequency, EBCT, choice of adsorption and reactivation technologies, interest rate and life of the facility, carbon loss rate, and GAC purchase cost. The study concluded that steel pressure GAC contactors seem more cost effective than concrete facility contactors for small GAC systems. The concrete facility contactors have significant economies for systems up to 30 mg/d and are cost effective for larger treatment systems. Carbon reactivation by infrared is cost effective for small systems, while fluid-bed reactivation is cheaper when spent carbon is above 3 billion lb/year. Multihearth reactivation is only slightly more expensive than the fluid bed and infrared systems. The GAC bed lives will probably be greater than two months with EBCT of 15 minutes or more. In such cases, total cost estimates for GAC systems range from ca. \$1.00/1000 gal (\$0.26/1000 L) for small (1 mg/d) systems to ca. \$.10/1000 gal (\$0.026/1000 L) for very large systems.

SUMMARY

Biohazardous organic compounds can be efficiently and preferentially removed from source waters by the use of GAC adsorption in treatment plants. The effluent waters contain no detectable levels of mutagens and carcinogens. This treatment is particularly efficient in the removal of VOCs, pesticides, and PCBs. Existing GAC technology is cost effective for both small and large water treatment systems.

REFERENCES

[1] Rosen, A. A. "The Foundations of Organic Pollutant Analysis", in *Identification and Analysis of Organic Pollutants in Water*, L.H. Keith, Ed., Ann Arbor Science Pub., Ann Arbor, Michigan, 3–14, 1976.

[2] Kopfler, F.C., Ed. "Drinking Water Disinfectants", in *Environmental Health Per-*

spectives, Department of Health and Human Services, Publ. No. NIH 86–218, Research Triangle Park, North Carolina, Vol. 69, 312 pp., 1986.

[3] Tabor, M.W., and J.C. Loper, "Separation of Mutagens from Drinking Water Using Coupled Bioassay/Analytical Fractionation", Int. J. Environ. Anal. Chem. 8, 197–215, 1980.

[4] Loper, J.C. "Mutagenic Effects of Organic Compounds in Drinking Water", Mutation Res. 76, 241–268, 1980.

[5] Loper, J.C., M. W. Tabor, L. Rosenblum, and J. DeMarco, "Mutagens of Chlorinated Drinking Water Removed by Treatment with Granular Activated Carbon", in *Water Chlorination: Environmental Impact and Health Effects*, Vol. 5, R. L. Jolley, R. J. Bull, W. P. Davis, S. Katz, M. H. Roberts. Jr., and V.A. Jacobs, Eds., Lewis Publishers, Chelsea, Michigan, pp. 1329–1339, 1985.

[6] Tabor, M.W. "The Role of Granular Activated Carbon in the Reduction of Biohazards in Drinking Water", in *Biohazards of Drinking Water Treatment*, R.A. Larson, Ed., Lewis Publishers, Chelsea, Michigan, pp. 213–233, 1988.

[7] Schalekamp, M. "Effectiveness of a European Surface Water Treatment System, Especially in Removal of Organics", in *Chemistry in Water Reuse*, Vol.2, W.J. Cooper, Ed., Ann Arbor Science. Ann Arbor, Michigan, pp. 465 et seq., 1983.

[8] Doull, J. "An Evaluation of Activated Carbon for Drinking Water Treatment", in *Drinking Water and Health*, Vol. 2, National Academy of Sciences Safe Drinking Water Committee, National Academy of Sciences, Washington, D.C., pp. 251–380, 1980.

[9] Cheremisinoff, P.N. and A.C. Morresi. "Carbon Adsorption Applications" in *Carbon Adsorption Handbook*, P.H. Cheremisinoff and F. Ellerbusch, Eds., Ann Arbor Science, Ann Arbor, Michigan, pp. 1–5, 1978.

[10] Clark, R.M., C.A. Fronk, and B.W. Lykins, Jr. "Removing Organic Contaminants from Groundwater", Environ. Sci. Technol. 22, 1126–1130, 1988.

[11] Gumerman, R.C., B.E. Burris, and S.P.Hansen. *Small Water System Treatment Costs*, Noyes Data Corp., Park Ridge, New Jersey, pp. 177 et seq., 1986.

[12] Adams, J. Q., and R.M. Clark. "Cost Estimates for GAC Treatment Systems", in J. Am. Water Works Assocn. 81, 32–42, 1989.

[13] Loper, J.C., M.W. Tabor, L. Rosenblum, and J. DeMarco. "Continuous Removal of Both Mutagens and Mutagen Forming Potential by a Full Scale Granular Activated Carbon Treatment System", Environ. Sci. Technol. 19, 333-339, 1985.

REVIEW OF WASTEWATER TREATMENT BY MAGNETIC PARTICLE TECHNOLOGY

B. A. BOLTO

CSIRO Bag 10, Clayton Victoria 3168, Australia

ABSTRACT

CSIRO research has made use of magnetic particle technology for a range of wastewater treatment techniques based on adsorption and coagulation processes. Contacting equipment for both types of reactions has been confirmed on the full-scale for water treatment. This approach to intensive processing offers lower capital and operating costs relative to conventional technologies. Pilot studies have been completed, or are well advanced, on a number of wastewater applications, such as metal recovery from electroplating rinsewaters, from sewage sludge and from hydrometallurgical effluents, and on the physicochemical concentration of sewege followed by anaerobic digestion or other physicochemical treatments. Laboratory tests have been carried out on some ten further applications. The approach is especially suited to the processing of slurries, sludges, muds and effluents containing suspended solids.

INTRODUCTION

CSIRO has for many years concentrated its water purification research on the use of finely-divided solid reagents to increase the intensity of processing. The old but severe problem of quickly separating very small particles from the treated waste liquor has been solved by utilizing a magnetic version of the reagent particles.

ADSORPTION PROCESSES

For adsorption processes magnetic fillers are incorporated into cross-linked polymer particles, which are employed in magnetized yet dispersed form in an agitated zone during the adsorption phase of the process. When adsorption is complete, transfer of the particles to a quiet settling zone results in magnetic flocculation and rapid settling of the floc [1]. Magnetic micro particles hence combine the rate advantage of finelydivided materials with the settling properties of much coarser particles. Subsequent regeneration is carried out by re-dispersion of the floc in an agitated zone for stripping of the adsorbed species with an appropriate chemical. The simple equipment required for this purpose makes for easy continuous operation on a large scale. This is facilitated by another characteristic of the magnetized floc: because of the high voidage, the particles may be directly pumped without significant attrition, an operation which is inconceivable with conventional ion-exchange resins.

Magnetic Resins

Magnetic micro resins were originally devised as a way of speeding up the reaction rates of inherently slow ion-exchange processes by providing larger interfacial areas [2]. They may be prepared in a variety of physical formats by incorporating ferromagnetic particles within synthetic polymer beads. The polymers are insolubilized by crosslinking, and may possess a diverse range of functional groups [1].

Typical dimensions of magnetic resins are 100 - 300 μm, versus conventional resins of 300 - 1200 μm. Shell resin formats have an active layer which is 24 - 40 μm thick. The settling rate of the resin is dependant on several factors: the type and amount of magnetic material present, the size and structure (whether homogeneous, composite or shell) of the resin beads, and the strength of the magnetic field. An acceptable method of magnetization is to pass the resin between the poles of a 1000 T horse shoe magnet. The settling curves for a series of homogeneous resins of varying magnetic content and bead size have been documented [2].

Contacting Equipment

For maximum efficiency, ion-exchange processes require countercurrent movement between the resin and the feed and regenerant solutions. Over the last twenty years a number of so–called continuous ion-exchange processes have been developed and some are in commercial use. However, most of these are not truly continuous since the water flow is stopped to permit resin transfer, which greatly adds to the complexity of such plants.

The development of magnetic resins, with their rapid settling and their ability to be transferred without interruption to the operation of the contact or regeneration stages, has stimulated a re-examination of the potential of continuous contacting. Our early studies were carriedout on the dealkalization reaction, using a magnetic weak acid shell resin in a simple fluidized bed adsorber to remove hardness and alkalinity from ground water and effluents [3].

Subsequent refinement of the equipment to allow large-scale operation made use of sieve plate columns (4) which contained a stirrer in each stage so that stable dual flow could be obtained, with resin downflow and the aqueous phase upflow, as illustrated in Figure 1. The results from pilot plant work showed that a much smaller resin inventory was required compared to conventional batch and also quasi-continuous contactors. Even versus a conventional continuous system the resin requirement is about one quarter, because of the faster reaction rates [3]. However, the normal technology cannot make use of resin of such small particle size. The contacting equipment for magnetic resins has been proven on the 1 ML/day scale in desalination studies.

The process has been applied, in the simpler column version, to the treatment of a sewage effluent [5]. The system was operated under extremely arduous conditions, with primary settled sewage which had been flocculated with lime and air stripped to remove ammonia. It had an average turbidity of 37 NTU, ranging from 5 to 135 NTU. The suspended solids included calcium carbonate and organic material, and the average COD was 150 mg/L. Some typical results for the removal of calcium bicarbonate from the wastewater are shown in Table 1.

Heavy Metal Ion Removal

Magnetic resins offer a new perspective in wastewater treatment in that they can cope with effluents containing high levels of suspended matter, and can easily be separated from non-magnetic slurries and even sludges.

Nickel Recovery from Rinsewater. The dealkalization sieve plate column was utilized in trials on nickel recovery from simulated and real electro-

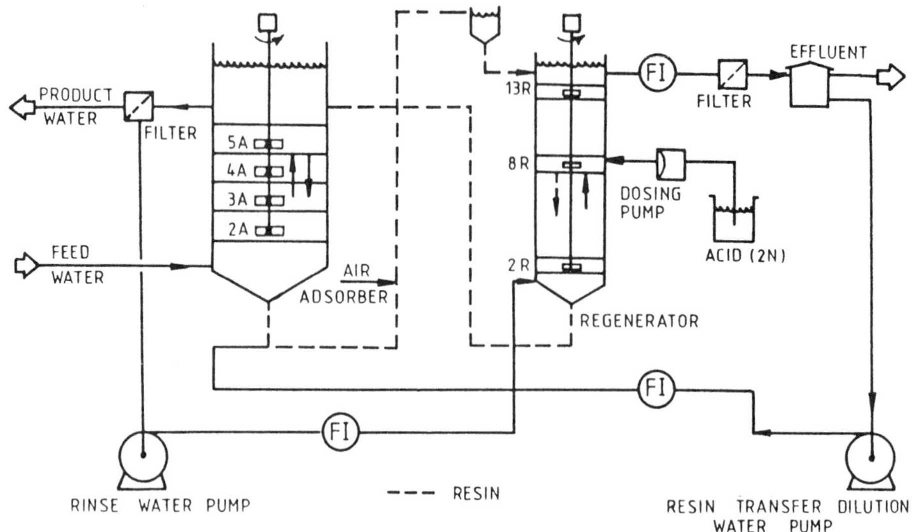

Fig. 1. Continuous magnetic ion exchange system for the dealkalisation process [4].

Table 1. Dealkalization of Sewage Effluent [5]

	pH	Concentration (mg/L)		Turbidity (NTU)
		Ca^{2+}	HCO_3^-	
Wastewater	11.2	66	198	135
Product	5.2	22	12	120
Regenerant effluent	2.2	1580	-	-

plating rinse waters [6]. The high capital cost, the need for prior clarification and the fouling problems which plague the operation of conventional ion exchange for the removal of heavy metal ions from industrial wastewaters were largely overcome by use of the system shown in Figure 2. A magnetic weak acid shell resin successfully brought the metal concentration in the rinsewater to below the limit for sewer discharge, while producing a concentrated regenerant effluent of 2 g Ni/L, suitable for recycle or further treatment and ultimate disposal. This liquor is still much more dilute than a plating bath, which typically contains 80 g Ni/L. However, water evaporating from the bath needs to be made up, and a further concentration factor of about 40 would yield a volume of liquor suitable for this. In experiments with real wastewaters the metal values in the concentrate could not be reclaimed because of the mixture of metals present. However, even where mixing cannot be avoided, the process offers the benefits of a major reduction in the size and cost of the plant needed to treat the rinse water by the conventional lime precipitation method, a reduced carryover of metal species from this smaller precipitation plant and reuse of the product water. The latter would require treatment on a carbon column to remove organics, but the water value would mean savings in both purchase cost and sewer charges.

The pilot plant is large enough to treat the full rinsewater flow from a medium-sized plating operation, and was used as a basis for cost-benefit estimates. Estimates of operating costs and potential savings are shown in Table 2, based on an assumed constant flow of 3 kL/hour and a metal content of 1.8 meq/L (53 mg Ni/L), of which 90% is removed with a 15% excess of regenerants. It can be seen that operating costs are approximately offset by savings in water and sewer charges, even without nickel recovery.

In an existing installation, capital costs can be offset only by the value of recovered metal salts. For a new installation, the capital cost of the ion-exchange plant can be set against the savings achieved by downsizing the clarifier (or eliminating it altogether if metal values are to be recovered). The potential savings shown in Table 2 show that the process offers an attractive payback time in cases where metal recovery is possible. Because of the use of smaller resin beads than in conventional continuous contactors, the process is faster and hence makes use of more compact and cheaper equipment.

Heavy Metals Uptake from Sewage Sludge. Magnetic resins can be recovered from sludges by the magnetic drum separators which are commonly used in mineral and coal industries [7, 8]. The engineering is an extension of the technology developed by CSIRO for the coagulation process to be described below. Operation of a pilot plant has demonstrated that the magnetic resin, when used as a 1 to 4% slurry of normal-sized beads (400 - 1000 μm), can be recovered with as little as 5 mL of leakage per kL of sludge.

The chemistry of heavy metal ion removal from sludges has been investigated in the laboratory with strong acid and chelating resins. Sludge is stirred with a small quantity of ion exchanger, together with a small dose of chlorine, at a pH of 2.5 to 3.0 for 30 - 60 minutes. The ion exchanger is recovered, regenerated with dilute sulphuric acid and recycled. The heavy metal ions in the acid effluent are precipitated with lime to give a small volume of metal hydroxide sludge which would be disposed of in a controlled landfill.

However, the original work has been based on sludges which have been spiked with metals, albeit with ample time for achieving equilibrium of the added metals with the biomass. It has been confirmed for anaerobic digester sludge acclimated with heavy metals. It will be essential to conduct investigations on real problem sludges before reasonable cost estimates can be made, but these investigations can be performed very economically in the laboratory, using conventional ion exchangers. If the results show promise, a simple pilot plant using the equivalent ion-exchanger manufactured in magnetic form can be operated to derive firm costings. The work so far indicates that the concentration of heavy metals in sludges can be lowered at an economic cost.

Decolorization of Pulp Mill Bleach Plant Effluent

Most of the color in kraft mill effluents is ascribed to oxidation and degradation products of lignin. The application of a magnetic weak base resin of the phenolic type to decolorize such effluents has been evaluated in the laboratory [9]. The resin, of size 70 - 300 μm, could reduce the color level from 3000 Pt-Co units to approximately 250 in a two-stage procedure. A continuous process was proposed which involves a co-current pipe reactor for the adsorption steps and counter-current upflow column for the regeneration (with alkali) and reactivation (with acid) steps. This method was adapted from a system originally developed for the dealkalization process [3, 10]. In the pipeline reactor the resin is fully entrained in the turbulent flow of feed water through the pipe. This is followed by a settler where the magnetized resin separates rapidly. In separating the resin from process streams at the various stages (using vacuum drum filters and magnetic drum separators), very high separation rates are achieved because of the magnetic properties of the resin. It is proposed that liquid wastes from the regeneration stages, after some concentration by evaporation, be incinerated in a recovery furnace. Compared with established methods such as lime treatment or fixed-bed ion exchange, the process appears to offer substantial cost economies. It is characterized by design simplicity and requires a comparatively small resin inventory due to the special properties of the magnetic micro-resin. It employs one fifteenth the volume of resin required by fixed-bed ion exchange [9]. A pre-clarification step is not required; this represents a major economic advantage.

Other Miscellaneous Applications

A variety of other avenues have been explored, such as the recovery of chromium and uranium oxyanions on magnetic weak and strong base resins, as well as the use of these adsorbents for the removal of nitrate, the isolation of valuable organic acids and the removal of colored materials from sugar syrups, fruit juices and stillage.

Further opportunities include zinc recovery from process and waste streams, and the recovery of protein from whey with a cellulosic magnetic strong acid resin. A magnetic chelating resin has been shown to have potential for mercury recovery from a mining effluent. Sodium chloride is used as the regenerant.

COAGULATION PROCESSES

Conventional coagulation processes for the removal of insoluble impurities and color involve formation of an adsorptive flocculant precipitate of aluminium or ferric hydroxide at a pH where the floc has an opposite charge to that of the material to be removed. The impurities thus bind to the floc and are removed through sedimentation. The flocs settle slowly, so a large sedimentation tank is necessary, and as settling is never complete, residual particles in the overflow must be removed by sand filtration. The facilities for the treatment of large volumes of water take up much land space, and there is a sludge disposal problem.

As a direct result of the magnetic ion-exchange work, magnetic adsorbents have been prepared which will very effectively remove particulate matter, colloids and colour in the one step [11]. Organic polymers of the quaternary ammonium type have been grafted onto magnetic polymeric cores to give structures akin to the weak acid shell resin described above, except that the polymer chains are now positively charged. Although physically present as a gel layer around the core, the grafted chains at the molecular level resemble threads attached to a smooth–surfaced sphere. The positively charged chains are very effective at adsorbing the usual impurities, which possess a negative charge. Unfortunately, although the system can be regenerated with brine, the cost of the strong base shell resin renders the method uneconomic relative to the alum process.

Cheaper magnetic coagulants have now been utilized, based on finely divided magnetite (Fe_3O_4) particles treated with alkali [12]. This has led to a completely new approach which overcomes many of the disadvantages of conventional coagulation methods. The "Sirofloc" process, as it is called, is faster and more efficient than the alum process, with greatly accelerated sedimentation of the magnetized coagulant and the virtual elimination of sludge. The particles may be reused repeatedly by regeneration with alkali. For the clarification step, a very small amount of acid is added, which gives clean magnetite a positive charge. The impurities in the water, in general, are negatively charged, and hence adhere to the surfaces of the magnetite. A coagulant aid in the form of a cationic polymer is added to bind the negatively charged impurity particles to the now negative surface of the magnetite. On alkali treatment, the magnetite attains a negative charge and the like-charged impurities are sloughed off.

Contacting Equipment

A 0.5 to 2% w/w slurry of 1-10 μm demagnetized particles is employed in a series of three stirred tanks [13, 14]. When removing clay and color bodies from surface or ground waters, contacting is carried out over 10-15 minutes at pH 5 to 6, followed by the addition of about 1 mg/L of cationic polyelectrolyte in the third tank. This reduces the amount of magnetite, which is necessary, and allows the use of higher pH levels. After magnetization the loaded magnetite is quickly and easily separated from the product water in a clarifier, the size of which is less than half that of equipment used in the conventional technology. A flowsheet is shown in Figure 3.

The flocculated, settled magnetite and the bound impurities are pumped from the bottom of the clarifier as a concentrated slurry; to regenerate the surface of the magnetite alkali is added. The impurities are released by the magnetite and are washed off and the cleaned magnetite is reused in the process. The separation of the magnetite from the wash liquor is achieved with magnetic drum separators. These are simple stainless steel hollow drums which rotate through a bath of the diluted magnetite and impurities. Permanent magnets are located inside the drum in the lower arc, and the magnetite is attracted to these regions as the drum surface passes the magnets. As the drum passes from a north to a south pole, the magnetite flips over to accommodate the change in field, which improves the washing efficiency. Raw water is used for washing and, depending on the flow scheme, the amount can be from 2 to 5% of the raw water flow. The magnetite is demagnetized merely by the turbulence encountered in its washing and transport back to the contact tanks.

The resultant wash water contains the original impurities concentrated 20 or more times; there is no gelatinous sludge. Disposal can often be back to the source stream or to the sea. Alternatively, the wash water can be further treated and recycled, but the need depends on the location and relative cost of the raw water.

Three full-scale "Sirofloc" plants of capacity 20 to 35 ML/day have been installed to treat waters ranging from complex anaerobic, highly colored ground water to surface waters having high levels of colour and moderate to low levels of clay [14].

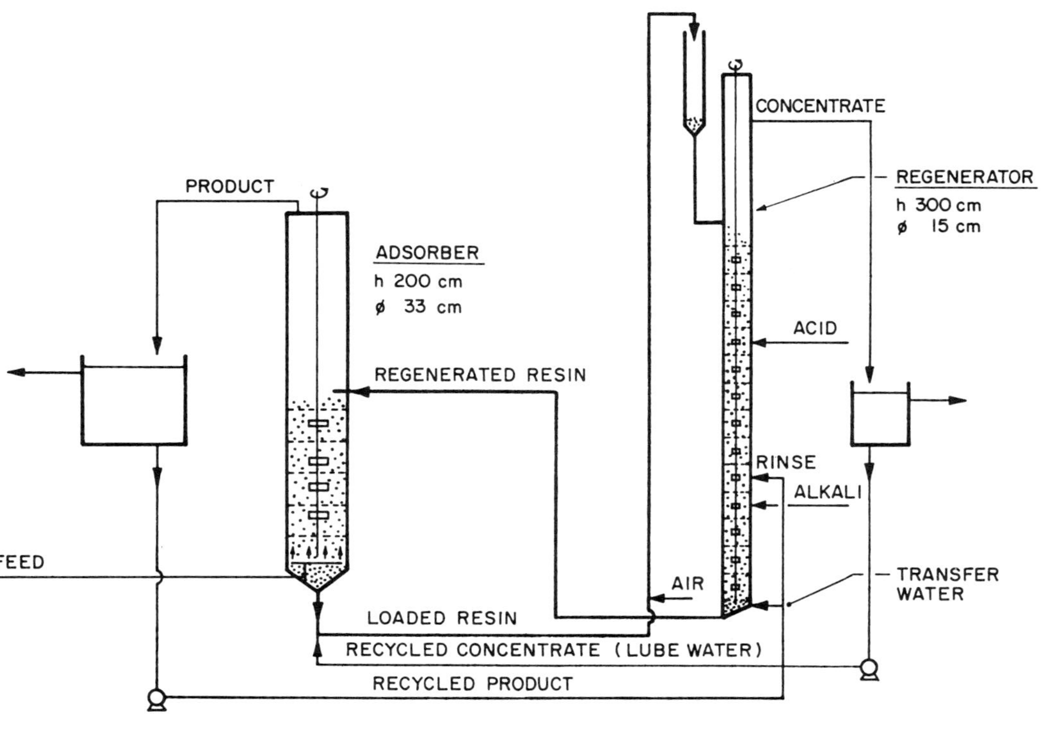

Fig. 2. Continuous magnetic ion exchange system for nickel recovery from electroplating rinse waters [6].

Table 2. Operating Costs (1988) and Potential Savings for Ni Recovery [6]

	Aust $/kL
COSTS	
A. Operating	
Acid	0.03
Alkali	0.05
Power (< 5 kW)	0.19
Labour ($< \frac{1}{2}$ h/shift	0.25
	–––
Total operating costs	0.52
B. Capital, on $35,000	1.46
SAVINGS	
A. Water Recovered, Metal Precipitated	
Purchase of water	0.43
Sewer charges	0.21
	–––
Total savings, metal not recovered	0.64
B. Water and Metal Recovered	
Alkali	0.05
Sludge haulage	0.03 to 0.10
Landfilling	0.17 to 0.64
Recovered metal	0.71
	–––
Additional savings from metal recovery	0.96 to 1.50

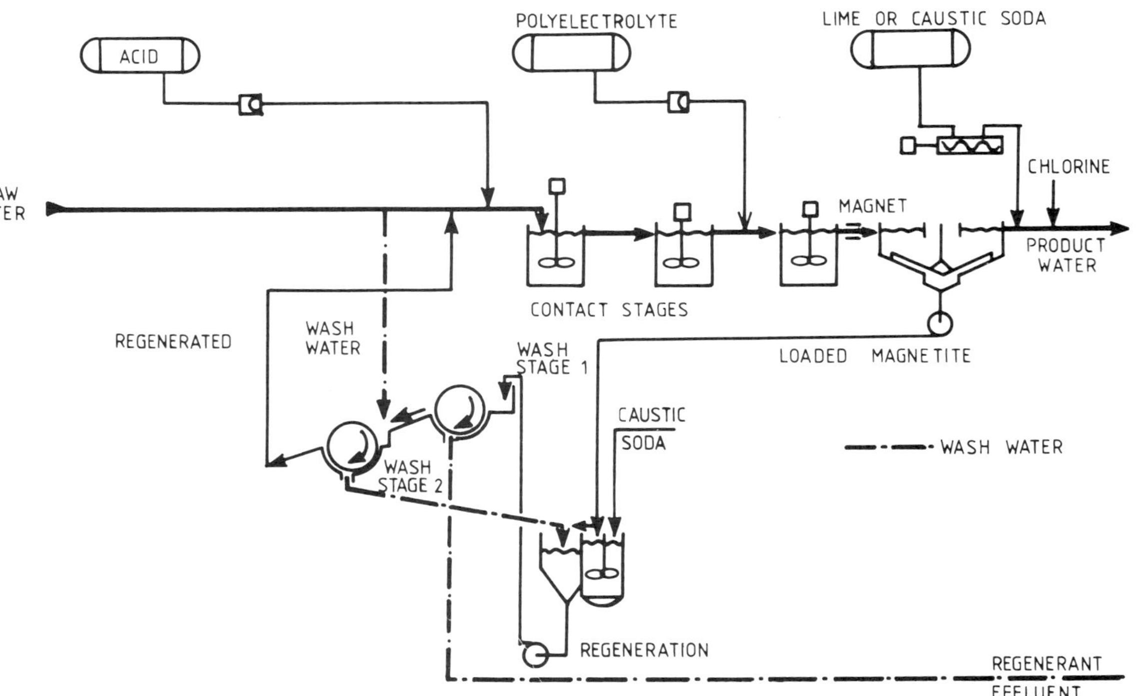

Fig. 3. Process flow sheet for water purification by a coagulation process based on magnetite [13].

Recovery of Metals from Hydrometallurgical Effluents

Magnetite particles may be adapted to the removal of heavy metals [14, 15]. Acidic waste streams are dosed with lime or alkali to selectively precipitate contaminant metal hydroxides to leave the highly valued metal in solution. Pilot-scale work confirmed that such effluents can be treated using a scheme similar to that used for the production of potable water, but employing acid for regeneration rather than alkali. The increase in processing intensity translates into significant capital and operating advantages in comparison with the existing technology. A 2 ML/day facility is now operating at an Australian zinc mine.

Chromate Removal from Cooling Tower Blowdown Water

Reduction of Cr(VI) species by ferrous ions, followed by precipitation of the hydrated metal oxides on magnetite, results in rapid and efficient removal of the chromium [16]. Bench-scale trials have shown that regeneration of the magnetite may be achieved by mechanical or chemical means. Again, a flow scheme based on Figure 3 is followed. The method should be competitive with the existing ferrous sulphate reduction-flocculation and ion-exchange processes.

Physicochemical/Biological Purification of Sewage

Sewage can be treated to secondary sewage treatment plant standards by a two-stage clarification with reusable magnetite particles [17]. The sewage organics, which are now concentrated 30-fold, can be digested in an anaerobic reactor or further treated by physicochemical means. The combination of these two processes has been explored in laboratory experiments, and a continuous pilot plant, demonstrating the first stage, is in operation. The research is aimed at providing sufficient design information to allow a reliable economic assessment of the process. At present the technique appears to offer considerable scope for a dramatic reduction in capital costs (of the order of 50%) when compared to conventional activated sludge systems.

Other Miscellaneous Applications

Many applications to industrial wastewaters treatment have been demonstrated on the laboratory scale [12, 15]. These include effluents from a steel cold rolling mill and a textile dyehouse, plus a paper machine white water.

Cold Rolling Mill Effluent. The method was tested on a light brown effluent of very high turbidity ($> 1,000$ NTU) and containing oil in emulsified form as well as droplets. The total solids content was 2.5 g/L. The liquor could be successfully treated at pH level of 4.0 with a dose of 12.5 g/L of magnetite and 5 mg/L of a cationic polyelectrolyte. The turbidity and color of the treated water were 3 NTU and 2 Pt-Co units, respectively, and visual observation indicated that there was also good oil removal. This performance was maintained over a period of seven cycles with no indication that there was any fouling of the magnetite surface.

Dyehouse Effluent. Successful color removal from synthetic dyehouse effluents has been demonstrated in the laboratory. Two effluents were studied: one, a wool reactive dye, Lanasol Red 6G, containing a small amount of cationic/nonionic detergent mix, had an absorption maximum at 510 nm, while the other was a mixture of Lanasol Red 6G and Maxillon Yellow with a considerable amount of cationic/nonionic detergent mix and an absorption maximum of 430 nm. When treated with 50 g/L of magnetite at the natural pH of the effluent (4.5), color removals of 80 to 100% were obtained. However, over a period of four cycles the color removal fell to 68% for the red effluent, and became very poor for the orange effluent. In all cases, regeneration with caustic soda produced a good release of color, but presence of surfactants appeared to cause a fall off in performance, indicating the need for more severe regeneration conditions.

Paper Machine White Water. Such waters are normally treated by the conventional alum process (average dose 120 mg/L) which has the disadvantage of considerably increasing the TDS and sulphate level of the water and, consequently, limiting the amount of water which can be recycled. Treatment of the effluent with both the conventional and the reusable magnetite processes showed that the latter process produced a superior quality product water and had the added advantage of not increasing the ionic content of the final water. Color removal was poor for both processes, probably because of the nature of the color, which arises from processing chemicals and wood extractives.

Algae Removal. Most algae carry a net negative surface charge, and thus can become bound to magnetite and removed from the raw water. The optimum treatment conditions vary with the raw water, but waters containing both high and low concentrations of algae can be successfully treated. In the case of seawater, spectacular removal was achieved at a neutral pH without the need to add polyelectrolyte. However, the addition to the water of 5 mg/L of phosphate ion as a nutrient resulted in no algae being taken up. Phosphate ions are strongly adsorbed on the magnetite surface and can effectively smother the surface of the magnetite [18].

Bacteria and Virus Removal. The subject has been actively studied with respect to magnetite particles [19 - 22]. Most bacteria have an isoelectric point within the pH range 2.5 to 4.5 and thus exhibit a negative surface charge at neutral pH. They are strongly bound to magnetite under appropriate pH conditions. Chlorinated hydrocarbons may be removed from water and wastewater by adsorbed bacteria [23]; denitrification has also been demonstrated. A similar approach is under study for manganese removal from surface and ground waters.

CONCLUSIONS

Magnetic particle technology has resulted in completely new contacting systems which are truly continuous in operation. This has been confirmed in pilot studies and in full-scale production plants, which show that the major benefit is reduction in size of the overall installation. The concept allows practical treatment of slurries, sludges, muds and effluents containing suspended solids.

REFERENCES

[1] Bolto, B.A., and Pawlowski, L., Wastewater Treatment by Ion Exchange, E & F Spon, London, pp. 10, 121, 1987.

[2] Blesing, N.V., Bolto, B.A., Ford, D.L., McNeil, R., MacPherson, A.S., Melbourne, J.D., Mort, F., Siudak, R., Swinton, E.A., Weiss, D.E. and Willis, D., Some ion-exchange processes for partial demineralization, Ion Exchange in the Process Industries, Soc. Chem. Ind., London, p. 371, 1979.

[3] Bolto, B.A., Dixon, D.R., Priestley, A.J. and Swinton, E.A., Ion exchange in a moving bed of magnetized resin, Prog. Water Tech. 9, 833, 1977.

[4] Swinton, E.A., Bolto, B.A., Eldridge, R.J., Nadebaum, P.R. and Coldrey, P.C., The present status of continuous ion exchange using magnetic microresins, Ion Exchange Technology, Soc. Chem. Ind., London, p. 542, 1984.

[5] Bridger, J.S. and Dixon, D.R., Continuous ion exchange using magnetic shell resins. IV. Pilot plant studies using unclarified feed, J. Appl. Chem. Biotechnol. 28, 17, 1978.

[6] Chin, W.J. and Eldridge, R.J., Treatment of metal finishing wastewaters by continuous moving-bed ion exchange with magnetic resins, CHEMCA 88 International Conf. for the Process Ind., Inst. Chem Eng. Aust., Vol 1, Canberra, p. 89, 1988.

[7] Swinton, E.A., Separation of ion exchangers from slurries by magnetic drum separators, Separation Process in Hydrometallurgy, Soc. Chem. Ind., London, p. 408, 1987.

[8] Swinton, E.A., Eldridge, R.J., Becker, N.S.C., and Smith, A.D., Extraction of heavy metals from sludges and muds by magnetic ion exchange, Sewage Sludge Treatment and Use Conference, EWPCA, Amsterdam, 1988.

[9] Bolto, B.A., Priestley, A.J. and Siudak, R.V., A continuous process for decolorising hardwood caustic bleach plant effluent using a magnetic weakly basic ion-exchange resin, Appita 32, 373, 1979.

[10] Priestley, A.J., Continuous ion exchange with magnetic shell resins. V. Process development, J. Chem. Tech. Biotechnol. 29, 273, 1979.

[11] Anderson, N.J., Bolto, B.A., Eldridge, R.J., Kolarik, L.O. and Swinton, E.A., Color and turbidity removal with reusable magnetic particles. II. Coagulation with magnetic polymer composites, Water Research 14, p. 967, 1980.

[12] Dixon, D.R. and Kolarik, L.O., Magnetic microparticles for treatment of natural waters and wastewaters, Chemistry for Protection of the Environment, ed. L. Pawlowski, A.J. Verdier and W.J. Lacy, Elsevier, Amsterdam, p. 179, 1984.

[13] Anderson, N.J., Bolto, B.A., Blesing, N.V., Kolarik, L.O., Priestley, A.J. and Raper, W.G.C., Color and turbidity removal with reusable magnetite particles. VI. Pilot plant operation, Water Research 17, p. 1235, 1983.

[14] Dixon, D.R., An alternative water treatment method using magnetite, GIT Supplement ACHEMA 2/88, p. 44, 1988.

[15] Anderson, N.J., Bolto, B.A., Dixon, D.R., Kolarik, L.O., Priestley, A.J. and Raper, W.G.C., Water and wastewater treatment with reusable magnetite particles, Water Sci. Tech. 14, p. 1545, 1982.

[16] Anderson, N.J., Bolto, B.A., Pawlowski, L., A method for chromate removal from cooling tower blowdown water, Nuc. Chem Waste Manag. 5, p. 125, 1984.

[17] Priestley, A.J., Sudarmana, D.L., and Woods, M.A., An alternative way to treat sewage, CHEMECA 88 International Conf. for the Process Ind., Inst. Chem. Engin. Aust., Vol 1, Canberra, p. 54, 1988.

[18] Dixon, D.R., Color and turbidity removal with reusable magnetite particles. VII. A colloid chemistry study of the effect of inorganic ions on the efficiency of clarification, Water Research 18, p. 529, 1984.

[19] McRae, I.C. and Evans, S.K., Factors influencing the adsorption of bacteria to magnetite in water and wastewater, Water Research 17, p. 271, 1983.

[20] McRae, I.C. and Evans, S.K., Removal of bacteria from water by adsorption to magnetite, Water Research 18, p. 1377, 1984.

[21] Atherton, J.G., and Bell, S.S., Adsorption of viruses on magnetic particles. I. Adsorption of MS2 bacteriophage and the effect of cations, clay and polyelectrolyte, Water Research 17, p. 943, 1983.

[22] Atherton, J.G., and Bell, S.S., Adsorption of viruses on magnetic particles. II. Degradation of MS2 bacteriophage by adsorption onto magnetite, Water Research 17, p. 949, 1983.

[23] McRae, I.C., Removal of chlorinated hydrocarbons from water and wastewater by bacterial cells adsorbed to magnetite, Water Research 20, p. 1149, 1986.

KINETICS OF THE REMOVAL OF HEAVY METAL HYDROXIDES BY DISSOLVED-AIR FLOTATION IN THE PRESENCE OF SODIUM OLEATE AS A COLLECTOR

E. KARLOVIĆ and D. MIŠKOVIĆ

Institute of Chemistry, Faculty of Sciences, Dr Ilije Djuričića 4, 21000 Novi Sad, Yugoslavia

ABSTRACT

The kinetics of removal of heavy metal hydroxides from their aqueous suspensions by dissolved-air flotation, both in the presence and absence of sodium oleate as a collector was studied. The systems considered were the hydroxides of copper and zinc, taken separately, and a mixture of hydroxides of copper, zinc, nickel, iron and chromium.

The results showed that the kinetics of precipitate flotation of metal hydroxides by dissolved-air techniques may be described by an empirical equation of the form $R = a \cdot t^b$, with the corresponding coefficients of determination in the range 0.70-0.98.

The metal recoveries were in the range 69–97%, and the flotation time was very short: 120 s. Addition of sodium oleate had a positive effect on the efficiency of metal hydroxide removal. This was confirmed by the higher values of the empirical kinetic coefficient a in the equation describing the flotation kinetics. In the range of pH 7-10, the hydrophobic nature of the surface of metal hydroxide particles was the main factor governing the kinetics of the flotation process.

Chemistry for the Protection of the Environment
Edited by L. Pawlowski *et al.*, Plenum Press, New York, 1991

INTRODUCTION

A variety of materials can be concentrated, as well as separated from each other, using foam flotation of surface-active components in a solution to concentrate preferentially at the solution/gas interface [1, 2, 3, 4]. With the use of surfactants as collectors, metal hydroxides can be separated from water suspensions by precipitate flotation [5, 6, 7, 8]. The role of pH in determining the form of the species present in solution and thereby its flotation as ions or precipitates is evident in the works of many authors [6, 9, 10, 11]. The applicability of dissolved-air precipitate flotation (DAF) of heavy metal hydroxides in the absence of a collector and frother has also been demonstrated [12].

The primary objective of this work was to study the kinetics of removal of heavy metal hydroxides from their aqueous suspensions by DAF techniques, both in the absence and presence of sodium oleate as a collector at different pH values.

EXPERIMENTAL

The systems studied were the hydroxides of copper and zinc, taken separately, and a mixture of the hydroxides of copper, zinc, nickel, iron and chromium.

Batch precipitate flotation with dissolvedair has been carried out on a conventional laboratory microflotation set-up, adopting the procedures described earlier [12]. The air flow rate corresponding to the saturation pressure of 300 kPa was dependent on the duration time of process, and ranged from 0.200 to 0.060 cm^3/s.

The suspensions of hydroxides of copper, zinc, nickel, iron, and chromium, each at the level of 80 mg/dm^3, and the hydroxides of copper and zinc when treated separately at the level of 400 mg/dm^3, were prepared from corresponding salts by precipitation with (5 mol/dm^3) NaOH.

The concentration of stock solution of sodium oleate in distilled water was $5 \cdot 10^{-4}$ mol/dm^3.

Concentrations of metal ions were determined by atomic absorption spectrophotometry using a Pay Unicam SP 191 instrument [13].

Efficiency of the flotation process, i.e., recovery R (%), was calculated according to the relation:

$$\%R = 100\frac{C_0 - C_t}{C_0}$$

were C_0 is the initial concentration of metal ion and C_t is its concentration in time t.

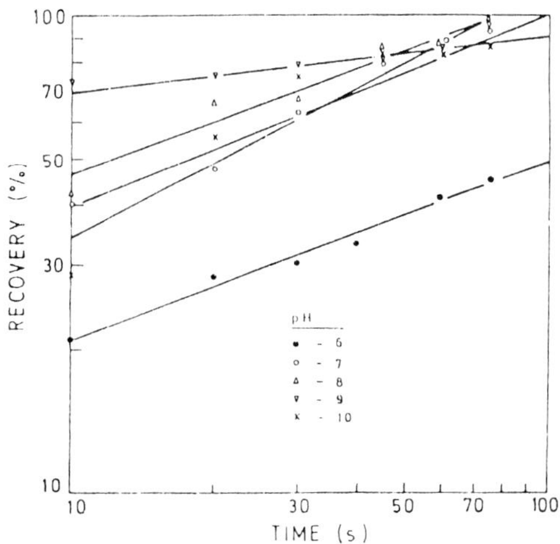

Fig. 1. Kinetics of $Cu(OH)_2$ flotation in the presence of sodium oleate ($20\ mg/dm^3$) at the saturation pressure of 300 kPa as a function of pH.

The kinetics of precipitate flotation of metal hydroxides using DAF techniques is described by an empirical equation of the form:

$$R = a \cdot t^b$$

were R is the total recovery (%), t the flotation time (s), and *a* and *b* the equation coefficients.

The dependence of ln R on ln t is represented by a straight line whose slope shows the increase of flotation recovery (coefficients *b*), and *a* is the intercept, which corresponds to the flotability of the substrate.

RESULTS AND DISCUSSION

Figure 1 shows the relationship between the recovery of $Cu(OH)_2$ and different times and pH values in the presence of sodium oleate ($20\ mg/dm^3$).

Separation of $Cu(OH)_2$ in the presence of sodium oleate is efficient in the alkaline range of pH, when recoveries of about 90% are achieved. The flotation period to 75 s was analyzed using the empirical equation of the form $\ln R = \ln a + b \cdot \ln t$. In Table 1 are listed the values of the coefficients *a* and *b*, and the corresponding coefficients of determination (r^2) of the equations describing the kinetics of flotational separation of $Cu(OH)_2$ in the presence of sodium oleate.

Table 1. Coefficients a and b and Determination Coefficient (r^2) of the Equation Describing Flotation Kinetics of $Cu(OH)_2$ in the Presence of Sodium Oleate (20 mg/dm^3) at Varying pH

	Coefficient		
pH	a	b	r^2
6.0	8.78	0.37	0.98
7.0	15.13	0.42	0.96
8.0	19.41	0.38	0.93
9.0	58.68	0.09	0.97
10.0	10.42	0.52	0.86

In the pH range of 6–9, the value of coefficient a increases with increasing pH, whereas at pH 10 this value is lower. This is in agreement with the results of flotation recovery (at pH 10 the recovery is lower). In view of the adsorption of the collector as a function of pH, the competition of the OH^- and oleate ions at pH above 9 is enhanced, resulting in a decrease in sorption of the collector, and thus, in a lower recovery.

Separation of $Cu(OH)_2$ at pH 8 in the absence of the collector at the saturation pressure of flotation pulp of 300 kPas, can be described by the following empirical equation:

$$R = 7.85 \cdot t^{0.52} \quad (r^2 = 0.86)$$

The maximum recovery of $Cu(OH)_2$ in the absence of the collector after 90 s of flotation is 72%. On the basis of kinetic coefficient a, which is higher in the presence of sodium oleate, we can conclude that separation of $Cu(OH)_2$ under these conditions is a prevailing physico-chemical process. On the other hand, pH does not substantially influence the coefficient b, which means that the hydrophobic nature of particles is the factor governing the kinetics of $Cu(OH)_2$ separation. The results of studying the effect of pH on kinetics of $Zn(OH)_2$ separation in the presence of sodium oleate (40 mg/dm^3) and saturation pressure of 300 kPa are shown in Table 2.

In the pH range of 8.0-9.5 the recovery after 120 s of flotation is almost 96%. In this interval, the values of coefficient a are the highest, reaching its maximum at pH 8.5. This is characterized by the maximum recovery of $Zn(OH)_2$ (96.9%). The lower flotation recovery (i.e., the lower values of coefficient a) at pH 10 is probably due to the adsorption of sodium oleate on the solid $Zn(OH)_2$ particles.

The kinetics of $Zn(OH)_2$ flotation at pH 8.5 in the absence of the collector and at the saturation pressure of flotation pulp of 300 kPa can be described by the following empirical equation:

Table 2. Coefficients a and b and Determination Coefficient (r^2) of the Equation Describing Flotation Kinetics of $Zn(OH)_2$ in the Presence of Sodium Oleate (40 mg/dm^3) at Varying pH

	Coefficient		
pH	a	b	r^2
7.0	7.95	0.22	0.84
8.0	12.21	0.44	0.70
8.5	44.31	0.16	0.66
9.5	19.30	0.36	0.77
10.0	5.72	0.61	0.80

$$R = 23.46 \cdot t^{0.23} \quad (r^2 = 0.89)$$

After 120 s of flotation, maximum flotation recovery of $Zn(OH)_2$ is 69%. A comparison of the values of kinetic coefficients in the equations describing the kinetics of $Zn(OH)_2$ separation at pH 8.5 in the presence and absence of sodium oleate indicates that the process is favored physicochemically in the presence of sodium oleate (both coefficient a and recovery are higher).

The results of the separation of mixtures of hydroxides of copper, zinc, nickel, chromium, and iron in the presence of sodium oleate (40 mg/dm^3) and at a saturation pressure of flotation pulp of 300 kPa are presented in Table 3.

The relatively high values of the coefficient of determination (r^2) for the equations describing kinetics of flotational separation show that the proposed empirical equation describes well the kinetics of separation of metal hydroxides from their mixture.

The values of coefficient a are higher for pH 6 and 7 than for pH 8 and above. At pH values below the pH of quantitative precipitation of hydroxides from an aqueous suspension, the positively charged ions formed by the metal ion hydrolysis also interact with the collector. The flotability of particles depends on the nature of the metal and its interaction with the collector. Thus, at pH 6 and 7, the maximum recovery (to 80%) is achieved in the flotation of hydroxides of copper, iron, and chromium, and somewhat lower recoveries are achived (about 50% at pH 6, and about 65% at pH 7) for the hydroxides of nickel and zinc.

In the pH range 8–10 the highest value of coefficient a corresponds to pH 8, and afterwards it shows a decrease. At pH 8 the highest value of coefficient a is observed for the hydroxides of nickel and zinc. As the

Table 3. Coefficients *a* and *b* and Determination Coefficient (r^2) for Equation Describing Flotation Kinetics of Metal Hydroxides from their Mixture in the Presence of Sodium Oleate (40 mg/dm^3) as a Function of pH

pH	Metal	Coefficient		
		a	b	r^2
6.0	copper	37.18	0.17	0.96
	nickel	32.33	0.08	0.94
	iron	32.83	0.19	0.91
	zinc	17.31	0.22	0.98
	chromium	30.98	0.21	0.96
7.0	copper	36.54	0.17	0.97
	nickel	40.00	0.10	0.99
	iron	31.53	0.19	0.95
	zinc	29.77	0.18	0.97
	chromium	26.71	0.23	0.98
8.0	copper	10.77	0.50	0.79
	nickel	22.80	0.32	0.81
	iron	8.69	0.54	0.81
	zinc	12.64	0.46	0.79
	chromium	10.52	0.50	0.79
8.5	copper	0.74	1.10	0.82
	nickel	9.36	0.52	0.88
	iron	0.81	1.08	0.81
	zinc	1.17	1.00	0.84
	chromium	0.13	1.50	0.79
9.5	copper	3.98	0.72	0.82
	nickel	10.97	0.53	0.91
	iron	1.89	0.93	0.87
	zinc	2.47	0.86	0.89
	chromium	0.02	2.06	0.81
10.0	copper	2.82	0.82	0.92
	nickel	10.89	0.50	0.95
	iron	2.23	0.87	0.91
	zinc	3.92	0.75	0.92
	chromium	0.02	2.01	0.83

Table 4. Coefficients a and b and Determination Coefficient (r^2) for Equation Describing Flotation Kinetics of Metal Hydroxides from their Mixture at pH 8 in the Absence of a Collector

pH	Metal	Coefficient a	b	r^2
	copper	6.14	0.60	0.87
8	nickel	15.76	0.38	0.90
	iron	4.31	0.68	0.83
	zinc	4.58	0.67	0.83
	chromium	9.27	0.49	0.82

pH quantitative precipitation of these hydroxides is between 9 and 10, it is obvious that at pH 8 both ionic and precipitation flotation are involved.

It is evident from Table 3 that in the pH range 8–10 the values of the coefficient b, are higher than at pH 6 and 7. The maximum value of this coefficient corresponds to pH 8.5. This means that the flotability of hydrophobic particles at this pH value is larger, and the difference in the rate of separation depends on the nature of hydrophobic particles and on their competition.

In the pH range of 8-10, the maximum recoveries for all the metals are very high: 94.5% (Cu), 93.9% (Ni), 93.8% (Fe), 92.9% (Zn), and 93.5% (Cr). The recoveries of all the investigated metal hydroxides at pH 8 are only 1-2% lower than those achieved at higher pH values, and the order of their separation is the same. For this reason, pH 8 was chosen to investigate the removal of the metal hydroxides from their mixture in the absence of the collector.

In Table 4 are given the values of coefficients a, b and r^2 in the equations describing the kinetics of flotation of metal hydroxides from their mixture at pH 8 in the absence of a collector.

As can be seen, the DAF technique enables efficient separation of metal hydroxides with no collector present in the mixture. The recoveries are only 2-10% lower then those obtained in the presence of a collector, whereas the coefficients b are approximately equal.

CONCLUSION

On the basis of the experimental results of separation of the metal hydroxides from their aqueous suspensions by DAF techniques we can conclude the following:

- The kinetics of precipitate flotation of metal hydroxides by dissolved-air techniques may be described by an empirical equation of the type: $R = a \cdot t^b$, where the coefficients of determination are in the range 0.70-0.98.

- The use of DAF techniques enables fast and simple separation of metal hydroxides precipitates. The flotation recoveries are in the range 69-98%, and the flotation time is very short: 120 s.

- Due to specific features of the DAF process, i.e., to forming fine air bubbles on hydroxide particles by their precipitation, it is possible to efficiently separate hydroxides of heavy metals from their aqueous suspensions with no collector and frother present in the system. The recoveries are in the range 50-90% and flotation time is 120 s.

- Addition of sodium oleate has a favorable effect on separation of metal hydroxides. By making precipitates hydrophobic, their recoveries increase for shorter flotation time. The greater flotability of particles is evident from higher values of the coefficient a. In the pH range 7-10, the hydrophobic nature of metal hydroxide particles is the main factor governing the kinetics of their separation.

REFERENCES

[1] Sebba, F., Ion Flotation, Elsevier Publishing Company, Amsterdam, New York, London, 1962.

[2] Lemlich, R. Adsorptive Bubble Separation Technique, Academic Press, New York, London, 1972.

[3] Wilson, D., Thackston E.L., Foam Flotation Treatment of Industrial Waters, Laboratory and Pilot Scale, US Environmental Protection Agency EPA, 600/2-80-138, June, 1980.

[4] Clarke, A.N, Wilson D.J., Foam Flotation, Theory and Application, Marcel Dekker Inc., New York-Basel, 1983.

[5] Skrylev, L.D., Mokrushin S.G., Zh. Prikl. Khim., **34**, 2403, 1961.

[6] Baarson, R.E., Ray C.L., Precipitate flotation, a New Metal Extraction and Concentration technique, American Institute of Mining, Metallurgical and Petroleum Engineers Symposium, Dallas, Texas, USA, 1963.

[7] Sheiham, I, Pinfold T.A., J. Appl. Chem., **18**, 217, 1968.

[8] Mahne, E.I., Pinfold T.A., J. Appl. Chem., **18**, 140, 1968.

[9] Rubin, A.J., Johanson D., Analyt. Chem., **39**, 298, 1967.

[10] Bhattachariya, D., Carlton A., Grives R., AIChE Journal, **17**, 419, 1971.

[11] Rubin, A., Foam Separation of Microcontaminants by Low-Flow-Rate Methods, Ph. D. Thesis, University of North Carolina, Chapel Hill, USA, 1966.

[12] Mišković, D., Karlović E., and Dalmacija B., The Investigation of Application of Dissolved Air Precipitate Flotation in the Absense of Collector and Frother for the Purification of the Wastewater Containing Metal Ions, in Studies in Environmental Science 23, Chemistry for Protection of the Environment, Elsevier, Amsterdam, 245, 1984.

[13] Chovanez, T., Ipari Vízvizsgàlatok, Müszaki könyvkiadò, Budapest, 1977.

APPLICATION OF MULTI-MEDIUM FILTRATION FOR WATER PREPARATION

S. MARIC[1], N. DRNDARSKI[2], N. JOVANOVIC[3], and S. HRISTOSKOVA[4]

[1]Institute of Chemistry, Faculty of Science 21000 Novi Sad, Dr. I. Djuricica 4, Novi Sad, Yugoslavia
[2]The Boris Kidric Institute of Nuclear Sciences Vinca Belgrade, Yugoslavia
[3]IHTM, Institute for Catalysis and Chemical Engineering, Belgrade, Yugoslavia
[4]Chemical-Biology Faculty, Plovdiv, NRB

ABSTRACT

Within a wider study of possible applications of some natural materials as filtration media, some physico-chemical properties of anthracite and quartz sand were investigated. The applicability of these materials was tested in the process of separation of suspended particles from kaoline suspensions (model system). They were examined under conditions of the single- and dual-medium open, fast, gravitational filtration. It was found that separation of suspended particles by filtration was almost complete. A significant increase was found in the active filtration period, as compared to classical sand filters.

Chemistry for the Protection of the Environment
Edited by L. Pawlowski *et al.*, Plenum Press, New York, 1991

INTRODUCTION

Description of filtration as a complex method of separation of a solid from liquid phase requires simultaneous consideration of a number of chemical, physico-chemical and hydraulic parameters. The adhesion forces between solid particles of the dispersed phase and the surface of the filtration medium depend on the adsorption characteristics of the surface and textural properties of the filtration medium.

Porosity of the filtration layer plays an important role in precipitation of suspended particles during the filtration cycle, whereas porosity of the filtration medium particles is of importance in filtrating colloidal systems of low concentrations [1 - 9].

In view of the above facts we have undertaken a study of the adsorption characteristics of some natural adsorbents, viz., anthracite and quartz sand.

These materials were characterized by determining their textural properties and tested under conditions of the single- and dual-medium filters in clarification processes of a model water system.

EXPERIMENTAL

All experiments were carried out using the apparatus shown in Fig. 1. Its main part was a glass tube (filtration column) 25000 mm high and 25 mm in diameter. The ambient temperature was 20 ± 0.5 °C.

Before filtration, the model water system was pretreated chemically by adding the coagulant and flocculant; the floccules formed were separated by sedimentation and the partially clarified water was fed into the filtration column. As coagulant, 51 mg dm^{-3} of $Al_2(SO_4)_3$ was used, and as flocculant, 2.0 mg dm^{-3} of anionic polyacrylamide A_3 [10].

Changes of the zeta potential of particles in the influent were measured by the electrophoretic method, using a zeta-meter ZM 77 (Zeta-meter Inc., New York).

Kinetics of the filtration cycle was monitored by measuring the head loss at different levels (depth) of the filtration medium. The efficiency of the filtration process was determined by measuring permanganate consumption, turbidity, pH and suspended matter content by standard methods [11].

Changes of aluminium concentration at different depths of the filtration medium were monitored by X-ray fluorescence spectrometry on a VRA-20 analyzer, VEB Carl Zeiss, Jena.

The pore volume distribution was investigated in the range of transient pores and macropores by the mercury method, using a Carlo Erba 2000 porosimeter.

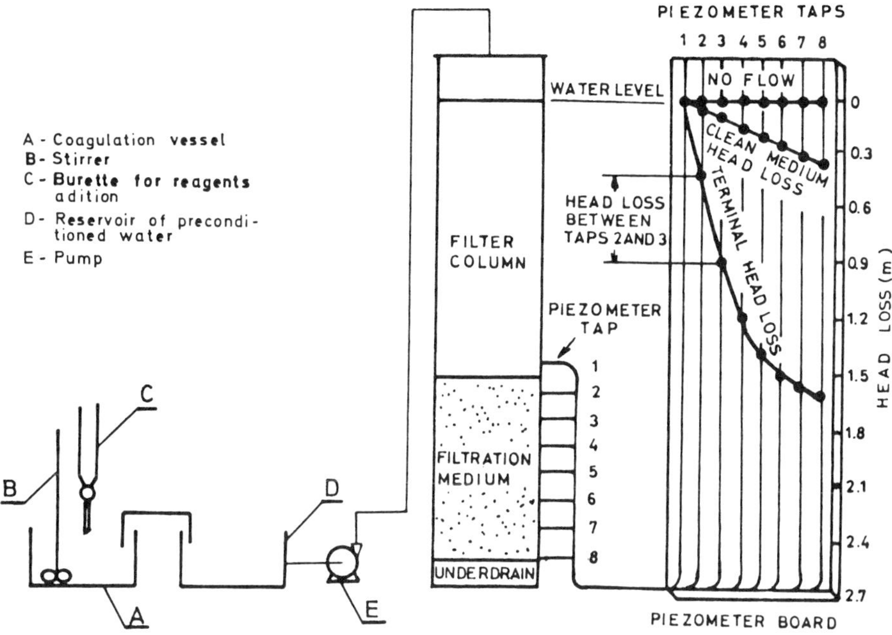

Fig. 1. Schematic diagram of the set-up for chemical conditioning of the influent and filtration.

Specific surface of samples was determined by the BET method, on the basis of low-temperature adsorption isotherms of nitrogen, employing a standard volumetric apparatus.

A JEOL ISM-35 scanning electron microscope was used to study the textural properties of the samples.

RESULTS AND DISCUSSION

Two types of filtration were investigated, viz., the single-medium filtration through quartz sand and dual-medium filtration using anthracite/quartz sand.

Characteristics of the Filtration Media

To analyze the microchanges occurring within the filtration system, the textural properties of the particle surface of all filtration media were investigated.

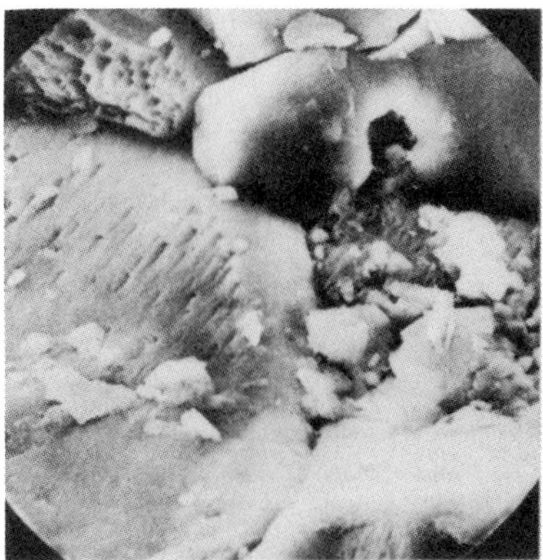
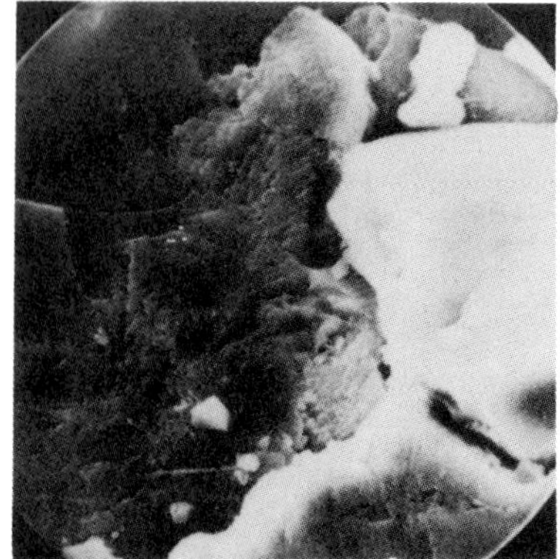

Fig. 2. SEM micrographs of a quartz sand sample (x 2000).

Fig. 2 presents the photographs of the quartz sand surface obtained by scanning electron microscopy (SEM) at magnification x 2000. The surface is layered in nature and shows a low roughness of its layers. The specific surface of quartz sand is low, 1.5 ± 0.2 m^2/g.

The surface of the anthracite sample is plain and smooth with a poorly developed porosity and low specific surface of 0.16 m^2/g. Some coalesced particles of the adsorbent of dimensions 7.7 x 10^{-4} x 1 x 10^{-2} mm can be discerned on the surface.

Fig. 3 presents the SEM micrographs of an anthracite sample.

Dual-medium Filtration through Anthracite-Quartz Sand

Separation of the suspended particles from the model water system preconditioned by coagulation was carried out on the dual-medium filtration bed anthracite-quartz sand. The filtration column was prepared by putting anthracite of particle dimensions 1.0 x 1.6 mm and about 400 mm thick as the first layer.

The results of physico-chemical investigations to characterize the efficiency of the dual-medium filtration of the model water system are shown in Table 1.

Fig. 4 presents the changes of physico-chemical parameters occurring during the filtration cycle, using anthracite and quartz sand.

This filter worked at a decreasing speed which was in the interval of 13.0 - 1.0 m/h.

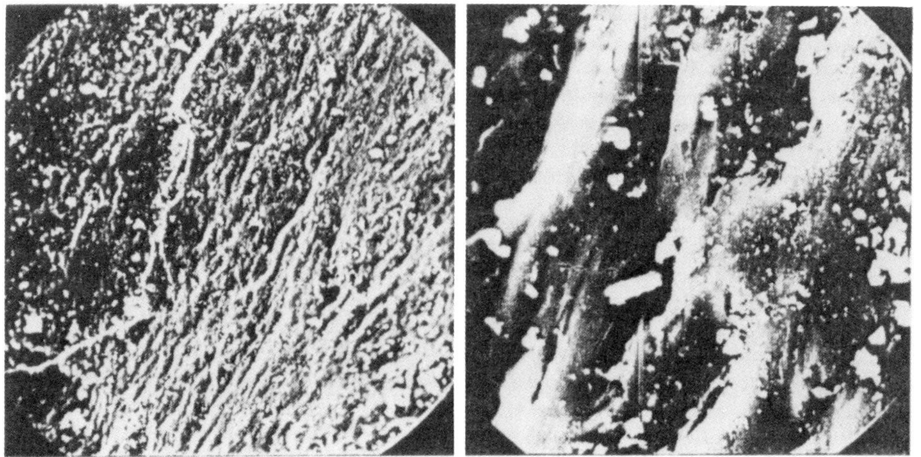

Fig. 3. SEM micrographs of an anthracite sample (x 320 and x 1300).

Table 1. Physico-Chemical Parameters of Dual-Medium Filtration Cycle

Parameter	Influent	Effluent
$KMnPO_4$, $mgdm^{-3}$	4.1	3.2
Turbidity, NTU	13.6	0.47
Dry matter, mg dm^{-3}	6.04	0.00
pH	7.30	7.60
$Al_2(SO_4)_3$: 51 mg dm^{-3}; A_3 : 2.0 mg dm^{-3}, t = 18.2°C		

Separation of the suspended particles took place on the filter surface and in other filter layers.

Detailed investigations of the motion and adhesion of the particles precipitated to the surface of the dual-medium filter were carried out by scanning electron microscopy and the obtained results are presented in Figs. 5 and 6.

In Fig. 5 the precipitate particles of very fine structure are visible, covering the anthracite grains. Cylindrical pores of 1.5 x 10^{-4} mm in diameter can also be noticed.

CONCLUSIONS

The purpose of this work was to characterize the influent (model water system) and some natural materials serving as filtration media for its filtration.

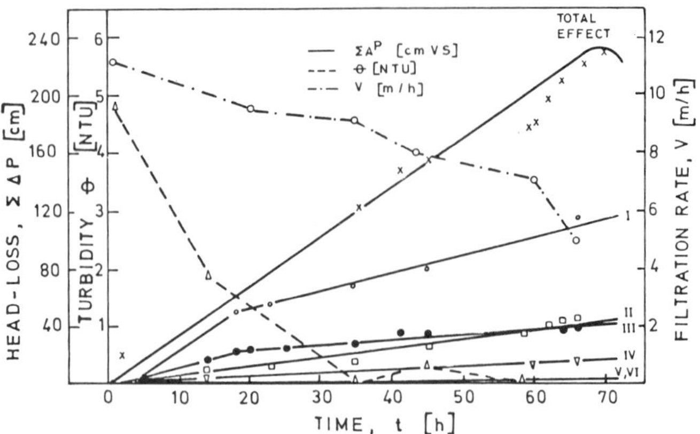

Fig. 4. Filtration cycle of the anthracite-quartz sand filtration bed.

Fig. 5. SEM micrographs of an anthracite sample taken from the surface of dual-medium filter (x 10000 and x 20000).

It was found that particles of clay and of other colloids are negatively charged; the measured zeta potential was - 36.4 mV, which corresponded to the turbidity of 105 NTU.

The studies of the textural properties of the filtration media showed that the specific surface of quartz sand was 1.5 ±0.2 m^2/g, and 0.16 m^2/g of anthracite.

It was concluded that the filtration using the dual-medium filter anthracite-quartz sand enables the removal of 100% of the suspended matter from the influent.

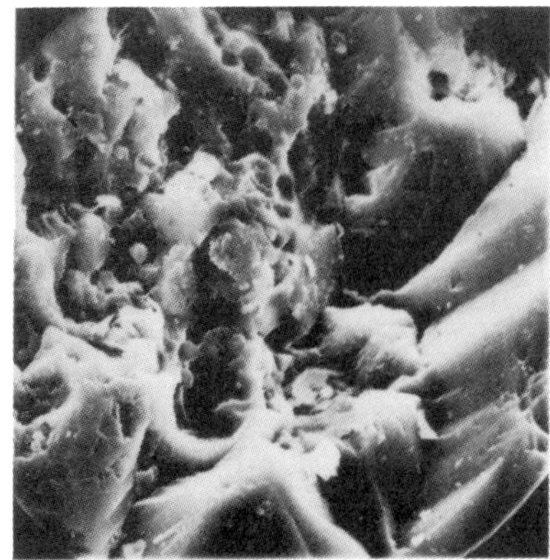

Fig. 6. SEM micrographs of quartz sand sample taken from the sixth level of the dual-medium filter (x 20000).

REFERENCES

[1] Ives K.J., Rapid filtration, Water Research, 4, 201-223, 1970.

[2] Kuljskyi L.A., Ispoljzovanie adgezionnyh i adsorbtsionnyh protsessov dlya udaleniya iz vody vzvesei i mikroorganizmov, Kiev, Naukova dumka, 1973.

[3] Zhurba M.G., Ochistka vody na zernistyh filjtrah, Ljvov, Vishcha Shkola, 1980.

[4] Nikoladze G.I., D.M.Mints, and A.A.Kastaljskii, Podgotovka vody dlya pidevogo i promyshlennogo vodosnabzheniya, Moskva, Vysshaya shkola, pp. 51-171, 1984.

[5] Beljankin D.S., B.V.Ivanov, and V.V.Lapin, Petrografija tekhnicheskogo kamnya, Izd. Akademii Nauk SSSR, Moskva, 1952.

[6] Logsdon G.S., V.C.Thurman, E.S.Fridt, and J.G.Stoecker, Evaluating sedimentation and various filter media for removal of giardian cysts, Journal AWWA, 12, 61-66, 1985.

[7] Craig K., Direct filtration: An Australian study, Jour. AWWA, 12, 56-61, 1985.

[8] Reed G.D., and P.C.Mery, Influence of floc size distribution on clarification. Jour. AWWA, 13, 75-80, 1985.

[9] Qureshi N., Comparative performance of dual- and mixed-media filters. Jour. AWWA, 9, 490-496, 1981.

NITRATES REMOVAL FROM SURFACE RIVER WATER BY MEANS OF A BIOSORPTION SYSTEM

B. DALMACIJA, Z. HAIN, D. MIŠKOVIĆ and M. KUKUČKA

Institute of Chemistry, Faculty of Sciences
I. Djuričića 4, 21000 Novi Sad, Yugoslavia

ABSTRACT

A study was carried out on the application of biologically activated carbon to remove nitrates from surface river waters. In one series of experiments the support in attached growth reactors was commercially available active carbon Karbozak H, while in the other series we used active carbon from agricultural wastes, produced in a pilot plant. The investigation was carried out using surface river water with about 116–117 mg $NO_3^- \cdot dm^{-3}$ added. Ethanol served as the source of carbon for denitrification bacteria. On the basis of the dissolved oxygen concentration in the influent, the optimal concentration of ethanol was found to be 60 mg $\cdot$ dm^{-3}.

The experiments were carried out at different hydraulic loads of the denitrification reactors. A high efficiency (near 100%) of nitrate removal was achieved.

The results showed that biologically activated carbons may be used for efficient removal of nitrates from water.

INTRODUCTION

Recently, an increase in the concentration of nitrates has been observed in both underground and surface waters. This increase is often due to an excessive use of fertilizers in agricultural production. When high doses

Chemistry for the Protection of the Environment
Edited by L. Pawlowski *et al.*, Plenum Press, New York, 1991

of fertilizers are used the soil quickly becomes rich in nitrates, which can easily reach waters. Untreated, and insufficiently treated, wastewaters represent another source of increased concentration of nitrates in water resources. Finally, the nitrates in water may come from wastes deposited on soil of inappropriate structure.

Removal of nitrates from drinking water is presently carried out by physico–chemical and biological denitrification processes. The experience gained in biological treatment of wastewaters and the encouraging results achieved in treating drinking water have directed investigations toward the latter procedures [1-5]. Biological denitrification is a process in which, in the absence of dissolved oxygen, bacteria (mainly heterotrophic) use nitrates and nitrites as final electron acceptors, so that they are transformed into molecular nitrogen. In the case of drinking water, it is necessary to introduce a source of carbon needed for bacterial development, and this can be either ethanol, acetic acid, glucose, or similar compounds.

Treatment of raw water aiming at nitrate removal is presently realized in denitrification reactors in which bacteria are attached to and supported by a granulated material [3]. When the size of the support particles is below 1 mm, the process can be carried out in a fluidized bed. As it is usually used to treat underground waters, an anaerobic reactor is the first unit in the treatment, which is followed by aeration, filtration and disinfection.

Here, we present the results of our studies of the use of an attached growth reactor with granulated active carbon as the supporting material for denitrification bacteria.

EXPERIMENTAL

The laboratory set-up used to study the efficiency of nitrates removal in a biosorption system with granulated active carbon is presented in a schematic form in Fig. 1.

In one series of experiments the commercial active carbon Karbozak H served as the support for denitrification, while in another it was active carbon from agricultural production wastes (CC–10), manufactured in a pilot plant [6, 7]. The first two columns of the set-up served as denitrification units, whereas the third column was aerated. Model water, prepared from surface river water (the Danube) by adding ca 116–175 mg $NO_3^- \cdot dm^{-3}$, was introduced at the bottom of the columns by means of pumps. About 60 mg$\cdot$ dm^{-3} of ethanol was added to the first column inlet.

In the first series of experiments, each column was filled with 1300 g of granular active carbon Karbozak H. The hydraulic retention time in each column was 7, 14, 25 and 39 minutes. In the second series each column

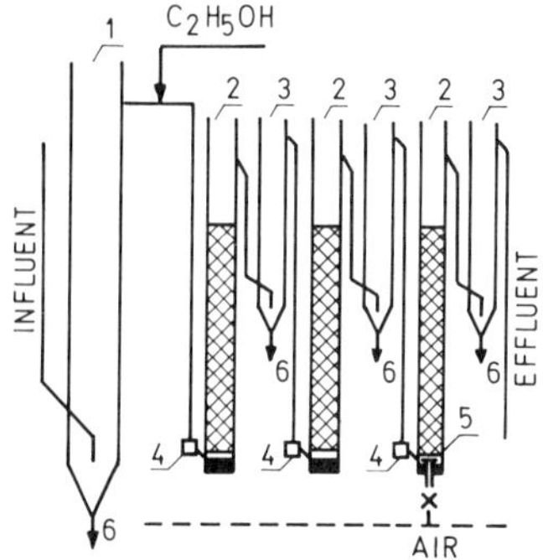

Fig. 1. Schematic diagram of the apparatus: 1 - primary settler, 2 - columns with activated carbon, 3 - settlers, 4 - pumps, 5 - sintered glass plate, 6 - taps for sludge removal.

contained 600 g of active carbon CC–10, and the experiments were carried out at hydraulic retention times (in each column) and 4, 6, 8.5, 17 and 42 minutes.

Standard analytical methods were used to determine the content of NH_4^+, NO_3^-, and NO_2^- in the influent and water after passing through each column [8].

In the second series, at the beginning of each experiment and during each hydraulic retention time, the iodine numbers of the active carbon and the concentration of microorganisms on its surface were determined [9]. The amounts of attached growth were determined in an indirect way by measuring concentration of the protein nitrogen [10].

RESULTS AND DISCUSSION

Table 1 presents average concentrations of the nitrates, nitrites and ammonia in the influent and water after passing each column, determined in both series of experiments for different hydraulic retention times.

The largest portion of nitrates is removed in the first denitrification reactor, after which further reduction of nitrates to elementary nitrogen takes place, as well as removal of part of the remaining nitrites. For the hydraulic retention time of 39 min the efficiency of nitrate removal with Karbozak H was 94.3%. At the same time the nitrite content in purified water was below 0.001 mg·dm^{-3}. The content of nitrites in purified water showed an increase at shorter retention times, reaching 17.24 mg·dm^{-3} at the hydraulic

Table 1. Average Concentrations of Nitrogen Substances

Retention time (min)	Influent	After col. I	After col. II	After col. III	Influent	After col. I	After col. II	After col. III
First series								
	NO_3^- (mg / dm^3)				NO_2^- (mg / dm^3)			
7	126.8	44.0	48.1	38.4	0.048	15.71	8.78	17.24
14	175.1	31.0	14.3	9.7	0.078	33.02	2.92	7.53
25	153.5	28.8	23.4	28.4	0.333	29.63	10.49	4.44
39	145.5	13.8	6.5	8.2	0.298	19.40	2.05	< 0.001
	NH_4^+ (mg / dm^3)							
7	0.00	0.48	0.96	0.55				
14	0.16	1.13	0.98	0.45				
25	0.15	0.88	0.27	0.70				
39	0.06	0.57	0.44	0.50				
Second series								
	NO_3^- (mg / dm^3)				NO_2^- (mg / dm^3)			
4	120.2	80.5	22.1	18.5	0.053	6.56	8.42	11.080
6	125.1	64.9	5.4	0.0	0.042	6.70	4.15	0.030
8.5	116.9	38.3	5.6	1.5	0.085	3.68	3.13	0.710
17	122.3	14.4	0.6	0.6	0.170	4.04	0.17	0.016
42	129.8	56.2	48.1	14.0	0.140	14.48	16.73	7.020
	NH_4^+ (mg / dm^3)							
4	0.04	1.32	1.16	1.07				
6	0.01	0.39	1.18	1.26				
8.5	0.03	1.14	1.04	1.01				
17	0.04	1.26	1.47	1.47				
42	0.15	0.96	0.95	2.56				

retention time of 7 min. When the active carbon CC–10 was used (second series), the highest efficiency (99.9%) of nitrate removal was achieved at the hydraulic retention time of 6 min.

It is evident from the above results that a better efficiency in nitrate removal and, what is much more important, a much shorter retention time are obtained using active carbon CC–10. Its advantage over commercial Karbozak H is also evident from the dependence of the efficiency of nitrate and total nitrogen compound on the nitrate load of active carbon (Figures 2 and 3).

In the experiments with Karbozak H, the efficiency of removal of total nitrogen compounds decreases for nitrate loads above 3 mg N·g^{-1}·day^{-1}; with the active carbon CC-10 this occurs above 4.5 mg N·g^{-1}·day^{-1}.

The explanation for a higher efficiency of CC–10 active carbon should

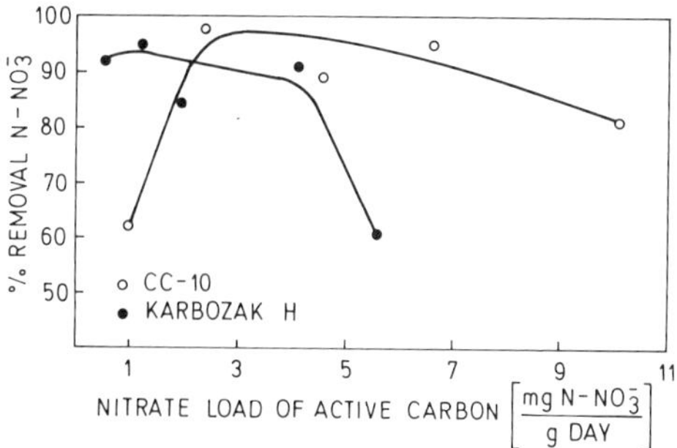

Fig. 2. Dependence of the efficiency of nitrate removal in denitrification columns on the nitrate load of active carbon.

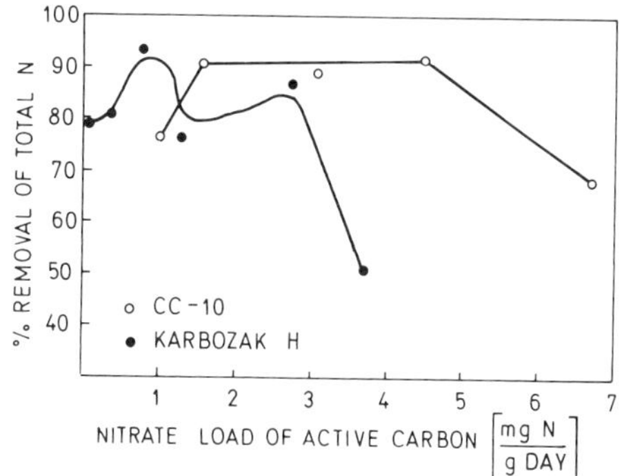

Fig. 3. Dependence of the efficiency of nitrogen compound removal in the set-up on the nitrate load of active carbon.

be sought in the favorable ratio of its micropores, mesopores and macropores (Table 2). It can be supposed that denitrification bacteria can colonize the carbon macropores, so that at higher hydraulic loads of the bioreactor their washout is hindered. On the other hand, the micropores retain ethanol needed for the denitrification process.

The advantage of a biosorption system when used for denitrification over the systems with inert support is in the ability of active carbon to adsorb on its surface both microorganisms and the source of carbon needed for the denitrification process. Thus, active carbon, by enabling simultaneous presence of microorganisms and substrate at increased concentrations, can accelerate the process of denitrification. At the same time, by their

Table 2. Porosity Characteristics of Active Carbons CC–10 and Karbozak H (KH)

	cm^3/g		%	
Parameter	CC - 10	KH	CC - 10	KH
Volume of micropores	0.40	0.50	10.72	31.85
Specific volume of mesopores	0.08	0.27	2.14	17.20
Volume of smaller macropores ($100 < D < 1.5 \cdot 10^4$ nm)	2.84	0.66	76.10	42.04
Volume of larger macropores ($D > 1.5 \cdot 10^4$ nm)	0.41	0.14	10.99	8.92
Total specific volume of pores (V^{Hg})	3.33	1.07	89.28	52.68

metabolic activity, microorganisms bring about regeneration of the active carbon surface so that new amounts of substrate can be adsorbed. This conclusion is supported by the dependence between the free surface degree and the amount of attached growth on the carbon surface (Fig. 4).

Free surface degree was calculated as the ratio of iodine number determined for different stages of the set-up operation and the iodine number of the original active carbon. It was established that an enhancement of the amount of biomass on the carbon surface is accompanied by an increase in the free surface degree.

On the basis of the above findings it can be concluded that the choice of active carbon to be used for denitrification processes in biosorption systems should be based on the determination of distribution and volume of pores in active carbon. Using active carbon of appropriate characteristics, it is possible to shorten significantly the hydraulic retention time, and thus, increase the load of the denitrification reactor.

CONCLUSIONS

Denitrification reactors in which denitrification bacteria are attached to granular active carbons make possible a very efficient removal of nitrate from surface river waters.

When Karbozak H active carbon is used, the efficiency of nitrate removal is 94.3% at the hydraulic retention time of 39 min. When, however, active carbon CC–10 is used, the highest efficiency (99.9%) of nitrates removal is obtained for the retention time of 6 min.

The higher efficiency of the CC–10 active carbon can be explained in terms of its more favorable porosity characteristics.

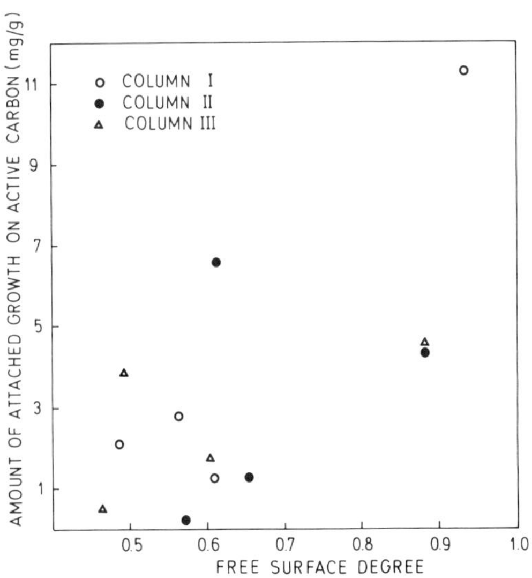

Fig. 4. Dependence of free surface degree on the amount of attached growth on active carbon.

REFERENCES

[1] De Renzo D.J., Nitrogen Control and Phosphorus Removal in Sewage Treatment, Noyes Data Corporation, Park Ridge, 3-432, 1978.

[2] Ginocchio C.J., Trinkwasser durch biologische Denitrifikation, Wasser, Energie, Luft, **75** (10), 250, 1983.

[3] Nitrates et eau potable, La technique de l'eau et de l'assainissement, **412**, 41 - 42, 1981.

[4] Popović M. and I. Popović - Brković, The occurrence of nitrates in water and their removal (in Serbo - Croat), Voda i sanitarna tehnika, **17**, (4, 5), 81 - 89, 1987.

[5] Dalmcija B., Z. Hain and D. Mišković, Vrbaški J., Application of biosorption systems for nitrates removal from surface waters (in Serbo - Croat), Proceedings "Water Protection 88", Dorjan, 1 - 3 June, 1988.

[6] Kukučka M., L. Santo, Ž. Živanov and D. Mišković, The possibility of producing active carbons from the agricultural waste materials, Carbon '88, Newcastle, 286 - 288, 1988.

[7] Kukućka M. and D. Mišković, The study of porosity characteristics of carbonaceous materials from agricultural production wastes (to be published).

[8] Standard Methods for the Examination of Water and Wastewater, Fourteenth edition. American Public Health Association, Inc., New York, 1975.

[9] Rice R.G., G.W. Miller, C.M. Robson, W. Kuhn, In Carbon Adsorption Handbook, P.N. Cheremisinoff, F. Ellerbusch, (Ed.), Ann. Arbor Sci., Ann Arbor, Mich., 1978.

[10] Lure Yu.Yu., Analiticheskaya khimiya promishlennykh stochnykh vod., "Khimiya", Moskva, 66 - 69, 1984.

REGENERATION OF THE EFFLUENT FROM SULPHUR DIOXIDE ABSORPTION IN WET DESULPHURIZATION OF FLUE GASES

I. TRZEPIERCZYŃSKA and M. A. GOSTOMCZYK

Institute of Environment Protection Engineering
Technical University of Wroclaw
Wybrzeże Wyspiańskiego 27, 50-370 Wroclaw, Poland

ABSTRACT

Regeneration of sorption effluences obtained from wet desulphurization of flue gases in calcium-based solutions involves oxidation of sulfites. The efficiency of calcium sulfite oxidation is responsible for the quality of gypsum obtained via this route and for its applications. It seemed, therefore, advisable to investigate this oxidation process with the aim to assess the contribution of iron (which is present in the reaction medium) to the efficiency of calcium sulfite oxidation. The process was conducted at the following parameters: suspension concentration, 1 to 3 wt.%; temperature, 293 K; air flow rate, 8.33×10^{-6} m^3/s. The efficiency of oxidation was found to depend on the pH of the solution, on the concentrate ion of the catalyst and on the duration of the process. The catalytic effect of iron and cobalt on the oxidation efficiency was compared, and its pH-dependence was quantified. Thus, in a neutral medium (pH = 7) the cobalt catalyst prevailed noticeably over the iron catalyst. In an acid medium (pH = 5.5) the efficiencies of oxidation following 1-hour aeration were identical for the two catalysts. After aeration times longer than 1 hour, the oxidation efficiency obtained in the presence of the iron catalyst was higher. The optimum oxidation parameters for a 1.0 wt.% calcium sulphite suspension over the iron catalyst were established. A maximum

Chemistry for the Protection of the Environment
Edited by L. Pawlowski *et al.*, Plenum Press, New York, 1991

efficiency of oxidation (82.5%) was achieved in the presence of a 0.5 wt.% Fe catalyst at pH = 5.5 and at the adopted oxidation time of 20 minutes.

INTRODUCTION

Of the various flue gas desulphurization methods, wet processes have received particular acceptance. Depending on the sorbent or treatment technology applied, wet processes produce sorption effluents, sludges and sediments which contain sulphites and sulphates. The most common sorbents of sulphur dioxide are calcium copmounds – also those contained in fly ash or bottom ash. The flue gas desulphurization process can be conducted so as to yield sulphite or sulphate sediments as end products. Sorption effluents are much easier to regenerate when they contain sulphites, because $CaSO_3 \cdot 1/2\ H_2O$ displays a lower solubility than does $CaSO_4 \cdot 2\ H_2O$. But the intermediate product obtained via this route requires further processing in order to improve those properties of the sediment that may unfavourably affect filtration, storage or the intended uses.

The oxidation of calcium sulphite in solution or suspension to gypsum as an end product and its utility in industrial-scale manufacture of building materials can make the desulphurization process a no-waste technology.

THEORETICAL BACKGROUND

The oxidation of the sulphites produced in the course of the process is one of the major stages in flue gas desulphurization. The oxidation of the calcium sulphites generated during SO_2 sorption is responsible for the quality and, consequently, for the possible uses of the gypsum obtained via this route. Many desulphurization technologies involve auto-oxidation reactions initiated by the oxygen included in the flue gas stream [1, 2]. In the presence of catalytic substances, the rate of auto-oxidation can be increased noticeably [3]. It is commonly accepted that the particulates with transition-metal-enrichment (V, Cu, Mn, Fe, Co, Cr), which are carried in the flue gas stream, account for a catalytic enhancement. Auto-oxidation with no catalytic enhancement runs slower. Among the many factors affecting the efficiency of oxidation, of particular importance are the presence of a catalyst in the reaction medium and the pH of the solution.

The oxidation of sulphites has been reported in specialized literature a number of times [4, 5, 6, 7]. In spite of this, the mechanism governing the reaction, or the factors affecting the reaction rate, or – last, but not least – the catalytic properties are still far from being sufficiently well described or understood.

Of the transition metals occurring in the sulphite oxidation medium, iron and its catalytic potential raise the most serious objections [4, 5].

As far as flue gas desulphurization is concerned, the catalytic properties of iron seem to be worth investigating. Of the transition metals accounting for dust particulate enrichment, the contribution of iron prevails over that of the remainder. Thus, Fe content measured in the particulates of the flue gas stream from the chemical plant CELWISKOZA (Jelenia Góra, Poland) amounted to 6.0 wt.%, whereas that of Mn, V, Cr, Cu, Ni and Co was 0.14, 0.04, 0.01, 0.01, 0.015, 0.009 wt.%, respectively.

To determine the contribution of iron to the oxidation of calcium sulphite in suspension, a series of experiments was run with $FeCl_3$ as a catalyst. The contribution of iron was compared to that of cobalt, a transition metal which is commonly regarded as a very good catalyst in the process of sulphite oxidation [4].

EXPERIMENTAL

Apparatus and Methods

Calcium sulphite in suspension was oxidized in a glass reactor with bottom supply of air through a sinter which covered the entire cross-section of the reactor (12.6 cm^2). The suspension 50 cm^3 in volume was prepared from laboratory sediments of $CaSO_3 \cdot 2\,H_2O$ (with $CaSO_3$ content amounting to 67-70 wt.%). Oxidation was conducted at 293 K and at an air flow rate of 8.33×10^{-6} m^3/s [7]. The concentration of the suspension (C_s) ranged from 0.1 to 5.0 wt.%. Catalyst concentration depended on the $CaSO_3$ content in the suspension. The oxidation process spanned a fixed period of time. At determined intervals, 5-cm^3-volume samples of the suspension were taken for analysis; pH variations in the solution were measured concurrently. Iodometry was used for determining the $CaSO_3$ content in the samples. On the basis of measured data, the efficiency of sulphite oxidation (U) was considered as a function of time (t), quantity of catalyst (C_c), and initial pH. The following equation was used to calculate the efficiency of oxidation:

$$U = \frac{(CaSO_3)i - (CaSO_3)f}{(CaSO_3)i} \times 100\%$$

where U is efficiency of oxidation (%), $(CaSO_3)i$ denotes initial $CaSO_3$ content in suspension (kg/m^3), and $(CaSO_3)f$ indicates final $CaSO_3$ content in suspension (kg/m^3).

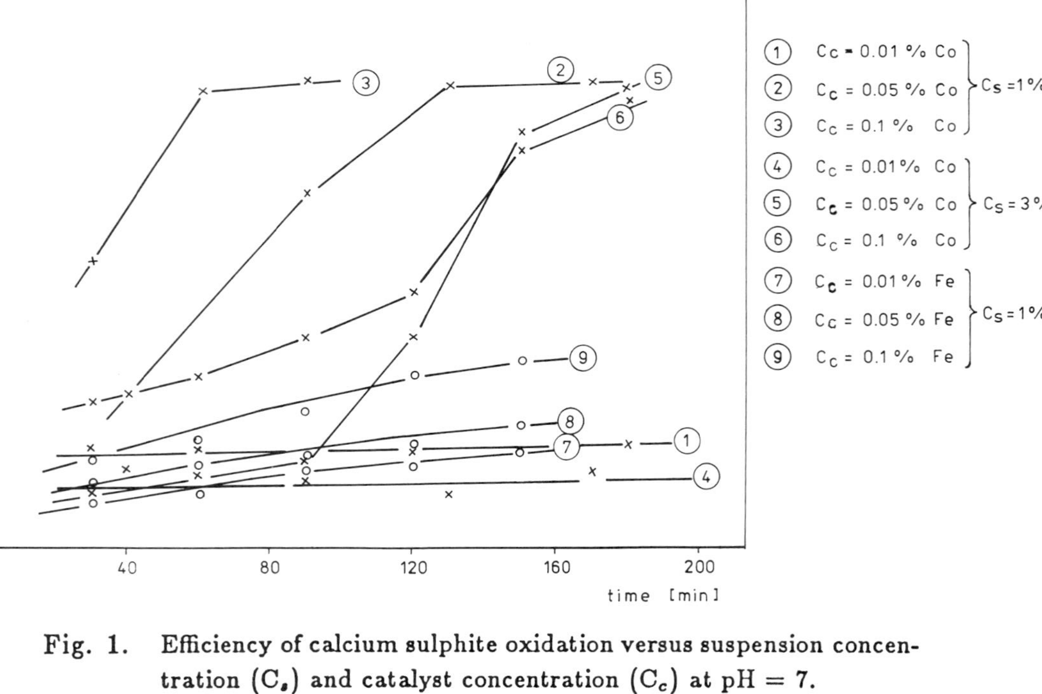

Fig. 1. **Efficiency of calcium sulphite oxidation versus suspension concentration (C_s) and catalyst concentration (C_c) at pH = 7.**

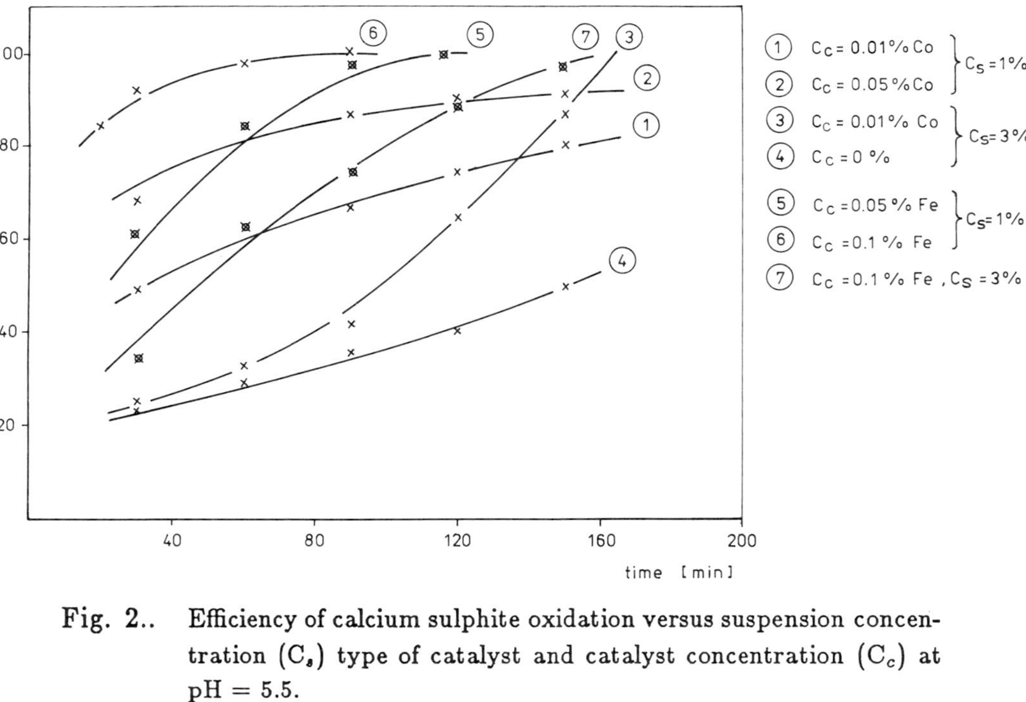

Fig. 2.. Efficiency of calcium sulphite oxidation versus suspension concentration (C_s) type of catalyst and catalyst concentration (C_c) at pH = 5.5.

DISCUSSION OF RESULTS

The results are plotted in Figs. 1 through 5.

Catalyst-enhanced oxidation of sulphites involved $CoCl_2$ and $FeCl_3$ with metal ion concentration varying between 0.01 and 0.1 wt.%, in proportion to the $CaSO_3$ content in the suspension. According to a previous study [7], the optimum pH for oxidation was found to be pH = 5.5.

As shown by the plots of Fig. 1, the presence of cobalt ions in the reaction medium has a substantial effect on the efficiency of sulphite oxidation . Addition of 0.05 wt.% of catalyst at pH = 7 increases the oxidation efficiency from about 20% for 0.01 wt.% Co concentration to about 100%. Complete oxidation is achieved after 130 minutes. Increasing the Co^{2+} content in the solution reduces the time required to achieve complete oxidation of sulphites. Addition of 0.1 wt.% of Co^{2+} leads to an almost complete oxidation after 60 minutes or so.

Addition of 0.01 to 0.05 wt.% of iron catalyst to the sulphite suspension at pH = 7 fails to improve the efficiency of oxidation. With a catalyst dose increased to 0.1 wt.%, oxidation efficiency rose by 10%. The plots also show that higher efficiencies are obtained at lower concentrations of the suspension.

Figure 2 relates the efficiency of oxidation to the duration of the process and to the concentration of the catalyst for suspension concentrations of 1 wt.% and 3 wt.%, and for pH = 5.5. Thus, oxidation efficiency increases with the increasing catalyst content, and is higher at a lower suspension concentration. Aeration conduced for 2 hours over the 0.05 wt.% Fe catalyst for the 1 wt.% suspension yielded 100% oxidation of sulphites. In the presence of 0.05 wt.% Co catalyst, the efficiency of sulphite oxidation amounted to only 90%. The two catalysts yield identical efficiencies after 1 hour of aeration. As shown by these data, the cobalt catalyst is far more effective than the iron catalyst provided that pH = 7 and the aeration time is shorter than 1 hour. At pH = 5.5, it is primarily the pH of solution that accounts for the rate of sulphite oxidation by influencing the solubility of the sediment particles.

As shown by the plots of Figs. 1 and 2, the time required for a complete oxidation of sulphites depends on the catalyst concentration and suspension concentration. To assess the oxidation rate which can be achieved within a time span acceptable in engineering practice, the oxidation process was conducted at pH = 5.5 over an Fe catalyst (0.1 to 1.0 wt.%) for $CaSO_3$ suspension concentrations higher or lower than those under investigation (0.1 to 5.0 wt.%). The adopted time span amounted to 20 minutes. The results are plotted in Fig. 3.

As shown by these plots, the increase in suspension concentration from 1 to 5 wt.% brings about a decrease in the efficiency of oxidation. As the catalyst concentration increases, so does the oxidation efficiency, which reaches its maximum at the Fe^{3+} concentration of 0.5 wt.%. From there, oxidation efficiency continues to decrease. This finding holds for suspension concentrations ranging between 1 and 3 wt.%. For suspension concentration of 5 wt.% no maximum was found to occur. This is likely to be so, because either the time span or the catalyst dose required to achieve this maximum is insufficient. The increasing efficiency of oxidation brings about a decrease in the pH of the solution. This phenomenon should be attributed to the mechanism governing the process [7]. The concentrations of the $CaSO_3$ suspensions adopted for the design of the boiler-room flue gas desulphurization in the chemical plant CELWISKOZA are low (they fall between 0.1 and 0.3 wt.%). More details are included in Reference [7]. In the study reported here, the oxidation process involved a 0.1 wt.% suspension concentration and a Fe catalyst with metal concentration ranging between 0.1 and 1.0 wt.% (which was in proportion to the $CaSO_3$ concentration in the suspension). According to the adopted design which included regeneration of the effluents from sorption with $Ca(OH)_2$, sulphites were oxidized in a weak acid and alkaline range of pH (from 6.5 to 9.0).

Figure 4 relates the efficiency of oxidation to pH and catalyst dose. Thus, in acidic solutions higher efficiencies are achieved with a catalyst concentration of 0.5 wt.%, and in alkaline solutions, with a catalyst concentration of 1.0 wt.%. Higher concentrations of ferric chloride bring about a decrease in the diffusivity of H^+ ions (as a result of the increased ionic force in the solution) and a decrease in the efficiency of the oxidation reaction. In alkaline solutions, the oxidation reaction is inhibited due to the decreased solubility of $CaSO_3$, so there is a manifestation of the catalytic effect which decreases with the decreasing pH of the solutions.

Oxidation in alkaline solutions calls for increased catalyst doses. And this should be attributed to the formation of sparingly soluble ferric hydroxide.

Figure 5 shows the effect of temperature and pH on the efficiency of oxidation. Thus, the oxidation of sulphites at increased temperature (319 K) and varying pH, which was carried out for 20 minutes, failed to lead to the expected efficiency variation. At increased temperature and in the presence of 0.5 wt.% Fe catalyst, oxidation efficiency increases in the acid medium and decreases in the alkaline one. Higher oxidation efficiencies in the alkaline medium can be achieved over a 1.0 wt.% Fe catalyst. In an acidic medium, the efficiency of oxidation decreases in the presence of the 1.0 wt.% Fe contact.

Summing up, at increased temperature the oxidation process is gov-

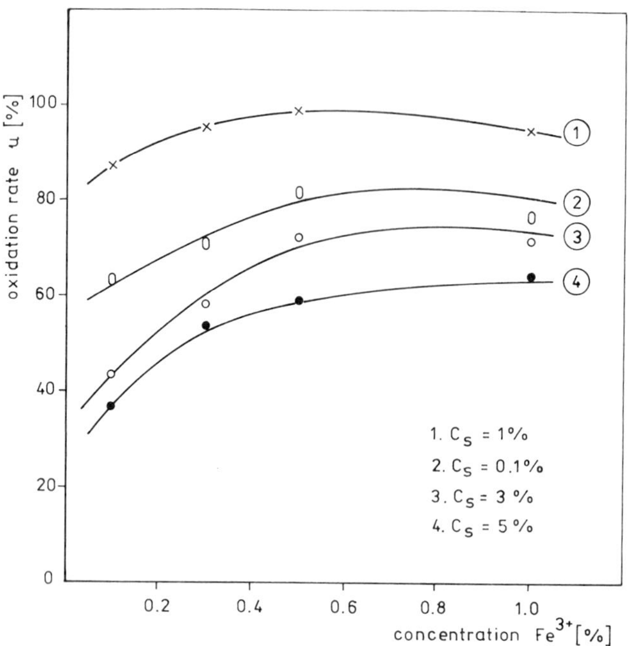

Fig. 3. Efficiency of calcium sulphite oxidation at suspension concentration ranging from 0.1 to 5.0 wt.% versus Fe^{3+} dose at pH = 5.5.

erned by the same accelerating and inhibiting mechanisms, which depend on the pH of the solution and on the concentration of the catalyst.

CONCLUSIONS

Analysis of the results obtained in this study enables the following conclusions to be drawn:

1. The efficiency of oxidation of calcium sulphite suspensions depends primarily on the type and concentration of the catalyst, as well as on the process conditions (pH of the solution, concentration of the suspension).

2. The oxidation efficiencies obtained as a result of 1-hour aeration of a 1.0 wt.% calcium sulphite suspension at pH = 5.5 are identical, irrespective of whether the 0,05 wt.% iron catalyst or the 0,05 wt.% cobalt catalyst has been applied.

3. The iron catalyst is more effective than the cobalt catalyst when the

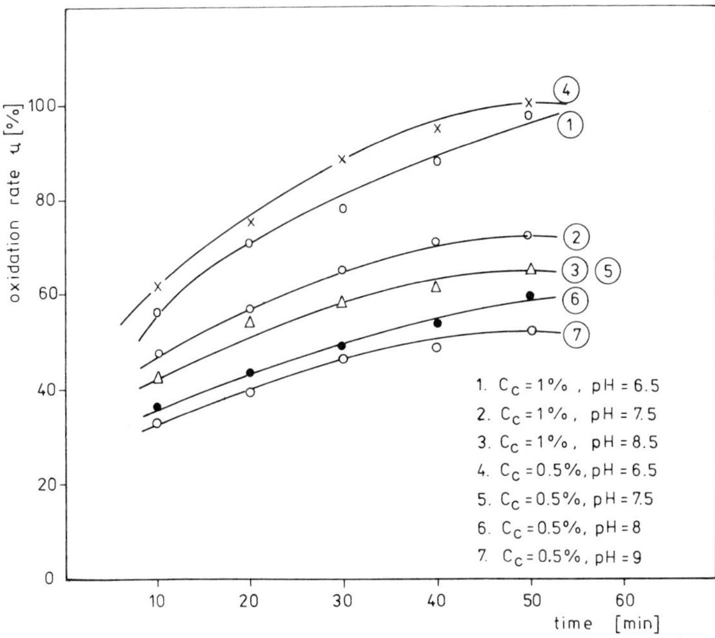

Fig. 4. Effect of catalyst quantity on the oxidation of 0.1 wt.% calcium sulphite suspension at pH = 6.5 to 9.0.

oxidation of calcium sulphite spans a longer period and is conducted in an acidic medium.

4. The cobalt catalyst prevails noticeably over the iron catalyst in neutral solutions.

5. The increase in catalyst concentration shortens the time required to achieve complete oxidation of sulphites; the increase in suspension concentration extends the required time, even if the same oxidation conditions are maintained.

6. The optimum oxidation parameters for suspensions obtained under industrial conditions (with concentrations ranging from 0.1 and 0.3 wt.%) and aerated for the adopted time of 20 minutes are as follows: pH, 5.5 to 6.5; C_c, 0.5 wt.% Fe. The efficiency of oxidation achieved at 298 K varies from 75.3 to 82.5%.

7. Oxidation of sulphites in solutions with pH higher than 6.5 requires increased catalyst concentration (up to 1.0 wt.%) and increased process temperature.

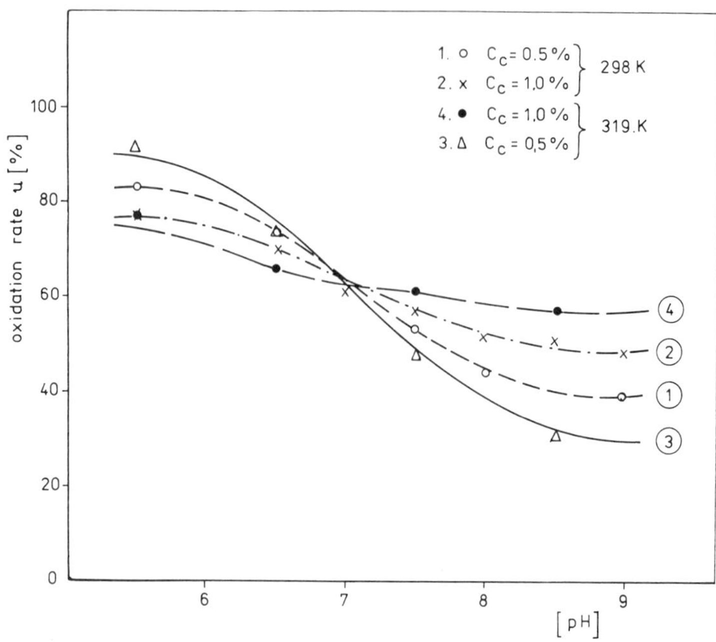

Fig. 5. Effect of temperature on the oxidation of 0.1 wt.% calcium sulphite suspension at pH = 6.5 to 9.0.

REFERENCES

[1] Merlet, H. (Saarberg-Hoelter-Lurgi GmbH), Flue Gas Desulfurization in the Federal Republic of Germany, Technical Solution and Operating Experience, Report, 1984.

[2] Idemura, H. New Thoroughbred FGD Plant (Chiyoda Thoroughbered 121 Plant), *Chem. Ecen. Eng. Rev.* **16** No. 11, 1984.

[3] Pasiuk-Bronikowska, W. and T. Bronikowski, Autooksydacja zwiazków siarki w roztworach wodnych, *Przem. Chem.* 5-6, 1982.

[4] Karlsson, H.T., and S. Bengtsson et al., Oxidation of sulfite to sulfate in FGD systems, *Coal Sci. Technol.* 9, 1985.

[5] Trześniowski, W.R., O utlenianiu i rozkladzie siarczynów amonowych, *Prace Naukowe Politechniki Szczecińskiej* No. 56, 1976.

[6] Ulrich, R.K. et al., Oxygen absorption into bisulfate solutions containing single and synergistic metal catalysts, *AIChE National Meeting*, Houston, 1983.

[7] Trzepierczyńska, I., Regeneracja roztworów posorpcyjnych w zmodyfikowanej metodzie dwualkalicznej odsiarczania spalin. *Report of the Institute of Environment Protection Engineering*, Technical University of Wroclaw, 1988. PWr/I-15/SPR-52/88.

PHYSICOCHEMICAL TREATMENT: OXIDATION – REDUCTION PROCESSES

CURRENT STATUS OF OZONE FOR MUNICIPAL WASTEWATER TREATMENT IN THE UNITED STATES

RIP G. RICE[a] and C. M. ROBSON[b]

[a] Rice International Consulting Enterprises
1331 Patuxent Drive, Ashton, MD 20861, USA

[b] Wentaga Environmental Service Inc. 1901 West Morris Street Indianapolis, IN 46721, USA

ABSTRACT

Although ozone technology has been applied for many years to the treatment of drinking water, its application for the treatment of municipal wastewaters is fairly recent. In the United States, pioneering efforts have been made over the past 20 years to develop the technology of wastewater disinfection with ozone. More than 45 sewage treatment plants in the U.S.A. have installed ozone systems, most for disinfection since the early 1970s, although only 25 of these plants use high purity oxygen for feeding the ozone generation equipment rather than air.

The use of ozone for pretreatment of sewage (by oxidation) preparatory to filtration and adsorption of organics by granular activated carbon has been shown to provide significant processing benefits, which include extension of the useful life of the GAC adsorbers before regeneration is required.

A new process for treating primary sewage sludge with ozone and oxygen under pressure, called the "Oxyozosynthesis" process, has operated successfully for ten years, and a second plant has come on-stream. Primary sludge is converted in a batch process to a light, solid material which is odorless and has been approved for land treatment in place of ocean disposal.

The cities of Denver, Colorado and El Paso, Texas have started wastew-

Chemistry for the Protection of the Environment
Edited by L. Pawlowski *et al.*, Plenum Press, New York, 1991

ater reuse programs, in which ozone plays an important role as a polishing and disinfecting agent.

INTRODUCTION

Early development of ozone treatment of drinking water was pioneered in Europe. In treating municipal wastewater with ozone, however, the U.S.A. and Japan have been the pioneers, with initial studies being undertaken in the late 1960s. The city of Indiantown, FL installed the first U.S. ozone disinfection plant for its sewage in 1975. As of mid-1989, there were at least 45 U.S. municipal wastewater treatment plants known to have installed ozone, mostly for disinfection. The city of Indianapolis, IN is operating the two largest sewage treatment plants in the world yet to use ozone, 120 and 125 million gallons/day (455,000 and 473,000 m^3/day).

A detailed survey of U.S. municipal wastewater treatment plants using ozone was conducted recently by Robson & Rice [1]. These authors found that only 25 plants actually are operating ozonating facilities today. There are many reasons why the other 20 plants were not operating their installed ozonation capabilities, but not all of these reasons bear on the ozonation systems themselves. Managements at many of the non-operational ozonation plants stated that costs proved to be higher than other disinfection options, some no longer were required to disinfect, and others had problems with the ozonation equipment. In many of those plants reporting equipment problems in 1983 [2], these have been overcome, and normal operations are reported today.

Only two European municipal wastewater treatment plants, both French, are known to be using ozone for treating sewage at full-scale plants, although a number of pilot plant studies have been reported in other countries. At Guéthary (southwest France on the Atlantic coast), the basic sewage treatment process is physical-chemical. In 1982, sand filtration plus ozone disinfection was added [3]. The second French plant to use ozone (at St. Michele-en-Grève) began operation in late 1983 [4].

In Japan and Korea, ozone treatment has found increased acceptance for treating effluents from "night soil" sewage treatment plants. Discharge requirements of effluents from these plants in 1981 were as follows:

BOD_5	:	< 10 mg/L
suspended solids	:	< 10 mg/L
COD	:	< 10 mg/L
color	:	not detectable

During the period 1978–1981, ozone was installed at 79 night soil treatment plants in Japan and Korea [5].

Most of the 25 operating U.S. municipal wastewater treatment plants

Table 1. Amount of ozone dosage to attain various levels of wastewater disinfection

Effluent Quality	Absorbed Ozone Dosage for 2.2 Total Coliforms per 100 mL	Absorbed Ozone Dosage for 70 Total Coliforms per 100 mL	Absorbed Ozone Dosage for 200 Fecal Coliforms per 100 mL
Filtered Secondary	35 to 40	15 to 20	12 to 15
Filtered Nitrified	15 to 20	5 to 10	3 to 5

use ozone for disinfection. However there are four notable exceptions. The plant at Chino Basin, CA uses ozone for removal of suspended solids; Marion, NY uses ozone for flotation removal of BOD after secondary treatment; and the Cleveland Westerly plant uses ozone to oxidize organics prior to sand filtration and GAC adsorption. In West New York, NJ and most recently in Hoboken, NJ, ozone is used to dewater sludge and convert it to a dry, off-white product which is quite acceptable for landfill.

Each of these innovative applications of ozone will be described briefly below, along with recent design experiences at the Vail, CO plant, which is the first "second generation" sewage treatment plant in the U.S. using ozone disinfection. Many of the design and equipment lessons learned in early ozone disinfection plants have been applied sussessfully to the Vail, CO plant.

DISINFECTION OF MUNICIPAL WASTEWATER

All but a few of the U.S. wastewater ozonation applications have been for effluent disinfection. The treatment plant effluents for which ozonation is applied normally are nitrified and highly clarified, frequently by tertiary filtration. Although a high level of indicator organism kill or removal already has been achieved in the upstream wastewater treatment process, some form of effluent disinfection is required to achieve common U.S. discharge standards, such as 200 fecal coliforms per 100 mL.

Table 1 provides general information on the ozone dosages required to achieve levels of effluent quality [1, 6].

Wastewater ozonation facilities can be classified on the basis of their feed gases: air and high purity oxygen. The use of air as the ozone feed

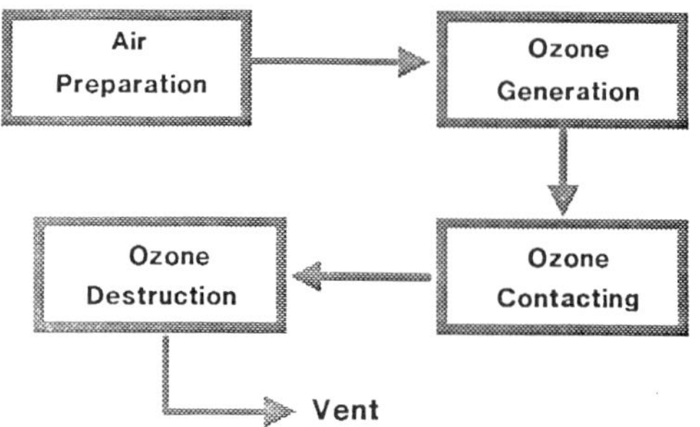

Fig. 1. Air feed process flow diagram.

gas is common throughout the world. In this type of system, ambient air is filtered and dried to a pressure through a corona discharge to generate an ozone-containing gas.

High purity oxygen for supplying ozonation equipment is generated by both cryogenic and pressure swing absorption (PSA) processes. Of the 15 U.S. wastewater treatment plants which installed oxygen for ozone generation, 11 installed cryogenic processes and 4 installed PSA systems. The selection of high purity oxygen production processes was based on the quantity of oxygen required for the ozonation/activated sludge process. The dividing point between PSA and cryogenic systems was at a wastewater flow of approximately 37,850 m^3/day (10 mgd) in late 1970s and early 1980s. This situation should change as larger capacity PSA and smaller capacity cryogenic systems become available.

In turn, high purity oxygen ozonation systems can be divided into "integrated" and "recycle" systems. The integrated system uses the oxygenated off-gas from the ozone contactor as the feed gas for the oxygen activated sludge system. The recycle system filters and dries the oxygen-enriched contactor off-gas prior to its being reused as feed gas to the ozone generators. These feed gas systems are illustrated in Figures 1-3.

The use of high purity oxygen as an ozone generation feed gas in the U.S. was a result of heavy promotion of oxygen activated sludge processes in the early 1970s. The potential availability of high purity oxygen at a wastewater treatment site then suggested the synergism of first using oxygen to generate ozone, and then reusing the oxygen-enriched off-gas from the ozone contactor in the activated sludge process. This application was particularly appealing for wastewater treatment facilities with stringent effluent standards requiring high levels of dissolved oxygen and low levels of residual chlorine.

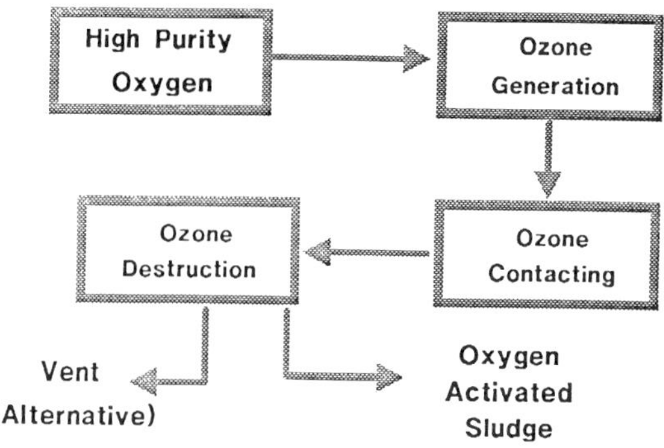

Fig. 2. Integrated high purity oxygen process flow diagram.

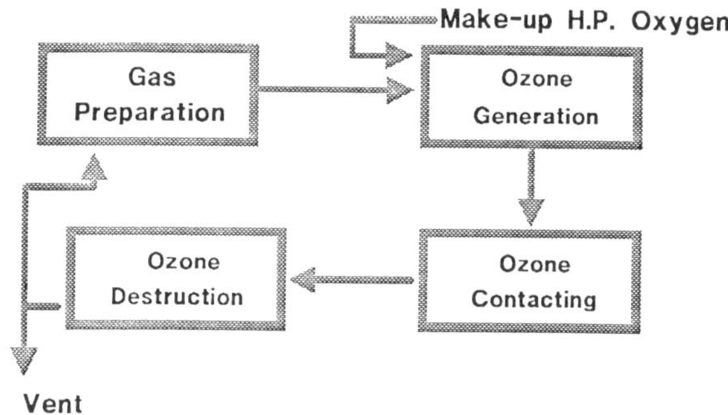

Fig. 3. Enriched oxygen recycle process flow diagram.

Disinfection at Vail, Colorado

Rakness et al. [7] reported the results of a six-months study to optimize process operation and performance characteristics of the ozone disinfection sytem, installed in 1982 at the 2.7 mgd (10 227 m^3/day) Vail, CO sewage treatment plant. Wastewater treatment preceding disinfection includes coarse screening, grit removal, flow measurement, flow equalization, activated sludge, and submerged contactor nitrification. The Vail plant includes microscreening prior to ozone disinfection.

Ozone Generation

Two 126-tube ozone generators are installed, each capable of producing 125 lbs/day of ozone requirements for the entire plant flow (2.7 mgd average flow, 5.4 mgd peak flow), with the second generator provided as backup.

Ozone Contacting System

Two ozone contacting basis are installed (one for each ozone generation system). Each basin is divided into two compartments, both adding ozone (through diffusers) countercurrent to the water flow. About 67% of the ozone generated is added to the first chamber and about 33% to the second. This type of contactor is very similar to many used in European drinking water treatment plants employing ozone. The majority of the ozone is added to the first chamber of the contacting basin to satisfy the immediate ozone demand of the wastewater and to build up a dissolved ozone residual. The second chamber requires a smaller amount of applied ozone to maintain the residual dissolved ozone level, and continues the disinfection process.

Both contactors are 6.1 m (20 feet) deep to maximize ozone transfer efficiency (Vail is 7,500 feet above sea level). The basins are covered and sealed to withstand pressurization while eliminating ozone leakage. Positive sealed access manholes were provided. Baffling was installed to prevent short-circuiting of the flow. A foam control demister was provided on the contactor exhaust gas line.

Ozone Piping and Connections

Type 304L stainless steel with welded connections has been installed at the Vail plant for ozone service. Schedule 40S is used for piping 1.5 inches in diameter or less, and Schedule 10S for piping $>$ 1.5 inches in diameter. Where welding was not possible, threaded or flanged connections were allowed. However, field-cut threaded connections had excessive leakage and had to be welded. Factory-cut threads, including threads on nipples, valves, and gas flow meters performed satisfactorily when attached to another factory-cut threading.

All ozone piping connections were required to be soap-tested at a pressure of 860 kPa (125 psig). It was also required that the ozone contact basin retain 5 kPa (20 inches of water column) pressure without any bubbles in the soap test.

Ozone Disinfection Evaluation

Disinfection requirements (200 fecal coliforms/100 mL) have been met consistently since startup of the Vail plant in October, 1982. Because the treated effluent is of such high quality (BOD_5 and Total Suspended Solids about 6 mg/L), disinfection is being attained at utilized ozone dosages as low as 1.5 mg/L, rather than requiring the 5.6 mg/L design dosages.

Ozone Transfer Efficiency

At the currently required applied ozone dosages of 1 to 3 mg/L (average 1.5 mg/L), the ozone transfer efficiencies measured during plant operation range from 98% to 94%, and average 97%. As a result of this high transfer efficiency, the applied and transferred ozone dosages at the Vail plant are essentially equal.

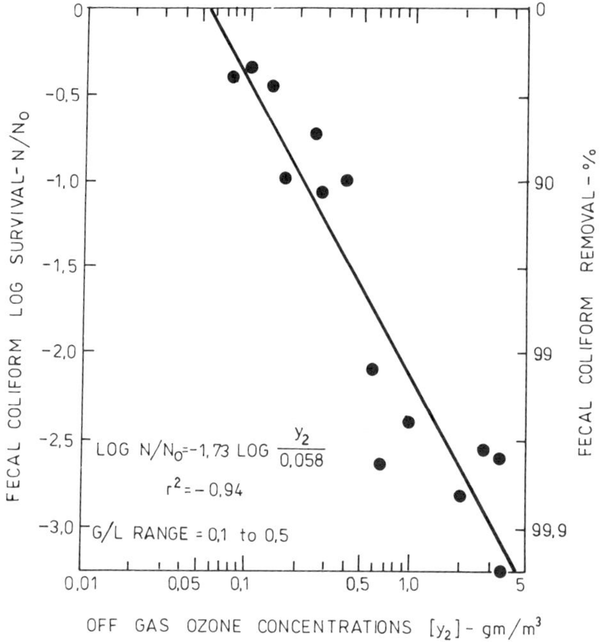

Fig. 4. Transferred ozone dose/disinfection response curve for Vail, CO disinfection system [7].

Ozonation Process Control

An additional and unique feature at the Vail plant is its measurement of the concentration of ozone in the contactor exhaust gases for controlling the ozonation process. This technique is based on work of Venosa & Meckes [8] and of Stover [9], who showed that coliform survival can be related to the ozone concentration in the contactor exhaust gases. This concept was tested at Vail by developing an exhaust gas ozone concentration/performance-response curve. Excellent correlation between log fecal coliform survival and exhaust gas ozone concentration has been obtained, as shown in Fig. 4.

This process control procedure is of specific advantage in controlling disinfection because sewage treatment plant effluents normally contain concentrations of ozone-demanding organic materials which quickly react with dissolved ozone. As a result, measurement of dissolvent ozone residuals is very difficult. In addition, relationships between the fecal coliform levels and levels of dissolved ozone in actual sewage treatment plant effluents still are controversial.

On the other hand, the amount of ozone measured in the contactor exhaust gases is immediately responsive to changes in wastewater quality (i.e., changes in ozone demand), as well as to changes in wastewater flow rate.

Summation

Vail municipal officials have chosen the ozone disinfection process for several reasons [10]. First, the effluent discharged is highly visible to the public (ski runs are located within plain view of the discharge). Second, the effluent discharges into a river used for fishing and human contact sports. Thus a "sparkling" effluent is desired, and ozone provides this. No longer is the Vail watewater discharge greenish, but is sparklingly clear and colorless.

Finally, Vail authorities are concerned with transmission of *Giardia lamblia* cysts. Although the current low dosages of ozone may be too low to guarantee total control, it is believed that ozone provides better protection against these organisms than does chlorine.

Disinfection at Indianapolis, Indiana

The two largest sewage treatment plants in the world to have installed ozone disinfection are in Indianapolis (Belmont (120 mgd) and Southworth (125 mgd) plants). The plants use the same wastewater treatment process which consists of trickling filter/activated sludge nitrification, mixed media filtration, and ozone disinfection. Both of these plants generate ozone from oxygen and began operating in late 1983. Many of the design features of the Vail, CO plant, with respect to the ozonation system, have been incorporated into these two plants.

Both plants discharge their total of 245 mgd (928,000 m^3/day) of effluent into the White River, which flows at an average rate of 35 mgd (132,575). Before the new plants began operating, the dissolved oxygen (DO) content of the White River dropped from about 6.5 mg/L to zero after passing the two sewage treatment plants, and many miles of river flow are required before the DO levels recover. The minimum DO requirement level is 4 mg/L. Since the new treatment process has been operating, the DO level of the river now is running at 8-10 mg/L [11].

Ozonation was selected over chlorination at Indianapolis for a number of reasons. First was the necessity to meet high residual DO levels in the White River. The city was also concerned about the impact of chlorinated organic materials on the river environment, and would have required dechlorination, then post-aeration to raise the DO level. Thus, the one-step ozonation process substitutes for the three steps of chlorination, dechlorination, then post-aeration [11].

Since the plants use pure oxygen for their activated sludge processing, ozone generation from oxygen is cost-effective when operated in an integrated mode, which includes reuse of the ozone contactor exhaust gases, which contain high levels of oxygen, and which are fed to the activated sludge system.

From a design engineering point of view, although unplasticized polyvinyl chloride (UPVC) piping had been designed into the ozone contactor underwater gas diffusion contacting system, during construction the Indianapolis engineers surveyed other operating ozone disinfection plants and decided to switch to Type 304L stainless steel. Supporting the porous plate diffusers with UPVC also was considered to be false economy. As the diffusers are tightened down, this fairly rigid plastic support can break. Stainless steel diffuser supports have been installed at both Indianapolis plants.

Contacting at Indianapolis was designed to be conducted in a four-chamber contactor, having counter-, counter-, co-, and countercurrent flows, respectively. Although the cocurrent chamber is believed to be less effective in achieving high ozone transfer efficiency, it does provide a longer detention time.

REMOVAL OF SUSPENDED SOLIDS

At Chino Basin, CA, a 5 mgd (18,940 m^3/day) secondary treatment plant has been using ozone for microflocculation and removal of suspended solids without adding chemicals since 1978 [12]. Treated effluents from this plant is recharged to groundwater. The State of California requires minimal addition of salts during wastewater treatment in order to protect the groundwater supplies in an aquifer basin having a long history of salt balance problems.

Ozonation at applied dosages of 10 mg/L reduces levels of suspended solids in secondary treated municipal wastewater without adding solids to the effluent, provides effective virus removal, and is cost-effective over other treatment processes considered. Alternative treatment procedures would add chemicals, which then would have to be removed (at considerable cost) in order to meet the groundwater recharge requirements.

Chlorine disinfection is still required in order to meet the California total coliform standard of $< 2.2/100$ mL (Most Probable Number - MPN).

FLOTATION REMOVAL OF BOD_5

In mid-1980 a 125,000 gal/day (473 m^3/day) sewage treatment plant went on-line in Marion, NY. This plants treats effluent to standards higher than secondary at a total yearly operating cost of $ 31,500. The operating and maintenance costs are modest because of the simple plant design, relying on low energy consuming aeration lagoons, and the ozonation facility [13].

Secondary treatment is provided by three aerated lagoons, which remove about 90% of the BOD_5. Most of the remaining 10% is treated by ozone flotation prior to discharge. In addition to lowering the BOD_5 content of the effluent to 5 mg/L, ozonation also destroys algae, aids in coagulation and removal of suspended solids (to a zero value), destroys bacteria and inactivates viruses, as well as provides a degree of protection against toxic substances which might enter the plant.

Ozone dosages required to accomplish these functions are on the order of 10 mg/L. The discharged wastewater meets or exceeds New York State water quality discharge levels, although the influent contains no oxygen and BOD_5 levels are 198 mg/L.

The total cost of the Marion wastewater treatment facility was $ 3,387,000, of which $ 1,009,110 represented the cost of the treatment plant itself, and $ 1,647,490 the collection system. Yearly operating and maintenance costs were budgeted (in 1980) at $ 14,400 for labor, $ 11,200 for administration, and $ 5,900 for equipment and chemicals. The annual cost to a typical household served by this plant is $ 200.

SLUDGE CONDITIONING

At the 10 mgd (37,879 m^3/day) wastewater treatment plant in West New York, NJ, a unique free-radical treatment process was installed in 1981 for converting primary sludge into a dry, free-flowing, cardboard-like material which is acceptable for landfill. The process has been named Hyperbaric Oxyozosynthesis, and the conditioning process is performed batchwise.

Into one of two 3,000 gallon (11.3 m^3) reactors are placed 1,500 gal (5.7 m^3) of comminuted primary sludge and the pH is adjusted to 3.5 with sulfuric or hydrochloric acid. Ozone is generated from high purity oxygen, and the mixture is compressed to 60 psig and injected into the reactor where the sludge, water and ozone/oxygen are mixed for about 90 minutes. Upon release of the pressure, much of the oxidized sludge undergoes flotation, and the foam is skimmed off and belt-pressed into the cardboard-like material which is sent to landfill. The liquid is returned to the front of the sewage treatment plant. Figure 5 shows a schematic diagram of the treatment process [14].

Before the Oxyozosynthesis was installed, West New York, NJ disposed of sludge by ocean dumping at an annual cost of $ 170,000. Total cost of the Oxyozosynthesis process is only $ 110,000 [15]. Thus, this city of 40,000 residents not only pays less for sewage treatment, but the sludge is no

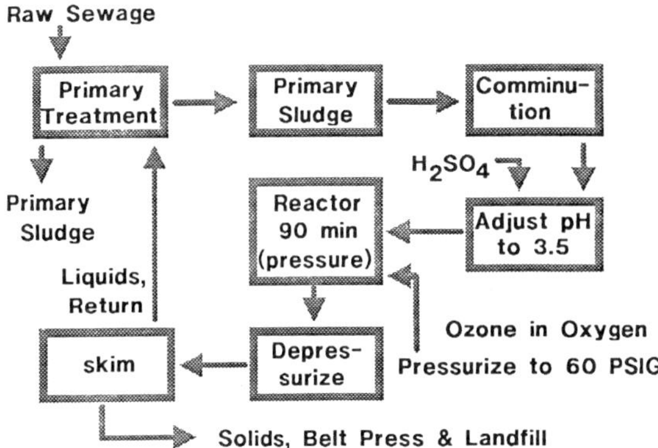

Fig. 5. West New York, NJ sewage treatment process, including Oxy-ozosynthesis.

longer polluting the Atlantic Ocean. In mid-1985, a second Oxyozosynthesis sludge treatment process was installed at Hoboken, NJ.

OXIDATION PRIOR TO GRANULAR ACTIVATED CARBON ADSORPTION

In the early 1970s, research personnel at the Cleveland Regional Sewer District began testing a 1-mgd (3,788 m^3/day) physical/chemical treatment process for the Westerly plant, which included granular activated carbon (GAC) adsorption of organics followed by ozone disinfection (Figure 6). Organics removal by the GAC column was found to be quite erratic, and caused wide swings in the disinfectant requirements. This indicated that ozone was reacting with materials which were not removed by GAC adsorption [16, 17]. In addition, the GAC column produced large quantities of sulfidic odors, which were quite objectionable. Also, reactivation of the GAC was required every thirty days or so.

In the effort to overcome this problem, the ozone treatment step was moved in front of the GAC column. This indeed solved the disinfectant variation problem, but also provided many additional process benefits. Sulfidic odors disappeared, and the useful life of the GAC was extended to over 21 months (Figure 7) [16, 18].

This performance was confirmed over a period of nearly two years in a 1-mgd (3,788 m^3/day) pilot plant, following which the modified treatment

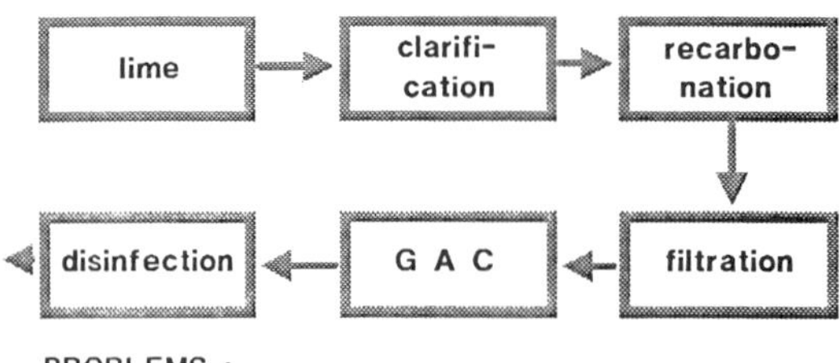

Fig. 6. Original design of the Cleveland Westerly plant.

process (Figure 8) was designed into the full-scale, 50-mgd (189,394 m^3/day) plant. However, this plant is still operating in the startup mode, due to problems with various system components, including the high purity oxygen ozonation system.

U.S. WATER REUSE PROJECTS INVOLVING OZONE

El Paso, Texas [20]

In June of 1985, the city of El Paso, Texas began operating the 10 mgd (37,879 m^3/day) Hueco Bolson Aquifer Recharge Project. This involves advanced treatment of sewage effluent for direct injection recharge of a fresh water aquifer for supplying 65% of El Paso's water supplies. In mid-1988, the plant was operating at 4 mgd (15,140 m^3/day).

The plant treatment scheme (Figure 9) involves two equal 5-mgd (18,940 m^3/day) lines. Equalization is followed by two-stage biological/physical PACT^R (Powdered Activated Carbon Treatment), which removes carbonaceous materials, nitrifies ammonia, and removes many organic materials in the first stage. Waste sludge is processed and powdered activated carbon is regenerated by wet air oxidation (236°C; 56.2 kg/cm^2).

In the second PACT stage, treatment for 1.25 h by anoxic detention, followed by 0.9 h aerobic detention provides denitrification and more organic removal. Methanol is fed to the second stage to be used as carbon source by the denitrifying bacteria.

Lime treatment follows PACT treatment to a pH of 11.1 (virus and

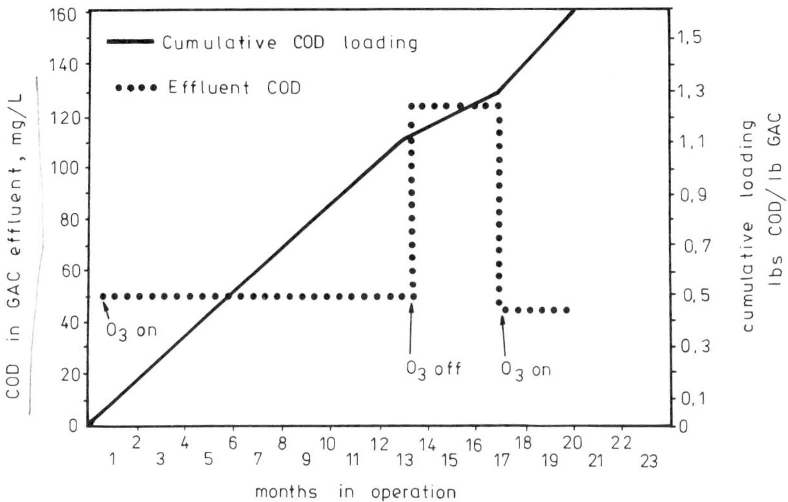

Fig. 7. Performance of GAC pilot plant unit with and without preozonation.

metal removals), then recarbonation with liquid CO_2, sand filtration and ozone disinfection in two 7.7 minute detention time stages, followed by an additional 7.7 minutes detention to allow completion of ozone reactions prior to filtration through granular activated carbon (GAC). GAC filtration is the final, polishing step prior to recharge. Pilot plant testing indicated that the primary GAC reactivation criterion will be trihalomethane formation potential. It is estimated that the GAC filter will last for two years prior to reactivation being required.

Just prior to storage in one of three 3.3 million gallon (12,500 m^3) reservoirs for a minimum of 8-h, the water is treated with 0.25 mg/L dosages of chlorine to minimize biological growths in the reservoirs. The 8-h storage allows sufficient time for the many water quality analyses to be conducted, prior to injection of the treated water into the aquifer recharge wells.

Anaerobic digestion is used as the sludge stabilization process (single, high rate, complete mix with dewatering on sand beds), with dried sludge sold as a soil conditioner. Methane gas produced in the plant will be utilized as an energy source.

Total costs for complete wastewater treatment at this plant were projected to be \$ 1.88/1,000 gal (\$ 0.50/m^3), including amortization of equipment. Costs associated with the ozone portion alone are \$ 1,016,000 for equipment and construction, and \$ 0.065/1,000 gal (\$ 0.0172/m^3).

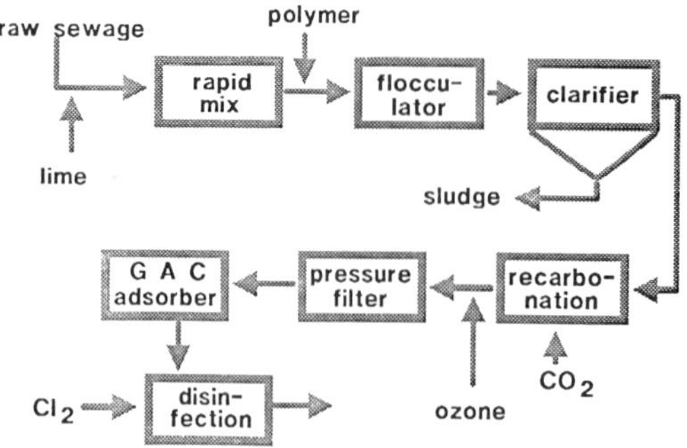

Fig. 8. Modified treatment process at Cleveland Westerly.

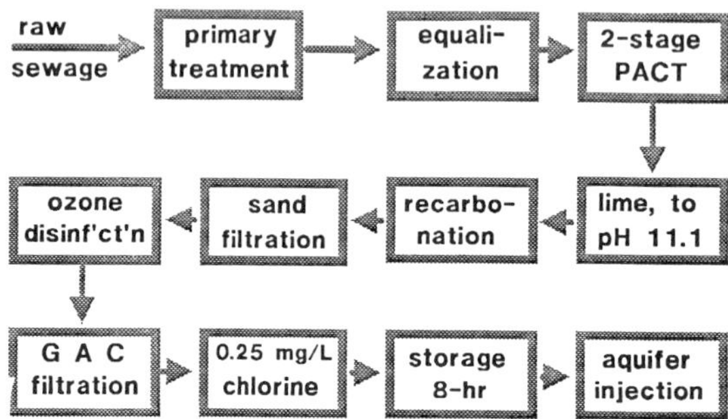

Fig. 9. Schematic of El Paso, TX wastewater reuse treatment process [20].

CURRENT STATUS OF WASTEWATER TREATMENT WITH OZONE IN THE USA

Junkins [2] and Robson & Rice [1] have documented many of the problems experienced at U.S. wastewater ozonation facilities. Many of these plants which incorporated ozonation and have encountered operational difficulties with their original ozonation facilities, have made or are making necessary modifications to the ozonation sytems to restore their operability.

These experiences, taken together with the insight of a growing number of U.S. engineers and scientists involved in the full-scale applications of ozone, have enabled a rapid growth of formal wastewater ozonation guidance documents. The Ozonation Manual for Water and Wastewater Treatment

[21] and the Handbook of Ozone Technology and Applications, Volume 1 [22] both were published during 1982. The U.S. Environmental Protection Agency [23] and the Water Pollution Control Federation [24] published wastewater disinfection design manuals with major portions dedicated to ozonation, in late 1986.

The experience of design engineers with previous involvement in the design and operation of "first generation" U.S. wastewater ozonation facilities has been well documented [25]. It is expected that the design of the Vail ozonation system will influence future U.S. wastewater ozonation facilities. Features of the Vail plant include a flexible, conservatively designed air preparation system, and a two-stage countercurrent flow contactor. Both of these features reflect previous European design practice for drinking water treatment.

The management and operational staffs of the majority of operating wastewater ozonation facilities remain enthusiastic about the improved quality (including aesthetics) of the ozonated wastewater. Therefore, ozonation appears likely to remain a part of many U.S. wastewater treatment facilities.

SUMMARY AND CONCLUSIONS

Most of the operating U.S. wastewater treatment plants using ozone installed it for disinfection. However, several other applications of ozone are being practiced, i.e., removal of suspended solids, flotation removal of BOD_5, sludge conditioning, oxidation prior to GAC adsorption, and reuse of treated wastewater for drinking purposes. About half of these facilities utilize high purity oxygen as the feed gas rather than air.

Operational experience with first generation U.S. wastewater ozonation facilities has identified a wide range of design and construction deficiencies. This knowledge is being incorporated into design manuals to be used for the design of future ozonation system. Despite problems, personnel involved in the operation of wastewater ozonation systems remain positive about the benefits of wastewater ozonation.

REFERENCES

[1] Robson, C.M. and Rice, R.G., Wastewater Ozonation in the U.S.A. - History and Current Status - 1988, in Ozone in Wastewater Treatment & Industrial Applications, Vol. 2, Proceeding, Ninth Ozone World Congress, New York, 1989, L.J. Bollyky, Editor (Norwalk, CT: Intl. Ozone Assoc., pp. 94-112, 1989.

[2] Junkins, R., Ozone Disinfection: An O&M Survey of Selected Water & Wastewater Treatment Plants, in Proc. Sixth Ozone World Congress (Norwalk, CT: Intl Ozone Assoc., May), pp. 131-133, 1983.

[3] Fressonet-Chambarlhac, B., Richard, Y., and Paré, M., Ozonation of Urban

Wastewater in a Tourist Zone, in Proc. Sixth Ozone World Congress (Norwalk, CT: Intl Ozone Assoc., May), p. 136, 1983.

[4] Ozone News, 12(1), 1984.

[5] Matsumoto, N., Experiences with Ozone in Japan, in Wasser Berlin '81. Proc. 5th Ozon-Weltkongress (Berlin, Federal Republic of Germany; Colloquium Verlag Otto H. Hess), pp. 401-408, 1981.

[6] Stover, E.L., Jarnis, R.N., and Long, J.P., High Level Ozone Disinfection of Municipal Wastewater Effluents, U.S. EPA Report No. EPA-600-2-81-040, 1981.

[7] Rakness, K.L., Stover, E.L., and Krenek, D.L., Successful Ozone Disinfection at Vail, Colorado, presented at 56th Annual Water Pollution Control Federation Conference, Atlanta, GA, Oct, 1983.

[8] Venosa, A.D, and Meckes, M.C., Control of Ozone Disinfection by Exhaust Gas Monitoring, in Municipal Wastewater Disinfection, A.D. Venosa & E.W. Akin, Editors (Cincinnati, OH: U.S EPA, Office of Research and Development), pp. 246-259, 1983.

[9] Stover, L.,Optimizing Operational Control of Ozone Disinfection, in Municipal Wastewater Disinfection, A.D. Venosa & E.W. Akin, Editors (Cincinnati, OH: U.S EPA, Office of Research and Development, pp. 260-276, 1983.

[10] Rakness, K.L., Process Applications, Inc., Ft. Collins, CO, Private Communication, 1985.

[11] Wells, D., City of Indianapolis, IN, Private Communication, Jan. 1985.

[12] Tate, C.H., Ozone in Advanced Wastewater Treatment - Chino Basin Municipal Water District Case Study, presented at Third Annual Conf., Nevada Water Poll. Control Assoc., Las Vegas, NV, Dec. 1977.

[13] Powell, L., Marion Facility Gets a Lot For its Money, Water & Sewage Works 127(7):46, 1980.

[14] Nebel, C., Sludge Treatment by the Oxyozosynthesis Process, in Proc Sixth Ozone World Congress (Norwalk, CT: Intl Ozone Assoc., May), p. 68, 1983.

[15] Peterson, S.A., and Mozell, E., Sludge Treatment by Oxyozosynthesis Stops Ocean Dumping, Public Works 115(1):42-43, 1984.

[16] Guirguis, W.A., Hanna, Y.A., Prober, R., Meister, T., and Srivastava, P.K., Reaction of Organics Non-Sorbable by Activated Carbon with Ozone, in Ozone/Chloride Dioxide Oxidation Products of Organic Materials, R.G. Rice & J.A. Cotruvo, Editors (Norwalk, CT: Intl. Ozone Assoc.), p. 611-631, 1976.

[17] Guirguis, W.A., Cooper, T., Harris, J. and Ungar, A., Improved Performance of Activated Carbon by Preozonation, J. Water Poll Control Fed. 50(2):308-320, 1978.

[18] Hanna, Y.A., Slough, J.W., and Guirguis, W.A., Ozonation As a Pretreatment Step for Physical/Chemical Treatment - Part II, presented at Intl. Ozone Assoc. Symp. on Advanced Ozone Technology, Toronto, Ontario, Canada, Nov. 1977.

[19] Miller, K.J., Total Water Management - The Denver Story, Public Works 114(2):38-39, 1983.

[20] Knorr, D.B., Status of El Paso, Texas Recharge Project, in Proceedings Water Reuse Symposium III, Vol. 1 (Denver, CO; AWWA Research Foundation), pp. 137-152, 1985.

[21] Masschelein, W.J., Editor, Ozonizaton Manual for Water and Wastewater Treatment, New York, NY: John Wiley & Sons, 1982.

[22] Rice, R.G., and Netzer, A., Editors, Handbook of Ozone Technology and Applications, Volume One, Ann Arbor, MI: Ann Arbor Science Publishers, Inc., The Butterworth Group, 1982.

[23] Stover, E.L., Haas, C.N., Rakness, K.L., and Scheible, O.K., Design Manual-Municipal Wastewater Disinfection, EPA Report No. EPA-600/1-88-021, Cincinnati, OH: U.S. EPA, Office of Research & Development, 1986.

[24] Wastewater Disinfection - Manual of Practice, Alexandria, VA: Water Pollution Control Federation, 1986.

[25] Rakness, K.L., Stover, E.L., and Krenek, D.L., Design, Start-up, and Operation of an Ozone Disinfection Unit, J. Water Poll. Control Fed. 56(11):1152-1159, 1984.

A REVIEW OF OZONE APPLICATIONS

IN MARINE AND FRESHWATER SYSTEMS

W. J. BLOGOSLAWSKI

International Ozone Association
Pan American Committee
83, Oakwood Avenue
Norwalk, Connecticut 06850, USA

ABSTRACT

The use of ozone to combat problems found throughout fresh and marine water treatment systems of the world is varied in application.

Presently, ozone is employed as a disinfectant in sewage treatment, eliminating the adverse impacts raw sewage or chlorine-treated sewage have on the delicate larval stages of marine species. The oxidant is also used to insure adequate disinfection in fresh water systems that are used to process and store fishery products, eliminating humic and fulvic acids and the subsequent formation of trihalomethanes that can be induced by the use of chlorine as a disinfectant. Ozone is also a major component of many depuration systems found throughout Europe which cleanse contaminated shellfish and insure a safe product for the human consumer. Finally, ozone is used in many marine aquaria and research facilities to provide suitable water for rearing and/or displaying marine animals.

This paper addresses the above-cited uses of ozone in marine and freshwater systems and reviews processes to enhance such applications.

Chemistry for the Protection of the Environment
Edited by L. Pawlowski *et al.*, Plenum Press, New York, 1991

INTRODUCTION

Large capacity ozone generators use dielectric-barrier discharge to produce a high voltage, silent corona around the high tension electrode and the grounding electrode. When air or oxygen is passed through the glowing discharge, molecular rearrangements via electrical excitement change oxygen into ozone. Only a few percent, normally 2–5%, of the carrier gas makes this change in spite of refinements to cool the reactor wall with water using heat exchangers.

At ambient temperatures, triatomic oxygen or ozone is an unstable gas which rapidly decomposes to oxygen in clean freshwater. The decomposition is accelerated by organic and inorganic compounds, silt, heavy metals, especially Mn and Fe, a pH above 9, and solution temperatures above 30°C. Since ozone decomposes autocatalytically, it cannot be stored in tanks but must be generated on site for immediate application. Some of ozone's aqueous uses include:

- Bacteria inactivation
- Elimination of humic and fulvic acid precursors of trihalomethanes and the improvement of coagulation of organics
- Reduction of off flavors, odors, and humic color in drinking water and in pulp wastes, prior to stream discharge
- Oxygenation of anoxic waters prior to sewage discharge
- Removal of substituted amines or phenols in sewage as a consequence of chlorination reactions
- Toxin inactivation and lysis of dinoflagellates and other microalgal cells prior to mariculture applications
- Production of ice free of spoilage bacteria for shelf-life extension of commercial fish

Other uses include disinfection of swimming pools, of marine and fresh water aquaria, and bottled drinking water. Ozone gas preserves cheeses, eggs, poultry, berries, fruit, some low fat meats and certain fish. Refrigeration ships employ ozone gas to reduce mold and fungi, in holds and cold food storage areas. Some of the ozone applications listed above will now be reviewed by section.

REVIEW

Seawater Disinfection

Ozone gas in an excellent means of seawater sterilization. It rapidly oxidizes organic compounds present in seawater reducing the biological and chemical oxygen demand of recycled seawater. In addition it kills bacteria, fungi, and viruses more rapidly than other oxidants [1]. An ozone disinfection system allows the mariculturist microbial control, providing an effective means of disease prevention, improved water quality, a method of rapid and effective depuration for contaminated animals, and a potential system for marine toxin inactivation.

In 1929, H. Violle [2] of the University of Marseille, France, noted that ozone could be considered an excellent sterilant for water supplies. He was the first investigator to report that seawater can be sterilized easily with ozone. In his paper, several experiments were described showing the disinfection of sealed tubes of seawater seeded with coliform bacteria and exposed to a stream of ozone. He found that ozone sterilized seawater seeded with around 1×10^6 cells/ml in a maximum of eight minutes.

In 1973 Blogoslawski *et al.* [3] constructed a pilot system to ozonize seawater. While monitoring this system, total counts of marine bacteria were observed to decrease by three logs as ozone dosage was increased. At a dissolved ozone concentration of 0.5 mg/liter, no marine microorganisms were detected.

In another study [4], ozone was found to be an effective control on bacterial populations in closed marine systems. It was noted that bacterial populations increased from one to three logs within 24 hours when ozone treatment was suspended but quickly fell back to their original levels when ozonization was resumed.

After examination of several seawater disinfection systems at a marine aquarium in Germany, ozone was selected as the most likely method to achieve disinfection of the water without causing harm to the animals being cultured [5]. Different methods of ozone-water contact systems were tried and it was found that sterilization could be achieved most effectively with a two-stage ozone process: First-stage ozonization served to remove large solids and proteinaceous organics by means of foam separation; the second ozone stage provided complete sterilization.

Mariculture Applications

Bivalve shellfish held in close contact in a mariculture facility are more susceptible to disease than shellfish in the wild. The most likely means of introducing and spreading a disease is through the water system. A mariculture facility which draws its water from a contaminated area must employ a disinfection system and ozone has been regarded as a method holding great promise [6].

Aside from the disinfection of water drawn from a source polluted by industrial and/or sewage contamination, ozone can be used in mariculture to cultivate food sources free from pathogenic bacteria. At the Monterey Abalone Farm, Hawaii, U.S.A., algal mats used as food for the abalone *Haliotis rufescens* are grown in large tanks of seawater sterilized by ozone. This ensures that the food source of the abalone contains no pathogenic material [7].

Toxin Inactivation

Undesirable water conditions for mariculture are not always created by man. In 1962, blooms of the dinoflagellate *Prorocentrum micans* prevented spawning in oyster hatcheries [8]. Blooms of toxic dinoflagellates occur throughout the world's oceans. These organisms, depending upon their species, either can be concentrated in shellfish tissues causing the mollusks to become poisonous to vertebrate consumers or can act in such a way that the toxin directly poisons fish.

In laboratory-scale studies, toxins extracted from the dinoflagellates *Gonyaulax catenella, Gonyaulax tamarensis* (var. *excavata*), and *Gymnodinium breve* were inactivated with ozone gas [9].

Should the water source for a mariculture facility originate from an area where a bloom of these toxic dinoflagellates occurred, many of the bivalves exposed to that water could be expected to become toxic or die. Ozonized seawater has been shown to inactivate the crude toxin from these blooms, as well as to detoxify shellfish (*Spisula solidissima* and *Mya arenaria*) which have absorbed the toxin [10, 11]. The routine use of ozone to sterilize seawater circulated to the culture species in a mariculture facility might reduce the public health threat and economic loss which would normally result from a toxic bloom.

Depuration

If shellfish held in a mariculture facility should become contaminated with pathogenic bacteria from domestic sewage spills, ozone can be used to depurate the bivalves quickly. In that way, shellfish which would have been lost to the retail market can be cleansed and sold.

It was reported in 1929 that oysters could be depurated by ozonized seawater without any accompanying change in taste or appearance of the oyster [2]. Later investigators expanded this finding to a pilot-scale study [12, 13]. Using contaminated oysters and mussels, it was found that bivalves held in ozonized seawater were rapidly and completely cleansed, while control animals held in raw seawater were not. A commercial depuration plant using ozonized seawater was established in 1963 [1]. It was such a success that ozone is now the disinfection method of choice in shellfish cleansing stations in France, having replaced chlorine. Investigation revealed that while the initial cost of equipment for ozone disinfection is higher than a comparable station for chlorine, in the long term, ozone is less expensive because the method does not require the additional purchase of any other material, whereas chlorine must be bought to replenish the supply of that oxidant in a chorine system [14]. An additional benefit of ozone treatment is that the food product has no chewy, chlorine-like taste or odor, while these undesirable characteristics have been noted in chlorine-treated shellfish [1].

Preservation of Fish

Ozonation of seawater inactivates the spoilage bacteria associated with it, so that ice prepared from ozonized water is disinfected. When this disinfected ice melts, the fish preserved with it are not contaminated.

In 1936, Salmon and LeGall [15] showed that storing freshly caught fish under ice which had been prepared from ozonized seawater extended the storage life of the flesh by more than five days. They stated that during the then normal French practice of cleaning and icing of fish, seawater was pumped aboard ship in the harbor areas and sent directly to ice-making machines. These harbor waters contain higher levels of bacteria then offshore waters; thus ice made from harbor waters will contain high bacterial levels. When this ice melts, the freshly cleaned fish are actually contaminated with high levels of bacteria.

In 1969, the Japanese research team of Haraguchi *et al.* [16] showed that soaking fresh jack mackerel (*Trachurus trachurus*) and shimaaji (*Caranx mertensi*) in 30% NaCl solution containing 0.6 mg/L of ozone for 30 to 60 minutes caused levels of viable bacterial counts on skin surfaces of the gutted fish to decrease to levels 1/100th of those of the control samples. The storage life of these fish was increased 1.2 to 1.6 days by applying such ozone treatments every two days.

At the National Marine Fisheries Service in Milford, CT, W. Blogoslawski [17] has shown that ice prepared from ozonized seawater and fresh water extends the storage life of fresh salmon by two to three days. Additional studies of sterilized ice demonstrated the preservation of fresh squid.

If this "sterilized ice" concept proves to be successful, the potentials for its adoption at ice makers throughout the world are significant. This is particularly true because of the recent dramatic increase in costs in areas where fish was traditionally cheaper than at present, and the desire to minimize losses due to spoilage.

Ozone Reactions in Seawater

The chemistry of ozone in seawater is complex, so care must be exercised during the use of this oxidant [18]. When added to seawater, ozone reacts rapidly with free bromide ions to form hypobromous acid and hypobromite ions. Bromamines may also be formed in proportion to the amino-nitrogen concentration. These products react further with organic compounds present in the water to form halogenated organic compounds [19, 20] which may prove harmful to larval stages of bivalves [21]. This problem may be avoided by maintaining an ozone dose that is sufficient to achieve bacterial disinfection of the seawater without producing the halogenated organic compounds in concentrations high enough to harm larvae.

Freshwater Disinfection – Sewage Treatment

Municipal waste treatment plants that discharge treated sewage into rivers and estuaries disinfect the effluent with biocides. Chlorine is the principal biocide currently used. Almost 1% to 2% of the entire United States production of chlorine is used for this purpose [22]. Chlorine-containing organics are produced in the chlorination disinfection process [20,23]. Therefore, discharged chlorinated wastewater effluents contain relatively stable chlorine-containing organics as well as a reactive chlorine residual. The toxicity of the chlorine residual in chlorinated effluents has been documented extensively in both fresh and marine waters [24]. Chlorine residuals from wastewater effluents consist principally of reaction products of chlorine with ammonia. When treatment effluents are discharged into marine systems, further reactions of the residuals with bromide ions produce new toxic components, some of which are yet to be identified. The chlorine-containing organic compounds in chlorinated wastewater effluents include such toxic chemicals as chlorophenols and mutagenic chemicals such as chlorinated pyrimidines [23]. The majority of the chlorine-containing, organic constituents in chlorinated effluents are unidentified.

Several municipal wastewater treatment plants disinfect effluents with ozone. If wastewater treatment plants using ozonation are located at estuarine and ocean-shore sites, the ozonated wastewater effluents may reduce marine pollution. Ozonation of fresh water sewage eliminates chlorine substituted phenols, amines, and trihalomethanes. Thus, ozone treatment provides effluent which is safer to the survival of developing marine organisms than the mutagenic pyrimidines or substituted organics which result from chlorine disinfection.

Pulp and Paper Treatment

Recent work in Poland [25] has indicated that ozonation following coagulation with alum or lime is very effective in removing COD, BOD_5, color and odor from wastewater effluent from pulp and paper-making processors. Thus, streams receiving the effluent are not subjected to low oxygen conditions which can cause fish kills.

Drinking Water

Disinfection of potable water supplies is achieved in many large cities with ozone. Cities such as Moscow (USSR), Montreal (Canada), Paris (France) and Los Angeles (USA) use ozone to reduce trihalomethanes while improving odor and taste. In addition, there are six ozone systems in operation in Poland for water conditioning (pers. comm., Bohdan Jasiński, Wroclaw).

CONCLUSIONS

The sections explored in this paper are examples of ozone's wide use as a marine and fresh water disinfectant.

Since ozone leaves no residual in fresh water, other disinfectants must be added to create a residual where required. In marine waters, ozone reacts with bromides creating an oxidant which can last 48 hours. While ozone is able to purify water quickly, cost considerations must also be taken into account. When comparing ozone, chlorine, or ultraviolet light treatment, ozone is always more expensive from a capital-cost standpoint, although ozone's operating costs compare favorably with other oxidants. Since ozone removes trihalomethanes (THM) more inexpensively than other treatments, it will likely become the disinfectant of choice where THM's are a problem.

ACKNOWLEDGEMENT

I am grateful to L.P. Tettelbach for completing the ozonized ice experiments.

REFERENCES

[1] Fauvel, Y. The use of ozone as a sterilizing agent in seawater for the depuration of shellfish. *International Commission for the Scientific Exploration of the Med. Sea, Monaco, Reports and Verbal Proc.,* **17** (3), 701–706, 1963.

[2] Violle, H. *Rev. Hyg. med. prev.,* **51**, 42, 1929.

[3] Blogoslawski, W.J., C. Brown, E. Rhodes and M. Broadhurst. Ozone disinfection of a seawater supply systems, in R.G. Rice and M.E. Browing, Eds., *Proc First International Symposium on Ozone for Water and Wastewater Treatment,* International Ozone Institute, New York, p. 674-687, 1975.

[4] Honn, K. V. and W. Chavin. *Mar. Biol.* **34**, 201, 1976.

[5] Sander, E. and H. Rosenthal. "Application of ozone in water treatment for home aquaria, public aquaria, and for aquaculture purposes", in W.J. Blogoslawski and R.G. Rice, Eds., *Aquatic Applications of Ozone,* Int. Ozone Institute, New York, 1975, p. 103.

[6] Blogoslawski, W.J. Influence of water quality on shellfish culture. C.M. 1983/F:8, Mariculture Cttee, Ref. Shellfish Cttee, ICES, Gothenburg, Sweden , **35**, 1983.

[7] Blogoslawski, W.J. Ozone as a disinfectant in mariculture. 3rd ICES Working Group on Mariculture, Actes de Colloques du C.N.E.X.O, Brest, France 371–381, 1977.

[8] Loosanoff, V.L., *Comm. Fish. Rev.*, 35, 1962.

[9] Blogoslawski, W.J. and M.E. Stewart, "Detoxification of marine poisons by ozone gas" in *Proc. Third Congress of the International Ozone Institute,* International Ozone Institute, Paris, 1977.

[10] Blogoslawski, W.J., *Jor. of Shell. Res.,* **7**, 4, 702-705, 1988.

[11] Blogoslawski, W.J., and M.E. Stewart and J.W. Hurst and F.G. Kern, III, *Toxicon,* **17**, 650, 1979.

[12] Salmon, A., J. Salmon, J. LeGall and A. Loir. *Annls. Hyg. publ. ind. Soc.,* **15**, 581, 1937.

[13] Salmon, J., J. LeGall and A. Salmon. *Annls. Hyg. publ. ind. Soc.,* **15**, **44**, 1937.

[14] Blogoslawski, W.J., Depuration and Clam Culture. From: Clam mariculture in North America. Ed. by J. Manzi and M. Castagna, Elsevier Publ., Amsterdam, Neth. 415–426, 1989.

[15] Salmon, J. and J. LeGall, *Rev. Gen. du Froid,* 317–322, 1936.

[16] Haraguchi, T., and U. Smidu and K. Aiso, *Bull. Jap. Soc. Sci. Fisheries* **9**, 915–919, 1969.

[17] Blogoslawski, W.J., *Ozonews,* **1**, 1982.

[18] Hoigne, J. The Chemistry of Ozone in Water. From: Process Technology for Wat. Treat. Ed. by S. Stucki, Plenum Publ., N.Y., 121–141, 1988.

[19] Helz, G.R., and R.Y. Hsu. *Limnol. Oceanogr.,* **23**, 858, 1978.

[20] Helz, G.R., R.Y. Hsu and R.M. Block. Bromoform production by oxidative biocides in marine waters, in R.G. Rice and J.A. Cotruvo, Eds., *Proc. Workshop on Ozone-Chlorine Dioxide Oxidation Products of Organic Materials,* International Ozone Institute, Cleveland, Ohio, p. 68, 1978.

[21] Stewart, M.E. and W.J. Blogoslawski, Effect of selected chlorine-produced oxidants on oyster larvae. From: *Water Chlorination: Chemical, Environmental Impact and Health Effects.* Ed. by R. Jolley, R. Bull, W. Davis, S. Katz, M. Roberts, Jr., and V. Jacobs, Lewis Publ., Michigan, 521–532, 1985.

[22] White, G.C. Current chlorination and dechlorination practices in the treatment of potable water, wastewater, and cooling water. From: *Water, Chlorination: Environmental Impact and Health Effects.* Ed. by R. Jolly, Ann Arbor Science Publ., March, 1–18, 1978.

[23] Glaze, W.H. and J.E. Henderson. *Jor. Wat. Pol. Con. Fed.,* **47**, 2511, 1975.

[24] Davis, W.P. and D.P. Middaugh. A review of the impact of chlorination processes upon marine ecosystems. From: *Water Chlorination: Environmental Impact and Health Effects.* Ed. by R. Jolley, Publ. Ann Arbor, Science Publ, Mich, 238–310, 1978.

[25] Sozanska, Z and M. Sozanski, Efficiency of ozonation as a unit process in the treatment of secondary effluents from the pulp and paper industry. From: *Ozone in Wastewater Treatment and Industrial Applications.* Ed. by J. Bollyky. Port City Press Publ., N.Y. 203–220, 1989.

FULL SCALE TREATMENT OF WASTEWATER EFFLUENT WITH HIGH ENERGY ELECTRONS

T. D. WAITE[1], W. J. COOPER[2], CH. KURUCZ[1],
R. NARBAITZ[2] and J. GREENFIELD[2]

[1]University of Miami
Coral Gables, Florida 33124, USA
[2]Florida International University
Miami, Florida 33199, USA

ABSTRACT

The use of high energy electrons for treatment of waste residuals is an emerging technology. Some work on treatment effects has been done at a laboratory scale, but virtually no data are available from full scale installations. The information that exists relates to domestic sludge irradiation only, and these data are quite meager.

The University of Miami and Florida International University have been awarded a grant from the U.S. National Science Foundation to rehabilitate and operate an electron beam irradiation unit located at the Central Wastewater Treatment Plant in Miami, Fla., USA. The unit is a 1.5 MeV, 50 Ma, insulated core transformer generator, designed to treat 120 gpm of digested sludge. The unit was constructed in 1983 at a cost of $2 million, and operated for one year. At that point the system was shut down, and remained inoperative for three years. We have now restored the system, and are testing the effects of electron irradiation on secondary effluent.

The data reported in this paper will describe the aqueous oxidants generated in the waste stream as a function of electron dose. Their characteristics and longevity will be described and compared to their effects on the chemical and biochemical composition of the waste stream.

Chemistry for the Protection of the Environment
Edited by L. Pawlowski *et al.*, Plenum Press, New York, 1991

INTRODUCTION

It has become clear over the past five years that water supplies around the world are becoming increasingly contaminated. Industrial and agricultural development has lead to the utilization of new compounds which in many cases are recalcitrant to environmental breakdown, and quite often are toxic to aquatic and human life. The result of these developments is a continued degradation of our global water supplies.

New technologies for water and wastewater treatment are therefore needed but very few have been developed. One process which has received recent attention world-wide is the use of high energy electrons for treating water and wastewater. This technology, which has been used successfully for years in medicine, as well as selected industrial processes, holds great promise for effectively treating many aqueous systems. While several sources of radiation are available for use on water and wastewater, and many have been tested, the electron beam has the advantage of not requiring a radioactive source. This fact alone has made this form of radiation much more acceptable to the general public.

In the case of electron beam radiation, electrons are generated by an electric current, and accelerated through an evacuated space under high voltage. After a short distance the electrons have achieved a sufficient velocity and, therefore, sufficient energy to pass through a thin window containing the vacuum, and enter into the medium outside. Their overall energy, however, is still small and the electrons are rapidly attenuated. At an acceleration voltage of approximately one to two million volts, the electrons will travel only three to four meters in air. If the electrons are impacted on water, their attenuation is more rapid and are only capable of traveling fractions of a centimeter. While this fact makes the process very safe and environmentally acceptable, it poses severe engineering problems for designing the treatment system.

One of the first attempts to utilize high energy electrons for wastewater treatment was reported by Trump [1]. His group developed a full-scale electron accelerator to treat municipal digested sludge. This pioneering work generated international interest in the feasibility of utilizing electrons, and since then many researchers have explored the possibility of using electrons for various types of water and wastewater treatment. In most cases, experiments have been run at a laboratory scale, utilizing existing laboratory scale accelerators. A review of these experiments has been given in Waite et al. [2].

In the past few years there has been extensive research in the radiation area. Miyata et al. [3] have reported on continuing ionizing radiation studies utilizing municipal wastewater. They are currently building a flow-through system for their electron accelerator which will treat approximately 10.8 cubic meters per hour. Gehringer et al. [4] have been evaluating the ability of high energy electrons to decompose chlorinated ethylenes in drinking water.

In Miami, Florida, U.S.A., The University of Miami and Florida International University have been awarded a grant by the National Science Foundation to evaluate the ability of a full scale electron beam process to treat water and wastewater streams. This plant has been operational for over a year and several papers have been given relative to its efficiency of treatment [5, 6]. This paper will describe the ability of a full scale electron beam irradiator to both inactivate microorganisms and destroy recalcitrant chlorinated organics in a secondary effluent.

METHODS AND MATERIALS

The electron accelerator utilized in this study is a 1.5 MeV, 50 Ma accelerator which is located at the Virginia Key Waste Water Treatment plant in Miami, Florida (USA). The system was originally designed to handle 120 gallons per minute of digested sludge, and was therefore located at the dewatering facility in the treatment plant. For the past year the machine has been utilized for research, and it has been reconfigured to handle different types of wastestreams as well as drinking water supplies. The machine can easily be operated in a research mode, as the current can be adjusted between zero and 50 milliamps. Actual absorbed dose is measured by sensitive electronic temperature devices, and for our flow system the dose is varied from zero to approximately 650 kilorads. Because this is a full scale unit, the data reported in this paper reflect the types of removal expected in a large operating facility.

During an experimental run, influent and effluent samples were collected at each of four beam currents. The four beam currents (0, 10, 30 and 50 milliamps) were set in random order, and each experiment was replicated a second time. Effluent samples smuthings mielling to allow for transit time of material between sampling ports. After each effluent sample was collected, the machine was changed to the next beam current, and the next influent sample was drawn three minutes later. This meant that there was approximately four to five minutes between consecutive influent (effluent) samples. In all experiments, the accelerating voltage of the machine was maintained at 1.5 million electron volts.

Both total coliform organism and total viable organism counts were

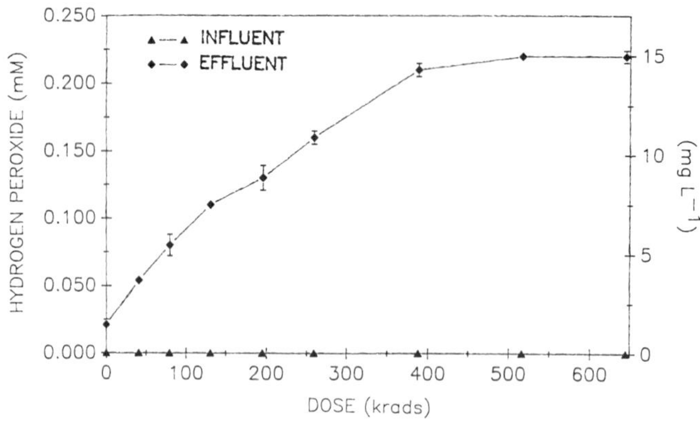

Fig. 1. Formation of peroxide as a function of total absorbed dose.

made on each influent and effluent sample collected. Total coliform organisms were enumerated, using the most probable number technique (MPN), after dilution in buffered sterile water. The bacterial counts were confirmed in BGB media and positive samples were inoculated in Eijhmans media to determine the occurrence of fecal coliforms.

Total viable bacteria were enumerated by dilution in buffered sterile distilled water with subsequent planting on nutrient agar. Cultures were incubated at room temperature for 48 hours, and colonies were then counted.

The disappearance of selected organics was monitored utilizing a Model 5890 Hewlett-Packard Chromatograph, equipped with a 7672A Autosampler. A 30 meter by 0.39 millimeter DB-624 column (J & W Scientific), and Electron Capture Detector was employed for the analyses. Helium carrier gas was supplied at 6 ml per minute, and nitrogen makeup gas at 40 ml per minute. The injector was maintained at 200° C and the detector at 250° C.

Hydrogen peroxide was also monitored at each dose during experimentation. Hydrogen peroxide was determined titrametrically by measuring the quantitative iodide oxidation in the presence of a molybdate catalyst.

RESULTS AND DISCUSSION

When wastewater is irradiated by high energy electrons, interactions occur which generate very powerful oxidants. The principal oxidant formed upon irradiation is the hydroxyl radical (OH·) and it is known to be a good oxidant of organic molecules. It is probably this hydroxyl radical that is

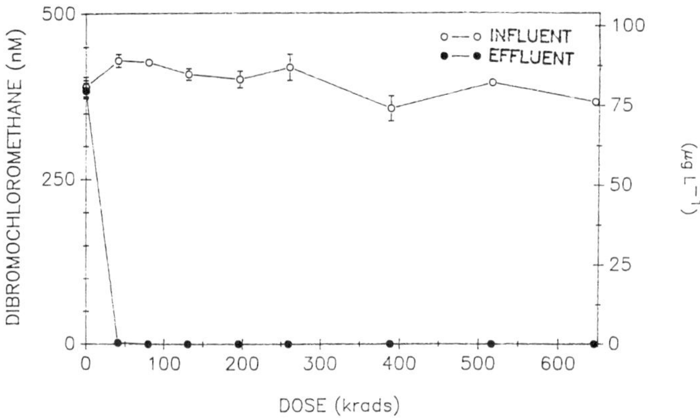

Fig. 2. Effect of irradiation on effluent.

responsible for most of the chemical reactions occurring upon irradiation. In addition to hydroxyl radical, a stable oxidant, which is most probably hydrogen peroxide, is formed in the wastewater. Figure 1 shows the formation of this oxidant as a function of total absorbed dose, and it can be seen that a residual greater than 0.2 mM is formed when the absorbed dose exceeds approximately 350 kilorads. There appears to be a limitation on generation of this residual oxidant, which probably reflects its interaction with hydroxyl radical. It has been observed that this residual can persist in secondary effluent for many days after radiation. It is clear from these data that any analyses of secondary effluent must take into consideration the presence of hydrogen peroxide so that erroneous data are not collected.

As noted above, because of the generation of hydroxyl radical by electrons, organic molecules are quickly destroyed. While many recalcitrant organics have been tested in our system, two examples of organic decomposition will be given here. Figure 2 shows the results of irradiation of an effluent which contains dibromochloromethane at a background concentration of approximately 400 nm. This compound was injected into the wastestream before it passed through the electron beam so these data represent steady state concentrations at various doses. It can be seen that dibromochloromethane is rapidly destroyed at even very low doses. In this case, less than 50 kilorads are required to completely destroy this particular compound in secondary effluent.

In contrast to dibromochloromethane, 1,1,1- trichloroethane is not as easily destroyed by electron irradiation. Figure 3 shows experimental data for this compound, and it can be seen that a dose in excess of 600 kilorads

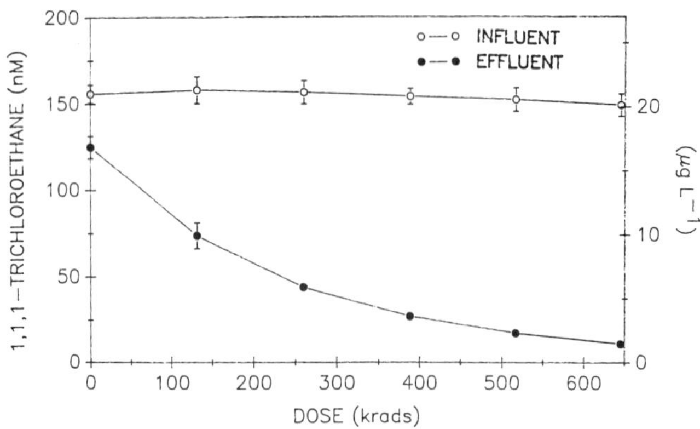

Fig. 3. Effect of irradiation on effluent.

is needed to completely remove the compound in the effluent. In this case, 1,1,1- trichloroethane was injected at approximately 150 nm during the test and even at 600 kilorads total absorbed dose, approximately 10 to 15 nm remained.

The observations shown in Figures 2 and 3 are in general agreement with reported rate constants for these types of compounds in the presence of hydroxyl radical, even though no rate constants exist for these actual compounds. In all of the decomposition studies to date, the relative rate of decomposition of organic molecules has been in agreement with rate constants of decomposition of those molecules in the presence of hydroxyl radical. This is further evidence that it is the hydroxyl radical that is most responsible for decomposition of recalcitrant organics in aqueous systems irradiated by high energy electrons.

In addition to destroying toxic organics, high energy electrons are capable of destroying microorganisms in wastewater. Trump [1] reported on studies of disinfection of sewage sludge, and the dose levels for inactivation of the microorganisms in that matrix are reasonably high. We have investigated the activation of microorganisms in secondary effluent (Figure 4). In this case, fecal coliform and total bacteria are compared in an unchlorinated secondary effluent as a function of total absorbed dose. It is seen that inactivation of these organisms is highly dependent on absorbed dose, and 3 to 4 orders-of-magnitude of inactivation can be expected when the total absorbed dose approaches 600 kilorads. It is also interesting to note that the total bacterial flora were inactivated at approximately the same level as the coliform for each absorbed dose. With other disinfectants, the total bac-

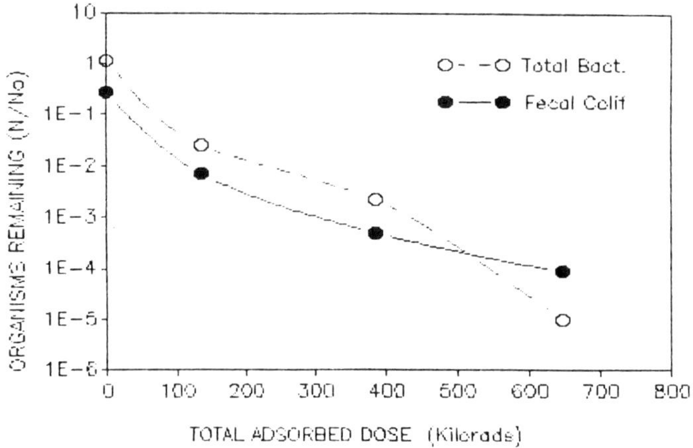

Fig. 4. Fecal coliform inactivation vs. total bacteria. Unchlorinated secondary effluent.

terial flora are inactivated much less efficiently than the coliforms. Because this is secondary effluent from a municipal waste treatment plant, the majority of the flora are indeed fecal coliforms; therefore, it would be expected that the majority of flora would be easily inactivated. However, the fact that the total flora are inactivated at approximately the same rate indicates that high energy electrons, or the radicals they produce, are indiscriminate disinfectants. If this observation can be verified in different aqueous systems, this would be a very important concept.

Figure 5 shows disinfection data for fecal coliform organisms as a function of radiation dose in two different wastestreams. In this case both secondary effluent and raw sewage were radiated and the inactivation kinetics are compared. Even though there is some scatter in the raw sewage data, it appears that the inactivation of coliform microorganisms will be similar in either secondary effluent or raw sewage. Therefore, the same inactivation dose relationships can be expected whether raw sewage or partially treated sewage is irradiated. Once again this points to the relative insensitivity of wastestream matrix to the efficiency of electron inactivation of microorganisms.

CONCLUSIONS

The data presented in this paper indicate that electron beam irradiation of wastewater will be an effective and efficient treatment process. We

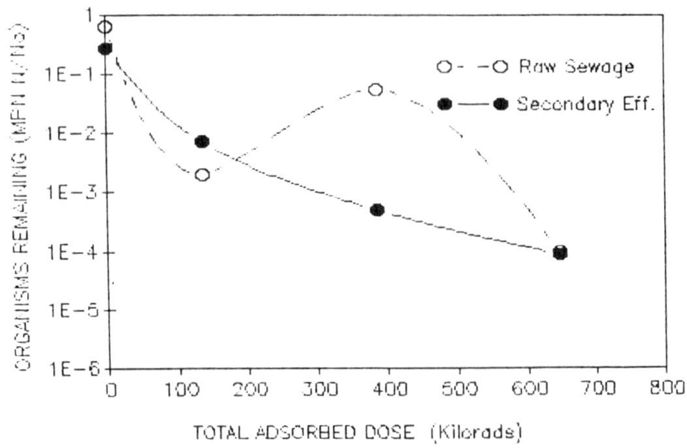

Fig. 5. Fecal coliform inactivation vs. radiation dose. Wastewater - Virginia Key Treatment Plant, Miami Fla.

have shown that highenergy electrons effectively destroy many toxic organic compounds while inactivating microorganisms, as modelled by intestinal coliforms. These data, which were collected at a full scale operating plant, show that the recalcitrant organics will probably be reduced by greater than 90% if not totally destroyed, and bacteria can be reduced by 3 to 4 orders-of-magnitude at approximately 600 kilorads total absorbed dose. In addition, there appears to be little effect of strength of sewage on electron beam efficiency, as similar data for disinfection were collected for both raw sewage and secondarily treated effluent.

REFERENCES

[1] Trump, J. G. High Energy Electron Radiation of Wastewater Liquid Residuals. Final Report, National Science Foundation, Washington, D.C., 1980.

[2] Waite, T. D., et al. Disinfection of Wastewater Effluents with Electron Radiation. Proceedings from National Conference on Environmental Engineering, ASCE, Austin, Texas, 1989.

[3] Miyata, T., et al.

[4] Gehringer, P., et al. Removal of Chlorinated Ethylenes afrom Drinking Water by Radiation Treatment. Proceedings from International Meeting on Radiation Processing, Noovdwijkerhout, The Netherlands, 1989.

[5] Waite, T., et al. The Use of High Energy Electrons for Treatment of Water and Wastewater. Proceedings from Joint CSCE-ASCE National Conference on Environmental Engineering, Vancouver, B.C., Canada, 1988.

[6] Cooper, W. C., et al. Removal of Halogenated Methanes, Ethanes, and Ethylenes in Oxygenated Secondary Wastewater Using High Energy Electrons. Proceedings from Division of Industrial and Engineering Chemistry, American Chemical Society Symposium, Atlanta, Georgia, 1989.

DESIGN PROCEDURE OF AN INDUSTRIAL-SCALE CHAMBER FOR WATER OZONATION

A. K. BIŃ[1], A. KONOPCZYŃSKI[2], J. RAABE[3]
and M. SZAFRAN[3]

[1]Institute of Chemical and Process Engineering
Warsaw University of Technology
ul. Waryńskiego 1, 00-645 Warszawa, Poland
[2]Institute of Precise Mechanics
Warsaw, Poland
[3]Department of Chemical Technology
Warsaw University of Technology
Warsaw, Poland

ABSTRACT

A design procedure for an industrial-scale ozonation chamber based on the liquid circulation is described.

INTRODUCTION

Ozonation of water is most frequently accomplished in contact chambers by means of dispersing ozone–containing air using finely porous ceramic diffusers. These provide good dispersion of gas into fine bubbles in water giving rise to high ozone transfer rates during its absorption coupled with chemical reactions. Usually the ozonated air is distributed within a contact

Chemistry for the Protection of the Environment
Edited by L. Pawlowski *et al.*, Plenum Press, New York, 1991

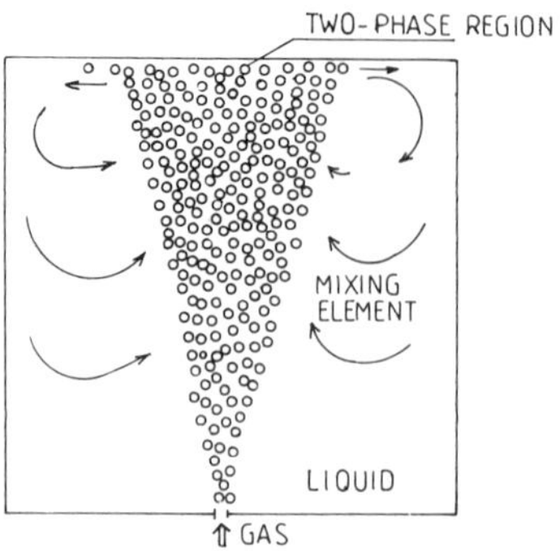

Fig. 1. Liquid circulation unit flow pattern.

chamber by means of discretely acting porous elements (domes or tubes). Typical design procedure of such chambers makes use of overall hydrodynamics and mass transfer characteristics referred to the total liquid phase volume in the chamber. This paper presents a more rigorous design procedure based on a concept of the liquid circulation unit and accounting for the effects of several factors on ozone transfer efficiency, suggested in the recent paper of Rakness et al. [1].

THEORETICAL BACKGROUND

When gas is sparged into a pool of liquid, a two-phase region is formed above each bubbling station. The density difference between this region and the surrounding liquid induces a flow of liquid in the two-phase region. This liquid flows upward in this region. Near the liquid surface, it turns and flows radially away from the two-phase region until it meets liquid flow from an adjacent bubbling station or the vessel walls. At this point the liquid flows downward into the vessel. The two-phase region and down-flowing liquid surrounding it form a liquid circulation unit. Mass transfer takes place only in the two-phase region. The component transferred into the liquid in the two-phase region is diluted in the liquid outside the region before it flows radially outward, then down and back into the two-phase region. The characteristics and number of liquid circulation units in a gas-liquid system depend on the number of bubbling stations and the distance between them, the gas flow rate per station, and the total liquid height in the vessel.

The liquid circulation unit, as introduced by Otero and Russell [2], is comprised of the two-phase region and the liquid mixing element (Fig. 1).

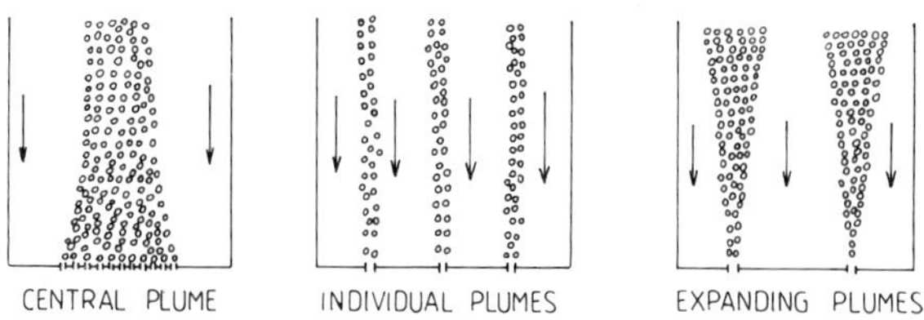

Fig. 2. Flow patterns for gas-liquid system.

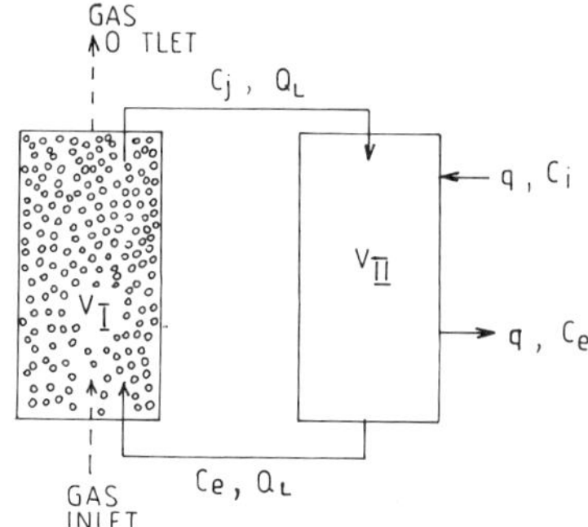

Fig. 3. Liquid circulation unit.

The fluid hydrodynamics and the mass transfer characteristics of the liquid circulation unit are a function of the system geometry and the operating conditions. Three distinct flow patterns for the tank-type systems (Fig. 2) have been identified by Otero et al. [3]. It appears that under typical practical conditions met in an industrial-scale chamber, cylindrical or expanding plume patterns are encountered most frequently.

Figure 3 demonstrates the concept of the liquid circulation unit for a continuous flow liquid system. In this case, liquid is introduced and removed for the mixing element at a flow rate, q. It is assumed that the liquid is well mixed and the gas is in plug flow. In the case of ozone absorption, the mass balance in the two-phase region can be written as

$$1.6576G\left|\frac{y_0}{1-y_0} - \frac{y_1}{1-y_1}\right| = (k_L a)V_I \Delta c_m \tag{1}$$

where G - air mass flow rate per porous disk, the numerical factor is the molar mass ratio of ozone and air, y_0, y_1 - ozone mole fractions in the inlet and the outlet gas, respectively, k_La - volumetric mass transfer coefficient in the liquid phase, V_I - volume of the two-phase region. The mean driving force

$$\Delta c_m = \frac{\Delta c_{inlet} - \Delta c_{outlet}}{\ln \dfrac{\Delta c_{inlet}}{\Delta c_{outlet}}} \tag{2}$$

where $\Delta c_{inlet} = c^0_{s,} inlet - c_e$; $\Delta c_{outlet} = c^0_{s,} outlet - c_j$
$c^0_{s,} inlet = c^o_s P_o y_o$; $c^0_{s,} outlet = c^o_s P_s y_1$; and c^o_s - Henry constant for ozone-water system, P_o, P_s - pressures above the bubbling station and at the liquid surface, respectively. Also

$$P_0 = P_s + \rho_L h_o g \tag{3}$$

were ρ_L - water density, g - gravitational acceleration, h_o - depth of the liquid layer in the chamber (above the disk). After substitution of the above given expressions into Equation 2 we get

$$\Delta c_m = \frac{c^o_s(P_o y_o - P_s y_1) - (c_e - c_j)}{\ln[\dfrac{c^o_s P_o y_o - c_e}{c^o_s P_s y_1 - c_j}]} \tag{4}$$

Usually, in the two-phase region, c_j=0 and c_e=0, since ozone is almost completely spent in chemical reactions (also it decomposes). For the mixing element region

$$Q_L c_j + q\, c_i = Q_L c_e + q\, c_e \tag{5}$$

The circulation flow rate, Q_L, between two regions can be taken from Smith and Goossens [4]:

$$Q_L = 0.16\, Q_G^{1/3} (h_o + 0.07)^{5/3} \tag{6}$$

where Q_G - volumetric gas flow rate per station related to the atmospheric conditions, the numerical value is the virtual origin correction for the two-phase jet. Equation 6 is valid over a wide range of the depth, h_o, from 0.03 to 15 m.

The volumetric mass transfer coefficient may be calculated from an empirical equation (c.f. Heijnen and Van't Riet [5])

$$k_L a_{20^0} = 0.63 u_G^{0.78} \tag{7}$$

where u_G - superficial gas velocity (in m/s) in the two-phase region. Equation 7 is originally valid for the system oxygen-water obtained for fine bubble diffusers, and has been established from our own considerations of the available experimental data collected, both in laboratory-scale bubble columns, and in commercial-scale aeration chambers equipped with porous dome elements (ceramic or Nokia polypropylene domes). For the latter, the data have been determined assuming of discretely operating bubbling stations created

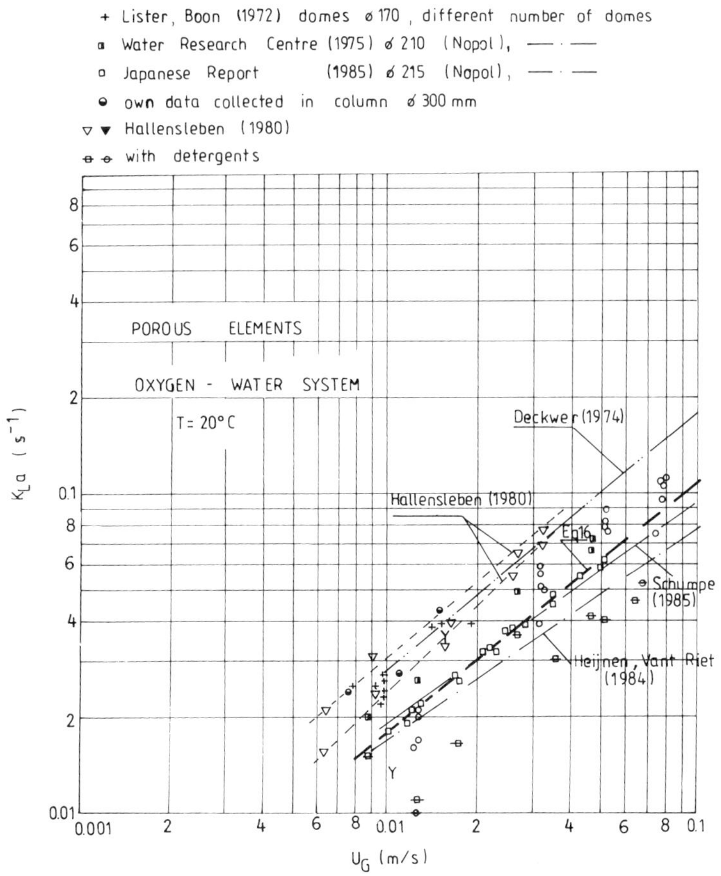

Fig. 4. $k_L a$ vs. u_G.

by porous domes and from calculations of the superficial gas velocity as that related to the two-phase region (either individual, cylindrical, or expanding plumes). The range of validity of equation 7 is for u_G from 0.01 to 0.1 m/s. Figure 4 shows experimental data for both laboratory bubble columns as well as commercial–scale chambers equipped with porous dome diffusers. It has to be pointed out that the $k_L a$ values depend much on the liquid phase properties (e.g.. ionic substances or contaminants dissolved in water). For the ozone-water system the values of $k_L a$ calculated for the oxygen-water system should be multiplied by the molecular diffusivity ratio of these two species raised to a power 0.5.

Although there is some experimental evidence [6] that the values of k_L in the ozone-water systems may be higher by a factor of about 2 as compared with the oxygen-water system at the same hydrodynamic conditions, the experimental data on k_La for the former do not confirm this conclusion. Also, the enhancement of k_L caused by usually unknown chemical reactions undergoing in practical situations is difficult to predict. Therefore, justified (and safer) to use the above outlined route of the k_La predictions based mainly on the available oxygen-water system data (the same approach is really assumed by Rakness et al. [1]).

Equations 1 and 4 can be solved simultaneously with the other values known to get a value of y_1, the outlet ozone concentration in the gaseous phase from an iterative process of computations. This provides a maximum value of the ozone transfer efficiency, TE_{max}

$$TE_{max} = 1 - y_1/y_0 \tag{8}$$

The maximum possible ozone concentration in the liquid phase results from

$$c_{max} = c_s^o P_o y_o \tag{9}$$

Now a linear dependence between TE and the bulk ozone concentration is established from the following considerations

$$\begin{array}{lll} at & c_b = 0 & TE = TE_{max} \\ at & c_b = c_{max} & TE = 0 \end{array} \tag{10}$$

Since ozone is partially decomposed in water, the following model is suggested to account for this effect [1]:

$$\ln(c_b/c_1) = -k_r T_d \tag{11}$$

where k_r - reaction constant, T_d - mean time of water in the ozonation chamber. The reaction constant, k_r, can be estimated from (in min^{-1}):

$$k_r = 5.2 \cdot 10^8 \exp(-5810.2667/T) \cdot 10^{0.17(pH-14)} \tag{13}$$

and T_d from the chamber geometry and the total water flow rate. A value of (c_b/c_1) is calculated from the expression valid for the perfectly mixed reactor:

$$c_b/c_1 = 1/(1 + k_r T_d) \tag{14}$$

A linear relationship between the bulk residual ozone concentration, c_b, and the transferred ozone dosage, c_1, is assumed as

$$c_b = (c_b/c_1)c_1 - c_d \tag{15}$$

where c_d - ozone residual suppression concentration which should be dependent upon the water quality.

The model is entered by assuming of TE<TE_{max} and then calculating c_1 from the known ozone dosage, OD:

$$c_1 = OD \cdot TE \tag{16}$$

Then c_b is estimated from equation 15 after an initial assumption of c_d. TE is now calculated from equation 11. Equation 5 is then used after establishing the ratio q/Q_L. Here Q_L should be large enough to provide an adequate mixing. The recommended value is to ensure at least ten turnovers through the two-phase region:

$$Q_L/q > 10 \tag{17}$$

and $q = Q/N$, were Q - total liquid flow through the chamber, N - number of the bubbling stations. When using equation 5 to get a value of c_e after taking c_j as equal to c_b ($c_i = 0$), it is expected that c_e should equal c_d, and this assumption enablesus to close the computational loop. The ozone transferred is calculated from

$$W_{OZ} = Q_N c_G TE \tag{18}$$

where Q_N - volumetric air flow rate at the standard conditions per single porous diffuser, c_G - ozone inlet concentration in the air (in g/Nm^3).

The outlined design procedure can easily be programmed on any computer. The initial data should contain the details of chamber geometry (length, width, and depth), water parameters (Q, T, pH, OD), porous element parameters (cross-sectional area, air flow rate per single element, Q_N, c_G, and number of elements, N). The computed output yields c_d, TE and W_{OZ}.

RESULTS AND DISCUSSION

Some typical characteristics of the ozonation process and the effects of the main process parameters (gas flow rate, ozone dosage, ozone concentration and water depth) are shown in Figures 5-7 for a porous ceramic disk diffuser of 300 mm in diameter immersed in an exemplary chamber of dimensions 3.2·3.0 m. The dependencies of the ozone transfer efficiency (TE) and the absorption rate (W_{OZ}) on the ozone dose and gas flow rate per single diffuser are shown in Figure 5. The transfer efficiencies decrease rapidly with an increasing OD, whereas the effect of Q_N is much less pronounced. The absorption rates of ozone drop only slightly with an increasing value of OD but differ much for the various gas flow rates. In Figure 6 the effects of the water depth in the chamber and the ozone concentration, c_G, on TE and W_{OZ} are demonstrated. An initially strong influence of h_0 on TE levels

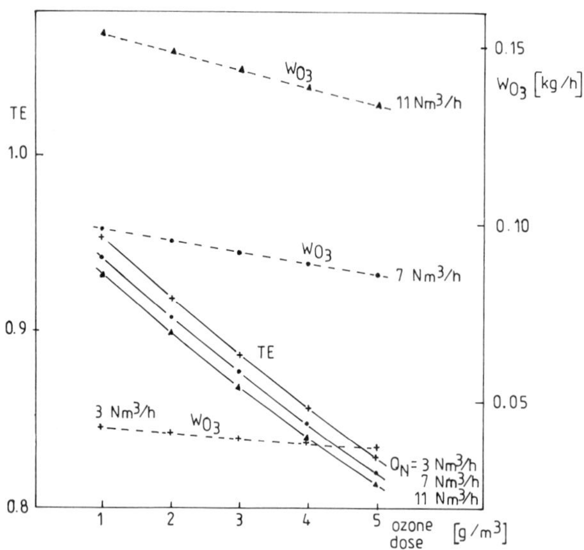

Fig. 5. Ozonation characteristics. Porous ceramic element ϕ 300 mm; $F=0.062\ m^2$; $C_G=15 g/Nm^3$; $h_o=5.5 m$; $q=0.023 m^3/s$; $T=10^oC$; pH=7.

off at h_0 higher than about 6 m. It is also evident that in order to attain the values of TE higher than 90% the applied water depths should exceed 5 m. As far as W_{OZ} is concerned, strong effect of c_G is seen, whereas h_0 variations affect the absorption rates of ozone only very slightly. The effects of Q_N and the ozone dose on TE are shown in Figure 7 . This figure is actually a reversed way of demonstrating the data given in Figure 5.

The overall chamber performance in terms of the absorbed ozone rate is easily obtained by multiplying the number of dome diffusers by W_{OZ} calculated from the above procedure for a single diffuser.

A part of the mass transfer consideration estimations of the degree of uniformity of gas distribution along the manifold and gas pressure drop in the piping system of the chamber can be made based on suggestions given by Bajura and Jones [7] and Riggs [8], supplemented by the pressure drop measurments carried out for ceramic dome diffusers. The results of such estimations show that good uniformity of gas distribution is achieved if the pressure drop in the lateral section (through the dome under normal "wet" operating conditions) is higher than about 300 Pa. The typical range of the "wet" pressure drop through the dome is from 1500 to 2500 Pa at the range of the air flow rates from 1 to 8 m^3/h per a single dome.

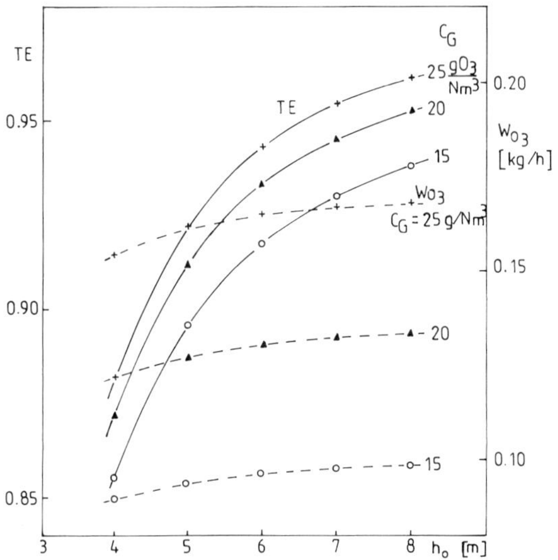

Fig. 6. TE and W_{O3} vs. h_0 for ceramic porous dome diffuser ϕ 300; Q_N=7 Nm^3/h; F=0.062 m^2; ozone dose=2 g/m^3; q=0.023m^3/s; T=10°C; pH=7.

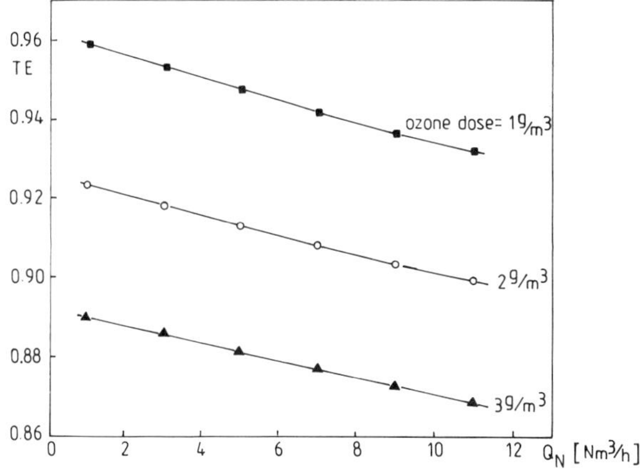

Fig. 7. TE vs. Q_N for different ozone dose; c_G=15gO_3/Nm^3; q=0.023m^3/s; h_o=5.5m; F=0.062 m^2; T=10°C; pH=7.

CONCLUSIONS

A design procedure for an industrial-scale ozonation chamber based on the liquid circulation unit concept is presented. The proposed procedure enables one to determine the main performance characteristics of the dome diffusers when applied in the ozonation process. The effects of the various relevant parameters on this performance are predicted and shown in a graphical form.

REFERENCES

[1] Rakness, K.L., R.C. Renner, B.A. Hegg, and A.G. Hill, "Practical design model for calculating bubble diffuser contactor. Ozone transfer efficiency", Ozone Sci. Engng. **10**, 173-214, 1988.

[2] Otero, Z., and T.W.F. Russell, "Design of commercial-scale gas-liquid contactors", AIChE J, **33**, 488-497, 1987.

[3] Otero, Z., J.N. Tilton, and T.W.F. Russell, "Some observations of flow patterns in tank-type systems", Int. J. Multiphase Flow, **11**, 583-589, 1985.

[4] Smith, J.M., and L.H.. Goossens, "The mixing of ponds with bubble columns", Fourth European Conference on Mixing BHRA Fluid Engineering, Noordwijkerhout, April 1982, Paper C2.

[5] Heijnen, J.J., and K. Van't Riet, "Mass transfer, mixing and heat transfer phenomena in low viscosity bubble column reactors", Chem. Engng J., **28**, B21-B42, 1984.

[6] Stanković, I., "Comparison of ozone and oxygen mass transfer in laboratory and pilot plant operation", Ozone Sci. Engng, **10**, 321-338, 1988.

[7] Bajura, R.A., and E.H. Jones, "Flow distribution manifolds", Trans. ASME, J. Fluids Engng, **98**, 654-666, 1976.

[8] Riggs, J.B., "Development of an algebraic design equation for dividing, combining, parallel, and reverse flow manifolds", Ind. Engng Chem. Res., **26**, 129-133, 1987.

OPTIMIZATION OF OZONE CONTACTORS IN A WATER TREATMENT PLANT USING MASS TRANSFER CORRELATIONS

I. STANKOVIĆ

Energoprojekt-Hidroinzenjering
Bulevar Lenjina 12, P.O. Box 20
11070 Beograd, Yugoslavia

ABSTRACT

In order to increase the removal efficiencies and the rates of ozone utilization in ozone contact columns, mass transfer of ozone to water with and without chemical reaction was investigated both in pilot plant bubble columns and in a Rushton type laboratory stirred reactor supplied with a variable turbine agitator. A comparison was made for different hydrodynamic conditions. The proposed mathematical models made it possible to simulate the mass transfer behavior of semi-industrial equipment by a laboratory apparatus. An engineering approach was developed for determination of the contactors mass transfer characteristics, in order to optimize the performance of a full scale water treatment plant.

INTRODUCTION

The new Belgrade water treatment plant, treating highly polluted Sava river water, has been in operation since 1987 with the nominal capacity of 2 m^3/sec, and with the possibility of future extension of up to 6 m^3/sec. The treatment process comprises preozonation, coagulation, floc-

Chemistry for the Protection of the Environment
Edited by L. Pawlowski *et al.*, Plenum Press, New York, 1991

culation, clarification, sand-anthracite filtration, granular activated carbon filtration, dosing of chemical (sulphuric acid, alu-sulphate, polyelectrolyte, lime, chlorine, fluorine), sludge treatment and granular activated carbon regeneration.

The pilot plant and laboratory investigations were carried out over a ten month period, prior to erection of the full scale plant, with the main task to optimize the treatment process in each stage [1].

Interfacial mass transfer plays a significant role in the design of ozonization contact columns. In order to examine the actual effectiveness of contact columns, it is essential to determine experimentally the mass transfer characteristics.

EXPERIMENTAL

The experimental ozonation system is presented schematically in Fig. 1.

Pilot plant measurements were performed in two bubble columns, made of Plexiglass, operating counter-currently and co-currently, respectively. The columns were 94 mm and 140 mm in diameter and 4300 mm in height. The capacities were 27 l and 60 l, and the distance between each sampling point was 900 mm. The liquid flow rates examined varied from 250 to 1500 l/h, and the gas flow rates from 25 to 250 Nl/h.

The laboratory experiments were carried out in a cylindrical stirred reactor of Rushton type, supplied with four baffles. Stirring was accomplished by a six-bladed turbine agitator, operating from 0 to 2000 rpm. The reactor was fitted with an air tight cover and a teflon ring around the stirrer axis. Gas was introduced through a sparger at the bootom. In all experiments the height of the liquid volume (5.92 lit) equaled the tank diameter (200 mm). The height of the reactor was 400 mm. Liquid flow rates examined varied from 1.5 to 3 lit/min., and gas flow rates from 35 to 80 Nl/h. A completely mixed systems was constructed to run in a continuous mode, with the liquid inlet and outlet at 10 cm above the bottom of the reactor.

Air was supplied by a compressor and introduced directly into the bubble columns or stirred tank. Ozone was generated by passing dried air through water cooled ozonizer.

The standard iodomeric method was used to analyze ozone concentration in the feed and exit gas, as well as in the liquid. Gas and liquid flow rates were measured by rotomaters, and gas volume by gasmeter. The dissolved oxygen concentration was measured by “HACH” dissolved oxygen meter.

LEGEND

FLOW METER
VALVE
PUMP
(T) THERMOMETER
(C) REVOLUTION COUNTER
(N) POWER REGULATOR
(M) MANOMETER
(A) AMMETER
(P) PRESSURE REDUCER
LIQUID PHASE
GAS PHASE

(1) PILOT PLANT OZONIZATION COLUMS
(2) LABORATORY STIRRED REACTOR
(3) OZONE GENERATOR
(4) STORAGE TANK
(5) COMPRESSOR
(6) AIR
(7) SAMPLING
(8) GASMETER
(9) OFF OZONE
(10) WATER INLET
(11) WATER OUTLET
(12) COOLING WATER INLET
(13) COOLING WATER OUTLET

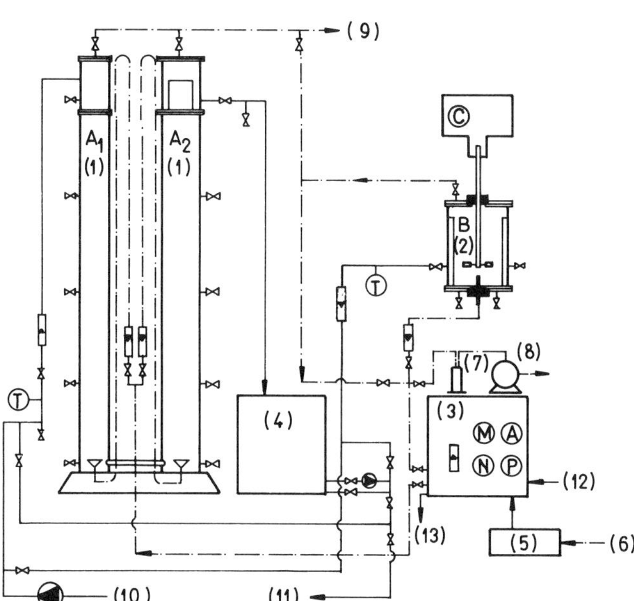

Fig. 1. The experimental system.

Table 1. Determination of Ozone Dosage in the First Column (A1)

L	G	C_{gin}	C_{Rout}	Ozone dosage
(l/h)	(Nl/h)	(mg/l)	(mg/l)	(mg/l)
1500	106.8	15.0	0.46	1.068
1250	106.8	15.0	0.60	1.282
1000	106.8	15.0	0.84	1.602
750	106.8	15.0	1.46	2.136

Clarified water: turbidity=1 NTU; $KMnO_4$ = 10.5 mg/l; pH = 7.1; t = 4.5°C.

RESULTS AND DISCUSSION

Ozone Dosages

Ozone dosages were calculated from the following relationship:

$$\text{Ozone dosage} = GC_{gin}/L \tag{1}$$

where:
G = gas flow rate
L = liquid flow rate
C_{gin} = ozone concentration in the inlet gas phase

Some typical results from the winter period are shown in Table 1.

The influence of ozone dosage on the ozone residual in water and on the ozone concentration in the off gas is shown in Fig. 2.

During the operation of the pilot plant, the ozone residual was adjusted to approximately 0.4 mg/l after the first column, and than maintained constant along the second column, by introducing the additional ozone dosage in order to compensate the ozone utilization due to decomposition and secondary reactions. Ozone distribution in both columns was calculated for different process conditions and is presented in Table 2.

MASS TRANSFER EXPERIMENTS IN BUBBLE COLUMNS

Physical Volumetric Mass Transfer Coefficient

The mass transfer mathematical model for a pilot plant bubble column was developed from concentration gradients along the column, during steady state operation, under plug flow assumption [2, 3].

The overall mass balance for physical absorption at any section of the counter-current bubble column has the following form:

$$GC_{gin} + LC_{Rin} = GC_{gout} + LC_{Rout} \tag{2}$$

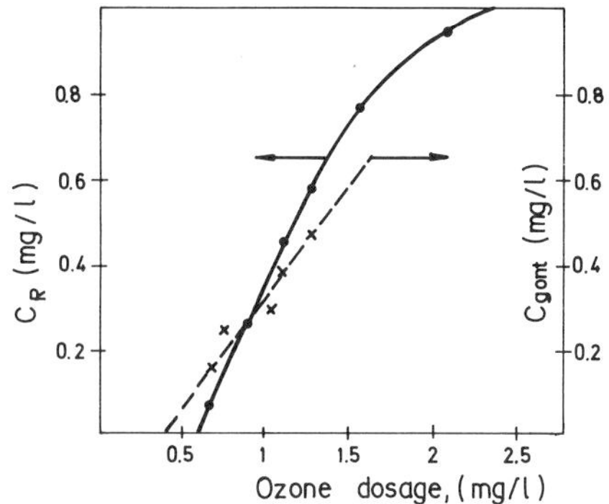

Fig. 2. Influence of ozone dosage on C_R and C_{gout}. Clarified water: turbidity= 0.8 NTU; SM> 3.6 mg/l; KM_nO_4=9.9 mg/l; UV(254 nm)=0.059 cm^{-1}; pH=7.1; t=20.5°C; G=50.1 Nl/h; C_{gin}=20.8 mg/l.

Table 2. Ozone Distribution in Ozonation Columns

L	C_R	G_1 A1	C_{gin}	C_{gout} A2	G_2 A2	OU/L A2	OZONE DOSAGE A1		A2		TOTAL
l/h	mg/l	Nl/h	mg/l	mg/l	Nl/h	mg/l	mg/l	%	mg/l	%	mg/l
1500	0.390	89.4	20.19	0.572	11.7	0.153	1.203	88.4	0.157	11.6	1.360
100	0.312	102.0	10.60	1.510	22.0	0.200	1.081	82.6	0.233	17.4	1.314
1000	0.328	55.8	20.09	0.890	13.4	0.257	1.121	80.6	0.269	19.4	1.390
500	0.365	55.8	9.83	0.986	18.3	0.324	1.098	75.3	0.360	24.7	1.458

Clarified water: turbidity=0.57 NTU; $KMnO_4$=17.36 mg/l; UV (254 nm)=0.084 cm^{-1}; pH=7.2; t=17°C.

where:
G = gas flow rate
L = liquid flow rate
C_g = ozone oxygen concentration in the air
C_R = free ozone residual

Assuming plug flow model:

$$U_L \, dC_R/dz = K_L^o a(P\,S\,C_g - C_R) \tag{3}$$

where:
V = liquid volume
U_L = superficial liquid velocity
z = vertical distance along the column

$K_L^o a$ = overall physical volumetric mass transfer coefficient
$PSC_g = C_R^*$ = gas solubility in water
S = distribution coefficient
P = absolute pressure

Ozonization experiments were performed with clarified water at t=4.5°C and pH=7. The ozone consumption for self decomposition was neglected.

The expression for overall physical volumetric mass transfer coefficient is derived after combination and integration of equations (2) and (3):

$$K_L^o a = \frac{\ln(\bar{P}SC_{gin} - C_{Rout})/(\bar{P}SC_{gin} - \bar{P}SLC_{Rout}/G + \bar{P}SLC_{Rin}/G - C_{Rin})}{H(PSL/G - 1)/U_L} \quad (4)$$

$\bar{P}$ is, the average absolute gas pressure and H the height of the integrated section.

The above equation holds for any section of a counter-current bubble column, providing the free residual ozone in the water exists between the two integrated points. In this work, equation (4) was applied for the bottom section of the first column. It was also assumed that the overall physical mass transfer coefficient and the effective interfacial area per unit volume of the bubble column were the same in all parts of the column.

According to Danckwerts [4] the necessary condition for no reaction in the liquid film is:

$$D_A k_1/(K_L^o)^2 \ll 1 \quad (5)$$

Diffusivity was calculated from the Vilke and Chang correlation at 20°C [5]: $D_A = 1.743\ 10^{-5}$ cm²/sec.

According to Hill and Spenser [6], the overall physical mass transfer coefficient is:

$$K_L^0 = 0.021 \text{cm/s}$$

The rate constant for ozone decomposition was calculated from Hewes and Davision's correlation [7], assuming the first order mechanism:

$$k_1 = 0.0033 sec^{-1} (pH = 8, t = 20°C)$$

Substituting into equation (5)

$$D_A\, k_1/K_L^O)^2 = 1.3\ 10^{-4} \ll 1$$

It was concluded that there was no appreciable enhancement of the mass transfer due to indirect oxidation by free radicals arising from ozone decomposition. Therefore, an expression for the overall physical volumetric mass transfer coefficient, given by eq. (4) for a bubble column, is valid both

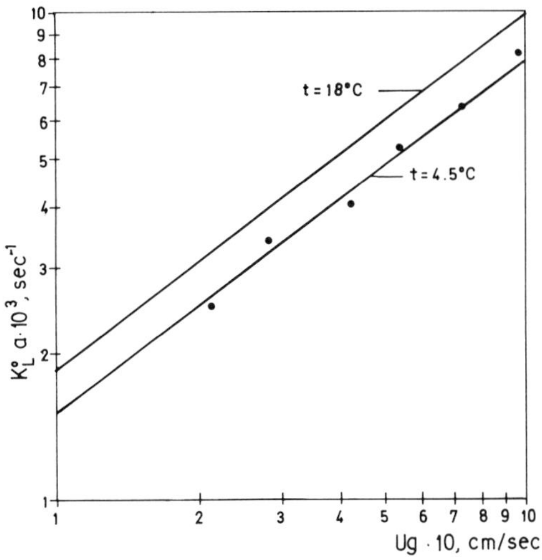

Fig. 3. Relationship between $K_L^o a$ and U_g for bubble column.

for clean and clarified water, provided the free ozone residual exists between the integrated points.

The temperature dependence of $K_L^0 a$ resulted from the temperature dependence of the diffusivity. Assuming that the effective interfacial area per unit volume of the column is not a temperature function, it follows from Higbie or Danckwerts model [4]:

$$K_L^o a\,(T_1)/K_l^o a(T_2) = \sqrt{D_A(T_1)/D_A(T_2)} \tag{6}$$

The plot of physical volumetric mass transfer coefficient versus gas superficial velocity (U_g) was found to be a straight line in log-log coordinates. This relationship, shown in Fig. 3, can be correlated by the following expression:

$$K_L^0 a \propto U_g^{0.72} \tag{7}$$

Volumetric Mass Transfer Coefficient for Absorption with Chemical Reaction

If the absorption was accompanied by direct chemical reaction, the mass transfer was enhanced, due to the fact that a significant fraction of ozone absorbed reacted in the liquid film.

The average overall mass transfer coefficient ($\overline{K_L a}$) and the actual ozone consumption for chemical reactions were determined measuring the inlet and outlet concentration of ozone in the gas phase. The free residual ozone at the outlet of the column was adjusted to equal zero.

Table 3. Clarified Water: Determination of $\overline{K_La}$, OU, and E in Bubble Column

L l/h	G Nl/h	C_{Rout} mg/l	C_{gin} mg/l	C_{gout} mg/l	OU mg/min	$\overline{K_La}$ sec^{-1}	E
1500	55.9	0	13.13	0.27	11.98	0.00696	2.08
1000	55.8	0	10.2	0.24	9.26	0.00689	2.06
500	55.8	0	6.9	0.19	6.27	0.00660	1.98

Turbidity = 0.7 NTU; SM=5.6 mg/l; UV=4.8 m^{-1}; KM_nO_4=10.54 mg/l; pH=7.2; t=17–18°C.

From the mass balance equation:

$$G\, dC_g/dz = K_L a A P S C_g \tag{8}$$

where A is the cross sectional area of the column.
After rearanging and integrating:

$$\overline{K_L a} = G \ln(C_{gin}/C_{gout})/V\bar{P}S \tag{9}$$

Ozone utilization rate was then computed from the balance equation.

$$OU = G(C_{gin} - C_{gout}) \tag{10}$$

Table 3 shows the results for $\overline{K_La}$, OU, and E, where E is, enhancement factor.

MASS TRANSFER EXPERIMENTS IN LABORATORY–STIRRED REACTOR

Physical Volumetric Mass Transfer Coefficient

An ozone mass balance was made on both the liquid and the gaseous phase at steady state conditions, under assumption of completely mixed gas-liquid system [2, 3].

The overall mass balance of ozone is:

$$G\, C_{gin} - G\, C_{gout} = L\, C_R + OU \tag{11}$$

The ozone mass balance in the gas phase is:

$$G\, C_{gin} - G\, C_{gout} = V\ K_L^o a(PSC_{gout} - C_R) \tag{12}$$

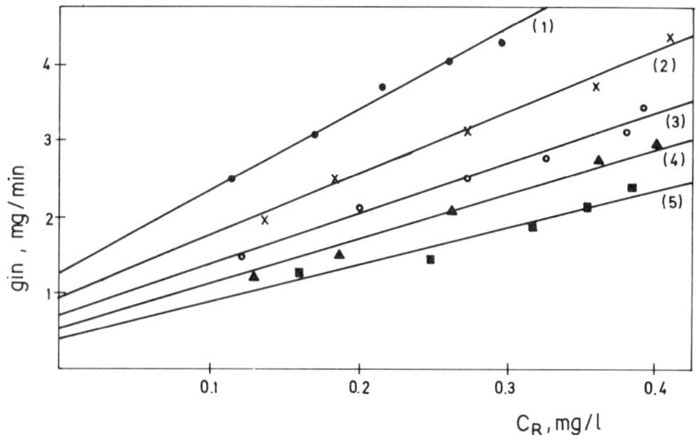

Fig. 4. Clean water: relationship between g_{in} and C_R for stirred reactor.

The solution of equations (11) and (12) leads to the following expression:

$$G\,C_{gin} = g_{in} = (OU + GOU/V\,K_L^o a\,PS) + (GL/V\,K_L^o a\,PS + G/PS + L)C_R \quad (13)$$

The proposed model is based on the linear relationship between the ozone supply rate (g_{in}) and the free ozone residual (C_R).

From the slope and the intercept of the straight line one can obtain:

$$K_L^o a = GL/V(\text{ tg } \alpha PS - LPS - G) \quad (14)$$

and

$$OU = mL/(\text{tg } \alpha - G/PS) \quad (15)$$

where m is the intercept.

For clean water the ozone utilization rate term (OU) refers to an ozone consumption for self decomposition.

Since there was no appreciable enhancement of the mass transfer due to ozone decomposition, the overall physical volumetric mass transfer coefficient was obtained from the slope of the straight line.

A linear plot of ozone supply rate versus free ozone residual is presented in Figs. 4 and 5. Results are summarized in Table 4.

A correlation between physical volumetric mass transfer coefficient and agitational speed was obtained as a straight line in log-log coordinates, as shown in Fig. 6.

The correlation is expressed as:

$$K_L^o a \propto n^{1.96} \quad (16)$$

Table 4. Clean Water: Determination of $K_L^o a$ and OU / t=18°C, pH=7.2/in Stirred Reactor

Fig.	N_o	L l/min	G Nl/min	n min^{-1}	$K_L^o a$ sec^{-1}	OU mg/min
	1	1.5	0.567	300	0.00135	0.234
	2	1.5	0.567	400	0.00217	0.249
3	3	1.5	0.567	500	0.00323	0.258
	4	1.5	0.567	600	0.00460	0.236
	5	1.5	0.567	700	0.00600	0.232
	6	1.5	1.217	300	0.00160	0.223
	7	1.5	1.217	400	0.00255	0.241
4	8	1.5	1.217	500	0.00410	0.249
	9	1.5	1.217	600	0.00570	0.247
	10	1.5	1.217	700	0.00730	0.242

Volumetric Mass Transfer Coefficient tor Absorption with Chemical Reaction

In the case of clarified water, the ozone utilization rate term (OU) from eq. (13) refers to an ozone consumption for direct chemical reactions with organic and inorganic materials and subsequent intermediates, and also for self-decomposition. No attempt was made to distinguish between the ozone utilization for chemical reaction, and for self-decomposition. The $K_L a$ obtained from the slope of the straight line corresponds to the overall volumetric mass transfer coefficient for absorption with chemical reaction.

The liquid flow rate in the stirred reactor was adjusted according to retention time in the pilot plant unit.

A linear relationship between ozone supply rate and free ozone residual for clarified water is presented in Fig. 7. Results are summarized in Table 5.

The ozone utilization rate and the overall volumetric mass transfer coefficient were also calculated from the inlet and outlet concentrations of ozone in that gas phase, applying equations (11) and (12). An assumption was made that the ozone concentration in the outlet gas equaled the ozone concentration averaged over the gas bubbles in the reactor, since the gas in the reactor was supposed to be perfectly mixed. Results are presented in Table 6.

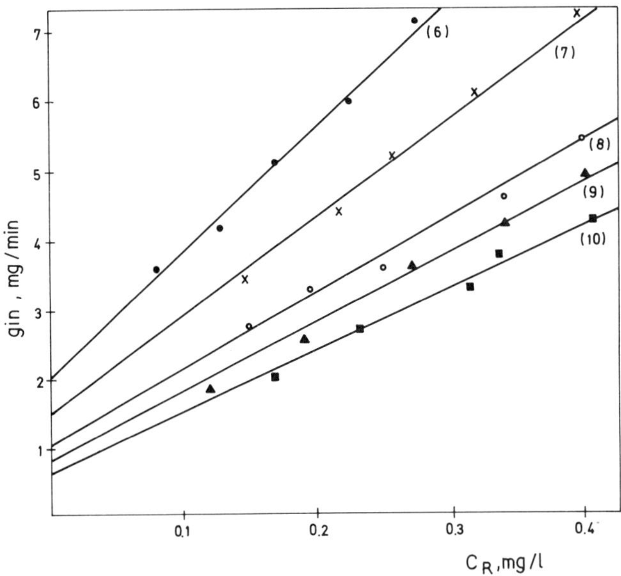

Fig. 5. Clean water: relationship between q_{in} and C_R for stirred reactor.

SCALE UP PROCEDURE

In the semi-industrial counter-current bubble column, a correlation between superficial gas velocity and the overall physical volumetric mass transfer coefficient was obtained.

In the laboratory–stirred reactor a correlation between overall physical volumetric mass transfer coefficient and agitational speed was obtained for different gas flow rates.

These two correlations enable one to establish a correlation between the superficial gas velocity of the semi-industrial unit, and the agitational speed of the laboratory–stirred reactor, producing the same overall physical volumetric mass transfer coefficients. The plot of U_g versus n is presented in Fig. 8 in log-log coordinates.

The correlation is expressed as

$$Ug \propto n^{2.5} \tag{17}$$

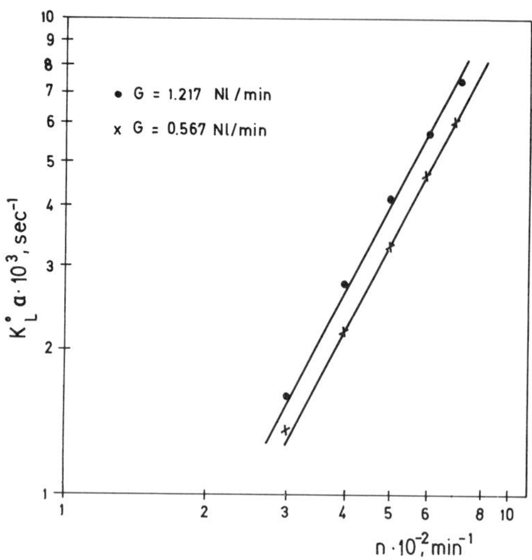

Fig. 6. Clean water: relationship between $K_L^o a$ and n for stirred reactor.

Table 5. Clarified Water: Determination of $K_L^o a$, OU and E in Stirred Reactor

N_o	L l/min	τ min	G Nl/min	n min^{-1}	$K_L^o a$ sec^{-1}	OU	E
1	1.5	4	0.617	300	0.00287	1.112	2.39
2	1.5	4	0.617	600	0.01012	1.217	2.41
3	1.5	4	1.233	300	0.00370	1.470	2.47
4	1.5	4	1.233	600	0.01249	2.442	2.40
5	3	2	0.617	300	0.00314	1.363	2.62
6	3	2	0.617	600	0.01100	1.523	2.63
7	3	2	1.233	300	0.00382	1.824	2.55
8	3	2	1.233	600	0.01394	2.793	2.68

turbidity= 1 NTU; UV=4.7 m^{-1}; KM_nO_4 =9.48 mg/l; t=12.7°C, pH=7.3.

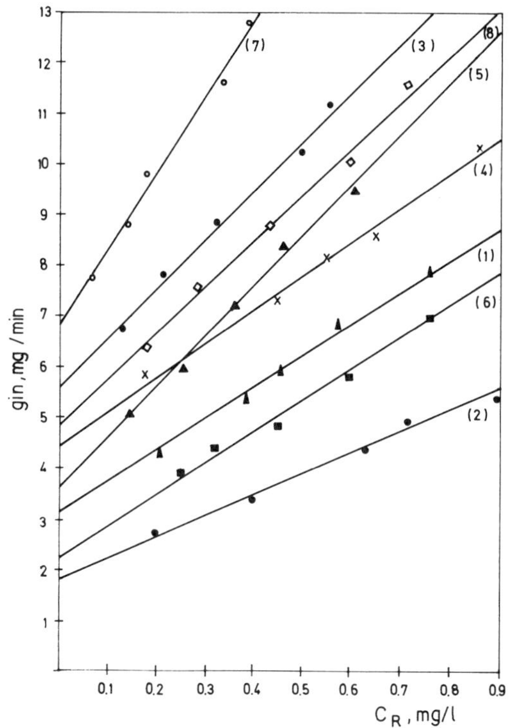

Fig. 7. Clarified water: relationship between q_{in} and C_R for stirred reactor.

Table 6. Clarified Water: Determination of $K_L^o a$, OU and E in Stirred Reactor

L l/min	τ min	G Nl/min	n min^{-1}	C_{gin} mg/l	C_{gout} mg/l	C_R mg/l	OU mg/min	$K_L a$ sec^{-1}	E
1.5	4	0.602	720	5.99	2.70	0.340	1.470	0.01600	2.54
2	3	0.602	720	5.99	2.32	0.216	1.777	0.01656	2.63
3	2	0.602	700	5.99	1.81	0.036	2.408	0.01665	2.77

turbidity=0.7 NTU; SM=3.6 mg/l; UV=5.9 m^{-1}; KM_nO_4 =9.9 mg/l; T=18.3°C, pH=7.1.

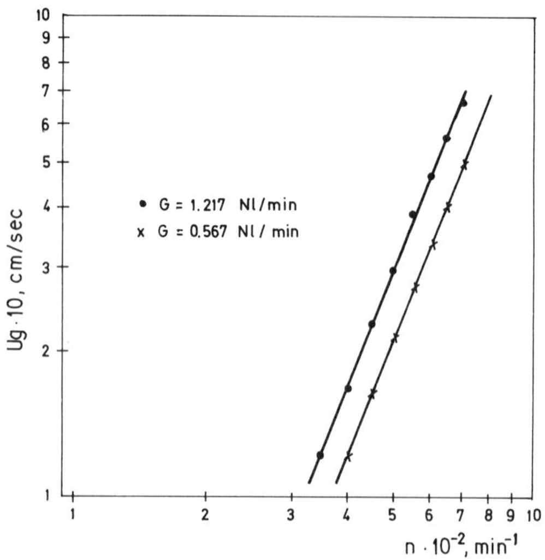

Fig. 8. Correlation between U_g (bubble column) and n (stirred reactor), producing the same $K_L^0 a$.

CONCLUSIONS

The basic task of this work was to estimate, for design purposes, how a large scale column will function using a laboratory reactor with the same detention time of the liquid phase.

It was concluded that there was no appreciable enhancement of the mass transfer due to indirect oxidation by free radicals arising from ozone decomposition. Therefore, an expression for overall physical volumetric mass transfer coefficient, given by eq. (4) for a bubble column, is valid both for clean and clarified water, provided the free ozone residual exists between the integrated points. A correlation between overall physical volumetric mass transfer coefficient and gas superficial velocity gives a straight line in log-log coordinates.

When absorption is accompanied by direct chemical reactions, an enhancement of the mass transfer is significant. The average values for overall volumetric mass transfer coefficient, the rate of ozone utilization, and the enhancement factor were determined for a bubble column.

A correlation was also established between overall physical mass transfer coefficient and agitational speed in a laboratory–stirred reactor. The overall volumetric mass transfer coefficient, rate of ozone utilization, and the enhancement factor for clarified water were computed from the linear plot of ozone supply rate versus free ozone residual.

A correlation between the superficial gas velocity of the pilot plant unit, and the agitational speed of the laboratory stirred reactor, producing the same overall physical volumetric mass transfer coefficients, made it possible to simulate the mass transfer behavior of semi–industrial equipment by a laboratory apparatus. The actual amount of ozone consumed by chemical reaction could be determined in a laboratory apparatus, and then employed to calculate what minimum applied dosage is necessary to effect the same degree of utilization in a large scale column.

REFERENCES

[1] Stanković, I. "Ozone Application in the New Belgrade Water Treatment Plant", 7th Ozone World Congress, Tokyo, Japan, 1985.

[2] Stanković, I. "Analysis of Ozonation and Aeration Contacting Systems Applied in Water Purification", Ph. D. Thesis, University of Belgrade, Yugoslavia, 1983.

[3] Stanković, I. "Comparison of Ozone and Oxygen Mass Transfer in a Laboratory and Pilot Plant Operation", Ozone Science and Engineering, Vol. 10, 1988, p 321–338.

[4] Danckwerts, P.V. "Gas Liquid Reactions", McGraw-Hill, 1970.

[5] Vilke, C.R. and P. Chang, A.I.Ch.E.J., 1–264, 1955.

[6] Hill, A.G. and H.T. Spenser, "Mass Transfer in a Gas Sparged Ozone Reactor", Proceedings of the first Symposium on Ozone for Water and Wastewater Treatment, 1973.

[7] Hewes, C.G and R.R. Davison, "Kinetics of Ozone Decomposition and Reaction With Organics in Water", A.I.Ch.E.J., Vol. 17, 17, pp 141, 1971.

PHOTOCHEMISTRY OF CHLOROORGANIC PESTICIDES IN THE UV AND VISIBLE REGION

J. ŁUBKOWSKI, T. JANIAK, J. RAK
and J. BŁAŻEJOWSKI

Department of Chemistry
University of Gdańsk
80-952 Gdańsk, Poland

ABSTRACT

Photochemistry of three chloroorganic pesticides, namely, lindane, pentachloronitrobenzene and methoxychlor, has been studied in methanolic media in the UV and visible regions. The 253.7 nm radiation initiates complex changes in the systems examined leading to the formation of chloride ions, oxidative species and numerous organic substances being the products of dechlorination, dehydrochlorination, degradation and isomerization of original molecules. The pesticides studied are unreactive towards singlet oxygen formed via sensitization by eosine in the visible region.

INTRODUCTION

Contamination of the environment by organic chemicals containing chlorine atoms presents a serious problem owing to the high toxicity of these derivatives [1-6]. The main source of the pollutants are chloroorganic pesticides which have been widely utilized for more than 40 years. Due to the high

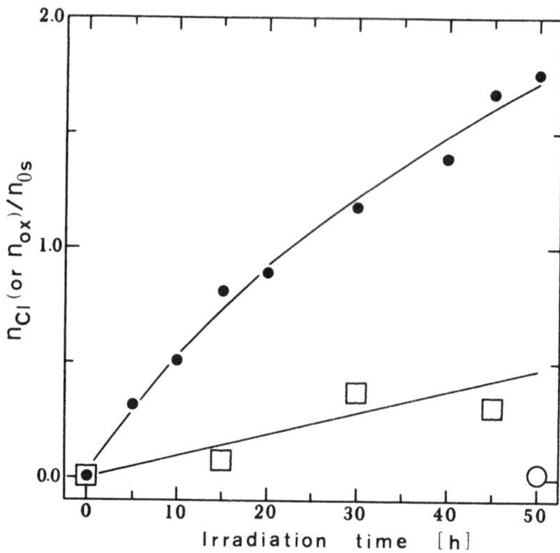

Fig. 1. The relative efficiencies of formation of chloride ions (n_{c1}/n_{os}) and oxidative species (n_{ox}/n_{os}, n_{os} = initial amount of substrate) versus time during exposure of γ-BHC with unfiltered radiation emitted by a medium-pressure mercury lamp (• and □, experimental points for n_{Cl}/n_{os}, and n_{ox}/n_{os}, respectively) and radiation of λ >300 nm (○, point for n_{Cl}/n_{os}, only).

stability of these compounds, they can remain unchanged in the environment for a long time. They can also be transferred to even more toxic oxygen-containing derivatives [7]. Therefore, the problem of degradation of these contaminants becomes immensely important. One of the factors influencing the environment is solar radiation, hence a knowledge of photochemistry of pesticides may provide information on the fates of these compounds in the environment.

The present paper is devoted to some complementary studies regarding photochemistry of γ-BHC (lindane, γ-hexachlorocyclohexane), PCNB (pentachloronitrobenzene), p,p'-methoxychlor (1,1'-(2,2,2-trichloroethylidene)bis[4-methoxy-benzene]) in the ultraviolet and visible regions. Despite studies undertaken in the past [8-12], several aspects of this problem still remain unexplained.

EXPERIMENTAL

The UV photolyses were carried out in a quartz cell, at room tem-

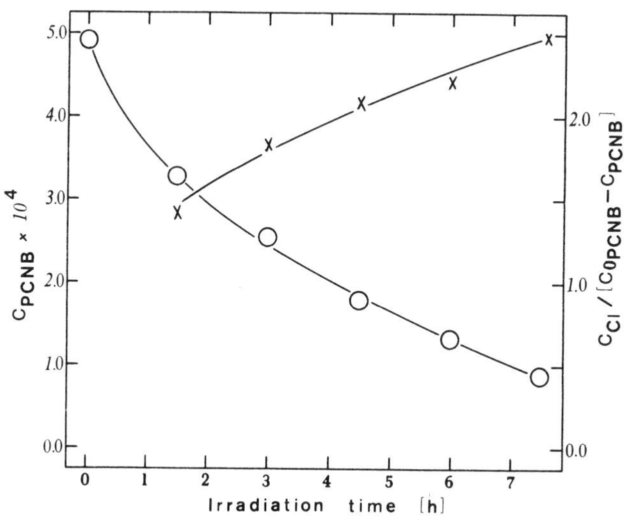

Fig. 2. Depletion of PCNB (o, C_{PCNB} in mol·dm^{-3}) and formation of Cl^- ions (x) (expressed as ratio of the amount of chloride ions formed to the amount of substrate depleted) versus exposure time.

perature, using a medium-pressure mercury lamp. The methanolic solutions (using Mallinckrodt, spectral grade solvent) of pesticides were exposed to either the whole radiation emitted by the lamp, or to a radiation of $\lambda > 300$ nm separated by a UG-11 glass filter (Schott u. Gen., Jena). Using malachite green leucocyanide actinometer, covering the region 250 - 320 nm [13], it was estimated that the mean radiation intensity of an unfiltered beam was ca. $4.0 \cdot 10^{-8}$ Einst s^{-1} at the entrance window of the cell. Combining the light source with a UG-11 filter, the incident radiation intensity was estimated to be ca. $2.6 \cdot 10^{-8}$ Einst s^{-1}. In a separate series of experiments, the methanolic solutions of pesticides containing dissolved eosine were saturated with oxygen (by bubbling O_2) and exposed in a Pyrex bulb to the visible light emitted from a 500 W halogen lamp.

Hexachlorocyclohexane isomers and p,p'-methoxychlor, all from P.O.Ch. (Poland), were of analytical grade (GC analytical standards) and were used as received. PCNB from Aldrich was subjected to vacuum sublimation prior to use. 1,1'-(dichloroethenylidene)bis[4-methoxy-benzene] (p,p'-DMDE) was synthesized as described in the literature [14]. Other pesticide reference standards, as well as standards of related compounds, were provided by the U.S. Environmental Protection Agency.

The extent of photolysis of pesticides was examined by determining

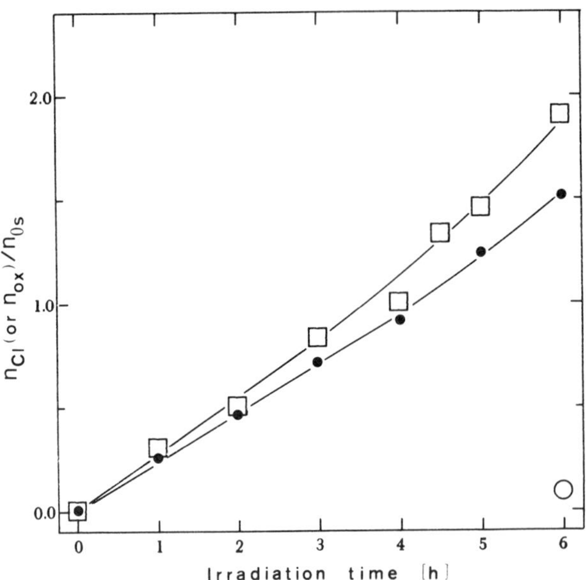

Fig. 3. n_{Cl}/n_{os} (•) and n_{ox}/n_{os} (□) vs. time during irradiation of PCNB (c_o=0.001 mol·dm^{-3}) with unfiltered radiation from a medium-pressure mercury lamp and above 300 nm (○, point only for n_{Cl}/n_{os}).

the amount of chloride ions (mercurometric method) and oxidative species (iodometric analysis) formed and by performing GC and HPLC analyses of organic reactants. The GC analyses were carried out on a Mera-Elwro model N504 (Poland) chromatograph using a 3.2 m long glass column (i.d.=0.004 m) packed with the 1.5% OV-17 + 1.95% OV-210 coated on a Chromosorb WAW (100/120 mesh). The column was treated with a silylating reagent (trimethylchlorosilane) and high concentrations of pesticides prior to use. The HPLC analyses were performed on an HPP 5001 chromatograph (Czechoslovakia) using a 0.15 m long column (diameter 0.0033 m) packed with Sepazon SCX C_{18}. As the mobile phase, a mixture of CH_3CN : H_2O of various v/v ratios was used.

RESULTS

Chloroorganic pesticides are poorly soluble in water [15], whereas, the compounds studied are relatively easily soluble in alcoholic media. Therefore, to study their photochemistry, we chose CH_3OH as a solvent which

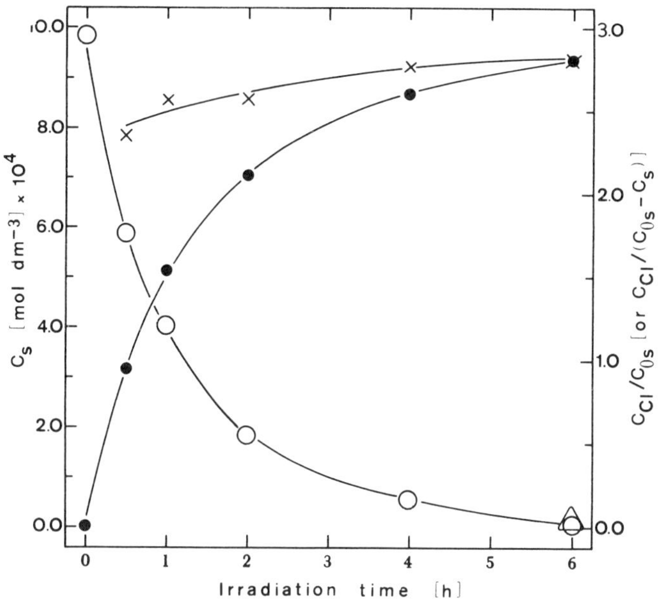

Fig. 4. Depletion of p,p'-methoxychlor (○) and formation Cl^- ions expressed either relative to the initial amount of substrate (•, C_{Cl}/C_{OS}) or as the ratio of the amount of chloride ions determined relative to the amount of substrate depleted (x, $C_{Cl}/(C_{OS} - C_S)$) versus irradiation time during irradiation with unfiltered radiation from a medium-pressure mercury lamp and above 300 nm (△, point only for C_{Cl}/C_{OS}).

exhibits some similarities to H_2O. All chloroorganic pesticides studied lose chlorine atoms upon photolysis. These species can be easily determined quantitatively as chloride ions. In experimental conditions chosen, i.e., in CH_3OH media, the photodecomposition of pesticides is always accompanied by the formation of oxidative species. Quantitative dependencies relating the amounts of both Cl^- ions and oxidative species formed to the initial amount of substrate are shown in Figures 1, 3 and 4. The exposure of the pesticides to ultraviolet light initiates various changes in parent molecules, which, however, are characteristic for individual compounds. They will be discussed subsequently.

The amount of chloride ions formed upon irradiation of γ-BHC relative to the initial amount of substrate varies almost linearly with time, reaching a value equal to unity after exposure for ca. 25 h (Fig. 1). This means that within this time an average Cl atom is detached from the original molecule. The amount of oxidative species produced during photolysis of γ-BHC, determined as the amount of molecular iodine released in acidic KI solutions, is roughly 5 times lower than the amount of Cl^- ions formed (Fig. 1). The material balances made on the basis of both above dependencies indicate that molecular chlorine is not very likely as a product of photolysis. The chromatographic analyses revealed at least 6 new substances in the photolyte. One of them was identified as α-BHC. Most reaction products have shorter retention times than those for substrate. This fact might suggest that they arise from lighter substances with a lower number of chlorine atoms in the molecule.

The existence of the absorption bands in PCNB and p,p'-methoxychlor molecules above 200 nm [16] results in the overall absorption of both compounds in the ultraviolet region being much stronger than in the case of the γ-BHC discussed above, the weak absorption of which is seen only at the short wavelength boundary of the ultraviolet region. In consequence, the irradiation time necessary to produce measurable changes in solutions of PCNB and p,p'-methoxychlor was almost one order of magnitude lower than in the case of γ-BHC, although the initial concentration of the latter compound was 10 times higher.

The extent of chemical changes vs. time in the methanolic solution of PCNB is shown in Figures 2 and 3. As can be noticed, the compound disappears gradually in time, reaching 0.18 of the initial concentration after 7.5 h exposure (Fig. 2). The dependencies n_{Cl}/n_{os} and n_{ox}/n_{os} are almost linear and identical in the first four hours of irradiation (Fig. 3). Both dependencies deviate from linearity upon prolonged exposures. These dependencies also indicate that the detachment of an average chlorine atom is accompanied by the formation of oxidative species capable of releasing one I_2 molecule. The average amount of chlorine atoms withdrawn from the molecule of substrate increases with the irradiation time (Fig. 2). It results from this experimental dependency that an average of 2 Cl atoms are detached from the PCNB molecule after ca. 4 h exposure. Chromatographic analyses revealed that the photochemical processes result in the formation of at least 6 new compounds. Of these we were able to prove only the presence of 1,2,4,5-tetrachloronitrobenzene, whereas, we did not confirm the presence of penta-, tetra- and tri-chlorobenzenes. The quantitative analyses revealed that tetrachloronitrobenzene accounts only for the depletion of less than 10% of PCNB. Radiation of over 300 nm causes minor changes in the methanolic solution of PCNB even after prolonged irradiation (Fig. 3).

The extent of chemical changes on irradiation of p,p'-methoxychlor dissolved in CH_3OH is shown in Figure 4. The characteristic feature of photolysis of the compound is its relatively fast depletion accompanied by the formation of large amounts of chloride ions and oxidative species. The character of experimental dependencies differs from that observed in the case of photolysis of γ-BHC or PCNB indicating much more pronounced ability of methoxychlor to undergo photochemical processes. Two points are worthy of mention. Firstly, the detachment of one average mol of chlorine atoms from the methoxychlor molecules is accompanied by the formation of oxidative species in an amount capable of releasing ca. 0.8 mol of I_2 molecules in acidic KI. Secondly, the ratio of amount of chlorine detached to the amount of substrate depleted increases with the irradiation time, reaching values equal to 2.8 on prolonged exposures. This indicates that dechlorination, dehydrochlorination or other changes in the original molecule are far advanced. p,p'-methoxychlor disappears, however, very slowly upon exposure of over 300 nm. Chromatographic analyses of the reactant mixture indicate the appearance of at least 5 products. Among them we were able to confirm only the presence of p,p'-DMDE.

We also investigated the possibility of singlet oxygen playing a role in the photochemically induced decomposition of γ-BHC, PCNB and p,p'-methoxychlor. For this purpose the oxygen saturated methanolic solutions of pesticides were exposed to visible radiation in the presence of eosine as a sensitizer. No changes in pesticide concentrations were observed, even after 50 h exposure. We did, however, monitor minor decomposition of the sensitizer after prolonged irradiation. Similar results were obtained when eosine was coated on an Amberlite (IRA-400) resin and irradiation was performed in a heterogeneous system. The above facts clearly indicate that all pesticides studied are completely unreactive to singlet oxygen.

DISCUSSION

The long wavelength absorption bands of γ-BHC are of either the $n \rightarrow \sigma^*$ or $\sigma \rightarrow \sigma^*$ type and fall in the vacuum ultraviolet region. Only long wavelength shoulders of these bands extend above 200 nm and cause relatively weak absorption at 253.7 nm. Above 300 nm the absorption of the pesticide is negligible. This means that solar radiation is practically notabsorbed by γ-BHC. The changes observed in Figure 1 are predominantly caused by 253.7 nm radiation emitted by a medium pressure mercury lamp. PCNB and p,p'-methoxychlor exhibit several bands in the 200-300 nm range [16, 17]. Some of them can be assigned to the transitions of the benzene chromophore ($\pi \rightarrow \pi^*$). Others are caused by the presence of Cl, NO_2 or OCH_3 groups which may contribute to the appearance of electron transfer transitions in which p-electrons of substituents are transferred to the π orbitals of the ring ($\pi \rightarrow \pi^*$). The absorption of both latter compounds is relatively strong in the 200-300 nm region. These pesticides must, therefore, absorb a fraction of the solar radiation.

The absorption of the UV radiation by the compounds studied produces electronically excited species. Complementary studies revealed that radiative processes contribute fairly insignificantly to the deactivation of the excited molecules. The rough estimation made on the basis of experimental results, presented in Figures 2 and 4 and actinometric measurements, revealed that quantum yield values for depletion of substrate never exceed 0.2 in the case of PCNB and are closet unity for p,p'-methoxychlor. This indicates that chloroorganic pesticides have a considerable photochemical reactivity.

It is difficult to outline the complete reaction mechanism solely on the basis of the results of the present work. Nevertheless, these results, together with certain information from literature, enable some general conclusions to be drawn. In all compounds studied, the C-Cl bond appears the weakest one. It is, thus, very likely that electronically excited molecules dissociate in the primary process to form Cl atoms and appropriate organic radicals. In the secondary processes both species may abstract hydrogen from solvent molecules to form HCl and appropriate dechlorinated molecules. In this way 2,3,5,6-tetrachloronitrobenzene is presumably formed. The absorption of UV radiation by molecules may also cause molecular elimination, i.e., detachment of, e.g., HCl or Cl_2 from parent molecules. Such processes are energetically constrained [18], although they can easily be realized in electronically or highly vibrationally excited states. The formation of p,p'-DMDE from p,p'-methoxychlor most probably proceeds by HCl elimination mechanism. In the case of γ-BHC isomerization to α-isomer is also feasible [19]. Lastly, photochemical processes may be accompanied by the destruction of the moiety of an original molecule. This may be expected particularly upon photolysis of p,p'-methoxychlor.

The results of the present studies clearly demonstrate that UV radiation causes withdrawal of chlorine atoms from molecules of chloroorganic pesticides. The photochemical processes are always accompanied by the formation of oxidative species whose nature is not known. It has, however, been revealed that these species disappear upon storage of photolyte. Interesting also is the stability of pesticides in the presence of singlet oxygen. This is a rather unexpected result since compounds containing double bonds usually undergo destruction in the presence of $^1O_2$ [20]. The present study confirms the very high stability of chloroorganic pesticides toward one of the most important factors influencing the environment, namely, solar radiation.

ACKNOWLEDGEMENT

We thank the U.S. Environmental Protection Agency (Research Triangle Park, NC 27711) for providing samples of chromatographic reference standards of pesticides and related compounds. This work was financed by the National Institute of Hygene (Warsaw) under CPBR 11.12 (contract No. B-42).

REFERENCES

[1] M. D. Reuber, Carcinogenicity of lindane, Environ. Res., 19 460-81, 1979.

[2] M. D. Reuber, Carcinogenicity and toxicity of methoxychlor, EHP, Environ. Health Perspect., 36 205-19, 1980.

[3] D. Y. Lai, Halogenated benzenes, naphthalenes, biphenyls and terphenyls in the environment: their carcinogenic, mutagenic, teratogenic potential and toxic effects, J. Environ. Sci. Health, C2 135-84, 1984.

[4] G. Renner, H. H. Otto and P. T. Nguyen, Fungicides hexachlorobenzene and pentachloronitrobenzene and their routes of metabolism, Toxicol. Environ. Chem., 10 119-32, 1985.

[5] H. Choudhury, J. Coleman, F. L. Mink, C. T. De Rosa and J. F. Stara, Health and environmental effects profile for pentachloronitrobenzene, Toxicol. Ind. Health, 3 5-69, 1987.

[6] J. Ashby and R. W. Tennant, Chemical structure, Salmonella mutagenicity and extent of carcinogenicity as indicators of genotoxic carcinogenesis among 222 chemicals tested in rodents by the U.S. NCI/INTP, Mutat. Res., 204 17-115, 1988.

[7] O. Hutzinger, M. J. Blumich, M. V. D. Berg and K. Olie, Sources and fate of PCDDs and PCDFs: an overview, Chemosphere, 14 581-600, 1985.

[8] D. G. Crosby and N. Hamadmad, Photoreduction of pentachlorobenzenes, J. Agr. Food Chem., 19 1171-4, 1971.

[9] R. G. Zepp, N. L. Wolfe, J. A. Gordon and R. C. Fincher, Light-induced transformation of methoxychlor in aquatic systems, J. Agr. Food Chem., 24 727-33, 1976.

[10] T. Bogacka, Kinetics of the breakdown of certain pesticides habitants, Rocz. Pańśtw. Zak l. Hig., 33 281-9, 1982.

[11] S. K. Chaudhary, R. M. Mitchell, P. R. West and M. J. Ashwood-Smith, Photodechlorination of methoxychlor induced by hydroquinone; rearrangement and conjugate formation, Chemosphere, 14 27-40, 1985.

[12] T. Baudemer and W. Thiemann, Degradation of organochlorine pollutants in water by ultraviolet radiation and addition of hydrogen peroxide, BBR, Brunnenbau, Bau Wasserwerken, Rohrleitungsbau, 37 413-17, 1986.

[13] J. F. Rabek, Experimental Methods in Photochemistry and Photophysics, Part 2, John Wiley and Sons, Chichester, p. 952, 1982.

[14] E. E. Harris and G. B. Frankforter, Concentrations of chloral and bromal with phenolic ethers in the presence of anhydrous aluminium chloride, J. Am. Chem. Soc., 48 3144-50, 1926.

[15] L. Weil, G. Dure and K. E. Quentin, Solubility in water of insecticide chlorinated hydrocarbons and polychlorinated biphenyls in view of water pollution, Z. Wasser Abwasser Forsch., 7 169-75, 1974.

[16] R. C. Gore, R. W. Hannah, S. C. Pattacini and T. J. Porro, Infrared and ultraviolet spectra of seventy-six pesticides, J. Ass. Offic. Anal. Chem., 54 1040-82, 1971.

[17] M. B. Abou-Donia, Ultraviolet spectroscopic studies of DDT - type compounds, Appl. Spectrosc., 29 261-4, 1975.

[18] J. Lubkowski, T. Janiak, J. Czermiński and J. Błażejowski, Thermoanalytical investigations of some chloroorganic pesticides and related compounds, Thermochim. Acta, 166 7-28, 1989.

[19] H. Steinwandter, Contributions to the conversion of HCH isomers by the action of uv light, I. Isomerization of lindane to α-HCH, Chemosphere, 5 245-8, 1976.

[20] L. B. Harding and W. A. Goddard, The mechanism of the ene reaction of singlet oxygen with olefins, J. Am. Chem. Soc., 102 439-49, 1980.

NEUTRALIZATION BY OZONE AND HYDROGEN PEROXIDE OF THIOPHENOL CONTAINING WASTEWATERS FROM PHARMACEUTICAL INDUSTRY

S. WIKTOROWSKI, R. TOSIK, and K. JANIO

Technical University of Lodz
Institute of General Chemistry
90-924 Łódź, Żwirki 36, Poland

ABSTRACT

Thiophenol – containing wastewaters from the pharmaceutical industry have an offensive odor, even if diluted a million times. They are very toxic if they contain 5-chloro- 2-nitroaniline or other refractory organic. Under optimal experimental conditions by using ozonation with hydrogen peroxide, it was found that:

- COD decreased nearly 50 percent,
- wastewaters were colorless and clear, and
- the offensive odor completely disappeared.

The removal of odor and color indicated that thiophenol and 5-chloro- 2-nitroaniline had been fully oxidized. This ozonation process combined with H_2O_2 is especially useful for offensive odor removal.

INTRODUCTION

In alkaline environment, as a result of ozone decomposition, hydroxyl radicals are generated; these free radicals are stronger oxidants than ozone. Their reaction is especially fast with aromatic hydrocarbons, alcohols and formic acid [1]. They also oxidize chlorohydrocarbons as well as many other

Chemistry for the Protection of the Environment
Edited by L. Pawlowski *et al.*, Plenum Press, New York, 1991

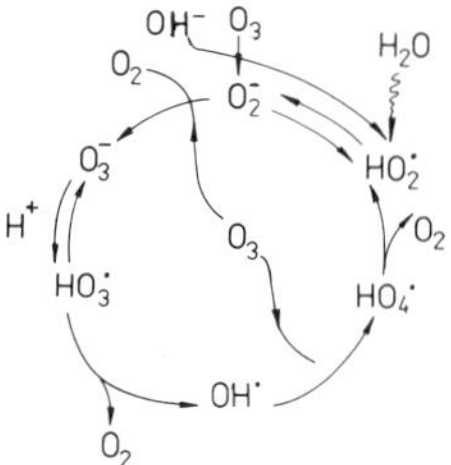

Fig. A .

chemical compounds that have a higher resistance to oxidizing with ozone [2].

Free radicals can be generated in reaction of ozone with hydrogen peroxide as well as due to catalytic decomposition of ozone or decomposition of hydrogen peroxide. The catalysts of ozone decomposition can also be: UV radiation, some metals and metal oxides or activated carbon [3]. Oxidation processes in which hydroxyl radicals take part are with regard to their effectiveness referred to as Advanced Oxidation Processes (AOP).

In water solution hydrogen peroxide reacts with ozone as a reductor, and also H_2O_2 causes ozone decomposition.

$$H_2O_2 + O_3 \longrightarrow HO\cdot + HO_2\cdot + O_2$$
$$HO_2\cdot + O_3 \longrightarrow HO\cdot + O_2$$
$$HO\cdot + O_3 \longrightarrow HO_2\cdot + O_2$$
$$HO\cdot + H_2O_2 \longrightarrow HO_2\cdot + H_2O$$

The rate of decomposition is higher when pH increases. This is so, since free radicals are also generated in the reaction of ozone with OH^- ions.

$$O_3 + OH^- \longrightarrow O_2^- + HO_2\cdot$$
$$O_2^- + O_3 \longrightarrow O_3^- + O_2$$
$$O_3^- + H_3O^+ \longrightarrow HO_3\cdot + H_2O$$
$$HO_3\cdot \longrightarrow O_2 + HO\cdot$$

Staehelin et al. [4] presented the following scheme [Fig. A] showing the decomposition of ozone in an alkaline environment.

It results from the above scheme that ozone decomposition is catalyzed by OH· radicals and O_2^- ions.

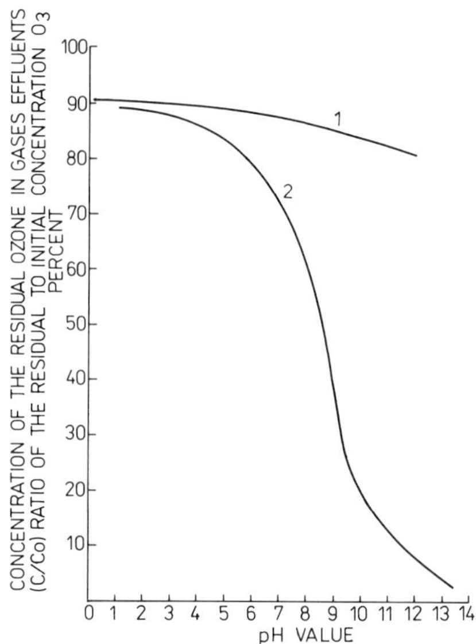

Fig. 1. Concentration of the residual ozone in gaseous effluents from the reactor as a function of pH; curve 1 ozonation of water; curve 2 ozonation of water with 0,25 mol/l H_2O_2.

SCOPE OF RESEARCH

This paper presents the results of studies on the possibilities of applying ozonation combined witch hydrogen peroxide as a method of rendering harmless pharmaceutical wastewaters containing thiophenol and 5-chloro-2-nitroaniline (CNA). These wastewaters have a very intensive offensive odor, which can be smelled even after considerable dilution (dilution many thousand times).

All experiments and observations were made using a barbotage reactor with fine-bubble dispersion, fed with an air-ozone mixture which contained c.55 mg/l O_3. The gas feed rate was about 15 l/hr. The volume of the ozonated solutions in the experiments was 200 cm^3. Hydrogen peroxide was added into the reactor before ozonation or was dosed continuously.

The ozonation materials were aqueous solutions of hydrogen peroxide, thiophenol, CNA as well as industrial wastewaters.

The concentration of ozone in the effluent gases and the redox potential of the solutions were monitored during the processes. The samples of the ozonized solutions were taken periodically for spectral analysis in the UV

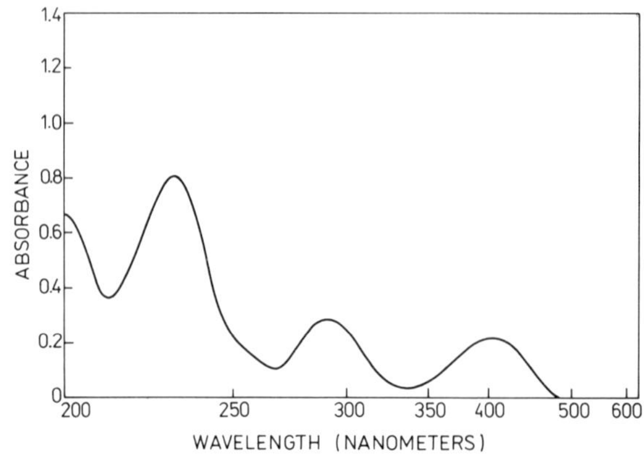

Fig. 2. Spectrophotometric absorbance curve of CNA, concentration 6 mg/l.

and VIS ranges in order to determine the progress of the oxidation reactions. After ozonation of wastewaters COD was determined.

EXPERIMENTAL RESULTS

OZONE DECOMPOSITION IN WATER SOLUTION CONTAINING HYDROGEN PEROXIDE

Ozone decomposition in aqueous solutions of hydrogen peroxide was analyzed at pH from 1 to 13; the initial concentration of H_2O_2 was 0,25 moles/l. pH was adjusted upwards with sodium hydroxide and downwards with hydrochloric acid.

The effect of pH on ozone decomposition in water and in water solution of hydrogen peroxide is illustrated in Fig. 1. It appears from Fig. 1 that ozone is effectively decomposed by hydrogen peroxide in an alkaline environment. It was found that the above mentioned decomposition was accompanied by a decrease of H_2O_2 concentration at approximately stoichiometric rate.

OZONATION OF THIOPHENOL AND CNA

7-10 mg/l thiophenol solutions as well as 6 mg/l and 60 mg/l CNA solutions were subjected to ozonation.

The efficiency of the reactions was measured by changes in light absorption for $\lambda=204$ nm and $\lambda=240$ nm wavelengths for thiophenol and $\lambda=288$

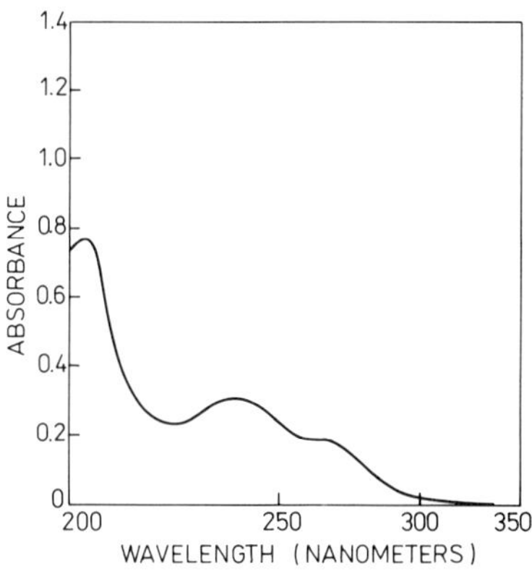

Fig. 3. Spectrophotometric absorbance curve of thiophenol solution, concentration 6 mg/l.

nm and λ=400 nm wavelengths in the case of CNA solution. The absorption maxima for CNA and thiophenol correspond to the above given wavelengths (Fig. 2, Fig. 3). Hydrogen peroxide was added to the reactor either before ozonation or continuously during the process.

It was found that even before ozonation, H_2O_2 reacted with thiophenol, but the reaction rate was relatively slow, and that without ozone the offensive odor was not eliminated. If air without ozone was fed through the reactor, thiophenol was slowly oxidated due to reaction with oxygen.

Effective oxidation of thiophenol and elimination of its odor were achieved as a result of ozonation combined with hydrogen peroxide added to the solution before ozonizing. The characteristic offensive odor of thiophenol disappeared after about 2 minutes of ozonation. The odor remained for a little longer if ozonation was without hydrogen peroxide or when ozonation was carried out with hydrogen peroxide, but the latter was introduced continuously during the process and not before it. The results of oxidation of thiophenol with air, hydrogen peroxide, air-ozone mixture and air-ozone with H_2O_2 are presented in Fig. 4.

The favourable effect of hydrogen peroxide on ozonation of thiophenol can be seen when the appropriate absorption spectra are compared (Fig. 5 and Fig. 6).

After a very short period of ozonation the λ=240 nm peak characteristic for thiophenol disappears, whereas absorption increases for λ=204 nm.

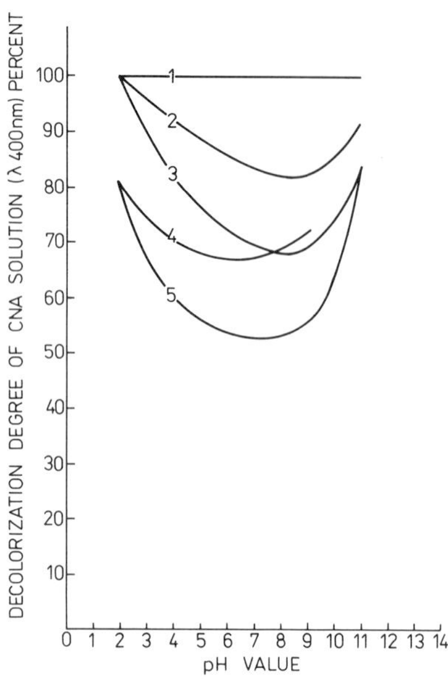

Fig. 4. Relationships of absorbance, measured for λ 240 nm, of thiophenol solution after treatment. Initial concentration of thiophenol 7,2 mg/l, pH 4,7; 1 - after reaction with H_2O_2; 2 - aeration of solution; 3 - ozonation of solution after addition of H_2O_2 -5 mmol/l; 4 - ozonation of solution with continuous dosing of H_2O_2 (amount of H_2O_2 was 25% mol in relation to O_3).

Especially significant changes in absorption take place during ozonation in the presence of hydrogen peroxide. These changes in absorption show that thiophenol oxidation products are formed, which have functional groups characterized by high absorption of light in the above wavelength $\lambda=204$ nm.

Unlike thiophenol, CNA is not oxidized by hydrogen peroxide, since in the mixture of these both components no change in light absorbance occurred in UV and VIS ranges of the spectrum within 20 minutes of observation. CNA solutions have a yellow color; they absorb light in the visible region of the spectrum $\lambda=400$ nm. Decolorization of CNA solution proceeds most favourably in acidic or alkaline solutions; less favourable results were achieved in the pH range from 4 to 9. If, however, during ozonation of CNA solutions hydrogen peroxide was continuously fed into the reactor at the rate of 25% moles in relation to the ozone flow, complete discoloring of the solution was obtained for the whole tested pH range from 2 to 13 (Fig. 7).

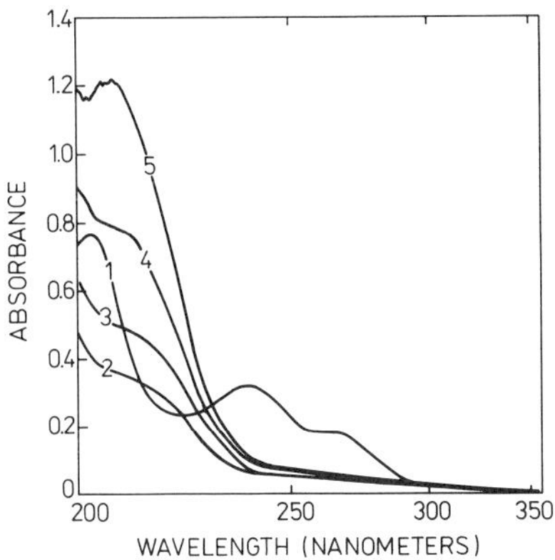

Fig. 5. Spectrophotometric absorbance curves of thiophenol solution after ozonation, measured in UV. Initial concentration of thiophenol 7,2 mg/l, pH 3,85; ozonation time: (1) - 0 minutes; (2) - 5 minutes; (3) - 10 minutes; (4) - 15 minutes; (5) - 20 minutes.

Ozone consumed during the ozonation of thiophenol and CNA solutions was determined by measuring the differences in the concentration of O_3 in gases fed into the reactor and in the effluent gases. The average concentration of ozone in the effluent gases from the reactor over 10 minutes of ozonation for different pH ranges from 1 to 11 pH is illustrated in Fig. 8.

In a solution containing 10 mg/l of thiophenol, ozone consumption was smaller than in water used for diluting thiophenol (Fig. 8 curves 1, 2). From the observations it can be concluded that thiophenol decreases the rate of ozone decomposition. Probably it reacts at a very fast rate with $HO^{\cdot}$ radicals, that catalyze ozone decomposition. In favourable conditions for ozone decomposition, i.e. at a high pH and in the presence of hydrogen proxide, the effect of trapping free radicals by thiophenol is less significant, and ozone consumption is very high (Fig. 8 curve 3).

The effect of hydrogen peroxide on ozone consumption, when ozonizing CNA solutions, was relatively small, and negligible in2 to 9 pH range (Fig. 8 curves 4,5).

In all likelihood, the reaction rate of $HO^{\cdot}$ radicals with CNA is much slower than with thiophenol, and the radicals mechanism seems to be of little importance.

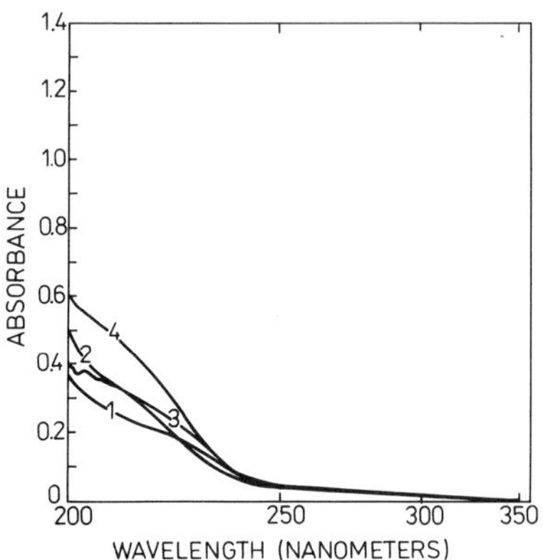

Fig. 6. Spectrophotometric absorbance curves of thiophenol solution with hydrogen peroxide after ozonation, measured in UV, after 5-time-dillution samples. Initial concentration 7,2 mg/l, pH 3,85, dose of H_2O_2 - 5 mmol/l; ozonation time: 1 - 5 minutes; 2 - 10 minutes; 3 - 15 minutes; 4 - 20 minutes.

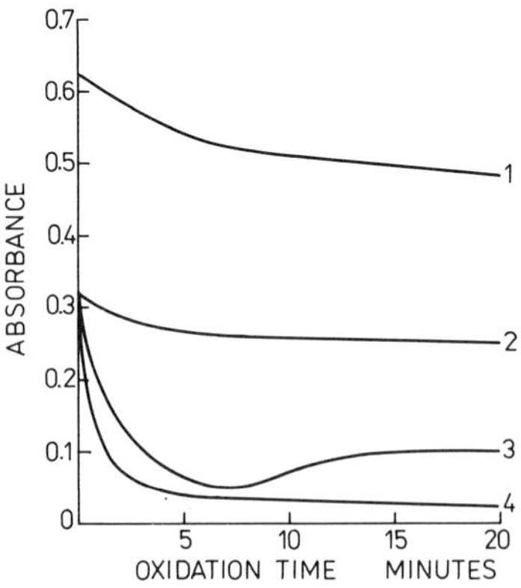

Fig. 7. Effect of pH on the degree of decolorization of CNA solutions after 10 minutes of ozonation; 1 - continuously dosing of H_2O_2 to CNA solution conc. 6 mg/l; 2 - solution of CNA conc. 6 mg/l without H_2O_2; 3 - solution of CNA conc. 5,2 mg/l with H_2O_2 added before ozonation; 4 - solution of CNA conc. 60 mg/l without H_2O_2; 5 - solution of CNA conc.; 60 mg/l with H_2O_2 added before ozonation.

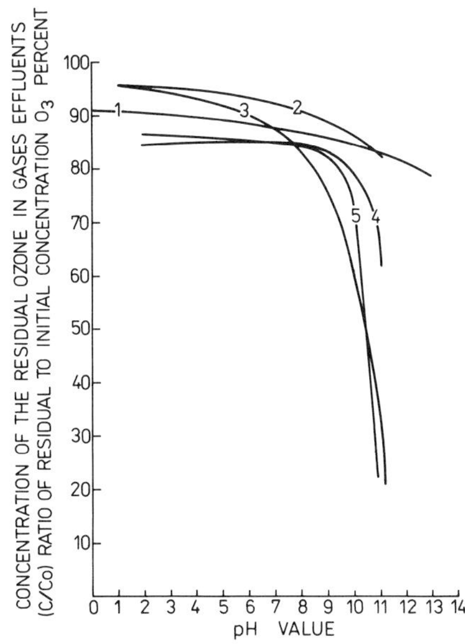

Fig. 8. Effect of pH of the ozonated environments on the concentration of ozone residual in effluent gases; 1 - water; 2 - solution of thiophenol conc. 10 mg/l; 3 - solution of thiophenol conc. 10 mg/l with addition 25 mmol/l H_2O_2; 4 - solution of CNA conc. 60 mg/l; 5 - solution of CNA conc. 60 mg/l with addition 25 mmol/ H_2O_2.

Table 1. Raw wastewaters characteristic

pH	COD mg/l O_2	BOD mg/l O_2	Suspension mg/l	Odor index	Color
8,8-9,6	7.700-11.700	600-800	800-2.000	10^3-10^6	yellow

Table 2. Results of analysis of ozonated wastewaters after initial decarbonization with lime and sedimentation. Raw wastewater analysis: COD 4100 mg/l O_2, Odor index 10^6.

No	pH	Ozone dose	Hydrogen peroxide dose	Ozone consumption	COD	Odor index
	value	mmol/l O_3	% mol	mmol/l O_3	mg/l O_2	
1	10	88	0	44	2300	200
2	10	100	5	58	1400	20
3	10	90	25	62	1800	40
4	10	105	125	95	2600	400

OZONATION OF THIOPHENOL AND CNA -CONTAINING WASTE WATERS FROM PHARMACEUTICAL INDUSTRIES.

Wastewater, the characteristics of which are presented in Table 1, and which had undergone initial lime treatment was decarbonized with lime and sedimentation settling for 2 hr before ozonation. The supernatant sludge-free liquid was ozonized for 1 hr, different amounts of hydrogen peroxide being added. The results are presented in Table 2.

It appeared that ozonation of wastewaters with small doses of H_2O_2 (5-25% in relation to ozone feed rate) gave the best results. In the conditions mentioned the odor of wastewaters after ozonation was the least intensive and least offensive, and COD reached minimal values. Decoloring of wastewaters when ozonizing them with small quantities of H_2O_2 was complete.

The lack of offensive odor and the removal of yellow color after ozonation indicate that thiophenol and CNA had been fully oxidized.

CONCLUSIONS

The authors have shown that in ozonation with H_2O_2 it is important to use adequate pH and adequate doses of H_2O_2.

Overdosing H_2O_2, especially in alkaline solutions can give bad results. COD decreased significantly, and the lowest COD value was reached when ozonizing wastewater with 5% H_2O_2 in relation to the quantity of ozone feed.

If the H_2O_2 dosage is appropriate and the pH sufficiently high it seems that it is possible to fully utilize ozone from an air-ozone mixture. This is important when taking into consideration the toxicity of ozone. It would also eliminate extra appliances for decomposition of residual ozone from effluent gases.

REFERENCES

[1] Z. Ignasiak, Człowiek i środowisko, 3/4/ pp.67. 1979.

[2] W.H. Glaze, Joon-Wun Kang, D.H. Chapin, Ozone: Science and Engineering, Vol. 9, pp. 335, 1987.

[3] J. Staehelin, R.E. Bühler, J. Hoigne, J. Phys. Chem. 88, pp. 5999, 1984.

[4] S. Razumowski, G. Zajkow, Ozone and its reactions with organic compounds. Elsevier; Amsterdam, NY, Oxford, Tokyo-1984.

ELECTROCHEMICAL PURIFICATION OF ALKALINE SOLUTIONS AFTER BLEACHING OF FABRICS

Z. GORZKA, A. SOCHA, K. JASIŃSKA
and M. KAŹMIERCZAK

Institute of General Chemistry
Technical University of Łódź
Żwirki 36, Poland

ABSTRACT

This paper contains results of investigations concerning the treatment of wastewaters from the process of cloth bleaching in cotton industry plants. It was found that the application of electrochemical oxidation allows about two-fold acceleration in liberation of organic compounds contained in lyes, as compared with natural sedimentation. The process of electrolysis brings about a two-fold increase in the amount of liberated organic compounds.

On average, from 1 dm^3 of wastewaters about 250 cm^3 of sediment or ca 20 g of dry substance is obtained. Investigations of electro-oxidation were carried out using a titanium electrode coated with a layer of $RuO_2 + TiO_2$ oxides, over the potential range 400–1000 mV. The progress in the reaction was determined from changes in COD values of lyes. Most advantageous results were obtained at a potential 600 mV and current density 1 A/dm^2. The results obtained indicate possible intensification of liberating organic substances from post-bleaching lyes. Sodium hydroxide solutions treated in this way can be recycled in the process.

Chemistry for the Protection of the Environment
Edited by L. Pawlowski *et al.*, Plenum Press, New York, 1991

INTRODUCTION

In the chemical treatment of cotton fibers called bleaching, organic substances contained mainly in the hemicellulose fibers get into the solution of sodium hydroxide. The contaminated lyes from the process of bleaching are drained into the sewage system. They are a great danger for the aqueous environment.

This paper presents an original method of wastewater treatment applying electro-oxidation of organic substances and sedimentation of the lyes formed while bleaching. Effectiveness and intensity of sedimentation in solutions after electrolysis and without it were compared.

METHODS OF MEASUREMENTS

The investigations were carried out on the solutions of polluted sodium lye obtained in the process of cotton bleaching. The process of anodic oxidation of wastewaters was carried out in an electrolyzer with separated and unseparated electrode spaces. The working volume of the anodic space was 300 cm^3. As anodes, sheet platinum and sheet titanium were used coated with ruthenium and titanium oxides with a surface of 160 cm^2. The cathode was a carbon cloth. The electrolysis was carried out at a temperature of 70°C. The progress of the reaction was determined by measurement of chemical oxygen demand (COD) values.

RESULTS AND DISCUSSION

Investigations of electrochemical oxidation of organic compounds contained in post-bleaching lyes started from determining the correlation between the current and the electrode potential. This gives basic information on the course of the electrode process. By means of the voltamperometric method the above correlation was determined for the oxidation reaction, using platinum and oxide electrodes (TiO_2, RuO_2). The electrode potential was measured in relation to a saturated calomel electrode (SCE). Examples of the correlation of the reaction current and the electrode potential is given in Fig. 1. The correlation shown in Fig. 1 indicates, that an increase in potential starting from 400 mV is accompanied by an increase in the current of oxidation of organic compounds. With further increases in potential the current value grows and reaches a practically constant value in the range 800 to 900 mV. Further increase in current at potentials higher than 900 mV results

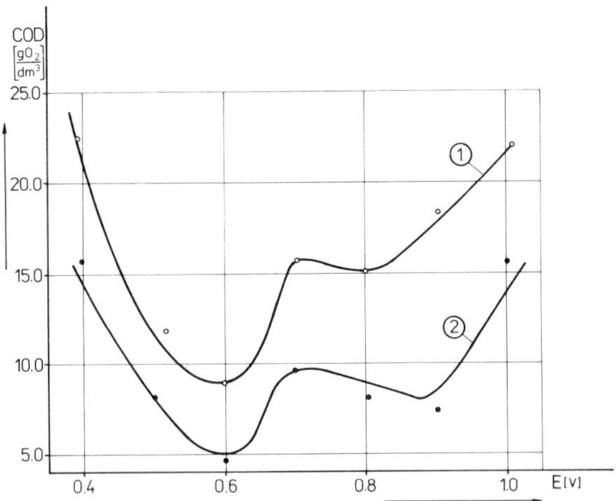

Fig. 1. The correlation between the current and potential in the oxidation of organic compounds contained in post-bleaching lyes; curve 1 - oxide electrode in the investigated solution, curve 2 - platinum electrode in the investigated solution, curve 3 - in the solution of NaOH

in continuous anodic oxidation of organic compounds and anodic oxidation of NaOH accompanied by the formation of oxygen.

Comparison of the changes in oxidation current on the platinum (curve 2) and oxide (curve 3) electrodes points to the fact that oxidation of organic compounds contained in wastewaters starts practically from the same potential, i.e., from about 400 mV. The observed current values are, however, 4O times higher when an oxide anode is employed.

In order to confirm the obtained data indicating a higher reaction rate when an oxide and not a platinum electrode is used, oxidation of organic compounds contained in lyes using the two electrodes was studied. The investigations were carried out at a constant potential 900 mV at which no oxygen was produced at the electrode and at 1300 mV at when the formation of oxygen resulted in the oxidation of organic compounds. The choice of such values was due to the necessity to test the effect of oxygen on the reaction by too high a value of the current at this potential.

The effects of these tests, reflected in the change in COD values of wastewaters before and after electrolysis and after 20 hours of sedimentation, are presented in Table 1.

Table 1. COD Values of Wastewaters after Electrolysis on Oxide and Platinum Electrodes and after 20 Hours of Sedimentation

Type of electrode	Oxide electrode		Platinum electrode	
Potential [mV]	900	1300	900	1300
COD after electrolysis [gO_2/dm^3]	14.96	18.44	21.12	21.12
COD after electrolysis and 20 h of sedimentation [gO_2/dm^3]	9.68	10.56	11.44	12.32

Reaction temperature 70°C, time of reaction 30 min.,
COD of raw wastewaters 29,04gO_2/dm^3, COD of raw wastewaters after 20 h sedimentation 16,72gO_2/dm^3.

The results in Table 1 indicate that application of anodic oxidation intensifies the process of liberating organic compounds contained in wastewaters. This is proved by lower COD value in the wastewaters subject to electrolysis and sedimentation as compared with those undergoing sedimentation without previous electro-oxidation. It was also found that the formation of sediment in wastewaters undergoing electrolysis and their complete sedimentation was twice as fast as compared with natural sedimentation. The above data indicate also that a greater decrease in COD and thus in the content of organic compounds is obtained when an oxide electrode is employed in the process.

Comparison of the effects of a decrease in COD after electrolysis at the above mentioned potentials and after sedimentation leads to the conclusion that the lower potential of electrolysis, i.e., 900 mV, is more advantageous for the process.

This fact does not take into account the effect of current value and hence the effect of reaction rate on the decrease in content of organic compounds. It seems that the higher potential value, i.e., 1300 mV, encourages anodic destruction of organic compounds contained in lyes. This was observed with the oxidation of hemicelluloses coming from coniferous and leafy trees for potential range 1.6 to 2.5 V [1]. Lower potential of oxidation, i.e., 900 mV, encourages polyaddition of dissolved hemicelluloses. With their oxidation by periodate, polyaldehyde and small amounts of formic acid were formed [2]. Therefore it was necessary to determine the correlation between changes in COD of the lyes and the electrode potential.Therefore measurements of the effect of the time of anodic oxidation on changes in COD values at a constant electrode potential were determined.

Table 2. COD Values of Solutions After Various Times of Electrolysis

time of electrolysis [min]	10	15	20	25
COD after electrolysis [gO_2/dm^3]	15.84	15.84	3.52	2.64
COD after electrolysis and 20 h of sedimentation [gO_2/dm^3]	11.44	10.56	2.64	1.32

Oxide electrode, potential 900 mV, reaction temperature 70C.
COD of raw wastewaters 17.6gO_2/dm, COD of wastewaters after 20 h of sedimentation 15.84gO_2/dm.

It is commonly known that the value of an electric charge passing through the system conditions the effect of the electrode reaction. The influence of reaction time on the effect of a decrease in COD values is presented in Table 2. Based on the data presented in Table 2 and analogous investigations for other portions of wastewaters it was found that the greatest decrease in COD of wastewaters is obtained after 20 to 30 minutes of the reaction. For further investigations the time of reaction of 20 minutes was chosen. As indicated by the data presented in Table 1, a decrease in the contents of organic compounds in wastewaters depends on the anode potential. In order to determine the above correlation, investigations into the changes in COD of wastewaters at different electrode potentials were carried out. The anode potential was changed within the range of 400 to 1000 mV during which an increase in the reaction current was observed but oxygen was not liberated on the electrode.

Examples of the correlation between the changes in COD of wastewaters and the electrode potential are shown in Fig. 2. It indicates that the greatest decrease in COD is obtained at a potential of about 600 mV. At this potential value the decrease in COD is about twice which that can be obtained with natural sedimentation of the wastewaters which have not undergone electrolysis. Also, the time of sedimentation in the wastewaters after electrolysis was half that of raw wastewaters. The sediments obtained from wastewaters undergoing electrolysis were less hydrated. On the average, from 1000 cm^3 of such wastewaters 250 cm^3 of sediment was formed by means of sedimentation. Application of a centrifuge to the sediment decreased its volume to about 50 cm^3. IR analysis showed that sediments obtained by natural sedimentation and those after electrolysis do not differ significantly. The above suggests a possibility of polyaddition of hemicellulose particles. Table 3 presents examples of the parameters of the reaction of anodic oxidation

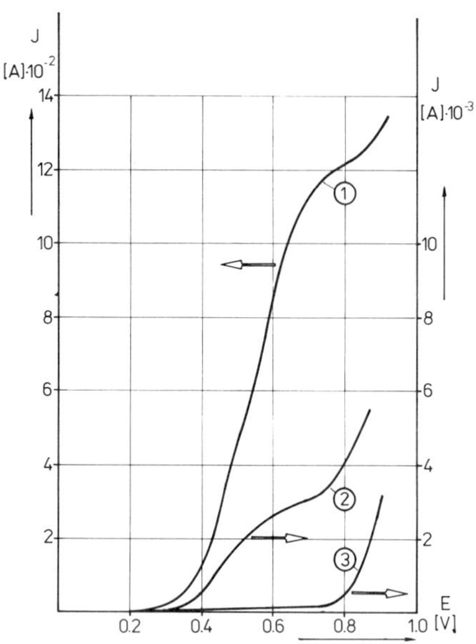

Fig. 2. The correlation between the changes in COD of wastewaters and the potential of an oxide electrode: curve 1 - COD of wastewaters after electrolysis, curve 2 - COD of wastewaters after electrolysis and 20 hours of sedimentation.

of wastewaters at the potential 660 mV. The COD of raw wastewaters was 22 gO_2/dm^3 and after 20 hours of natural sedimentation, 17.47 gO_2/dm^3. After electrochemical oxidation, COD of the wastewaters decreased to 13.2 gO_2/dm^3 and after 20 hours of sedimentation to 7.94 gO_2/dm^3. From the data presented in Table 3 it follows that for an additional decrease in COD of about 10 gO_2/dm^3, as compared with natural sedimentation an electric charge of about 0.51 Ah must be used for every 300 cm^3 of the wastewaters. Therefore, an additional decrease in COD value as compared with natural sedimentation requires about $5x10^{-4}$ kWh/g COD.

CONCLUSIONS

Based on the investigations it was found that the method of electrochemical oxidation increases and accelerates the effect of removing organic compounds contained in post-bleaching wastewaters. On average, from the wastewaters subjected to electrolysis, twice the organic compounds are liberated in half the time as compared with natural sedimentation of raw wastewaters. It was shown that the above process is most efficient at the potential

Table 3. Change in Current Intensity I [A], Voltage Between Electrodes U [V] and Electric Charge Q [Ah] in the Reaction of Anode Oxidation of Wastewaters at the Potential 660 mV

time [min]	I [A]	U [V]	Q [Ah]
1	1.45	3.02	0.02
5	1.34	2.93	0.12
10	1.29	2.90	0.23
15	1.15	2.84	0.33
20	1.21	2.88	0.43
25	0.99	2.78	0.51

of about 600 mV. Mean current density registered during the reaction is about $1A/dm^2$ and a voltage drop between electrodes is about 3V. In order to decrease the COD value below that obtained with natural sedimentation about $5x10^{-4}$ kWh/g COD must be used. Thus the application of electrochemical oxidation of organic compounds contained in post-bleaching lyes leads to an increased and accelerated liberation of the above compounds in wastewaters. This fact indicates the possibility of repeated or multiple recycling of sodium hydroxide solutions in the process of cloth bleaching and thus a considerable decrease in the amount of impurities discharged to the sewers and hence into the environment.

REFERENCES

[1] Antipova I.A., Yanilkin V.V., Medvedeva S.A., Babkin V.A., Gorokhova V.G., Nikiforov V.P., Lanina N.A., Khim. Drev. 1, 82, 1985.

[2] Kin Z., Hemicelulozy, Chemia i wykorzystanie, PWRiC Warszawa, 1980.

PHYSICOCHEMICAL TREATMENT: MEMBRANE PROCESSES

MEMBRANE OPERATIONS FOR WATER RENOVATION IN THE TEXTILE INDUSTRY

K. LIPIŃSKI, A. R. SZANIAWSKI

Water Engineering Institute
Technical University of Szczecin
Al. Piastów 50A, 71–310 Szczecin, Poland

D. SZANIAWSKA

Chemical Engineering
and Physical Chemistry Institute
Technical University of Szczecin
Al. Piastów 42, 71–065 Szczecin, Poland

ABSTRACT

Membrane operations currently represent the most suitable techniques for treatment of textile wastes. The application of such techniques results in potentially significant reduction in freshwater demand and waste treatment requirements and the conservation of energy and materials.

For the treatment/purification of textile wastes, polymeric membranes in the form of tubular, spiral-wound and hollow-fiber were used. Dynamic membranes formed on metal, carbon and ceramic microporous supports were also successful.

Conventional textile waste utilization techniques based on coagulation and

Chemistry for the Protection of the Environment
Edited by L. Pawlowski *et al.*, Plenum Press, New York, 1991

sedimentation do not provide the complete removal of many dyes used in the textile industry.

Adsorption methods involve a high capital expenditure and are troublesome in service.

Radiation treatment of textile wastes is very expensive.

Previous studies and results of investigations by the authors permitted the development of our own alternative concept for water renovation in the textile plants. The proposed membranes were dynamic (alternative I) and polymeric (alternative II).

The research group at the Water Engineering Institute of TU Szczecin is investigating the possibilities of domestic dynamic membranes production. The research is directed toward manufacture of predictable, reproducible and highly stable membranes. Dynamic membranes obtained in preliminary tests had good performance during synthetic waste purification.

INTRODUCTION

The dyeing operations in the textile industry require large quantities of water. In these processes, the amount of auxiliary non-consumable materials exceeds that of the consumable components. Thus, recycling of water and process materials is motivated by the potentially significant reduction in freshwater demand and waste treatment requirements and the conservation of energy and materials. One of the main pollution products of the textile industry are the dyehouse effluents. These create difficulties on discharge because of their poor biodegradability and the high levels of surface-active agents, color and salts.

The problem of dyehouse wastewaters utilization may be solved with different techniques without recovery of water and auxiliary chemicals, as well as with those which allow recovery of both water and auxiliary chemicals.

The most suitable and universal are methods in which membrane operations are employed.

At the Water Engineering Institute of Szczecin Technical University, the concept of water renovation technology at textile industry plants was developed. The concept was based on the application of polymeric as well as dynamic membranes.

The domestic prepared dynamic membrane are now investigated.

THE CONVENTIONAL METHODS OF DYEHOUSE WASTEWATERS UTILIZATION

The dyehouse wastewaters utilization techniques based on coagulation and sedimentation were realized with ferrous sulphate or aluminium sulphate [1,2]. The effluent after sedimentation was pumped to a lagoon and aerated by surface aerators.

The value of BOD after aeration was below 450 mg/l. This method did not provide the complete removal of many kinds of dyes used in the textile industry. The removal of dyes is usually incomplete since they are stable to light and oxidizing agents and are resistant to aerobic digestion.

The adsorption techniques of dye removal from dyehouse wastewater involves the use of activated carbon and peat [3,4]. In fact, granular activated carbon has proved to be so effective in removing toxic and non-toxic dissolved organic matter that it is widely used to remove taste and odor from municipal drinking water supplies. In industrial applications, granular activated carbon is packed into one or more large vessels or "adsorbers". The wastewater is pumped through the carbon and as it passes through it, most of the contaminants are left behind on the activated carbon, allowing the water to be recycled due to the BOD and COD removal of 95% and 75%, respectively, and 100% color removal. Once saturated, the carbon can be removed from the adsorber and processed in a high-temperature furnace which drives off and destroys the adsorbed material. Its adsorptive capacity restored, the carbon can then be replaced in the adsorber and rused again.

However, peat is only a fraction of the price of activated carbon and its ability as an adsorbent medium indicates that it could be an effective and economic material for color removal from effluents. Several dyestuffs have been tested using peat as an adsorbent and basic dyes show a considerable affinity for peat, which adsorbs up to its own weight of dyestuff. Methods of regeneration are not well established although the combustion characteristics of peat and adsorbed dye have been determined, with the possibility of burning the peat to produce low pressure steam.

In general, adsorption techniques are expensive to install and reasonably difficult to operate.

The other techniques of textile wastewaters treatment are a combination of sorption and biological regeneration [5]. Organic matter contained in wastewater is adsorbed on an adsorbent (e.g., activated carbon) contained in a fixed bed.

The exhausted adsorbent is regenerated by circulating, in an upflow mode, a liquid stream containing an aerobic biological culture.

The resultant bio-oxidation of the eluted organic matter continues to take place until the adsorbent is reactivated. The treatment of dyehouse wastewater in a four-column pilot gave the following results: COD reduction – 49.0–81.1%, TOC reduction – 47.8–81.7%, color reduction – 99.4–99.5% and BOD reduction – 95.0%. The radiation treatment of textile wastewaters involves the high pressure (10.5–14.0 MPa) radiolytic oxidation of textile effluents [6]. High-pressure radiolysis has been shown to reduce pH and color from pure dyes in aqueous solution from wool dyeing process wastes. Cost of this process seems to be prohibitive for large volume textile wastes.

The radiational-chemical and radiational-biological techniques are not as expensive [7]. In the first case free chlorine, sodium hypochlorite, hydrogen peroxide and ozone are employed as chemical oxidizing agents. The concurrent application of ozonation and atomic radiation will provide COD and BOD reduction and waste decoloration in the range 42-98% and 12-100%, respectively, depending on the pH of textile wastewaters.

MEMBRANE TECHNIQUES FOR TEXTILE WASTEWATERS UTILIZATION

Textile wastewaters may be treated by two basic membrane operations: ultrafiltration (UF) and reverse osmosis (RO).

Satisfactory results in industrial UF and RO runs have been obtained not only with polymeric membranes in tubular, spiral-wound and hollow-fiber configuration, but the dynamic membranes deposited on microporous metal, carbon and ceramic supports have also yielded satisfactory effects. Reverse osmosis carried out with polymeric cellulose acetate tubular membranes and aromatic polyamide hollow-fiber membranes has been reported to reduce the concentration of organic matter (COD) by 80%. The average surfactant and color reduction was 98-99% at 85-95% water recovery [8,9].

The spiral-wound module with polyamide membranes has given excellent rejection performance for the treatment of textile dyehouse effluents [10]. The treated effluent, at 85-95% water recovery, was of excellent quality due to the COD, TDS, TOC and colour removals of around 98-99%.

The dynamic membranes are particularly suitable for the treatment of textile wastewaters, i.e., high strength, high fouling and hot industrial effluents.

The dynamic membranes are formed in the process of filtration of solutions containing the appropriate components (additives) through microporous supports. The microporous supports have a nominal pore size ranging from 3 nm (30 Å) to 5 μm and supports alone have no ability to reject microparticles or ions of microsolutes [11].

As a result of deposition of the dispersed substance on the support surface facing the solution, the semipermeable layer is formed. The principle of dynamic membrane operation is shown in Fig. 1.

A wide assortment of additives has been shown to form membranes in this way: hydrous oxides Zr(IV), Th(IV), Fe(III), synthetic organic polyelectrolytes (polyvinylopirolidon), natural polyelectrolytes, including humic

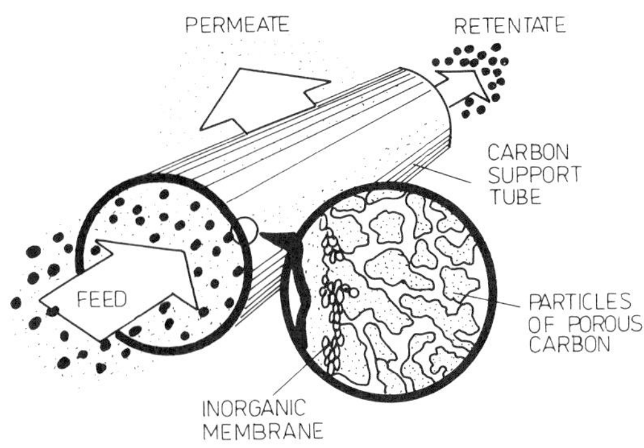

Fig. 1. The principle of dynamic membrane operation.

acid and clays, and material in waste streams, such as sewage or pulp mill wastes [12].

Ultrafiltration dynamic membranes used for the treatment of textile wastewaters are formed by deposition of hydrous Zr(IV) oxide on microporous support tubes. They exhibit salt rejection in the range 5–20%. The dynamic Zr(IV) oxide ultrafilters are very useful for renovation of dye process wash water, wool scouring [13], and textile size recovery [14].

The dynamic reverse osmosis membranes are prepared by sequential deposition of hydrous Zr(IV) oxide followed by poly(acrylic acid) on a suitable microporous support under pressure and cross–flow conditions. Although not competitive with the conventional RO membranes for desalination, the resulting RO membranes exhibit salt rejection in the range 80-90%. The treatment of textile wastewaters by dynamic RO membranes at 70-97% water recovery provide COD and BOD reduction in the range 71–97% and 74–99%, respectively, and color removal in the range 89–100% [15-18].

THE CONCEPT OF WATER RENOVATION IN TEXTILE PLANTS

In the textile industry, discharge from the dyehouse occurs in three forms: the wash and rinse waters discharged continuously during the dyeing run and the dye drop from the dye pad following each run. To achieve the maximal dye recovery, the residual dye pad solutions ought to be recycled to the dye baths. Usually after several cycles, the residual dye pad solutions must be removed from the dyehouse. On the other hand, the whole amount of the wash and rinse water must be purified for reuse.

The schematic diagram of such a general recovery system is presented in Fig. 2.

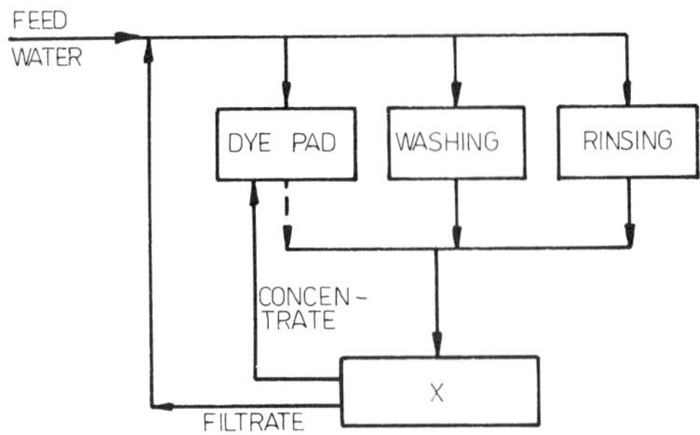

Fig. 2. General schematic diagram of recovery system.

Considering previous published studies and the results of investigations of the authors it is clear that only membrane techniques will permit reaching the target X (see Fig. 2) [19,20].

Taking into account the nature of dyehouse wastewaters, the authors propose the following possible membrane operations:

- reverse osmosis on dynamic with membranes simultaneous removal of organic substances and salts (at least 80% removal),
- ultrafiltration on dynamic membranes or tubular polymeric membranes followed by conventional reverse osmosis,
- ultrafiltration on capillary membranes, subsequent to pre-treatment operations removing suspended matter and colloids, followed by reverse osmosis.

The recovery of the whole amount of dyes and auxiliary chemicals is very hard to realize. Therefore, the authors' concept of the recovery system is based on two assumptions: the recovery of water for reuse will be as high as possible, and the utilization (or maximal recovery) of residual dyes and auxiliary components will provide the minimum of pollution to the environment.

Taking into consideration the above assumptions, there are two possible alternatives of this concept. The concept of the recovery system according to the alternative I (Fig. 3) is based on RO dynamic module, while the proposal for alternative II (Fig. 4) is the employment of conventional (polymeric) UF and RO modules.

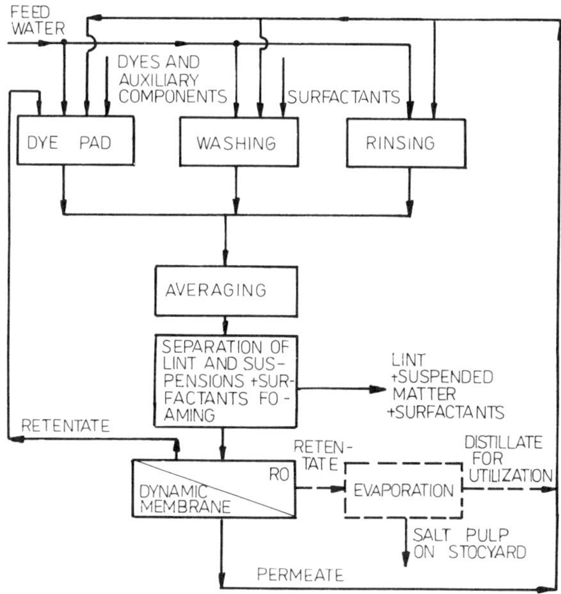

Fig. 3. The conceptual schematic diagram of the recovery system for dyehouse wastewaters (alternative I).

The wastewater pre-treatment depends on the applied filtration process. When the ultrafiltration on dynamic or tubular membranes is employed, there is no need for pre-treatment. The application of capillary ultrafiltration modules involves the pre-treatment directed toward removal of suspended matter and colloids.

The recovery system designed and constructed according to alternative I will permit the recycling of hot wastewaters, while the other (alternative II) will have to be fed with cooled–down wastewaters.

In general, our developed recovery systems provide maximal or partial (depending on requirements) recovery of water, dyes and auxiliary chemicals.

FORMATION OF DOMESTIC DYNAMIC MEMBRANES – PRELIMINARY WORK

It is well known that the advantages of dynamic membranes are high-temperature stability and accommodation of highly permeative fluxes. They are therefore suitable for ultrafiltration because high temperature during operation, cleaning or sterilization is required in ultrafiltration applications. It

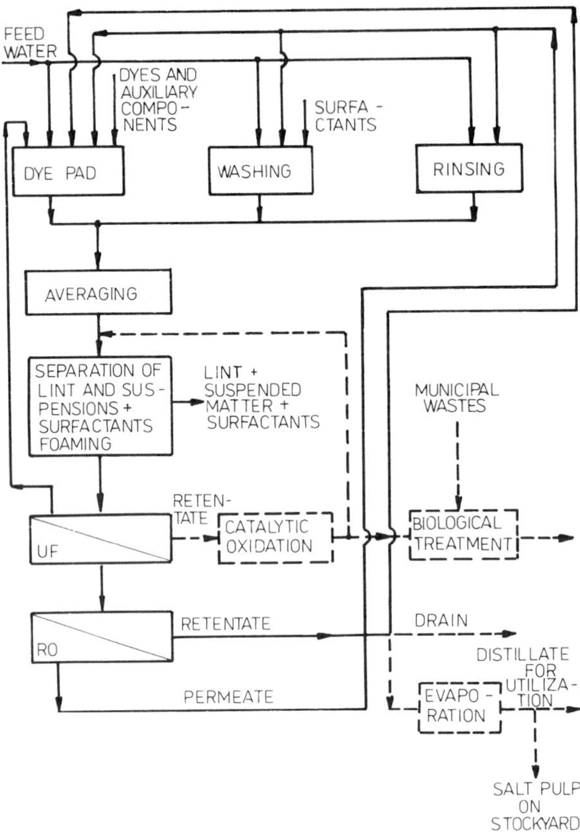

Fig. 4. The conceptual schematic diagram of the recovery system for dye-house wastewaters (alternative II).

can be supposed that an inorganic colloid has a larger resistance to high temperature than polymers such as cellulose acetate and polyamide. At the Water Engineering Institute of TU Szczecin, preliminary investigations were undertaken to prepare domestic dynamic membranes [21]. Ceramic and nickel tubes were used as the supports for the dynamic membranes.

The ceramic support tubes, having outer diameters of 15.0 and 20.2 mm, a length of 650 mm and wall thicknesses of 1.5 and 4.0 mm, were manufactured at AGH (Mining and Metallurgical Academy) in Kraków.

The nickel support tubes had outer diameters of 9.0 and 10.7 mm, a length of 240 mm and wall thicknesses of 1.8 and 2.2 mm and were produced by "Baildon" Steel-Works in Katowice.

Chemical reagent zirconyl chloride octahydrate ($ZrOCl_2 \cdot 8H_2O$) was employed as Zr(IV) colloid (additive).

The dynamic membrane formed for 30 min. at an operating pressure of 1.0 MPa, and $ZrOCl_2 \cdot 8H_2O$ concentration, 0.382 g/l gave an average COD reduction around 70% for synthetic waste, according to Weinberger [22]. The synthetic waste had the following composition (in mg/l) : peptone-150, urea-50, CH_3COONa-30, NaCl-30, HCl-7, $MgSO_4 \cdot 7H_2O$-50, $Na_2HPO_4 \cdot 12H_2O$-63, $NaHCO_3$-168, $CaCl_2$-7, soaps-2, starch-100.

DISCUSSION

In the paper it has been pointed out that membrane operations are the best techniques for treatment of textile wastes.

At the Technical University of Szczecin, the concept of textile wastewaters purification for auxiliary chemicals and water reuse based on polymeric as well as dynamic membranes has been worked out.

Dynamic membranes formed on domestic microporous ceramic and nickel supports exhibited good performance during synthetic waste purification.

REFERENCES

[1] Turner M.T., Waste water treatment and re-use within the textile industry, Water Services, 8, 1978.

[2] Earle A.P., Textile wastes treatment, Water and Wastes Eng., 2, 57, 1970.

[3] Rizzo J.L., Chemical Eng. Prog. Symp. Series, 67, 107, AICHE, 466, 1971.

[4] Mc Kay G., Peat-an adsorbent/filtration medium for waste-water treatment, Water Services, 6, 357, 1980.

[5] Rodman C.A., Shunney E.L., Novel approach removes colour from textile dyeing wastes, Water and Wastes Eng.,9, E-18, 1971.

[6] Garrison A.W., Radiation used to treat textile wastes, Water and Wastes End., 9, 65, 1971.

[7] Perkowski et al., Wysokoefektywne metody oczyszcznia ścieków i odnowa wody, cz.II, Kraków, 247, 1985.

[8] Slater C.S., et al., Membrane processes used for the treatment of industrial effluents, Desalination, 24, 155, 1978.

[9] Mapelli P., et al., Membrane processes used for the treatment of industrial effluents, Desalination, 24, 155, 1978.

[10] Treffry-Goatley K., et al., Reverse osmosis treatment and reuse of textile dyehouse effluents, Desalination, 47, 313, 1983.

[11] Dytnerskij J., Obratnyj osmos i ultrafiltracja, Izd. Chimia, Moskwa, 1978.

[12] Johnson J.R., Jr, et al., Hyperfiltration XXI, Dynamically formed hydrous Zr(IV)oxide-polyacrylate membranes, J. Electroanl. Chem., 37, 267, 1972.

[13] Beaton N.C., Textile Inst. and Ind., 11, 361, 1975.

[14] Grizzle T., Ultrafiltration applications in the textile industry, prepared for Gaston County Dyeing Machine Company, USA, 1981.

[15] Carre Inc.(USA) bulletin, Recent applications in dynamic membranes, 1980.

[16] Carre Inc.(USA) bulletin, Information package, 1982.

[17] Carre Inc.(USA) bulletin, Hyperfiltration of textile process water for reuse, 1980.

[18] Brandon C.A., Jernigan D.A., Closed cycle textile dyeing: Full scale renovation of hot wash water by hyperfiltration, Dasalination, 39, 301, 1981.

[19] Lipiński K., Szaniawski A., Oczyszczanie i zagospodarowanie ścieków z aparatów farbiarskich, raport nr 12–1793 dla COBR MW Polmatex-Cenaro, Polit. Szcz., Szczecin, 1986(unpubl.).

[20] Lipiński K., et al., The possibilities of applying the dynamic membrane in Polish industry, Proc. of Int. Conf. on Particle Technology in Relation to Filtration and Separation, Antwerp, 4. 85, 1988.

[21] Lipiński K. et al., Badania instalacji laboratoryjnej z modu lem jednorurowym dla różnych typów rur nośnych i membran dynamicznych, raport nr. 03.01.88.10 dla Polit. Wroc l., Polit. Szcz., Szczecin,1988(unpubl.).

[22] Weinberg L. W., Powers T. W., The detergents and water quality standards, Journal American Oil Chemistry Society, 11, 41, 736, 1964.

RECOVERY OF Cr(VI) WITH EMULSION LIQUID MEMBRANES (ELM) IN MECHANICALLY STIRRED CONTACTORS

E. SALAZAR, M. I. ORTIZ and A. IRABIEN

Departamento Ingenieria Química
Facultad de Ciencias
Universidad del País Vasco
Apdo. 64448080 Bilbao, Spain

ABSTRACT

Chromic acid, coming from sulfuric baths employed in electroplating processes, is one of the most toxic elements to the environment. These processes usually operate at atmospheric conditions and present a Cr concentration of about 125g/l, which leads to a concentration of 5g/l Cr in the waste effluents. In this work the viability of the recovery of chromium by ELM in a mechanically stirred tank was investigated, analyzing the influence of membrane formation and operation variables on membrane stability by means of factorial design of experiments.

INTRODUCTION

Two main problems are related to industrial wastewater disposal:

i. The environmental damage produced by the discharge of toxic elements.

ii. The loss of raw materials which after recovery could be recycled to the industrial processes.

Dischare wastewater from sulfuric baths used in metal electroplating processes contains hexavalent chromium, usually present as chromic acid. When the chromic bath is depleted, waste residuals have a concentration of 5 g/l Cr(VI) in relation to an Cr(VI) initial concentration of 125 g. [1]

Hexavalent chromium is one of the most toxic metals to human health. As a reference the maximum value proposed by the World Health Organization for Cr(VI) content in drinking water is 0.05 mg/l [2], therefore it is necessary to reduce the discharge of wastes containing this metal to natural streams.

The separation-concentration of Cr(VI) by emulsion liquid membranes has been reported in several works [3,4,5], starting from an initial concentration of chromium lower than that used in electroplating baths: Cr(VI)= 14-100 mg/l and Cr(VI)=100-800 mg/l.

In this work the viability of the recovery of Cr(VI) by ELM was studied, starting from aqueous solutions with a concentration of 5 g/l to obtain a final Cr(VI) concentration in the stripping solution of 125 g/l in order to be recycled to the industrial process. Two different aspects have been analyzed:

i. The influence of the experimental variables on the stability of the membrane by means of a factorial design of experiments.

ii. Under the conditions selected from the previous point the influence of the concentration of the stripping agent on the yield of chromium recovery.

EXPERIMENTAL

Analar–grade reagents were used to prepare aqueous solutions the CrO_3 (Merck), H_2SO_4 (Merck), Na_2SO_4 (Probus) and NaOH (EKA Nobel). The components of the organic phase were: Kerosene (Petronor S.A.) as diluent, Aliquat 336 (Hankel Co) as metal extractant, n-Decanol (Fluka) as stabilizer and SPAN-80 (Sigma) as surfactant. The employed kerosene is normally used for industrial requirements.

Emulsions were prepared at a high stirring speed, 8000 rpm, with a stirrer Ultraturrax T25 (IKA), obtaining a distribution of drop diameters of 1-5 μm, and the emulsion container was submerged in a water bath in order to dissipate heat generation during stirring.

Both phases (aqueous and membrane) were put into contact in a glass made stirred tank, provided with four stainless steel baffles and a plastic turbine type impeller connected to a stirrer Heidolph RZR-2000 speed controller (0-700 rpm).

Samples were taken from the botton of the tank, and after separation of the two phases, the aqueous one was prepared for analysis. Figure 1 is a scheme of the experimental apparatus.

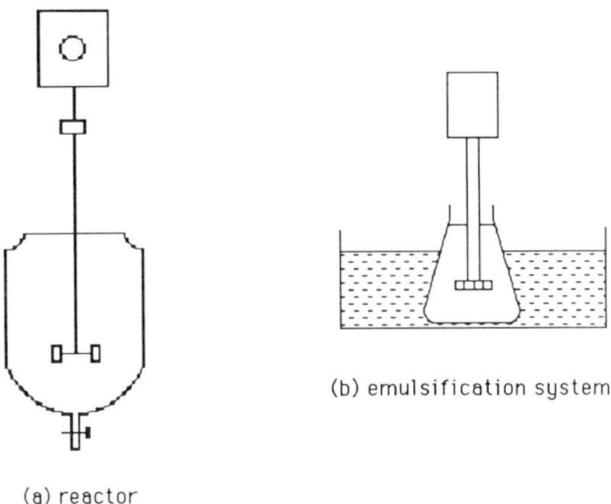

Fig. 1. Experimental apparatus.

In experiments performed for the evaluation of membrane stability the aqueous phase was titrated with hydrochloric acid to determine the concentration of sodium hydroxide as a result of membrane breakage.

In extraction experiments chromium (VI) was analyzed after dilution of the aqueous samples in a Perkin Elmer 1100 B absorption spectrophotometer.

RESULTS AND DISCUSSION

Membrane Stability

The evaluation of the influence on membrane stability of the variables surfactant (SPAN-80) concentration, inner aqueous phase/organic phase volumetric ratio, emulsification time and stirring speed in the tank reactor was analyzed through a factorial design of experiments, where variables were changed according to Table 1.

The ratio of external aqueous phase volume to that of inner aqueous phase volume was kept constant at to 10/1.

Assuming that chromium should reach a concentration in the inner phase of 125 g/l and according to the reaction stoichiometry, 5N NaOH was used as the stripping agent in all cases. Experimental results, are plotted in Figure 2, show that membrane breakage is strongly dependent on stirring time, thus making high contact times ineffective. On the other hand, it is

Table 1. Factorial Design of Experiments

variables	range		
	minimum	medium	maximum
SPAN-80 (%volume)	1	3	5
O/IA	0.5/1	1/1	2/1
t_s (min.)	5	10	15
ω (rpm)	200	300	400

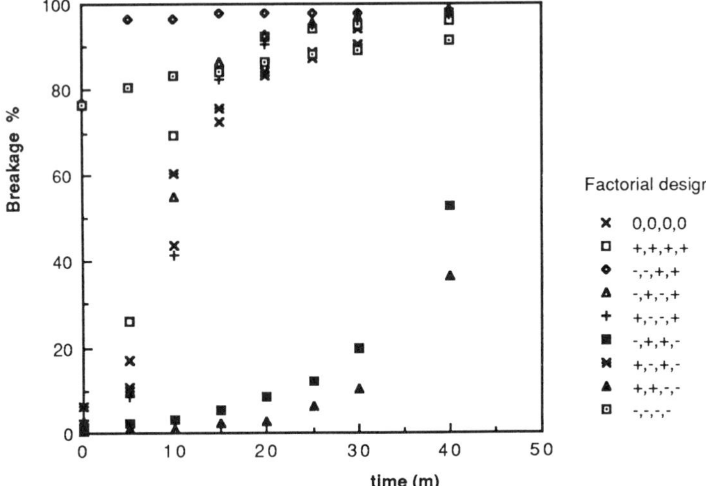

Fig. 2. Membrane breakage vs time.

also observed that experimental results are close to a mean value with a certain degree of deviation.

Separation-Concentration Process:

After results were obtained in the study of membrane stability, several experiments were carried out to analyze the influence of the stripping agent, sodium hydroxides on Cr(VI) recovery.

The following variables were kept constant:

external aqueous phase composition:

Cr=5 g/l H_2SO_4=5 g/l NA_2SO_4=40 g/l

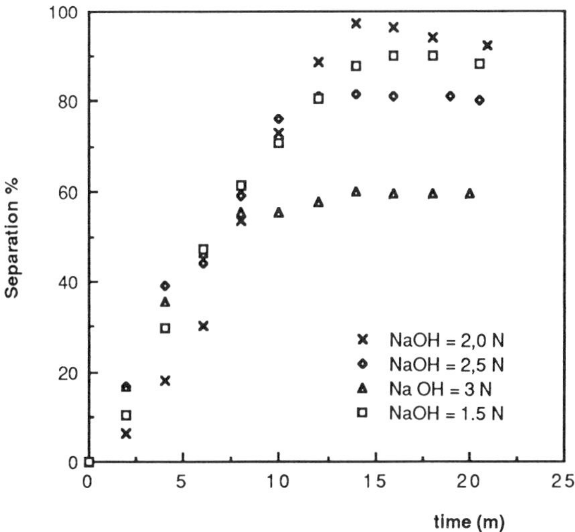

Fig. 3. Chromium separation vs time.

organic phase (% volume):

Aliquat 336=3 SPAN-80=5 n-Decanol=5

operation variables:

O/IA=1/1 EA/M=5/1 t_s=10 min. ω=300 rpm

Sodium hydroxide concentration varied from 1.5 to 3.0 N. Figure 3 presents the experimental results.

Although differences are observed in maximum extraction yields with different concentrations of sodium hidroxide, analysis of the Cr(VI) contained in the stripping phase gave a similar value in all cases. Therefore it was suggested reaching a maximum concentration of chromium in the inner part of the membrane depended on the water transport through the organic phase.

CONCLUSIONS

i. From the analysis of membrane stability (50% breakage in 10 min.) it has been concluded that the separation-concentration process for Cr(VI), employing NaOH 5N as stripping agent in a stirred tank, is only viable at very short contact times, $t_r \leq 5$ min.

ii. It is very likely that membrane instability is mostly related to the surfactant behavior (SPAN-80) which allows the transfer of water through the organic wall.

iii. The maximum concentration of Cr(VI) attainable in the stripping phase seems to be limited by water permeation through the membrane, reaching a maximum value Cr=30 $\pm$ 5 g/l.

REFERENCES

[1] "Ion Exchange Membranes", Ed. D.S. Flett; Ellis Horwood Ltd., London 1983.

[2] Peters, R.W., *Environmental Progress* 6, 2, M4, 1978.

[3] Fuller, E.J., Li, N.N; *Jour. Mem. Sci.,* **18**, 251-271, 1984.

[4] Hochhauser, A.M., Cussler, E.L., *AIChE Symp. Ser.* **71** (152) 136 1973.

[5] Weiss, S., Castañeda, A., ISEC'88, 6-24, 1988.

PHENOL RECOVERY WITH SUPPORTED LIQUID MEMBRANES: EXPERIMENTAL STUDY

A. M. URTIAGA, M. I. ORTIZ and
A. IRABIEN

Departamento Ingeniería Química
Facultad de Ciencias
Universidad del Páis Vasco
Apdo. 64448080 Bilbao, Spain

ABSTRACT

Phenol is considered to be one of the most hazardous wastes in the aqueous media. Phenol is used in the production of phenolic resins, which are formed by condensation of phenol with formaldehyde or acetaldehyde. The resulting aqueous effluents contain mainly phenol, formaldehyde and low molecular weight resins. Presently, a series of conventional processes are used for the phenol recovery from the aqueous effluents, solvent extraction being a widely extended method.

Supported liquid membranes offer an attractive alternative to solvent extraction. In this system, the organic extractive phase is immobilized in the microporous walls of a polymeric hollow fiber which is separating two aqueous phases. The aqueous effluent circulates through the bore of the fiber, while the stripping solution circulates along the external wall of the fiber. This configuration allows the coupling of the extraction and stripping operations in a single step.

In this work, aqueous solutions with a phenol concentration of 5 g/l have been treated. The experimental module is formed by one or three hollow fibers of polypropylene (Accurel; ENKA, A.G.) of 0.6 mm inner diameter. The fibers were inserted into a hollow glass tube of 4 mm inner diameter. The solvent used is kerosene, which is pumped through the bore of the fiber to impregnate it. The stripping solution is sodium hydroxide (1M). Phenol is a weak acid and can pass through

Chemistry for the Protection of the Environment
Edited by L. Pawlowski *et al.*, Plenum Press, New York, 1991

ping solution is sodium hydroxide (1M). Phenol is a weak acid and can pass through the organic membrane in its undissociated form, leading to sodium phenolate in the phase containing sodium hydroxide.

The phenol concentration can be reduced from 5 g/l to 90 ppm, the point at which experiments were stopped. The final concentration of sodium phenolate was 40 g/l.

The experimental results show the feasibility of the simultaneous separation and concentration of phenol in a supported liquid membrane system using kerosene as liquid membrane and microporous polypropylene as supporting material.

INTRODUCTION

Phenol is considered to be a hazardous waste in aqueous media.

Phenol effluents are mainly due to chemical and oil-refining industries. The production of phenolic resins, which are formed by condensation of phenol with formaldehyde or acetaldehyde, gives aqueous effluents containing mainly phenol, formaldehyde and low molecular weight resins. Presently, a series of conventional processes are used for phenol recovery from the aqueous effluents, solvent extraction being a widely extended method.

Liquid membranes offer an attractive alternative to solvent extraction and the application of emulsion liquid membranes to phenol recovery from waste effluents has been previously studied [1,2].

Recently Sengupta [3] proposed a method for the separation of phenol from aqueous solutions by hollow–fiber–contained liquid membranes.

In a supported liquid membrane (SLM) system, the organic extractive phase is immobilized on a microporous support which separates two aqueous phases. The solid substrate can be in the form of a flat sheet or a hollow fiber with porous walls which when inserted in modules allows the design of reactors with high surface to volume ratios. The aqueous effluent containing phenol circulates through the bore of the fiber while the stripping solution is pumped along the external wall of the fiber. Phenol in the inner phase diffuses through the liquid membrane and reacts with the stripping solution. The acid base reaction involved is:

$$C_6H_5OH + NaOH \quad \rightarrow \quad C_6H_5ONa + H_2O$$

Phenol is a weak acid and can pass through the organic membrane in its undissociated form to give sodium phenolate in the external phase.

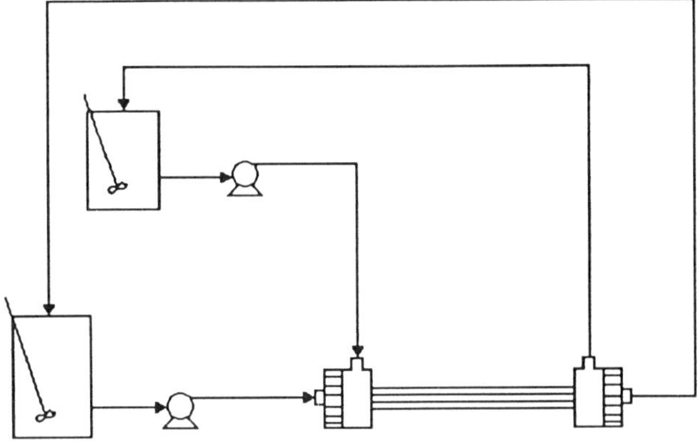

Fig. 1. Schematic diagram of experimental set-up.

Sodium phenolate is a strong electrolyte and remains dissociated, $C_6H_5O^-$ and Na^+. These ions are not soluble in the organic phase. In this way, phenol diffuses continuously due to the reaction in the outer phase.

In this work, aqueous solutions with a phenol concentration of 5 gr/l have been treated, leading to final concentrations of 90 ppm. A first order kinetic law for the transport of phenol with chemical reaction through SLM has been proposed and the feasibility of the simultaneous recovery and concentration of phenol has been stated.

EXPERIMENTAL METHOD

Figure 1 shows a schematic diagram of the experimental setup. The permeation module is formed by one or three hollow fibers of polypropylene (ACCUREL; ENKA, A.G.) inserted into a hollow glass tube (inner diameter 4 mm). The hollow fibers are firmly epoxy-bound to the end of the glass shell, with the ends of the fibers exceeding the shell. There are two lateral openings in the glass shell to circulate the stripping solution along the external wall of the fibers while the feed solution goes through the lumen of the fibers. Inner and outer fluxes come from magnetically stirred tanks where they are recirculated by a peristaltic pump (Tokyo Rikakai, MP-3).

The characteristics of the support material are as follows:

Diameter (mm)		Pore Dimensions (mm)	Porosity %	Tortuosity
Inner	Outer			
0.6	1.0	0.2	75	1.67

The organic solvent used as the liquid membrane kerosene, which is impregnate the fiber by being pumped through the fiber bore. Due to the hydrophobic nature of the material, fibers are inmediately impregnated, and the pumping is maintained for one hour.

The feed flow is 225 ml of an aqueous solution with a phenol concentration of 5 gr/l, which through the fiber with a lineal velocity of 8 cm/s. The stripping solution is NaOH (1M) flowing with a linear velocity of 2 cm/s through the glass shell.

At different times 1 ml samples of the inner phase where taken out and analyzed after dilution in a UV PYE-UNICAM spectrophotometer; when reaction was completed, the stripping phase was also analyzed in order to check the mass balance.

RESULTS AND DISCUSSION

The experimental results have been fitted to a first order kinetic law as follows:

$$ln\frac{C_0}{C} = \frac{a\,Kw}{V}t$$

where C_0 is the initial concentration in the feed, C is the feed concentration at time t, a is the mass transfer area of the membrane (cm^2), Kw is the overall mass transfer coefficient (cm/s), V is the volume of the feed tank (cm^3) and t is the transfer time (s).

The experimental results for one- and three-fiber modules are shown in Figure 2. The rate constant is calculated from the slope of the regression lines. The values of the mass transfer coefficient obtained as KW.a are:

$$(Kw.\,a)_{1\,fiber} = 1.37m^3/hr.$$

$$(Kw\,a.)_{3\,fibers} = 4.81cm^3/hr.$$

Taking into account that the transfer area of a single fiber is 3.82 cm^2, the resulting overall mass transfer coefficients are as follows:

$$Kw_1 = 9.97 \cdot 10^{-5} cm/s$$

$$Kw_2 = 11.7 \cdot 10^{-5} cm/s$$

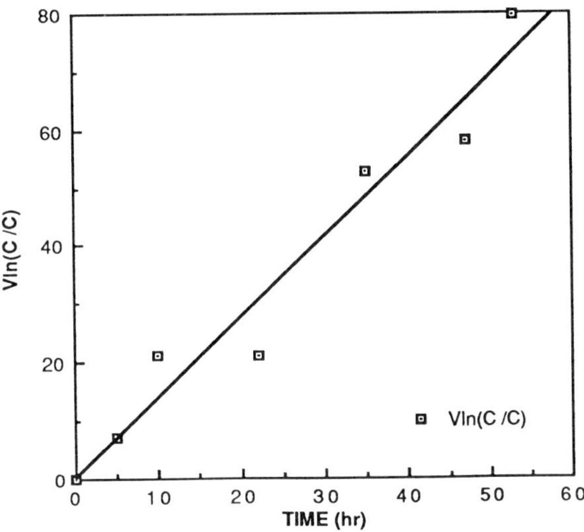

Fig. 2. Experimental results.

The analysis of the caustic stripping phase was carried out when the reaction was completed, leading to a concentration of 40 gr/l in sodium phenolate.

Experimental results show the feasibility of the simultaneous separation and concentration of phenol from diluted effluents.

REFERENCES

[1] Cahn, R.P. and Li, N.N. "Separation of Phenol from Waste Water by the Liquid Membrane Technique", *Sep. Sci.*, **9(6)**, 505, 1974.

[2] Kim, K., Ihm, S. "Simulation of phenol removal from water by liquid membrane emulsion" *Ind. Eng. Chem. Fund.*, **22**, 167, 1983.

[3] Segunpta, A., Basu, R. and Sirkar, K.K. "Separation of Solutes from Aqueous Solutions by Contained Liquid Membranes" *AICHE Journal,* **34(10)**, 1698, 1988.

NEW ROD–TYPE MEMBRANES MADE OF VARIOUS POLYMERS FOR ORGANIC DYE SEPARATION

K. MAJEWSKA-NOWAK, J. WIŚNIEWSKI
and TOMASZ WINNICKI

Institute of Environmental Protection Engineering
Technical University of Wrocław
Wybrzeże Wyspiańskiego 27, 50–370 Wrocław
Poland

ABSTRACT

The objective of this study was to optimize the casting parameters for polysulfone rod–type membranes, which are to be used for the decolorization of organic dye solutions. The study has led to the following findings: membranes which were cast from a 12.5 wt.% polysulfone solution and were gelated in a bath containing 1.5 wt.% of dimethylformamide display very good transport and separation properties. They retain organic dyes of molecular weights higher then 600 with an efficiency of 96 to 99%, at a volume flux approaching 2.4 m^3/m^2d and a pressure of 0.15 MPa.

A preliminary test was carried out to investigate the casting of rod–type poly(vinyl chloride) and polyacrylonitrile membranes, as well as their utility in the decolorization of dye solutions. It was found that rod–type poly (vinyl chloride) or polyacrylonitrile membranes display similar transport and separation properties as do polysulfone membranes.

The mean pore radius of rod–type membranes has been established and the mechanisms which are likely to govern the separation of organic dyes by these membranes have been suggested.

Chemistry for the Protection of the Environment
Edited by L. Pawlowski *et al.*, Plenum Press, New York, 1991

INTRODUCTION

It is a well-established fact that dye effluents discharged without appropriate treatment bring about a number of undesirable changes in the recipient stream. Increased color concentration makes the water unfit for domestic or industrial uses. It also reduces light transmittance, thus limiting aquatic plant growth and self-purification processes [1].

Dye effluent treatment by conventional methods was found to be insufficient. That is why environmental scientists and engineers have directed their attention to pressure membrane techniques. The application of membrane systems not only enables high removal efficiences, but also allows reuse of water and some of the valuable waste constituents (especially dyes), which have been recovered from the treated stream [2].

Of the various types of membrane modules, the tubular module seems to be the most appropriate for ultrafiltration of dye wastewater. This is so because the module has a low vulnerability with regard to mechanical pollutants, thus simplifying the pretreatment process. The module is also easy to rinse [3].

A typical design of a tubular membrane module consists of a membrane which has been placed inside a porous supporting tube [2,4]. In the system described in this paper the membrane has been cast directly on the surface of a porous bar. The bar acts not only as a membrane support, but also as an element of permeate transport. This module is called a rod–type membrane module.

EXPERIMENTAL

Casting Procedure

The membranes were formed on the surface of poly(methyl methacrylate) bars (30 cm long and 10 mm in diameter). The forming process was carried out in the apparatus which is shown in Fig. 1. A removable cylinder liner (3) spreads the casting solution (2) as a film of fixed thickness over the surface of a moving bar (1). The membranes were cast in liners with the following gap widths: 0.4 0.7, and 0.8 mm. The bar moved at three different speeds: 2.5 cm/s 5.0 cm/s, and 10.0 cm/s.

The casting process involved three polymeric solutions: Polysulfone (PS), poly(vinyl chloride) (PVC) and polyacrylonitrile (PAN). Thus, PS membranes were cast from solutions containing 10.0 11.25 or 12.5 wt.% of PS–P3500 (Union Carbide) in dimethylformamide (DMF) and were seasoned in a gelating bath with DMF content varying from 0 to 1.5 wt.%. PVC

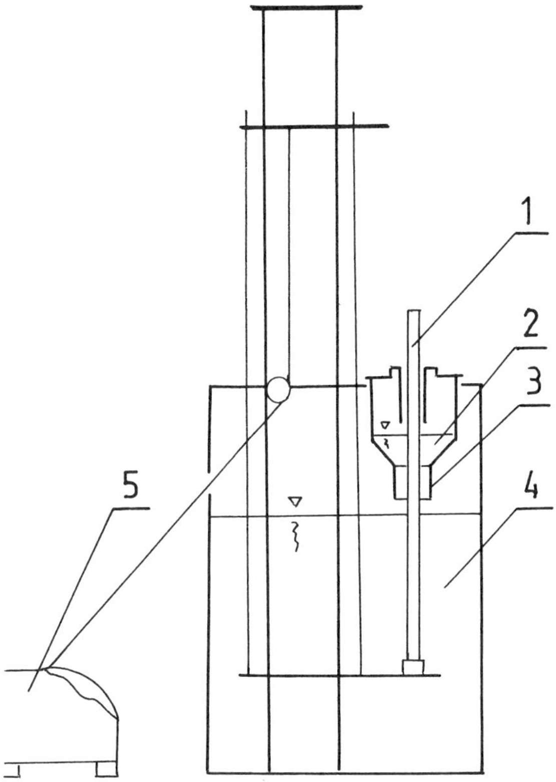

Fig. 1. Apparatus for the casting of rod-type membranes: 1 – porous bar, 2 – casting solution, 3 – membrane-forming liner, 4 – gelating bath, 5 – driving force.

membranes were prepared from a 14 wt.% solution of PVC-S64 (made by Zak lady Azotowe, Tarnów) in DMF and were gelated in distilled water. The casting solution for PAN membranes contained 14 wt.% of PAN fiber (Chemitex-Anilana, Lódź) in DMF. The gelating bath consisted of distilled water.

Experimental Systems

Figure 2 gives the schematic of the experimental system which enables investigation of the transport and separation properties of rod-type membranes. A plunger pump (5) sends the solution from the feeding tank (6) to the membrane module (1) through an air accumulator (4). The module contains a porous bar which is covered with a membrane (2) of an effective

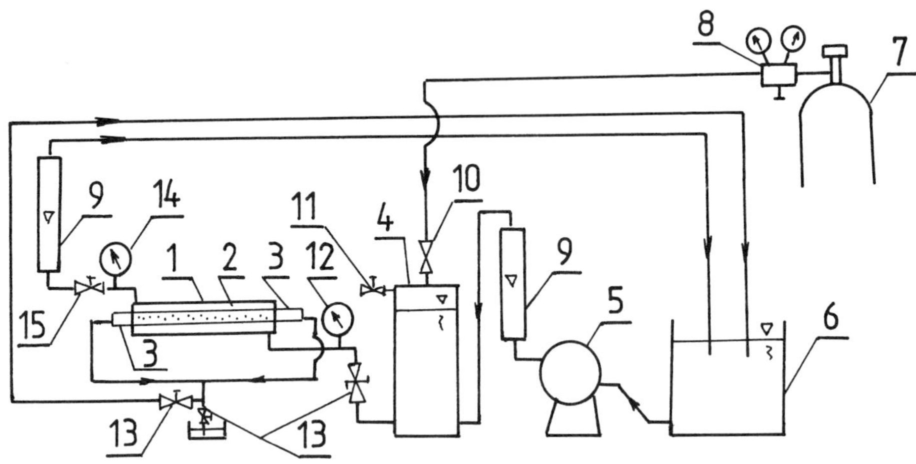

Fig. 2. Apparatus for the investigation of rod-type membranes: 1 – membrane module, 2– rod-type membrane, 3 – outlet for permeate, 4 – pressure accumulator, 5 – plunger pump, 6 – feed tank, 7 – gas cylinder, 8 – pressure reducing valve, 9 – rotameter, 10 – non-return valve, 11 – ventilating valve, 12 – contact manometer, 13 – cut-off valve, 14 – manometer, 15 – control valve.

surface area amounting to 81.2 cm^2. A control valve (15) keeps the pressure of the module at the level desired. Stub pipes (3) pass the permeate to the feeding tank (6) in order to maintain a constant concentration of the feed.

The device for measuring the mean pore radius of rod-type membranes is shown in Fig. 3. The membrane (2) (with an effective surface area of 49.4 cm^2) is placed in the pressure housing (1) which is fed with nitrogen from a gas cylinder (4). Gas pressure control in the apparatus is provided by a valve (5). A rotameter (3) enables measurement of the gas flow rate through the membrane.

Methods of Investigation

Table 1 gives the characterization of the three organic dyes which were used in this study. Dye concentration in aqueous solutions amounted to 100 g/m^3.

Ultrafiltration of model dye solutions was carried out under steady conditions at a pressure of 0.15 MPa and a linear velocity of fluid flow in the module of 0.38 m/s. Permeate flow through the membrane as well as dye

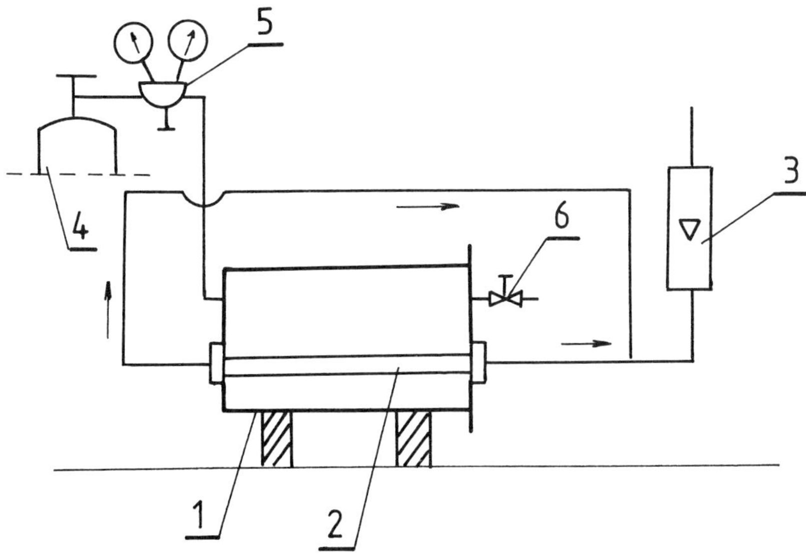

Fig. 3. Apparatus for measuring the mean pore radius of rod-type membranes: 1 – pressure housing, 2 – rod-type membrane, 3 – rotameter, 4 - gas cylinder, 5 – reducing valve.

concentrations in the permeate and in the feed were measured throughout. On the basis of measured data, the volume flux of the dye solution and the dye retention coefficient were calculated.

Pore radius was determined by the gas permeability method under steady conditions in the pressure range of 0.02 to 0.2 MPa.

RESULTS AND DISCUSSION

Effect of Casting Parameters on the Transport and Separation Properties of PS Membranes

On the basis of preliminary investigations the optimum operating parameters of the device for PS membrane casting have been established. Thus, the optimum gap width and the optimum bar moving velocity amount to 0.7 mm and 5 to 10 cm/s, respectively. The PS membranes obtained via this route display a high hydraulic permeability, as well as an appropriate selectivity to high-molecular-weight organic dyes.

To improve the transport properties of PS membranes, the polymer

Table 1. Characterization of the Experimental Dyes

Brand name	Molecular weight	Type of dye	Apparent radius of dye particle, nm[5]
Direct Meta Black, C.I. 30235	781.2	triazo direct	0.58
Helion Grey C.I. 27720	617.0	diazo direct	0.54
Methyl Orange C.I. 13025	327.0	mono-azo acidic	0.44

concentration in the casting solution was varied. When PS concentration decreased, there was a considerable reduction of the membrane thickness - from 207 μm (average membrane thickness obtained with a 12.5% PS casting solution) to 130 μm (average membrane thickness obtained with a 10.0% PS casting solution). The decrease of PS concentration in the casting solution from 12.5 to 10% brought about a 3–fold increase in hydraulic permeability (Fig. 4) at a slight deterioration of the separation properties with regard to high-molecular-weight organic dyes (Direct Black, Helion Grey). But there was a noticeable decrease in the efficiency of Methyl Orange retention (from about 70% for membranes cast with the use of 12.5% PS solution to about 30% for membranes cast with the use of 10% PS solution).

These changes in the membrane properties should be attributed to the aforementioned thickness variations, as well as to the changes that occur in the internal structure. With the decreasing polymer concentration, the rate of non-solvent penetration into the layer of the cast film increases, thus contributing to the increase in porosity and pore size [6].

Successive investigations on the choice of optimum casting conditions for rod-type PS membranes aimed at determining the effect of the solvent which was present in the gelating bath on the membrane properties. In this study, the membranes were cast from a 12.5% PS solution and were gelated in distilled water with DMF addition.

When DMF concentration in the bath increased, membrane thickness decreased rapidly – from 230 μm (average thickness of water–gelated membrane) to 141 μm (average thickness of membranes gelated in 1.5% DMF solution). The presence of DMF exerts a substantial influence on the transport and separation properties of the membranes (Fig. 5). Addition of DMF in amounts of 1.5 wt.% makes the hydraulic permeability increase to a value

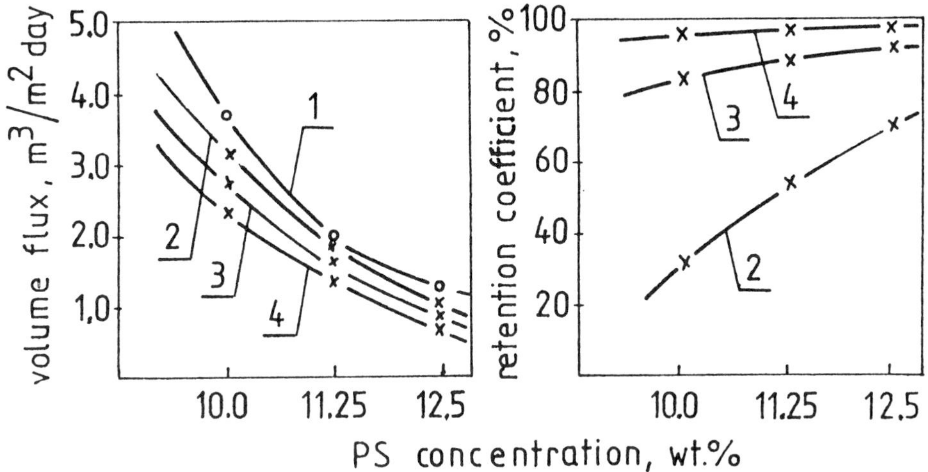

Fig. 4. Volume flux and retention coefficient versus PS concentration in the casting solution: 1 – water, 2 – Methyl Orange, 3 – Helion Grey, 4 – Direct Meta Black (membranes were cast at a gap width of 0.7 mm and a bar moving velocity of 10 cm/s.

which is over three times as high as that of the water–gelated membrane. It should be noted that the presence of DMF in the gelating bath has a favourable effect on the separation properties of the membranes. This effect becomes particularly distinct in the case of Methyl orange. The retention coefficient increases from 73% (for water-gelated membranes) to 85% (for 1.5% DMF-gelated membranes).

In this particular case, the variations in the internal structure of the membrane are associated with the decreasing difference in the chemical potential between solvent and non-solvent [7,8]. Slow precipitation of the polymer accounts for the formation of spongy structures with predominance of small size pores. As a result, hydraulic permeability increases and membrane selectivity improves.

Assessing the Utility of PVC and PAN Membrane in the Separation of Organic Dyes

Rod–type PS membranes yield high efficiencies of organic dye removal from aqueous solutions. But taking into account the limited availability of

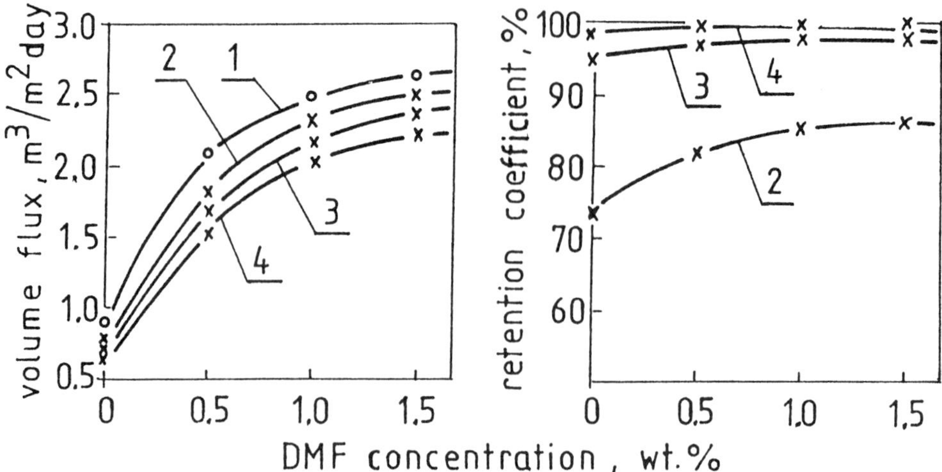

Fig. 5. Volume flux and retention coefficient versus DMF concentration in gelating bath for PS membranes: 1 – water, 2 – Methyl Orange, 3 – Helion Grey, 4 – Direct Meta Black (membranes were cast at a gap width of 0.7 mm and a bar moving velocity of 5.0 cm/s).

polysulfone on the Polish market, it seemed advisable to investigate the possible applications of two domestic polymers – poly(vinyl chloride) and polyacrylonitrile – for the same purpose. The membranes were cast from solutions with 14 wt.% concentration of the polymer, which had been heated to the temperature of 313 K. The casting process involved a bar moving velocity of 2.5 cm/s and a gap width of 0.4 mm. Average membrane thickness amounted to 94 μm and 105 μm for PVC and PAN, respectively.

The results of preliminary investigations on the transport and separation properties are listed in Table 2. As shown by these data, both PVC and PAN membranes display very high (above 95%) coefficients of retention to high-molecular-weight dyes (Direct Black, Helion Grey). PVC membrane also exhibit a comparatively high coefficient of retention (71.9%) to Methyl Orange. These results - along with the water flux values (1 to 2 m^3/m^2d at 0.15 MPa) – quantify the similarity of the properties of PVC or PAN membranes with those of PS membranes. Only the relative permeabilities of PVC and PAN membranes (the quotient of dye solution flux and water flux) were found to be considerably lower than that of the PS membrane. This phenomenon should be attributed to the high values of the dye retention coefficients.

Table 2. Volume Fluxes and Retention Coefficients of Organic Dye for PVC and PAN Membranes

Testing solution	Type of membrane			
	PVC		PAN	
	Volume flux, m^3/m^2d	Retention coefficient, %	Volume flux, m^3/m^2d	Retention coefficient, %
Water	1.38	–	2.28	–
Direct Meta Black	0.59	99.6	1.31	99.2
Helion Grey	0.67	97.9	1.59	95.0
Methyl Orange	0.74	71.9	1.73	48.6

Membrane Structure and Membrane Properties

To define the mechanism of dye particle retention by rod–type membranes, their mean pore radius was determined and compared with the apparent radius of the dye particles. The pore radius of the membrane was established by measuring the permeability of the membrane to gas [8,9]. The advantage of using the gas permeability method is that the mean pore radius can be determined without establishing the porosity of the membrane. Rod-type membranes are firmly attached to the support, which practically eliminates porosity determinations.

The calculated results are given in Table 3. As shown by these data, the casting parameters have an influence on the membrane structure. Thus, to obtain a fine-pore PS membrane of high hydraulic permeability, it is advisable to gelate the cast film in distilled water with DMF. A fine-pore structure is also typical for PVC and PAN membranes – and this account for their excellent separation properties.

It is interesting to investigate how the pore radii of the membranes compare with the apparent radii of the dye particles (Table 1). The apparent radius of dye particles was calculated, assuming that no aggregates were formed. It should be noted that dye particles have the shape of a dipole with a minimum length of 2 nm and a minimum width of 0.6 to 0.8 nm [5]. Even though the dye particles are noticeably smaller in size than the pores of the investigated membranes, the efficiency of dye retention remains very high. Taking this into account, it seems advisable to postulate the following mechanism: Dye separation by ultrafiltration is not only governed by the

Table 3. Mean Pore Radii of Rod-Type Membrane

Membrane	Mean pore Radius, nm
PS–10[a]	30.5
PS–11.25[a]	26.1
PS–12.5[a]	24.3
PS–12.5[a]–0[b]	23.5
PS–12.5[a]–0.5[b]	18.1
PS–12.5[a]–1.0[b]	16.2
PS–12.5[a]–1.5[b]	14.7
PVC–14[a]	10.4
PAN–14[a]	12.6

Note: a = polymer concentration in the casting solution;
b = DMF concentration in the gelating bath.

molecular sieve principle. The efficiency of retention may be influenced by the phenomenon of dye adsorption in the membrane pores, as well as by the electrostatic interaction between the dye particle and the membrane material. The latter process is of particular importance when azo dyes and polysulfone membranes are involved. Negatively charged dyes are repelled by the membrane which is also negatively charged. As a result, the efficiency of dye retention increases. This finding was substantiated by Liu Tinghui and co-workers [10].

Retention efficiency may also be influenced by the ability of the dyes to form aggregates. It is anticipated that, when this aggregability is high, the dye particle size will approach 100 nm [11]. In this particular case, dye separation will proceed via the molecular sieve mechanism.

CONCLUSIONS

Our attempts to optimize the casting process have shown that the transport and separation properties of rod-type PS membranes are effectively controlled by varying the concentration of the polymer in the casting solution and the concentration of the solvent (DMF) in the gelating bath. Thus, the best properties to yield an efficient ultrafiltration of organic dyes have been obtained with a 12.5% PS casting solution and a 1.5% DMF gelating bath. The rod-type PS membranes formed via this route enable 98 to 99% retention of organic dyes with molecular weight above 600 at a volume flux approaching 2.4 m^3/m^2d, and a pressure amounting to 0.15 MPa.

Preliminary tests of rod-type PVC or PAN membranes with regard to their transport and separation properties have supported their utility when applied to the separation of organic dyes. Both the membranes display transport and separation properties similar to those of the PS membranes.

REFERENCES

[1] B. Koziorowski, Industrial Wastewater Treatment, Wydawnictwo Naukowo-Techniczne, Warszawa, pp. 456–462, 1980 (in Polish).

[2] C.A. Brandon, Reuse of total composite wastewater renovated by hiperfiltration in textile dyeing operation, *Ind. Water Eng.*, **12**, 14–16, 1976.

[3] M. Bodzek, O. Kominek and J. Zieliński, Application of reverse osmosis and ultrafiltration to the technology of water and wastewater treatment, in: A. Machalski (Ed.), Nowa Technika i Inżynierii Sanitarnej, Wodoci agi i Kanalizacja, Arkady, Warszawa, pp. 27–34, 1981.

[4] R. Matz and Y. Meittis, Design and operating parameters for tubular ultrafiltration membrane modules, *Desalination*, 24 , 281–294, (1978).

[5] T. Hori, M. Mizuno and T. Shimizu, Dye diffusion in water swollen cellulose membranes and in bulk water, *Colloid. and Polym. Sci.*, 258, 1070–1076, (1980).

[6] H. Strathmann and K. Kock, The formation mechanism of phase inversion membranes, *Desalination*, **21**, 241–255, (1977).

[7] H. Strathmann, K. Lock, P. Amar and R.W. Baker, The formation mechanism of asymmetric membranes, *Desalination*, **16**, 176–203, 1975.

[8] H. Yasuda and J.T. Tsai, Pore size of microporous polymer membranes, *J. Appl. Polym. Sci.*, **18**, 805–819, 1974.

[9] E.J. Hopfinger and M. Altman, Experimental verification of the dusty-gas theory for thermal transpiration through porous media, *J. Chem. Phys.*, **10** 2417–2428, 1969.

[10] Liu Tinghui, T. Matsuura and S. Sourirajan, Effect of membrane materials and average pore size on reverse osmosis separation of dyes, *Ind. Eng. Chem. Prod. Res. Dev.*, **22** 77–85, 1983.

[11] L.Ya. Kukushkina, E.V. Migalati, Nikoforov et al., Ultrafiltracjonnoje vydelenije krasitielej is vodnych rastvorov, *Zh. Prikl. Chim.*, **50**, 1847–1852, 1977.

MEMBRANIZED CONDUCTING POLYMER ASSISTED PHOTODEGRADATION FOR ENVIRONMENTAL PROTECTION

L. CAMPANELLA, C. MORGIA

Department of Chemistry
University "La Sapienza"
Rome, Italy

ABSTRACT

Photodegradation of chlorophenols when catalyzed by membranized conducting polymers is studied. The influence of additions to the membrane composition of an oxidase enzyme and of a protecting redox mediator is investigated, aimed at realizing a natural photodegradative process based on membrane containing conducting polymers operating in sunlight and in the presence of humic matter.

INTRODUCTION

The direct or mediated adsorption of UV-light can be applied to the degradation of some polluting organic compounds in order to lower their environmental impact when disposed. Chlorophenols, due to their high toxicity, ability to be bioaccumulated and easy degradability, are among the compounds to which photodegradation can be successfully applied. Photodegradation is a clean, simple and cheap process. In previous papers we exposed and discussed the results we obtained when we applied catalysis from conducting polymers [1-8] to the photodegradation of some organic pollutants; the membranization of the catalyst and the possible mechanism of the reaction were also presented [9-12]. Substantially the study evidenced that this catalysis presents some advantages over that of the traditional inorganic semiconductors, especially with reference to recovery and cost aspects.

Two innovations are presented here; an important result is obtained by using polyphenylacetylene as catalyst instead of polybenzylpropargilamine. The former needs a doping process with iodine, which is released during the photodegradation reaction. The latter is simply doped by treatment with an acid [5, 13, 14, 15], due to the presence in its molecule of NH groups able to be promoted and activated. A further improvement can be obtained by immobilizing in the catalyst membrane an oxidase enzyme and an opportune redox mediator (hydroguinone, humic acids) in solution also able to protect the enzyme molecules.

EXPERIMENTAL

The preparation of the polymer and of the membrane containing PVC as supporting material and buthylphtalate as plasticizing agent is already described in the literature [2, 5, 13]. The experimental conditions require the exposure of the chlorophenol (orto, meta, para) solutions to a 40 W light filtered by borosilicate glass. The volume of the solution is 50 ml and the reaction is followed by spectrophotometric measurements performed with a Beckman 40 DU Model instrument, while the treated solution is homogenized by magnetic stirring. The solutions were analyzed by HPLC (Perkin Elmer Series 10-liquid chromatography) to try to interprete the transformations occurring during the photodegradation. Laccase and peroxidase were the enzymes used (500 units); purified humic acids (0.2 mg/L) and hydroquinone (1 mg/L) the mediators dissolved in solution.

A membrane with a surface of 24.6 cm^2 containing PVC, DBF and PBPA was first prepared and used as a catalyst for photodegradation of parachlorophenol. The reaction is slow to begin and reaches an extent of 60% during the first 96 hrs. For the two other isomers, values of about 60% were observed for photodegradation an improvement in comparison with the results obtained when the catalyst was a membrane based on polyphenylacetylene as the conducting polymer, for which the best reaction obtained generally did not exceed 50% in two days of reaction; this means that the latter reaction is faster than the former one.

In order to gain rapidity we activated PBPA by immersion of the membrane containing it for 24 hrs in acid solution (HCl 1+1 by volume). The results we obtained show that the introduction in to the membrane, which acts as catalyst, of the conducting polymer activated in acid solution

Table 1. Photodegradation Values Catalyzed by Polymeric Conducting Membranes of PBPA

a

MEMBRANE	MEMBRANE (cm^2) AREA	TIME (h)	% DEGR.
PVC-DBF-PBPA (unactivated)	24.63	96	60
PVC-DBF-PBPA (activated)	"	72	60

b

COMPOUND	SUP. MEMBRANE (cm^2) AREA	TIME (h)	% DEGR.
o-Cl-phenol	24.63	72	40.0
p-Cl-phenol	"	72	60.0
m-Cl-phenol	"	48	70.0

a) unactivated b) activated by immersion in acid solutions

presents some advantages; degree of photodegradation is increased by almost 20% (Tab. 1).

The absence of the iodine as dopant also has another important advantage. As degradation can in this case be considered as an oxidation it is possible to immobilize an oxidase enzyme in the membrane with the hope it can act as a further catalyst for the reaction; this would not be possible with iodine–doped polymer due to the denaturating action on the enzyme of this substance. So a double catalysis (by conducting polymer and by enzyme) system was experimented with.

The enzyme was first added to the solution in order to explore its efficiency; it was successively immobilized in the membrane containing the other catalyst, represented by the polymeric conductor activated by immersion in hydrochloric acid solution. Two different enzymes were used: laccase and peroxidase.

In the described experimental conditions the three considered chlorophenols were temptatively degraded with a membrane as catalyst, constituted by PVC, DBF, PBPA and enzyme (500 units).

The first enzyme we used was a peroxidase, constituted by a large number of enzymes able to oxidize a certain number of substrates in the presence of hydrogen peroxide.

Among these substrates we find guaiacol and pirogallol, but generally most of the phenols and aromatic amines can be used in the study of the catalytic activity of the enzyme. In the presence of the enzyme the rate

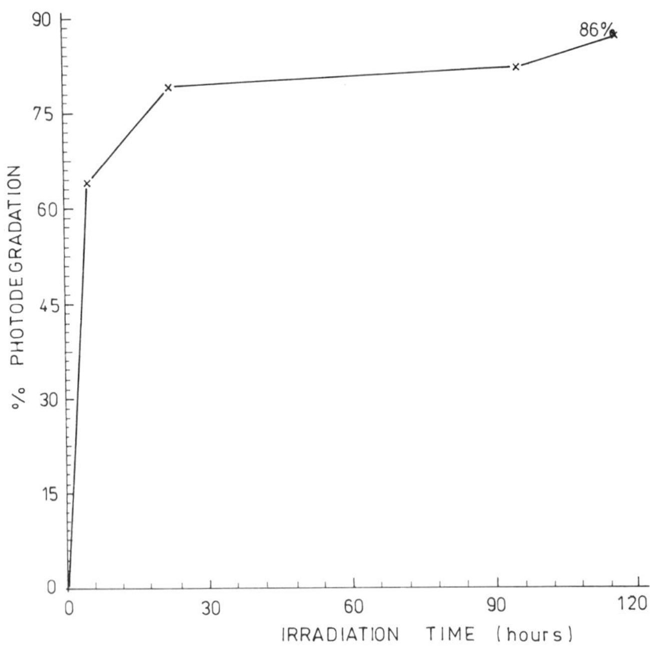

Fig. 1. Photodegradation of m-chlorophenol catalyzed by a polymeric conducting enzymatic membrane.

of the reaction is sharply increased; in a few hours (1.5 - 2.5) degradation degrees about 60% are obtained. If the experiment is prolonged beyond the maximum that can be obtained, not only is no progress observed, but sometimes a regression occurs for all three isomers (Fig. 1).

In order to interpret the observed behavior it was assumed that with the passage of time the activity of the enzyme is lowered by the peroxyradicals formed in the degradation reaction and by its possible photooxidation. Accordingly, a stabilizing agent of the enzyme, which at the same time could be oxidized by photoreaction, was introduced into the system, thus protecting the enzyme from possible damage and to act as redox mediator. Hydroquinone was experimented with first; it behaved as hydrogen peroxide in our previous studies.

Then because hydroquinone groups are present in humic acids, because these substances account for the presence in natural waters of radicals such as $\cdot OH$, $RO_2\cdot$ and, on the other hand, because humic acids occur naturally in soil so that they can be imagined as present in the natural photodegradation process, hydroquinone was replaced by humic acids.

After purification, humic acids were dissolved in basic solution and

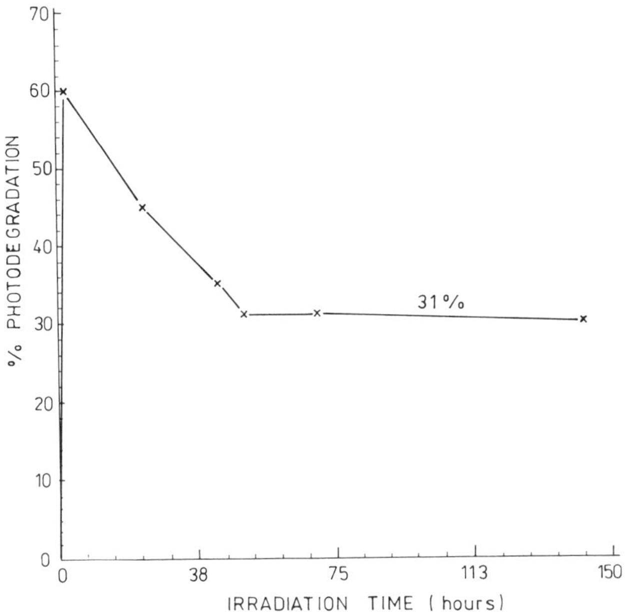

Fig. 2. Photodegradation of m-chlorophenol catalyzed by a polymeric conducting enzymatic membrane containing a protecting enzymatic mediator.

opportunely mixed with the phenolic solutions to be tested so that they were practically neutral.

In these conditions chlorophenols were temptatively photodegraded by a catalyzing membrane containing laccase as the immobilized enzyme. The behavior of the reaction with the passage of time is shown in Figure 2. Degradation degrees of about 80% and higher can be attained in a short time (Fig. 2).

In order to investigate the function of the enzyme, some complementary experiments were performed reaching the following conclusions:

1) by comparing the reaction behavior at different concentrations of the compound to be degraded (fixed enzyme concentration) and at different concentrations of the enzyme (fixed chlorophenol concentration) it can be concluded that no reaction between enzyme and chlorophenol occurs, at least not in a short time;

2) by spectrophotometric investigation of peroxidase and chlorophenol solution behavior in light and darkness, two peaks are observed at 270 and 400 nm with the values of the maximum higher when in darkness; all the

Table 2. **Photodegradation Data in Different Conditions (Conducting Polymer, Enzyme, Mediator Possibly Present in the Membrane Acting as Catalyst)**

COMPOUND	CONDUCTING POLYMER	ENZYME	MEDIATOR	TIME (h)	% DEGR
o-Cl-fen	PBPA	peroxidase		1.5	40
p-Cl-fen	"	"		1.5	35
m-Cl-fen	"	"		2.0	60
o-Cl-fen	PBPA	laccase		20	45
p-Cl-fen	"	"		24	45
m-Cl-fen	"	"		25	40
o-Cl-fen	PBPA	"	hydroquin	50	60
p-Cl-fen	"	"	"	60	70
m-Cl-fen	"	"	"	60	70
o-Cl-fen	PBPA	"	humic ac.	80	80
p-Cl-fen	"	"	"	90	80
m-Cl-fen	"	"	"	90	100

values are constant till 70 hrs, when an increase in absorbance at 400 nm (due to peroxidase) both in light and in darkness is observed. It can be concluded that the enzyme does not undergo any degradation due to temperature and that no new peaks appear in the spectrum;

3) the presence of the solvent used, THF, does not denaturate the enzyme even if the solvent remains partially bound in some way to the enzyme;

4) on following the enzymatic reaction by spectrophotometry the growth of the peak at 270 nm and a decrease of that at 400 nm are observed at the same time.

A new peak at 668 nm with a chromatic variation of the solution appears: the modification of the spectrum is probably due to the formation of a dimer (H_2A_2) which evidently is formed neither in presence of light nor when the enzyme is dispersed in solution or immobilized in the membrane with conducting polymer. The presence of the dimer is therefore able to modify the enzymatic action, although this is insufficient to explain why the reaction stops at a certain moment; the dimer formation is expected to behave as an inhibiting step of a radical–forming process, we observed the contrary unless we admit that depletion of oxygen in the reaction room plays a role.

To determine the behavior of peroxidase was to be ascribed to its particular nature or to its general enzymatic action, we experimented also

Table 3. Photodegradation of Oil Waste Water

% DEGR.	TIME (h)
0	0
30	24
45	48
60	96

with of laccase (paradiphenoloxidase) which, having an operating optimum pH of 5.5, looked very advantageous to use considering the experimental pH very near to the optimum. The results of the experiments are shown in Table 2 compared with the previous ones: the yields of the reactions are as high as in the previously described tests for the first few hours, but in this case the reaction continues for longer times; after about 20 hours the reaction tends to stop high acidity and a chloride concentration not always compatible to the enzyme's nature.

The application to oil mill waste waters is suggested because of the problem of the disposal of these waters and of the recovery of the wastes.

The control of the results of this application was performed by COD measurements and satisfying results were obtained (Tab 3), particularly if the complex polluted matrix to be degraded was considered in comparison with some inhibition generally observed in softer conditions for enzymatic processes (Tab. 3).

REFERENCES

[1] Campanella, L., Salvi, A.M., Morgia, C., Sbardellati, R., Inquinamento, 12, 32, 1987.

[2] Campanella, L., Salvi, A.M., Sammartino, M.P., Tomassetti, M., La Chimica e l'Industria 68, n. 12, 71, 1986.

[3] Campanella, L., Mazzei, F., Morgia, C., Sammartino, M.P., Tomassetti, M., Baroncelli, V., Battilotti, M., Colapicchioni, C., Giannini, I., Porcelli, F., in Acta of National Meeting on "Biocatalizzatori", Roma, 27-29 October 1987, E.N.E.A. Ed., 257, 1987.

[4] Campanella, L., Morgia, C., Tomassetti, M., Ortaggi, G., in Acta of National Meeting IX A.I.C.A.T., Palermo, Dic. 1987, 23, 1987.

[5] Campanella, L., Morgia, C., Salvi, A.M., Chimica Oggi, 65, May 1988.

[6] Campanella, L., Morgia, C., et al., Analysis 16, 120, 1988.

[7] Campanella, L., Morgia, C., Tomassetti, M., Ortaggi, G., Thermochimica Acta, Submitted, 141, 27, 1989.

[8] Campanella, L., Mazzei, F., Morgia, C., et al., Electrofinn Analysis, an International Conference an Electroanalytical Chemistry, June 1988, Trukur Abu (Finland).

[9] Mill, T., Mabey, W., in Environmental Exposure from Chemicals 1, n. 8, C.R.C., Ed. 1986.

[10] Mabey, W.R., et al., Laboratory Protocols for evaluating the fate of organic chemicals in air and in water, EPA, Final Report, EPA 600/3-82-002, Washington, Dic., 1982.

[11] Pelizzetti, E., Barbeni, M., Pramauro, E., Serpone, N., Borgarello, E., Janiesau, M.A., Hidaka, H., La Chimica e l'Industria, 67, n. 11, Nov. 1985.

[12] Pelizzetti, E., Minero, C., La Chimica e l'Industria, 69, n. 10, Oct. 1987.

[13] Furlani, A., Paolesse, R., et al., in Recent developments in ion exchange, P.A. Williams and M.J. Hudson Ed., 315, 1987.

[14] Campanella, L., Ferri, T., et al., in Recent developments in ion exchange, P.A. Williams and M.J. Hudson Ed., 315, 1987.

[15] Campanella, L., Morgia, C., Mazzei, F., Tomassetti, M., Cordatore, C., in Acta of National Meeting X A.I.C.A.T., pISA, Dec. 1988, in press.

MISCELLANEOUS METHODS FOR REMOVAL OF POLLUTANTS

THE LAND DISPOSAL OF WASTES:

A STUDY OF SOME APULIAN SOILS

D. PETRUZELLI*, A. LOPEZ*, L. LIBERTI** and
G. PETIO***

* Istituto di Ricerca sulle Acque
C.N.R., 5, Via De Blasio, 70123 Bari, Italy
** Istituto Chimica Applicata
Università di Bari
200, Via ReDarid, 00198 Roma, Italy
*** Istituto di Ricerca sulle Acque
C.N.R., 1, Via Reno, 00198 Roma, Italy

ABSTRACT

The increasing interest in water reuse by land application of wastewaters and/or land disposal of solid wastes is posing much concern on underground water resources quality, and this calls for a deeper understanding of the physico-chemical interactive phenomena during percolation of the liquid-phase through the natural porous media (soil, subsoil).

Among other processes responsible for attenuation (dispersion, retention, degradation) of a persistent chemical, in a fluid phase leaching along the flow path of a porous material, sorption and ion exchange phenomena play a relevant role in determining transport and availability of a pollutant substrate.

After a review of the models available for predicting the fate and transport of contaminants in the subsurface environment the paper will furnish an example of application of lab-scale physico-chemical characterization of the natural porous material (solid-liquid interaction isotherms, cation exchange capacity, geochemical

Chemistry for the Protection of the Environment
Edited by L. Pawlowski *et al.*, Plenum Press, New York, 1991

and pedological properties) to the solution of field problems.

With reference to solid-waste landfill disposal, the paper assesses the transport of heavy metals such as Cr^{3+}, Cd^{2+}, Cu^{2+}, Pb^{2+} on typical soils and subsoils of the Apulia region (S.E. Italy).

INTRODUCTION

The assessment of attenuation by dispersion, retention, and degradation in soil of chemicals still persisting in partially treated wastewater has led to much work on the artificial recharge of groundwaters and operations for counteracting seawater intrusion into coastal aquifers [1 - 4]. Analogous problems arise from landfill disposal of solid wastes, both of industrial and municipal origin. Accordingly, Italian legislation for the preservation of natural bodies of water and for landfill disposal (law 31976 [5] and DPR 915/82 [6], respectively) is rightly restrictive, since insufficient knowledge is available concerning the transport and fate of chemicals in natural porous materials, e.g., soil and subsoil.

The present paper is concerned with the land disposal of both solid and liquid wastes, and deals with physicochemical aspects of heavy metal percolation in soil.

Following a brief theoretical introduction to the problem, it reports a lab-scale characterization and quantification of the interactive phenomena (equilibrium isotherms and other physicochemical parameters) at the liquid-solid interface. Based on these data, it is possible to evaluate the average fronts of migration of heavy metal ions in different soil systems, thus helping to assess the environmental impact and eventual compliance with current legislation.

THEORY

Processes regulating the transport and fate of a persistent molecule passing through a natural porous material can be classified in three disciplinary areas, viz., hydrogeologic, abiotic and biotic [7].

Hydrogeologic processes are based on physical transport phenomena such as advection, mechanical dispersion, and molecular diffusion [8, 9]. Such phenomena are described in terms of Darcy's Law [8]:

$$V = -K_{sat} \cdot i \, , \qquad (i)$$

the dispersion law [8]:

$$D_x = \alpha \cdot V + D_0 \, , \qquad (ii)$$

and Fick's Law of molecular diffusion [9]:

$$J = D_0 \cdot \frac{dC}{dx} \,. \qquad (iii)$$

Abiotic interactions of a pollutant solute in a percolating liquid phase may be classified in terms of:

a) Sorption phenomena based on surface electrostatic interaction (adsorption) and/or partition between the bulk of the phases involved (absorption) [10].

b) Ion exchange as the result of electrostatic interaction of an ionic solute with charged functional groups in the solid phase, e.g., carboxylate, phenate, aluminate, (hydr)oxide, etc. These functional groups may be associated with the soil organic fraction (humic and fulvic acids, macromolecular structures) and/or with the soil mineral fraction (zeolites, oxides, etc.) [11 - 13].

c) Chemical transformation of the pollutant by, e.g., hydrolysis or redox reactions [7, 14 - 16].

Biotic phenomena depend upon biochemical transformation of organic substrates, and only play a relevant role in the presence of appropriate microorganisms, and where nutrient availability, contact times, and suitable redox conditions, etc., are assured.

The transport and fate of a chemical substance in a natural porous medium are highly complex processes, involving the interaction of all the above phenomena: the net result is the algebraic sum of the contributions of each.

A undimensional general transport equation for a chemical substance in such a medium can thus be described in the following terms [8]:

$$\frac{dc}{dt} = \underbrace{v\frac{dC}{dx}}_{\text{(advection)}} + \underbrace{D_x\frac{d^2C}{dx^2}}_{\text{(dispersion)}} - \underbrace{\frac{\rho}{\theta}\frac{dS}{dt}}_{\text{(sorption)}} \pm \underbrace{\frac{dC}{dt}}_{\text{(reaction)}} \qquad (iv)$$

The last term on the right-hand side of the equation may be positive, negative, or zero, depending on whether the substance is formed, transformed, or conserved in the system.

If the interactive phenomena at the liquid-solid interface are assumed to be rapid with respect to the percolation of the fluid, i.e., the Local Equilibrium Assumption [11] in which $\frac{dS}{dt} = \frac{dS}{dC} \cdot \frac{dC}{dt}$, then equation (iv) can be rearranged in the following form:

$$R\frac{dC}{dt} = -v\frac{dC}{dx} + D_x\frac{d^2C}{dx^2} \qquad (v)$$

where the "retardation factor" term:

$$R = 1 - \frac{\rho}{\theta}\frac{dS}{dC} \qquad (vi)$$

represents the retardation of the migration of the pollutant with respect to the front of migration of the solvent (water). In this context, the retardation factor R can be represented as $\frac{v}{v_c}$, where v is the filtration velocity of the water and v_c that of the solute. Physically, retardation results from interaction of the pollutant at the liquid-solid interface during percolation of the solution in the natural porous material.

Quantitative evaluation of this is obtained by experimental determination of the equilibrium isotherm $(\frac{dS}{dC})$ of the pollutant on the solid phase.

Under steady-state conditions in an isotropic porous material Darcy's Law (see equation (i)) can be combined with the retardation factor relations (equations (v) & (vi)) to obtain the average migration distance Z of the solute in the porous medium [14 - 15]

$$Z = \frac{K_{sat} \cdot i \cdot t}{Rn_{\varepsilon}} . \qquad (vii)$$

In other words, from experimental determination of the equilibrium isotherm for the particular system, and from physicochemical (CEC, %O.C.) and geotechnical (K_{sat}, ρ, θ, n_{ε}) characterization of the solid phase, it is possible to evaluate the retardation factor R, and from this the average migration distance Z, of the chemical in the porous medium.

EXPERIMENTAL

Four typical Apulian soils (from the region of Puglia, in S.E. Italy) were examined in this study:

Terra rossa (red soil)	ex Minervino,	district of Lecce.
Staturo sand	ex Altamura,	district of Bari.
Lucera clay	ex Lucera,	district of Foggia.
Monte Marano sand	ex Gravina,	district of Bari.

Table 1 reports the chemical and mineralogical composition of these different materials. In Table 2 are given their geotechnical properties as reported in the literature, and their carbon contents determined as %O.C. by standard soil analysis techniques [24]. All soils sieved < 100 U.S. mesh, and air-dried.

Equilibrium isotherms were determined with a series of synthetic solutions containing Cd^{++}, Cr^{+++}, Cu^{++} and Pb^{++} as their nitrate salts, having

Table 1. Chemical and Mineralogical Composition of the Different Soils Investigated

	1	2	3	4
Illite - Muscovite	oxides:		46.00	28.00
Kaolinite	goethite,		4.30	6.00
Chlorite	ematite,		4.50	13.00
Montmorillonite	magnetite,		8.30	
Quartz	gibbsite,	40		6.00
Feldspar	quartz.		6.40	18.00
Plagioclase			3.90	5.50
Phyllosilicate		20		19.00
Mica				
Calcite		20	22.50	4.50
Dolomite			4.10	
SiO_2	35.2	50	45.70	48.70
Al_2O_3	31.1	20	11.10	16.10
Fe_2O_3	14.8	5	3.50	3.80
TiO_2	2.0		0.50	0.70
MnO	0.1		0.02	0.03
MgO	1.3		2.50	1.30
K_2O			2.00	2.10
Na_2O			0.80	0.60
$CaCO_3$		20		21.10
$MgCO_3$				2.10

1: Terra Rossa (red soil)

2: Staturo Sand

3: Lucera Clay

4: Monte Marano Sand

an initial concentration of 10 mg/l (ppm) of the metal ion. Batchwise equilibrations were run following the ECI (Environmental Conservative Isotherm) technique proposed by Griffin and Roy [14], in which variable quantities of the solid phase (20 - 50 mg) were contacted with constant volumes (20 ml) of the liquid phase, giving different ratios of soil to solution. Ratios in the range 1:100 to 1:10000 were used, depending upon the retention capacity of the solid phase. Sample mixtures for equilibration were stirred on rotating (30 rpm) drums in an environment thermostated at $20^o \pm 2^oC$ for 48 hrs. Preliminary kinetic experiments indicated 48 hrs as the optimum equilibration time; samples at 56 and 64 hrs showed variations of less than 5% in the measured equilibrium concentration. After equilibration and phase sep-

Table 2. Geotechnical and Physicochemical Characteristics of the Soils Investigated

	ρ_r (g/cm^3)	ρ_{dry} (g/cm^3)	K_{sat} (cm/s)	θ (cm^3/cm^3)	n_e (cm^3/cm^3)	CEC (meq/100g)	%O.C.
Terra Rossa	2.70	1.50	10^{-7}	0.44	0.05	7.5	0.176
Staturo Sand	2.58	1.66	10^{-5}	0.36	0.25	6.0	0.037
Monte Marano Sand	2.68	1.45	10^{-4}	0.46	0.30	2.0	0.02
Lucera Clay	2.66	1.73	10^{-7}	0.35	0.08	11.5	0.5

See definition of symbols at end of paper.

aration using 0.45 μm filters, the metal ion concentration in the filtrate was determined by flame atomic absorption on a Model 951 spectrophotometer (from the Instrumentation Laboratory, Wilmington, MA, U.S.A.). Each sample was run in triplicate, and blanks were determined to allow for any adventitious metal uptake from plastic bottles, filters, and all other material in contact with the equilibrating solutions.

RESULTS AND DISCUSSION

The simplest landfill for disposal of solid wastes of industrial origin is the so-called II Category, type B, according to current Italian legislation [6]. For this type of landfill, artificial impermeabilization by the use of clay or plastic liners is not required provided that at least 100 cm of free material is present between the bottom of the landfill and the maximum freatic level of the aquifer. A necessary condition for the disposal of solids in these landfills is that the leachate should have a maximum concentration of heavy metals less than ten times the limit set for liquid effluent discharge to bodies of water accessible to the public [15]. Experience indicates that in most cases this limit is not normally reached, and in the present work a concentration of 10 ppm was taken as a convenient maximum figure [18 - 23]. Figs. 1 - 4 show the experimental equilibrium isotherms for the ions Pb^{2+}, Cd^{2+}, Cu^{2+} and Cr^{3+} on the four soil samples investigated. The interpolated curves may generally be described by non-linear Freundlich isotherms, with the exception of that for Cd^{2+} uptake on the Monte Marano sand which can be represented by a linear Freundlich equation. The derived constants are reported in Table 3, together with the R and Z values calculated from them using equations (vi) and (vii) (assuming a ten-year operational life for the landfill).

In general, it appears that the attenuation capacity of the different soils towards the ionic species considered follows the order:

Clay > Terra rossa > Sand

while the mobilities of the various ionic species in the solid phase are:

$$Cu^{2+} > Cd^{2+} > Cr^{3+} > Pb^{2+}$$

This is clearly seen in Figure 5 where a correlation between R and Z is shown.

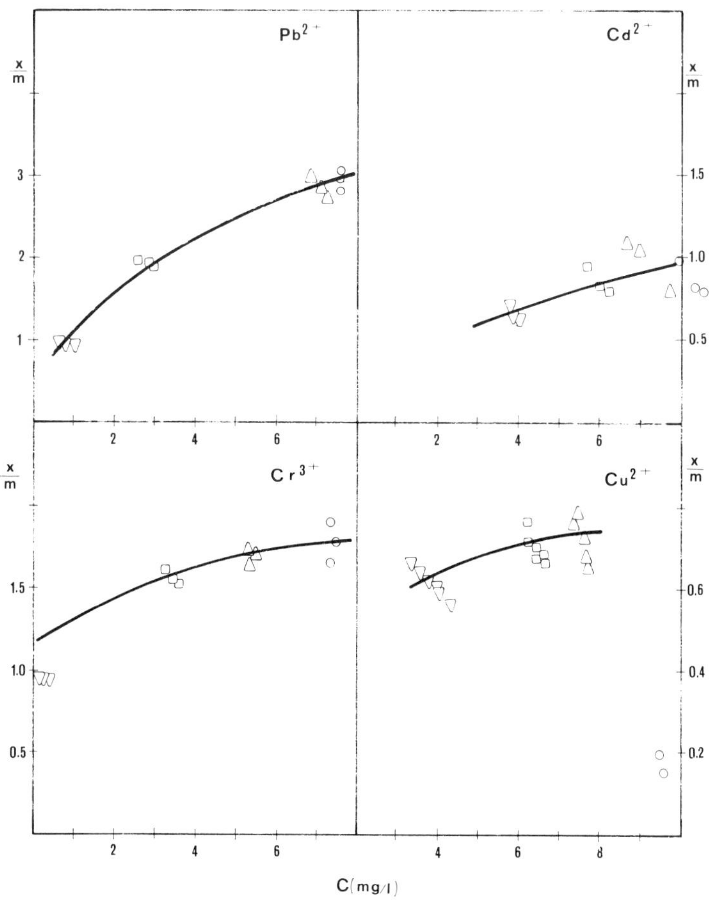

Fig. 1. Experimental equilibrium isotherm relative to Terra Rossa (red soil) from Minervino, Lecce (Italy).

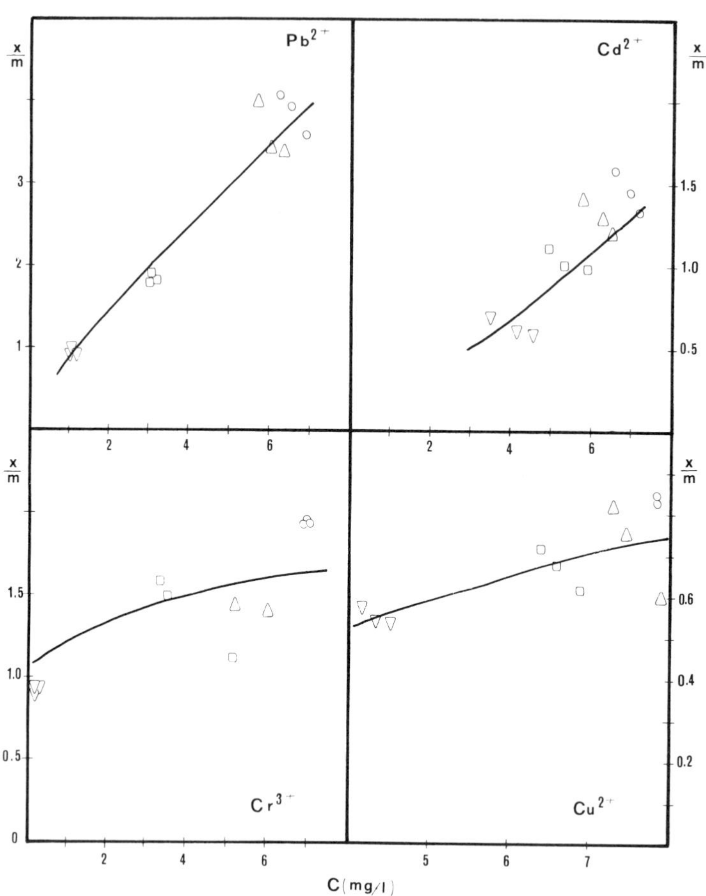

Fig. 2. Experimental equilibrium isotherm relative to Staturo Sand from Altamura, Bari (Italy).

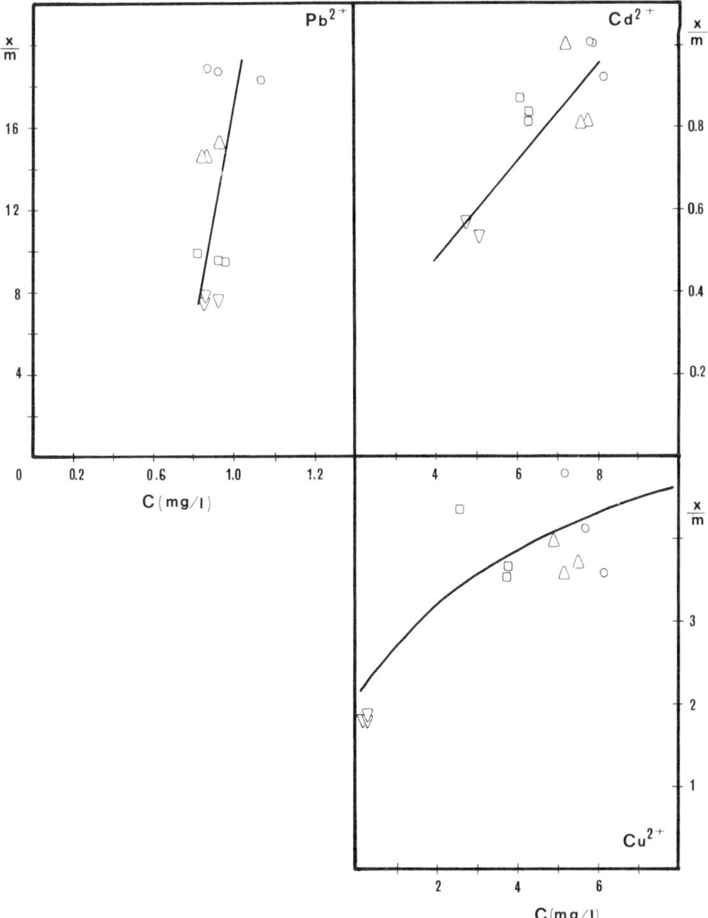

Fig. 3. Experimental equilibrium isotherm relative to Monte Marano Sand from Gravina, Bari (Italy).

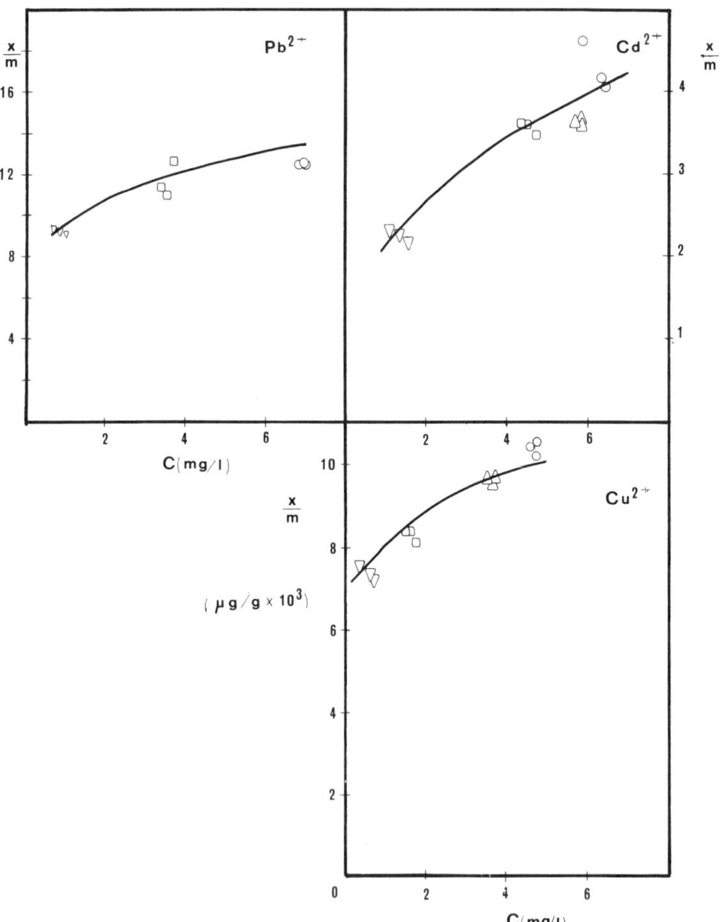

Fig. 4. Experimental equilibrium isotherm relative to Lucera Sand from Lucera, Foggia (Italy).

Table 3. Retardation Factor R, Migration Distance Z and Experimental Equilibrium Isotherms on Different Soils Investigated

Soils	Me^{n+}	R	Z (cm)	Experimental isotherm (Freundlich non-linear)
Terra Rossa	Pb^{2+}	2104	0.30	$S_{Pb} = 1084\ C_{Pb}^{0.5}$
	Cd^{2+}	637	0.99	$S_{Cd} = 394\ C_{Cd}^{0.42}$
	Cr^{3+}	1145	0.55	$S_{Cr} = 1290\ C_{Cr}^{0.16}$
	Cu^{2+}	458	1.38	$S_{Cu} = 492\ C_{Cu}^{0.18}$
Staturo Sand	Pb^{2+}	3775	3.30	$S_{Pb} = 797\ C_{Pb}^{0.82}$
	Cd^{2+}	1487	8.52	$S_{Cd} = 137\ C_{Cd}^{1.18}$
	Cr^{3+}	1253	10.10	$S_{Cr} = 1209\ C_{Cr}^{0.16}$
	Cu^{2+}	618	20.41	$S_{Cu} = 254\ C_{Cu}^{0.53}$
M. Marano Sand	Pb^{2+}	172557	0.60	$S_{Pb} = 13538\ C_{Pb}^{1.34}$
	Cd^{2+}	7109	14.80	$S_{Cd} = 122\ C_{Cd}$ (*)
	Cu^{2+}	2822	37.30	$S_{Cu} = 2527\ C_{Cu}^{0.25}$
Lucera Clay	Pb^{2+}	10700	0.04	$S_{Pb} = 9740\ C_{Pb}^{0.16}$
	Cd^{2+}	3572	0.10	$S_{Cd} = 2099\ C_{Cd}^{0.35}$
	Cu^{2+}	8519	0.05	$S_{Cu} = 8119\ C_{Cu}^{0.14}$

(*) eq. Freundlich (linear).

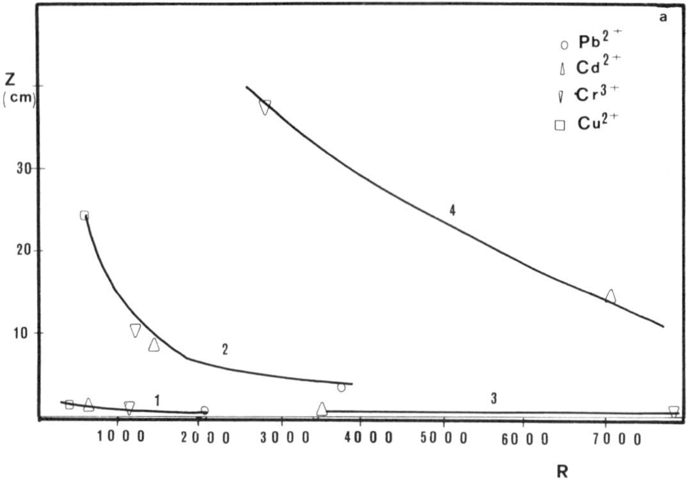

Fig. 5. Migration distance Z vs. retardation factor R for different soils investigated.

1: Terra Rossa (red soil);

2: Monte Marano Sand;

3: Lucera Clay;

4: Staturo Sand.

Generally speaking, migration distances of ions in the different soils are inversely proportional to the retardation factors over the range 500 to 3000; for values of R greater than 6000, no significant reduction of migration of Cu^{2+} on either Monte. Marano or Altamura sands, the calculated migration distances after ten years (37 and 70 cm, respectively) are still significantly less than the 100 cm enforced by Italian legislation. [6]

CONCLUSION

This work confirms that the required 100 cm of free material between the base of the landfill and the maximum excursion of the freatic level of the aquifer is adequate for groundwater quality protection, at least for the range of soils and heavy metal ions investigated.

From the experimental point of view, the characterization of liquid-solid interactions by means of equilibrium studies of this type appears to provide a relatively straightforward and versatile method of estimating the attenuation properties of soils for pollutants. However, it is necessary to remember that in real life, particularly in underground systems, the problems are much more complex. Besides, showing pedological inhomogeneities and structural variation in the components of the solid phase may well be variation, in both space and time, of the measured parameters, anything more, an approximation to the environmental behavior of a real systems.

ACKNOWLEDGEMENTS

The authors are indebted to Prof. Michele Maggiore, Department of Geophysics, University of Bari, for supplying the soil and subsoil samples.

Technical assistance of Mr Luigi Valente is gratefully acknowledged.

REFERENCES

[1] Consorzio "Villoresi"; Artificial recharge of the aquifer in the Milan area. Appendix V, Report 8, ICID Symposium 1978.

[2] Argo, D.G., Mc Cline, N.M., Groundwater recharge operations at Water Factory 21, Orange County, CA. Ch. 12 in T. Asano Ed. "Artificial Recharge of Groundwater" Butterworths Pub. Co., London, 1985.

[3] Idelovich, E., Michail, M., Groundwater recharge for wastewater reuse in the Dan's Region Project: Summary of five years experience, 1977-81; Ch. 15 in T. Asano Ed. "Artificial Recharge of Groundwater" Butterworths Pub. Co., London, 1985.

[4] Bouwer, H., Rice, R.C., Renovation of wastewater at 23rd Ave. Rapid infiltration project, J.W.P.C.F. 56, 1 76-83, 1983.

[5] Italian Law n° 319, 10.05.1976 "Norme per la tutela delle acque dall'inquinamento".

[6] D.P.R. 915/76 of Republic of Italy. "Regolamentazione in materia di smaltimento di rifiuti solidi".

[7] Scheunert, I., Klein, W., Predicting the movement of chemicals between environmental comparts, In P. Sheehan Ed. "Appraisal of tests to predict environmental behaviour of chemicals" J. Wiley and Sons, 1985.

[8] Bear, J., Dynamics of fluids in porous media, Elsevier Pub. Co. N.Y., 1972.

[9] Crank, J., The mathematics of diffusion, Clarendon Press, Oxford, 1975.

[10] Schwarzemback, R.P., Westall, J., Sorption of hydrophobic trace organic compounds in groundwater systems, Wat. Sci. Technol. 17, 9, 39 - 55, 1985.

[11] Valocchi, A.J., Roberts, P.V., Parks, G.A., Street, R.L., Simulation of the transport of ion exchange solutes using laboratory determined chemical parameter values, Groundwater 19, 6, 600 - 607, 1981.

[12] Rubin, J., Transport of reacting solutes in porous media: Relation between mathematical nature of the problem formulation and chemical nature of reactions, Wat. Resources Res. 19, 1231 - 1238, 1983.

[13] Anderson, A., Rubin, J., Adsorption of inorganics at solid-liquid interface, Ann Arbor Sci. Pub., Ann Arbor MI, 1984.

[14] Roy, W.R., Krapac, I.G., Chou, S.F.J., Griffin, R.A., Batch type adsorption procedures for estimating soil attenuation of chemicals, U.S. EPA Rep. 600, October 1986.

[15] Griffin, R.A., Roy, W.J., Feasibility of land disposal of organic solvents. Preliminary assessment., Rep. Env. Inst. Waste Management., Univ. Alabama, February 1986.

[16] Rao, P.S., Davidson, J.M., Estimation of pesticide retention and transformation parameters required for non-point sources pollution models, In M.R. Overcash Ed. Environmental impact of non-point source pollution, Ann Arbor Sci Pub. Ann Arbor, p.23, 1980.

[17] Water Research Institute, Metodi analitici per i fanghi, Quad. IRSA n° 64, Roma, 1984.

[18] Beretta, G.P., Diffusione degli inquinanti e risanmento di falde compromesse dallo smaltimento di rifiuti solidi, Ingegneria Ambientale 15, 3 - 4 147, 1986.

[19] Danke, J.T., Reardon, E.J., Migration of a contaminant in groundwater at a landfill (V): A case study, J. Hydrol. 63, 109 - 30, 1983.

[20] Egbocka, B.C.E., Cherry, J.A., Forwalden, R.N., Frind, E.O., Migration of a contaminant at a landfill. A case Study (III), J. Hydrol., 63, 51 - 80, 1983.

[21] Hoeks, J., Pollution of soil and groundwater from land disposal of solid wastes, Inst. Land and Water Manag. Res. Tech. Bull. Wageningen, The Netherlands, 1976.

[22] J. Hoeks; The mobility of pollutant in soil and groundwater near waste disposal sites; Ibid. Tech. Bull. 1977.

[23] Hoeks, J., Measures to control groundwater pollution near waste disposal sites, Ibid. Tech. Bull., 1981.

[24] Metodi normalizzati per l'anlisi del suolo, S.I.S.S., Edagricole, Bologna, Italy, Cod. 2674, 1 Ed, 1985.

LIST OF SYMBOLS

v	=	velocity of the fluid in the porous media,
K_{sat}	=	Darcy's permeability coefficient of the porous media,
i	=	Darcy's hydraulic gradient,
D_x	=	dispersion coefficient of the porous media,
D_0	=	liquid-phase molecular diffusion of the pollutant substrate,
J	=	molecular flux of the pollutant substrate,
C	=	liquid-phase concentration of the pollutant substrate,
x	=	linear coordinate,
t	=	time,
S	=	solid-phase concentration of the pollutant substrate,
R	=	retardation factor,
ρ	=	specific gravity of the porous material,
θ	=	volumetric water content (porosity) of the porous material,
Z	=	migration distance of the pollutant substrate in the porous material,
n_ε	=	effective porosity of the porous material.

PHYSICOCHEMICAL TREATMENT OF MUNICIPAL-INDUSTRIAL WASTEWATER

M. ROŠ, B. MEJAČ

Boris Kidrič Institute of Chemistry
Hajdrihova 19, P.O. Box 30
61115 Ljubljana, Yugoslavia

ABSTRACT

A new wastewater treatment plant is planned for the town of Ljubljana. According to the plan, the first phase is designed to solve the problem of primary treatment only; the second phase with a complete treatment with a biological unit is to follow later. Wastewater from pulp and paper industries (the latter is located about 25 km away from the Ljubljana treatment plant) is to be drained through it as well. As the dry summer months result in a severe pollution overloading of the recipient, the investor decided that until the biological unit has been built, the recipient water during the critical periods (extending over 1 month per year at the most) is to be preserved by simultaneous primary and physicochemical treatment.

Physicochemical treatment of municipal wastewater and combined municipal and industrial wastewater have been studied. It has been found that by adding 50 g/m^3 of $Al_2(SO_4)_3$ and 1 g/m^3 of nonionic polyelectrolyte after neutralization with $Ca(OH)_2$, a 60-63% decrease of pollution, as COD (the value of COD = 500 - 900 mg/L), can be achieved. With the same additives, treatment of combined municipal and manufacturing wastewaters is only 32-36% effective according to COD value; The resulting sludge was very difficult to settle, and the supernatant was turbid. For complete clarification 150-175 mg/L $Al_2(SO_4)_3$ and $Ca(OH)_2$ with 1 mg/L of nonionic polyelectrolyte has to be added: In this case 45% of COD efficiency is reached.

INTRODUCTION

Municipal wastewater treatment is relatively rarely done by way of chemical and physicochemical methods, because biological treatment is generally more suitable and efficient. However, there are some specific cases and conditions in which coagulation and flocculation stand as sensible alternatives or supplements to biological purification [1, 2] provided that the municipal water in question contains mostly undissolved suspended solids and dispersed matter [3, 4]. Nevertheless, this does not apply to those municipal wastewaters with adjoined larger quantities of industrial wastewater containing mostly dissolved substances [5].

The municipal wastewater of Ljubljana, a city with approximately 300,000 inhabitants, proved to be composed of a considerable volume of dispersed and emulated wastes, drained in by either domestic wastewater or a large number of adjoined industrial wastewaters.

PROBLEM STATEMENT

During the first stage of the two-phase construction of the central wastewater treatment plant only mechanical treatment will be carried out by way of settlers, thickeners, and anaerobic stabilization of primary sludge. With regard to specific composition of the treated wastewater, it is possible to reduce its organic load by by sedimentation by approximately 35% in the case of COD bases and by approximately 25% in the case of BOD_5; nevertheless, a large part of its existing organic load appears in such a form that it may be removed by coagulation and flocculation.

The untreated wastewater, which is being drained into the Ljubljanica river and further on into the Sava river, has a very negative impact on the two recipients. The state of condition will not improve considerably until after the first stage of construction of the central wastewater treatment plant has been completed. The fact remains that the two recipients exhibit considerable self-purification capacity; nonetheless, this holds true only duing suitable hydrologic conditions prevailing for a short period of a year. On the other hand, the extremely unsuitable water conditions, statistically proved to last for approximately 1 month a year, make the situation alarming. Attempts to overcome this problem by various measures have proved unsuccessful (aeration of the recipient, production decrease in the case of excessive polluters, to mention just a few). Until the secondary stage of the wastewater treatment plant has been completed, the problem should be restrained by additional chemical treatment, i.e., by a combination of primary sedimentation by coagulation and flocculation.

Table 1. Hydrologic Conditions of the Two Recipients (Rounded-off Values)

		The Ljubljanica river	The Sava river
average flow rate per year	m^3/s	57	99
average low flow rate	m^3/s	10	26
lowest low flow rate	m^3/s	4.5	14
highest flow rate (100 years)	m^3/s	358	1610

Table 1 bellow gives hydrologic conditions of the Sava river and the Ljubljanica river, which receive the municipal wastewater of the city. Fig. 1 provides a scheme of the city map indicating the main polluters, the anticipated location of the wastewater treatment plant, and the two rivers the city pollutes.

The wastewater treatment plant will receive and treat two kinds of pollution. The first load of wastes originates in the central part of the city of Ljubljana, with accompanying industrial facilities, and the second load in the town of Medvode, which is located approximately 25 km upstream from the site of the anticipated plant. Municipal wastewaters will be drained into the existing central sewage pipeline. The most important organic waste in Medvode originates in pulp and paper industry manufacturing wastewater. Table 2 shows the main pollution parameters of the two cities' wastes and total waste rate to load the central wastewater treatment plant.

One of the major reasons for such a disposition, i.e., for the collective treatment of the municipal wastewater of Ljubljana and combined wastewaters of Medvode (municipal and the pulp and paper industry wastewater), is to preserve the water quality of the Sava river in this region, for it is the main feeding source of ground water, which is one of the sources of the city's drinking water supply.

The aim of this assignment was to test the possibilities and efficiency of physicochemical treatment of municipal wastewater either with adjoined pulp and paper industry wastewater or without it.

MATERIALS AND METHODS

The preliminary tests were carried out by a standardized device called a jar test. Coagulation lasted for 5 to 10 minutes at 80 RPM, and flocculation

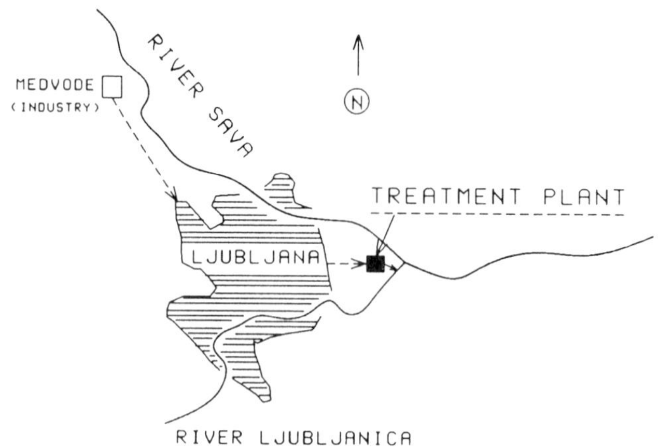

Fig. 1. The river basin scheme with main pollution sources and location of the wastewater treatment plant.

Table 2. Flow Rates and Waste Load for the Two Main Collectors

Wastewater	Q (m^3/d)	COD (kg/d)	BOD_5 (kg/d)
Collector Ljubljana (D) (town + industry)	135000	115000	50000
Collector Medvode (I) (industry predominantly)	16000	60000	13000
Combined (total) (T)	151000	175000	63000

was kept on for 15 to 20 minutes at 40 RPM. The laboratory tests were done by means of a completely mixed reactor with spade mixer; coagulation lasted 5 minutes at 130 RPM, whereas flocculation continued for 15 minutes at 15 RPM.

The municipal wastewater (D) was collected at the entry of the future wastewater treatment plant, and the pulp and paper wastewater (I) was collected at the point where it is drained off the factory.

The combined type of water (T) for the test purposes was synthetically prepared in a laboratory by mixing the two wastewater samples in the same ratio as shown in Table 2.

Standard inorganic coagulants, i.e., $FeCl_3$ in a 42% solution, $Al_2(SO_4)_3.18H_2O$ and $FeSO_4.7H_2O$, were used, and in order to adjust the pH, NaOH and $Ca(OH)_2$ were applied. All data regarding the quantity of the additives are expressed as anhydrous salts. All three kinds of polyelectrolytes, the cationic, the nonionic and the anionic were tested; we applied an anionic Lamfloc 7985, a cationic Lamfloc 818, and a nonionic Lamfloc 3085 (Fratelli Lamberti, Lambratte, Italy). The COD tests were carried out according to the Standard Methods [6].

RESULTS AND DISCUSSION

We made use of elimination jar tests to determine the quantity and the appropriate kind of the necessary chemical additives, the pH level, and the removal efficiency. We also determined appearance quality of the treated wastewater, the size and consistency of flocks and measured the settling velocity. Each test resulting in a clarified or at least opalescent supernatant liquid was considered successful.

In treating the municipal wastewater, the applicabilities of aluminium sulphate and ferric chloride generally proved equivalent. Either of the two chemical additives was to be applied to the wastewater with COD ranging from 600 to 900 mg/L, namely, in the quantity ranging from 75 to 90 mg/L of the tested wastewater. Approximately 50 mg/L of the sodium hydroxide was added in order to adjust the pH level. The resulting flocks settled very slowly and incompletely. However, by adding only 1 mg/L of polyelectrolyte, the settling velocity noticeably increased and the flocks turned out to be larger and more consistent, with the supernatant liquid being clarified. The removal efficiency rate amounted to 71 - 77% (considering removal of settling and non-settling suspended solids). Most favorable results were achieved by using the nonionic type of polyelectrolyte, followed by cationic and anionic types.

The tests where calcium hydroxide served as a neutralising agent did give inferior results of removal efficiency; nevertheless, the wastewater did not require polyelectrolyte to be clarified. In this case the flocks were relatively small, and the removal efficiency ranged between 60 and 62%. By adding 1 mg/L of nonionic polyelectrolyte the removal efficiency improved to 62 - 68%, the resulting flocks being easily settled and more consistent. Application of other kinds of polyelectrolyte proved less efficient. The tests where ferrous sulphate and calcium hydroxide were used did give the analogue results, but this procedure required oxidation of Fe^{2+} into Fe^{3+} by means of an air blowing process. Chemical additives were approximately similar and the settling velocity increased if the nonionic type of polyelectrolyte was was added. The resulting removal efficiency was 60 to 62%.

Treatment of the combined wastewater (T), which comprised a considerable part of matter originating from pulp production (lignin sulphonates) differed from treatment of the previously described wastewater sample (D) in many aspects. Namely, at least twice as much coagulant and flocculant was added to get the water sample clarified, and the removal efficiency, expressed by the COD parameter, was considerably lower. Quantity of the additive was increased due to dispersed properties of the calcium lignin sulphonates, whose zeta-potential is -49 mV. Lower removal efficiency (at already clarified supernatant) was caused by a relatively smaller portion of suspended solids in the combined wastewater sample (T), based on the fact that the organic load appearing in the manufacturing wastewater is mostly dissolved.

Aluminium sulphate and ferric chloride proved to be the most suitable inorganic coagulants, whereas the results obtained with ferrous sulphate were inferior and it was necessary to add polyelectrolyte. The best results were obtained with nonionic types of polyelectrolytes, which increased the size of the resulting flocks and their settling velocity.

To clarify the water sample whose COD ranged from 1000 to 1300 mg/L, the addition of 150 - 175 mg/L of the aluminum sulphate was required, whereas the pH level adjustment was maintained by adding 150 - 175 mg/L of calcium hydroxide. The achieved removal efficiency was approximately 45%. By way of a mass balance sheet it could be concluded that in this case a portion of dissolved lignin sulphonates was also removed. If the quantity of aluminum sulphate and calcium hydroxide was 100 - 125 g/L, the total removal efficiency was approximately 45.9%, the supernatant being turbid. With the addition of only 75-90 mg/L the removal efficiency was reduced to approximately 35% (COD basis).

Tests in a Completely Mixed Reactor

The main purpose of these tests was to determine the quantity of the settled solids resulting from the clarifying process and, their settling and thickening capacity. We used the two chemical substances that have proved the most suitable in the previous jar tests, i.e., aluminum sulphate and calcium hydroxide. We found a considerable statistical scatter of the results due to the fact that an equal COD value does not necessarily imply that the distribution of the removable suspended solids and dissolved substances in raw wastewater is equal as well. Nevertheless, the removal efficiency results did not essentially digress from those of the jar tests.

The results are collected in Table 3. To allow comparison, it includes the results of the primary settling (mechanical treatment), i.e., without coagulation and flocculation. Curves given in Fig. 2 show the volume of the

Table 3. Results of Treatment of Individual Kinds of Wastewaters (D in T)

	Wastewater D					Wastewater T			
Raw wastewater									
COD (mg/L)	637.	637.	637.	1170.	1170.	1015.	1015.	1074.	1074.
pH	7.6	7.6	7.6	7.4	7.4	7.2	7.2	7.0	7.0
Additives (mg/L):									
Al-sulphate	-.	75.	90.	-.	150.	175.	175.	150.	175.
Ca-hydroxide	-.	75.	90.	-.	150.	175.	175.	150.	175.
Polyelectrolyte	-.	1.	1.	-.	1.	1.	1.	-.	-.
Treated water:									
COD (mg/L)	417.	185.	162.	947.	616.	555.	549.	714.	670.
pH	7.7	8.0	7.8	7.4	7.8	7.5	7.8	7.4	7.4
Removal effic (%)	34.5	71.0	74.6	19.1	47.3	45.3	45.9	39.2	42.9

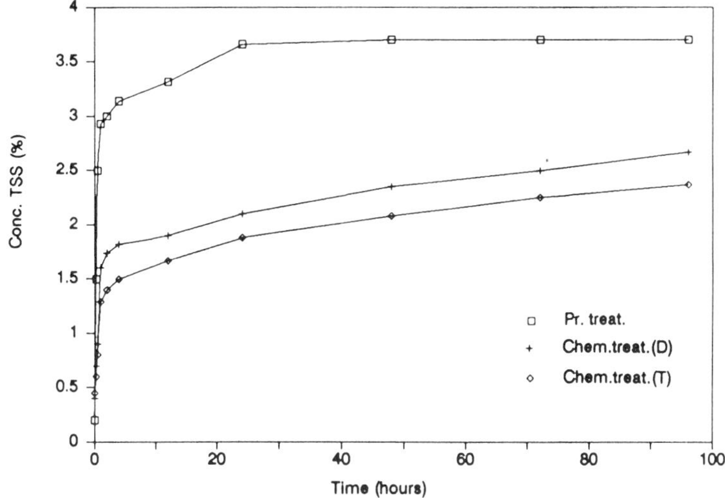

Fig. 2. Settled sludge volume.

settled sludge in mL per L of the treated water, whereas the curves in Fig. 3 show the quantity of TSS in sludge (expressed in %).

CONCLUSIONS

The two-phase construction of a new central wastewater treatment plant is to be in Ljubljana, a town with approximately 300,000 inhabitants. In addition to its municipal wastewater with accompanying industrial facilities, the wastewater of Medvode (a town with several thousand inhabitants),

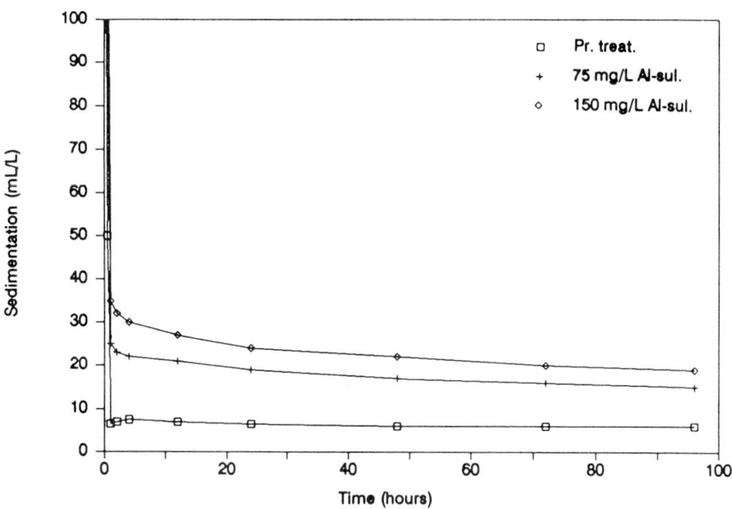

Fig. 3. Concentration of TSS in sludge during settling and thickening (expressed in %).

whose central sewage pipeline is loaded with large quantities of pulp and paper wastewater, will feed the plant. The first stage of the central wastewater treatment plant anticipates mechanical treatment only; nonetheless, during the extremely unfavorable hydrologic conditions of the recipient (lasting for approximately 1 month a year), its wastewater should be additionally physicochemically treated by flocculation.

In our laboratories we have carried out all the necessary tests on municipal wastewater of Ljubljana and on the same water with adjoined wastewater of Medvode. We found out that physicochemical treatment of the municipal wastewater of Ljubljana, the samples of which were taken from the existing central sewage pipeline (wastewater D), functioning as pretreatment during extremely unfavorable hydrologic conditions of the two recipients, was acceptable, for the waste rate was decreased by about 75%. A result of such treatment is formation of voluminous sludge, the volume of which is 2.5 to 3 times larger than that resulting from the primary treatment alone. The combined wastewater (wastewater T), comprising lignin sulphonates and other substances originating in the pulp and paper industry wastewater (wastewater I), requires excessive quantities of chemical additives, causing a 3 to 3.5 times larger quantities of sludge. The removal efficiency rate (COD basis) does not exceed 45%, even though the suspended solids are removed. Therefore we suggest that the pulp and paper wastes be treated separately.

REFERENCES

[1] Balmer, P.: Experience with the precipitation of municipal sewage with lime. Publ. Chalmers Tek. Hoegsk. Geteborg Inst. Vatenfoersoeynings Avloppstek **3**, 1984, 84 (abstr.): Chem. Abstr. **105**, 120100r, 1986.

[2] Karlson, J.: Chemical precipitation: Alternative or complement to biological treatment. Schriftenr. Ver. Wasser-Boden-Lufthyg, **62**, (1985), 277 (abstr.): Chem. Abstr. (1986), **104**, 10068h.

[3] Ademoroti, C.M.A.: Water purification by fluidized bed technique. Wat. Res., **20**, (1986), 1105-1109.

[4] Farooq, S. and Bari, A.: Physicochemical treatment of domestic wastewaters. Environ. Technol. Letters, **7**, (1986), 87-90.

[5] Rehbun, M. and Streir, S.: Physicochemical treatment of strong municipal wastewater. Wat. Res., **8**, (1977), 195-201.

[6] Standard Methods for the Examination of Water and Wastewater (1985), 16th ed., APHA, AWWA WPCF, Washington.

ACIDOGENIC FERMENTATION OF THE ORGANIC FRACTION OF MUNICIPAL SOLID WASTES

M. BECCARI, L. CAMPANELLA, M. MAJONE,
E. ROLLE and O. TODINI

Department of Chemistry
University "La Sapienza"
P. le A. Moro 5, 00185, Rome, Italy

ABSTRACT

The production of volatile fatty acids ($C_2 - C_6$) by acidogenic fermentation of the organic fraction of municipal solid wastes was investigated in batch tests. Substrate conversion was about 20% but a mild alkaline pretreatment doubled it. Volatile fatty acids were practically the only product: acetic acid was generally prevalent but higher acids always represented more than 50% of the total yield. In addition, characteristic kinetic constants were calculated by batch experimentation. A preliminary process modelling is also presented.

INTRODUCTION

Increasing shortage of fossil fuels has stimulated great interest in the production of liquid fuels and of valuable chemicals from ubiquitous and renewable feedstocks by biotechnological processes. In this context, the acidogenic fermentation of the organic fraction of municipal solid wastes may be an attractive alternative for treating wastes before disposal as well as for recovering valuable products from them. In fact, through this process volatile fatty acids (VFAs) may be produced. VFAs are recoverable by solvent extraction and then they can be decarboxylated and dimerized by electrochemical processes to obtain alkanes and olefins. Estrification using novel techniques could be another interesting method for obtaining valuable chemicals from these acids [1, 2, 3].

The primary objectives of the work were:

– to assess the biotreatability of this substrate by determining the yields of its bioconversion to VFAs and the related distribution of the various acids, both with and without a previous mild alkaline treatment;

– to analyze the kinetics of the process by determining which equations, plus their related constants, may describe rates for biomass growth, substrate consumption and product formation;

– to model the process and to simulate its performance mathematically in a continuous-flow stirred-tank reactor with recycle (CFSTR) for various operating conditions.

EXPERIMENTAL

The composition of the organic fraction of municipal solid wastes that was used in this work is shown in Table 1. Biotreatability tests were performed directly on this substrate (MSWOF) as well as on a previously treated one (pMSWOF). Pretreatment was carried out [4], under continuous stirring, with a 0.5% w/v NaOH solution (16 ml per gram of substrate) for 72 hours at 25°C, followed by neutralization with dilute HCl. The liquid phase of this pretreated substrate was obtained by a 0.45 μm filtration and was also used for the biotreatability tests. Other biotreatability tests were performed on the liquid phase obtained from treating MSWOF with distilled water, the other conditions being the same as for pretreatment. Furthermore, a glucose solution was used as reference substrate. Blanks were performed with no substrate at all and the obtained acid production was accounted for when calculating true substrate conversion percentages.

The tests were performed in 200 mL assay bottles submerged in thermostatic water baths and periodically stirred; incubation temperature was 25°C. Assay bottles were filled with the seed inoculum, constituted by anaerobically digested sewage sludge (non acclimated), and with the substrate diluted in distilled water in order to obtain the final chosen COD concentration. Tests on glucose, MSWOF and pMSWOF were conducted at 10 gCOD/L (the COD added with the inoculum being 10% of the COD added with the substrate). Tests on the liquid phases from MSWOF and pMSWOF were conducted at 6.0 gCOD/L (the COD added with the inoculum being 28% of

Table 1. MSWOF Characterization

pH	6.82	
water content	45.70	%
on dry basis:		
COD	1.02	gCOD/g
Volatile suspended solids	79.14	%
total N	4.69	g/kg
NH_4-N	0.84	g/kg
organic N	3.85	g/kg
total P	1.94	g/kg

the COD added with the substrate). The anaerobic conditions during the filling operation were maintained by continuously flushing nitrogen gas. No nutrient solution was added. The initial pH ranged from 6 to 7, and tests were performed in the presence of sodium salt of bromoetanosolfonic acid ($5.0 \cdot 10^{-4}$ mol/L) which inhibits methanogenesis [2]. At regular time intervals, samples were collected taking care to maintain anaerobic conditions in test bottles. The samples were centrifuged for a period of 30 minutes at 4000 rpm and the supernatant was filtered on 0.45 μm filters. Both pH and VFAs content were determined on the filtrate. VFAs analysis was performed by gas chromatography with a flame ionization detector on a separation column Supelco SP1200 according to [5].

Kinetic tests were performed in batch only on pMSWOF. Filling and sampling procedures and operating conditions were as stated above except but for the initial COD, which was 6 g/L (the COD added with the inoculum being 20% of the COD added with the substrate). Total COD for both liquid and solid phases was determined by the dichromate titrimetric method, according to [6]. Moreover, in order to measure biomass concentration another aliquot was sampled on which dehydrogenase activity was determined according to [7] and [8].

RESULTS AND DISCUSSION

Biotreatability

Fig. 1 shows MSWOF, pMSWOF and glucose conversion percentages to total VFA vs time. Conversion percentages were calculated by expressing VFAs concentrations in terms of gCOD/L. Conversion yields for MSWOF

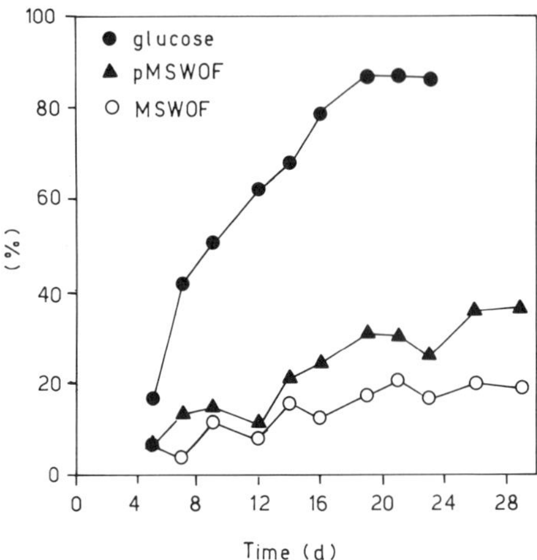

Fig. 1. Conversion percentage to total VFAs vs. time for MSWOF, pMSWOF and glucose.

and pMSWOF are 18 and 35% in a month, respectively. Thus, pretreatment doubles the conversion yield. Glucose fermentation is faster and brings about a high final yield (90%). As shown in Fig. 2, conversion yields obtained by using the soluble fractions from MSWOF and pMSWOF as the only substrate are also high and comparable to glucose conversion. This result shows that conversion to VFAs is actually the only path through which soluble substrate degradation occurs. Moreover, since yields for MSWOF and pMSWOF are higher than the initial soluble fraction of both substrates (10 and 17%, respectively), hydrolysis of the particulate substrate contributes to acid production.

Table 2 shows the distribution of the various acids (expressed as percentages of the total yield at the highest conversion) for each substrate. Acid distribution for MSWOF and pMSWOF is very similar: in both cases acetic acid is prevalent but the percentage of higher acids (more easily recoverable in the next phase of the process) is about 50%. Acid distribution for the soluble fractions from both substrates is better as C_4 and higher acids turn out to be more than 50% of the total. This different behavior could be the result of a higher partial pressure of H_2 in the fermentation of the soluble substrates. As hydrogen is a side product of the fermentation step, higher and quicker conversion produces higher partial pressure which orients acid productions to the highest acids [9]. During fermentation pH decreases to values within the range 5.5 – 6.0. Reproducibility of replicates for both total yields (see Fig. 3) and acid distribution is fair enough.

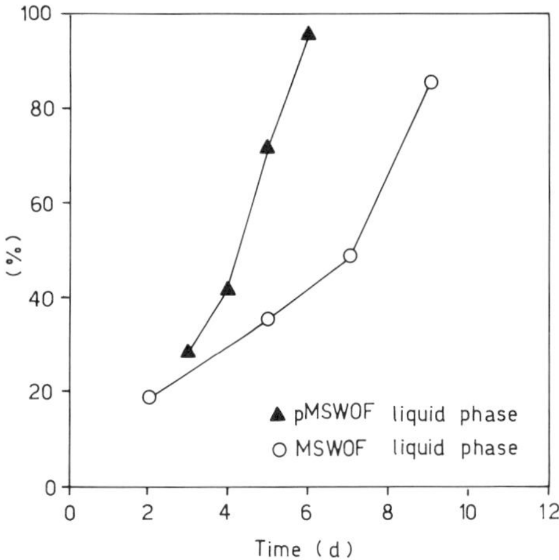

Fig. 2. Conversion percentage to total VFAs vs. time for MSWOF, and pMSWOF liquid phase.

Kinetic Analysis

Since pretreatment had been effective for increasing the conversion yield, kinetic tests were performed by using only pMSWOF as substrate. In Table 3, solid phase and liquid phase total concentrations and biomass and VFA concentrations are shown as functions of time. According to the kinetic modelling suggested by Eastman and Ferguson [10], concentrations for all the components of the system were measured directly or converted in terms of COD so that both reactants and products can be compared directly. Soluble and particulate substrate concentrations are then calculated by subtraction. From Table 3, the total COD concentration of the system turns out to be constant during the test, which indicates strict anaerobic conditions and negligibility of gaseous products in the material balance. Figures 4 and 5, based on the data in Table 3, show that:

– the biomass remains in a lag phase for about two days, during which the soluble substrate is consumed for the adjustment of the enzymatic pool, then it starts to grow in a typical Monod way;

– the particulate substrate decreases slowly and regularly in time;

– the soluble substrate is quickly consumed and after 5 – 6 days it presents a low, stable concentration. In the meantime, acid production goes on. This indicates that the biomass can use the soluble substrate at a rate at least equal to that of particulate substrate hydrolysis.

The best interpretation of the experimental data regarding particulate

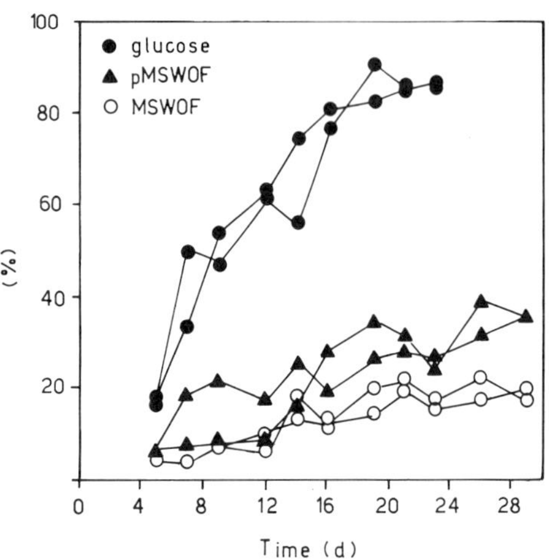

Fig. 3. Reproducibility of conversion percentage tototal VFAs vs. time for MSWOF, pMSWOF and glucose.

Table 2. Acid Distibution in Total VFAs Production (at the Highest Conversion for Each Substrate)

Acid distribution (% of total VFAs)	MSWOF	pMSWOF	MSWOF liquid phase	pMSWOF Liquid phase
acetic	48	48	29	15
propionic	39	30	17	18
i-butyric	–	–	–	10
butyric	8	14	34	18
i-valeric	1	6	1	11
valeric	2	2	9	15
caproic	2	–	10	13

hydrolysis is obtained by assuming a kinetic expression as:

$$-\frac{dF}{dt} = K_h F \, (^*) \qquad (1)$$

(* an alphabetically ordered list of symbols is reported below.)

The hydrolysis constant K_h has been calculated by the linear regression of experimental data according to the integration of (1). A value of $K_h = 5.2 \cdot 10^{-4}\ h^{-1}$ was obtained with a linear correlation coefficient of 0.971. Since K_h is very small a linear dependence on time could also be used. Other different forms for (1) do not improve the obtained correla-

Table 3. Experimental Results from the Kinetic Test on pMSWOR

time (h)	total concentration (gCOD/L)			concentration (gCOD/L)			
	liquid phase	solid phase	total	biomass	soluble substrate	particulate substrate	VFAs
19.1	1.24	6.05	7.29	1.17	1.20	4.88	0.04
49.9	1.27	6.23	7.50	1.20	1.01	5.03	0.26
69.9	0.94	6.63	7.57	1.62	0.58	5.01	0.36
93.9	0.75	6.65	7.40	1.80	0.30	4.85	0.45
115.1	0.65	6.75	7.40	—	0.14	—	0.51
140.9	0.67	6.81	7.48	1.88	0.06	4.93	0.61
165.4	0.65	6.75	7.40	1.91	0.02	4.84	0.63
187.7	0.79	6.69	7.48	1.96	0.09	4.73	0.70
236.1	0.87	6.59	7.46	1.98	0.09	4.61	0.78
264.4	0.91	6.67	7.58	2.01	0.11	4.66	0.80
285.2	0.89	6.45	7.34	2.04	0.09	4.41	0.80
310.3	0.95	6.43	7.38	2.14	0.13	4.29	0.82
333.0	0.97	6.43	7.40	2.09	0.00	4.34	0.97
354.0	1.04	6.42	7.46	2.11	0.07	4.31	0.97
402.2	1.00	6.30	7.30	2.09	0.01	4.21	0.99
426.6	1.05	6.36	7.41	2.14	0.00	4.22	1.05

tion. Furthermore this expression shows that the hydrolysis rate does not depend on biomass activity. A blank experiment, performed on pMSWOF in distilled water without any biomass, showed that hydrolysis proceeds at a similar rate (K_h resulted $4.2 \cdot 10^{-4}$ h^{-1}); thus biomass influence on hydrolysis can actually be considered negligible.

To interpret the biomass growth, soluble substrate decrease, and VFA production, the following expressions can be assumed:

$$\frac{dX}{dt} = \mu X \qquad \mu = q\frac{S}{K+S} \tag{2}$$

$$-\frac{dS}{dt} = \frac{\mu}{Y}X + \frac{dF}{dt} \tag{3}$$

$$\frac{dP}{dt} = \alpha(\frac{1}{Y} - 1)\mu X \tag{4}$$

from which q, K, Y and α are to be calculated.

The mathematical form of the previous expressions allows calculation with the integral method for Y and α, while the differential method is to be applied for calculating q and K.

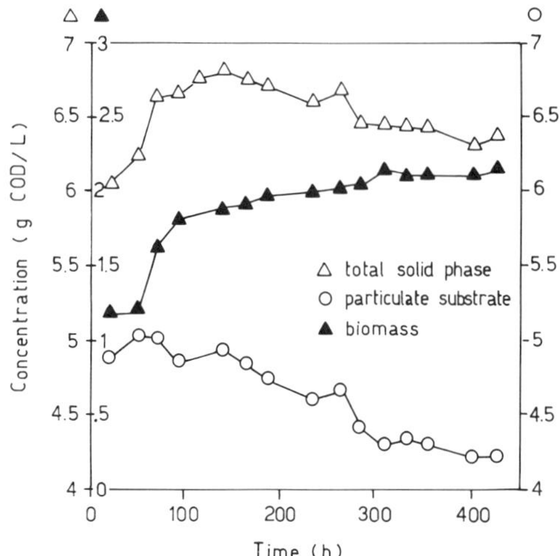

Fig. 4. Kinetic test on pMSWOF: total concentration of solid phase and concentrations of biomass and particulate substrate vs. time.

By introducing (3) into (2) and by integrating:

$$X = X_0 + Y[(S_0 - S) + (F_0 - F)] \tag{5}$$

Linear regression of the experimental data according to (5) gives $Y = 0.46$, with a correlation coefficient of 0.974. In the same way, by introducing (3) into (4) and by integrating:

$$P = P_0 + \alpha(1 - Y)[(S_0 - S) + (F_0 - F)] \tag{6}$$

from which linear regression gives $\alpha = 0.89$, with a correlation coefficient of 0.969. Such a value indicates that VFA production accounts for about 90% of the total formed product. The remaining 10% should be gaseous products (i.e., molecular hydrogen) which were not measured experimentally. In terms of overall material balance, they correspond to a negligible 2% of the total COD.

For the differential analysis of the experimental data, (3) can be transformed into:

$$\frac{X}{\frac{dX}{dt}} = \frac{K}{q}\frac{1}{S} + \frac{1}{q} \tag{7}$$

Linear regression of the experimental data according to [7] gives $K = 2.83$ gCOD/L and $q = 2.7 \cdot 10^{-2}\ \mathrm{h}^{-1}$, with a correlation coefficient of 0.978. Values obtained for the constants lie within the range usually found for the acidogenic phase in anaerobic processes.

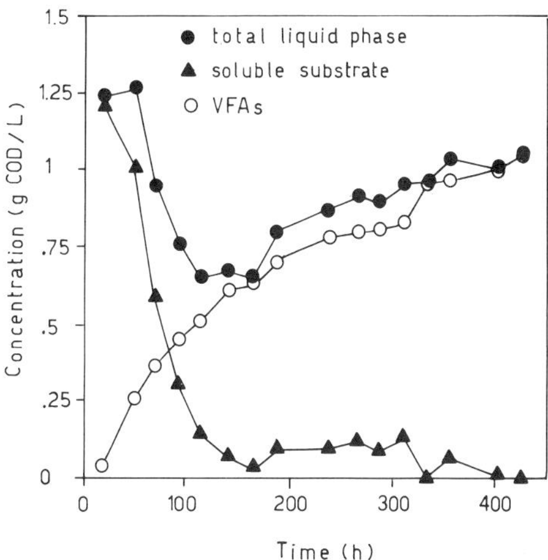

Fig. 5. Kinetic test on pMSWOF: total concentration of liquid phase and concentrations of VFAs and soluble substrate vs. time.

By assuming the expressions (1), (2), (3) and (4) and the derived constants, process performance can be calculated for a continuous-flow stirred-tank reactor with recycle (Fig. 6).

The calculation of the concentration inside the reactor for all process variables (F, X, S, P) was performed as a function of the recycle ratio within the range 0.0 – 0.5, and of the hydraulic residence time (HRT) within the range 4 – 40 days (2 days being the wash-out time with no recycle). Influent concentrations for particulate and soluble substrate were assumed to be 25 and 5 gCOD/L, respectively. The chosen thickening coefficient C was 2.5. As an example, Figure 7 shows the conversion percentage of the particulate substrates vs. HRT. The soluble substrate concentration is always very low and product concentration varies from 2 to 10 g/L.

CONCLUSIONS

Acidogenic fermentation of MSWOF shows a good yield of VFAs (90% of total product), but a low substrate conversion (about 20% of initial COD). A mild alkaline pretreatment at room temperature doubles conversion, and hence acid production, making the process more interesting for industrial application. As the soluble fraction of the substrate is completely converted, hydrolysis of the particulate controls the performance of the overall process.

The batch kinetic study shows that the hydrolysis rate is well fitted

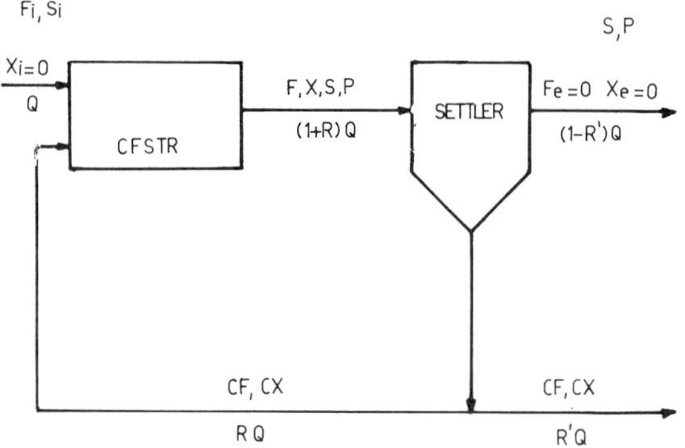

Fig. 6. Flowsheet of CFSTR with recycle.

by a first-order expression with respect to the particulate substrate with no evident dependence on bacterial activity. Hydrolysis rate constant K_h is low, confirming the importance of this step in the process also from a kinetic point of view. Other constants in the kinetic model lie within the normal range for anaerobic processes.

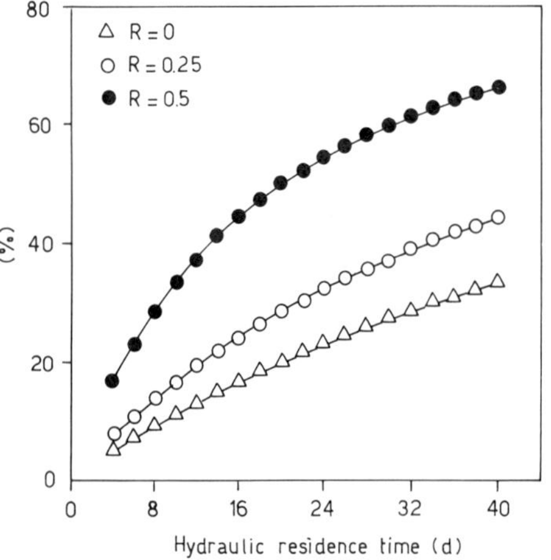

Fig. 7. Particulate substrate conversion percentage vs. hydraulic residence time in a CFSTR at different recycle ratios.

A kinetic modelling allows a simulation of process performance for a continuous-flow stirred-tank reactor with recycle under different operating conditions. With such a model a complete economical and technical assessment will be made after further research has allowed improvement of conversion yield by optimizing substrate pretreatment, and oriented acid production to the highest acids by controlling hydrogen partial pressure in the gaseous environment of the process.

ACKNOWLEDGEMENTS

The research was supported by a Ministry of Education grant (National Research Project).

LIST OF SYMBOLS

F - particulate substrate concentration (ML^{-3}),
K - Michaelis-Menten half velocity coefficient (ML^{-3}),
K_h - hydrolysis rate constant (T^{-1}),
P - total VFAs concentration (ML^{-3}),
q - maximum specific rate of biomass growth (T^{-1}),
Q - volumetric flow rate (L^3T^{-1}),
R - recycle ratio (L^3T^{-1}/L^3T^{-1}),
R' - waste recycle ratio (L^3T^{-1}/L^3T^{-1}),
S - soluble substrate concentration (ML^{-3}),
t - time (T),
Y - net biomass yield coefficient (MM^{-1}),
X - biomass concentration (ML^{-3}),
α - VFAs yield coefficient (MM^{-1}),
μ - specific biomass growth rate (T^{-1}).
Subscripts:
$_0$ - initial time,
$_i$ - influent,
$_e$ - effluent.

REFERENCES

[1] Levy, P.F., Sanderson, J.E., Kispert, R.G. and Wise, D.L., Biorefining of biomass to liquid fuels and organic chemicals, Enzyme Microb. Technol., 3, 207-215, 1981.

[2] Levy, P.F., Sanderson, J.E. and Wise, D.L., Development of a process for production of liquid fuels from biomass, Biotechnol. and Bioeng. Symp. n. 11, 239-248, 1981.

[3] Wise, D.L., Leuschener, A.P. and Levy P.F., Suppressed methane fermentation of selected industrial wastes: a biologically mediated process for conversion of whey to liquid fuel, Developments in Industrial Microbiology, 26, 194-207, 1985.

[4] Datta, R., Acidogenic fermentation of corn stover, Biotechnol. and Bioeng., 23, 61-77, 1981.

[5] Ottenstein, D.M. and Bartley, D.A., Improved gas chromatography separation of free acids C_2-C_5 in dilute solution, Anal. Chem., 7, 952-955, 1971.

[6] Istituto di Ricerca Sulle Acque, Quad. Ist. Ric. Acque, 64, Metodi analitici per i fanghi, vol. 3, Parametri chimico-fisici, Consiglio Nazionale delle Ricerche, Roma 1985, in italian.

[7] Lenhard, G., A standardized procedure for the determination of dehydrogenase activity in samples from anaerobic treatment systems, Wat. Res., 2, 161-167, 1968.

[8] Klapwijk, A., Drent, J. and Steenvoorden J.H.A.M., A modified procedure for the TTC-dehydrogenase test in activated-sludge, Wat. Res., 8, 121-125, 1974.

[9] Mc Inerney, M.J. and Briant, M.P., Reviev of methane fermentation fundamentals in Wise D.L., Fuel gas production from biomass, 1 CRC Press Inc., 19-46, 1981.

[10] Eastman, J.A. and Ferguson, J.F., Solubilizatiom of particulate organic carbon during the acid phase of anaerobic digestion, J. WPCF, 53 (3), 352-366, 1981.

RECENT ADVANCES IN OZONE TREATMENT OF DRINKING WATER

R. G. RICE

Rice International Consulting Enterprises
1331 Patuxent Drive, Ashton, MD 20861, USA

ABSTRACT

Ozonation has been used to treat drinking water since 1906 starting in Nice, France. Today, more than 3 000 plants are using ozone, for a variety of purposes throughout the world. After a brief review of the classical disinfection and oxidation uses of ozone in drinking water treatment, some of the newer applications of ozone will be discussed. These include the use of multiple stages of ozonation, the once-through use of oxygen to produce ozone, new concepts in disinfection of cysts, viruses and bacteria (Ct values), Advanced Oxidation Processes (ozone coupled with UV radiation, hydrogen peroxide, or elevated pH), and the impacts of production of Assimilable Organic Carbon (AOC) on downstream processing.

INTRODUCTION AND HISTORICAL DEVELOPMENT

In 1906, the city Nice, France installed ozonation to disinfect raw water supplies coming into this community. By 1916, 49 water treatment plants had adopted ozonation, mostly in France, and by 1940, this number had grown to 114. By 1977, at least 2 000 plants had installed ozone facilities. Today, at least 3 000 plants are known to be using ozone for one or more

Chemistry for the Protection of the Environment
Edited by L. Pawlowski *et al.*, Plenum Press, New York, 1991

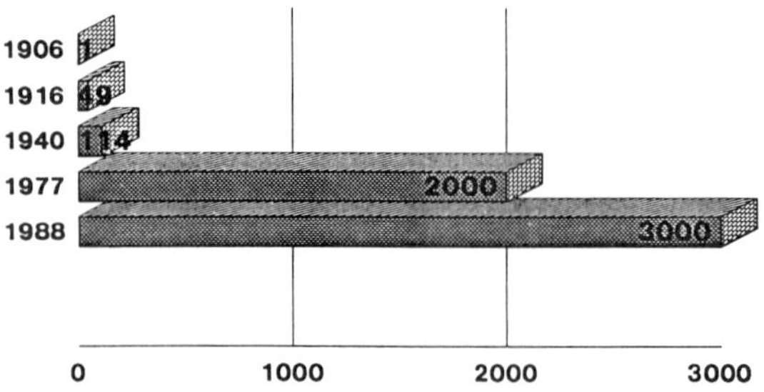

Fig. 1. Worldwide growth of ozone drinking water installations.

applications (Figure 1), and some 1500 of these are located in France.

In the United States, the longest continuously operating ozone drinking water treatment plant is at Whiting, Indiana, close to Chicago. This plant installed ozone in 1940. Our second longest operating plant is at Strasburg, PA, which installed ozone in 1973. Since 1977, 38 additional U.S. water treatment plants have installed ozone, and well over 100 more are under design. New drinking water treatment regulations being promulgated by the U.S. Environmental Protection Agency are encouraging U.S. municipalities to install ozonation, for disinfection of microorganisms, as well as to avoid the production of halogenated organic byproducts of chlorine disinfection.

MAJOR USES OF OZONE

Table 1 lists the major uses of ozone in drinking water treatment [1]. The oldest and best known application is disinfection.

Disinfection

In the 1960s, French public health scientists [2,3] determined that when the immediate ozone demand of a water has been satisfied and a residual concentration of dissolved ozone of 0.3 mg/L is attained and maintained for a minimum of three minutes, 99.9% inactivation of Poliomyelitis viruses types I, II, and III can be guaranteed. In practice, these conditions have been expanded to provide 0.4 mg/L residual ozone concentration held for at least four minutes. Some plants even maintain the 0.4 mg/L residual ozone for up to 10-12 minutes. Recent data from studies sponsored by the U.S. EPA show that such ozonation conditions also can guarantee inactivation of Giardia cysts [4].

Table 1. Major Uses of Ozone in Drinking Water Treatment

Disinfection

- Bacteria kills
- Virus Inactivation
- Cyst Inactivation

Oxidation

- Of Inorganic Materials
  - Iron, Manganese, Nitrate Ion, Sulfide Ion, Cyanide Ion, Arsenic
- Of Organic Materials
  - Taste, Odor, Color, Phenols, Detergents, Pesticides, THM precursors, Algae control, Biofouling control, Promotion of biodegradation,
- Microflocculation, Turbidity Control

Oxidation of Inorganic Materials

Ozone oxidation of iron, nitrite, sulfide, and cyanide ions proceeds rapidly to produce innocuous or easily filtered materials. Divalent manganese is readily oxidized by ozone past the insoluble tetravalent stage to the soluble permanganate stage. However, passage of the ozonized solution through granular activated carbon (GAC) filters quickly converts the soluble permanganate ions to insoluble manganese dioxide, which is easily removed by filtration. In fact, many ground water treatment plants with manganese problems deliberately convert all soluble divalent manganese to permanganate by ozonation, in order to avoid precipitating MnO_2 in the ozonation system, and having to maintain the equipment more often as a result.

Trivalent arsenic (soluble) is easily oxidized to the soluble pentavalent form. However, in the presence of calcium, magnesium, or ferric ions, the pentavalent arsenate ions are easily precipitated.

Oxidation of Organic Materials

Because of the varied geometric nature of molecular structures which cause different contamination problems, different organic molecules react with ozone at widely varying rates. These range from seconds (for phenol, cresols, and aliphatic non-halogenated olefins) to hours (for tetrachloroethylene, nitrobenzene, lower molecular weight aliphatic alcohols, oxalic acid, urea). In addition, even those organics which do oxidize readily are rarely completely mineralized to carbon dioxide and water, especially under ozonation conditions normally practiced in drinking water treatment plants (applied ozone dosages: 1 to 5 mg/L; short ozone contact times: 5 to 15 minutes).

Under such ozonation conditions, most organic molecular structures are only partially oxidized. Carbonyl groups (>C=0) are introduced into the molecular structures, forming aldehydes, ketones, and carboxylic acids, for the most part. Thus the partially oxidized organic materials are more polar than were the original organics. As a consequence of their increased polarity, the partially oxidized organics also are more readily biodegradable. Advantage can be taken of deliberately producing biodegradable organic materials to reduce the levels of dissolved organic materials. After ozone oxidation, the water can be passed through a biological filtration step. Thus, the ozonation step produces biodegradable dissolved organic carbon (DOC), and biodegradation mineralizes the easily degradable organic materials.

Ozone For Turbidity Control

Colloidal sized particles are suspended in water because of their small size, but also because of positive or negative charges on their surfaces. Likecharged particulates remain suspended unless their charges are neutralized. Ozone is capable of neutralizing some charged particles, but not all. Consequently, not all colloidal, charged, turbidity-causing particulates can be treated effectively with ozone. Furthermore, only sufficient ozone should be added to neutralize the surface charges. If excess ozone is added, the particles may simply be recharged, and resuspend themselves.

Microflocculation With Ozone

The term "microflocculation" usually refers to the enhancement of flocculation or coagulation of soluble organic materials (micropollutants) caused by ozone oxidation and changes made in the molecular structures of the soluble organics. Because of the microflocculation effect, it is possible for clear solutions to become cloudy when ozonized.

Reckhow et al. [5] have discussed microflocculation phenomena in detail, and offer six possible mechanisms for the observed effects:

1. **Organics are oxidized to form carboxyl groups.** Trivalent cations (iron, aluminum) present in the raw water can bridge carboxyl and phenolic groups on one organic polymeric molecule with those of

another, increasing molecular weight, and causing insolubility. Alternatively, the polar groupings produced upon ozonation can be more easily adsorbed on the surface of other polar materials (i.e., turbid particulates, added flocculants, etc.).

2. **Carboxylic acid groups directly precipitate with calcium.** With high natural hardness, or with lime softening, this effect can be maximized.

3. **Organics adsorbed to natural particulates are liberated during ozonation.** The oxidized organics thus become accessible to flocculants, whereas without oxidation, the adsorbed organics are inaccessible.

4. **Metastable organic intermediates which may form during ozonation (ozonides, organic peroxides, organic free radicals) may undergo condensation or polymerization reactions.** The products formed may be more easily removed by association with flocculants, or act as bridging polyelectrolytes themselves.

5. **Ozone-induced rupture of organo-metallic complexes releases oxidized metals, such as Fe(III) and Mn(IV).** The released polyvalent metals then act as coagulants or adsorbents for dissolved, polar organics.

6. **Algae can be lysed by ozone, liberating biopolymers.** The liberated nucleic acids, proteins, polysaccharides may act as natural organic coagulants.

Excessive ozonation can be as detrimental to "microflocculation" removal of organic materials as small dosages can be beneficial. For example, Farvardin and Collins [6] showed that there is an optimum preozonation dosage (OPD) for aiding the coagulation of humic substances. With ozone dosages beyond the OPD, the benefit is reduced, then eliminated, and further ozonation becomes detrimental to the coagulation-flocculation removal process.

CHEMISTRY OF OZONE – REACTION MECHANISMS

Although ozone is relatively stable in air (half-life in hours), it is much less stable (and more reactive) in aqueous solution (half-life in minutes). Consequently, ozone cannot be stored, and must be generated and used onsite. Ozone is only partially soluble in water, about 10-20 times more than is oxygen. Its solubility is governed by Henry's law; thus the higher the concentration of ozone in the gas mixture applied to water, the greater will be the amount of ozone which dissolves in water, and is available to perform useful work.

In aqueous solution, ozone can react directly as the O_3 molecule (at lower pH ranges below ~ 7), or as the hydroxyl free radical, which is formed with increasing rapidity above pH ~ 8. Direct reactions of the ozone molecule occur within seconds with such solutes as sulfite ion, nitrite ion, olefinic aliphatic hydrocarbons, phenols, polyaromatic hydrocarbons, some organic amines, and sulfides. However, some chlorinated solvents react much slower, even though they might be unsaturated. Compounds such as saturated hydrocarbons, benzene itself, or tetrachloroethylene require hours and even days to undergo complete reaction with ozone by the direct reaction mechanism. On the other hand, these same compounds oxidize much faster with hydroxyl free radicals.

Hydroxyl free radicals can be promoted and quenched by some natural water ingredients. Thus in waters containing high levels of alkalinity, at neutral pH, bicarbonate ions will quench the hydroxyl free radicals as they form, thus promoting direct reactions of molecular ozone. On the other hand, addition of hydrogen peroxide, in weight ratios to ozone of 0.125 to 0.8, accelerates the formation of the reactive hydroxyl free radicals by decomposing ozone, thus minimizing direct reactions and maximizing free radical formation.

At pH 10.3, bicarbonate ion converts into carbonate ion, which is 28 times as effective a free radical scavenger as is the bicarbonate ion [8]. Consequently, if ozone is added to water containing carbonate ions above pH 10.3, the ozone molecule will decompose rapidly to hydroxyl free radicals, which then are quenched rapidly by the carbonate ions. The net result is the addition of considerable ozone, but with little oxidative effect.

POSITIONS OF OZONE IN WATER TREATMENT PROCESSING

Single-Stage Ozonation

Early French application of ozone was for disinfection. For this purpose, ozone normally is installed after filtration associated with conventional treatment (chemical addition, flocculation, sedimentation, filtration). It is at this point in the treatment process that the water is cleanest, and applied ozone will not be "wasted" in coping with extraneous ozone-demanding materials. Some raw waters contain iron and manganese, and oxidation of these impurities produces considerable amounts of insoluble materials, which must be removed by filtration. In such instances, ozone oxidation is placed advantageously in an early point in the treatment process, i.e., before filtration. Other raw waters may be highly turbid, or contain high levels of algae, or contain dissolved organic impurities which become insoluble upon oxidation (microflocculation). With these types of raw waters, an initial stage of low-level ozonation (up to 1 mg/L applied ozone dosage) results in considerable improvement in the water quality. However, post-filtration disinfection still must be practiced.

Multiple-Stage Ozonation

From the above discussion, it should be clear that two stages of ozonation during water treatment can be very effective in overcoming many raw water quality problems, and at the same time provide higher quality finished water. For example, preozonation for turbidity control, microflocculation, or for iron and manganese oxidation can be followed by an intermediate ozonation (prior to filtration) for oxidation of more refractory organics (color, taste and odor, detergents, some pesticides, THM precursors, etc.). Alternatively, preozonation can be followed by post-filtration ozonation for the same oxidation purposes as well as primary disinfection. Such postozonation can be followed by GAC filtration for adsorptive removal of non-oxidizable organics and/or for biological removal of oxidized organics. As of 1989, there are an estimated 300 water treatment plants which have installed two-stage ozonation.

In 1980, the world's first three stage ozonation drinking water treatment process was installed in the Méry-sur-Oise plant (270,000 m^3/day) near Paris [9]. Preozonation at the rate of 0.75 mg/L is applied to the raw water, which then is stored 2+ days in a reservoir. During this time, nitrification of some ammonia and mineralization of some DOC occurs.

Water then enters the treatment plant where it is ozonized before the rapid mix at a level of 2.25 mg/L applied ozone dosage. This provides considerable oxidation of dissolved organics, and replenishes the DO level. Following filtration, ozone disinfection is practiced with an average applied ozone dosage of 1.5 mg/L, before the treated water is sent to GAC filtration. After GAC filtration, the water is treated with sodium hypochlorite solution. Before water is distributed, it is dechlorinated (sodium bisulfite) to a free chlorine residual of about 0.3 mg/L.

The total amount of ozone applied today at Méry-sur-Oise in three stages is 4.50 mg/L. It is of interest to observe that the original ozone treatment process installed at Mèry (conventional treatment, followed by post-filtration ozone disinfection) required an average of 4 mg/L of ozone. A much higher water quality is attained, and at only a 6% increase in operating costs.

Table 2. Ct values for 99.9% inactivation of *Giardia lamblia* cysts (mg/l-min) [4]

Disinfectant	pH	0.5°C	5°C	10°C	15°C	20°C	25°C
Free Chlorine	6	171	122	91	61	46	30
2 mg/L	7	261	186	140	93	70	47
	8	377	269	201	134	101	67
	9	521	371	279	186	139	93
Ozone	6-9	3	2	1.67	1.4	1	0.67
Chlorine Dioxide	6-9	54	36	27	18	14	9.3
Chloramine (preformed)	6-9	3,300	2,200	1,850	1,500	1,100	750

NEW DEVELOPMENTS IN OZONE TECHNOLOGY

Disinfection

The original French viral inactivation conditions (0.4 mg/L held for at least 4 minutes contact time) have demonstrated at least 99.9% (3-logarithms) inactivation of polioviruses types I, II, and III [2,3]. Recent studies in the United States have extended this "Ct Value" philosophy to *Giardia* cyst organisms [4, 10].

The U.S. EPA is in the process of promulgating sweeping regulatory actions for the treatment of drinking water to cope with not only microorganisms such as *Giardia* cysts, enteric viruses, *Legionella* bacteria, and heterotrophic plate count organisms, but also to minimize the formation of "disinfection byproducts", currently defined as byproducts of chlorination. Since it is well-known that cyst organisms are much more difficult to inactivate than are viruses or bacteria (with ozone, chlorine, or chlorine dioxide), the indicator of disinfectant performance efficiency is being measured by the "Ct value".

Table 2 shows typical Ct data for the 3-log inactivation of *Giardia* cyst organisms by means of chlorine, chlorine dioxide, ozone, and preformed monochloramine, at temperatures ranging from 0.5°C to 25°C. It can be seen immediately that the very low Ct values for ozone prove it to the most effective disinfectant of the four.

The conditions for oxidation of organics with ozone are optimized, it is usually found that the Ct values required to attain 3-logs of *Giardia*inactivation are attained simultaneously. This means that in many cases, primary disinfection credit can be taken for ozone simultaneously while attaining the amount of organics oxidation desired.

Advanced Oxidation Processes

In the 1970s, it was found [11,12] that when ozonation is conducted simultaneously with UV irradiation (254 nm), although no dissolved ozone could be measured, nevertheless, oxidation alone. Subsequently, the ozone/UV process has been commercialized by Westgate Research Corporation, now ULTROX International [13,14].

Later studies showed than the same rapid oxidation effect could be obtained by ozonating aqueous solutions containing trace quantities of hydrogen peroxide [15, 16, 17], or simply by raising the pH of the aqueous solution above 8. All of these processes deliberately produce the hydroxyl free radical, (OH), by decomposition of molecular ozone. The hydroxyl free radical is an even more powerful oxidizing agent than is ozone, having an oxidation potential of 2.80 V, compared with 2.07 V for molecular ozone. However, hydroxyl free radicals have very short half-lives, on the order of microseconds, compared with much longer times for the ozone molecule.

Many organic compounds which normally are stable to direct reaction with the ozone molecule can be oxidized rapidly by the hydroxyl free radical. Thus, chlorinated solvents such as trichloroethylene (TCE) and tetrachloroethylene (PCE) can be destroyed by hydroxyl free radicals. Glaze et al. [18] and Aieta et al. [19] describe laboratory and pilot plant experiments, respectively, of the ozone/hydrogen peroxide advanced oxidation process for destroying TCE and PCE in groundwaters contaminated annualized treatment costs for removing TCE and PCE from ground waters by air stripping, air stripping with gas phase GAC adsorption, liquid phase GAC adsorption, and by the ozone/peroxide advanced oxidation process [19]. Although the ozone/peroxide process is approximately 20% more costly than air stripping, it is significantly less expensive than the other two alternatives.

At the present time, these "advanced oxidation processes" have not been well studied, and their mechanisms are not yet fully elucidated, although all involve generation of the hydroxyl free radical, the formation of which is unique to ozone and the combination of UV radiation with hydrogen peroxide.

Glaze et al. [20] compare the generation of hydroxyl free radicals from the three ozone sources described above, along with the process involving UV decomposition of hydrogen peroxide. As shown in Table 4, although the stoichiometric yield of hydroxyl free radicals is greatest from the photolysis of hydrogen peroxide, ozone photolysis yields more radicals in water treatment practice because of the higher molar extinction coefficient of ozone compared to that of hydrogen peroxide.

McGuire & Davis [23] have presented the potential cost advantages of the ozone/hydrogen peroxide process for the full-scale treatment of sur-

Table 3. Comparison of annualized treatment costs for removing TCE & PCE

Cost Type	Air Stripping	Air Stripping With Gas-Phase GAC Adsorption	Liquid Phase GAC Adsorption	Peroxide-Ozone AOP
Capital cost (annualized $ 10 yrs, 8%)	48,400	108,000	192,500	35,000
O & M costs - $	30,200	73,300	13,900	63,900
GAC replacement cost - $	—	105,100	210,400	—
Total annualized costs - $	78,600	291,400	416,800	98,900
Cost per 1 000 gal - $ (based on 2 000 gal flow)	(0.075)	(0.277)	(0.397)	(0.094)

face waters in a different fashion. The capital cost for Metropolitan Water District of Southern California to achieve a 20 μg/L THM level at its five surface water plants with GAC adsorption is estimated at $ 1.3 billion. Conventional ozone treatment (4 mg/L applied ozone dosage, 10 minute contact time) would cost approximately $ 300 million to attain the same 20 μ/L THM level. Assuming that the indicated cost benefits of the ozone/peroxide process carry through to full-scale, up to $ 150 million in capital costs of ozonation equipment could be saved. Over $ 100 million could be saved by halving the applied ozone dosage, and an additional $ 40 million could be saved by halving the contact time.

Table 4. Theoretical amounts of oxidation and UV required for formation of hydroxyl radicals in ozone-peroxide-UV systems [20]

System	Moles of Oxidant Consumed per Mole of OH Formed		
	O_3	UV[a]	H_2O_2
Ozone - Hydroxide Ion[b]	1.5	—	—
Ozone - UV	1.5	0.5	(0.5)[c]
Ozone - Hydrogen Peroxide[b]	1.0	—	0.5
Hydrogen Peroxide - UV	—	0.5	0.5

a Moles of photons (Einsteins) required for each mole of OH formed.

b Assumes that superoxide formed in the primary step yields one OH radical per O_2^-, which may not be the case in certain waters.

c Hydrogen peroxide formed *in situ* [21,22].

Assimilable Organic Carbon (AOC), Its Measurement and Significance in Water Treatment

This parameter, relatively new to water treatment professionals, was defined originally by Dutch water treatment specialists [24] to quantify the amount of dissolved organic carbon in water which can be consumed by specific strains of microorganisms. The most significant consequence of the AOC concept is that at its higher levels, microorganisms in a water (say in the distribution system) can proliferate. On the other hand, low concentrations of AOC mean lower quantities of food available for microorganism regrowth. Consequently at low AOC levels, regrowth in the distribution system will be minimized.

Dutch water treatment scientists maximize the production of AOC within the treatment plant, then provide for biological treatment **within the water treatment plant** to assure biological destruction (assimilation) of the biodegradable materials so that they will not be released from the plant and passed into the distribution systems [25, 26]. The optimum treatment to maximize formation of AOC currently used in Dutch water treatment plants is ozonation; the optimum biological treatment to assure minimizing the level of AOC **within the water treatment plants** is GAC adsorption (with BAC activity).

In France, Hascoët et al. [27] have adopted the term "Biodegradable Dissolved Organic Carbon", BDOC, for AOC, and have conducted similar preozonation studies, followed by GAC filtration, with similar results.

Evolution of AOC during and after Ozonation

Measurement of AOC made after dual media filtration, after ozonation, and after GAC filtration show that ozonation increases the colony counts (indicating the presence of AOC) by an order of magnitude. Following filtration through GAC, the colony count is lowered to its original preozonation level [25].

Under drinking water ozone oxidation conditions, only a fraction of the initial DOC (10-20%) is converted to AOC. On the other hand, ozonation in conjunction with hydrogen peroxide, UV radiation, or at elevated pH (where hydroxyl free radicals predominate) may result in higher AOC yields.

For those treatment plants which cannot afford to provide both a sand and GAC filter, Bablon et al. [28] recommend a dual media filter to maximize high rate biological filtration. The recommended filtration system is 40 cm of GAC on top of 60 cm of sand. Such a dual media filter provides good biological removal of ozonized organic materials, as well as being better able to remove ammonia (by nitrification). This dual media filter is more effective in removing organics biologically than is a sand filter, but not as effective as a sand filter followed by a GAC filter. For medium quality raw waters, ozonation plus dual media filtration probably is sufficient for optimum organics removal, and is far more economical than installation of individual sand and GAC filters in series.

Effects of Chlorination on AOC Removal

The effect of postchlorination on the AOC produced by ozonation depends on the influence of this treatment step on:

a) the microbiological quality of the produced water;
b) the formation of halogenated organic compounds;
c) the formation of mutagenic compounds;
d) the effect on regrowth in the distribution system.

The effect of chlorination on regrowth depends on the net result of retardation of bacterial multiplication by free chlorine and the simultaneous production of assimilable organic compounds. Further research is needed to quantify these effects [25]. However, the Amsterdam Water Works has concluded that the negative effects of postchlorination are greater than its positive effects on water quality. Consequently, postchlorination has been stopped at the Leiduin and Weesperkarspel water treatment plants [29], which use the process combinations of ozone/GAC as terminal treatments.

Ozone Generation from Oxygen

General Considerations

For many applications, pure oxygen is a more attractive ozone feed gas than air for a number of reasons:

1. Higher production density (more ozone produced per unit area of dielectric).

2. Lower specific energy consumption (energy supplied per unit area of dielectric).

3. Higher gas-phase ozone concentrations are made possible by the use of oxygen (6-12% versus 1-4%).

4. Essentially double the amount of ozone can be generated per unit time from oxygen than from air (for the same power expenditure); ozone generation and contacting equipment can be halved in size when using oxygen, to generate and contact the same amount of ozone.

5. Smaller gas volumes are handled using oxygen than air, for the same ozone output; thus costs for ancillary equipment are lower with oxygen feed gas than with air.

6. Air preparation equipment is eliminated.

7. Ozone transfer efficiencies are higher, due to the higher concentration of ozone in oxygen.

8. Oxygen feed is better suited to intermittent (frequent stop/start) operation; air preparation equipment requires lengthy startup periods to reach optimum dew point levels.

However, the economic implications of these advantages must be weighed against the capital expenditure required for on-site oxygen production, or operating costs associated with purchase of liquid oxygen produced off-site.

Oxygen Production for Ozone Generation at Water Treatment Plants

There are currently two primary methods for producing oxygen on-site for ozone generation, pressure swing adsorption of oxygen from air, and cryogenically (liquefaction of air, followed by fractional distillation separation of oxygen from nitrogen). A membrane separation method is being developed, which currently produces oxygen-enriched air (30-40% oxygen).

In addition, purchased liquid oxygen (LOX) can be added to dried air to produce oxygen-enriched air (as at the Tailfer plant serving Brussels, Belgium) [30].

Gaseous oxygen produced on-site by pressure swing adsorption typically is 93 to 95% pure [31], while liquid oxygen produced cryogenically generally is 99.5% pure [32]. Consensus among suppliers of ozone in the United States used to be that oxygen becomes cost-effective over dried air for generating ozone at production rates above 3,500 lbs/day. However, liquid oxygen storage capacity is being installed to generate ozone at small water treatment plant (much less than 100 lbs/day). The number of drinking water treatment plants currently known to be using oxygen is only a fraction of those using air (11 of about 3,00), 10 of these 11 plants have installed oxygen capability since 1980.

OXIDATION PRODUCTS OF OZONATION [33, 18]

In general, ozone oxidation products of organic materials can be classified into the primary categories of aldehydes, ketones, and acids. Within those categories, lower molecular weight aldehydes (C_1-C_3 and C_7-C_{12}) and acids have been isolated in the highest yields. Aromatic compounds produce oxidized aromatic derivatives as initial ozonation products, prior to forming aliphatic aldehydes, acids, and ketones upon continued ozonation.

Some volatile organic compounds containing carbon-halogen bonds) can be destroyed by ozone coupled with UV radiation or with H_2O_2, at least to the point of yielding chloride ion quantitatively. Organic oxidation products which have been identified following ozonation of such compounds include halogenated acids.

Some amino acids can be oxidized to destruction with ozone; others only partially to destruction. Still others can form small amounts of polymers upon ozonation. If amino acids are not oxidized to destruction by ozone, and the ozonized water is chlorinated, there is a possibility of forming acetonitrile or chloroacetonitriles, from acetaldehyde produced by ozonation of other organic materials.

Many synthetic organic chemicals (SOCs – pesticides) react with ozone, at widely differing rates. Intermediate oxidation products have not been identified for the majority of ozone-reactive pesticides. In a few cases, primary ozone oxidation products have been shown to be at least as toxic as the starting materials (e.g., heptachlor forms heptachlorepoxidel malathion and parathion form their respective oxons).

Secondary oxidation products can be produced by further reaction of the primary oxidation products. In particular: acetaldehyde can form acetonitrile upon chloramination; nitrites or nitrophenols can increase the yield of chloropicrin upon subsequent chlorination; THMFP levels can be increased by ozonation, particularly if chlorine is added immediately after ozonation, without an intervening biological filtrator step.

Application of low levels of ozone (up to 1 mg/L) to source waters can result in small increases in mutagenicity (measured by the Ames test) [34, 35, 36,37]. Other oxidants (chlorine, chlorine dioxide) produce greater increases in mutagenic activities. When higher ozone doses are applied (4 mg/L) any mutagenicity formed is destroyed. In any event, passage of the ozonized waters through a GAC column removes all traces of mutagenic activity which may form upon ozonation.

It is clear that the best technological water treatment "insurance policy" known to date to minimize oxidized organic materials, TOXFP, AOC, and secondary/daughter treatment step, which can involve GAC filtration. However, it is critically important that no free chlorine be present in the water until after such GAC (biological) treatment step.

SUMMARY

Although ozone has been used for treating water since the early 1900s, it has only been since World War II that its use has become extensive throughout the world. In the past 20 years, considerable ozone research activity has been conducted under a variety of drinking water treatment conditions. These studies provide deeper understanding of the ability of this versatile material to oxidize pollutants and to disinfect ubiquitous microorganisms.

In the early days, only a single stage of ozonation was employed in water treatment processes; today two-stage ozonation processes are routine. Many single-stage ozonation plants have been retrofit with a second stage (usually preozonation). Three French treatment plants are now using three stages of ozonation. Costs for applying ozone three times is only slightly higher than the cost of the original single stage ozonation process. At the same time, the quality of water produced is considerably higher than when treated by the original process.

In the United States, new regulations being promulgated by the U.S. Environmental Protection Agency are encouraging American water utilities to adopt ozonation, both for primary disinfection (inactivation of *Giardia* cysts and enteric viruses), as well as for avoiding the production of halogenated organic byproducts. Although ozonation cannot produce any chlorinated organic byproducts directly, if bromide ion is present in the raw water, brominated organic derivatives can be produced upon ozonation, as well as during subsequent chlorination.

Advanced oxidation processes (ozonation coupled with the addition of hydrogen peroxide, UV radiation at 254 nm, or at elevated pH ranges) form the reactive hydroxyl free radical, which is capable of reacting with compounds which normally are unreactive (or only very slowly reactive) to molecular ozone. The combination of ozone with hydrogen peroxide (the "Peroxone" process) shows potential to reduce the amount of ozone required by 33% to 50% for the oxidation of such refractory materials as trichloroethylene, tetrachloroethylene, geosmin, and 2-methylisoborneol at lower cost than by ozone alone.

Organic oxidation products formed during ozonation of natural products generally are aldehydes, ketones, and carboxylic acids. Most of these are readily biodegradable, and are capable of being mineralized (to CO_2 and water) if a biological filtration step is provided after ozone oxidation. These same aldehydes, ketones, and carboxylic acids also are produced in natural waters treated with potassium permanganate, chlorine dioxide, and even chlorine itself.

In recent years, the use of pure oxygen as a feed gas for generating ozone has increased in large drinking water treatment plants. Small scale ozonation systems, which must operate intermittently, also can benefit from oxygen feed gas, since the startup time can be much shorter using oxygen than when using dried air to feed the ozone generator.

REFERENCES

[1] Rice, R. G., Robson, C. M., Miller, G. W., Hill, A. G., "Uses of Ozone in Drinking Water Treatment", *J. Am. Water Works Assoc.* 73(1):44-57, 1981.

[2] Coin, L., Hannoun, C., Gomella, C., "Inactivation of Poliomyelitis Virus by Ozone in the Presence of Water", *La Presse Médicale* 72(37):2153-2156, 1964.

[3] Coin, L., Gomella, C., Hannoun, C., Trimoreau, J.C., "Ozone Inactivation of Poliomyelitis Virus in Water", *La Presse Médicale* 75(38):1883-1884, 1967.

[4] U.S.Environmental Protection Agency, "National Primary Drinking Water Regulations; Filtration and Disinfection; Turbidity, *Giardia lamblia*, Viruses, *Legionella*, and Heterotrophic Bacteria; Final Rule", *Federal Register* 54(124):27485-27541, 1989.

[5] Reckhow, D. A., Singer, P.C. and Trussell, R.R., "Ozone as a Coagulant Aid", in *Ozonation: Recent advances and Research Needs*, AWWA Seminar Proceeding No. 20005, Am. Water Works Assoc. Pub., Denver. CO, 17-46, 1986.

[6] Favardin, M.R. and Collins, A.G., "Preozonation as an Aid in the Coagulation of Humic Substances – Optimum Preozonation Dose", *Water Research* 23(3):307-316, 1989.

[7] Hoigné, J., "Mechanisms, Rates and Selectivities of Oxidation of Organic Compounds Initiated by Ozonation of Water", in *Handbook of Ozone Technology and Applications*, R.G. Rice & A. Netzer, Editors, Ann Arbor Science Publishers, Inc., Ann Arbor, MI, 341-379, , 1982.

[8] Yurteri, C., and Gurol, M., "Ozone Consumption in Natural Waters: Effects of Background Organic Matter, pH, and Carbonate Species", *Ozone: Science & Engineering* 10(3):277-290, 1988.

[9] Rapinat, M., "Recent Developments in Water Treatment in France", *J. Am. Water Works Assoc.* 74(12):610-617, 1982.

[10] Wickramanyake, G.B. and Sproul, O.J. "Ozone Concentration and Temperature Effects on Disinfection Kinetics", *Ozone:Science & Engineering* 10(2):123-135, 1988.

[11] Garrison, R.L., Mauk, C.E., Prengle, H.W., Jr., "Advanced Ozone Oxidation System for Complexed Cyanides", in *Proc. First Intl. Symposium on Ozone for Water and Wastewater Treatment,* R.G. Rice & M.E. Browning, Editors (Norwalk, CT: Intl. Ozone Assoc.), pp. 551-577, 1975.

[12] Prengle, H.W., Jr., "Evolution of the Ozone/UV Process for Wastewater Treatment", presented at IOA/EPA Colloquium on Wastewater Treatment & Disinfection with Ozone, U.S. Environmental Protection Agency, Municipal Environmental Research Laboratory, Cincinnati, OH, Sept. 15, 1977.

[13] Leitis, E., Zeff, J.D., Smith, M., "Chemistry and Application of the UV-Ozone Purification Process", Final Technical Report, OWRT Contract 14-34-001-9436 (Washington, DC: Dept. of Interior, Office of Water Research & Technology), 1981.

[14] Zeff, J.D., Leitis, E., Nguyen, D., "UV-Oxidation Case Studies on the Removal of Toxic Organic Compounds in Ground, Waste and Leachate Waters", presented at the PACHEC 1988 Conference, Acapulco, Mexico, Oct. 19-23, 1988.

[15] Nakayama, S., Esaki, K., Namba, K., Taniguchi, Y., Tabata, N., "Improved Ozonation in Aqueous Systems", *Ozone: Science & Engineering* 1(2):119-131, 1979.

[16] Duguet, J.P., Brodard, E., Dussert, B., Mallevialle, J., "Improvement in the Effectiveness of Ozonation of Drinking Water Through the Use of Hydrogen Peroxide", *Ozone: Science & Engineering* 7(3)241-258, 1985.

[17] Bollyky, L.J., "Pilot Plant Studies for THM, Taste and Odor Control Using Ozone and Ozone-Hydrogen Peroxide", in *The Role of Ozone in Water and Wastewater Treatment, Proceedings of the Second International Conference,* D.W. Smith & G.W. Finch, Editors (Kitchener, Ontario, Canada: TekTran International, Ltd.,) p. 211, 1987.

[18] Glaze, W.H., Kang, J.-W., "Advanced Oxidation Processes for Treating Groundwater Contaminated With TCE and PCE: Laboratory Studies", *J. Am. Water Works Assoc.* 80(5):57-63, 1988.

[19] Aieta, E.M., Reagan, K.M., Lang, J.S., McReynolds, L., Kang, J.-W., Glaze, W.H., "Advanced *Oxidation Processes for Treating Groundwater Contaminated With TCE and PCE: Pilot-Scale Evaluations", J. Am. Water Works Assoc.* 80(5):64-72, 1988.

[20] Glaze, W.H., Kang, J.-W., Chapin, D.W., "The Chemistry of Water Treatment Processes Involving Ozone, Hydrogen Peroxide and Ultraviolet Radiation", *Ozone: Science & Engineering* 9(4):335-352, 1987.

[21] Peyton, G.R., Glaze, W.H., "Mechanism of Photolytic Ozonation", in *Photochemistry of Environmental Aquatic Systems,* ACS Symposium Series 327, R.G. Zika & W.J. Cooper, Editors (Washington, DC: Am. Chemical Soc.,) pp. 76-88, 1986.

[22] Taube, H., "Photochemical Reactions of Ozone in Solution", *Trans. Faraday Soc.* 53:656, 1957.

[23] McGuire, M.J., Davis, M.K., "Treating Water with Peroxone: A Revolution in the Making", *WATER/Engineering & Management* 135(5):42-49, 1988.

[24] Van der Kooij, D., Visser, A., Hijnen, W.A.M., "Determining the Concentration of Easily Assimilable Organic Carbon in Drinking Water", *J. Am. Water Works Assoc.* 74(10):540-545, 1982.

[25] Van der Kooij, D., "The Effect of Treatment on Assimilable Organic Carbon in Drinking Water", in *Treatment of Drinking Water for Organic Contaminants*, P.M. Huck & P. Toft, Editors (New York, NY: Pergamon Press), pp. 317-328., 1987.

[26] Van der Kooij, D., Hijnen, W.A.M., Kruithof, J.C., "The Effects of Ozonation, Biological Filtration and Distribution on the Concentration of Easily Assimilable Organic Carbon (AOC) in Drinking Water", in *Proceedings, 8th Ozone World Congress, Volume 1* Zürich, Switzerland: Intl. Ozone Assoc.,) pp. D-96 – D-113, 1987.

[27] Hascoët, M.-C., Servais, P., Billen, G., "Use of Biological Analytical Methods to Optimize Ozonation and GAC Filtration in Surface Water Treatment", in 1986 Annual Conference Proceedings (Denver, CO: Am. Water Works Assoc.,) pp. 2050222, 1986.

[28] Bablon, G.P., Ventresque, C., Ben Aïm, R., "Developing a Sand-GAC Filter to Achieve High-Rate Biological Filtration", *Am. Water Works Assoc.* 80(12):47-53, 1988.

[29] Schallart, J.A., "Disinfection and Bacterial Regrowth; Some Experiences Before and After Stopping the Safety Chlorination by the Amsterdam Water Works", in *Proceedings, Internat. Workshop on Water Disinfection,* Mülhouse, France, April 9-10, pp. 228-236., 1986.

[30] Masschelein, W.J., "The Use of Oxygen-Enriched Process Gas", in *Ozonization Manual for Water Wastewater Treatment,* W.J. Masschelein, Editor (New York, NY: John Wiley & Sons), pp. 13–17, 1982.

[31] Namba, K., Honda, T., "A High Efficiency Ozone Generation System", in *Proceedings, 8th Ozone World Congress, Volume 1* (Zürich, Switzerland: Intl. Ozone Assoc.), pp A11–A18, 1987.

[32] Geering, F., "Experiences With Ozone Treatment of Water In Switzerland", in *Proceedings, 8th Ozone World Congress, Volume 1* (Zürich, Switzerland: Intl. Ozone Assoc.), pp B59–B75, 1987.

[33] Rice, R.G. and Gomez-Taylor, M., "Occurrence of By-Products of Strong Oxidants Reacting with Drinking Water Contaminants – Scope of the Problem", *Environmental Health Perspectives* 69:31-44, 1986.

[34] Van Hoof, F., Janssens, J.G., Van Dyck, H., "Formation of Mutagenic Activity During Surface Water Preozonation and Its Removal in Drinking Water Treatment", *Chemosphere* 14(5):501-510, 1985.

[35] Kool, H.J., van Kreijl, C.F., Hrubec, J., "Mutagenic and Carcinogenic Properties of Drinking Water" in *Water Chlorination: Chemistry, Environmental Impact and Health Effects, Vol. 5* , R.L. Jolley, R.J. Bull, W.P. Davis, S. Katz, M.H. Roberts, Jr., V.A. Jacobs, Editors (Chelsea, MI: Lewis Publishers, Inc.,) pp. 187–205, 1985.

[36] Bourbigot, M.M., Hascoët, M.C., Levi, Y., Erb, F., Pommery, N., "Role of Ozone and Granular Activated Carbon in Removal of Mutagenic Compounds", *Environmental Health Perspectives* 69:159-163, 1986.

[37] Cognet, L., Courtois, Y., Mallevialle, J., "Mutagenic Activity of Disinfection By-Products", *Environmental Health Perspectives* 69:165-175, 1986.

THE COMPOSITION OF "SCALING" ON SEAWATER RO MEMBRANES

G. PEPLOW

Department of Chemistry
University of Malta
Msida, Malta

ABSTRACT

The commissioning of the seawater Reverse Osmosis Plant at Ghar Lapsi, Malta, was a major asset for the production of drinking water for the Island. The feedwater, product and brine samples from the Plant, were analyzed periodically to monitor its initial performance.

The results indicated loss of product water flow and high salt passage which were somewhat higher than expected. The cleaning solutions used to regenerate the hollow fiber permeators were analyzed and found to contain trace metals, which are potential foulants.

Chemical tests on fouled membrane strands confirmed the presence of some trace metal deposits. An XRF analysis on some fouled membrane strands indicated the presence of other trace elements, namely, V, S, Al and Sr, but excluded the presence of Zn, Co, Mn, Ni and Cd which were part of the research programe.

Chemistry for the Protection of the Environment
Edited by L. Pawlowski *et al.*, Plenum Press, New York, 1991

SEMIPERMEABLE MEMBRANES FOR SEA WATER REVERSE OSMOSIS

The membranes which are of interest in this report are cast from an aromatic polyamide polymer (aramid) originated by Richter and Hoehn [1]. Several criteria were established to develop a membrane which would withstand the rigors of sea water desalination by replacing cellulose acetate polymers because they resist mechanical change as well as chemical and biological attack [2]. During extensive research programes involving more than 100 different polymers, cellulose acetate and cellulose esters showed considerable promise in initial tests but frequently deteriorated during extended testing. Furthermore, some tests were carried out on permeators employing membranes which have been in operation at pressures of 5,000 kpa and higher, providing fresh water from the sea for four years and longer [3]. The polyamide membrane responded favorably in these experiments to the following requirements [4].

- high permeability to water for achieving high production rates
- low permeability to salts to obtain acceptable product water quality
- small thickness dimensions to maximize water flow but strong enough to withstand high pressures
- minimal changes in transport and mechanical properties after long exposures to high pressure
- resistance to chemical and biological attack

fabrication of shape to offer high surface to volume rations.

These requirements were met by the design of the B-10 permeator membrane manufactured by Du Pont commercially as far back as 1974. The membrane consist of a very fine hollow fiber. A bundle of these fibers make up the complete permeator to allow a large surface area exposed to feed water for permeation.

The basic unit of the membranes consists of a polyethylene oxide-isophthalamide copolymer in which all the amide cross links are in the side chains:

~CH_2 — CH

CH_2

O O

~ C — [benzene ring] — C — N — CH_2 — CH_2

NH

CO

[benzene ring] — C ~

O

Each permeator contains about 2,300,000 fibers which are formed on a forming machine [5].

The fibers are asymmetric, meaning that they consist of a very thin dense "skin" (0.1 - 1.0/μm) on the surface, and a thick, porous layer of the same polymer.

The increase in salt concentration at the membrane surface is called concentration polarization. Unless this concentration polarization is difused back into the bulk of the salt water, precipitation of salts and hence scaling on the membrane surface will occur. The performance of the permeator is shown to strike a balance at a pressure of 5,500 kPa, arbitrarily set as a guide. Increase in feedwater pressure increases product flow in a linear fashion; however, barely any improvement in salt rejection properties is effected. On the other hand, decrease in feedwater pressure increases the salt passage exponentially with a more linear decrease in product water flow.

FEEDWATER TREATMENT

For optimum reverse osmosis performance, water fed to the permeators must be pre-treated to remove gross amounts of solids and prevent formation of solid matter inside the permeator by precipitation or biological growth. Therefore, upstream of the permeator, pre-treatment consists of:

(a) filtration to remove large particles;

(b) adjustment of solubility parameters to prevent precipitation of sparingly soluble salts (scaling) and/or oxides (fouling);

(c) coagulation of collodial matter;

(d) chemical treatment to prevent biological growth.

The membranes also require periodic cleaning programes usually every two or three months to complement the pre-treatment procedures. When pre-treatment is inadequate, cleaning will be less effective in restoring the reverse osmosis performance and the need for cleaning will increase. The product flow rate will normally decrease when fouling sets in. This is due to compacting of the fiber bundles. Other performance parameters influenced by the pre-treatment are the bundle pressure drop and the salt passage.

At constant operating conditions and proper pre-treatment, the bundle pressure drop will remain essentially unchanged with time. Thus, an observed increase in bundle pressure drop usually indicates fouling.

The B-10 permeators are treated with tannic acid [6] and polyvinyl methyl ether [7], before being released for commissioning plants. The tannic acid is adsorbed on the membrane surface and diffusive salt rejection is enhanced.

Treatment with the ether also enhances salt rejection by reducing coupled flow. This happens when the residual ions at the surface move along with the oncoming feed flow. The ether creates enough polarity to destabilize any static hydrated ions to an appreciable distance prependicular to the membrane surface.

MEMBRANE FOULING DUE TO THE PRECIPITATION OF TRACE METALS PRESENT IN SEA WATER

The possibilities of metal ions combining with various anions present in sea water may induce precipitation of such salts in the brine. Such precipitations may be catalyzed by oxidation with dissolved oxygen. Most of the studies have been carried out in laboratory or pilot membrane systems [8]. The oxidation of iron and manganese is given by:

$$4Fe(HCO_3) + O_2 + 2H_2O \rightarrow 4Fe(OH)_3 + 8CO_2$$

$$4Mn(HCO_3)_2 + O_2 + 2H_2O \rightarrow 4Mn(OH)_3 + 8CO_2$$

Such oxidations to insoluble M(III) hydroxides are dependant on pH, oxygen and M^{2+} concentrations.

Surface water analysis has shown the existence of several trace metals in particulate matter. Therefore even if plant feed water is acidified to pH 7.5 and filtered through cartridge filters of 1μm, such particles will still exist to affect membrane performance. Apart from depositing as scale, potential exists for some of the metals to be eventually oxidized to higher states to form soluble or insoluble hydroxides, carbonates, sulphates, etc.

At the working pH of 7.0 - 7.5 of the feedwater some metal ions have low pMe^{+Z} solubility, (where $[Me^{+Z}]$ is the metal concentration of oxidation state +z). Reverse osmosis membrane guidelines indicate a limit of 50 μg/l Fe^{2+} at a pH of 7.5.

In the presence of dissolved oxygen, Fe^{2+} can be oxidized to Fe^{3+} to form amorphous $Fe(OH)_3$ precipitates. Such guidelines are based on the balance of pH, dissolved oxygen and Fe^{2+} concentrations (9).

Due to the complex nature of sea water, potential precipitates may include carbonates and bicarbonates.

EXPERIMENTAL: DETERMINATION OF METALS IN FOULED MEMBRANE STRANDS

Method A

About 25gms of strand were weighed accurately in a $150cm^3$ beaker and treated with 30cm 6M HNO_3. The solutions was covered with a watch-glass and left on a hot plate at 60°C for 3 days.

The strands were carefully removed from the beaker by means of a teflon rod, rinsing thoroughly with DI water. The acid solutions was then boiled until no brown fumes were given off. The solution was then transferred to a 50 cm^3 volumetric flask and analyzed for various metals on the PE303 against standards. A blank was prepared in a similar way excluding the strand samples.

Method B

A few strands of the fibers were analyzed qualitatively by X-ray fluorescence and indicated possible presence of the following elements on their surface:

Ti	Sr
Cu	Si
[Pb] ?	Ca
V	K
Cr	S
Fe	

Some elements which are also of interest in this study were not detected, namely, Co, Mn, Ni, Zn and Cd.

These results supported the theory that fouling was also due to trace metal deposits. In order to confirm these result chemical tests were conducted to complement the XRF results.

Method C

Studies on fouled membrane strands by X-Ray fluorescence indicated the presence of vanadium. A colorimetric method of analysis was conducted on ashed residues of these strands to confirm the XRF observations.

Five separate 20gm samples of the strands were accurately weighed and placed in 30cm^3 Ni crucibles. Three of the samples were spiked with 20 μg, 50 μg and 75 μg of standard V solutions, respectively. A sixth crucible was used as a blank. About 2gms of NaOH pellets were added to each crucible and the mixture fused for 10 minutes over an electric burner. On cooling, the residues were treated with 15cm^3 of DI water and left to dissolve overnight. The solutions were then transferred to 25cm^3 volumetric flasks, rinsing the crucibles with 5cm^3 of DI water. The flasks were cooled in ice and made up to the mark with dropwise addition 50% H_2SO_4, keeping the flasks in ice to prevent overheating.

The set of six solutions were then treated with 5% $KMnO_4$ solutions dropwise until a permanent pink color was obtained. This was followed by 10cm^3 concentrated HCl, 4cm^3 of N-Benzoyl phenylhydroxylamine solution (NBPHA) 0.5% in $CHCl_3$ and 10cm^3 $CHCl_3$.

The mixtures were shaken to extract the vanadium complex as a pink/purple $CHCl_3$ layer into 25cm^3 volumetric flasks. The aqueous layer was again extracted with 10cm^3 $CHCl_3$ and pooled with the previous extract. The volumetric flasks were made up to the mark with $CHCl_3$.

The absorbance was read on the Shimadzu 210 UV spectrophotometer in the 0.02 absorbance range.

The experiment was repeated to obtain concentration curves at a lower concentration, using three spiked standards of strands with 5 μg, 10 μg, and 20 μg vanadium, and another experiment using 4 μg, 8 μg, 12 μg, 16 μg, and 20 μg vanadium.

RESULTS: PRESENCE OF TRACE ELEMENTS IN FOULED MEMBRANE STRANDS

The solution obtained by means of Method A was analyzed for the elements Fe, Zn, Cu, Ca, Al, Sr, Cr and Mn on the PE303 and IL55 atomic absorption spectrophotometers.

Table 1. Trace Elements in Fouled Membranes

Fe	mg/kg	0.54
Zn	mg/kg	0.16
Cu	mg/kg	0.15
Ca	mg/kg	1.99
Al	mg/kg	0.18
Sr	mg/kg	0.08
Cr	mg/kg	0.013
Mn	mg/kg	0.11

Table 2a. Test A: Vanadium Strands Range 0.0 - 3.75 mg/kg

Sample No	Amount of V Added mg/kg	Amount Detected mg/kg
1	1.00	1.67
2	2.50	3.34
3	3.75	4.51
4	-	0.56
5	-	0.73

Vanadium Content = 0.712 mg per kg of strands

Table 1 shows the results of the elements found, calculated in terms of the weight of membranes.

These results continue to support the previous results for the presence of such metals in the membranes. The values for iron and calcium once again predominate. However, zinc also gives a positive value although the element could not be detected by the XRF analysis.

PRESENCE OF VANADIUM IN MEMBRANE STRANDS

The colorimetric determination for canadium by Method C was conducted three times with the following results:

Test A (Table 2a) was performed by using a standard calibration curve of 0 – 4 mg/kg vanadium. The results for the duplicate unknowns show that the vanadium content was below 1mg/kg, lower than the lowest standard used. The standard calibration curve obtained was quite satisfactory.

A new set of calibration standards containing 0 – 1.0 mg/kg was used in Test B (table 2b) to correspond with the concentration of the vanadium

Table 2b. Test B: Vanadium Strands Range 0.0 - 1.0 mg/kg

Sample No	Amount of V Added mg/kg	Amount Detected mg/kg
1	0.25	0.94
2	0.50	1.31
3	1.00	1.80
4	-	0.69
5	-	0.81

Vanadium Content = 0.76 mg per kg of strands

Table 2c. Test C: Vanadium Strands Range 0.0 - 1.0mg mg/kg

Sample No	Amount of V Added mg/kg	Amount Detected mg/kg
1	0.2	1.01
2	0.4	1.14
3	0.6	1.34
4	0.8	1.58
5	1.0	1.84
6	-	0.78
7	-	[1.08]

Vanadium Content = 0.77 mg per kg of strands

in the strands. The calibration curve for such a low concentration was very satisfactory. A third experiment using a total of five standards was more reliable (Test C; Table 2c). The mean value of vanadium in the strands was therefore found to be 0.747 mg/kg.

CONCLUSION

The study leaves no doubt as to the presence of metallic fouling on the membranes. The XRF analysis indicating presence of vanadium was confirmed by AAS. The absence of Zn and Mn in the XRF experiment was probably due to the fact that only a very small surface area of the strands was scanned. Therefore, this limitation may not have offered a thoroughly representative analytical area. The presence of Zn and Mn were confirmed by the AAS experiment.

The amounts of metal fouling found on the strands suggest a more efficient treatment technique is required to control scaling of such metals. Lowering the pH to 6.5 – 7.0 does not apparently provide suitable conditions to stabilize the particulate or ionic metal species. Lowering the pH further is precluded due to potential corrosion problems.

Perhaps an exercise for further work is to study the possibility of dosing the feedwater with organic microgranulates or chelating agents which occlude metal species from the sea water.

REFERENCES

[1] Richter, J. W. and Hoehn, H. H., US Pat. 3,567,632.

[2] Shields, C. P. *Proc. 6th Intl. Symp.; Fresh Water from the Sea,* 3, 295, 1979.

[3] Kellar, R. A. *World Water*, 2, 44, 1979.

[4] Beasley, J. K. *Desalination*, 22, 181–187, 1977.

[5] McGinnis, P. R. and O'Brian, G. J., US Patent, 3690,000.

[6] Ganci, J. B., US Patent 3,853,755.

[7] Ganci, J. B., Jensen, J. H. and Smith, F. H., US Patent, 3,808,303.

[8] Jackson, J. M. and Ladott D. *Desalination*, 12, 361–378, 1973.

[9] Water Quality and Treatment, 3rd Ed., A.W.W.A. McGraw Hill N.Y., 83, 1971.

RECOVERY OF CADMIUM BY CRYSTALLIZATION OF CADMIUM CARBONATE IN A FLUIDIZED-BED REACTOR

C. DOTREMONT, D. WILMS, D. DEVOGELAERE
and A. VAN HAUTE

Catholic University of Leuven
Department of Chemical Engineering
de Croylaan, 26, B 3030 Leuven, Belgium

J. VAN DIJK

DHV Consulting Engineers
Postbus 85, 3800 AB Amersfoort, The Netherlands

ABSTRACT

Heavy metals can be recovered from spent plating baths by growing crystals of metal carbonate in a fluidized-bed reactor. The optimal conditions for crystallizing cadmium carbonate have been investigated on a laboratory scale pellet reactor initially seeded with quartz sand. With a C_t/Cd feeding ratio of 1.6 mol/mol, a Cd-load of 0.41 kg Cd per square meter reactor cross-section per hour and a pH of 7.9, the effluent cadmium concentration is below 1 ppm. The pellets, with a diameter of 1 mm, consist mainly of cadmium carbonate with a small amount of cadmium hydroxide; by dissolving them in a strong acid, a concentrated solution of cadmium ions is obtained.

INTRODUCTION

There are a number of situations in water and wastewater treatment where the dissolved component has to be removed. This is usually done by precipitation. The sludge produced in this manner has no useful application and must be transported to and disposed off in a waste landfill: a troublesome and expensive solution.

Examples of such situations are :

- removal of calcium from drinking water (partial softening), by precipitating $CaCO_3$ with lime,
- removal of phosphates from wastewater by precipitating iron-, aluminium- or calcium–phosphate,
- removal of heavy metals, e.g., from electro-plating baths, by precipitating metal hydroxides.

The necessity of forming a sludge can in many cases be circumvented by transforming the component that has to be removed into a crystalline compound with a sufficiently low solubility (e.g., $CaCO_3$, $Ca_3(PO_4)_2$, metal carbonate). By making the solution slightly supersaturated with respect to this compound (one adds resp. NaOH, $Ca(OH)_2$ and Na_2CO_3), and this within an environment where a high concentration exists of crystals of this compound, (heterogeneous) crystallization will take place, but no homogeneous nucleation. These pure crystals can easily be separated from the liquid phase. Moreover, in most cases they find a useful application. The best environment for such a controlled crystallization is a fluidized–bed reactor: such a reactor provides conditions of a high concentration of crystal surface and of excellent mixing.

The process of fluidized-bed reactor crystallization was developed by DHV. The first reactors were designed for the softening of drinking water: almost 40 full scale plants are now in operation, mainly in the Netherlands [1]. The largest softening plant, situated at the Municipal Water Works of Amsterdam, consists of 10 pellet reactors with a height of 8 meters, a diameter of 3 meters and a capacity of 750 m^3/h each. No sludge is produced and the $CaCO_3$ pellets are sold to the livestock feed industry.

Removal of phosphates from wastewaters is a second application [2]. The first full scale phosphate removal plant is in operation at the domestic wastewater treatment plant (12,000 p.e.) of Westerbork, the Netherlands. No sludge is produced and the $Ca_3(PO_4)_2$ pellets are sold to the phosphate processing industry.

A third application is the removal and recovery of heavy metals from, e.g., spent electro-plating baths [3].

This paper will deal with the crystallization of cadmium carbonate. With the help of a laboratory scale reactor and by using pure solutions of

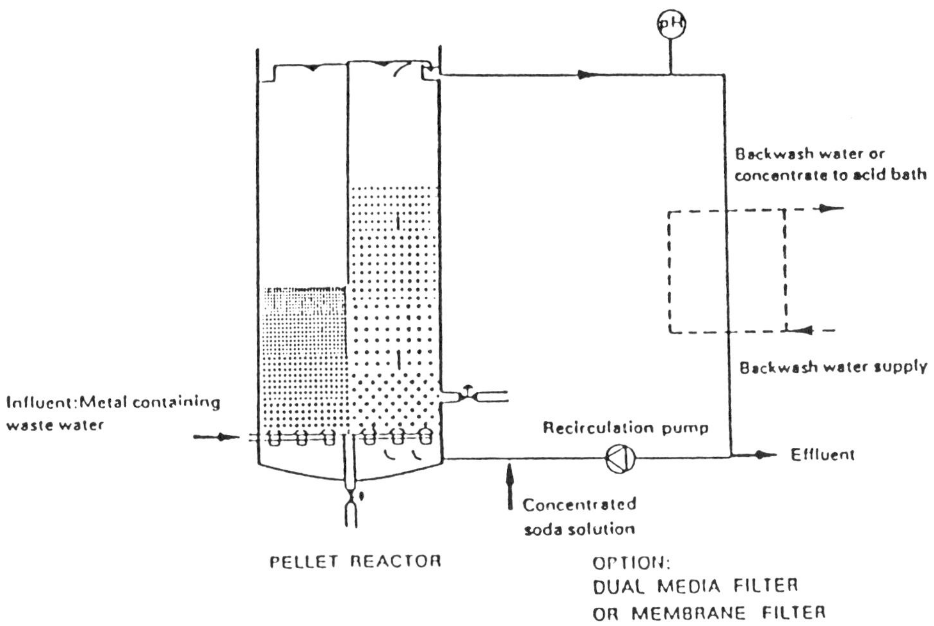

Fig. 1. Principle of the pellet reactor system.

cadmium nitrate, optimal conditions for a number of parameters have been studied at the University of Leuven. At present, DHV is working with real baths in full scale pellet reactors.

SYSTEM DESCRIPTION

Principle of the Pellet Reactor System

The pellet reactor system is shown in Fig. 1. The reactor consists of a cylindrical vessel, partially filled with a suitable seed material, e.g., filter sand. The fluid velocity in the reactor is high (75 – 125 m/h); as a result the pellet bed is kept in a fluidized state.

The heavy meta–containing wastewater is injected at the bottom of the pellet reactor: here this influent is mixed with a recirculation stream, to which a concentrated solution of sodium carbonate is being added. An important parameter is the ratio of the number of moles of C_t (this is the sum of the molar concentrations of CO_2, bicarbonates and carbonates) that is being fed in unit time, over the number of moles of heavy metal introduced in unit time with the influent. In order to obtain a sufficiently low metal concentration in the effluent, an overdose of C_t in relation to the metal dose is normally needed.

The mixture of recirculation and influent streams at the bottom of the reactor is supersaturated with respect to the metal carbonate and this supersaturation acts as a driving force for crystallization of the metal carbonate onto the pellets; on the other hand, the supersaturation may not be so high that it would lead to spontaneous nucleation: therefore, the higher the metal concentration in the influent, the more effluent is to be recirculated to provide an adequate dilution.

Because the crystallization reaction is very fast and because high hydraulic loads can be applied, the pellet reactor system is compact and has therefore relatively low investment costs. The process is pH-controlled and very stable thanks to the buffering action of the carbonate system and the dampening effect of the recirculation system; hence the operation is easy and operational costs are low.

Pellet Handling and Reuse

The pellets are growing and have to be removed periodically, because their specific surface would become too small. After release of the bigger pellets, which automatically concentrate at the lower part of the reactor, new seeding material (usually sand) is added.

The metal carbonates are pure and can be re-used by dissolving them in a strong acid: the carbonate escapes as CO_2, and a pure concentrated metal solution is obtained, while the seeding sand can be re-used in the reactor. The pure concentrated metal solution can be re-used in the metal finishing, the metal processing or the chemical industry.

Postfiltration

Depending on the type of metal and wastewater, small quantities of suspended solids (carry-over) can form in the pellet reactor. For one part, this carry-over consists of amorphous metal hydroxide, for an other part, of small fragments of pellets formed as a result of erosion in the fluidized bed. If the amount of carry-over becomes too high, a dual-media or a membrane filter should be incorporated in the recirculation system. The dual-media filter has to be backwashed regularly. The effluent is used as backwash water. By mixing backwash water with the acid wastewater and - if necessary - by adding some concentrated acid, the carry-over dissolves. This solution is then returned to the reactor to be treated once again so that the metal is removed as metal carbonate pellets and so that no waste remains.

EXPERIMENTAL STUDY OF THE CRYSTALLIZATION OF CADMIUM CARBONATE

In order to evaluate the influence of a number of process parameters, such as pH, feeding ratio of C_t/Cd, cadmium-load, recirculation ratio and total mass of pellets in the reactor, a great number of experiments have been carried out with a laboratory scale fluidized-bed reactor having an internal diameter of 2.0 cm and a height of 2.40 m. The system did not contain a filter to remove any suspended solids from the recirculation or the effluent stream. The initial seeding material consisted of quartz-sand with a diameter of 0.2 to 0.3 mm; during the experiments the size of the carbonate pellets was kept constant between 1.0 and 1.1 mm. The fixed-bed height of the pellets was 1.00 m. In all experiments the spent cadmium-bath was simulated by a solution of pure $CdSO_4$ with a concentration of 10.0 mmol/l; the Na_2CO_3 had a concentration which varied between 10 mmol/l to 70 mmol/l. In order to install a desired pH of the reactor's effluent, a solution of H_2SO_4 with a concentration of 1 mmol/l was fed as well. The three solutions were dosed by means of three peristaltic pumps equipped with an adjustable pumping rate.

Influence of pH

In a first series of experiments, the effluent pH has been varied between 7.52 and 10.56 by dosing appropriate amounts of H_2SO_4, keeping all other parameters constant:

- molar ratio of the feeding rate of Na_2CO_3 and $CdSO_4$: 1.60 mol/mol, giving an effluent C_t between 2.2 and 3.1 mmol/l;
- Cd-load: 0.41 kg Cd/m^2h;
- hydraulic load : 27.7 m/h;
- fixed-bed height : 1.0 m; expanded-bed height: 2.40 m.

Prior to taking samples of the effluent for analysis, the reactor was run several hours at constant pumping rates till a steady state was obtained (constant effluent pH). Table 1 shows the results of a number of such experiments. The first four parameters have been measured directly, the others have been obtained by calculation and mass balances.

There is only a small difference between the C_t of the effluent and the C_t in the mixture at the bottom of the reactor; whereas the C_t/Cd feeding ratio is only 1.60 mol/mol, this ratio is considerably higher at both the bottom and the top of the reactor (e.g., 18 and 257 at pH 8). Total, soluble and amorphous Cd in the effluent as a function of its pH are shown in Figures 2 and 3.

The total cadmium concentration in the effluent is minimal around pH 8 -

Table 1. Characteristics of the Effluent and of the Liquid Phase at the Bottom of the Reactor

	effluent			bottom			Formation in one pass	
pH	sol. Cd (ppm)	total Cd (ppm)	C_t (mmol/l)	sol. Cd (ppm)	total Cd (ppm)	C_t (mmol/l)	pellets (ppm)	amorphous Cd (ppm)
8.02	1.257	1.889	2.529	15.325	15.935	2.647	14.046	0.0217
8.01	1.261	1.718	2.449	15.329	15.770	2.569	14.052	0.0157
8.00	1.013	1.440	2.281	15.089	15.502	2.407	14.062	0.0147
7.79	1.822	2.339	2.249	16.863	17.361	2.379	15.022	0.0186
8.35	0.207	0.576	2.470	14.927	15.283	2.593	14.707	0.0131
8.64	0.116	0.528	2.497	14.969	15.366	2.619	14.838	0.01457
9.07	0.187	0.410	2.494	15.167	15.379	2.617	14.969	0.0079
9.34	0.324	0.660	2.605	15.045	15.369	2.725	14.709	0.0118
9.82	0.272	0.782	2.706	14.881	15.374	2.818	14.592	0.0174
10.22	0.094	1.409	2.839	14.962	16.231	2.954	14.822	0.0453
10.30	0.090	1.786	2.944	14.833	16.470	3.054	14.685	0.0579
7.52	2.181	3.216	2.187	16.821	17.819	2.321	14.603	0.0372
10.29	0.330	1.957	2.856	15.065	16.639	2.969	14.679	0.0556
10.56	0.034	2.111	3.103	12.019	14.037	3.1885	11.926	0.0594

Cd-load 0.41 $kg/m^2.h$, C_t/Cd feeding ratio 1.6 mol/mol.

10; at higher pH values it increases steeply.
This increase can be explained by:
- spontaneous nucleation of $CdCO_3$ as a result of a high supersaturation in the mixing point, and/or
- precipitation of $Cd(OH)_2$ (s), the solubility of which is exceeded above pH = 10.
- the formation of soluble cadmium hydroxide complexes above pH = 10.

Influence of the C_t/Cd Feeding Ratio

In three series of experiments, the first with a constant effluent-pH of 7.3, the second with a pH of 7.65 and the third with a pH of 8.0, the feeding C_t/Cd ratio was varied between 1.30 and 17 mol/mol (the effluent-C_t changed in the same time from 1.12 to 50 mmol/l), whilst the other process parameters (Cd-Load, hydraulic load, fixed- and expanded-bed height—were kept at the same values as in the previous runs.

There is no big difference in Cd effluent concentration between the three experiments. Even at small C_t effluent concentrations there is a good Cd elimination. The cadmium concentration does not change much at C_t values between 1 and 12 mmol/l, whereas at higher C_t concentrations amorphous cadmium is formed due to spontaneous nucleation of $CdCO_3$.

To reduce cost of chemicals, one would operate the crystallization

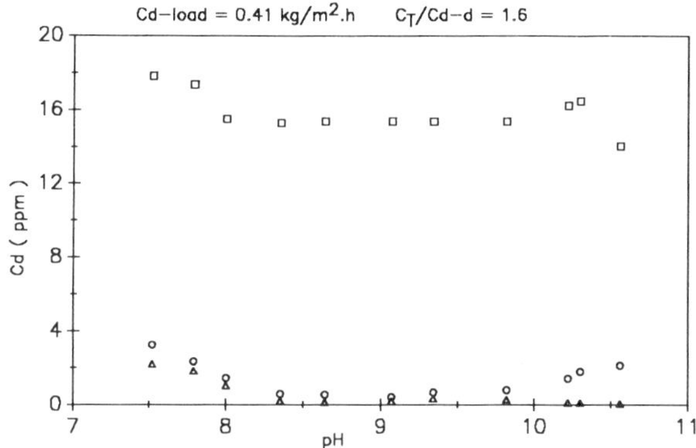

Fig. 2. Effluent Cd–conc. vs pH; o cadmium not filtrated, △ cadmium filtrated, □ cadmium not filtrated at the bottom of the reactor.

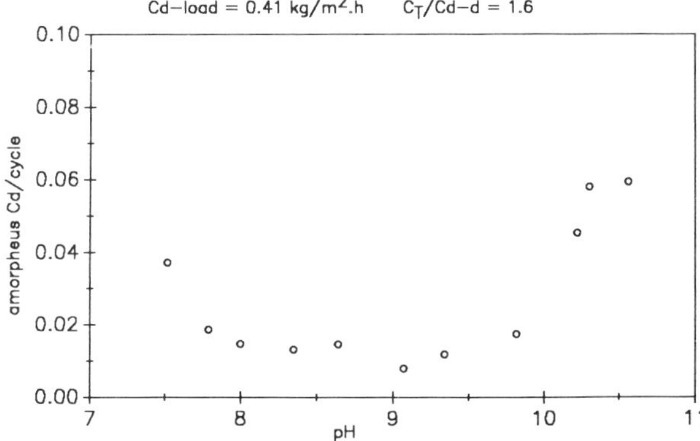

Fig. 3. Amorphous Cd vs pH.

reactor at the lowest dose of Na_2CO_3, at which equally low concentrations of cadmium in the effluent can be obtained; this seems to be a feeding ratio C_t/Cd around 1.6 mol/mol, corresponding to an effluent C_t of 2.3 mmol/l at pH

Influence of the Cd-load

In a new series of experiments, the cadmium load was varied from 0.14 to 3.74 kg Cd/m^2h. All other process variables were kept constant: effluent-pH 7.9, feeding ratio C_t/Cd 1.6 mol/mol, hydraulic load varied between

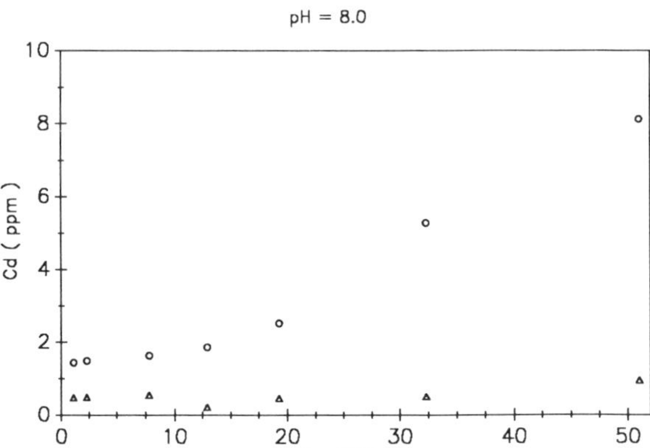

Fig. 4. Effluent Cd–conc. vs effluent C_t; o cadmium not filtrated, △ cadmium filtrated.

29 and 38 m/h, fixed-bed height 1.0 m, expanded bed height 2.40 m. The effluent cadmium concentration as a function of the cadmium load is shown in Figure 5. One observes a steady increases of the effluent concentration with the applied load, especially at loads higher than 1 kg/m^{2}h. The cadmium concentration in the filtered effluent samples increase as well, probably due to the colloidal character of part of the carry-over.

Influence of the Hydraulic Load

In order to study the influence of the hydraulic load, the recirculation ratio was changed; as a result the liquid velocity through the reactor varied between 18 and 77 m/h. The cadmium load was kept constant at 0.43 kg/m^{2}h, the fixed-bed height at 1.0 m and the effluent-pH and -C_t at 7.9 resp. 21 mmol/l. The results are represented in Figure 6. The hydraulic load seems to exert only a minor influence on the cadmium concentration in the effluent; the slight increase of undissolved cadmium at high recirculation rates may be due to pellet erosion.

Influence of the Pellet Mass in the Reactor

The crystallization capacity (defined as the amount of cadmium carbonate that can crystallize in unit time per unit cross-section of the reactor) will evidently strongly depend on the fixed-bed height: indeed, for uniform-sized pellets, this is proportional to the total pellet mass or the total available crystal surface in the reactor. One can assume that if there is more crystal

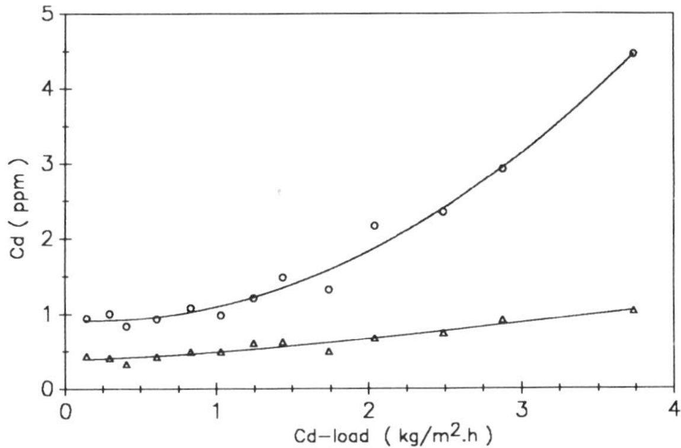

Fig. 5. Effluent Cd-conc. vs Cd-load; o Cadmium not filtrated, △ Cadmium filtrated.

surface on which the metal carbonate can crystallize, then the system can handle a higher supersaturation.
In order to investigate the specific influence of the pellet mass, the fixed bed was decreased from experiment to experiment by withdrawing an upper part of the bed. The first run was with a fixed-bed height of 1.030 m, the last run with a height of 0.135 m. The other process parameters were kept constant: effluent-pH 7.9, C_t/Cd 1.6 mol/mol, Cd-load 1.00 kg/m^2h and hydraulic load 31.2 m/h. The resulting effluent cadmium concentrations are depicted in Figure 7. One observes that the same effluent quality is obtained as long as the fixed-bed is higher than 0.35 m.

APPLICATIONS IN PRACTICE

The pellet reactor system for heavy metal recovery has wide fields of application in:
- the metal finishing industry
- the chemical industry
- the metal processing industry.

In the metal finishing or plating industry the heavy metals can be recovered from all kinds of spent concentrate, passivating and drag-out baths and from rinsing water.
There is some full scale experience and a lot of pilot-plant experience in treating these wastewaters.
In the chemical industry all kinds of heavy metal–containing wastewaters

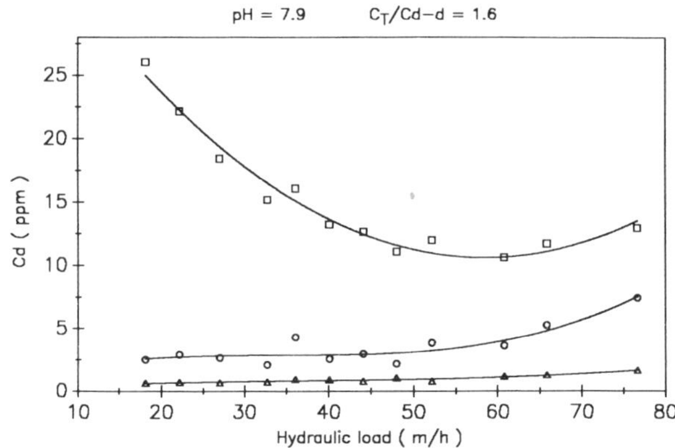

Fig. 6. Cd–conc. vs hydraulic load; o cadmium not filtrated, △ cadmium filtrated, □ cadmium not filtrated at the bottom of the reactor.

can be treated. Pilot-plant experience has been obtained on a number of different wastewaters of different chemical industries.

The pellet reactor for heavy metal recovery can also be applied in the metal processing industry. Possible applications include a process step in a wet metal recovery process from ores or the treatment of pickling baths in the steel and stainless steel industry.

COSTS

The pellet reactor system is compact and has relatively low investment costs. The crystallization process is stable and process operation is easy so process control is relatively simple and cheap. Depending on the degree of automation, the system needs half an hour or two hours attendance a day in the case of full automation or no automation, respectively.

The main operational costs are the costs of chemicals and personnel. Sodium carbonate and caustic soda are used depending on the metal and the wastewater. The costs of the chemicals amount to 700–1,500 US$ per ton recovered metal and resemble those of a hydroxide plant. The costs of energy are relative low because there is only some consumption of electricity by the pumps.

Generally the heavy metals can be reused in the same industry. Depending on the metal this results in savings on the purchase of heavy metals of 1,000-30,000 US$ per ton, metal.

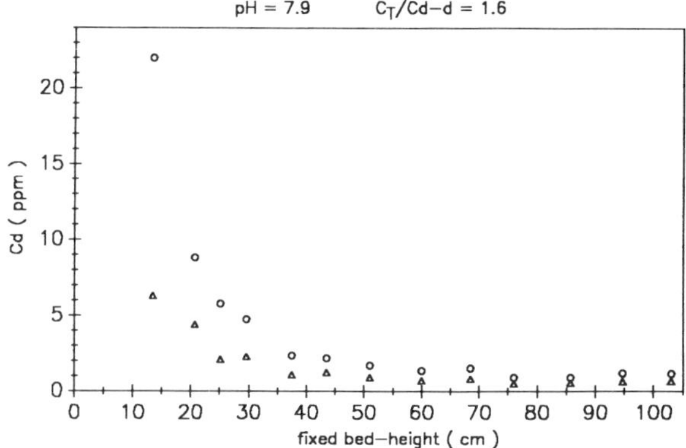

Fig. 7. Effluent Cd vs fized-bed height; o cadmium not filtrated, △ cadmium filtrated, □ cadmium not filtrated at the bottom of the reactor.

There is no sludge or waste production, thus no disposal problem. The savings on the disposal of chemical waste amount to 1700-8500 US$ per ton of metal depending on the metal content of the waste. (The disposal costs for chemical waste in Western Europe amount to 170 US$ per ton of chemical waste).

In general the pellet reactor system pays for itself in a few years.

REFERENCES

[1] Van Ammers M., Van Dijk J.C., Graveland A., Nühn P.A. (1986). State of the art of pellet softening in the Netherlands. Water Supply, 4, 223-235.

[2] Van Dijk J.C., Braakensiek H. (1984). Phosphate removal by crystallization in a fluidized bed. Water Science and Technology, 17, 133-142.

[3] Schöller M. Van Dijk J.C., Wilms D.A. (1987). Recovery of heavy metals by crystallization in the pellet reactor. In: Environmental Technology K. De Waal and W. Van den Brink (Eds). M. Nijhoff Publ., Dordrecht, 294-303.

ENVIRONMENTAL MANAGEMENT IN THE NITROGEN FERTILIZER INDUSTRY IN THE ARAB WORLD: AN EGYPTIAN CASE STUDY

A. HAMZA

High Institute of Public Health
Alexandria University, Egypt

ABSTRACT

The nitrogen fertilizer industry has grown rapidly in the Arab world recently. At the same time there has been a tendency towards the construction of larger complexes for the production of various nitrogenous products using state-of-the-art technologies. These trends were dictated by the global economic pressure to improve cost-effectiveness for this highly competitive industrial sector.

This paper presents a review of the production technology and pollution abatement in the fertilizer industry in the Arab world. The paper also presents a case study of production practices at an Egyptian fertilizer plant where the release of raw process effluents causes environmental problems for the aquatic life and poses health risks for the community.

A long-term study was undertaken to identify sources and loads of major pollutants in the plant and to examine treatment possibilities for process effluents; emphasis was placed on the urea process condensate as it has the greatest potential for releasing harmful pollutants to the environment. The study centred on the development of a multitreatment system which comprises stripping, enzymatic hydrolysis and biological nitrification to produce effluent suitable for irrigation and/or discharge to the Mediterranean. Fish bioassay indicated possible release of the treated effluent after dilution without harmful effects on aquatic life in Abu Kir Bay.

Chemistry for the Protection of the Environment
Edited by L. Pawlowski *et al.*, Plenum Press, New York, 1991

INTRODUCTION

The nitrogen fertilizer industry has grown rapidly in the Arab world during the past two decades. At the same time, there has been a noticeable tendency towards the construction and operation of integrated fertilizer manufacturing complexes. The primary production facility is typically a large ammonia (NH_3) plant which utilizes natural gases. Integrated with it, are various production systems for the manufacture of urea, ammonium nitrate or ammonium phosphates.

This integration of large scale production systems in the Arab states, and the acquisition of state-of-the-art technologies have produced considerable economies in production costs and permitted incorporation of advanced waste control technologies.

The demand for nitrogen fertilizers will continue its rapid growth in the foreseeable future to meet the needs of the agricultural development plans of the Arab countries. Equally the scarcity of water supplies and the rising costs of wastewater will promote improved water management practices particularly in the integrated fertilizer complexes of the Gulf states.

This paper reviews pollution control practices in the fertilizer industry in selected Arab states. It also presents the results of a long-term study to identify pollution sources from a fertilizer plant in Alexandria, and to assess various options for treatment of heavily polluted process wastewater.

POLLUTION ABATEMENT IN THE NITROGEN FERTILIZER INDUSTRY IN THE ARAB WORLD: A BRIEF OUTLOOK

The petrochemical company in BAHRAIN produces NH_3 and methanol, each with a capacity of 100 tons/day (t/d). The natural gas from Khuff is utilized as a feedstock and fuel at an average rate of 95,000 cubic metres per hour (m^3/h); sea water is supplied at an average of 18,500 m^3/h for cooling process water [1]. The basic source of pollution is the purge stream from the recycled gases separated after the synthesis reaction. It contains NH_3 and methanol residues, as well as carbon monoxide (CO), carbon dioxide (CO_2), hydrogen (H_2), and nitrogen (N_2). The NH_3 content of the purge gases is virtually eliminated by water scrubbing, which is followed by cryogenic liquification of the purge stream and fractional distillation. Effluents of NH_3 processing consist mainly of process condensates, demineralizer effluent and boiler blowdown; the combined stream is presently discharged without pretreatment to the Gulf.

There are four nitrogen fertilizer plants in EGYPT. The Abu Kir Fertilizers and Chemical Industry (AKFCI) in Alexandria utilizes offshore gases

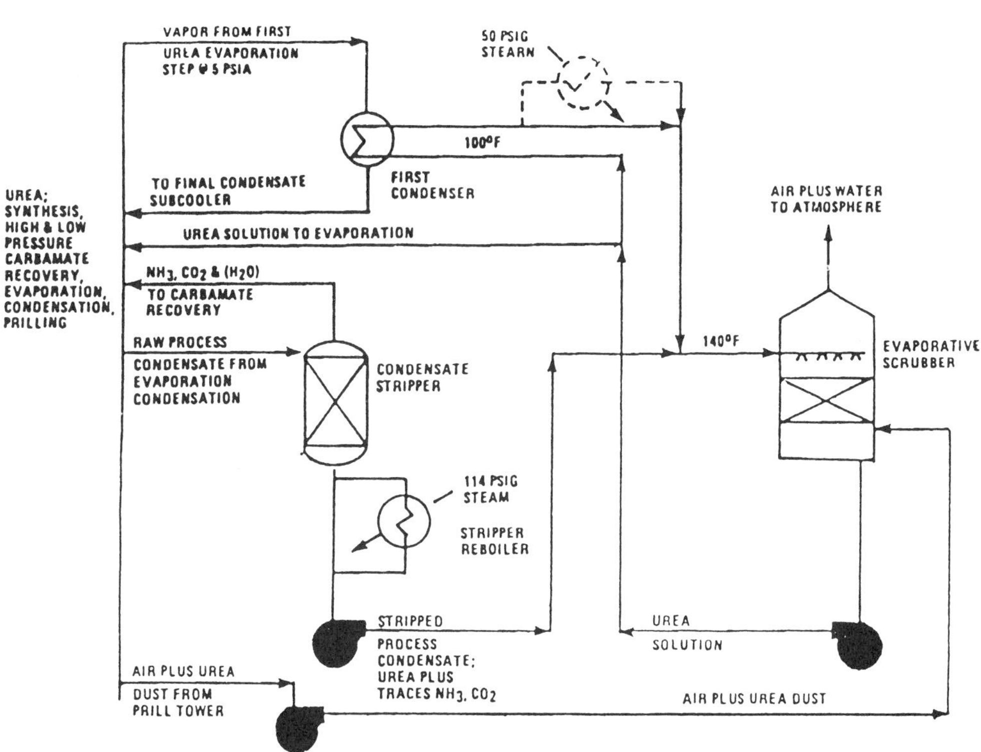

Fig. 1. Evaporative scrubbing system for urea production plant.

to produce 500,000 t/y urea. Construction is underway for a new facility to produce ammonium nitrate at the same plant. The Talkaha I plant has a capacity of 380,000 t/y of nitrogen fertilizers and Talkaha II plant produces 570,000 t/y urea. The Suez plant produces 250,000 t/y nitrogen fertilizers. The installed capacity for NH_3 production in Egypt will increase from 1.136 million t/y in 1985 to 1.466 million t/y in the year 2000 [2]. During the same period wastewater discharge from the fertilizer plants will increase from 37.6 million m^3/y to 57.9 million m^3/y [3]. Though air emissions from the urea prilling towers and NH_3 synthesis are controlled, some plants are not equipped with urea hydrolysis units and hence process condensates are discharged with high concentrations of urea (3–5 g/l) which causes adverse effects on the receiving water streams. Plans are underway to install waste recovery systems especially in Talkaha plants which discharge their effluents into the River Nile. A modified neutralization system for the production of ammonium nitrate has reduced pollution load and increased NH_3 recovery by about 15 t/d. This is equivalent to an increased fertilizer production by about 10,000 t/y [4].

The three urea plants in IRAQ are designed to produce 800,000 t/y; a new plant is planned in Baji with a rated capacity of 560,000 t/y [5]. Air pollution emissions are effectively controlled by wet scrubbing (Fig. 1); most of the residual urea in the process effluents is recovered by hydrolysis before final discharge to the water stream.

The Petrochemical Company of KUWAIT produces 990,000 t/y NH_3, of which about 65% is used for the production of 800,000 t/y urea. The purged gases that accumulate in the synthesis loop (methane, argon, H_2 and N_2) are saturated with NH_3. In the past, gas scrubbing and the discharge of NH_3-saturated effluent caused severe environmental problems, and a net annual loss of 3600 tons of NH_3. Presently about 99.5% of the lost NH_3 is being recovered and the clean washing water is recycled to the scrubber so as to alleviate pollution problems from this source. A new system that recovers urea dust from the prilling towers has also been installed. Previously, the process effluent contained about 1% each urea and NH_3, which rendered it unfit for either forestry or disposal into the sea. At present residual urea is hydrolyzed to NH_3 and CO_2. The effluent is then subjected to stripping to achieve concentration of maximum 60 mg/l each of NH_3 and urea; which makes the water suitable for agricultural purposes [6].

The two fertilizer plants in QATAR has a total capacity of 59,000 t/y NH_3 and 660,000 t/y urea. A gas sweetening plant has been installed recently to remove sulphur from offshore feedstock gases. Both urea dust from the prill towers and NH_3 gases are emitted in minor amounts while H_2S removed from natural gas is flared with the SO_2 emissions.

The Al Jubial Fertilizers Company in SAUDI ARABIA produces

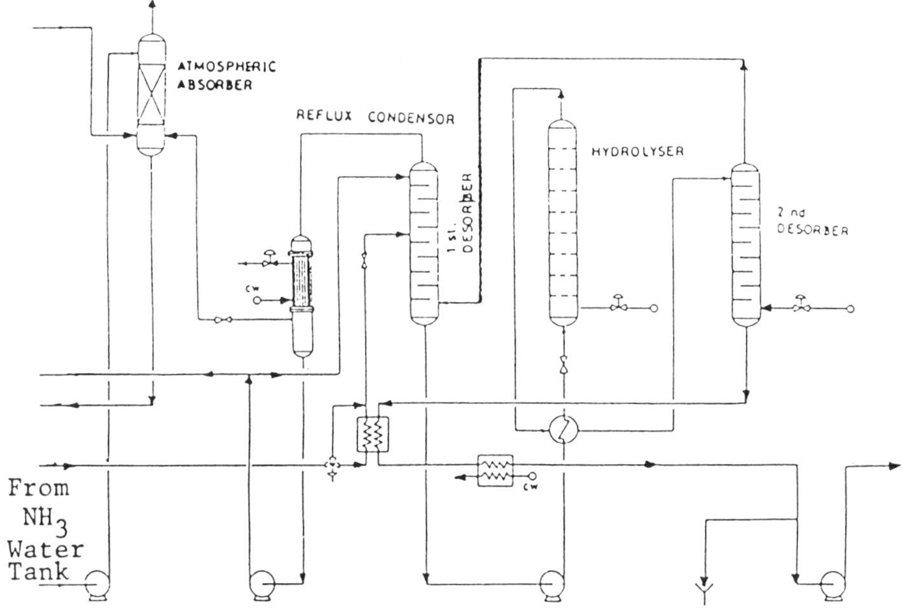

Fig. 2. Hydrolysis-desorption of residual urea in process condensate.

500,000 t/y urea and 300,000 t/y NH_3. Production from this plant meets the needs of the local market while another 500,000 t/y plant will be commissioned soon to cater for the export market. The plant employs a cryogenic system for the separation of hydrogen and recovery of NH_3 from the purge gas which is the cause of NO_x emissions in the primary reformers; urea emitted from prill towers is controlled by wet-scrubbers which give particulate emission of 0.08–0.24 kg/t of product, and NH_3 at a rate of 0.3 kg/t of product. Process condensates are treated in hydrolysis - desorption system where NH_3 as well as the urea hydrolysis products ($NH_3 + CO_2$) are recovered and recycled to the synthesis process.

The Ruways fertilizer plant in the UNITED ARAB EMIRATES produces 495,000 t/y urea and 30,000 t/y NH_3; both products are destined for the export market. The effluent consists of oily waste and process condensates with high levels of NH_3 and urea. The oily effluent is pretreated in an oil separator, and the process condensate is treated in hydrolysis-desorption unit as shown in Fig. 2, then subjected to aeration in oxidation pond before discharge into the Gulf. The urea dust from the prill tower is subjected to wet scrubbing and air emissions usually contain less than 30 mg/m^3 urea and 20 mg/m^3 NH_3 [6].

MANAGEMENT OF UREA FERTILIZER WASTEWATER: AN EGYPTIAN CASE STUDY

The AKFCI plant in Alexandria commenced production in 1979. The plant consumes 1.2 million m^3/d of natural gases supplied from Abu Kir offshore field. Production processes are depicted in Fig. 3.

The process consists of:

1. Ammonia Production

Natural gas has 95% methane, 3.2% ethane and about 2.8% mixture of propane, butane, isobutane, CO, S and N_2. Residues of S are removed by passing the gas through a zinc oxide layer which forms zinc sulphide. The gas is then subjected to a process of reformation in the presence of catalyst where the hydrocarbon is converted to H_2 and CO_2 by the following reactions:

$$CH_4 + H_2O \rightarrow CO_2 + 3H_2 \qquad \text{(endothermic)}$$
$$CO + H_2O \rightarrow CO_2 + H_2 \qquad \text{(exothermic)}$$

The overall reaction is endothermic (heat absorbing) and takes place in two stages:

- First in the primary reformer where indirect heat is supplied via natural gas burners.
- Second in the secondary reformer where indirect heat is supplied via burning part of the H_2 with a calculated quantity of air. The amount of air added is determined by the quantity of N_2 required for NH_3 production.

The outlet of the secondary reformer contains, apart from H_2 and N_2, CO_2, CO and small quantities of methane. Since CO is a poison for the downstream catalyst, the gases are passed through two shift converters containing catalyst where most of CO is converted to CO_2 with evolution of heat:

$$CO + H_2O \rightarrow CO_2 + H_2$$

The outlet gases after shift conversion contain about 0.35% CO. These gases are then sent to the CO_2 absorption section where CO_2 is absorbed preferentially in K_2CO_3 solution and the CO_2 recovered again from the K_2CO_3 solution by steam stripping. The recovered CO_2 is sent to the urea plant.

The gas coming out of the CO_2 absorption system still contains small quantities of CO and CO_2 which have to be removed. This is achieved in the methanator where the following reactions take place:

$$CO + 3H_2 \rightleftharpoons CH_4 + H_2O$$

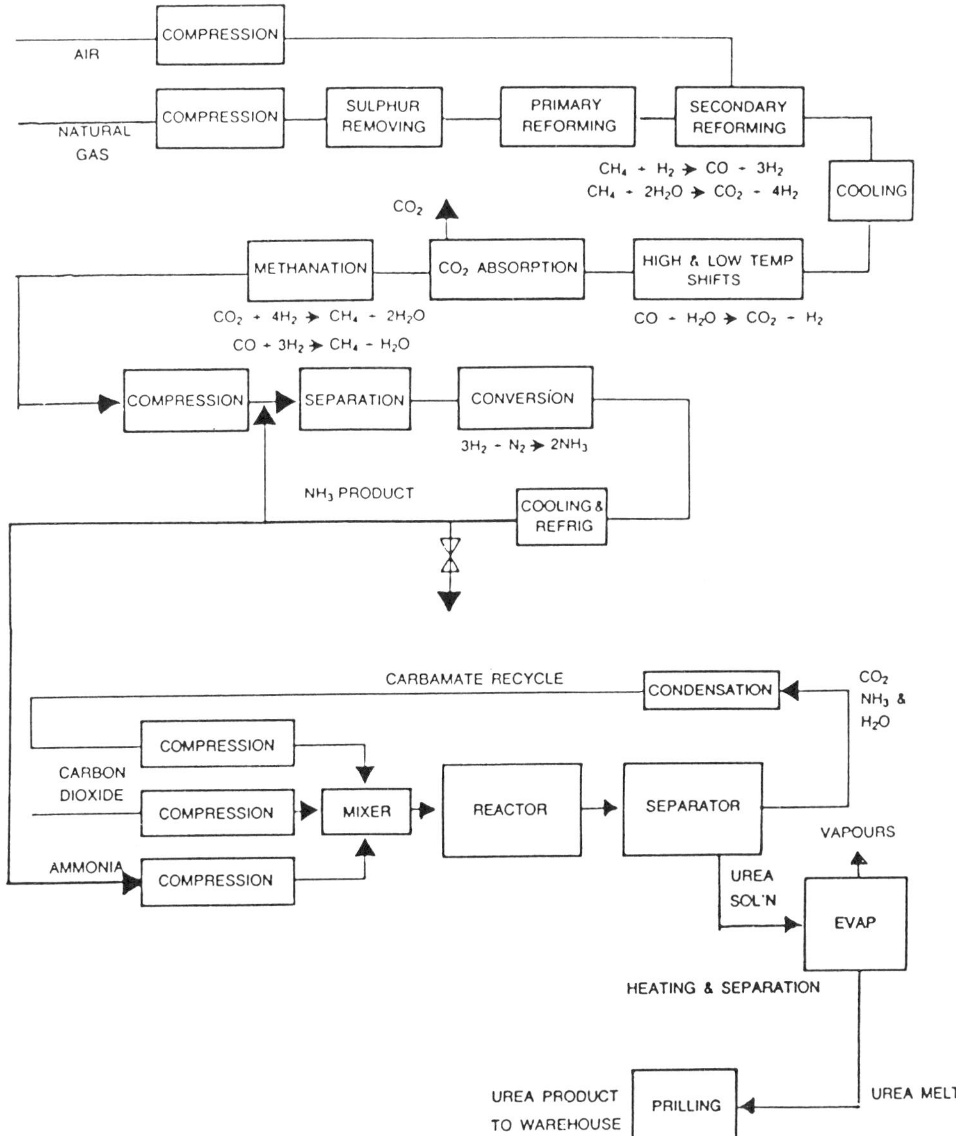

Fig. 3. Ammonia and urea processing at Abu Kir fertilizers.

$$CO_2 + 4H_2 \rightleftharpoons CH_4 + 2H_2O$$

Both reactions are exothermic and the outlet gas contains less than 10 ppm $CO + CO_2$. The synthesis gas leaving the methanator consists of H_2 and N_2 in the ratio of 3:1. This is compressed to a pressure of 220 kg/cm^2 and passed through an NH_3 synthesis converter containing an iron catalyst where NH_3 is produced at a temperature of 400–500 ^oC. The conversion per pass being 28%, the unconverted gas is recycled.

The produced gaseous NH_3 is cooled and liquefied by chilling. All the heat evolved in the various stages of the process is recovered to produce steam which is used either as a prime mover or as process steam.

2. Urea Production

Urea is produced by reacting liquid NH_3 and gaseous CO at about 190^oC and 150 kg/cm^2 pressure according to the following reactions:

$$2NH_3 + CO_2 \rightarrow NH_2COONH_4 \quad (1)$$
$$NH_2COONH_4 \rightleftharpoons NH_2COONH_2 + H_2O \quad (2)$$

In the first reaction, where the CO is converted to ammonium carbamate, the reaction is exothermic and goes to completion.

In the second reaction, ammonium carbamate decomposes to form urea and water. This reaction is endothermic and does not go to completion.

The carbamate formation of reactions (1) occurs in the high pressure condenser and the heat of formation evolved is used to produce low pressure steam which is utilized elsewhere in the process. The urea formation takes place in the reactor, also called the autoclave, where about 60% of the CO_2 added is converted to urea. The solution from the autoclave containing a mixture of urea, water and unconverted carbamate is fed to the high pressure stripper where the carbamate is decomposed back again to NH_3, CO and recycled in the system.

The solution leaving the high pressure stripper containing urea, water and undecomposed carbamate is expanded and heated with steam whereby the carbamate is decomposed to NH_3 and CO_2 which is recycled back to the synthesis section and the solution led to the urea solution tank.

The urea solution, which has a concentration of about 72% is further concentrated to 99.5% by two stage vacuum evaporation. The concentrated urea melt from the second stage evaporation is fed to the prilling equipment on top the prilling tower.

The prilling equipment distributes the urea melt in fine droplets over the cross section of the prill tower. During their fall in the tower, the droplets solidify and cool, the heat being carried away by air aspirated through holes at the bottom of the tower and via belt conveyors, are transported, after screening and further cooling, to storage.

3. Water Treatment Systems

The raw water from the nearby Rakta canal is withdrawn at an average rate of 1200 m^3/h. The treatment system consists of precoagulation with ferrous sulphate and polyelectrolyte followed by chlorination and rapid sand filtration. The treated water is stored in two storage tanks (2700 m^3 each) and used for domestic, processing and cooling purposes. The boiler water is further treated in anion/cation exchange units at a rate of 360 m^3/h.

Cooling water is recycled to five cooling towers at a rate of 25000 m^3/h while makeup varies from 400 to 600 m^3/h according to the rate of buildup of dissolved solids and other contaminants. A flowsheet of the water treatment system is shown in Fig. 4.

MATERIALS AND METHODS

A survey of the process effluents at AKFCI was carried out using an advanced mobile laboratory (Medicoach, USA) which has transmitters for Turbidity (Turb.), Dissolved Oxygen (DO), pH, and Chemical Oxygen Demand (COD). Other parameters monitored in the processing effluents were Biochemical Oxygen Demand (BOD_5), Total Solids (TS), Suspended Solids (SS), Volatile Solids (VS), Chlorides (Cl), Sulphates (SO_4), Alkalinity (Alk), Phosphates (PO_4), Total Hardness (T. Hard), NH_3 and Dissolved Organic Carbon (DOC) from which urea was calculated by multiplying by a factor of 5.

Trace metal analyses were performed using Jarell Ash Atomic Absorption Spectrophotometer model 850. Chemical and trace metal analyses were in accordance with the Standard Methods [7].

Treatability studies were limited to the process condensate from the urea plant which contains abnormally high concentrations of NH_3 and urea due to exclusion of the hydrolysis-desorption unit from the production system. The treatment train shown in Fig. 5 consisted of an NH_3 stripping and recovery unit (ASRU), enzymatic hydrolysis unit (EHU) and nitrification unit (NU). The ASRU encompasses a 220 cm packed column of 20cm internal diameter. The column is sealed from outside air and filled with a plastic media (Filter Pack, Mass Transfer) with a specific surface area of 118 m^2/m^3 and a volume void ratio of 0.93. Most of the NH_3 discharged to the air stream from the stripping column was removed in the adsorption unit which consisted of a 150 cm glass tube of 6 cm internal diameter. The

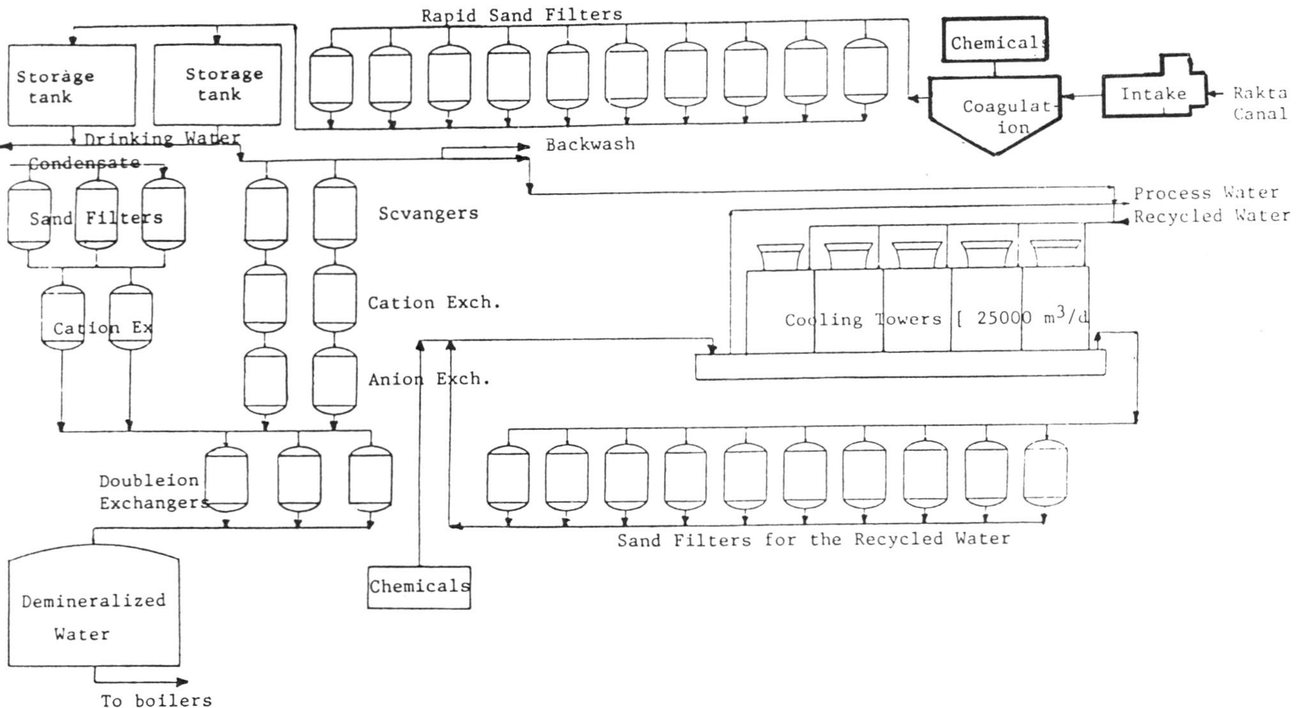

Fig. 4. Water treatment system at Abu Kir fertilizer plant.

Process
Condensate
C
D
Neutralizat-
ion tank
Stripped
gases
Vent
Stripping
Column
Plastic
Packing
Air
Dil.
NH3absorption
Tube
Ammonia Stripping/Absorption Unit
A
Enzymatic
Hydrolysis
Unit
A. Influent to hydroly
B. Distribution Head
C. pH Monitor-Controller
D. Chemical Mixing
E. Sludge Outlet
F. Hydrolysis Compartement
G. Filtration Compartement
H. Overflow Outlet
I. Backwash Effluent
J. Backwash Influent
Nitrification Unit
L. Primary Clarification
M. DO Monitor
N. Aeration Compartement
O. Secondary Clarification
P. Sludge Outlet
Q. Air Diffusers
R. Recirualation Pump
S. Return Sludge
T. Sludge Wasting
U. Final Effluent

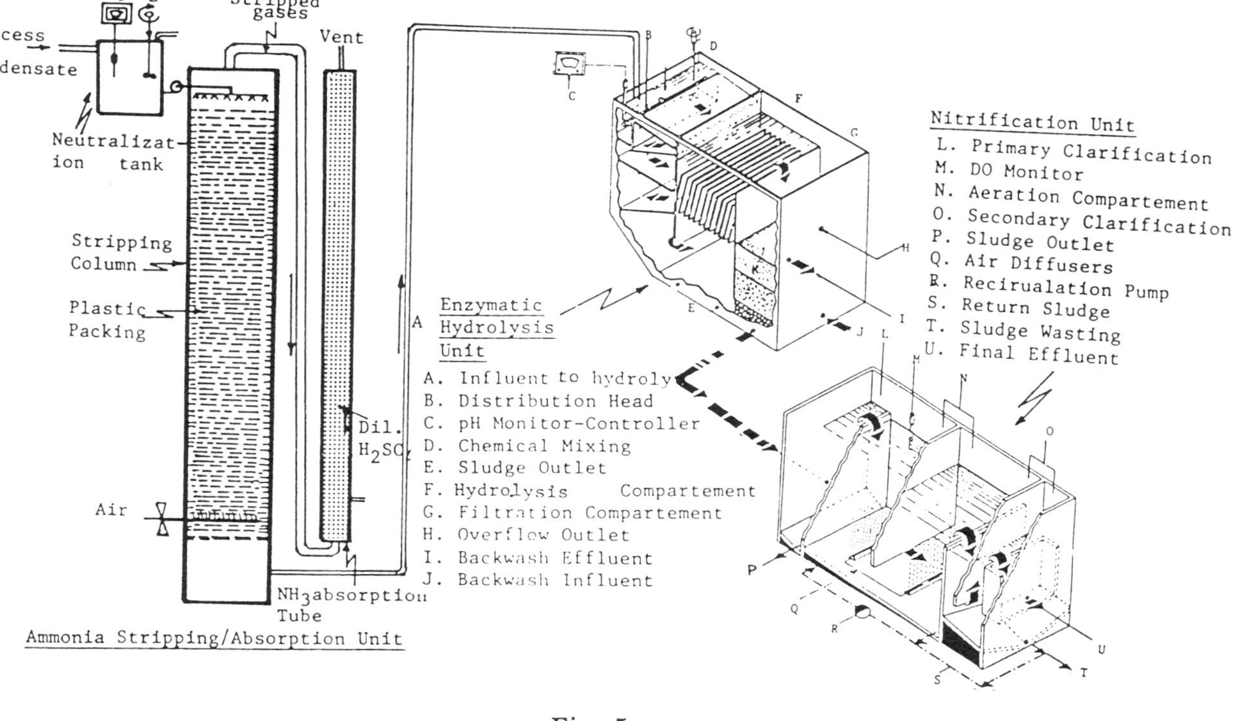

Fig. 5.

absorption medium was dilute sulphuric acid which effectively trapped the NH_3 gas to form ammonium sulphate. The absorbent was periodically replaced with fresh acid when the ammonium sulphate concentration reached a level of 25–30%. The stripped effluent was fed to the EHU at a controlled rate using a Masterflex multihead pump model 7536-60. The EHU included neutralization (3.6 l), hydrolysis (76 l) and filtration sections (30 l). The filter contained a bottom layer (20 cm) of 1.1 mm sand and a top layer which consisted of 95% sand 0.45 mm and 5% powdered activated carbon. The effluent of the EHU was continuously fed to the NU at predetermined flow rates. The NU has an aeration section of 40 l volume followed by a clarification section of 10 l. Air flow to the aeration unit was monitored by YSI DO meter model 57. The pH was adjusted during various treatment stages by a pH monitor/controller (Coliparmer Model 5651-00).

RESULTS AND DISCUSSION

Survey of AKFCI industrial effluents

The water and wastewater sources selected for this survey study, were as follows:

1. Raw water from Rakta Canal;
2. Treated water;
3. Recycled water from the cooling towers;
4. Effluent from the ammonia plant;
5. Effluent from the urea plant;
6. Combined effluent discharged to the Abu Kir Bay.

Raw water withdrawn from Rakta Canal is heavily polluted as it receives appreciable quantities of the polluted effluents from upstream industrial plants. The AKFCI is located at the end of the canal. The poor quality of the raw water requires excessive use of chemicals to achieve the desired water quality needed for the various processing operations.

Cooling water constitutes 75–80% of the total water use. It includes water coming in contact with the gases and water used for indirect cooling. Indirect cooling and part of the direct contact water are passed through the cooling towers and the rapid sand filters, and recycled after makeup with fresh water.

The cooling water overflow is presently discharged with the process effluents to the adjacent Abu Kir Bay. The results of the analytical survey indicate that the cooling water effluent has a comparable quality to that of the Rakta Canal water so it is advisable to divert the overflow to an upstream location in the canal rather than its discharge in the Mediterranean. This will

Table 1. Trace metal analysis of the AKFCI combined effluent (μg/l)

	Cr	Pb	Fe	Zn	Cd	Ni	Mn	Cu
x	36.2	83.8	220	421	10.1	82.1	119	26.3
R	15- 75	65- 120	140- 230	250- 650	6- 17	65- 130	80- 174	5- 20
DS	23.3	21.3	61.0	148	4.9	45.3	41.5	5.4

$n = 15$

increase the canal flow by 1500 m^3/d to meet the increased water demands for the various industrial and agricultural projects in the area.

The NH_3 and urea processing effluents constitute 15–20% of the combined effluent. They contain abnormally high concentrations of NH_3 and urea. The decision to exclude the hydrolysis-desorption unit during the installation of the plant was merely based on economic reasons without due regard to the devastating environmental consequences due to continued disposal of about 8–10 t/d urea in effluents discharged in the Bay. Suspended solids in the combined effluent consist mainly of inorganic matters which do not settle readily, thus reducing the transparency of the water. Suspended solids in the wastewater may also plug the fish gills and form deposits on the Bay bottom which impede the normal growth of flora and fauna and therefore hinder the development of aquatic life.

The high COD in the process effluents is attributed to the organic pollutants in the raw water and the added organic load from the residual urea in the process condensate.

The abnormally high level of NH_3 emitted with the combined effluent to Abu Kir Bay has an appreciable adverse effect on fish life. Factors which may increase NH_3 toxicity are high pH, elevated temperatures, CO_2, and bicarbonate alkalinity. Other problems created by excessive discharge of NH_3 and urea include eutrophication or uncontrolled algal growth, odour problems and lower DO levels.

Trace metal analysis of the combined effluent given in Table. 1 indicates normal concentrations of the tested metals; their presence at these low levels cannot be attributed to release from industrial sources such as spent catalysts but may emanate from equipment wear and corrosion of pipes and storage tanks.

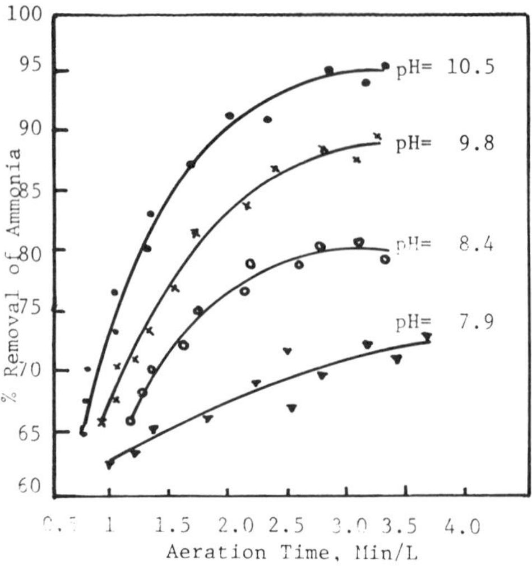

Fig. 6. Effect of aeration time on ammonia stripping.

TREATMENT OF UREA PROCESS CONDENSATE

1. Ammonia Stripping

Stripping is considered an easy and simple method for removal of NH_3 from the urea process condensate. Fortunately, the problem of reduced removal efficiency at low temperatures ($< 10^oC$) was not encountered in this study as the ambient air temperature was always in the range of $18 - 20^oC$.

As the pH of the influent was in the range of 7.0–8.5, it was presumed that both ammonium ion (NH_4^+) and dissolved NH_3 gas were present in solution. At pH=7 only NH_4^+ may be present while at pH=12 only NH_3 exists which speeds up the process of gas liberation and removal [8].

The pH was adjusted in the alkaline range using spent caustic from a nearby paper plant. Removal efficiency of NH_3 at various pH and aeration periods are depicted in Fig. 6.

Results indicate an optimum removal efficiency of about 95% at pH=10.5 using an aeration period of 3 min/l. The plastic packing used in ASRU was preferred over wood chips as it does not suffer from delignification at high pH values. The effluent of ASRU contained residual NH_3 in the range of 14–25 mg/l while residual urea levels were greater than 3 g/l; this is expected as stripping has little or no effect on the removal of organic nitrogen. As a result of mixing with other process and utility effluents the NH_3 concentration may decrease by a third to a quarter to about 4–6 mg/l in the combined effluent discharged into the Bay. Available acute and chronic NH_3 toxicity data for salt water organisms are very limited. Mean LC_{50} values for marine invertebrate species range from 0.94 to 18.3 mg/l NH_3 [9].

Based on the results of this phase of the study the final effluent discharged from the plant contains an NH_3 loading of 0.12–0.16 kg/t product which compares favourably to the USA limitations of 0.27 kg/t for the prilled urea plants. On the other hand, organic nitrogen release is about 1.12 kg/t compared to a maximum USA limit of 0.46 kg/t [10, 11].

The release of excessive concentrations of urea has promoted eutrophication and created unfavourable conditions to the aquatic life in the Abu Kir Bay. The reduced water quality is also causing severe operational problems in the adjacent power plant. A study has been just commenced to assess means of protecting water intake to the power plant; a top priority issue is to study methods of eutrophication control caused by excessive urea release from AKFCI.

To solve this problem it is suggested to use the stripped effluent which contains high urea levels for irrigation purposes in the nearby agricultural areas surrounding the AKFCI plant. The effluent is rich in organic nutrients needed for soil enrichment, while residual NH_3 may be detoxified by combination with carbon skeletons without straining carbohydrate metabolism. Preliminary investigations carried out during the course of this study indicated the suitability of the stripped effluent for irrigation of various crops (cotton, wheat, and vegetables). If it is mandated that AKFCI effluent is to be discharged to the Bay, urea process condensate must be treated either by desorption hydrolysis which is still being considered by AKFCI management as economically unjustifiable or by enzymatic hydrolysis and nitrification as suggested below.

2. Urea Removal

The NH_3-stripped effluent was neutralized to pH 7.0–7.3, and then subjected to enzymatic hydrolysis in an EHU using powdered soyabean as a source of urease. The hydrolysis reaction is represented by the equation

$$NH_2CONH_2 + H_2O \xrightarrow{\text{urease}} 2NH_3 + CO_2$$

The hydrolysis was carried out under the following conditions: temperature (29–31°C), pH (7.3–7.8) using periods up to 20 days. Because the separation and purification of the enzyme is difficult and expensive and extensive purification often reduces enzyme activity, the enzyme used was in

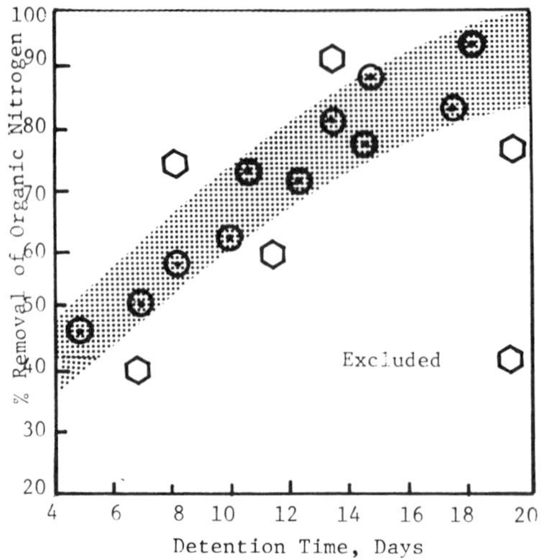

Fig. 7. Effect of hydrolysis time on removal of organic nitrogen.

the crude form with a considerable content of inert material (soya powder). The results of urea hydrolysis were erratic as shown in Fig. 7. This irregular performance was attributed to the high substrate concentration in the influent, use of impure enzyme, use of comparatively low temperature for hydrolysis and the predominance of alkaline conditions which may inhibit enzyme activity. The residual organic nitrogen of EHU was in the range of 220–250 mg/l using a detention period of 16 days. This corresponds to an average urea content in the effluent of 470–535 mg/l.

As urea hydrolysis results in stoichiometric formation of NH_4^+; such high NH_4^+ levels may cause acute fish toxicity, loss of equilibrium, hyperexcitability and oxygen uptake. Thus NH_4^+ was removed in the last stage of treatment by biological nitrification.

Biological nitrification is achieved by two autotrophic bacteria as follows:

- Oxidation of NH_4^+ to nitrite by Nitrosomonas

$$NH_4^+ + 1.5\ O_2 \rightarrow 2\,H^+ + H_2O + NO_2^-$$

- Oxidation of nitrite to nitrate by Nitrobacter

$$NO_2^- + 0.5\ O_2 \rightarrow NO_3^-$$

The overall reactions is represented by the following equation

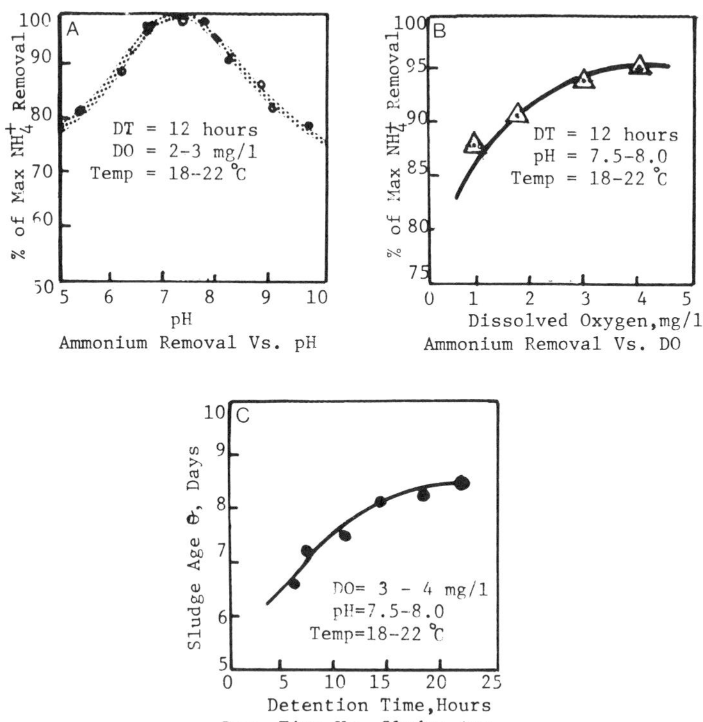

Fig. 8. Effect of operating conditions on the nitrification of process condensate

$$NH_4^+ + 1.86\ O_2 + 1.98\ HCO_3^- \rightarrow$$
$$\rightarrow 0.021\ C_5H_7NO_2 + 0.98\ NO_3 + 1.04\ H_2O + 1.88\ H_2CO_3$$

Since nitrification reduces the HCO_3^- level and increases the H_2CO_3 level, it tends to depress pH, which can be offset by stripping of CO_2 during aeration [12, 13].

Nitrification was achieved by an attached growth biological system. The influent of the EHU was combined with the biological solids returned from the final clarifier and biocell (packed aeration section recycle) to form a mixed liquor which was pumped to the biocell at a controlled rate. This mixture when passed through the biocell was able to create an attached bacteria growth. This system provided nitrification of residual NH_4^+, and oxidation of biodegradable organics. Preliminary investigation indicated optimum operation at mixed liquor volatile suspended solids MLVSS of 3–4 g/l. The results illustrated in Fig. 8 indicate a marked decline of the nitrification rate as the

Table 2. Values of LC_{50} for urea process condensate before and after air stripping

Species	Stage	Age days	Length cm	96 LC_{50} %Vol/Vol Influent 650 mg/l NH_3	Treated 8-10 mg/l NH_3
Cyprinus Carpio	Hatched embryo	2	0.5-0.6	–	6.3
	larval stage	7	–	1.9	6.3
Mugil Capito	Fry	45	2.0-3.0	1.1	8.9
Mugil Cephalus	Fry	30	1.9-2.5	1.5	9.2
Sparus Aurata	Fingerlings	120	7.2-9.0	1.5	7.6

Note: Fish toxicity is rated:
very high, when 96 LC_{50} 0-5%Vol/Vol,
high, when 96 LC_{50} 6-12%Vol/Vol.

pH moves towards the acidic or the alkaline range; optimum pH was about 7.5. The concentration of DO was a major factor in controlling the nitrification process. Though levels less than one mg/l DO may support the growth of nitrifiers, it was evident that optimum nitrification required a minimum DO of 3 mg/l. However higher DO concentrations were not warranted as shown in Fig. 8B. The amount of oxygen required for nitrification is estimated at 4.6 kg O_2/kg NO_3^- while oxygen needed for endogenous respiration of the sludge is 0.1 kg O_2/kg MLVSS. It should be noted that the above nitrification conditions are valid only for the range of temperature used in this study (18–22°C).

Other control parameters evaluated in the treatability study were detention time (DT) and sludge age (Θ_c). Optimum detention time was 13 hours. Sludge age defined as

$$\Theta_c = \frac{X_a V}{X}$$

where:
Θ_c = sludge age, days,
$X_a V$ = average aeration basin MLVSS,
X = sludge wastage, mass per day, and
V = basin volume.

As shown in Fig. 8C Θ_c increased as DT increased with an optimum value for Θ_C in the biocell of 8.2 days.

Fish bioassay was carried out according to the Standard Methods [7].

The results of Table 2 indicate that the raw process condensate is extremely toxic for all tested fish species due to its high NH_3 content. To overcome this toxicity, a dilution factor of at least 100 times may be required. The treated effluent with much lower levels of NH_3 still exhibits appreciable toxicity as the effluent requires 20 time dilution to avoid the toxic and pathological damage to the tested fish species. Such dilution can be achieved by mixing the stripped condensate with other NH_3-free process and utility effluents, and by dilution when mixing with sea water in the Abu Kir Bay.

REFERENCES

[1] Environmental Protection Committee, Assessment of Source of Air, Water and Land Pollution in the State of Bahrain, Manama, Bahrain 1985.

[2] UNIDO, Industrial development review series: Egypt, United Nations Industrial Development Organization, IS, 637, May 1986.

[3] UNDP, Industrial water and wastewater production, United Nations Development Programme, Report No 10 UNDD-EGY/73/024, March 1981.

[4] Salem, M.M., Recycling of industrial wastes in the fertilizer industry, Proceedings of Int. Seminar on Clean Technology and Pollution Treatment, UNEP Industry and Environment, Cairo, March 1988.

[5] ESCWA, A framework for a masterplan for the development of technological capabilities in the oil refining, petrochemicals and fertilizer Industry, Economic and Social Commission for Western Asia, E/ESCWA/ID/85, 1985.

[6] Hamza, Abatement of pollution in petroleum refining and downstream industries in the Arab world, Submitted for Publication, J. of Industrial Cooperation in the Gulf, 1989.

[7] Standard Methods for the examination of water and wastewater, APHA-AWWA-WPCF, 15th ed., American Public Health Association, Washington D.C., 1980.

[8] EPA, Process design manual for nitrogen control, US Environmental Protection Agency, Technology Transfer, October 1975.

[9] WHO, Ammonia - Environmental Health Criteria, World Health Organization, EHC 54, 1986.

[10] WHO, Compendium of environmental guidelines and standards for industrial discharge, World Health Organization EFP/83. 49, 1983.

[11] EPA, Effluent guidelines and standards, fertilizer manufacturing point source category, Federal Register CRR-40-part 418, 385-403, July 1983.

[12] Culp, R.L. et al., Handbook of advanced wastewater treatment, Van Nostrand Reinhold Environmental Engineering Series, 1978.

[13] Bress, D., Eliminating effluents from urea and ammonium nitrate plants, Proceedings of the Fertilizer Institute Environmental Symposium, New Orleans, Louisiana, 123, 1976.

CHANGES IN WATER CHEMISTRY

IN THE SOUŠ RESERVOIR

L. MACEK, M. MACH and A. GRÜNWALD

Technical University of Prague
Faculty of Civil Engineering
Department of Sanitary Engineering
Praha, Czechoslovakia

ABSTRACT

The Souš Reservoir is situated in the Protection Region of the Jizera Mountains, and since 1974 it has been used as one of the most important sources of drinking water for the Jablonec - Liberec conurbation. The reservoir was built in 1912–15 and reconstructed in 1924–27. In the last decade the region in which the reservoir is located was strongly affected by acid depositions. Forest death, soil acidification and erosion as well as changes in water quality were observed there. This paper deals with the evaluation of changes of the water quality in the Souš Reservoir during the last 10 years, i.e., with pH and aluminum content. The changes were evaluated on the basis of monthly averages and smooth averages.

INTRODUCTION

Ecological effects of acid deposition have not been fully recognized yet, but they are numerous. Among them are acidification of natural water sources and surface waters, leaching of soil, damage to vegetation and corrosion of materials. The term "acid deposition" is now used for precipitation with large amounts of hydroxide ions (H_3O), increased concentrations of sulphate and nitrate anions as well as for precipitation with a high content of ammonium, various heavy metals and other trace elements and organic micropollutants.

The principal precursors of acid deposition are emissions of sulphur dioxide (SO_2) and nitrogen oxides (NO_x), but acidity is affected also by a variety of other emissions, such as hydrochloric acid, ammonia, volatile organic compounds and alkaline dust. These pollutants are both natural and man-made. Natural emissions include biogenic emissions from terrestrial, tidal and nutrient-rich oceanic areas and non-tropogenous emissions from natural combustion, geothermal activity, lighting and airborne soil and water aerosols.

The form of acidity in depositions is not quite clear; generally two types of acidity are recognized, strong and weak acids. A large array of proton sources occurs in depositions. Detailed chemical analyses of acid precipitation show the presence of mineral acids H_2SO_4 and HNO_3, weak acids such as organic acids and H_2CO_3 and Brönsted acids (e.g., dissolved Fe, Al and NH_4^+).

The most serious effects of acid depositions on catchment water quality are considered to be decreased pH and alkalinity and increased base cation and aluminum concentration. In Czechoslovakia, according to the observations near Dubá in 1981, the average year values of pH ranged from 4.0 to 4.5. The extreme values in 1981 were 3.7 and 6.1. Lower pH values were observed in the north-westerns areas of Czechoslovakia [1,2].

The Souš Reservoir is situated in the north of the Czechoslovak Republic, in the Jizera Mountains, north-east of the town Jablonec n. N. The Jizera Mountains are in the region strongly affected by acid depositions. The depositions of SO_2 in the Krkonoše Mountains, east of the Jizera Mountain, are, according to Kurfürst [3], caused by power stations: half of the pollution is from Czech sources, the rest from East Germany and Poland. Weiss [4] has found the long-term average of average day concentrations of SO_2 in the Jizera Mountains to be about 70 $\mu g.m^{-3}$, but the average day concentrations during winter are sometimes higher than 600 $\mu g.m^{-3}$.

The Jizera Mountains are one of the most humid regions in Czechoslovakia. The annual average precipitation ranges from 1,200 to 1,600 mm. The surface water quality of this area was studied by Mach et al. [5] in eight small streams. The lowest pH value of 3.04 and the highest concentrations of aluminum were determined during spring thawing. The highest pH values occurred in summer, with a maximum of 7.62. Rivers rising in peat-bogs, such as the Kamenice River, are more acid. This phenomenon is caused not only by acid depositions, but also by an increased content of humic acids.

The Souš Reservoir was built in 1912-1915 on the Černá Desná River, reconstructed in 1924–27, and since 1974 it has been used as a source of drinking water for the Jablonec-Liberec conurbation. It is 770 m above sea level with water intakes at 754 and 760 m above sea level. The maximum depth of the reservoir is about 17 m. The basin of the reservoir is at present deforested. The forests seriously damaged by acid depositions were cut down in 1988. Sheet and especially rill erosion in the places of the transport of wood constitute a great problem to this area. The erosion increases the amounts of mineral and suspended solids entering the reservoir.

EVALUATION OF ACIDIFICATION OF THE SOUŠ RESERVOIR

The results of weekly analyses done by the laboratory of the Souš Water Treatment Plant were used for evaluation of the acidification of the water in the Souš Reservoir. Monthly averages of pH in this reservoir in the period 1975–85 are presented in Fig. 1. It can be seen that the values of monthly pH averages oscillated from 3.4 to 5.0. The lowest value was determined in 1981, the highest in 1982. The average monthly aluminum concentrations of this period are shown in Fig. 2. Aluminum concentration increased after 1982. The lowest concentration of Al was 0.41 $mg.l^{-1}$ by the end of 1982, the highest 1.47 $mg.l^{-1}$ in 1984.

Smooth monthly averages were calculated from the monthly averages using the following equation:

$$x_i^S = \frac{1}{N} \sum_{i=i-N/2}^{I+N/2} \overline{x_i} \qquad (1)$$

where:

x_i^s is the smooth monthly average in month i

N is the amount of monthly averages (N=12)

$\overline{x_u}$ is the average value in month i

The smooth averages are shown jointly with monthly averages of pH and aluminum concentrations in Fig. 1 and 2. It is evident that the smooth monthly averages better express the development of acidification.

While in 1975–81 pH was decreasing continuously, in the following year it suddenly increased from 3.97 (yearly average in 1981) to 4.64 (1982). Comparing the smooth monthly averages of pH with the smooth monthly precipitation totals (Fig. 3), it can be found that increased pH occured in the period with the lowest precipitation of the whole studied period 1975–85. The plot of the empirical autocorrelation function of monthly averages of pH in the Souš Reservoir is shown in Fig. 4. The values of the autocorrelation

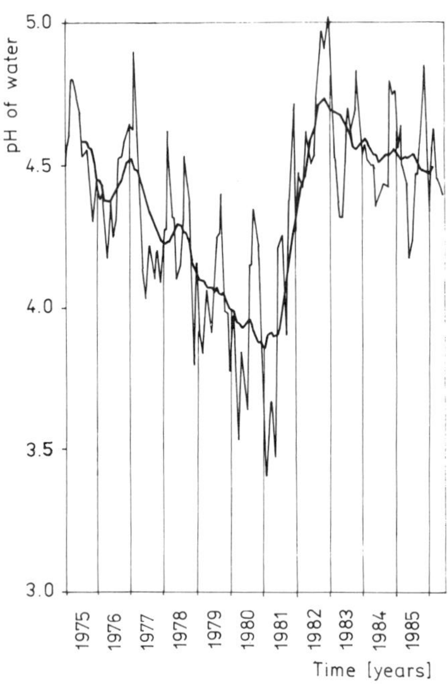

Fig. 1. Monthly averages (fine line) and smooth monthly averages (thick line) of pH in water of the Souš Reservoir during 1975–1985 period.

coefficient are calculated from the equation [6]:

$$r_\tau = \frac{\sum_{i=1}^{m}(x_i - \overline{x}_i \cdot (x_{i+\tau} - \overline{x}_{i+\tau})}{\sigma_i \cdot \sigma_{i+\tau} \cdot (m-1)} \tag{2}$$

where:
x_i are the random realizations of the process from 1 to m
$x_{i+\tau}$ are the random realizations of the process from $1+\tau$ to $m+\tau$
τ is time parameter of the autocorrelation function ($\tau = 0$ to $\tau = m - n$)
$\overline{x}_i$, $\overline{x}_{i+\tau}$ are the mean values of the series of variables x_i, $x_{i+\tau}$
$\overline{\sigma}_i$, $\overline{\sigma}_{i+\tau}$ are the standard deviations of the series of variables x_i, $x_{i+\tau}$
m is the total of variables, n=12

The critical value of the confidential interval (r_α) for levels of importance $\alpha = 0.02$ were calculated using the tables of normal distribution. It can be seen from Fig. 4 that in 1981 the seasonal component of pH fluctuation in the reservoir was absolutely eliminated.

From Fig. 1 to 4 it is evident that the effects of quantity and quality of precipitation on acidification increase are insignificant; pH increase in 1982

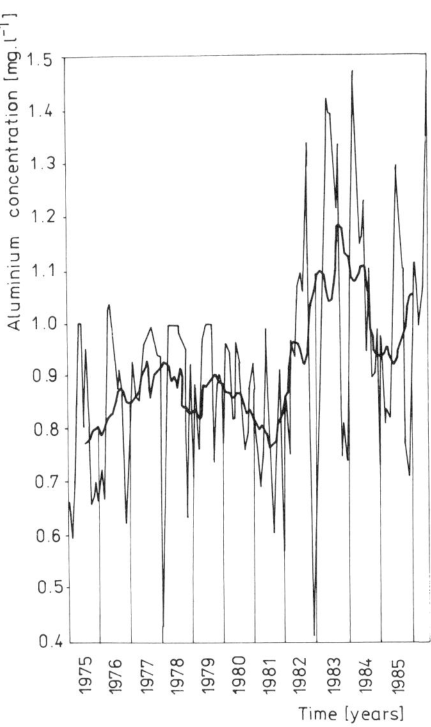

Fig. 2. Monthly averages (fine line) and smooth monthly averages (thick line) of concentration of aluminum in water of the Souš Reservoir during 1975–1985 period.

with lower precipitation in the studied period was caused especially by lower acid depositions in catchment area and by lower run-off of acid water from peat-bogs.

CONCLUSIONS

Gradual development of acidification can be deduced from calculated monthly averages of pH of water and aluminum concentration in the Souš Reservoir in the Jizera Mountains. The danger is that after exhausting the buffering capacity of the basin this process could be intensified with all negative effects on the environment. A close relationship was found between yearly precipitation totals and pH of water in the reservoir. Lower precipitation totals were accompanied by an increase of pH of water and vice versa.

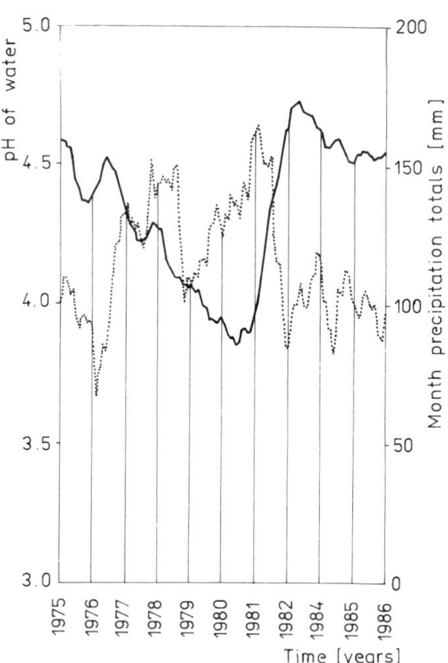

Fig. 3. Smooth monthly averages of pH in water of the Souš Reservoir (thick line) and smooth monthly totals of precipitation (dotted line) during the 1975–1985 period.

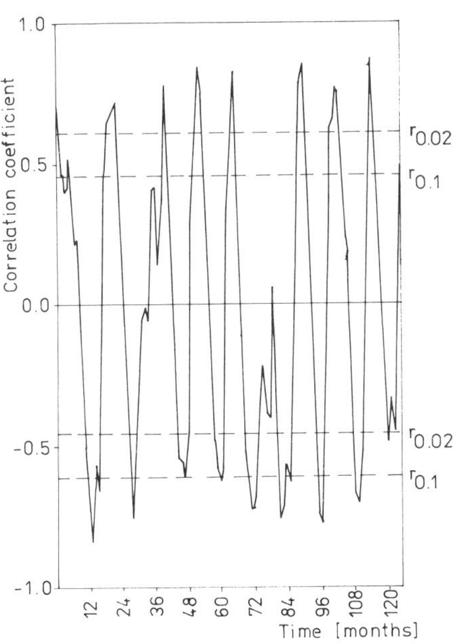

Fig. 4. Plot of empirical autocorrelation function of monthly averages of pH in water of the Souš Reservoir (N=12) during the 1975–1985 period.

REFERENCES

[1] Richter, V.:Czech hydrometeorologic institute and observations of water quality. VTEI 26, 2, p. 55, (in Czech), 1984.

[2] Jiřele, V.: Secondary affects of air pollution on water sources. Conf. on biosphere, ČSVTS Praha, (in Czech), 1984.

[3] Kurfürst, J.: Evolution and prognosis of air pollution in middle Europe. Conf. Vodohospodářské dusledky imisní kalamity v Jizerských horách. ČSVTS Ústí n. L., p. 35-39, (in Czech), 1988.

[4] Weiss, K.: Imission situation of the Jizera Mountains and the Frýdlant area. Conf. Vodohospodářské dusledky imisní kalamity v Jizerských horách. ČSVTS Ústí n. L., p. 40-43, (in Czech), 1988.

[5] Mach, M. et al.: Study of the metal content in the waters of the Protected Region of the Jizera Mountains. Technical Papers, Tech. Univ. of Prague, Fac. of Civil Eng., V, No. 8, p. 111-122, 1988.

[6] Votruba, L. and Broža, V.: Water Management in Reservoir. SNTL Praha, 2. edition, p. 143-147, (in Czech), 1980.

METAL LIXIVIATION OF STEEL FOUNDRY DUST

A. ANDRÉS, J. VIGURI, P. BILBAO and A. IRABIEN

Departamento de Ingeniería Química
Facultad de Ciencias Universidad del País Vasco
Apdo. 644. 48080 Bilbao Spain

ABSTRACT

Solid waste, steel foundry dust, comes from gas filtration units in the steel industry. Taking into account two main manufacturing processes leading to common steel and special steel, two different byproducts have been investigated: dust P1 from special steel manufacture and dust P2 from common steel manufacture.

A method based on leaching procedures is used in the toxicity evaluation of solid wastes. The U.S. EPA Toxicity Characteristic Leaching Procedure (TCLP) tries to simulate a weakly acidic medium usually found in sanitary landfills, and Distilled Water Extraction Procedures try to characterize the leaching behavior of solid wastes. Both methods have been applied in this work for characterization of steel foundry dusts.

The toxic metals under consideration are chromium, cadmium and lead. The results for special steel foundry dusts using TCLP show a mean value of: $Pb = 32 \pm 2\ ppm$; $Cr = 0.02\ ppm$; $Cd = 10 \pm 1\ ppm$, and for common steel foundry dust: $Pb = 355 \pm 8\ ppm$; $Cr = 0.02\ ppm$; $Cd = 25 \pm 2\ ppm$.

The Distilled Water Extraction Procedure leads to a different final pH value depending on the dust; special steel dust, 11.8 and common steel, 3.2. The pH difference seems to be responsible for the obtained values for metal leaching; special steel dust: $Pb = 0.7 \pm 2\ ppm$, $Cd = 0.1 \pm 0.04\ ppm$ and $Cr > 10\ ppm$; common steel dust: $Pb = 2.8 \pm 0.3\ ppm$, $Cd = 27.5 \pm 2\ ppm$ and $Cr = 0.06 \pm 1\ ppm$.

Chemistry for the Protection of the Environment
Edited by L. Pawlowski *et al.*, Plenum Press, New York, 1991

INTRODUCTION

Steel foundry dusts are the solid wastes exhausting from filtration units in steel industries.

Two different raw materials are usually employed in the steel manufacturing processes: (i) pig iron from blast furnaces and (ii) scrap iron. In the first case the whole installation includes an oxygen convertor, whereas with scrap iron only furnaces are used. Even if the latter is the most commonly established installation, both procedures lead to the emission of high volumes of steel foundry dusts [1].

The resulting solid waste is a dusts (60% is smaller than 5μm) with variable composition of heavy metals (Cd, Pb, Cr ...), depending on the following characteristics: (a) composition of the raw material, (b) oxygen utilization and (c) furnace work conditions [2].

In general terms two different types of steel foundry dust can be distinguished: (i) that generated in the special steels manufacture processes, where the raw material is previously classified and mixed with different alloys depending on the desired characteristics and (ii) dust produced in the manufacture of common steel with a higher content of carbon and where the scrap materials coming from galvanoplastic processes have a higher content of Zn and Pb. A typical description of the main components of both types of steel foundry dust is shown in Tables 1 and 2 [3].

Only in recent years has attention been paid to the toxicity of this industrial waste, which usually has no further utilization because of the environmental degradation caused by its storage and/or disposal [4].

Because of the small size of steel dusts they present a high surface value for leaching and they are also easily dispersed by wind action causing environmental problems in the nearest zones; therefore it is convenient to dispose of this waste in controlled landfills. In this case and in order to avoid pollution of groundwaters from leachates as well as atmospheric pollution it is desirable to perform inertization processes [6]. Taking into consideration the recycling technologies for this type of waste, three main groups can be distinguished: (i) classical technologies for the recovery of zinc and lead which are very dependent on the amount of zinc in the dust to be treated (Waelz furnace) [5]; (ii) new plasma based technologies which demand a big effort in research and development but which as yet have clear results; and (iii) processes tending to reuse dust in the same foundry installation by means of a physico-chemical treatment [6].

From an analysis of the different alternatives the USEPA (American Agency for Environmental Protection) considers the solidification/inertization process the most promising technology in order to control the environmental impact of waste disposal [7, 8]. The toxicity of two different steel foundry

Table 1. Analysis of the Composition of Dust Coming from the Manufacture of Special Steels in %

Zn	10	–	20
Pb	2	–	4
Cu	0.2	–	0.8
Sn	0.04	–	0.1
Cd	0.04	–	0.06
S total	0.06	–	1.2
F	1	–	2
Cl	0.1	–	0.3
C	0.2	–	0.6
FeO	26	–	30
MnO	4	–	5
CaO	5	–	12
MgO	1.2	–	4.5
BaO		$<$	0.01
Al_2O_2	0.4	–	0.8
SiO_2	1.5	–	6.0
Na_2O	0.5	–	0.2
K_2O	0.8	–	2.5

dusts has been evaluated: dusts coming from the manufacture of special steels – P1, and dusts coming from the manufacture of common steel – P2, using two different leaching tests in order to get consistent values of the metal levels in the leachates.

EXPERIMENTAL

Different leaching procedures have been developed for industrial wastes, and comparisons have been reported [9]; the leaching tests employed in this work to characterize foundry dusts are TCLP (Toxicity Characteristic Leaching Procedure) and a second one based on extraction with distilled water (Equilibrium Leach).

The TCLP test a simulation of the waste leaching procedure in a sanitary landfill. It is very appropriate to analyze the mobility of organic and/or inorganic compounds which are usually present in solid, liquid or multiphase wastes.

In this extraction procedure which has been standardized by the EPA (USe Federal Register, June 13, 1986), high liquid/solid (waste) ratios (20/1)

Table 2. Analysis of the Composition of Dust Coming from the Manufacture of Common Steels in %

Zn	20	–	35
Pb	5	–	10
Cu	0.1	–	0.4
Sn	0.06	–	0.1
Cd	0.05	–	0.07
S total	0.08	–	1.4
F	1.5	–	3.0
Cl	0.2	–	0.5
C	0.5	–	0.8
FeO	20	–	22
MnO	3	–	4
CaO	4	–	5
MgO	2	–	3
BaO		$<$	0.01
Al_2O_2	0.2	–	0.4
SiO_2	1	–	2
Na_2O	1.5	–	3.0
K_2O	1.5	–	2.5

are stirred in standard equipment for 18 hours. The employed extractant (liquid) is selected as a function of the alcalinity of the solid waste; after the extraction has been completed the liquid phase is filtered off and samples taken for analysis [10].

In the Equilibrium Leach test a waste sample is put into contact with distilled water for 7 days in rotary equipment keeping low liquid/solid ratios (4/1). The filtered liquid phase is analyzed to determine metals concentration (Pb, Cd, Cr and Zn). The analytical determination is performed in an atomic absorption spectrophotometer Perkin Elmer 560 with a hollow cathode lamp. Previous to the analysis, samples are diluted to reach the linearity range of analysis for each metal: $Pb \leq 20\ ppm$; $Cd \leq 2\ ppm$; $Cr \leq 5\ ppm$ and $Zn \leq 1\ ppm$.

LEACHING RESULTS

The results obtained from samples treated according to the TCLP test expressed as concentration of metals in the leachate from the filtration unit are shown for P1 samples in Table 3 and for P2 samples in Table 4.

Table 3. Results from P1 Samples

Sample	Pb (ppm)	Cr (ppm)	Cd (ppm)
1.0	27.9	0.01	10.5
2.0	30.2	0.02	9.5
3.0	29.7	0.02	10.0
4.0	28.2	0.01	10.0
5.0	31.8	0.02	9.8
6.0	31.6	0.02	10.5
Mean value	32 ± 2	0.02	10 ± 1

Table 4. Results from P2 Samples

Sample	Pb (ppm)	Cr (ppm)	Cd (ppm)
7.0	365	0.02	25
8.0	358	0.01	24
9.0	350	0.02	25
10.0	355	0.02	27
11.0	358	0.02	26.5
12.0	347	0.01	24
Mean value	355 ± 8	0.02	25 ± 2

Table 5. Zn Concentration in Leachates from P1 and P2 Samples

Waste	Zn (ppm)
Special steel dusts (P1)	3000
Common steel dusts (P2)	4500

For both types of steel dusts the final concentration of lead and cadmium is much higher than the maximum values allowed by the USEPA, which are 5 ppm for lead and chromium and 1 ppm for cadmium.

These results lead to the conclusion that these wastes have to be considered as toxic wastes rather than inert, following the indications given by the USEPA.

Besides the previous results, zinc concentration has also been analyzed in the filtered liquids, obtaining data shown in Table 5.

This metal has to be considered from a different point of view because it not important to decide whether a final waste is toxic or not but to study the recycling possibilities by conventional procedures (Waelz furnace) or new technologies (plasma).

Table 6. Results from the Equilibrium Test

Waste	pH	Pb (ppm)	Cd (ppm)	Cr (ppm)
Special steel dusts (P1)	11.8	0.7 ± 0.2	0.1 ± 0.04	>10
Common steel dusts (P2)	3.2	2.8 ± 0.3	27.5 ± 2	0.06

When the steel dusts are treated following the considerations of the equilibrium test, results shown in Table 6 for metals concentration and final pH have been obtained.

From the obtained results the following conclusions can be drawn:

(i) There is not a simple relationship between lead, cadmium and chromium leaching from steel foundry dusts, using the TCLP test and the equilibrium test, probably due to the pH evolution in the system.

(ii) The final pH varies greatly with the type of dust, P1 samples having a pH value of 11.8, whereas P2 samples have a pH value of 3.2 (data from equilibrium test), as well as with the test used.

(iii) This final pH seems to influence dramatically the solubility of Pb and Cr in the leachate, having no influence in the case of Cd. For this reason the final concentration of lead is higher in P2 samples than in P1, and just the contrary takes place for Cr, which is much more soluble in basic media leading to higher concentrations in P1 samples than in P2.

REFERENCES

[1] Apraiz, J., Fabricación de acero, Ed. Urmo. Bilbao, 1980.

[2] De Lora, F., Miro, J., Técnicas de Defensa del Medio Ambiente, Ed. Labor S.A., Madrid, 1978.

[3] ASER, Nueva solución a los resíduos sólidos polucionantes de acerias, V Reunión de la comisión técnica de acerías, Madrid, 1986.

[4] UNESID, La siderurgia Española. Proceso Siderúrgico, Madrid, 1986.

[5] Maczekard, H., Kola, R., Recovery of Zinc and Lead from Electric - Furnace. Steelmaking Dust at Berzelius, Journal of Metals, Enero, 1980.

[6] McAloon, T.P., So little Time, So Many Choices, I&SM, 14 - 20 August, 1987.

[7] Van Note, K., Miller, C.C., Lichter, J,C., Foundry waste stabilization: Laboratory testing and conceptual design, Transactions of the American Foundry Society, 93, 395 - 404, 1985.

[8] Wiles, C., Apel, L., Critical characteristics and properties of hazardous waste solidification / stabilization, EPA Contract 68 - 03 - 3186, 1986.

[9] Baldi, M., Chierico, G., Riganti, V., Leaching of heavy metals from solid wastes in landfill site, Chemistry for protection of the Environment 1987, Studies in Environmental Sciences 34, L. Pawlowski, E. Mentasil, W.J. Lacy and C. Sarzanini (Eds.) Elsevier, 99 - 109, Amsterdam, 1987.

[10] U.S. EPA and ENVIRONMENT CANADA, Test Methods for Evaluation of Solid Waste; Physical, Chemical Methods, Environmental Protection Service, Burlington, Ontario, Canada, April 1988.

REMOVAL OF CHLOROPHENOXY ACID DERIVATIVES FROM WASTEWATERS

P. ANIELAK, K. JANIO and J. JANKOWSKI

Institute of General Chemistry
Technical University of Łódź
Żwirki 36, 90-924 Łódź, Poland

ABSTRACT

Strongly basic anion exchangers (Wofatit SBK and Wofatit SBW) and non-ionic macroreticular adsorbents (Amberlite XAD type) were used for the removal of chlorophenols and chlorophenoxyalkanoic acids from industrial effluents. Static experiments showed high affinity of the adsorbents for 4-chloro-2-methylphenoxyacetic acid (partition coefficients ranged from 1,800 to 5,800). Its adsorption decreased with an increasing pH value of the solution.

Results of column tests indicated that the anion exchangers as well as non-ionic adsorbents might be used for purification of strongly polluted wastewater containing chlorophenoxy acids. High degree of removal of the pollutants (from 98 to 100%) was achieved in a wide range of pH values.

Spent anion exchangers might be regenerated with methanol solution of HCl. Methanol or aqueous solutions of NaOH were used as regenerants for exhausted nonionic adsorbents. Those regenerants permitted a high efficiency of regeneration.

INTRODUCTION

Derivatives of chlorophenoxy acids [1] are widely applied active substances for a large group of herbicides and growth regulators. Synthesis of these herbicides is accompanied by production of large volumes of wastewater particularly heavily loaded with organic substances (COD 5-10 g O_2/l). These wastes contain unreacted chlorophenols, chlorophenoxy acids and a large amount of sodium chloride. Herbicides which contain chlorophenoxyalkanoic acids are less toxic to aquatic organisms than chlorinated hydrocarbons [2]. However, usable preparations are 10-100 times stronger poisons than the pure compounds. Additionally the herbicides contain dibenzo-p-dioxins, which are strongly toxic and resistant to biochemical degradation.

Currently employed methods of neutralization are based on precipitation and chemical oxidation. They are not very effective and they can hardly be applied to recovery of chemicals. A novel method of neutralization seems to be the electrochemical oxidation of organic compounds existing in wastewater [3]. The highly polar character of reagents prompted the use of ion exchange and adsorption methods for purification of wastewaters.

Early equilibrium studies on ion exchange of chlorophenols and chlorophenoxy acids on strongly basic anion exchangers showed a great affinity of those resins to these compounds [4]. Selectivity coefficients ranged from 10 to 50, depending on the degree of crosslinking of the anion exchanger. Dynamic investigations showed a high efficiency in ion exchange columns for resins with low and standard degrees of crosslinking [5]. Even so, the column efficiency is dependent on the salinity of the untreated solution but this does not exclude the practical application of this process. Highly effective regeneration of anion exchangers, in particular for resins with a low degree of crosslinking, was obtained using aqueous or mixed electrolyte solutions [5].

In this paper the investigations on the application of synthetic adsorbents and ion exchangers for neutralization of wastewaters polluted with chlorinated phenoxyalkanoic acids are presented.

EXPERIMENTAL

The following styrene-divinylbenzene copolymers were examined:

1. Nonionic macroreticular adsorbents, Amberlite XAD-4 and a copolymer produced in Professor B. Kolarz Laboratory (Institute of Organic Technology, Techn. University of Wroclaw, Poland) and named "PW" [6]. Their specific surface areas were 1065 and 612 m^2/g, respectively.

2. Strongly basic anion exchangers containing ionogenic trimethylammonium groups (Wofatit SBW - type I) and dimethylethanolammonium groups (Wofatit SBK - type II).

Adsorbate stock solution was prepared by dissolving 2 mmol (401 mg) 4-chloro-2-methylphenoxyacetic acid (MCPA) in one liter of water. The pH of this solution was 3.1. The ion exchange experiments were carried out with model wastewater containing sodium 2,4-dichlorophenolate (15 mmol/l), sodium 2,4-dichlorophenoxyacetate (5 mmol/l) and sodium chloride (2 g/l). The pH of this solution was 8.5, and COD was about 5 g O_2/l.

Concentration of organic substances in the investigated solutions was determined by the measurement of optical absorption at wavelength 290 nm. COD was determined by the Jeris method [7].

In the batch adsorption experiments 200 mg of adsorbent were weighed into stoppered volumetric flasks and 100 ml portions of different concentrations of MCPA were added. Then hydrochloric acid, sodium hydroxide, and methanol were added. The samples were equilibrated at 298 K with continuous agitation. After reaching equilibrium the concentration of the supernatant liquid was determined.

The fixed, bed experiments were carried out on bench scale in glass columns of 16 mm diameter containing 20 ml of anion resin, and in 11 mm diameter columns containing 10 ml of adsorbent material, respectively. Flow rates ranged from 10 to 90 BV/h. Column effluent was fractionated. Concentration of the organic compounds was determined in individual portions. Operating capacity of the resins was investigated in succeeding working cycles. Saturation was realized with a flow rate of 15 BV/h, desorption at 5 BV/h. Regenerative agent dose was about 100 g per liter of exhausted resin. 0.5 M hydrochloric acid in a water-methanol (1:9 v/v) mixture was used to regenerate spent anion exchangers. Exhausted adsorbents were regenerated using 0.1 M sodium hydroxide aqueous solution.

RESULTS AND DISCUSSION

Adsorption Studies

Adsorption isotherms of MCPA onto the investigated resins are given in Fig. 1. Both adsorbents show high adsorption capability in aqueous solution (curves 1 and 2). These experimental curves were compared with well-known theoretical isotherms. Experimental isotherms can be satisfactorily characterized by either Langmuir or Freundlich equations. Correlation coefficients calculated for linear forms of the equations were higher than 0.99. Constant a_m in the Langmuir equation was considered as a measure of the completed monolayer adsorption. Calculated values of a_m were 2.24 and 1.28 mmol/g, partition coefficients at infinite dilution were 5,800 and 1,800 for Amberlite XAD-4 and PW sorbent, respectively.

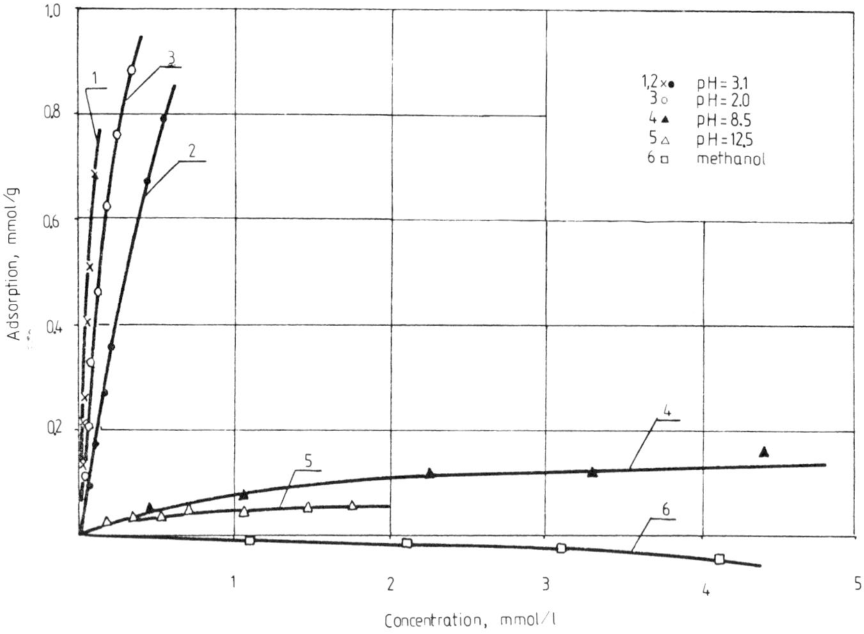

Fig. 1. Adsorption isotherms of MCPA on Amberlite XAD-4 (1) and PW adsorbent (2-6) from aqueous solutions.

Isotherms 3 to 5 in Fig. 1 illustrate the adsorption of MCPA on PW adsorbent from acids or basic aqueous solutions. Adsorption decreases with increasing pH value. It should be noted that MCPA has an acid characteristic with an ionization constant pK_a equal to 3.28. Its adsorption is strongest in that pH region which corresponds to the lowest degree of dissociation. Calculated monolayer capacities were 1.38, 0.38, and 0.13 mmol/g at the pH 2, 8.5, and 12, respectively.

Inspection of curve 6 in Fig. 1 clearly shows that methanol decreases MCPA adsorption. The reason for this is its high affinity for the resins. Adsorption then reaches a negative value. This means that methanol as well as aqueous solutions of NaOH can be used as regenerants for MCPA desorption.

Table 1 contains results of column tests showing the influence of flow rate of influent on breakthrough time and working adsorbent capacity. The column has reached a break point when the concentration of solute in the effluent has risen to the standard value 10% of c_o. Analysis of data summarized in Tab. 1 shows that high efficiency of the column process can

Table 1. Flow Rate Effect on Breakthrough Time and Breakthrough Capacity

Volume flow rate, BV/h	Breakthrough time, h		Breakthrough capacity, mmol/g	
	XAD-4	PW	XAD-4	PW
10	20.7	12.7	1.43	1.56
20	8.1	4.8	1.13	1.10
40	4.5	2.5	1.19	1.17
90	0.99	0.93	0.59	1.0

Influent concentration of MCPA = 2 mmol/l

be reached at a flow rate of about 10 BV/h for both adsorbents. Residual concentration of solute in column effluent at the optimum rate was 1 to 2% of c_o and the operating capacities of both adsorbents reached 1.4 mmol/g. An increase of the flow rate of the influent causes a decrease of operating capacity.

Working capacity of adsorbent bed and desorption ratio has been examined in three succeeding cycles (Fig. 2, Tab. 2). The results obtained indicated that the desorption ratio had been reaching a very high and almost constant value (0.93 or 0.97). The used regenerant level was about 100 g NaOH per liter of adsorbent. The operating capacity of the adsorbent bed decreased in succeeding cycles because of incomplete desorption of MCPA. Compared with aqueous regenerants, desorption efficiency may be improved by using methanol. Methanol not only shows high affinity for the resins, but also dissolves 60 times more solute than water.

Ion Exchange Studies

As follows from the data presented in Fig. 3, both the investigated anion exchange resins can be used for removing chlorophenols and chlorophenoxyacetates from their aqueous solutions. Residual concentrations of these compounds in the column effluent are too small to be determined by standard spectrophotometric methods. Absorbance of wastewater after neutralization was lower than absorbance of distilled water used as a reference material. Data presented in Tab. 3 also confirm these conclusions. Residual COD determined in purified wastewater was caused by the oxidation of chlorides present in column effluent (about 1.9 g Cl/l). Flow rates through the ion exchange resin for both anion resins was about 20 BV/h, , giving a high efficiency in the column process. Higher flows gave disadvantages in the shape of the breakthrough curves and to a decrease of the working exchange capacity. As can be seen from the Fig. 3, type II anion resin with higher crosslinking (Wofatit SBK) was more sensitive to changes in flow rate than

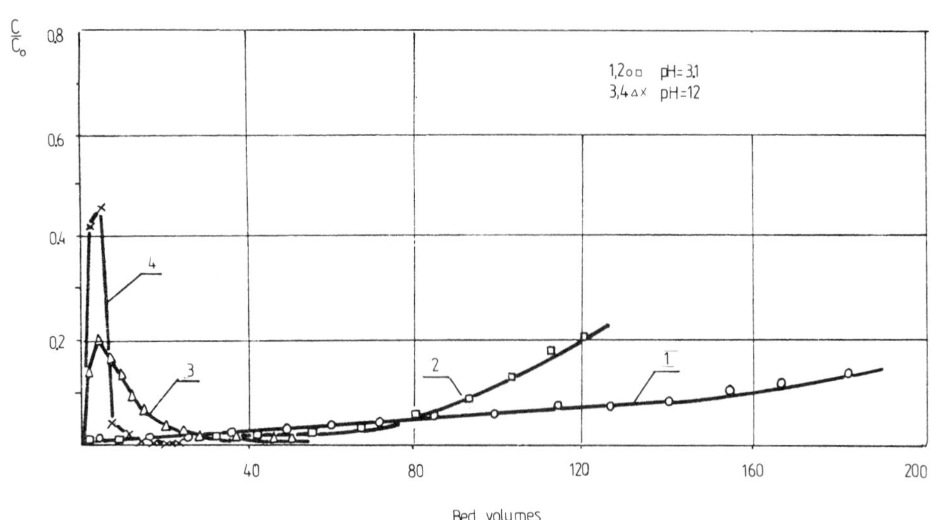

Fig. 2. Adsorption (1,2) and desorption (3,4) of MCPA in a dynamic system (cycle No 1). Adsorbents: Amberlite XAD-4 (1,3), and PW sorbent (2,4); c_o=2 mmol/l.

Table 2. Changes of Working Capacity of Adsorbent Bed and Desorption Ratio with the Number of Cycles

Cycle No	Working capacity of bed, mmol/BV		Desorption ratio	
	XAD-4	PW	XAD-4	PW
1	3.43	2.94	0.93	0.98
2	3.20	2.81	0.91	0.96
3	3.07	2.70	0.96	0.98

that resin (Wofatit SBW) with lower crosslinking. Working capacities of both the resins were similar at low flow rates (0.7 mmol/ml). Investigations into the change in working exchange capacity in successive cycles (5 cycles) showed that at a regenerant level of 100 g HCl per liter of anion resin, the value remained constant.

SUMMARY

The results of the studies may be summarized as follows.

1. Strongly basic anion exchangers and nonionic macroreticular adsor-

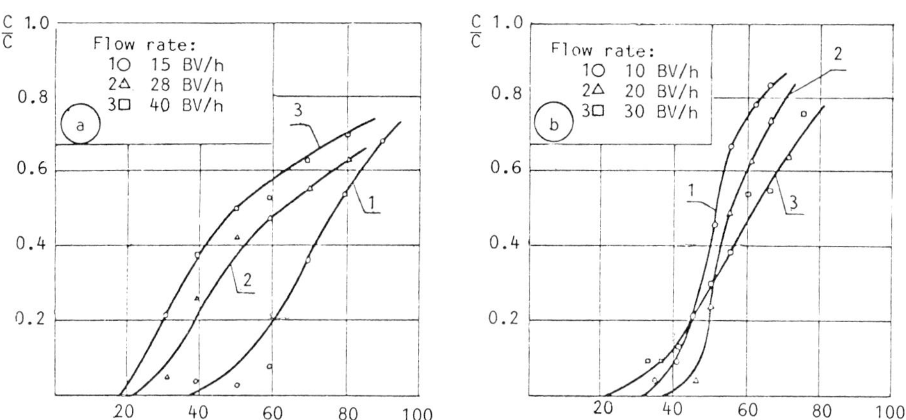

Fig. 3. Breakthrough curves of chlorophenols and chlorophenoxyacetates in synthetic wastewater. Ion exchange resins: Wofatit SBK (a); Wofatit SBW (b).

Table 3. Results of Treatment of Model Wastewater by Strongly Basic Resins in Dynamic System

Parameters	Wofatit SBW	Wofatit SBK
COD of influent wastewater, mgO_2/l	5,100	4,930
Organic matter content in treated wastewater (spectrophotom.), mg/l	0	0
COD of effluent wastewater, mgO_2/l	560	610

bents show a high affinity for chlorophenoxy acid derivatives. Adsorption of MCPA depends on pH value of the solution.

2. Column tests indicate that the investigated resins may be used for purification of strongly polluted wastewaters containing chlorophenoxy acids. High removal efficiency of pollutants has been obtained.

3. Spent anion exchangers may be regenerated using a methanol solution of HCl, and nonionic adsorbents - with methanol or an aqueous solution of NaOH. High regeneration efficiency was achieved.

4. Methanol can be recovered by a simple distillation of the spent regenerant.

5. Using the data, conditions for a highly efficient neutralization procedure, in a wide range of pH values, can be recommended for a pilot plant scale installation.

REFERENCES

[1] Fedtke C., Biochemistry and physiology of herbicide action, Springer-Verlag, Berlin 1982.

[2] Edwards C. A., Environmental pollution by pesticides, Plenum Press, London 1973.

[3] Socha A., Gorzka Z., Koziol H., Kaźmierczak M., Chemistry for protection of the environment, 5th International Conference, Leuven 1985.

[4] Jankowski J., Ph.D. Dissertation, Techn. University, Lódź 1975.

[5] Janio K., Jankowski J., Physicochemical methods for water and wastewater treatment, International Conference, Lublin 1976.

[6] Kolarz B., Chemia Stosowana, 19, 71, 1975.

[7] Jeris J.S., Water and Wastes Eng., 4, 89, 1967.

COLOR REMOVAL FROM A TEXTILE DYEHOUSE EFFLUENT BY FLY ASH

A. M. ESTEVES COELHO

National Laboratory of Civil Engineering
Chemistry Division
Av. do Brasil, 101 – 1799 Lisbon, Portugal

ABSTRACT

Fly ash is a residue obtained during the combustion of pulverized coal at power plants.

Nowadays in Portugal the annual production of fly ash amounts to 260,000 t and in the coming years is expected to double.

The present study reports use of Portuguese fly ash in the decoloration of a textile effluent from a wool-polyester blend dyed with dispersible and metallized dyes. This effluent was filtered through a column, 200 cm high and with 10 cm internal diameter, containing 6.5 kg of fly ash.

The main conclusions were that color intensity and COD of raw effluent decreased and the percent removal achieved was 80% in color and 28% in COD.

INTRODUCTION

When pulverized coal is burnt in power plants, the inorganic coal ash undergoes complex reactions and gives rise to microsized glassy spheres which are swept along in the gaseous stream. These particles are then collected in electrostatic precipitators and this residue is called "fly ash" (Fig. 1).

Chemistry for the Protection of the Environment
Edited by L. Pawlowski *et al.*, Plenum Press, New York, 1991

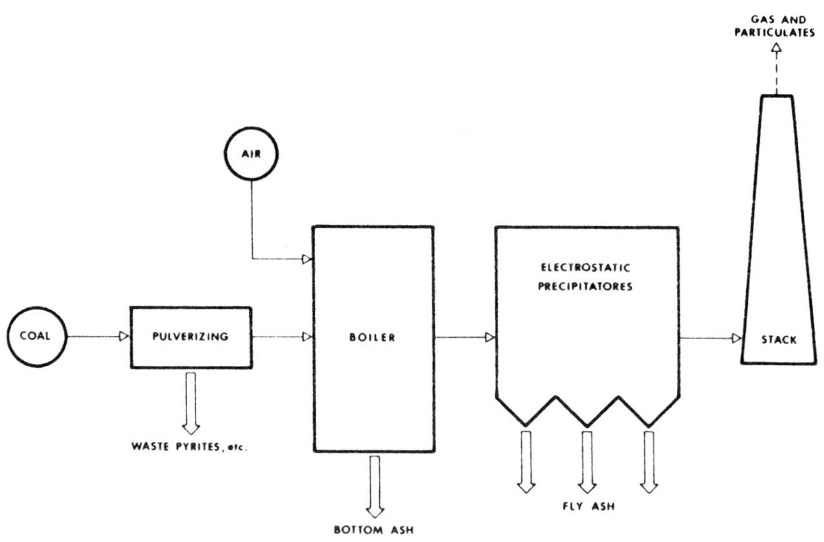

Fig. 1. Solid waste production points for a low sulfur coal-fired power plant.

As dependence on coal as a source of energy has developed, fly ash production has increased in many countries. Amounts of fly ash produced depend on the characteristics of the coal and on the capacity of power plants. In Portugal there is a coal power plant with 1200 MW capacity and another of the same capacity is being constructed. The power plant burns betuminous coal from U.S.A., United Kingdom and Poland, with about 14% ash and 1% sulfur, and produces 260,000 t of fly ash annually [1, 2].

This great amount of residue must be disposed of in a manner which is both environmentally acceptable and economically feasible.

For that reason studies of utilization of fly ash have been made on industrial effluents for removal of detergents, oils, phenols, copper and color [3, 4, 5, 6, 7].

With the purpose of studying the applicability of Portuguese fly ash for treatment of industrial wastewater, the textile industry was selected. In the textile industry the operations which give rise to significant liquid wastes are desizing, scouring, bleaching, mercerizing, dyeing, printing and special finishing. This industry is mainly concentrated in the north of Portugal, in Oporto and Braga districts, corresponding to 84% of the overall textile production in the country [8]. The textile industry converts natural and syn-

Table 1. Raw Effluent Quality

pH	5,13
Conductivity (20°C S/m)	0,17
COD (mg/l)	6000
Colour — Absorbance (600 nm)	1,24
Colour — Times of dilution	4500

thetic fibers into fabrics and other textile products. In Portugal raw material consumption in 1981 was mainly cotton (50%) and synthetic fibers (35%) and industrial processes make use of a wide variety of dyes and chemicals such as acids, bases, salts, detergents, wetting agents, etc.

Although environmental pollution is a subject of concern in Portugal, only a few factories have effluent treatment systems. The others discharge their wastewaters directly into municipal sewers, soil and rivers [9].

As color from dyes is aesthetically objectionable particularly in drinking and recreational waters, and fly ash is an abundant residue, it seemed of interest to study its applicability to color removal of textile dyehouse effluents.

EXPERIMENTAL CONDITIONS AND RESULTS

An effluent from dyeing a wool-polyester blend (45% + 55%) with dispersive and metallized dyes was selected. The raw effluent had a deep blue color and an odor of aromatic compounds. The following parameters were determined: pH, conductivity, chemical oxygen demand (COD) and color (Table 1).

To observe the effectiveness of dye removal by means of fly ash, a Perspex column 200 cm high and with 10 cm of internal diameter was constructed. Half of the column was filled with 6.5 kg of fly ash (Table 2) and the column was fed with raw effluent during 22 days, using a total volume of 21 dm^3 of effluent (Fig. 2).

The filtered effluent was collected everyday except on week-ends and determinations of pH, conductivity, COD and color were carried out on all samples (Table 3, Fig. 3).

Table 2. Fly Ash Characteristics

Chemical composition (%)		
	SiO_2 -	51,8
	Al_2O_3 -	27,8
	Fe_2O_3 -	7,3
	TiO_2 -	1,8
	CaO -	2,5
	MgO -	1,1
	Na_2O -	0,6
	K_2O -	2,0
	SO_3 -	0,8
	L.I. -	3,9
Specific gravity		2,32 g/cm^3
Permeability		$8,9.10^{-7}$ cm/s
Specific surface (Blaine)		3500 cm^2/g

The ultraviolet and visible spectra of raw effluent and filtered samples were recorded (Figs. 4, 5, 6).

All determinations except for color followed Standard Methods 1980.

Colour was determined by spectrophotometry, using a spectrophotometer Pye Unicam SP8-100, and by a dilution method. In the spectrophotometric method, concentration c of dye is directly related with its absorbance A on a fixed wavelength by the mathematical expression $A = \varepsilon cl$, designated as Beer's law (ε is characteristic of the absorbing molecule or ion and l is the path length of the cell utilized).

In the dilution method a known volume of effluent is diluted with water until it is tintless and the ratio of the final volume to the sample volume is called "times of dilution".

CONCLUSIONS

The pH of the effluent filtered through fly ash changes from acidic to basic. Conductivity increases in the first filtered samples, decreasing later to a value near that of the raw effluent.

The organic compounds determined by COD reached a 60% removal in first samples of filtered effluent, decreasing then to 28%.

It was observed that color can be completely removed from the first 7 dm^3 of effluent passed through fly ash; after this volume, however, fly ash starts to become exhausted and color removal decreases gradually to 80% (Fig. 7).

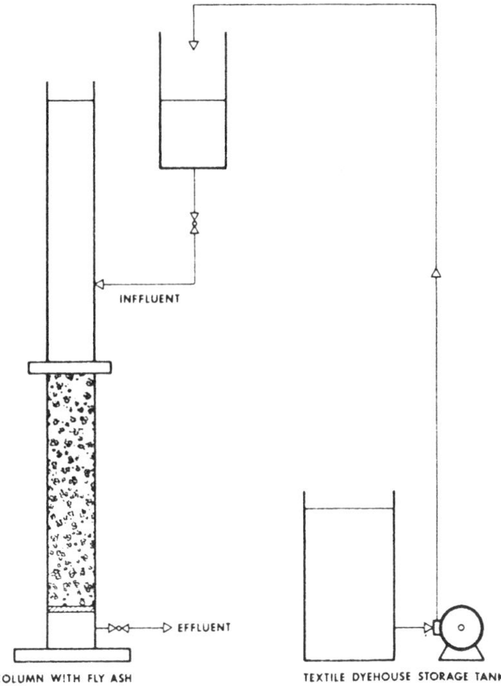

Fig. 2. Laboratory experiment flow diagram.

In the spectrum of raw effluent we can see two broad bands with a central maximum of absorbance at 600 nm and 370 nm.

In the spectrum of filtered effluent the band of maximum absorbance at 600 nm vanished and two other bands at 370 nm and 290 nm appeared.

The band at 290 nm is a characteristic of benzonoid compounds and the band at 370 nm probably corresponds to the polynuclear aromatic compounds such as naphthalenes and anthracenes.

Comparing the characteristics of the raw effluent and the filtered effluents, it can be concluded that fly ash treatment gives rise to highly alkaline effluent, a slight increase of dissolved solids, a decrease of some organic compounds and a decrease of dye concentration. It was also observed that the filtered effluent is odorless.

From the technical point of view fly ash could be used in color removal from textile effluents with the additional interest of removing other organic compounds; however, it seems that significant amounts of fly ash would be required to achieve a 100% color removal.

Table 3. Effluent Filtered by Fly Ash Column

SAMPLING DAYS	1	4	5	7	8	9	10	11	12	13	15	18	22
pH	9,50	9,56	9,60	9,90	9,90	9,21	9,58	9,45	10,0	9,52	9,51	9,90	10,26
Conductivity ($20^{o}C$ S/m)	0,70	0,37	0,29	0,28	0,29	0,29	0,29	0,32	0,34	0,35	0,33	0,21	0,20
COD (mg/l)	-	2400	-	3600	-	-	-	-	-	-	-	-	4300
Colour – Absorbance (600 nm)	0,0	0,0	0,0	0,0	0,0	0,10	0,25	0,26	0,26	0,24	0,22	-	0,24
Colour – Times of dilution	0	0	0	0	0	60	-	-	-	-	-	-	22
Colour removal (%)	100	100	100	100	100	92	80	79	79	81	82	-	81
COD removal (%)	-	60	-	40	-	-	-	-	-	-	-	-	28

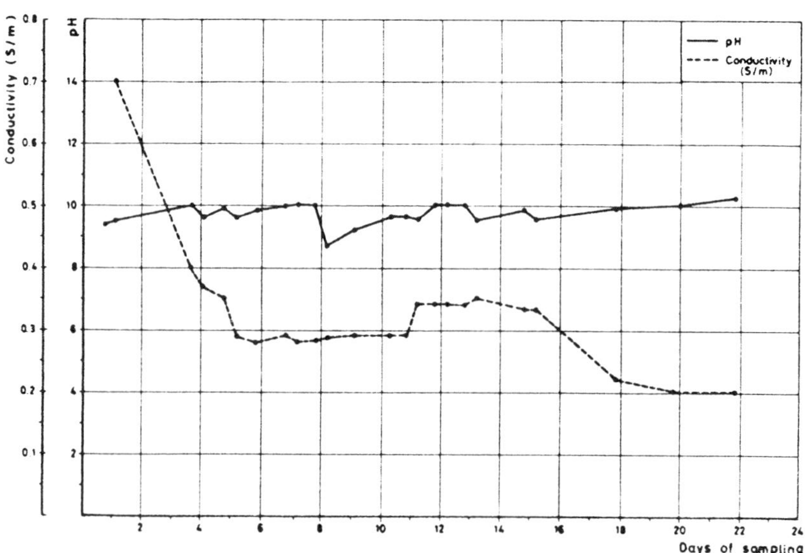

Fig. 3. Conductivity and pH of Effluent Filtered by Fly Ash

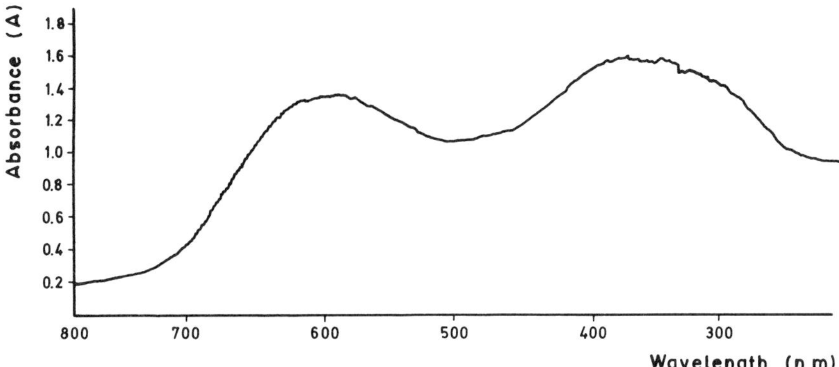

Fig. 4. Visible and ultraviolet spectrum of raw effluent.

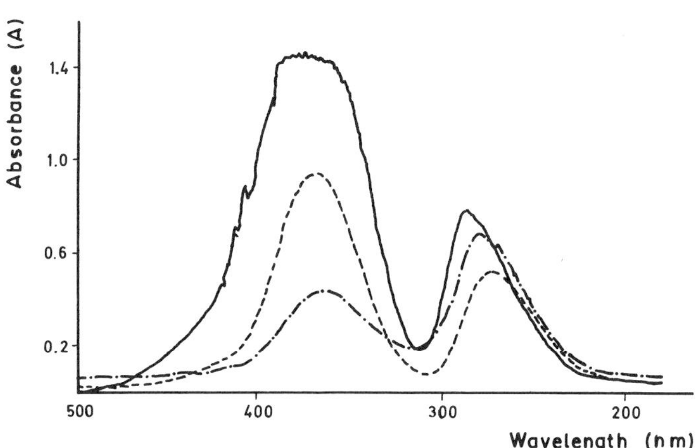

Fig. 5. Visible and ultraviolet spectrum of the filtered effluent sampled on the first —, fourth - - - , and fifth days — o — o —o.

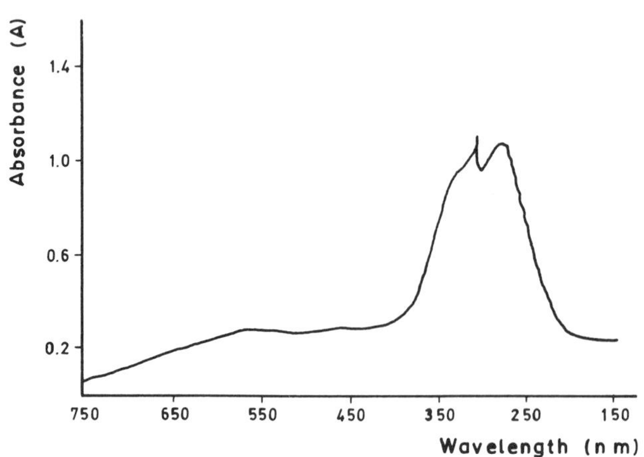

Fig. 6. Visible and ultraviolet spectrum of the filtered effluent sampled on the thirteenth day.

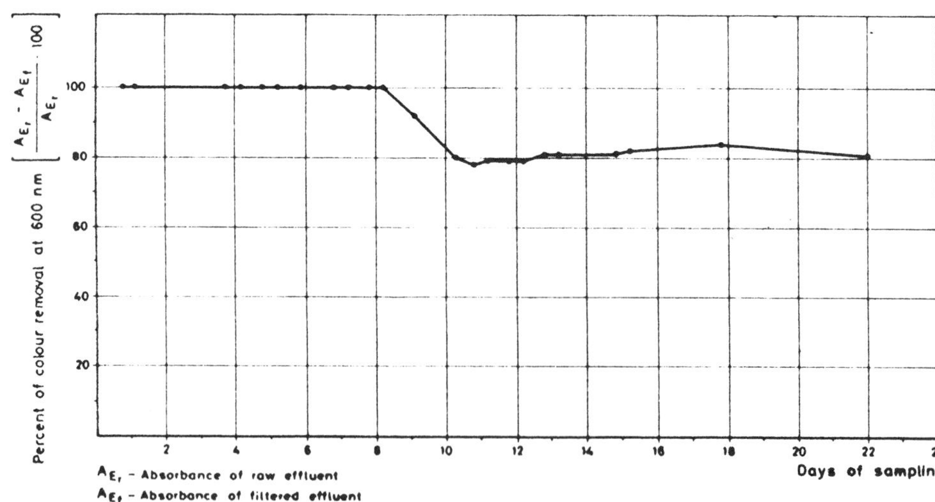

Fig. 7. Colour removal of raw effluent by fly ash.

REFERENCES

[1] Branco, J., Cinzas de carvão, suas caracteristicas e campos de utilização. Colóquio sobre Poluição Industrial, Instituto Superior Técnico, 1985.

[2] Cabaço, R., Aroso, M., Utilização das cinzas volantes de Sines em betumes e argamassas, Ingenium, Revista da Ordem dos Engenheiros n° 19, pp. 32 - 48, 1988.

[3] Bhargana, R., Mathor, R.P., Khanna, P., Removal of detergent from wastewater by adsorption on fly ash, Indian J. Environ. HLTH. vol 16, n° 2, pp. 109 - 120, 1974.

[4] Dubey, A.P., Rajev, Shukla, N.P., Monotoring and treatment of effluent from hydrogenated vegetable oil industry, Report not published.

[5] Karl Lorenz, Phenolwasser - Nachreinigungsverfshren mittels Winklergasstaub und Winklerashe, Gesundh Ing., Heft 11/12 (75. Yahrg), pp. 189 - 194, 1954.

[6] Tien - Yung J. Chu, Steiner, G.R., McEntyre, C.L., Removal of complex copper - ammonia ions from aqueous wastes with fly ash, 32nd Ind. W.Water Conference, pp. 359 - 376, 1977.

[7] Mac Donald, D.G., Bakhshi, N.N., Rao, D.N., Colour removal from pulp mill effluent using fly ash. CPAR Project report 611-1 Dep. of Chemistry and Chemical Engineering, University of Saskatchewan, 1977.

[8] Cartaxo, L., Pinelas, R., Almeida, M.F., Determinação das cargas poluidoras brutas produzidas pelos sectores de actividade industrial em Portugal Continental, Direcção dos Serviços de Controle de Poluição - 21.7/Adm, 1985.

[9] Santos, M.A., Utilização da sagua na industria - Inquérito na Bacia Hidrogrufica do Rio Ave, Laboratório Nacional de Engenharia Civil ITH 14, 1984.

DRY FLUE GAS DESULPHURIZATION

AT LOW TEMPERATURES

F. CORTABITARTE, J. R. VIGURI, M. I. ORTIZ,
and J. A. IRABIEN

Departamento Ingeniería Química
Facultad de Ciencias
Universidad del País Vasco
Apdo. 644-48080 Bilbao. Spain

ABSTRACT

Although wet desulphurization processes have been developed mainly due to their technological feasibility in comparison to the high temperature required in dry processes, a new type of in-duct dry-sorbent injection technology at low temperature provides low-cost retrofit SO_2 control options for existing coal-fired power plants. This technology involves injection of a dry sorbent, typically hydrated lime, in conjunction with flue gas humidification.

Some workers have dealt with the study of the variables having influence on the yield of solid utilization concluding that sorbents are the most reactive at or below the local dewpoints and they have proposed some hypotheses to explain the reaction mechanism under experimental conditions.

Although some efforts have also been directed to the kinetic modelling of the reaction between SO_2 and lime at low temperatures in recent years, they have not succeeded in interpreting the influence of experimental variables by means of a kinetic model with constant parameters in the whole researched range of variables. In this work, experiments at different initial concentrations of sulfur dioxide have been performed in a fixed bed with variable heights of the solid, evaluating the influence of both variables on sulfation rates. Through experiments with different

Chemistry for the Protection of the Environment
Edited by L. Pawlowski *et al.*, Plenum Press, New York, 1991

influence of both variables on sulfation rates. Through experiments with different reagent solids (i) commercial lime (Se=8m^2/g), and obtained lime under controlled conditions (Se=18.5m^2/g) the evaluation of the characteristics of the solid on its maximum utilization has been also performed

INTRODUCTION

Acid rain and acidification of our environment have emerged as a serious global problem during recent decades. The far most dominating air pollutants contributing to acid rain are nitric oxides and sulfur dioxide. The global annual manmade emissions of sulfur are currently about 75 to 100 million tons; about 70 percent of this originates from combustion of fossil fuels.

Efforts to control sulfur emissions to the environment include several types of flue gas desulfurization (FGD) systems. From the point of view of the phases present in the sulfation medium, FGD processes can be classified as wet processes giving a wet by-product which is difficult to be disposed of and dry processes giving a dry reaction product for easy disposal or reuse.

Emerging in-duct dry-sorbent injection technology may provide low-cost retrofit SO_2 control options for existing coal-fired power plants [1,2]. The technology involves injection of a dry sorbent, typically hydrated lime, in conjunction with flue gas humidification. Since the waste is dry, spent sorbent can be recycled and allow more complete sorbent utilization.

In-duct desulfurization processes currently under active development include the Coolside process (Consolidation Coal Company) and the HALT process (Dravo Lime Company) [3]. In the first one, hydrated lime is injected dry in the duct downstream of the air preheater, and the flue gas is subsequently humidified to a close approach to adiabatic saturation by finely atomized water sprays. With the injection of sorbent upstream of the water sprays, sorbent particles may interact with evaporating droplets. An alternate process configuration is used in the HALT process; it envolves injection of sorbent downstream of the water sprays after droplet evaporation.

Some workers have dealt with the study of the variables having influence on the yield of solid utilization, concluding that sorbents are at the most reactive at or below the local dewpoints and they have proposed some hypotheses to explain the reaction mechanism under the experimental conditions [4,6].

Klingspor and coworkers [7,8] fitted their experimental data to a shrinking core model only after a certain lime conversion had been reached and Ruiz Alsop [9] employed a modified shrinking core model changing the values of model parameters within relative humidity.

This paper describes the influence of the process variables on the rate of sulfation when a humidified flue gas containing a mixture of SO_2, CO_2, N_2 and air flows through a fixed bed containing hydrated lime as solid sorbent diluted in inert silica sand.

From experiments performed at different initial concentrations of SO_2 kinetic data expressed as C_{SO_2} vs. time for different heights of bed have been obtained, allowing the evaluation of these variables on the rate of sulfation.

EXPERIMENTAL PROCEDURE

Experiments have been performed in a fixed bed made of glass where the sorbent (calcitic solid) is dispersed in inert silica sand; the entire bed is supported on a 3.6 cm diameter fritted glass plate contained in the glass cylinder. The simulated flue glass passes down through the sand bed at a face velocity of up to 2.5 cm/s. The simulated flue gas is generated by mixing gases from cylinders in appropriate amounts using calibrated rotameters operated at slightly elevated pressure to obtain stable readings.

Prior to the addition of the SO_2 to the gas mixture, the latter is passed through an humidification system where the gas is contacted with water vapor produced at, steady rate in two absorbers of 250 ml each. Both flasks, containing small glass spheres in order to improve the contact between gas and liquid phases, are submerged into a water-bath kept at constant temperature (by means of a controller). After humidification of the flue gas, and addition of the corresponding amount of SO_2, the gas is conducted to the reactor through a system of preheated pipes reaching the desired temperature before the entrance to the fixed bed. A schematic diagram of the experimental apparatus is shown in Fig. 1. The exact wet bulb temperature was measured just downstream of the reactor using a wetted wick thermometer. The difference between dry and wet bulb temperatures is the "approach to saturation", Δts, which together with the pressure measurement (U-tube manometer) establishes the relative humidity at test conditions.

When the reacted flue gas goes out of the reactor it is passed through a refrigerated pipe in order to eliminate the water contained to prevent damage to the analysis system which is a hot wire detector. It has been checked that the amount of SO_2 retained in the condensed water is always less than 0.3% of total amount of SO_2 in the gas phase.

Prior to testing, the fixed bed reactor is by-passed, and the initial test gas SO_2 concentration is measured in the gas chromatograph.

Two different types of sorbents have been utilized in the experiments: commercial hydrated lime (Dolmitas del Norte, S.A.) having a BET specific surface (Se) of around $8 m^2/g$ and the following chemical composition deter-

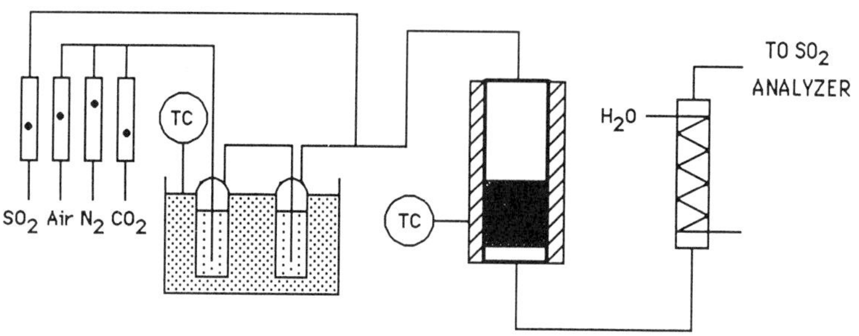

Fig. 1. Schematic diagram of the experimental apparatus.

mined from thermogravimetric decomposition runs, 88% $Ca(OH)_2$, 6.21% $CaCO_3$ and the rest CaO and some impurities; and a hydrated lime obtained in the laboratory under controlled conditions having a BET specific surface Se=18.5 m^2/g and the following composition, 78.4% $Ca(OH)_2$, 17.1% $CaCO_3$ and the rest CaO and some impurities, and mean particle diameter is smaller than 40 μm in both cases.

RESULTS AND DISCUSSION

The range of experimental variables was selected according to the characteristics of composition of flue gases exhausting from combustion plants: the following table shows the range of variation of process variables:

variable	maximum level	minimum level
SO_2 (ppm)	6000	3000
H_2O(%)	16	4
CO_2(%)	15	5
O_2(%)	4	0
$T^{e.}$ ($^{\underline{o}}$C)	80	50

Data from previous experiments were used to make sure that the linear gas velocity was high enough to avoid the external mass transfer the controlling step; therefore in subsequent experiments the gas velocity was u=5 cm/s or Qg=3 l/min. It was also checked that the ratio of sorbent/inert solid had no influence on the obtained results (1/30) and that there was no

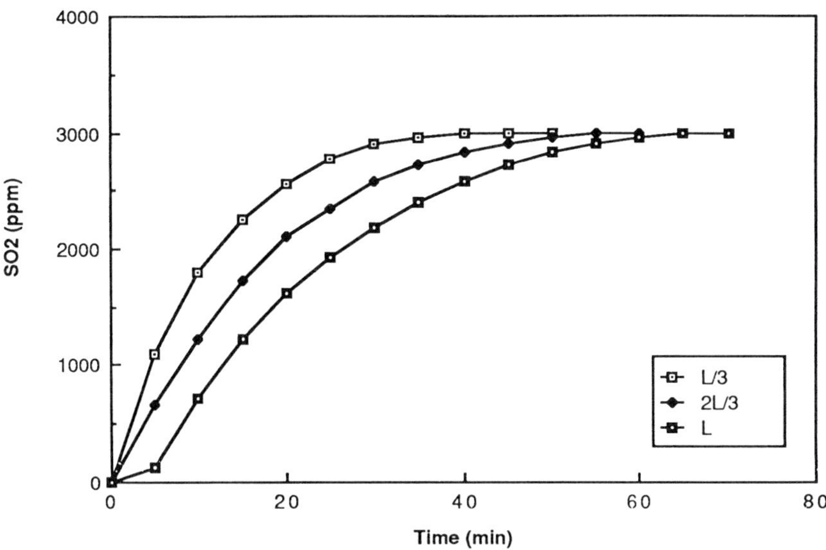

Fig. 2. Experimental results at initial SO_2 concentration of 3000 ppm (time as independent variable).

adsorption due to the inert silica sand. The time delay before analysis was measured. The first kinetic experiments were performed with the commercial lime but the results showed that this solid was not appropriate because in all cases the conversion obtained for the sorbent was less than 5%.

Therefore another experiment was planned, working with the hydrated lime obtained under controlled conditions (Se=18.5 m^2/g) and the following values for the variables: SO_2, 300ppm; CO_2, 12%; O_2, 2%; H_2O, 12%; $T^{e,}$ 52ºC and u=5cm/s.

Under these conditions 20% of conversion for the solid sorbent was obtained, so the same conditions were applied in the following experiments.

In order to evaluate the influence of the initial SO_2 concentration, runs at 3000 and 6000 ppm were performed at different bed heights ($Ca(OH)_2$/inert: 3/90, 2/60 and 1/30). The obtained results expressed as SO_2 concentration versus reaction time are shown in Figs. 2 and 3 and expressed as SO_2 concentration vs bed height in Figs. 4 and 5.

By calculating the maximum solid conversion from a mass balance it was seen to be dependent on the initial concentration of SO_2 and independent of the height of the bed, as shown in Fig. 6.

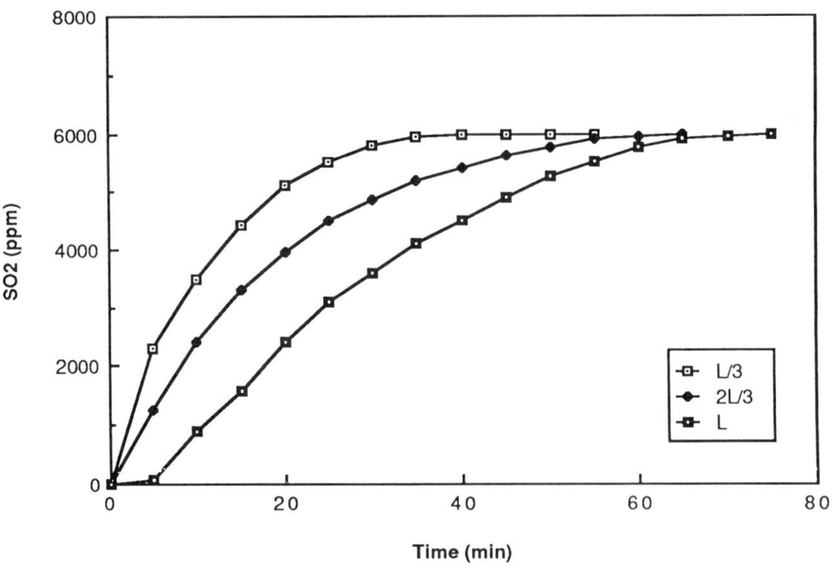

Fig. 3. Experimental results at initial SO_2 concentration of 6000 ppm (time as independent variable).

KINETIC ANALYSIS

The equations that describe the absorption of a component from a moving gas stream by a fixed solid in a paced bed consist of two partial differential equations obtained from material balances in the gaseous and solid phases.

For the gas phase:

$$U(\partial C/\partial z) + \epsilon(\partial C/\partial t) + r = 0 \tag{1}$$

where z and ϵ are the height and porosity of the bed, respectively, and r the rate of sulfation (mol/l bed.min). For the solid phase:

$$\partial x_B/\partial t = \partial/d_B \cdot r \tag{2}$$

where ∂x_B is the solid conversion, the stoichiometric coefficient and d_B the molar density of the solid in the bed (moles of B/l bed volume).

From the experimental curves $C_{SO_2^{-t}}$ and $C_{SO_2^{-z}}$ the partial derivatives $\partial C_A/\partial t$ and $\partial C_A/\partial z$ have been obtained and after substitution in equation (1) reaction rate values have been calculated.

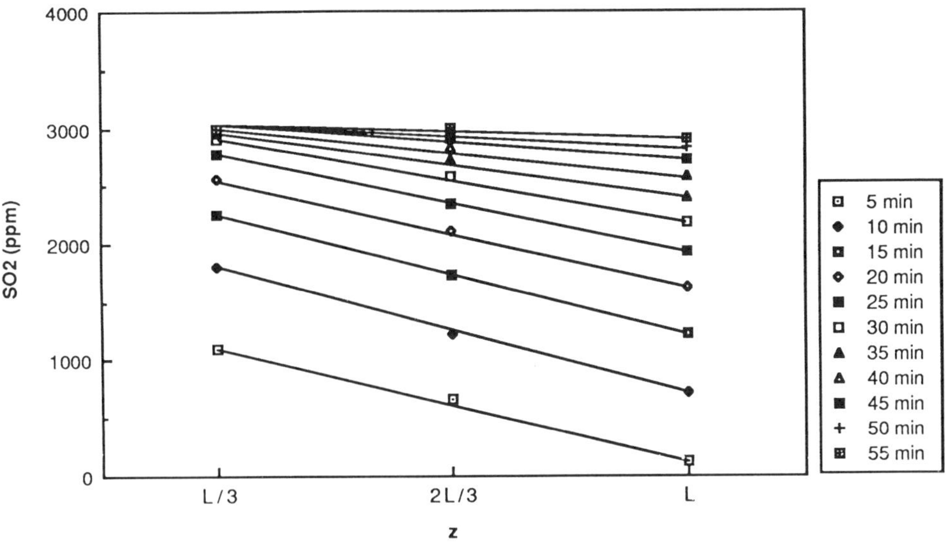

Fig. 4. Experimental results with Z as independent variable (C_{SO_2}=3000 ppm).

Taking into account an adsorption kinetic model as controlling step, where the available surface follows the shrinking core model:

$$r = \frac{K \cdot \epsilon \cdot C}{1 + K_A C}(1 - x_B/x_M)^{2/3} \tag{2}$$

x_M is the maximum conversion in the experimental conditions which after rearranging into a linear form leads to:

$$\frac{(1 - x_B/x_M)^{2/3}}{r} = 1/K\epsilon \cdot 1/C + K_A/K\epsilon \tag{3}$$

A linear plot is shown in Figure 7, from which:

$$K = 3.81\,10^2 \quad (min^{-1})$$

and

$$K_A = 2.09\,10^4 (l/mol)$$

DISCUSSION

The first modelling effort for this reaction was the work of Klingspor and coworkers [7,8]. They used an integral shrinking core model with only

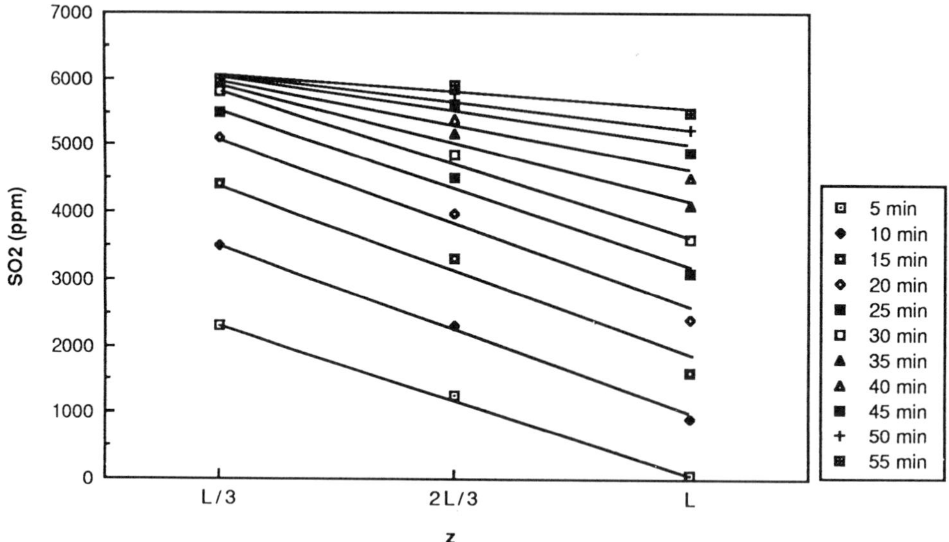

Fig. 5. Experimental results with Z as independent variable (C_{SO_2}=6000 ppm).

reaction kinetics to explain the dependence of reaction rate on the $Ca(OH)_2$ conversion reached. They attributed the sharp decrease on the reaction rate observed at initial times to a decrease in surface roughness, but did not attempt to correlate this decrease in surface roughness with $Ca(OH)_2$ conversion. Their model neglected the effects of SO_2 diffusion through the product layer and also the SO_2 concentration and $Ca(OH)_2$ concentration profiles in the fixed bed reactor.

On the other hand, Ruiz Alsop [9] interpreted the data obtained in a fixed bed reactor by means of a modified shrinking core model. The variables under study were the relative humidity of the gas (19 to 74%) and the initial concentration of SO_2. She showed good agreement between simulated curves and her experimental data by changing the values of model parameters within relative humidity attributing this fact to a change in controlling mechanism. She never considered the height of the bed as an experimental variable in order to evaluate its influence.

In this work the height of the bed (Z) is considered as an experimental variable in order to evaluate properly its influence on the reaction rate and it is shown that in the experimental range of variables under study sulfation rates follow anadsorption kinetic model, where the available surface follows the shrinking core model.

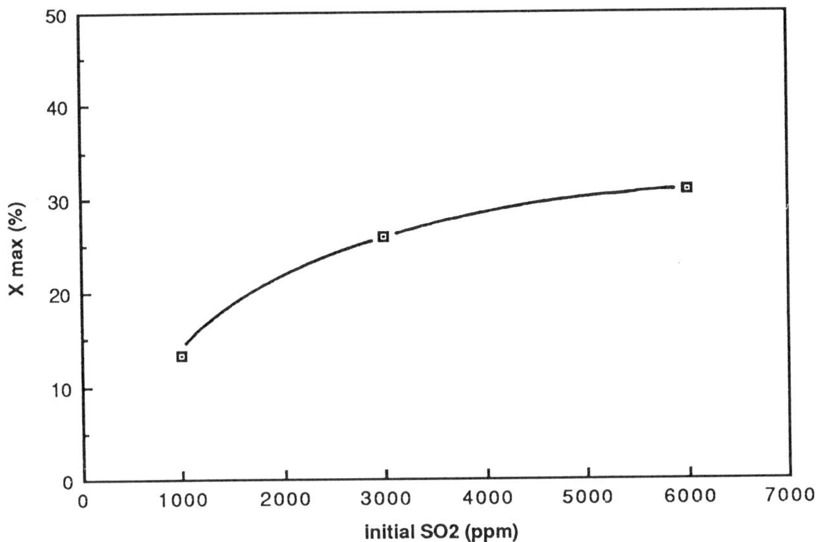

Fig. 6. Maximum solid conversion versus initial SO_2 concentration.

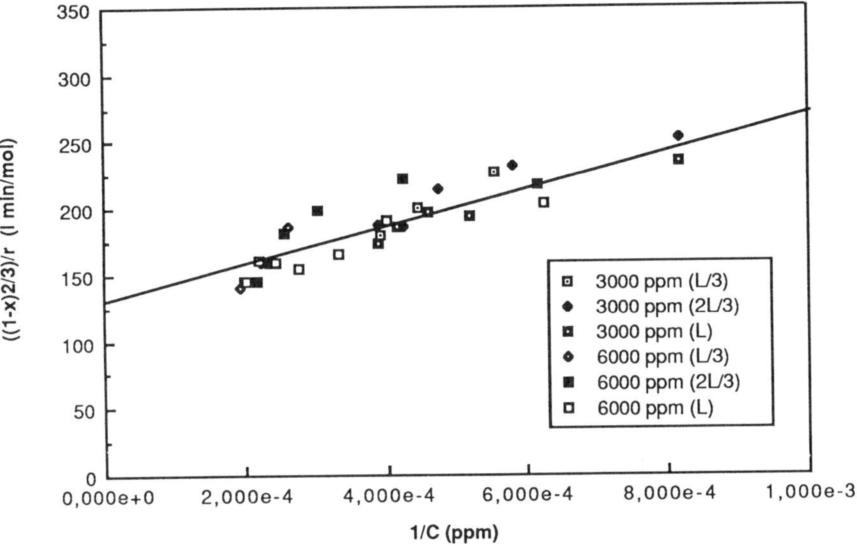

Fig. 7. $(1\text{-}X_B/X_M)^{2/3}/r$ versus $1/C$.

CONCLUSIONS

1.- In flue gas desulphurization processes by in-duct dry-sorbent injection technology and with a process temperature above the dew point the maximum sorbent utilization depends strongly on the characteristics of the initial solid employed, mainly its specific surface. The obtained results show that with commercial lime ($Se=8m^2/g$) it is not possible to obtain conversion for the solid sorbent higher than 5%, whereas when hydrated lime obtained under controlled conditions is employed ($Se=18m^2/g$) utilization conversion of the solid can be as high as 30%.

2.- Evaluating the influence of the initial concentration of SO_2 and height of the fixed bed on the sulfation rates from experimental results it is shown that under the investigated range of variables reaction rates, can be fitted by an adsorption model.

3.- The obtained results in this work show the inconsistency of previously reported kinetic efforts trying to find a suitable kinetic model for the reaction between SO_2 and lime at low temperatures because they have neglected the influence of height of the fixed bed on reaction rates.

ACKNOWLEDGEMENTS

This work has been financially supported by the Spanish CICYT under Project N^O. PA86-0147

One of the authors (F. Cortabitarte) wishes to thank the Basque Government for the award of a Scholarship.

REFERENCES

[1] Yoon, H.; Theodore, F.W.; Burke, F.P.; Koch, B.J.; Corder, W.C. "Low Capital Cost, Retrofit SO_2 Control Technologies for High Sulfur Coal Applications". Proceedings of the 79th Annual Meeting of the Air Pollution Control Association, Minneapolis, MN. (June 1986).

[2] Statnick, R.M.; F.P. Burke and B.J. Koch and D.C. McCoy and H. Yoon "Status of Flue Gas Sorbent Injection Technologies", Proceedings of the Fourth Annual Pittsburgh Coal Conference, Pittsburgh, PA, (Sept. 1987).

[3] Stouffer, M.R; H. Yoon and F.P. Burke "An Investigation of the Mechanisms of Flue Gas Desulfurization by In-Duct Dry Sorbent Injection" *Ind. Eng. Chem. Res.*, **28**, 20-27, 1989.

[4] Chen, S.L.; J.A. Cole and S.B. Greene and J.C. Keamlich and B.J. Overmol "Low temperature capture of SO_2 using calcium-based sorbents/kinetics of sulfur and nitrogen reactions" Unpublished report.

[5] Jorgensen; Chang, J.C.S., "Evaluation of Sorbent and Additives for Dry SO_2 Removal" *Environmental Progress*, **6(1)** 26-32, 1987.

[6] Ruiz-Alsop, R.N. and G.T. Rochelle "Effect of Delisquescent Salt Additives on the Reaction of SO_2 with $Ca(OH)_2$. Reprinting of Papers presented at the 189th National Meeting of the American Chemical Society, Miami Beach, Fla., 88–97, April 1985.

[7] Klingspor, J.; H.T. Karlsson and I. Bjerle "A kinetic study of the dry SO_2-Limestone reaction at low temperature" *Chem. Eng. Commun.*, **22**, 81–103, 1983.

[8] Klingspor, J.; A. Stromberg and H.T. Karlson and I. Bjerle "Similarities between lime and limestone in wet-dry scrubbing" *Chem. Eng. Process.,* **18(5)**, 239–247, 1984.

[9] Ruiz-Alsop, R.N., "Effect of relative humidity and additives on the reaction of sulfur dioxide with calcium hydroxide" (Univ. Texas, Austin TX, USA) 1986, 218 pp. Avail Univ. Microfilms Int., Order No DA 8700274 Diss. Abstr. Int B **47(9)**, 3874–5, 1987.

INDEX